The Blood–Cerebrospinal Fluid Barrier

The Blood–Cerebrospinal Fluid Barrier

Edited by

Wei Zheng

Adam Chodobski

CRC Press is an imprint of the
Taylor & Francis Group, an **informa** business
A TAYLOR & FRANCIS BOOK

CRC Press
Taylor & Francis Group
6000 Broken Sound Parkway NW, Suite 300
Boca Raton, FL 33487-2742

First issued in paperback 2019

ISBN-13: 978-0-415-32074-0 (hbk)
ISBN-13: 978-0-367-39310-6 (pbk)

Library of Congress Card Number 2004057098

Library of Congress Cataloging-in-Publication Data

The blood-cerebrospinal fluid barrier / edited by Wei Zheng and Adam Chodobski.
p. ; cm.
Includes bibliographical references and index.
ISBN 0-415-32074-7 (alk. paper)
1. Blood-brain barrier. 2. Choroid plexus. 3. Cerebrospinal fluid.
[DNLM: 1. Choroid Plexus--blood supply. 2. Blood-Brain Barrier. 3. Cerebrospinal Fluid.
4. Choroid Plexus--physiology. 5. Choroid Plexus--physiopathology. WL 307 B655 2004]
I. Zheng, Wei, Ph.D. II. Chodobski, Adam. III. Title.

QP375.5.B57 2004
612.8'24--dc22 2004057098

Visit the Taylor & Francis Web site at
http://www.taylorandfrancis.com

and the CRC Press Web site at
http://www.crcpress.com

Preface

The first account of the cerebroventricular system was provided by Galen of Pergamon (ca. A.D. 129–210). It was not until the seventeenth century, however, that the physician Thomas Willis came up with an explanation of choroid plexus (CP) function that was surprisingly close to the current concept of cerebrospinal fluid (CSF) production. By that time the anatomy of the ventricular system, the CP, and the meninges had been well described. Despite some progress in understanding the workings of the CP–CSF system, it took researchers almost three more centuries to realize that the CP is not only the major source of CSF, but that it is also the main site of the blood–CSF barrier.

The tight and adherens junctions connecting the choroidal epithelial cells form a barrier between the blood and the CSF. The CP shares the properties of the blood–CSF barrier with the arachnoid membrane. However, unlike the latter barrier, the blood–CSF barrier in the CP plays an active role in maintaining homeostasis of the brain microenvironment and in regulating various central nervous system (CNS) functions. Rapid development of experimental methodologies, including molecular techniques, which has taken place over the last 20 years, has facilitated advancements in research on the blood–CSF barrier and greatly enhanced our knowledge of the CP-CSF system. The CP now emerges as an organ possessing multifaceted functions, including broad enzymatic activities and the ability to synthesize proteins and transport peptides, nutrients, xenobiotics, and their metabolites. Increasing evidence indicates that the CP/blood–CSF barrier plays an important role in neuroendocrine signaling, neuroimmune and neuroinflammatory responses, drug metabolism, and protection from chemical-induced neurotoxicity. The critical role of the CP in brain development is also well recognized, and ample data have been obtained to suggest that the CP/blood–CSF barrier is involved in the aging of the CNS and the progression of some neurodegenerative diseases.

The lack of a comprehensive and up-to-date overview of the CP and its barrier function has prompted us to assemble this book. The book is comprised of four sections. In the **first section**, the reader will find a description of the ultrastructure and molecular features of the blood–CSF barrier and an analysis of the morphological changes in the CP that occur in various CNS disorders. In this section, the ontogeny of the CP is also discussed. The **second part** of the book focuses on various physiological functions of the CP/blood–CSF barrier. These include the molecular mechanisms of CSF formation and the peptide-mediated regulation of this process, transport of nutrients and essential elements, transport and metabolism of drugs, production of polypeptides, and neuroendocrine signaling. The **third section** analyzes the involvement of the CP/blood–CSF barrier in various CNS disorders, such as cerebral ischemia, Alzheimer's disease, metal-induced neurodegenerative diseases, and neuro-AIDS. It also discusses aging of the CP-CSF system and the role of the CP in immune signaling. In addition, this section contains chapters on the molecular mechanisms of hydrocephalus and the clinical aspects of CNS disorders involving the CP and CSF circulation. The **last part** of the book describes the methodologies currently employed in blood–CSF barrier research, such as the *in situ* CP perfusion technique, choroidal cell culture systems, and *in vivo* methods of CSF production measurement.

We believe it is time to take the blood–CSF barrier out from the shadow of the blood–brain-barrier and recognize its importance in controlling the microenvironment of the neuropil and functioning of the CNS. It is with this intention that we wish to provide the reader with an expert opinion on recent advancements in both basic and clinical research on the CP/blood–CSF barrier. We hope that this book will not only benefit those whose primary interest is in the CP-CSF system, but that it will also serve as an important source of reference to clinicians, neuroscientists, and toxicologists in academia, the pharmaceutical industry, governmental agencies, as well as graduate students.

We wish to thank our colleagues for the time and effort they have given in the preparation of the chapters for this book. In addition, we thank Dr. David S. Miller, National Institute of Environmental Health Sciences, National Institutes of Health, Research Triangle Park, NC, who kindly provided the images of the choroid plexus for the book cover. We would also like to thank our families for their continuous support, without which our work could not have been completed.

Adam Chodobski
Wei Zheng

Introduction

Wei Zheng

The brain has four major fluid compartments: the blood that flows through entire brain structures; the interstitial fluid (ISF) bathing neurons and neuroglia; the cerebrospinal fluid (CSF), which circulates around brain ventricles and spinal cord; and the intracellular fluid within brain cells. There is no major diffusional barrier between the ISF and CSF. Thus, materials present in either of these two fluid compartments are free to exchange and reach their destination cells. However, for a bloodborne substance to enter the ISF or CSF, it must first pass across the cell layers that safeguard the milieu of the central nervous system (CNS). Such cell layers possess unique cell types that join tightly together through intercellular connections, forming two major barrier systems designed to protect the brain microenvironment from fluctuations in concentrations of ions, metabolites, nutrients, and unwanted materials in the blood. The barrier located in the cerebral endothelia that separates the systemic circulation from the ISF compartment is defined as the blood–brain barrier, and the barrier in the choroidal epithelia that separates the systemic circulation from the CSF compartment is known as the blood–CSF barrier.

Despite the existence of two barrier systems in the brain, the focus of brain barrier research in the last century seems to be rather unbalanced, favoring the blood–brain barrier. This is not entirely surprising, taking into consideration the blood–brain barrier's intimate contact with neuronal constituents, its large surface area that covers nearly every part of brain organization, and its practical association in the development of CNS-effective therapeutic drugs. Nonetheless, inadequate research efforts on the blood–CSF barrier may somewhat reflect the general underestimation of the function of this barrier, and, to a certain extent, the insufficient understanding of the function of the choroid plexus, the tissue that comprises the blood–CSF barrier.

The choroid plexus is developed primarily from spongioblasts. In the human brain at autopsy during which the CSF is usually completely drained, the choroid plexus can be seen to extend along the floor of the lateral ventricles, hang down from the roof of the third ventricle, and overlie the roof of the fourth ventricle. The size of the tissue in one ventricle when it is in a dry, condensed state appears to be nothing more than that of an index finger. These autopsy impressions, which have over the years become the popular perception among many neuropathologists, can be dreadfully misleading and may lead to a misjudgment of the function of this tissue. Recent advancement in technology has enabled a micro-video probe to be inserted directly into the lateral ventricles. The vivid image illustrates that the live choroid plexus pervades the ventricles, stretching in concert with the heart pulse.

To better understand this, it may be helpful to make an analogy between the choroid plexus and a fishnet. The fishnet occupies barely one corner of a fishing boat; yet upon spreading in the water, it extends to cover a fairly large area. Likewise, in the life situation where the brain ventricles are full of the CSF, the choroid plexus expands to fill nearly all the cerebral ventricles. Unlike the fishnet, however, the plexus tissue possesses well-developed brush-type borders (i.e., microvilli) on the apical epithelial surface. These brush borders further protrude into the CSF and increase the choroidal epithelial surface area, enabling rapid and efficient delivery of the CSF as well as other materials included with the CSF secretion.

The brush borders of the choroidal epithelia also function to trap, take up, and enzymatically degrade the metabolites spilled into the ISF. A common notion regarding the brain's clearance of metabolites is that the metabolites are released into the ISF between neurons and glial cells; they are further drained into the CSF, where the bulk flow of CSF originating from the choroid plexus carries the metabolites to the subarachnoid space and further up to the dural sinuses for elimination. One should keep in mind that brain metabolites in the ISF would ultimately enter the CSF in the ventricles, if they are not transported to the bloodstream en route to the blood–brain barrier. Once in the ventricle, the metabolites first encounter countless brush borders of the choroid plexus. Clearly richer plexus borders would retain metabolites more so than do ependymal lining cells, which possess fewer foldings and microvilli, and, accordingly, the brushes of choroidal epithelia literally filter the CSF. Thus, from a neuroanatomical viewpoint, it is tempting to believe that the choroid plexus serves as a primary site for efflux or clearance of neuronal metabolites.

The lack of understanding of the blood–CSF barrier has finally come to an end in the face of a growing body of evidence that suggests a critical role of this barrier in monitoring the CNS homeostasis (Davson and Segal, 1995). In addition to its role in the movement of materials between the systemic circulation and the CSF, the choroid plexus avidly participates in various aspects of brain function, including involvement in the early stages of brain development, neuronal functional maturation, brain immune function, and neuroendocrine regulation. More recently, a great research effort has been devoted to the understanding of the pathophysiological influence of the blood–CSF barrier in neurological disorders, including its involvement in chemical-induced neurotoxicity, aberrant brain development, and, possibly, the initiation of neurodegenerative diseases (Strazielle and Ghersi-Egea, 2000; Zheng, 2001; Zheng et al., 2003). It is quite possible to say that the cerebral microcapillary injury (i.e., dysfunction of the blood–brain barrier) could be the cause of numerous brain homeostasis disorders; yet attention must be equally given to the blood–CSF barrier, without which, however extensive the research would be, it will remain difficult to explain the ultimate alteration in CNS milieu.

As a barrier between the blood and CSF, several unique anatomical and physiological features of the choroid plexus make the tissue a focal point of brain homeostasis research. First, the choroid plexus possesses a fairly large surface area despite its relatively small weight. The mature choroid plexus is typically less than 5% of brain weight (Cserr et al., 1980). For the adult human brain weighing 1.3 to 1.5 kg, the choroid plexus weighs about 65 to 75 g. The rat lateral ventricle choroid plexus weighs typically ~3 mg. Although the tissue weight is relatively small, the total apical surface area of the choroidal epithelium approximates 75 cm^2, about one half that of the blood–brain barrier (155 cm^2) (Keep and Jones, 1990). A recent study by Speake and Brown (2004) has estimated that the membrane area of a choroidal epithelial cell is at least ten times greater than what had been estimated by histological studies. This translates into a total surface area of the blood–CSF barrier that is within the same order of magnitude as that of the blood–brain barrier.

The surface area of the choroid plexus can be further increased by the unique structure of the microvilli. As mentioned earlier, the choroidal epithelial cells possess the apical microvilli and multiple basolateral infoldings (Johanson, 1995). This feature shall be expected with respect to its physiological role in the secretion of CSF and in the delivery of substances (e.g., drug molecules) to the cerebral compartment. From a pharmacological and toxicological point of view, the extended surface area increases the chances of the tissue being exposed to drugs or toxicants from either side of the barrier.

Second, unlike the endothelial cells in the blood–brain barrier, the endothelia in the choroid plexus are extremely fenestrated and quite leaky, possibly lacking tight junctions between adjacent cells. Large molecules such as proteins can readily pass from the blood through the fenestrated capillary and then into the connective tissue. Yet, most of these materials are prevented from further entering the CSF by the tight junctions between the epithelial cells. To produce the CSF and nurture the brain, the fenestrated phenotype of choroidal endothelial cells allows the materials to enter and, subsequently, enrich in the choroidal epithelia; from there the tissue may select the needed substances to be delivered to the CSF. The leaky choroidal endothelia also allows xenobiotics and toxicants to gain access to the choroidal epithelia, some of which ultimately build up in the epithelia, either due to strong protein binding or a weak back-flux to the blood.

Third, in comparison to other brain regions, the choroid plexus has fast blood flow. The blood supply of the choroid plexus is derived from two posterior choroidal arteries, which branch from the internal carotid arteries. On the basis of data from experimental animals, the blood flow rate to the choroid plexus is about 4 to 6 mL/min per gram of tissue (Maktabi et al., 1990), whereas the average blood flow to brain is about 0.9 to 1.8 mL/min per gram of tissue. The rapid blood flow to the choroid plexus warrants an efficient influx of chemicals to the cerebral compartment or the efflux in the opposite direction. Understandably, both the rich blood supply of the choroid plexus and the leaky endothelial layer help increase the chance for the choroidal tissue to be insulted by toxic materials in the blood circulation.

Fourth, the tight junctions between the epithelial cells in the blood–CSF barrier seem less effective, or somewhat "leakier," than those between endothelial cells of the blood–brain barrier (Davson and Segal, 1995; Johanson, 1995). The electrical resistance is the parameter commonly used to reflect the tightness of the barrier in the sense of the barrier's resistance to ionic conductivity. When the cells are grown on a permeable membrane, the cells with barrier characteristics produce tight junctions and form a cellular barrier between the two compartments. The trans-endothelial or epithelial electrical resistance (TEER) can then be determined to estimate the tightness of the barrier. The TEER value for an established endothelial barrier (i.e., the blood–brain barrier) usually exceeds 1,300 ohms-cm^2, whereas the value for an established epithelial barrier is about 80 to 200 ohms-cm^2. Thus, the limited tightness of the choroidal barrier may provide a pathway for materials in the CSF to enter the brush borders of the choroid plexus, serving as a cleansing mechanism in the CNS. Some investigators have referred to the choroid plexus as a "kidney" to the brain (Spector and Johanson, 1989). The real kidneys remove unwanted materials from the blood to urine; the choroid plexus removes the materials from the CSF to the blood, should the CSF be analogously viewed as the "blood" to supply nutrients to neurons and glial cells in the CNS.

Finally, the unique anatomical location of the choroid plexus within the cerebral ventricles may be associated with certain chemical-induced neurotoxicities. For example, the hippocampal formation forms the wall of a certain part of the lateral ventricles. As discussed earlier, the live choroid plexus is loosely floating in the CSF and fills the spaces of all ventricles. The choroid plexus in the lateral ventricle is closely adjacent to the hippocampal formation and other neuronal structures. It remains unknown whether the close contact between the choroid plexus and some brain structures makes it easier for substances derived from the choroid plexus to come into contact with brain tissue and therefore affect certain brain functions. Nonetheless, it will be an interesting subject for future investigation.

Because of these anatomical and physiological characteristics, the choroid plexus is clearly a vital physiological compartment in the brain. This book, the first of its kind exclusively devoted to the blood–CSF barrier, does not attempt to be nor is it suitable for encyclopedic collection of all aspects of advancement in this area. Nonetheless, it is my belief that we must establish between the blood–brain barrier and blood–CSF barrier a cohesive relationship, which will allow, at the minimum, a clearer conception, so that before we can explain brain homeostasis in any meaningful sense, we must first be aware what causes and manages it.

REFERENCES

Cserr, H.F., Bundgaard, M., Ashby, J.K., and Murray, M. (1980). On the anatomic relation of choroid plexus to brain: a comparative study. *Am. J. Physiol.* 238, R76–81.

Davson, H. and Segal, M.B. (Ed.). (1995). *Physiology of the CSF and Blood-Brain Barriers.* CRC Press, New York.

Johanson, C.E. (1995). Ventricles and cerebrospinal fluid. In: *Neuroscience in Medicine,* Conn, P.M. (Ed.). Lippincott, Philadelphia. pp. 171–196.

Keep, R.F. and Jones, H.C. (1990). A morphometric study on the development of the lateral ventricle choroid plexus, choroid plexus capillaries and ventricular ependyma in the rat. *Develop. Brain Res.* 56, 47–53.

Maktabi, M.A., Heistad, D.D., and Faraci, F.M. (1990). Effects of angiotensin II on blood flow to choroid plexus. *Am. J. Physiol.* 258, H414–H418.

Speake, T. and Brown, P.D. (2004). Ion channels in epithelial cells of the choroid plexus isolated from the lateral ventricle of rat brain. *Brain Res.* 1005, 60–66.

Spector, R. and Johanson, C.E. (1989). The mammalian choroid plexus. *Sci. Am.* 261, 68–74.

Strazielle, N. and Ghersi-Egea, J.F., (2000). Choroid plexus in the central nervous system: Biology and physiopathology. *J. Neuropathol. Exp. Neurol.* 59, 561–574.

Zheng, W. (2001). Neurotoxicology of the brain barrier system: New implications. *J. Toxicol. Clin. Toxicol.* 39, 711–719.

Zheng, W., Aschner, M., and Ghersi-Egea, J.F. (2003). Brain barrier systems: A new frontier in metal neurotoxicological research. *Toxicol. Appl. Pharmaco.* 192, 1–11.

Editors

ADAM CHODOBSKI, PH.D.

Dr. Chodobski was born in Warsaw, Poland, and received his master's degree in biomedical engineering from the Polytechnic University in Warsaw. He continued with doctoral studies at the Medical School of Warsaw, where he received his Ph.D. in neuroscience and the Polish Ministry of Health's Award for Outstanding Research in Science and Medicine. Dr. Chodobski then pursued his postdoctoral training at the Howard Florey Institute of Experimental Physiology and Medicine in Melbourne, Australia, before joining the faculty of Brown University School of Medicine in 1991.

Dr. Chodobski is currently an associate professor of clinical neuroscience at Brown University School of Medicine. In 2000, Brown University awarded him an honorary master's degree in recognition of his scientific accomplishments. He is an active member of several societies and serves on the Advisory Committee of the International Workshop on Choroid Plexus. In 1999, he, together with his wife Dr. Joanna Szmydynger-Chodobska, founded the Gordon Research Conference Series on "Barriers of the CNS."

For many years Dr. Chodobski's research interest has been in neuropeptides. His focus has been primarily on vasopressin and its role in the regulation of fluid homeostasis in the central nervous system, with the emphasis on the choroid plexus being a source and target for the polypeptides. Currently, Dr. Chodobski and his group are working on the molecular and cellular mechanisms of cerebral edema and the mediatory role of the choroid plexus-CSF system in brain inflammatory response.

WEI ZHENG, PH.D.

Dr. Zheng is an associate professor of health sciences at Purdue University. He is a member of the Environmental Health Sciences Review Committee of the National Institute of Environmental Health Sciences. He also serves as a member of an Advisory Committee of the International Workshop on Choroid Plexus based in Lyon, France, and as a neurotoxicology consultant to both pharmaceutical industry and law firms for brain bioavailability of neurotoxicants or drugs. He is active in the Society for Neurosciences and the Society of Toxicology. In addition, Dr. Zheng is the chief executive officer of Life Plus, LLC, a toxicology and testing devices company based in Purdue Technology Park in West Lafayette, Indiana.

Raised in the picturesque city of Hangzhou, China, Dr. Zheng received his bachelor's degree in pharmacy and master's degree in pharmacology at Zhejiang University College of Pharmacy in Hangzhou. He completed his doctoral degree in pharmacology and toxicology at the University of Arizona College of Pharmacy in Tucson, Arizona. After his postdoctoral training in 1993, he joined the faculty of the School of Public Health and College of Physicians and Surgeons at Columbia University as an assistant professor and was promoted to associate professor in 2000. He joined the faculty of Purdue University in 2003.

Dr. Zheng is a recognized scholar in the field of neurotoxicology of brain barriers. During his tenure at Columbia and now at Purdue, his research group conducted a pioneer investigation exploring contributions of brain barriers, especially the blood–CSF barrier, in chemical-induced neurodegenerative disorders. His group also developed an immortalized choroidal epithelial cell line for toxicological study and for drug screen. He has published more than 70 original research papers, book chapters, and review articles, and more than 100 abstracts.

Contributors

Susanne Angelow, Ph.D.
Institut für Biochemie
Westfälische Wilhelms-Universität
Münster
Münster, Germany

Milton W. Brightman, Ph.D.
National Institutes of Health
Bethesda, MD

Peter D. Brown, Ph.D.
School of Biological Sciences
University of Manchester
Manchester, UK

Shushovan Chakrabortty, Ph.D.
Center for Interventional Pain Medicine
University of Michigan
Ann Arbor, MI

Ruo Li Chen
Institute of Gerontology and Centre for
Neuroscience Research
King's College London
London, UK

Adam Chodobski, Ph.D.
Department of Clinical Neurosciences
Brown University School of Medicine
Providence, RI

Marek Czosnyka, Ph.D.
Department of Academic Neurosurgery
University of Cambridge
Cambridge, UK

Sarah L. Davies
School of Biological Sciences
University of Manchester
Manchester, UK

John E. Donahue
Department of Pathology
Rhode Island Hospital
Providence, RI

John A. Duncan, III, M.D., Ph.D.
Department of Clinical Neurosciences
Brown University School of Medicine
Providence, RI

Katarzyna M. Dziegielewska, Ph.D.
Department of Pharmacology
University of Melbourne
Parkville, Australia

Carl Joakim Ek, Ph.D.
Department of Pharmacology
University of Melbourne
Parkville, Australia

Steven R. Ennis, Ph.D.
Department of Neurosurgery
University of Michigan
Ann Arbor, MI

Hans-Joachim Galla, Ph.D.
Institute of Biochemistry
Westfalische Wilhelms Universitat
Münster, Germany

Jean-François Ghersi-Egea, Ph.D.
INSERM U433
Faculté de Médecine Laennec
Lyon, France

Julie E. Gibbs, Ph.D.
Center for Neurosciences
King's College London
London, UK

Miles Herkenham, Ph.D.
Section on Functional Neuroanatomy
National Institute of Mental Health
National Institutes of Health
Bethesda, MD

N. Higgins, Ph.D.
Department of Academic Neurosurgery
University of Cambridge
Cambridge, UK

Chizuka Ide, M.D.
Department of Anatomy and Neurobiology
Kyoto University Graduate School of Medicine
Kyoto, Japan

Yutaka Itokazu, M.D.
Department of Anatomy and Neurobiology
Kyoto University Graduate School of Medicine
Kyoto, Japan

Conrad E. Johanson, Ph.D.
Department of Clinical Neurosciences
Brown University School of Medicine
Providence, RI

Richard F. Keep, Ph.D.
Department of Neurosurgery and Molecular and Integrative Physiology
University of Michigan
Ann Arbor, MI

Kazushi Kimura, Ph.D.
Department of Anatomy
Mie University Faculty of Medicine
Tsu City, Japan

Masaaki Kitada, Ph.D.
Department of Anatomy and Neurobiology
Kyoto University Graduate School of Medicine
Kyoto, Japan

G. Jane Li, Ph.D.
School of Health Sciences
Purdue University
West Lafayette, IN

Simon Lowes, Ph.D.
Department of Drug Metabolism and Pharmacokinetics
AstraZeneca Pharmaceuticals
Cheshire, UK

Naoya Matsumoto, M.D., Ph.D.
Department of Anatomy and Neurobiology
Kyoto University Graduate School of Medicine
Kyoto, Japan

Ian D. Millar
School of Biological Sciences
University of Manchester
Manchester, UK

David S. Miller, Ph.D.
Laboratory of Pharmacology and Chemistry
National Institute of Environmental Health Sciences
National Institutes of Health
Research Triangle Park, NC

S. Momjian, Ph.D.
Department of Academic Neurosurgery
University of Cambridge
Cambridge, UK

B. Owler, Ph.D.
Department of Academic Neurosurgery
University of Cambridge
Cambridge, UK

Alonso Pena
Department of Academic Neurosurgery
University of Cambridge
Cambridge, UK

Carol K. Petito, M.D.
Department of Pathology
University of Miami School of Medicine
Miami, FL

John D. Pickard, M.D., Ph.D.
Department of Academic Neurosurgery
University of Cambridge
Cambridge, UK

Lise Prescott, M.D.
Centre Hospitalier de Verdun
Verdun, Québec, Canada

Jane E. Preston, Ph.D.
Institute of Gerontology and Centre for Neuroscience Research
King's College London
London, UK

John B. Pritchard, Ph.D.
Laboratory of Pharmacology and Chemistry
National Institute of Environmental Health Sciences
National Institutes of Health
Research Triangle Park, NC

Zoran B. Redzic, M.D., Ph.D.
Department of Pharmacology
University of Cambridge
Cambridge, UK

Samantha J. Richardson, Ph.D.
Department of Biochemistry and Molecular Biology
University of Melbourne
Parkville, Australia

Norman R. Saunders, Ph.D.
Department of Pharmacology
University of Melbourne
Parkville, Australia

Malcolm B. Segal, Ph.D.
Centre for Neuroscience Research
GKT School of Biomedical Sciences
King's College London
London, UK

Gerald D. Silverberg, M.D.
Department of Neurosurgery
Stanford University School of Medicine
Stanford, CA

Tracey Speake, Ph.D.
School of Biological Sciences
University of Manchester
Manchester, UK

Edward G. Stopa, M.D.
Department of Pathology
Brown University School of Medicine
Providence, RI

Nathalie Strazielle, Ph.D.
Research and Development in
Neuropharmacology
Faculté de Médecine Laennec
Lyon, France

Joanna Szmydynger-Chodobska, Ph.D.
Department of Clinical Neurosciences
Brown University School of Medicine
Providence, RI

Sarah A. Thomas, Ph.D.
Center for Neurosciences
King's College London
London, UK

Charles E. Weaver, M.D., Ph.D.
Department of Clinical Neurosciences
Brown University School of Medicine
Providence, RI

Michael R. Wilson, Ph.D.
Department of Academic Anaesthetics
Imperial College School of Medicine
Chelsea and Westminster Hospital
London, UK

Jianming Xiang, Ph.D.
Department of Neurosurgery
University of Michigan
Ann Arbor, MI

Wei Zheng, Ph.D.
School of Health Sciences
Purdue University
West Lafayette, IN

Contents

SECTION THREE Blood–CSF Barrier in Diseases

SECTION FOUR Current In Vivo *and* In Vitro *Models of the Blood–CSF Barrier*

Section One

Ontogenesis and Morphology of the Choroid Plexus

1 Development of the Blood–Cerebrospinal Fluid Barrier

Carl Joakim Ek, Katarzyna M. Dziegielewska, and Norman R. Saunders
University of Melbourne
Parkville, Australia

CONTENTS

1.1 INTRODUCTION

The choroid plexus is a richly vascularized tissue situated in the roof of each of the four brain ventricles. In the adult it is specialized for secretion, as one of its main functions is to produce cerebrospinal fluid (CSF); but in order to maintain homeostasis of the brain environment it is also a barrier interface between blood and CSF. One of the earliest uses of the term "blood–brain barrier" was that of Stern and Gautier (1922). They proposed the term to describe the mechanisms they considered would be necessary to maintain the physiological integrity of the elements of the nervous system. They argued that the normal functioning of the nervous system would require multiple control mechanisms, but were careful to point out that the term blood–brain barrier (*Barrière hématoencéphalique*) did not prejudge either the morphological structure or mode of action of barrier mechanisms. The choroid plexuses were only mentioned in passing as one of the structures likely to be involved in overall brain barrier mechanisms. In more modern times, the choroid plexus is often more specifically referred to as the blood–CSF barrier (see Figure 1.1). It is becoming increasingly apparent that this distinction is particularly important in the developing brain. The function of the choroid plexus in development might be quite different from that of the adult brain given that both the size of the choroid plexus and the size of the CSF-filled brain ventricles in relation to brain size are much larger in development than in the adult (Johanson, 1995). A primitive ventricular system is first formed when the neural tube closes and amniotic fluid is trapped within the central canal. This closure is associated with a subsequent rise in the intraventricular fluid pressure and increase in CSF protein concentration and coincides with a start of rapid brain enlargement. Desmond (1985) and Desmond and Jacobson (1977) showed that the ventricular fluid might provide an essential pressure force for normal brain expansion and morphogenesis. These two studies showed that the drainage of the embryonic ventricular system in the chick, which lowers the CSF pressure, resulted in smaller tissue volume in all parts of the CNS and only about 50% of the normal cell number in the brain.

Numerous studies have shown that the concentration of protein in CSF in the developing brain is very high compared with that in the adult (see Figure 1.2). Studies in species in which it has been possible to obtain CSF samples from very early in development [e.g., sheep (Dziegielewska et al., 1980) and opossum (Knott et al., 1997)] confirm that the concentration actually increases in the early stages after neural tube closure, after reaching a peak around the time when the neocortex first begins to differentiate into the first identifiable layer of its structure. That the pressure and composition of the ventricular fluid is important for brain development raises interesting questions about developmentally regulated changes at both the blood–CSF and CSF–brain interfaces, since these two interfaces control the composition of this fluid. The choroid plexus differentiation has not started yet in any of the ventricles at the time of neural tube closure, so some other tissue must form the fluid inside the central canal (Catala, 1998).

The two structural components of the choroid plexus, the central fibrovascular stroma and the overlaying epithelium, have different developmental origin. The stroma arises from the mesenchyme, whereas the epithelial cells derive from the neuroepithelium lining the ventricles. Wilting and Christ (1989) showed, using grafted chick tissue, that prospective epithelial cells are destined to form a choroid plexus before plexus morphogenesis starts and that choroid plexus differentiation is not due to induction by the underlying vasculature. The choroid plexus appears early in brain development in all mammalian species studied and is believed to start producing CSF at an early stage. It is therefore likely to be important in regulating the environment and the nutritional needs for the growing brain, especially since there are few blood vessels in the brain at these early stages in development (see Figure 1.3).

1.2 MORPHOLOGICAL DEVELOPMENT OF THE CHOROID PLEXUS

Choroid plexus tissue can be found in each of the four brain ventricles, the two lateral ventricles, the 3rd ventricle, and the 4th ventricle. The different choroid plexuses in young and adult animals have been shown to have similar structural appearances (Tennyson, 1971; Davis et al., 1973; Gomez and Potts, 1981). This may

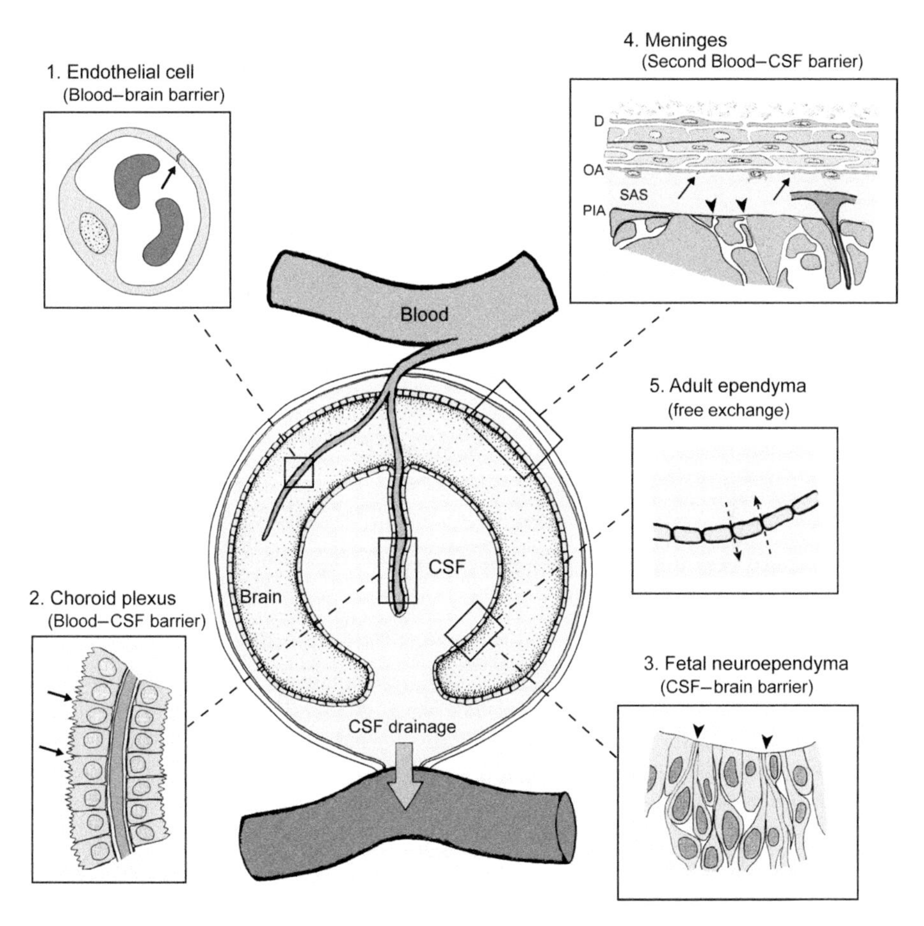

FIGURE 1.1 The brain barriers in the developing and adult animal. The central drawing is a schematic diagram of the three main compartments within the central nervous system (blood, CSF, and brain). (1) The blood–brain barrier, which is located between the lumen of cerebral blood vessels and the brain parenchyma (arrow points to the location of tight junction). (2) The blood–CSF barrier, which is situated between the lumen of choroidal blood vessels and the CSF (arrows point to the tight junctions between epithelial cells). (3) The CSF–brain barrier, which is situated at the inner ventricular surface and the outer pia-glia limitans. This has only been shown to be a functional barrier in the early developing brain (Fossan et al., 1985) where the neuroependymal cells are connected to each other by strap-junctions (arrowheads). (4) The meningeal barrier, which is located between the CSF filled subarachnoid space (SAS) and the overlaying blood circulation. The cells in the outer surface of the arachnoid have tight junctions that are believed to form this barrier (arrows). The dura mater (D) has fenestrated blood vessels which are leaky, whereas the blood vessels in the arachnoid and on the pial surface have functional barrier characteristics similar to the blood vessels inside the brain. Similar to the ventricular CSF–brain interface, a functional barrier seems to exist on the outer pial surface of the brain (arrowheads) during early development but disappears later (OA = outer arachnoid; PIA = pia arachnoid). (5) The mature ventricular surface is made up of a flat layer of ependymal cells (connected to each other by gap junctions), which do not seem to impede the exchange of molecules between CSF and brain. (Adapted from Ek, 2002.)

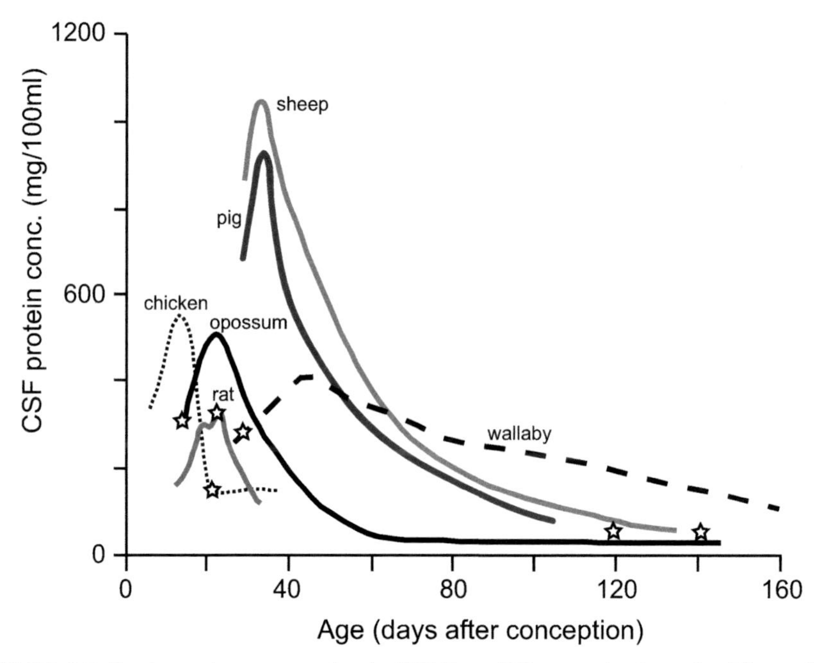

FIGURE 1.2 Total protein concentration in CSF from different animal species. The peak of protein concentration occurs early in development in all animal species and is not related to the time of birth (stars). For instance, in the opossum and tamar wallaby the protein peak is after birth, in the rat it is just around birth, and in the sheep and pig it is well before birth. (Modified from Saunders, 1992.)

be the reason why the choroid plexus is often referred to as one entity and the results from studies of one choroid plexus are often generalized for all choroid plexuses. Some authors do not even mention which choroid plexus has been used in their studies, and since the timing of choroid plexus development in each ventricle is different, this can be of significance in developmental studies. Most commonly, the telencephalic (lateral ventricular) plexuses are studied and to a lesser extent the myencephalic (4th ventricular) plexus, whereas the diencephalic (3rd ventricular) plexus is rarely used in studies, possibly because it is much smaller than the other plexuses. The different choroid plexuses appear during development in the same order in each mammalian species studied: first the myelencephalic choroid plexus, then the telencephalic plexuses, and last the diencephalic plexus. Data from several studies concerning the appearance of the different choroid plexuses in mammalian species and chick are summarized in Table 1.1. We have used our own material to confirm the data from rat and mouse.

Although the timing of appearance and the rate of maturation are different for each choroid plexus, all choroid plexuses appear to go through similar morphological changes during development. The choroid plexus forms as an invagination of the neuroependyma, and at the same time the neuroependymal cells differentiate into more specialized epithelial cells. Kappers (1958) distinguished three stages in the

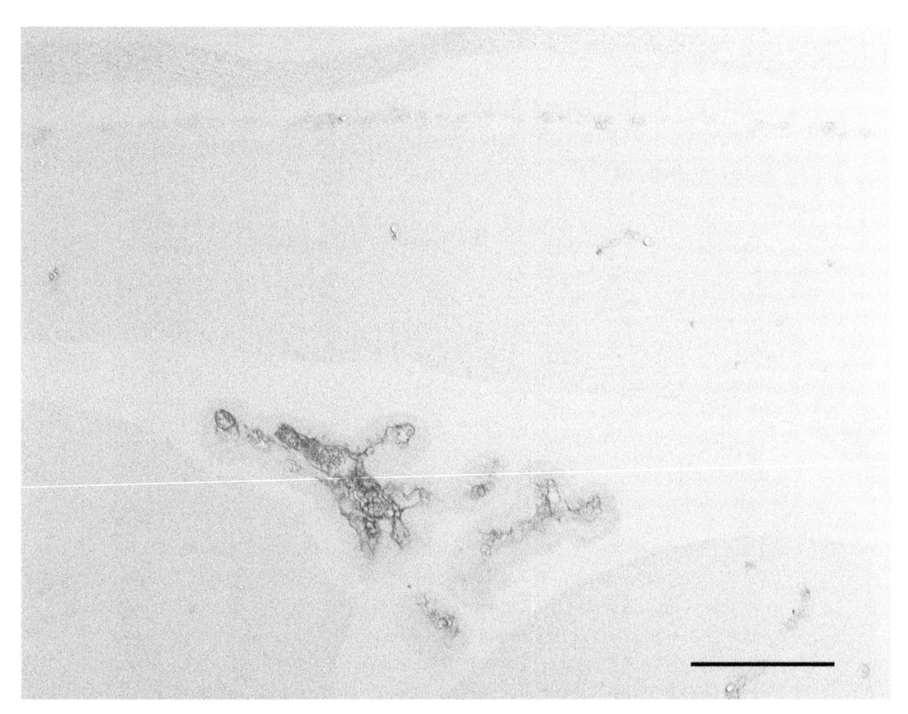

FIGURE 1.3 This micrograph illustrates the high vascularity of the choroid plexus compared to the neocortex in early development. Section is from a P7 opossum and has been stained with antibodies against plasma proteins. (Scale bar is 100 μm.)

development of the choroid plexus based on the maturation of the epithelial cells. However, Shuangshoti and Netsky (1966) and Tennyson and Pappas (1968) later characterized four stages, the difference in classification being that Shuangshoti and Netsky (1966) and Tennyson and Pappas (1968) divided the last stage (stage III) into two separate stages based on timing when the epithelium is poor or rich in glycogen and also on reposition of the nucleus. In the classical three-stage maturation process, the epithelial cells form a pseudostratified epithelium in stage I, in stage II the epithelial cells form a simple high columnar epithelium, and in stage III the epithelial cells flatten and widen forming a squamous epithelium (see Figure 1.4). The ultrastructure of the choroid plexus in development has been examined by transmission electron microscopy (TEM) in several mammalian species, including the rat (Cancilla et al., 1966; Keep et al., 1986; Keep and Jones, 1990), mouse (Sturrock, 1979; Zaki, 1981), rabbit (Tennyson and Pappas, 1964; Tennyson and Pappas, 1968; Tennyson, 1971), opossum (Ek et al., 2003), and human (Shuangshoti and Netsky, 1966; Tennyson, 1971). The fine structure of the choroid plexus appears to go through similar structural changes in all these species. Structural changes associated with this maturation of the epithelial cells from stage I to stage III include an increase in the mitochondrial content, formation and elongation of villi and cilia on the apical membrane, increase in basal infoldings of epithelial cells, the endoplasmic reticulum changing from tubular cisternae running in all directions to

TABLE 1.1
Timing of the Appearance of Different Choroid Plexuses in Different Species During Brain Development

Species	MCP	TCP	DCP	Gestation (days)
Mouse[3,4]	E11	E11	E14.5	21
Rat	E13	E13	E15–E16	21
Sheep[5]	E21	E24	E24–E33	150
Rabbit[6]	E14	E14	—	30–32
Chick[1,2]	<E5	E5	E6	23
Opossum[9]	<P0	<P0	P2–P3	14
Dunnart	<P0	P1	P3	10–11[10]
Human[7,8]	Stage 18–19	Stage 20	Stage 21	40 weeks

Data from [1]el-Gammal (1981), [2]el-Gammal (1983), [3]Sturrock (1979), [4]Zaki (1981), [5]Jacobsen et al. (1983), [6]Tennyson and Pappas (1964), [7]O'Rahilly and Müller (1990), [8]Otani and Tanaka (1988), [9]Ek et al. (2003), [10]Selwood and Woolley (1991). According to Butler and Juurlink (1987), stage 18–19 corresponds to E45–E48, stage 20 to E50, and stage 21 to E54 in human development. MCP, Myencephalic choroid plexus; TCP, Telencephalic choroid plexus; DCP, Diencephalic choroid plexus.

flattened parallel cisternae, and apical-luminal polarization of the cells, along with structural changes of the nucleus (see Figure 1.4). These changes are probably a reflection of progressive specialization of the choroid plexus as a highly secreting epithelium during development. Several studies have shown that CSF production increases during development (Bass and Lundborg, 1973; Evans et al., 1974; Johanson and Woodbury, 1974). This is not due just to an increase in overall mass of the choroid plexus, but also to an increase in production per weight of choroid plexus tissue (Johanson and Woodbury, 1974). That one of the main functions of the choroid plexus is to produce CSF is today unquestioned. During development there is an increase in mitochondrial content, mainly as a result of an increase in the number of mitochondria, but also due to an increase in the size of individual mitochondria. The increase in mitochondrial content of the epithelial cells is probably a reflection of an increase in CSF production of the choroid plexus, which is a secretory process and therefore is dependent on energy. The production of CSF is also dependent on fluid transfer between the blood vessels and into the epithelial cells. The increase in basal infolding of the epithelial cells may be a reflection of this increase in fluid transfer. As well as producing CSF, the choroid plexus is also responsible for a constant regulation of the composition of the CSF. Several studies have shown that the regulation and transfer of ions by the choroid plexus rise during development (Jones and Keep, 1987; Parmelee and Johanson, 1989; Parmelee et al., 1991; Preston et al., 1993). The formation of microvilli and cilia on the apical surface is an indication of an increase in such exchange mechanisms between the choroidal

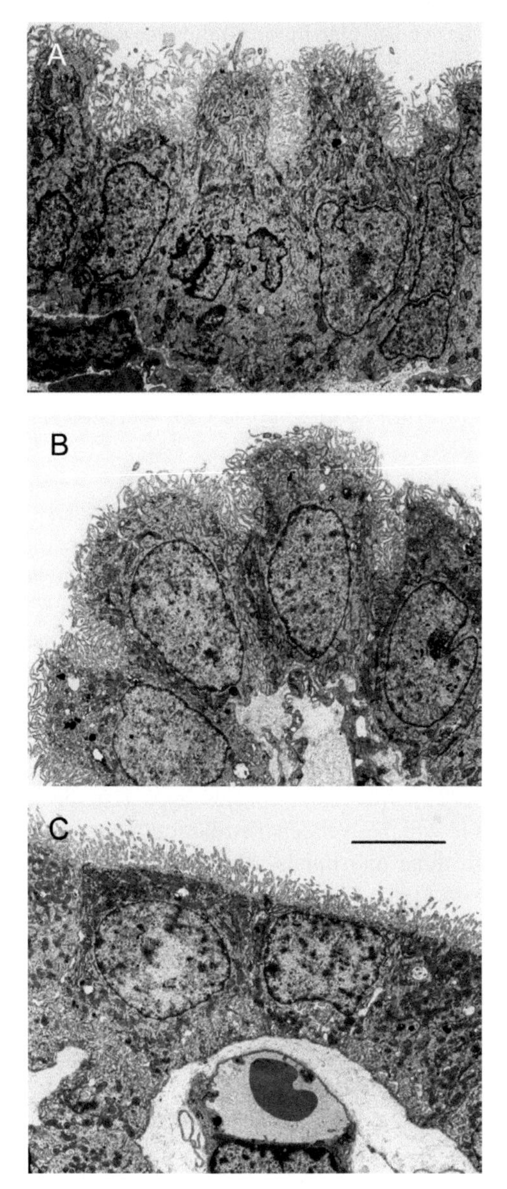

FIGURE 1.4 Electron micrographs showing the structural changes occurring between stages I and III in choroid plexus development in the opossum. The epithelium changes from a pseudostratified epithelium with tall and thin epithelial cells (A), to a columnar epithelium (B), and finally to a squamous epithelium (C). Other changes in the epithelial cells are an increase in mitochondrial content and microvilli, nucleus change from lobulated to spherical form, and a higher density of organelles toward the apical (CSF) surface. (Scale bar is 5 μm.) (From Ek et al., 2003. *Journal of Comparative Neurology* © 2003 Wiley-Liss.)

epithelial cells and the ventricular fluid. Studies have also shown a distinct change in the structure of the nucleus of the epithelial cells with age from being irregular in shape with large invaginations to being almost spherical with a smooth membrane (Dohrmann and Herdson, 1969; Ek et al., 2003). The lobulated nucleus along with a high abundance of rough endoplasmic reticulum and polyribosomes in the interlobulated cytoplasm is an indication of greater interaction between the nucleus and cytoplasm in younger animals. The main change in the core of the plexus is that the extracellular matrix becomes thicker and richer in connective tissue.

Added confirmation that the changes in the ultrastructure of the choroid plexus are a reflection of its functional development comes from studies of the aging choroid plexus. Some morphological and functional changes in development appear to be reversed in aging rats; for instance, there is a decrease in the number of microvilli in older animals while at the same time the CSF secretion rate is reduced (Serot et al., 2001; Preston, 2001).

The choroidal epithelial cells in most species have a stage during which they store large amounts of glycogen. The glycogen appears early in the epithelial cells (around E13 in mice and E15 in rats), and the amount increases until birth but is lost rapidly afterward. At E21 (birth) in rats, glycogen has been shown to occupy about 40% of the volume in the epithelial cells of the telencephalic plexus and 18% in the myelencephalic plexus (Keep et al., 1986; Keep and Jones, 1990). The favored hypothesis is that this glycogen provides an energy source for the growing brain by increasing the amount of glucose either in the CSF or in the choroidal capillaries. Alternatively, it has been suggested that the glycogen is involved in the formation of glycosaminoglycans of the choroidal basement membrane (Kappers, 1958; Sturrock, 1979; Zaki, 1981). Marsupial species appear to be different from all the eutherian mammals in which glycogen distribution in the developing choroid plexus has been studied. In three marsupials—the short-tailed South American opossum (*Monodelphis domestica*), the North American opossum (*Didelphis virginiana*), and the stripe-faced dunnart (*Sminthopsis macroura*)—we have not been able to find such a stage of glycogen storage in the epithelial cells. This may be because the brain in marsupials develops under different oxygen tension (almost all choroid plexus development occurs after birth) than in eutherian mammals, and thus anaerobic metabolism of glycogen may be less important in species born at such an immature stage of brain development.

So are there ultrastructural differences in the different choroid plexuses during development? Few studies have examined the different choroid plexuses in a comparative way during development using the electron microscope, and especially the diencephalic plexus seems to have been omitted in most studies. Keep, Jones, and Cawkwell (1986) and Keep and Jones (1990) made quantitative measurements of some morphological features of the developing telencephalic and myelencephalic ventricular choroid plexuses in the rat. They stated that the morphological development of the two choroid plexuses appeared similar, although the timing of morphogenesis was different. Tennyson (1971) studied the same two choroid plexuses of the rabbit and at some stages of human development. That study reported some ultrastructural differences between the two choroid plexuses, with the most noticeable being that the glycogen-rich stage of the epithelial cells was much more prominent in the telencephalic plexus than the myelencephalic plexus, similar to what was found in the rat (Keep et al., 1986; Keep and Jones, 1990). Our studies in the opossum that used choroid plexus tissue from all four brain ventricles did not reveal any significant morphological differences between the different plexuses when the different timing of development of the plexuses was taken into count (Ek et al., 2003). The only other comparative ultrastructural studies of the developing choroid plexuses were presented by el-Gammal (1981, 1983) who studied the surface changes of the lateral and third ventricular choroid plexus in embryonic chicks using the scanning electron microscope. These studies showed that the morphogenesis of telencephalic plexuses was distinct in regard to proximal to distal differentiation, whereas the cells in the diencephalic plexus appeared to differentiate from several unrelated sites within the choroid plexus. To our knowledge this distinction between the two plexuses has never been reported for any mammalian species.

Not only does it seem that each individual choroid plexus goes through similar developmental stages, but cells within each choroid plexus also go through a similar developmental process. The choroid plexuses show distal to proximal differentiation of the epithelial cells in that the cells at the root of the choroid plexus are less mature than the ones in the more distal parts of the plexus villi (see Figure 1.5). This is in

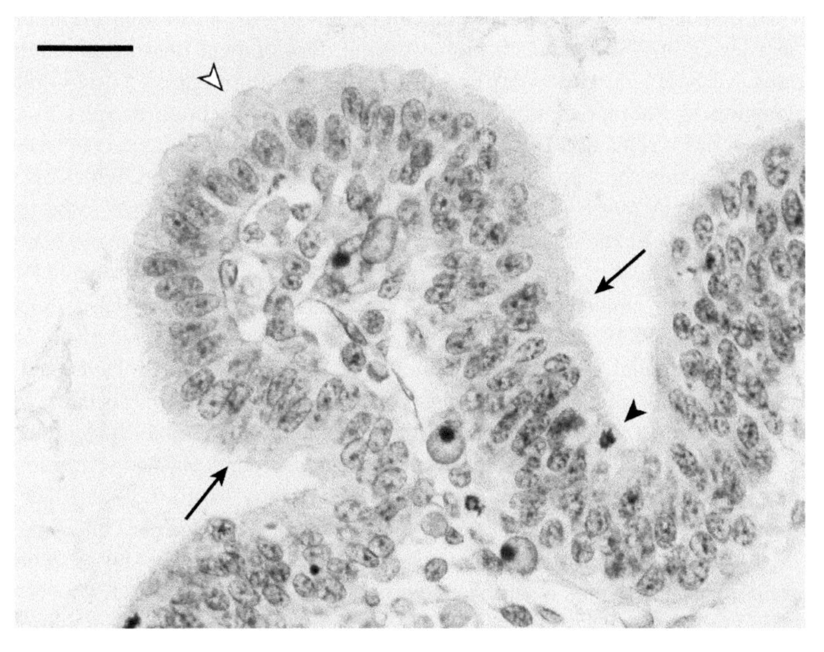

FIGURE 1.5 Micrograph of the telencephalic choroid plexus just after its first appearance in a stripe-faced dunnart (*Sminthopsis macroura*). Note that the epithelial cells in the root of the plexus have a pseudostratified appearance (arrows) whereas the cells in the most distal parts have started to differentiate into a columnar epithelium (unfilled arrowhead). The distal cells are the oldest epithelial cells since the choroid plexus grows from the root where mitotic figure can be found (one can be seen at the right hand side of the root, filled arrowhead). Scale bar is 20 μm.

agreement with the belief that the choroid plexus grows from the proximal parts (most mitotic figures are found in the root) where the cells show a pseudostratified appearance until later in development (Knudsen, 1964; Tennyson and Pappas, 1964; Ek et al., 2003).

1.3 THE BLOOD–CSF BARRIER IN DEVELOPMENT

The choroidal blood–CSF barrier in the developing brain has probably been characterized better than any other of the blood-CNS interfaces. The choroid plexus is one of the circumventricular organs and is in many ways specialized for transfer of blood solutes and fluid. The tissue is rich in blood vessels that have fenestrations, and even large tracers seem to extravasate into the perivascular space (as opposed to cerebral vessels, which are tight). However, the epithelial cells have complex tight junctions in between them at their apical side that are believed to form the structural basis for the choroidal blood–CSF barrier. These junctions have been shown to stop the movement of many tracers in adults (Becker et al., 1967; Brightman and Reese, 1969; Milhorat et al., 1973). There is a general belief that the brain barriers are not present, or at least to a lesser extent functional, in the developing animal, although this belief has been repeatedly challenged. This belief seems to stem from the finding that the transfer of water-soluble molecules from blood to CSF and brain and the protein levels in CSF are much higher during development than in the adult (see Figures 1.2 and 1.7). However, there is today much information to show that the explanation for the apparently higher permeability of the brain barriers in young animals is not a reflection of poorly developed barrier (tight junction) functions.

Tight junctions between the epithelial cells of the choroid plexus, which resemble the *zonulae occludentes* in the brain endothelium, have been shown to be present early during development in mice (Zaki, 1981), rats (Dermietzel et al., 1977), sheep (Møllgård and Saunders, 1975; Møllgård et al., 1979), rabbits (Tennyson and Pappas, 1968), humans (Tennyson and Pappas, 1968; Bohr and Møllgård, 1974; Saunders and Møllgård, 1984), chicks (Delorme, 1972), and opossums (Ek et al., 2003). Changes in epithelial tight junction complexity during development have been studied in the rat by Tauc et al. (1984) using TEM. They showed that as early as E14 the junctions are structurally similar to those in the mature animal, forming continuous belts around the epithelial cells. Freeze-fracture studies on sheep embryos have shown that complex tight junctions are present as early as the E30 and that there are only minor changes even as late in fetal life as E125 (Møllgård and Saunders, 1975; Møllgård et al., 1976; Møllgård et al., 1987). There was no evidence for discontinuous strands or a change in the proportion of complex strands between E30 and E125 (sheep are born around E150) (Møllgård et al., 1979). An even more direct way to study the functional state of the tight junctions of the epithelial cells is to use tracers that are administered systemically and to investigate whether they can penetrate the epithelial tight junctions. The most commonly used tracer is horseradish peroxidase (HRP), a 40,000 Da protein that can be easily visualized for both light and electron microscopy. Tauc et al. (1984) administered HRP intravenously and found that it was stopped at the epithelial tight junctions at least as early as E14 in the rat. Ek et al. (2003) used a much smaller molecule, a 3000 Da biotin-dextran, and showed that the tight junctions impede the paracellular movement of this tracer already at birth in the opossum (comparable to an E14 rat embryo in regard to cortical and choroid plexus development). Studies in this species have suggested that there might be a transcellular pathway through the epithelial cells that decreases during development (Ek et al., 2001; 2003). The blood–CSF barrier of the chick has also been extensively studied during development. Wakai and Hirokawa (1981) reported that a proper functional barrier to HRP is established at E10, and a similar study by Bertossi et al. (1988) showed comparable results, except that the timing of the apparent junction tightening was slightly later. This would suggest that the avian choroidal barrier is functional somewhat later in development than the mammalian barrier. However, some methodological problems with this work, especially that of Wakai and Hirokawa (1981), have been identified (Saunders 1992). In some systems HRP has given anomalous results exhibiting permeability across epithelial interface where smaller tracers have not; this was attributed to a possible damage on cell membranes by enzyme activity of HRP (Mazariegos et al., 1984).

The concentration of proteins in CSF reaches a peak early in development; thereafter it gradually decreases with age (see Figure 1.2). This generally occurs during fetal development in eutherian mammals, reaching a total protein concentration that is 10 to 40 times the adult level. This has been interpreted by some (e.g., Adinolfi, 1985) as a reflection of immature barriers, but with further studies it has become clear that this is probably not the case (Dziegielewska et al., 1991; Habgood et al., 1992). If the reduced permeability of proteins with age were a reflection of tight junction development between the epithelial cells, the structure of the tight junctions would be expected to change. The reported minor changes in tight junction complexity during development (Møllgård et al., 1979) are unlikely to account for the dramatic changes in permeability of proteins that have been reported (Dziegielewska et al., 1980). When endogenous albumin, exogenous albumin from other species, and chemically modified derivates of albumin were injected into sheep fetuses, native unmodified protein reached the highest concentrations in the CSF. Similar studies in neonatal rats (Habgood et al., 1992) and opossums (Knott et al., 1997) produced comparable results (see Figure 1.6). The albumins were identified immunocytochemically within choroid plexus epithelial cells early in development. This included one study of albumin at the EM level (Balslev et al., 1997). Thus, a possible transporting system seemed to be able to distinguish between different kinds of albumin. It was also found that the greatest number of choroidal epithelial cells containing albumin coincided with the peak in CSF protein concentration in opossums (see Figure 1.7). The authors suggested that the route for albumin is intracellular through the epithelial cells of the choroid plexus in a process that can distinguish between native and modified albumins. This selective transporting mechanism seems to be present only during early development, since in the more mature brain there was no difference in the penetration between different species of albumin (Knott et al., 1997). Balslev et al. (1997) showed using EM immunocytochemistry that albumin can be found in the tubulo-cisternal endoplasmic reticulum (TER) of the choroidal epithelial cells after administering it into the blood in sheep. They found no albumin passing the interepithelial cleft and proposed that the TER is the main route of transfer for albumin from blood into CSF in the immature animal.

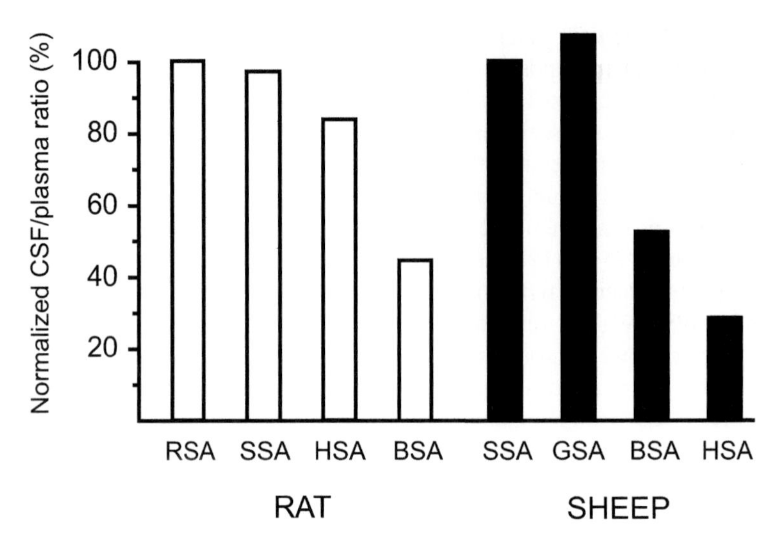

FIGURE 1.6 Steady-state CSF/plasma ratios for different albumin species in P2 rat and E60 sheep normalized to the naturally occurring albumin ratios. Note that in the rat sheep albumin (SSA) appear to be transported into the CSF at the same rate as rat's own albumin (RSA), whereas both human (HSA) and bovine (BSA) albumins are discriminated against. (From Habgood et al., 1992.) In the E60 sheep, goat albumin (GSA) reaches similar levels in CSF to sheep's own albumin while bovine and human albumin are discriminated against. (Adapted from Dziegielewska et al., 1991.)

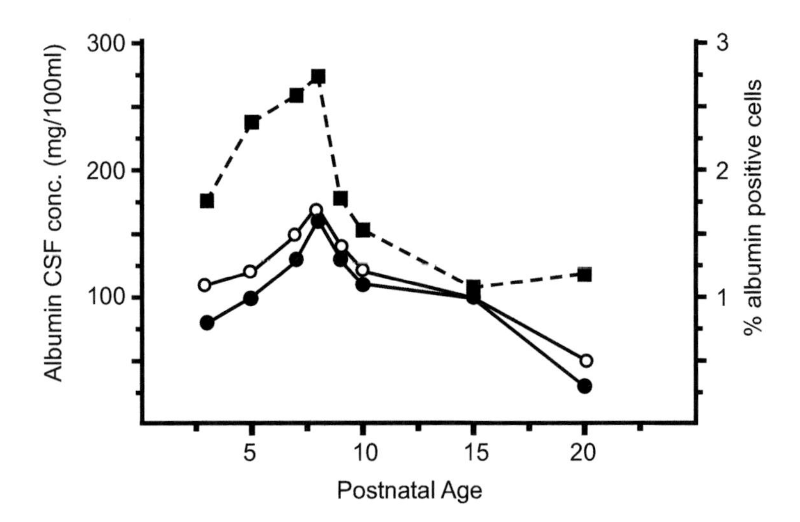

FIGURE 1.7 Relation between concentration of albumin in the CSF of opossum (left y-axis) and proportion of choroidal epithelial cells that are immunoreactive for the protein (right y-axis). Filled squares (■) are concentration of albumin in mg/100ml, open circles (❍) are % of albumin-positive cells in the myencephalic plexus, and filled circles (●) are % of albumin-positive cells in the telencephalic plexus. (Adapted from Knott et al., 1997.)

1.4 CORRELATING TIGHT JUNCTIONAL STRUCTURE WITH PERMEABILITY OF BLOOD–CSF BARRIER

Claude and Goodenough (1973) made a quantitive study of structural features of tight junctions (e.g., strand number and depth) in freeze-fracture replicas of different epithelia. They compared this with results from measurements of transepithelial resistance published by others and showed that there seemed to be a trend that epithelia with more complex tight junctions had a higher resistance, suggesting that it is the complexity of the tight junctions that dictates the tightness of the epithelia as a whole. However, there are important exceptions, and a major problem with this hypothesis is that there is little or no evidence that molecules actually traverse tight junctions. Indeed, as will be reviewed below, there is now evidence that small, lipid-insoluble molecules do not enter the brain and CSF via tight junctions, at least not in the developing brain. Direct measurements of the permeability of the choroidal blood–CSF barrier in development are difficult because of the technical problems inherent in such experiments. It is important to remember that in the adult, blood-borne solutes can reach the CSF (and brain) by several pathways (blood–CSF and blood–brain interfaces), and it is difficult to know which pathways are used by a certain molecule and the quantitative contribution of a particular pathway. Permeability studies from blood into brain or CSF therefore measure the sum of all these pathways and not the permeability of a specific interface. However, in the developing brain the situation may not be the same. A developmentally specific barrier at the CSF–brain interface within the brain and possibly also over the external surface of the brain has been demonstrated that prevents the passage of proteins between the cells of the neuroependyma early in brain development (Fossan et al., 1985; Møllgård et al., 1987). This impermeability appears to be due to a membrane specialization of adjacent neuroependymal cells lining the cerebral ventricles (strap junctions) (Møllgård et al., 1987). This makes it possible to treat the CSF and brain extracellular fluid in the developing brain as effectively separate compartments, as was done by Dziegielewska et al. (1979) with respect to permeability data obtained for a wide range of fetal sheep ages. Although Dziegielewska et al. (1979) obtained some preliminary electron microscopical evidence that the intercellular spaces between neuroependymal cells were narrow, a clear morphological basis for this assumption was not available at the time of these studies. However, the conclusion made by Dziegielewska et al. (1979) that permeability across the blood–CSF barrier could be accounted for by unrestricted (i.e., large pore) diffusion is now well supported by the morphological findings of Fossan et al. (1985) and Møllgård et al. (1987).

In addition, blood–CSF permeability studies, in which proteins and the much smaller BDA 3000 have been measured quantitatively and the route into the brain demonstrated by immunocytochemistry, suggest that the predominant route for these molecules to penetrate into the immature brain is via the choroid plexus and CSF (Ek et al., 2001; 2003). This is perhaps not surprising since the brain is poorly vascularized at the stage when blood–CSF transfer appears high (Figure 1.3). Immunocytochemical examination of the distribution of proteins and BDA-3000 showed that these molecules had an intracellular distribution in the ventricular zone, presumably as a result of uptake from CSF (Ek et al., 2001). This is an important finding for the interpretation of blood–brain barrier permeability studies. It illustrates the importance of combining quantitative in vivo permeability and morphological methods when studying brain barrier mechanisms. In studies of adult brain barrier permeability to small lipid-insoluble molecules such as sucrose and inulin, it has generally been assumed that these molecules are distributed in the extracellular space of the brain. They are of course used to estimate extracellular space volumes elsewhere in the body, but this is not applicable in the brain because the slow penetration and sink effect of CSF secretion result in very low volumes of distribution in the brain that are much smaller than other independent estimates of brain extracellular volume (Davson and Segal, 1969). Nevertheless it has generally been assumed that molecules such as sucrose and inulin have an extracellular distribution in the brain and that brain/plasma ratios can be used as index of blood–brain barrier permeability. This approach has been extended to similar permeability studies in fetal and newborn animals, and it has been assumed that the much larger brain/plasma ratios that are characteristic of immature animals reflect much greater permeability of the blood–brain barrier to these small molecules. This can now be seen to be an inappropriate interpretation. If a small molecule such as BDA-3000, which appears to have similar permeability characteristics to inulin (5000 Da), enters the immature brain via the choroid plexuses and CSF and has a cellular distribution in brain parenchyma as a consequence of uptake from CSF, then brain/plasma ratios clearly do not reflect permeability across the cerebral endothelial interface (blood–brain barrier). The situation in the adult brain is not clear, as similar experiments with BDA-3000 have not yet been done. If it turns out that such small molecules also enter the adult brain via the choroid plexus epithelial cells rather than across the cerebral vessels, this would suggest that the choroid plexuses may have a far greater role to play in overall brain homeostasis than currently thought. However, this is not a new idea; it was suggested by Stern and Gautier (1922) as an important mechanism for the adult brain and by Klosovskii (1963) for the fetal brain many years ago.

In the sheep, rat, and opossum, the steady-state CSF/plasma concentrations of various lipid-insoluble molecules have been shown to decrease significantly during development (see Figure 1.8) (Ferguson and Woodbury, 1969; Dziegielewska et al., 1979; Habgood et al., 1993; Ek et al., 2001). This reduction in steady-state ratios could be explained by an increase in the CSF sink. The continuous production of CSF flushes out solutes in the CSF, an effect that has been named the CSF sink (Oldendorf and Davson, 1967; Davson and Segal, 1969). Although the CSF secretion rate increases dramatically during development in rats and sheep (Johanson and Woodbury, 1974; Evans et al., 1974) when the rate is related to the total CSF volume and calculated as a turnover rate of CSF per minute, it is not much higher in adults compared to young animals (Saunders, 1992). It therefore seems unlikely that the decreases in steady-state ratios can be explained by an increase in CSF sink. Structural investigations of the fetal sheep tight junctions of both cerebral endothelial and choroidal epithelial cells have reported that they closely resemble those of the mature sheep (Møllgård and Saunders, 1975; Møllgård et al., 1979). In the rat, the exact onset of the structural characteristics for the brain barriers is still disputed, but nearly all studies show that the brain barriers are structurally mature at birth (Tauc et al.,

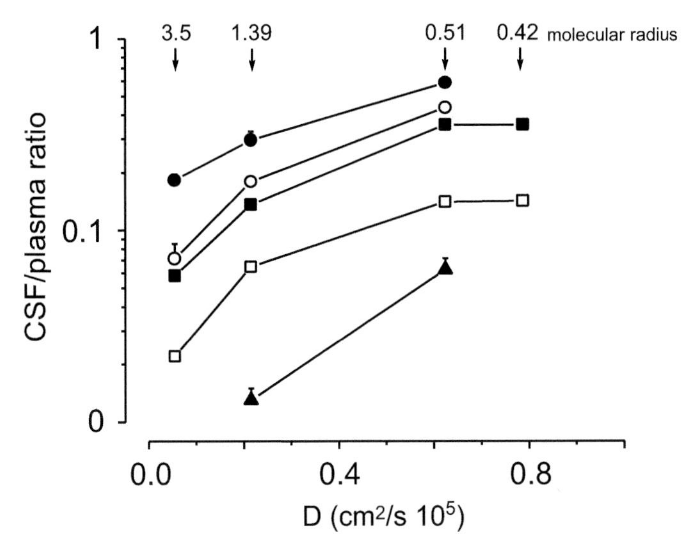

FIGURE 1.8 Steady-state CSF/plasma ratios (y-axis) versus the diffusion coefficients (D) (x-axis) for molecules of different molecular radii (albumin 3.5, inulin 1.39, sucrose 0.51, L-glucose 0.42) in P5–P7 (●), P10–P13 (○), P15–P17 (■), P32–P37 (□), and P65 (▲) opossums, showing an inverse relationship between diffusion coefficient and the CSF/plasma ratio. Note that with increasing age there is a near-parallel decrease in these curves. This implies that the transfer of water-soluble molecules from blood to CSF is by unrestricted diffusion through large aqueous "pores." Furthermore, it suggests that during development there is a decrease in the number of these "pores" rather than a decrease in pore diameter with age, which would alter the slopes of these curves (Habgood et al., 1993). (Adapted from Ek et al., 2001.)

1984; Kniesel et al., 1996). In the opossum, Ek et al. (2003) showed that the epithelial tight junctional structure does not seem to change during development and that these junctions were impermeable to a 3000 Da biotin-dextran (see Figure 1.9). Hence it appears that the apparent decrease in permeability to lipid-insoluble molecules during development from blood to CSF in the rat, sheep, and opossum is not just a reflection of either maturation of the tight junctions or an increase in CSF sink, but is due to some yet-undetermined mechanism.

1.5 SUMMARY AND CONCLUSION

This chapter presents available information on the structure and function of the developing choroid plexuses of the mammalian brain. Where possible, the relation between structure and function in vivo has been stressed. The fundamental basis for all brain barrier mechanism lies in the diffusion restraint provided by tight junctions between the cells that form the various barrier interfaces. These intercellular junctions in effect translate the permeability properties of cell membranes of single cells into similar properties across the extensive structural interfaces illustrated in Figure 1.1. In addition, specialized permeability properties of the tight junctions (often

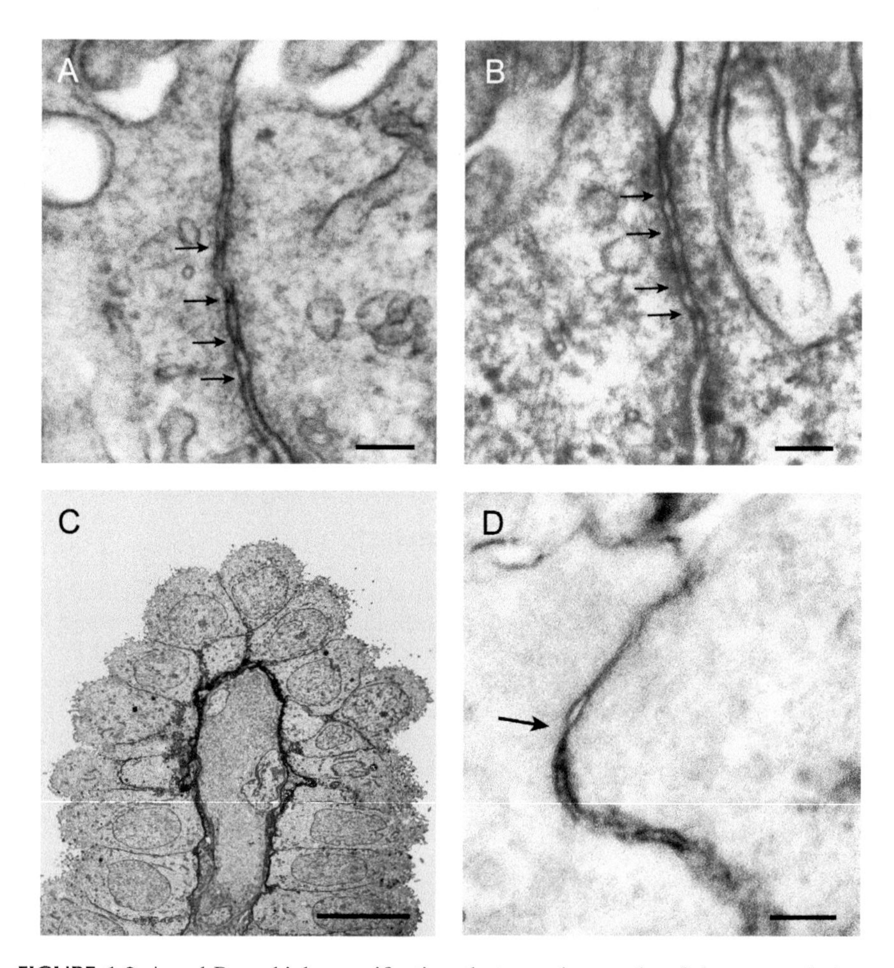

FIGURE 1.9 A and B are high-magnification electron micrographs of the most apical part of the intercellular cleft between two epithelial cells in the opossum at birth (A) and 2 months (B). These sections show that several contact points (arrows) can be seen within the tight junction, which form the structural basis for the choroidal blood–CSF barrier and that the number of these contact points does not seem to change with age. C and D illustrate the distribution of a 3000 Da biotin-dextran (BDA-3000) in the choroid plexus of a P13 (C) and P1 (D) opossum 30 min after an intraperitoneal injection. BDA-3000 is stained with a HRP/Streptavidin complex followed by DAB, which gives an electron-dense reaction product. As seen in C, the reaction product is most abundant in the perivascular space and in between epithelial cells, showing that the tracer readily leaked out of the choroidal blood vessels. However, as seen in D, which shows the most apical part of the interepithelial cleft, the tight junctions appear to stop the movement of the tracer, showing that these junctions are functional at this early stage of development. (Scale bars are 100 nm in A, B, and D, and 10 μm in C.) (From Ek et al., 2003. *Journal of Comparative Neurology* © 2003 Wiley-Liss.)

referred to as the paracellular pathway) themselves have long been thought to be important for some functional characteristics of barrier interfaces, not only in brain barriers but also in epithelial and endothelial interfaces throughout the body. It is widely believed that the paracellular pathway is an important route for the transfer of water, ions, and small lipid-insoluble molecules across the different brain barrier interfaces. Conflicting experimental evidence has generated much discussion about the structural and functional state of tight junctions in cerebral blood vessels and of the epithelial cells of the choroid plexuses in the developing brain. The evidence summarized in this chapter suggests that tight junctions in the blood–CSF barrier of the immature brain are not permeable to either proteins or smaller water-soluble molecules and that these molecules enter the immature brain predominantly via the choroid plexuses and CSF. It is proposed that this route of entry for at least these classes of compound is functionally important at early stages of brain development during the initial vascularization of the immature brain. As the brain becomes vascularized, there is presumably a change in the relative importance of entry from blood into the brain across the blood–CSF and blood–brain barriers, in which the latter becomes quantitatively more important, as is thought to be the case in the adult brain. The contribution of the paracellular pathway to water and ion transfer across barriers in the immature brain is unclear, and it remains to be shown whether the choroid plexuses are also the predominant route of entry for metabolically important molecules, such as glucose and amino acids, in the immature brain. The significance of the high concentration of protein in the CSF of developing brains (see Figure 1.2), together with the presence of a diffusion barrier at the CSF/brain interface (see Figure 1.1) could provide a driving force, due to high colloid osmotic pressure, involved in the expansion of the brain ventricular system necessary for normal brain development. Individual proteins present in the CSF could also have specific or nonspecific nutritive properties and/or provide signals that are important for the division, differentiation, and migration of cells in the ventricular zone, since these cells have a unique, direct contact with the CSF.

1.6 ACKNOWLEDGMENT

Supported by NIH Grant RO1NS043949.

REFERENCES

Adinolfi, M. (1985) The development of the human blood-CSF-brain barrier. *Dev. Med. Child Neurol.* 27: 532–7.

Balslev, Y., Dziegielewska, K.M., Møllgård, K. and Saunders, N.R. (1997) Intercellular barriers to and transcellular transfer of albumin in the fetal sheep brain. *Anat. Embryol.* 195: 229–36.

Bass, N.H. and Lundborg, P. (1973) Postnatal development of bulk flow in the cerebrospinal fluid system of the albino rat: clearance of carboxyl-[^{14}C]inulin after intrathecal infusion. *Brain Res.* 52: 323–32.

Becker, N.H., Novikoff, A.B. and Zimmerman, H.M. (1967) Fine structure observations of the uptake of intravenously injected peroxidase by the rat choroid plexus. *J. Histochem. Cytochem.* 15: 160–5.

Bertossi, M., Ribatti, D., Nico, B., Mancini, L., Lozupone, E. and Roncali, L. (1988) The barrier systems in the choroidal plexuses of the chick embryo studied by means of horseradish peroxidase. *J. Submicrosc. Cytol. Pathol.* 20: 385–95.

Bohr, V. and Møllgård, K. (1974) Tight junctions in human fetal choroid plexus visualised by freeze-etching. *Brain Res.* 81: 314–8.

Brightman, M.W. and Reese, T.S. (1969) Junctions between intimately apposed cell membranes in the vertebrate brain. *J. Cell Biol.* 40: 648–77.

Butler, H. and Juurlink, B.H.J. (1987) *An Atlas for Staging Mammalian and Chick Embryos.* Boca Raton, FL: CRC Press.

Cancilla, P.A., Zimmerman, H.M. and Becker, N.H. (1966) A histochemical and fine structure study of the developing rat choroid plexus. *Acta Neuropathol.* 6: 188–200.

Catala, M. (1998) Embryonic and fetal development of structures associated with the cerebrospinal fluid in man and other species. Part I: The ventricular system, meninges and choroid plexuses. *Arch. Anat. Cytol. Pathol.* 46: 153–69.

Claude, P. and Goodenough, D.A. (1973) Fracture faces of zonulae occludentes from "tight" and "leaky" epithelia. *J. Cell Biol.* 58: 390–400.

Davis, D.A., Lloyd, B.J.J. and Milhorat, T.H. (1973) A comparative ultrastructural study of the choroid plexuses of the immature pig. *Anatom. Rec.* 176: 443–54.

Davson, H. and Segal, M.B. (1969) Effect of cerebrospinal fluid on volume of distribution of extracellular markers. *Brain* 92: 131–6.

Delorme, P. (1972) Ultrastructural differentiation of intercellular junctions in the endothelium of capillaries in the embryonic telencephalon of the chicken. *Z. Zellforsch. Mikrosko. Anat.* 133: 571–82.

Dermietzel, R., Meller, K., Tetzlaff, W. and Waelsch, M. (1977) In vivo and in vitro formation of the junctional complex in choroid epithelium. A freeze-etching study. *Cell Tissue Res.* 181: 427–41.

Desmond, M.E. (1985) Reduced number of brain cells in so-called neural overgrowth. *Anat. Rec.* 212: 195–8.

Desmond, M.E. and Jacobson, A.G. (1977) Embryonic brain enlargement requires cerebrospinal fluid pressure. *Dev. Biol.* 57: 188–98.

Dohrmann, G.J. and Herdson, P.B. (1969) Lobated nuclei in epithelial cells of the choroid plexus of young mice. *J. Ultrastruct. Res.*, 29: 218–23.

Dziegielewska, K.M., Evans, C.A., Fossan, G., Lorscheider, F.L., Malinowska, D.H., Mollgard, K., Reynolds, M.L., Saunders, N.R. and Wilkinson, S. (1980) Proteins in cerebrospinal fluid and plasma of fetal sheep during development. *J. Physiol.* 300: 441–55.

Dziegielewska, K.M., Evans, C.A., Malinowska, D.H., Møllgård, K., Reynolds, J.M., Reynolds, M.L. and Saunders, N.R. (1979) Studies of the development of brain barrier systems to lipid-insoluble molecules in fetal sheep. *J. Physiol.* 292: 207–31.

Dziegielewska, K.M., Habgood, M.D., Møllgård, K., Stagaard, M. and Saunders, N.R. (1991) Species-specific transfer of plasma albumin from blood into different cerebrospinal fluid compartments in the fetal sheep. *J. Physiol.* 439: 215–37.

Ek, C.J. (2002) Transfer mechanisms into the developing brain, PhD thesis, University of Tasmania, Australia.

Ek, C.J., Habgood, M.D., Dziegielewska, K.M., Potter, A. and Saunders, N.R. (2001) Permeability and route of entry for lipid-insoluble molecules across brain barriers in developing *Monodelphis domestica. J. Physiol.* 536:841–53.

Ek, C.J., Habgood, M.D., Dziegielewska, K.M. and Saunders, N.R. (2003) Structural characteristics and barrier properties of the choroid plexuses in developing brain of the opossum (*Monodelphis domestica*). *J. Comp. Neurol.* 460: 451–64.

el-Gammal, S. (1981) The development of the diencephalic choroid plexus in the chick. A scanning electron-microscopic study. *Cell Tissue Res.* 219: 297–311.

el-Gammal, S. (1983) Regional surface changes during the development of the telencephalic choroid plexus in the chick. A scanning-electron microscopic study. *Cell Tissue Res.* 231: 251–63.

Evans, C.A., Reynolds, J.M., Reynolds, M.L., Saunders, N.R. and Segal, M.B. (1974) The development of a blood-brain barrier mechanism in foetal sheep. *J. Physiol.* 238: 371–86.

Ferguson, R.K. and Woodbury, D.M. (1969) Penetration of ^{14}C-inulin and ^{14}C-sucrose into brain, cerebrospinal fluid, and skeletal muscle of developing rats. *Exp. Brain Res.* 7: 181–94.

Fossan, G., Cavanagh, M.E., Evans, C.A., Malinowska, D.H., Møllgård, K., Reynolds, M.L. and Saunders, N.R. (1985) CSF-brain permeability in the immature sheep fetus: a CSF-brain barrier. *Brain Res.* 350: 113–24.

Gomez, D.G. and Potts, D.G. (1981) The lateral, third, and fourth ventricle choroid plexus of the dog: a structural and ultrastructural study. *Ann. Neurol.* 10: 333–40

Habgood, M.D., Knott, G.W., Dziegielewska, K.M. and Saunders, N.R. (1993) The nature of the decrease in blood-cerebrospinal fluid barrier exchange during postnatal brain development in the rat. *J. Physiol.* 468: 73–83.

Habgood, M.D., Sedgwick, J.E., Dziegielewska, K.M. and Saunders, N.R. (1992) A developmentally regulated blood-cerebrospinal fluid transfer mechanism for albumin in immature rats. *J. Physiol.* 456: 181–92.

Jacobsen, M., Møllgård, K., Reynolds, M.L. and Saunders, N.R. (1983) The choroid plexus in fetal sheep during development with special reference to intercellular plasma proteins. *Dev. Brain Res.* 8: 77–88.

Johanson, C.E. (1995) Ventricles and cerebrospinal fluid. In P.M. Conn (ed.). *Neuroscience in Medicine.* Philadelphia: J.B. Lippincott Company.

Johanson, C.E. and Woodbury, D.M. (1974) Changes in CSF flow and extracellular space in the developing rat. In A. Vernadikis and N. Weiner (eds.), *Drugs and the Developing Brain.* New York: Plenum Press.

Jones, H.C. and Keep, R.F. (1987) The control of potassium concentration in the cerebrospinal fluid and brain interstitial fluid of developing rats. *J. Physiol.* 383: 441–53.

Kappers, J.A. Structural and functional changes in the telenchephalon choroid plexus during human ontogenesis. Paper presented at Ciba Foundation Symposium on the Cerebrospinal Fluid, Boston, 1958.

Keep, R.F. and Jones, H.C. (1990) A morphometric study on the development of the lateral ventricle choroid plexus, choroid plexus capillaries and ventricular ependyma in the rat. *Brain Res. Dev. Brain Res.* 56: 47–53.

Keep, R.F., Jones, H.C. and Cawkwell, R.D. (1986) A morphometric analysis of the development of the fourth ventricle choroid plexus in the rat. *Brain Res.* 392: 77–85.

Klosovskii, B.N. (1963). *The Development of the Brain and Its Disturbance by Harmful Factors.* Oxford: Pergamon.

Kniesel, U., Risau, W. and Wolburg, H. (1996) Development of blood-brain barrier tight junctions in the rat cortex. *Brain Res. Dev. Brain Res.* 96: 229–40.

Knott, G.W., Dziegielewska, K.M., Habgood, M.D., Li, Z.S. and Saunders, N.R. (1997) Albumin transfer across the choroid plexus of South American opossum (*Monodelphis domestica*). *J. Physiol.* 499: 179–94.

Knudsen, P.A. (1964) Mode of growth of the choroid plexus in mouse embryos. *Acta Anat.* 57: 172–182.

Milhorat, T.H., Davis, D.A. and Lloyd, B.J.J. (1973) Two morphologically distinct blood-brain barriers preventing entry of cytochrome c into cerebrospinal fluid. *Science* 180: 76–8.

Mazariegos, M.R., Tice, L.W. and Hand, A.R. (1984) Alteration of tight junctional permeability in the rat parotid gland after isproterenol stimulation. *J. Cell Biol.* 98: 1865–77.

Møllgård, K., Balslev, Y., Lauritzen, B. and Saunders, N.R. (1987) Cell junctions and membrane specializations in the ventricular zone (germinal matrix) of the developing sheep brain: a CSF-brain barrier. *J. Neurocytol.* 16: 433–44.

Møllgård, K., Lauritzen, B. and Saunders, N.R. (1979) Double replica technique applied to choroid plexus from early foetal sheep: completeness and complexity of tight junctions. *J. Neurocytol.* 8: 139–49.

Møllgård, K., Milinowska, D.H. and Saunders, N.R. (1976) Lack of correlation between tight junction morphology and permeability properties in developing choroid plexus. *Nature* 264: 293–4.

Møllgård, K. and Saunders, N.R. (1975) Complex tight junctions of epithelial and of endothelial cells in early foetal brain. *J. Neurocytol.* 4: 453–68.

Oldendorf, W.H. and Davson, H. (1967) Brain extracellular space and the sink action of cerebrospinal fluid. Measurement of rabbit brain extracellular space using sucrose labeled with carbon 14. *Arch. of Neurol.* 17: 196–205.

O'Rahilly, R. and Müller, F. (1990) Ventricular system and choroid plexuses of the human brain during the embryonic period proper. *Am. J. Anatomy* 189: 285–302.

Otani, H. and Tanaka, O. (1988) Development of the choroid plexus anlage and supraependymal structures in the fourth ventricular roof plate of human embryos: scanning electron microscopic observations. *Am. J. Anat.* 181: 53–66.

Parmelee, J.T., Bairamian, D. and Johanson, C.E. (1991) Response of infant and adult rat choroid plexus potassium transporters to increased extracellular potassium. *Brain Res. Dev. Brain Res.* 60: 229–33.

Parmelee, J.T. and Johanson, C.E. (1989) Development of potassium transport capability by choroid plexus of infant rats. *Am. J. Physiol.* 256: R786–91.

Preston, J.E. (2001) Ageing choroid plexus-cerebrospinal fluid system. *Microsc. Res. Tech.* 52: 31–7.

Preston, J.E., Dyas, M. and Johanson, C.E. (1993) Development of chloride transport by the rat choroid plexus, in vitro'. *Brain Res.* 624: 181–7.

Saunders, N.R. (1992) Ontogenic development of brain barrier mechanism. In M.W.B. Bradbury (ed.) *Handbook of Exp. Pharmacology: Physiol. and Pharmacology of the Blood-Brain Barrier* Berlin: Springer-Verlag.

Saunders, N.R. and Møllgård, K. (1984) Development of the blood-brain barrier. *J. Dev. Physiol.* 6: 45–57.

Selwood, L. and Woolley, P.A. (1991) A timetable of embryonic development, and ovarian and uterine changes during pregnancy, in the stripe-faced dunnart, Sminthopsis macroura (*Marsupialia: Dasyuridae*). *J. Reprod. Fertil.* 91: 213–27.

Serot, J.M., Foliguet, B., Bene, M.C. and Faure, G.C. (2001) Choroid plexus and aging in rats: a morphometric and ultrastructural study. *Eur. J. Neurosci.* 14: 794–8.

Shuangshoti, S. and Netsky, M.G. (1966) Histogenesis of choroid plexus in man. *Am. J. Anat.* 118: 283–316.

Stern, L. and Gautier, R. (1922) Recherches sur le liquide céphalo-rachidien. II Les rapports entre le liquide céphalo-rachidien et les éléments nerveux de l'axe cérébro-spinal. *Arch. Int. Physiol.* 17: 391–48.

Sturrock, R.R. (1979) A morphological study of the development of the mouse choroid plexus. *J. Anat.* 129: 777–93.

Tauc, M., Vignon, X. and Bouchaud, C. (1984) Evidence for the effectiveness of the blood-CSF barrier in the fetal rat choroid plexus. A freeze-fracture and peroxidase diffusion study. *Tissue Cell* 16: 65–74.

Tennyson, V.M. (1971) The differences in fine structure of the myelencephalic and telencephalic choroid plexuses in the fetuses of man and rabbit, and a comparison with the mature stage. *Acta Neurol. Latinoam.* 1: 11–52.

Tennyson, V.M. and Pappas, G.D. (1964) Fine structure of the developing telencephalic and myelencephalic choroid plexus in the rabbit. *J. Comp. Neurol.* 123: 379–412.

Tennyson, V.M. and Pappas, G.D. (1968) The fine structure of the choroid plexus adult and developmental stages. *Prog. Brain Res.* 29: 63–85.

Wakai, S. and Hirokawa, N. (1981) Development of blood-cerebrospinal fluid barrier to horseradish peroxidase in the avian choroidal epithelium. *Cell Tissue Res.* 214: 271–8.

Wilting, J. and Christ, B. (1989) An experimental and ultrastructural study on the development of the avian choroid plexus. *Cell Tissue Res.* 255: 487–94.

Zaki, W. (1981) Ultrastructure of the choroid plexus and its development in the mouse. *Z. Zellforsch. Mikrosk. Anat.* 95: 919–135.

2 Normal and Pathologically Altered Structures of the Choroid Plexus

Lise Prescott
Centre Hospitalier de Verdun
Verdun, Canada

Milton W. Brightman
National Institutes of Health
Bethesda, Maryland, U.S.A.

CONTENTS

The primary emphasis of this review is the extracellular matrix and its possible roles in the functioning of the normal and altered plexus. This aspect has received only limited attention, as attested to by the frequent allusions, in this account, to the epithelial stroma of other organs. The discussion also concerns the structural peculiarities of the apical microvilli and how these features affect the cells' functions, especially vesicular uptake and movement of solutes. The participation of the choroid plexus as both target and mediator of systemic and local pathological processes is a reminder of how similar to one another the choroid plexus and kidney are in their structural responses to pathological events. These correlations continue to be validated by current reports.

2.1 VASCULATURE AND EPITHELIUM

Unlike the continuous endothelium of brain parenchymal vessels, that of the choroid plexus (CP), like most of the circumventricular organs, is fenestrated (Becker et al., 1967; Brightman, 1967, 1968). A developmental increase in the number of endothelial fenestrae in the capillaries of the CP stroma has been related to the maturation of cerebrospinal fluid (CSF) secretion (Keep and Jones, 1990). These endothelial "holes" are round, about 60 nm in diameter, but do not perforate the endothelium because they are bridged by a thin diaphragm. The diaphragm is permeable to water, but not to solutes (Pino, 1985). It is likely that solutes cross the fenestrated capillary wall by vesicular transport or by passing through intercellular junctions that are patent. The junctional route has been demonstrated to be available in choroid plexus papillomas (see Figure 2.1) (Prescott and Brightman, unpublished results; Brightman, 1977). Some of these junctions are short, straight, and patent to circulating probe molecules, such as horseradish peroxidase (HRP). The short, open junctions are encountered more frequently in tumors than in normal CP. A complete, periendothelial basal lamina surrounds the fenestrated capillary and comprises the inner border of the stromal space, the outer border of which is the subepithelial basal lamina (Figure 2.2A). The morphology of the fenestrated endothelium in normal CP is, therefore, very similar to that of choroid plexus papillomas.

2.1.1 JUNCTIONAL STRUCTURES OF THE BLOOD–CSF BARRIER

The CP projects into the cerebral ventricles and is thereby situated between blood and CSF. The plexus regulates the exchange of fluid and solutes between blood and CSF in part by acting as a passive, paracellular barrier between the two compartments. The structural locus of the CP barrier to solutes, such as the glycoprotein horseradish peroxidase (HRP) (MW 43,000), is the tight junction between its epithelial cells (see Figure 2.2B) rather than its endothelial cells (Becker et al., 1967; Brightman, 1967, 1968). Because the cleft is occluded, solutes cannot bypass the epithelial cells to reach the ventricle, so the cells have the opportunity of actively selecting, modifying, or rejecting solutes.

Like the endothelial tight junctions of the brain (Bodenheimer and Brightman, 1968), the tight junctions of the CP epithelium cannot be penetrated by HRP molecules infused into the cerebral ventricles (Brightman, 1968; Brightman and Reese, 1969). The CP epithelial barrier to colloidal La^{3+} is not as tight as the endothelial barrier of the vessels in the cerebral parenchyma. The paracellular passage of the exogenous, circulating La^{3+} colloid is blocked by the first or second tight junction

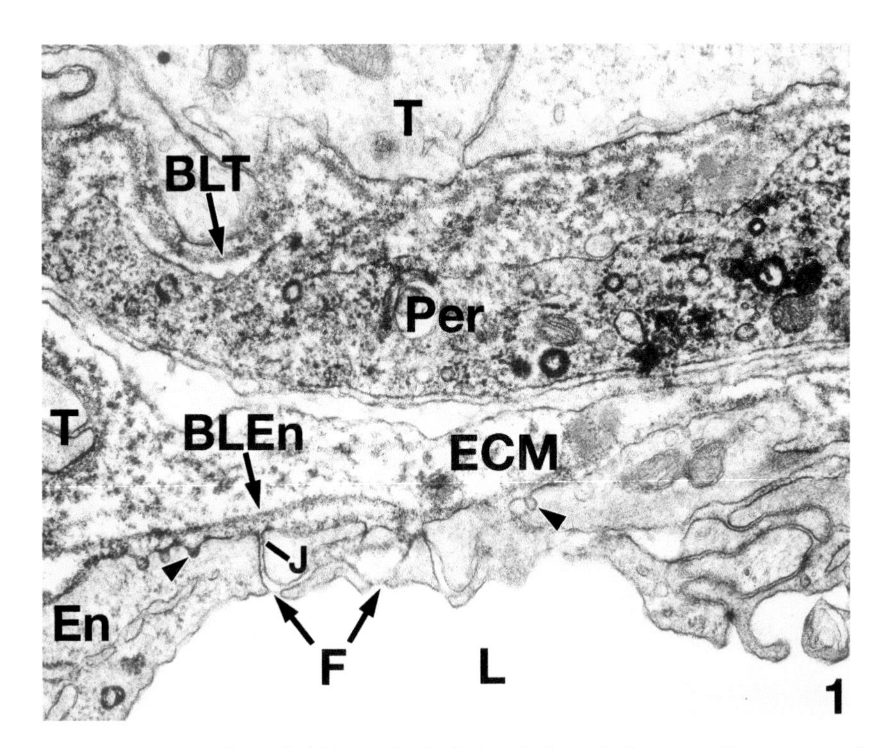

FIGURE 2.1 Endothelial cell (En) of virally induced choroid plexus papilloma cannot be distinguished from En of normal choroid plexus (compare with Figure 2A). The fenestrae (F) are bridged by a thin diaphragm. The intercellular cleft is short and straight to end in a junction (J) that may be patent. The serpentine junction at the right margin of the field is usually tight. Endocytotic or transcytotic pits (arrowheads) are more numerous than those of parenchymal barrier vessels. Circulating horseradish peroxidase (HRP), flushed from the vessel lumen (L) by fixative, had crossed the vessel to enter its basal lamina (BLEn), then into the perivascular, extracellular matrix (ECM), from which it was endocytosed by a pericyte (Per). Some HRP inundated basal lamina (BLT), subtending tumor cells (T).

between brain endothelial cells, but is able to penetrate several tight junctions between some choroid epithelial cells (Brightman and Reese, 1969). It was inferred from this penetration that either the epithelial junctions were not as occlusive as the endothelial junctions or that they did not form the continuous, uninterrupted belt or zonule expressed by the endothelium. This inference was borne out by the passage of ionic La^{3+}, smaller than the colloidal form, through the entire series of tight junctions within a junctional complex, so that some ions passed through the junctional paracellular clefts between CSF and CP stroma in either direction (Castel et al., 1974).

Adherens junctions, situated directly beneath the tight junctions, also tether adjacent choroidal epithelial cells to each other and ependymal cells to one another. Twenty-four hours after phorbol ester, a carcinogen that activates macrophages, was infused into the lateral cerebral ventricles of the rat, the proteins cadherin and β-catenin, a protein associated with cadherin, were diminished in both choroidal

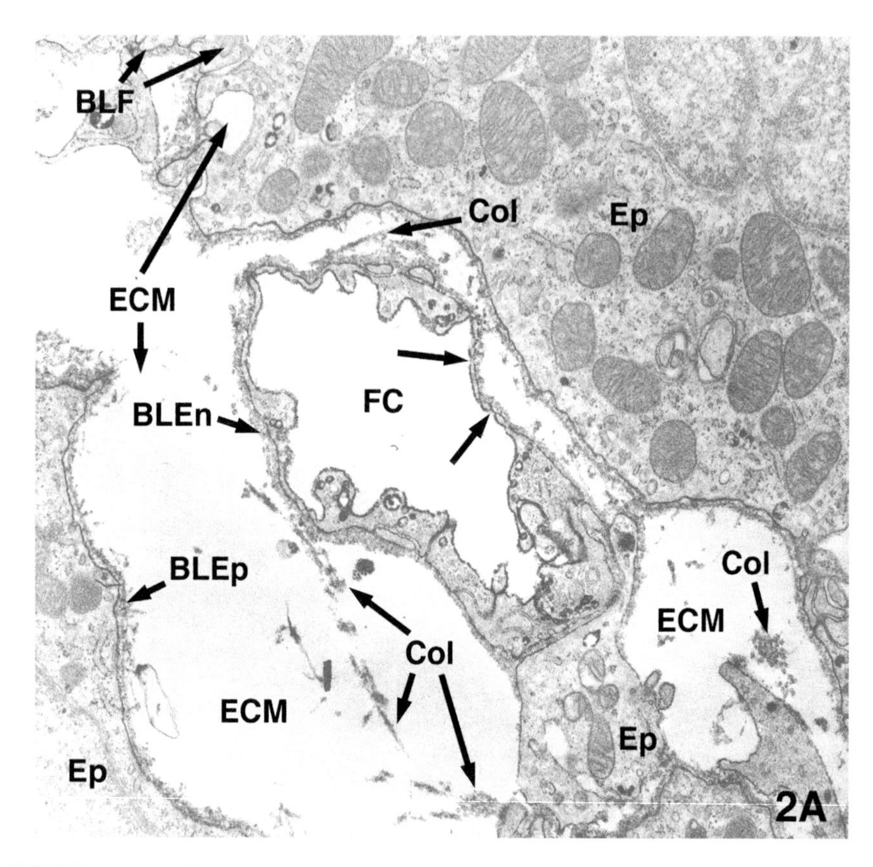

FIGURE 2.2 (A) Choroid plexus epithelial cells (Ep) with basolateral folds (BLF) surround capillary (FC) with fenestrae (arrows) lying in extracellular matrix (ECM) or stromal space. The stromal space is bordered on one side by an endothelial basal lamina (BLEn) and on the other side by an epithelial basal lamina (BLEp). Collagen fibrils (Col) lie embedded within the ECM. Horseradish peroxidase, given intravenously to this mouse, exuded from the FC into ECM and thence infiltrated all basal laminae and, eventually, the intercellular clefts of the epithelium. (B) View of two adjacent, normal, choroid plexus epithelial cells. Circulating horseradish peroxidase has crossed fenestrated capillaries to enter the extracellular matrix (ECM) between epithelium and capillaries, then flowed between two of the cells as far as the apical tight junction (TJ) isolating the cleft from the ventricle into which project many microvilli (V). Some endocytotic or transcytotic pits (arrowheads) have imbibed HRP, others (P) have not. The cells contain typical Golgi apparatus (G) and mitochondria (M). (From Brightman, 1968.) *(continued)*

epithelium and in ependyma. The α–catenin, another component of the adheren's junction, was unaffected (Lippoldt et al., 2000). The immunoreactivity of the occludin component of the tight junctions of the choroidal epithelium, after exposure to phorbol ester, was markedly reduced, although the ZO-1 protein, associated with tight junctions, was unchanged (Lippoldt et al., 2000).

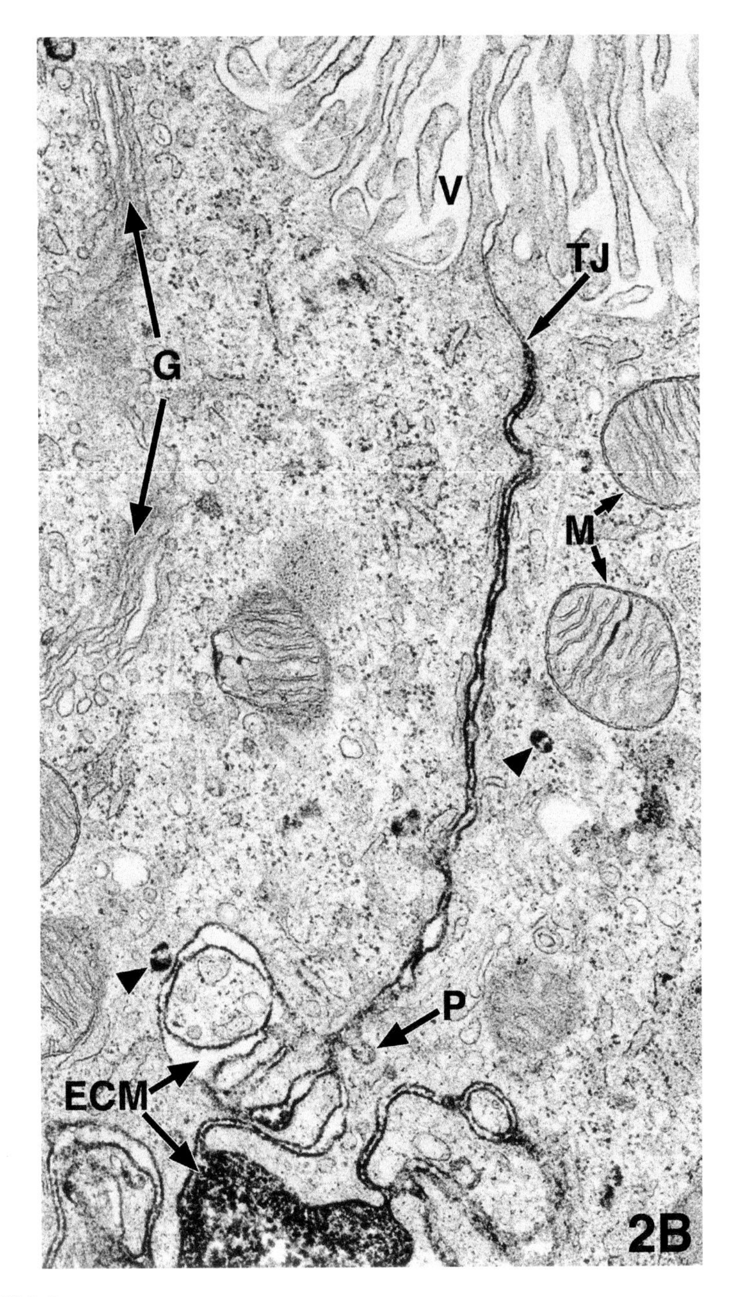

FIGURE 2.2

2.1.2 Endocytosis

The incorporation of solutes for either their processing within the choroidal epithelial cell or their transfer to the CSF is by way of coated vesicles. Coated pits are microindentations of the plasmalemma that arise at all surfaces—apical and basolateral—of the epithelial cells. The coat consists of the protein clathrin, which encases the Ω-shaped pit (see Figure 2.3) that will pinch off as a coated vesicle which, in various cell types, is 100 to 150 nm wide (Bishop, 1997). Clathrin is a requisite for endocytosis in many cell types. The formation of the endocytotic pit and its invagination, neck formation, and scission from the cell membrane as a vesicle depend on clathrin (see DiPaolo and DeCamilli, 2003). Endocytosis of some ligands and of some virus is also carried out by noncoated vesicles (Bishop, 1197), including, perhaps, some of those arising from the basolateral surface of the choroid plexus epithelium.

Although the numerous, slender microvilli projecting from the apical surface into the ventricles add a great deal of surface area available for fluid secretion, and enzyme and carrier activities, only a small fraction of the apical surface is available for endocytosis and exocytosis (Brightman, 1975). It is only the relatively small intervillous portion of the apical surface of choroidal epithelium, that portion between the bases of adjacent microvilli, that can form pits that become endocytotic vesicles (see Figure 2.3). The cell membrane that covers the microvilli is apparently incapable of invaginating as endocytotic pits. Much of the basolateral surface, in contrast, can form them (see Figure 2.2B). Apical and basal pits eventually pinch off as endocytotic or transcytotic vesicles. When ferritin (445 kDa) is perfused through the cerebral ventricles, it is not internalized by the expanse of microvillous plasmalemma, but rather by the small area presented by the intervillous plasmalemma (see Figure 2.3).

The same depot, the lysosomes, accepts solutes internalized from both poles of the choroid plexus epithelial cell. When ferritin was infused intraventricularly, followed by HRP injected intravenously in the same mouse, the ferritin was endocytosed by coated vesicles budding from the apical pole of the cells (see Figure 2.3). Ferritin was then carried to lysosomes by the vesicle, which fused with the lysosomes' membrane to release the ferritin molecules into the lysosomes. HRP was endocytosed by coated vesicles, originating from the basolateral cell membranes and carried to the same lysosomes (see Figure 2.4) (Brightman, 1975). The lysosome is a membrane-bounded organelle containing acid hydrolases and disposes of its cargo by enzymatic digestion. Acceptance by the same lysosome of unneeded solutes entering from different regions of the cell implies an economy of intracellular storage and disposal space.

2.1.3 Transcytosis

Unlike endocytosis, which is performed by clathrin-coated vesicles, the vesicular transport of solutes across cells from one pole to the opposite one and the eventual exocytosis of that vesicle's cargo is referred to as transcytosis. In the CP epithelium, transcytosis takes place in either direction (Villegas and Broadwell, 1993). A special

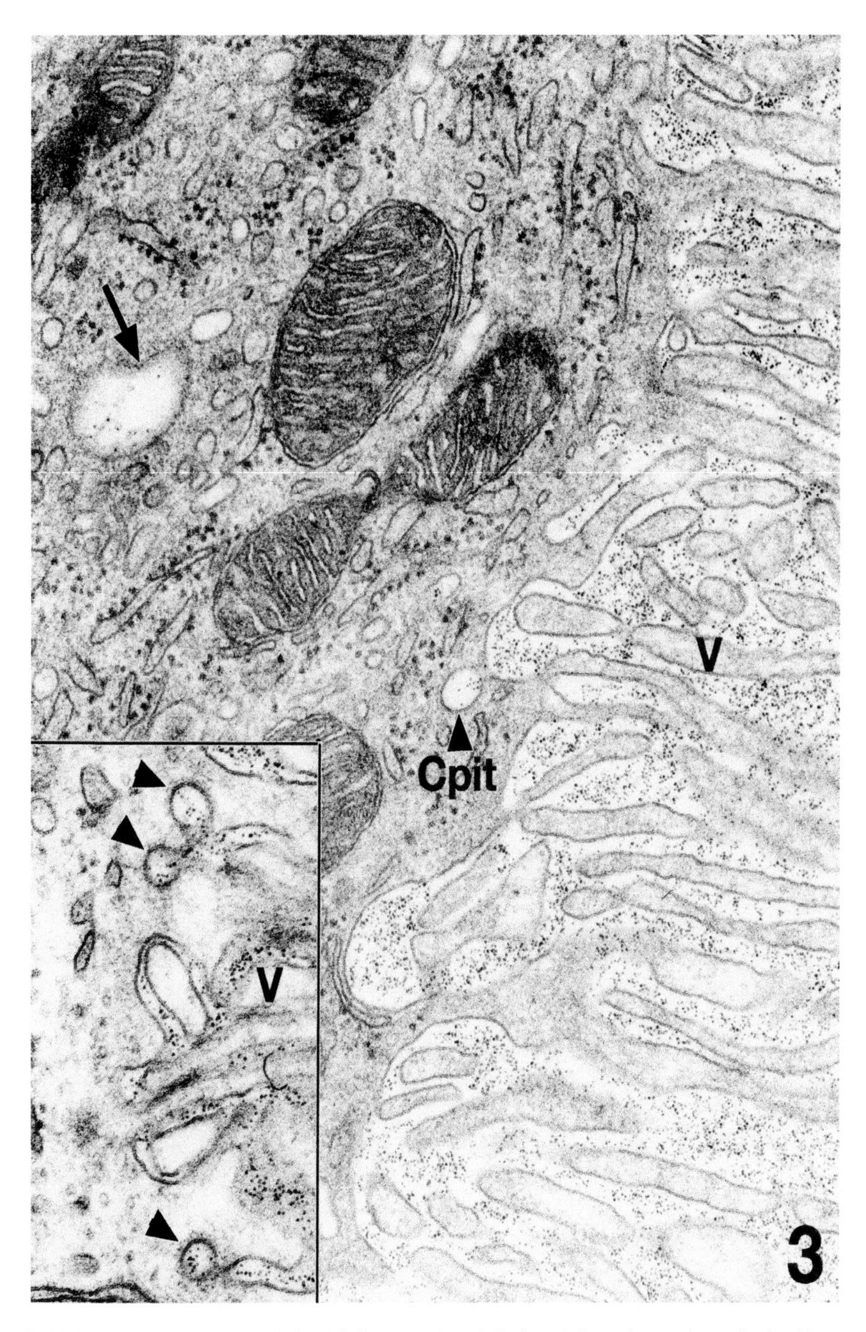

FIGURE 2.3 Clathrin-coated pits (Cpit; arrowheads in inset) invaginate the apical cell membrane of a normal choroid plexus epithelial cell. The pits contain ferritin (small dots), perfused through the cerebral ventricles, into which project many microvilli (V). A few ferritin molecules lie within a "vacuole" (arrow), which may actually be an invaginated pocket of the ventricle. The coated pits form only between microvilli, not in them. (From Brightman, 1975.)

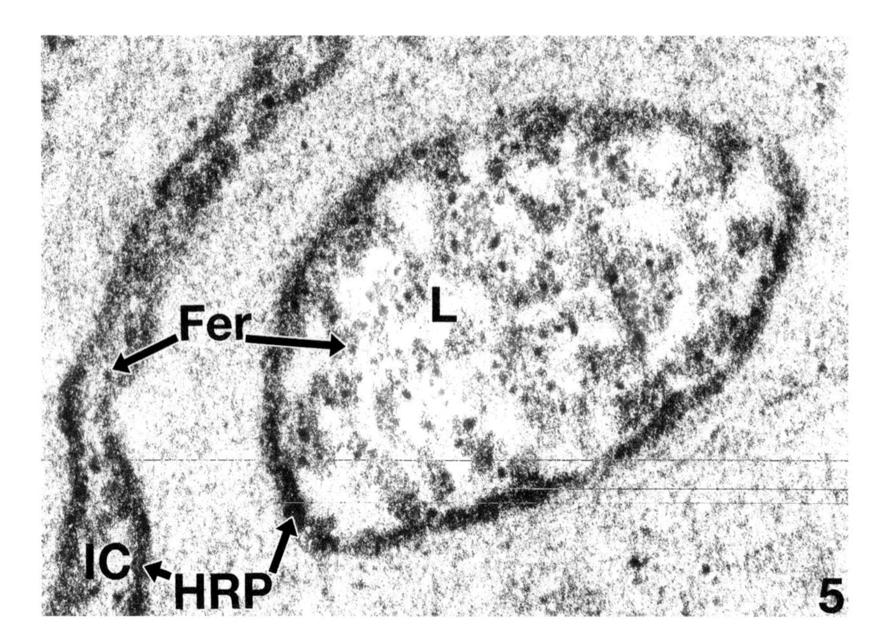

FIGURE 2.4 Horseradish peroxidase (HRP) has been injected intravenously, then ferritin (Fer) infused intraventricularly in a mouse. Both probe molecules entered the intercellular clefts (IC) and were vesicularly transferred to the same lysosome (L) within a choroid plexus epithelial cell. The lysosome membrane thus fused with the membrane of vesicles that had been derived from both apical and basolateral poles of the cell. (From Brightman, 1975.)

type of vesicle, the caveola, 50 to 80 nm wide, in various cell types (Bishop, 1997), does the transporting across the cell. The vesicle in Figure 2.4 labeled with HRP that had been injected intravenously was transcytosed by CP epithelial cells from their basolateral surface and had been conveyed to a point very near the apical surface facing the ventricle. This vesicle may have been a caveola. The caveola originates as a microdomain within the plane of the plasmalemma, which is rich in sphingomyelin and cholesterol. The caveola also contains a 22 kDa scaffolding protein, caveolin-1, which regulates a number of functions, including the internalization of the caveolar vesicle. In order for subsequent transcytosis to take place, the neck of the incipient vesicle must be separated from the cell membrane. This scission is mediated by tyrosine phosphorylation of caveolin-1 and dynamin. Thus, caveolar fission is enhanced by phosphatase inhibition and diminished when kinases are inhibited (Minshall et al., 1993), The importance of caveolin-1 is further illustrated by the augmented transcytosis of serum albumin across pulmonary endothelium in which caveolin is overexpressed by the induction of diabetes in rats (Pascarlu et al., 2004).

The electric charge on a solute molecule is another important determinant of whether or not it can pass across a cell layer. Native, anionic ferritin, infused ventriculocisternally, is prevented from entering the paracellular cleft between adjacent CP epithelial cells. The ferritin is available only to the coated endocytotic pits at the cells' apical surface, and only a few molecules are carried to lysosomes: dense

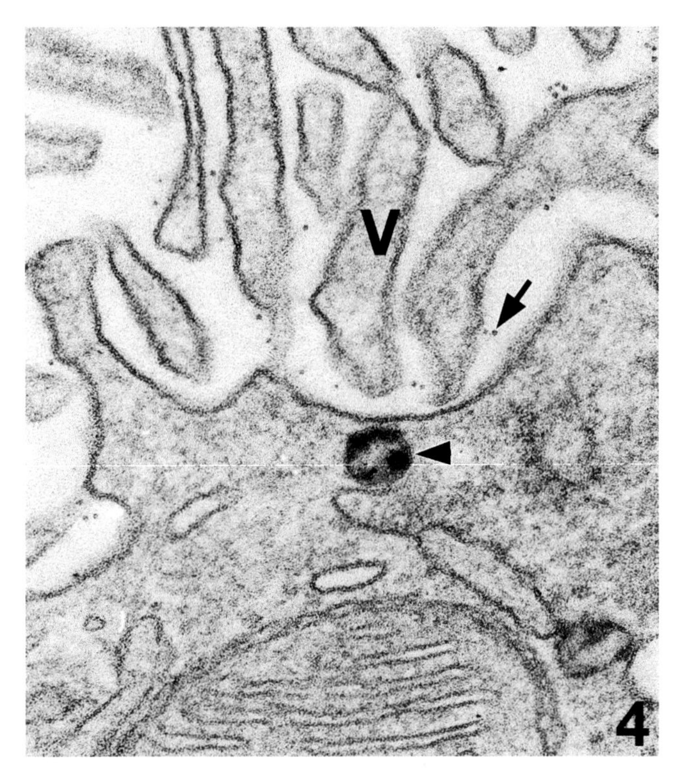

FIGURE 2.5 A vesicle (arrowhead), probably transcytotic, contains horseradish peroxidase (HRP) that it has endocytosed from the basolateral surface of the cell. The vesicle has moved across the cell almost to the apical plasma membrane facing the ventricle (V) into which ferritin (Fer) had been infused. The arrow points to a single ferritin molecule. (From Brightman, 1975.)

bodies and multivesicular bodies (Brightman, 1968). If, however, the ferritin is rendered strongly anionic, it can be endocytosed from the CSF by apical coated vesicles, which then transport it to the basolateral, extracellular clefts in addition to lysosomes (Van Deurs et al., 1981). Cationization of various molecules also greatly enhances their transport across the blood–brain barrier (Pardridge et al., 1990). The augmented uptake of cationic ferritin by the CP epithelium implies that, like brain endothelium (Schmidley and Wissig, 1986; Vorbrodt, 1989), there is a net negative charge on the luminal or ventricular face of the cells' apical cell membrane. In the CP endothelium, this anionic charge is concentrated on the diaphragm of the fenestrae in the stromal microvessels (Dermietzel et al., 1983). Most of the cationic ferritin accumulates at the diaphragms leaving only a fraction to cross the endothelium to enter the perivascular space.

A third vesicular mechanism of transcytosis across an epithelial or endothelial cell layer is used when the solute is a ligand for a receptor on the cell's plasma membrane. The ligand is not engulfed by a vesicle that is routed to the lysosomal compartment for enzymatic destruction. Instead, the ligand binds to its specific receptor, and the resultant receptor-mediated transcytosis ferries it across the cell from one extracellular compartment to another. Endothelial receptor-mediated transcytosis has been demonstrated with transferrin (Broadwell et al., 1996), the iron-transporting metalloprotein, which has receptors on the surface of blood vessels throughout the brain. A variant of receptor-mediated transcytosis is adsorptive transcytosis, which takes place when a protein or peptide is conjugated to a lectin, which then binds to its receptor, a specific oligosaccharide on the cell's surface. The binding induces pit and vesicle formation, internalization of the vesicle with its attached lectin-peptide, and transcytosis across endothelium or epithelium (Villegas and Broadwell, 1993).

2.1.4 Microvilli

The configuration of the CP is faithfully preserved by critical point drying of the tissue before it is examined with scanning electron optics. In such tissue, the apical surface of CP epithelium forms a bleb-like protrusion, and it is from this domed surface that slender, numerous microvilli arise (Peters, 1974). The form and dimensions of the microvilli are of particular interest because these pleomorphic extensions of the apical surface of the epithelium (see Figure 2.2B, Figure 2.3, and Figure 2.5) are the site of metabolic enzymes and leukocyte adhesion molecules. The microvilli are also susceptible to toxins and to pathogens. Unlike the ependyma, which bears a tuft of kinocilia as well as microvilli upon its apical, ventricular surface, the apical pole of a choroidal epithelial cell has few or no cilia but is thrown into slender folds extending into the ventricle as microvilli (see Figure 2.2, Figure 2.3, and Figure 2.5). The CP microvilli greatly increase the surface area of the apical face so that, in the mature human plexus, the total apical surface area is about 200 cm^2. In rodents, the total apical surface area, including the microvilli, of the CP in the lateral ventricle of a 30-day-old, young adult rat is 75 cm^2 (Keep and Jones, 1990). It would appear, therefore, that the apical surface area is, in terms of brain weight, proportionately much greater in the rodent than in the human choroid plexus. During development of the rat, the microvilli increase in number and, especially, in length so that the apical surface density, i.e., the surface area per unit volume of epithelial cell, of the epithelium is increased fourfold per unit volume of cell. By the time the animals have reached early maturity, the apical surface density has enlarged to a value of 3.27 ± 0.30 $\mu Mm^2/\mu Mm^3$ (Keep and Jones, 1990). The increase in the area of the basolateral surface is not as great as that of the microvilli.

The large surface area provided by the microvilli is of no advantage for some functions. The microvilli are entirely free of any formed organelle, such as a ribosome. They are sensitive to a variety of unrelated noxious agents and to environmental conditions such as changes in gravity and head position (Davet et al., 1998; Masseguin et al., 2001). They are incapable of forming endocytotic or exocytotic pits and vesicles and so do not participate in vesicular traffic, including that of receptor-mediated transfer of ligands. Their thickness and contours appear to be highly labile and, while a microvillus may respond momentarily to a stimulus by expanding its tip, this terminal club may collapse into the more usual thin microvillus, approximately 50 nm wide. This width is near the 20 to 30 nm range purported to be the threshold of a cell process before it can indent as an endocytotic pit (Brightman, 2002); the CP microvilli may usually be too thin to form vesicles. Receptor-mediated endocytosis, therefore, may not be performed by these microvilli. It would be of considerable interest to see whether they lack clathrins and caveolins, deficiencies that would account for the absence of the two main types of vesicles.

Nevertheless, adhesion molecules and the enzymes involved in the secretion of CSF, e.g., Na^+/K^+-ATPase, are located on the microvilli. The localization of the enzyme was convincingly demonstrated at the electron microscopic level with the specific, irreversible, binding of radiolabeled ouabain to the enzyme (Quinton et al., 1973). The apical localization of this ATPase in CP epithelium and in the pigmented epithelium of the retina is atypical. The common distribution is basolateral. The polarity of ATPase distribution depends on sorting signals that regulate the placement of its subunits into a given sector of plasmalemma (Rizzolo, 1999). Another enzyme involved in CSF secretion, carbonic anhydrase, has also been identified on CP microvilli (Masuzawa et al., 1981). Neutral amino acid transporters may be harbored by microvilli, although the transporters have only been demonstrated at the apical surface (Ross and Wright, 1984), which may or may not include the microvilli. The expanded surface area presented by the microvilli would provide a high density of enzymes and transporters on the CSF surface of the choroidal epithelium. Because of their content of these enzymes, it is surmised here that, if only the microvilli were injured or sloughed off and the rest of the cell remained intact after exposure to noxious agents, CSF could not be secreted by these focally disrupted cells.

2.2 EXTRACELLULAR MATRIX

The complex, hydrated gel or "ground substance" of the ECM consists of, among other components, a meshwork of proteoglycans and sulfated glycosaminoglycans. These molecules are, in turn, part of large proteins common to the ECM of various organs—heparan sulfate proteoglycans (HSPGs). The HSPGs are polyanionic molecules consisting of a linear protein core covalently bound to heparan sulfate chains that include heparin regions (Nader et al., 1987). A component of the perivascular basement membrane (summarized by Taipale and Keski-Oja, 1997) and developing choroid plexus (Handler et al., 1997) is the HSPG perlecan. As its name implies, perlecan has a beaded structure and a very large, 467 kDa protein core (Murdoch and Iozzo, 1993). During murine development, perlecan, which may be involved in the trafficking of transthyretin, to which it binds in vitro (Smeland et al., 1997), also maintains the structural integrity of basement membranes and is expressed in the choroid plexus and in organs other than brain during development (Handler et al., 1997). HSPGs, including perlecan, are not only within the matrix's basal laminae, but also at the external surface of cell membranes.

Beneath the choroidal epithelium lies an extracellular compartment bounded by basal laminae, one being the subepithelial or outer lamina, the other being peri-endothelial or the inner lamina. The basal lamina, discernible only by electron microscopy as an electron-dense band about 50 to 60 nm thick at the base of CP epithelium and ependyma and as a cover over astroglia, has come to be regarded as an interconnected system throughout the brain (Mercier et al., 2002). Although the designation "basal lamina" is still used interchangeably with "basement membrane," the latter term refers to a composite structure that includes the basal lamina, collagen fibrils and filaments, and other components visualized, collectively, by light microscopy. The epithelial basal lamina of different organs consists of three predominant molecular networks: collagen type IV, laminin, and fibronectin (Courtoy et al., 1982). During development, a particular isoform of collagen type IV is common only to the basement membrane of organs that filter out solutes: choroid plexus, renal glomerulus, pulmonary alveolus, and peritoneum (Saito et al., 2000). These molecular networks are embedded in the ECM.

Like the stromal, i.e., ECM, space of the choroid plexus, the perivascular space around the parenchymal barrier vessels of the brain may be regarded as a two-component system. The first or outer border is the basal lamina covering the astroglia. The second or inner border is a perivascular basal lamina. The CP stromal space likewise lies between a perivascular and a subepithelial basal lamina and is occupied by extracellular matrix in the normal CP (see Figure 2.2). In the CP tumor the stromal space lies between a perivascular and a subtumor basal lamina (see Figures 2.1 and 2.6).

2.2.1 Extracellular Matrix Effect on Epithelial Permeability

The ECM regulates many cell functions primarily through signals carried by integrins, which are glycoprotein, transmembrane, adhesion receptors that bind to ECM ligands or counterreceptors and are then linked to the actin cytoskeleton by way of adaptor proteins (van der Flier and Sonnenberg, 2001). Adhesion between epithelial cells and ECM is effected through the binding of their clustered integrins with ECM components, such as fibronectin. By signaling actin and intermediate filaments within the cytoplasm, the shape and movement of the cells can be modified.

The assembly of ECM fibrin (Danen et al., 2002) and laminin-1 (Aumailley et al., 2000) is also mediated through integrins. The details of how ECM, integrins, and cytoskeleton interact are beyond the scope of this discussion, but this brief summary illustrates the complexity of the relation between the cell and its matrix. The composition of the ECM itself affects the adhered cell's properties.

One of these properties is the permeability of intercellular tight junctions. The role of the ECM on the permeability of brain endothelial cells has been demonstrated in vitro by varying the composition of the matrix. When the cells were maintained on an ECM comparable to that found in brain—one containing laminin, collagen type IV and fibronectin at a certain ratio—the transendothelial resistance (R) was 2.3- to 2.9-fold higher than when the cells were grown on rat-tail collagen. It was concluded that the ECM influenced the permeabillity of the tight junctions (Tilling et al., 1998). The diffusion of albumin across endothelial monolayers, maintained on millipore filters coated with collagen type I, another constituent of basement membranes, reduced the diffusion of albumin across the layer by 30%. When a matrix of fibrin was used instead of collagen, the permeability was unaffected (summarized by Lum and Malik, 1994). The paracellular permeability of epithelium is also regulated by its tight junctions, the patency of which may, like endothelium, depend on the composition of the extracellular matrix laid down by the CP, a relationship yet to be fully explored.

The closeness of the contact between an epithelial cell layer and its ECM may be another factor in determining the permeability of the layer to the paracellular passage of ions across it. Epithelial cells with continuous tight junctions have a high R in vitro. The high R is also partly due to the intimate contact between the cells and their substrate. This close proximity of a few nanometers between the epithelial cell's basal cell membrane and the substrate impedes the extracellular movement of electrolytes, thereby increasing the trans-epithelial R (summarized by Lo et al., 1999). It may be surmised that, in situ, a pathological event, such as the release of collagenases or heparanases by macrophages during inflammation, may digest the ECM to the degree that the cells no longer make close contact with the changed ECM. The permeability across the epithelium would then be determined primarily by the tight junctions, while the ECM influence would be diminished; solutes would flow more readily through the digested ECM.

It is emphasized that the polarity and permeability of the choroidal epithelium depend on the presence of its ECM, but other constitutive functions do not. When dissociated CP epithelial cells are grown upon matrigel, a complete ECM derived from a tumor, the cells maintain their polarity and function. However, when the matrigel is replaced by collagen I as a single-component matrix, the cells lose their polarity and organization as a layer but still retain, at a high level, their transthyretin mRNA. The cells also perform the same pattern of protein synthesis as the cells that had been grown on matrigel (Thomas et al., 1992). The ECM may be vital for the expression of some of the choroid plexus's functions, but not all of them.

2.2.2 Depletion of Extracellular Matrix

The loss of ECM constituents can affect the function of the overlying epithelium. In addition to perlecan, a prominent HSPG of brain and kidney is agrin (~300 to 400 kDa). When agrin is stripped of its heparan sulfate polysaccharide side chain, the filtering capacity of its basement membrane is markedly diminished. The resulting albuminuria is manifested as a human disease (Raats et al., 1999). During cerebral ischemia in primates, a diminution in perivascular collagen IV, laminin, or fibronectin—all major components of basal lamina—is associated with the loss of endothelial integrity, which becomes apparent as focal bleeding in the brain (Hammann et al., 1995). The CP is more vulnerable to ischemia, coupled with hypoxia, than is the ependyma or subependymal cells (Rothstein and Levison, 2002). If, however, the ECM is also affected, stored growth factors may be released to enhance angiogenesis and, perhaps, neural repair. The predominant mechanism whereby the ECM is stripped of components is enzymatic.

Enzymatic degradation of the basal lamina system and other constituents of the ECM is achieved when activated leukocytes and tumor cells release matrix metalloproteinases (MMPs) in addition to heparanase and heparatinase. The MMPs are a family of Zn^{2+} endopeptidases, such as collagenases and stromelysin, that sculpt the ECM during embryogenesis, wound healing, neovascularization, and cell migration

(Pagenstecher et al., 1998). This remodeling of the matrix is controlled by limiting the MMPs' activity with a concurrent release of tissue inhibitors of MMPs (TIMPs). These counterbalancing processes take place during inflammation. In murine experimental autoimmune encephalomyelitis, one member of the MMP inhibitor, TIMP-3, is contributed primarily by the CP (Pagenstecher et al., 1998). In the CSF of patients with viral meningitis, the effects of several MMP isoforms may be modulated by TIMP-3 (Kolb et al., 1998). These enzymes, responsible for the remodeling of the choroid plexus's own ECM during inflammation, may thus indirectly affect the epithelium's junctions and permeability.

The ECM components of the choroid plexus that are freed by MMP activity might act as antigens. The released constituents include collagen IV, the alpha chains of which are very similar to those of the kidney and pulmonary alveolus (Saito et al., 2000). To approximate the products released from a damaged basement membrane, that of the CP has been solubilized and injected with complete Freund's adjuvant into rabbits. The consequences are the deposition of immunoglobulins in the basement membrane of the hosts' CP and renal glomerulus together with the development of hemorrhagic pneumonitis (McIntosh, 1974). These changes are comparable to those in the human autoimmune condition, Goodpasture's syndrome. The identity and relative importance of the components derived from the choroidal ECM that incite these immunological responses have yet to be determined.

2.2.3 Extracellular Matrix as Depot

The concept of basement membrane as a storage depot for growth factors that are releasable by enzymatic activity grew out of the observations on the distribution of the endothelial mitogen, fibroblast growth factor-2 (FGF-2), in various tissues. Despite the wide distribution of the mitogen in the ECM of normal tissues, the low rate of endothelial proliferation was taken to mean that FGF-2, was sequestered within its basement membranes away from its receptors (Vlodavsky et al., 1991). The notion of the ECM as an available depot was concordant with the earlier finding that an angiogenic factor was, indeed, stored within the basement membrane of the cornea (Folkman et al., 1988). The perlecan, associated with the plasma membrane of brain-derived endothelial cells, binds the ligand, FGF-2, and being on the cell membrane, is near the ligand's receptors. This perlecan is able to mediate uptake of the factor by the cells (Deguchi et al., 2002). However, the growth factors are often harbored within ECM as well as at the cell surface.

Before growth factors, cytokines, and other molecules sequestered in ECM can act on their target cells, they must first be released from the matrix. The release of bound factors is achieved by proteolytic activity. About 70% of FGF-2 bound to perlecan is liberated by the heparanase secreted by proliferating endothelial cells (Elkin et al., 2001). Collagenase, plasmin, and the MMP stromelysin also free bound factors (Whitelock et al., 1996). After FGF-2 is added to endothelial cell cultures and binds to the ECM secreted by the cells, most of the factor is rapidly released by the addition to the culture medium of heparan sulfate or heparitinase (Bashkin et al., 1989). The extrapolation to in vivo events, such as injury or inflammation, was made when it was noted that biologically active FGF-2 could be released from basement membrane stores through the action of heparanase secreted by platelets and leukocytes (Ishai-Michaeli et al., 1990), thereby promoting angiogenesis and endothelial cell differentiation (Knox et al., 2002). Other HSPGs, syndecan-1 and glypican-4, also apparently situated at cell surfaces rather than in basement membranes, are expressed in the developing brain at the ventricular sites where brain cells proliferate (Ford-Perress et al., 2003). If, in the CNS, growth factors can be stored in its extracellular matrices, angiogenesis, brain cell differentiation, and migration may be regulated as it is in other organs (Brightman and Kaya, 2000).

Within the CP itself, the capacity of its injured epithelium to regenerate is still open to question. The combination of ischemia and hypoxia leads to a patchwork of cell loss that is quickly reversed by what appears to be the migration of surviving epithelial cells into the denuded areas (Johanson et al., 2001). Although some cells disappeared, no mention was made of changes in the underlying ECM. One possible change would be the actual augmentation of ECM components such as perlecan. This possibility is suggested by the increased expression of the mRNA for perlecan in brain injured by a stab wound; the level of its expression may be regulated by the pro-inflammatory cytokine interleukin-1α (Yebenes et al., 1999). Augmentation of HSPG components could also presage a heightened capacity to store growth factors, such as epidermal growth factor (Raab et al., 1997). Ischemia, however, does not necessarily lead to mitosis of CP epithelium; no mitotic figures were found among the epithelial survivors which, it was inferred, migrated to the gaps left behind by the dead cells (Johanson et al., 2001). This conclusion can be tested by the administration of tritiated thymidine to rule out a mitotic contribution to the reconstruction of the epithelium.

Advantage might be taken of the binding and releasing properties of HSPGs. In that connection, it has been suggested that exogenous perlecan might be administered as a vehicle for the introduction of growth factors into injured areas of brain. The subsequent release, on demand, by proteinases secreted by incoming activated leukocytes might be useful for the repair of injured choroid plexus and brain tissue (Brightman and Kaya, 2000). A similar suggestion has been made to administer exogenous heparan sulfate so as to enhance the growth of neuronal precursors (Ford-Perriss et al., 2003).

2.3 PATHOLOGY OF THE CHOROID PLEXUS

2.3.1 Choroid Plexus Tumors

The basal lamina acts as a barrier to the migration of cells during normal development but serves as a guidance track for the migration of tumor cells. In the benign type of tumor, the CP papilloma, the basal lamina subtending CP epithelium maintains its position. In CP papillomas, the basal lamina, according to the figures of Furness et al., (1990) and a report on human papillomas (Carter et al., 1972), extends linearly from its usual perivascular site along the bases of tumor cells (see Figure 2.6). Like the pattern in mechanically or thermally injured brain (Anders and Brightman, 1986), the basal lamina of the tumor is, in places, redundant, several sheets of basal lamina

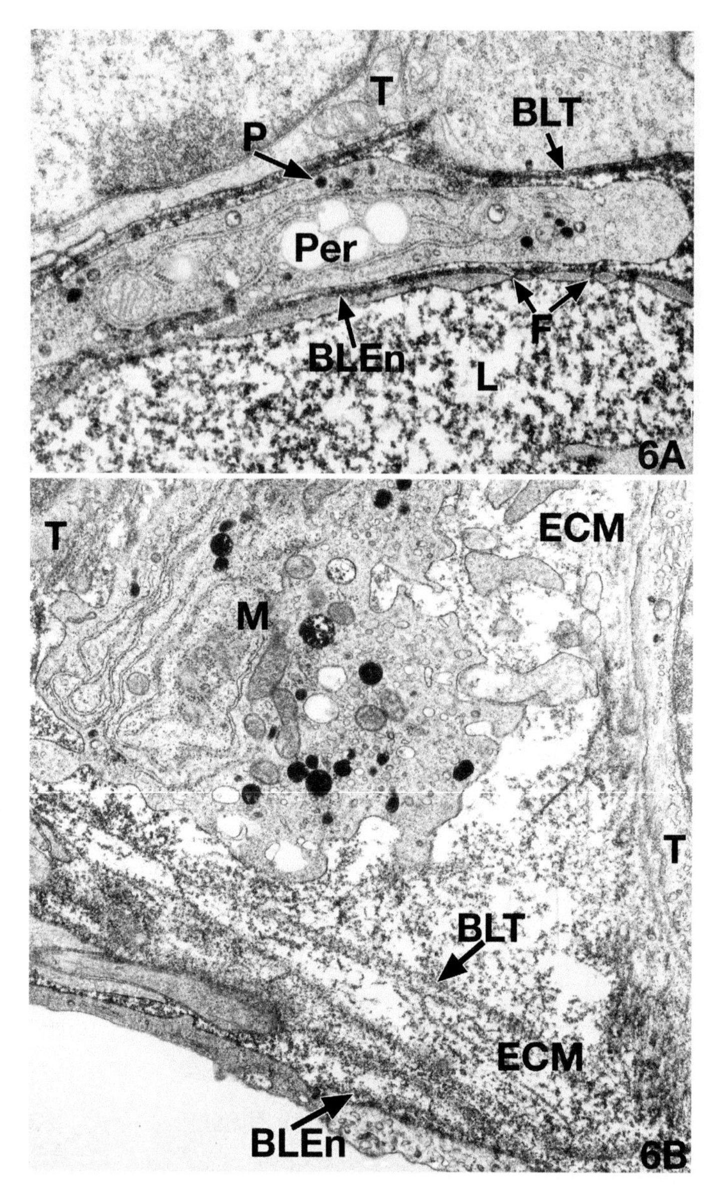

FIGURE 2.6 Basal laminae and extracellular matrix of a virally induced choroid plexus papilloma in a hamster that had subsequently been given horseradish peroxidase (HRP) intravenously. (A) HRP in the lumen (L) of a capillary, the endothelium of which is interrupted by fenestrae (F) and surrounded by its basal lamina (BLEn). The BLEn forms one border of the stromal space or extracellular matrix, occupied by a pericyte (Per) with endocytotic pits (P). The other border of the space is the basal lamina (BLT) subtending tumor cells (T). (B) Redundant basal lamina in this region of the tumor includes the peri-endothelial BLEn, which lies alongside several, approximately parallel, basal lamina (BLT) around tumor cells (T). This part of the tumor's stroma or extracellular matrix (ECM) is sufficiently large to accommodate a cell, which is probably a macrophage (M).

having been secreted more or less in parallel (see Figure 2.6B). In malignant neoplasms, only short linear fragments of basal lamina course between the tumor cells. There are fewer physical, laminar barriers to the invasive migration of the malignant cells (Barsky et al., 1983).

The fine structure of choroid plexus papillomas, induced by the intracerebral inoculation of simian vacuolating virus-40 into neonatal Syrian hamsters, as described here and that appears, spontaneously, in patients (Carter et al., 1972; Milhorat, 1976), is similar to normal CP (Prescott and Brightman, unnpublished observations; Brightman, 1977). One salient feature already alluded to is the short, straight junction that is patent from end to end (see Figure 2.1) between some of the endothelial cells. Such junctions may also be found in normal CP vessels. Circulating horseradish peroxidase (HRP) rapidly passes across these stromal vessels, presumably through the open junctions, to inundate the subepithelial matrix of the CP, where it coats collagen fibers and is endocytosed by perivascular cells (see Figures 2.1 and 2.6) and by the tumor cells.

A striking feature of the tumor's extracellular space is that it may, rarely, invaginate the tumor's cell plasma membrane so deeply as to approach the nuclear membrane, especially if the tumor cell has a lobulated nucleus into which an extracellular lacuna can extend (see Figure 2.7). In such tumor cells, the environment may be brought very close to the cell nucleus. More commonly, the highly tortuous, narrow sectors of the extracellular spaces in the CP papilloma interdigitate closely

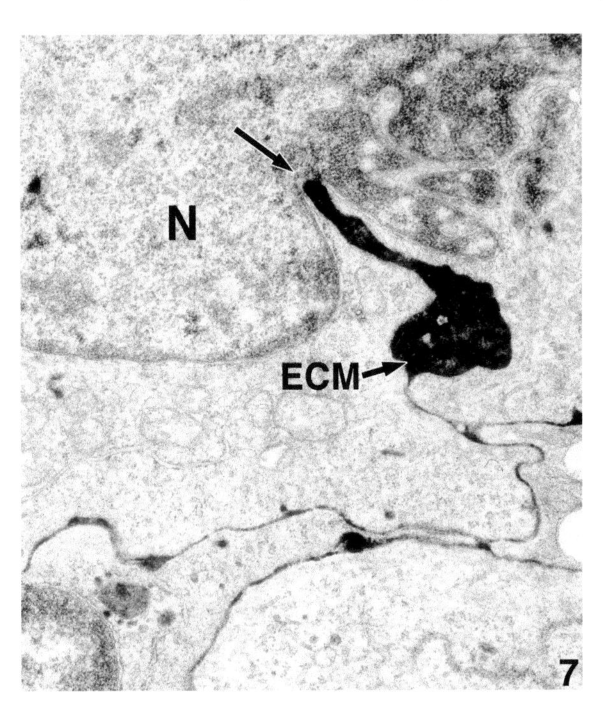

FIGURE 2.7 The extracellular matrix (ECM) of this region of the choroid plexus papilloma impinges on the lobulated nucleus of (N) a tumor cell.

and elaborately to accommodate the highly folded parts of adjacent cells. This interdigitaton is even more elaborate than that between the basolateral surfaces of normal CP epithelial cells. The intercellular clefts commonly dilate to form numerous, small cisterns or lacunae (see Figure 2.7). The stromal perivascular space of the CP papilloma is, in some regions, greater and more variable in dimensions and extent than that of the normal CP and can accommodate migrating macrophages (see Figure 2.6B).

2.3.1.1 Extracellular Matrix of Tumors

The extent and continuity of the basal lamina in tumors have been correlated with the degree of the tumor's malignancy. As derivatives of the neuroectoderm, the CP epithelium and the ependyma are separated from the underlying mesenchyme by a basal lamina. Thus, the formation of a basal lamina connotes an epithelial phenotype. The counterpart of the subepithelial basal lamina in normal CP (see Figure 2.2) is the basal lamina subtending tumor cells (see Figures 2.1 and 2.6). A feature that distinguishes between the epithelium and ependyma is the organization of the ECM when the two cell types become pathological. In the relatively benign CP papillomas, the basement membrane, as delineated by immunostaining of its laminin content, consists of the normal perivascular and subepithelial portions according to the published figures (Furness et al., 1990). In ependymomas, the basement membrane is restricted, as an oval line, around blood vessels without branches among the tumor cells, a distribution that implies a more permissive extracellular space for the spread of neoplastic ependymal cells.

In addition to its capacity to form neoplastic growths, the CP itself is used as an invasive route for other tumor cells. Human meningioma cells, injected into frontal lobe white matter or subdural space of the immunoincompetent, athymic mouse, spread along the leptomeninx and into the CP and ventricles, presumably from its tela choriodea extension (McCutcheon et al., 2000). Human, malignant astrocytoma cells, in contrast, spread throughout much of the brain by following the myelinated axons of fiber bundles as well as basal lamina. When xenografted into the cerebral cortex of rats, the astrocytoma cells migrate along the basal lamina around parenchymal vessels, at the glia limitans of the brain's subpial surface and between the ependymal and subependymal interface (Bernstein et al., 1989). The cells' widespread dissemination along the basal lamina route reflects the basal lamina as a ubiquitous extracellular structure that permeates the CNS (Mercier et al., 2002).

2.3.2 Toxins

The CP epithelium together with its basal lamina is viewed as a filter, much like the renal glomerulus. As a protector of the brain, the CP becomes a target for a number of toxic metals and metalloid ions. Cd^{2+} imbibed at a low concentration is filtered by the murine CP, which shows minor structural changes; but at 100 ppm, the cytoplasm of the CP epithelial cells becomes vacuolated, microvilli are lost, and the apical surface may rupture. A less severe response is apical blebbing, which also follows the administration of Cd^{2+} (Valois, 1989). A caveat in the interpretation of pathological tissue fixed in aldehydes is, of course, that the precise form taken by the blebs or protrusions may be a second-order change in the morphology of an injured portion of the cell. But it is unlikely that the apical eruptions as blebs are fixation artefacts. Blebs contain cytoplasmic organelles and are enclosed by cell membrane. Some of these protrusions pinch off from the cell and are released as globules into the ventricular CSF. The induction and release of blebs, large enough to be discernible by light microscopy, have been observed in live specimens (Gudeman et al., 1989) and, accordingly, are real events.

A common mechanism—destabilization of the cytoskeleton—may account for blebbing and other alterations brought about by a variety of toxins. The microvilli of CP are pleomorphic and irregular in dimensions and arrangements (see Figure 2.2, Figure 2.3, and Figure 2.5). Microfilaments containing F-actin course through the apical cytoplasm (McNutt, 1978), and are primarily responsible for the shape, contours, and movements of all cell types. The intravenous injection of cytochalasin D, which prevents the addition of actin molecules to actin filaments, and of colchicine, which inhibits the polymerization of tubulin into microtubules, results in several structural responses. These reactions are contraction of the cells accompanied by compaction of microfilaments and microtubules at the base of apical protrusions or blebs (Koshiba, 1987). The cytoskeleton of CP epithelium might also respond to prolonged changes in gravity. The CP microvilli of rats maintained in a head-down position become longer and thinner (Davet et al., 1998; Masseguin et al., 2001). It is reasonable to conclude that treatments that disrupt the cytoskeleton may lead to blebbing and other distortions of the microvilli.

A more specific morphological response to a toxin follows the intraperitoneal injection of the lysosomotropic drug chloroquine into neonatal rats (Kaur, 1997). The result is a massive accumulation of lysosomes in CP epithelial cells and in ependymal cells. Lamellar bodies and cytoplasmic vacuoles also arise in ependymal cells, but not in CP epithelial cells (Kaur, 1997). Apical blebbing, apparently, does not take place.

Some toxins act directly on the CP to produce inflammation. The choroid plexitis associated with the systemic administration of the cytotoxin cyclophosphamide is manifested by bleeding and exudation of serum contents into the stroma of the CP, in addition to epithelial necrosis (Levine and Sowinski, 1973). Those stromal vessels that are spared retain their fenestrae.

There appear to be paradoxical results following administration of cyclophosphamide in humans. This agent is used in the treatment of nephritis associated with systemic lupus erythematosis (SLE) (Flanc et al., 2004). Before treatment, the choroid plexus may be swollen, image-enhanced by radiocontrast material, and hemorrhagic, but after an initial administration of cyclophosphamide with methylprednisolone, followed by repeated intravenous injections of 750 mg of cyclphosphamide per month for 6 months, the CP appeared normal (Duprez et al., 2001). Whether this dosage and regimen are comparable to those used for rat, the results were remarkably different.

2.3.3 Inflammation

Inflammation that begins in the choroid plexus usually spreads to the leptomeninges. The pial core of the CP stroma is responsible for the extension. This meningeal pathway is particularly well documented in bacterial infections and other inflammatory processes. The subcutaneous inoculation of mice with *Listeria monocytogenes* leads to the development of plexitis and meningitis when, presumably, infected mononuclear phagocytes, i.e., epiplexus Kolmer cells, migrate to the leptomeninx (Prats et al., 1992). The vectorial spread (summarized by Tuomanen, 1996) is inferred from the predilection of some pathogens for the CP. For example, the CP is a selective target for Sendai and lymphocytic choriomeningitis viruses and trypanosomes (Levine, 1987). CP was the only tissue in brain affected by streptococci injected intravenously into pigs. Within the plexus, the epithelium's microvilli were disrupted, but the vessels were unaffected and pathogen gained access to the ventricles (Williams, 1990). A route for viral entry into brain is directly across the barrier vessels throughout the brain's parenchyma, the interstitial fluid of which is confluent with the ventricular CSF (Brightman, 1968; Brightman and Kaya, 2000). If the *streptococcus* followed this route, diffuse lesions would develop early in the infection, but this sort of distribution does not occur (Tuomanen, 1996). A systemic infection that reaches the CP may spread not only to the meninges but also into the brain parenchyma. Thus, a high level of bacteremia with, e.g., *L. monocytogenes*, sustained for a sufficient interval in mice, produces not only a severe choroid plexitis but a meningoencephalitis as well (Berche, 1995).

The migration of leukocytes through the CP and along the pia-arachnoid is mediated by a cell surface property common to many cell types: intercellular adhesion molecules (ICAMs). One such molecule, ICAM-1, enables the transitory adhesion of leukocytes to endothelium or epithelium, a preliminary event in the migration of leukocytes across cell layers. ICAM-1 and other adhesion molecules are upregulated in the CP of mice with experimental autoimmune encephalomyelitis; the enhanced expression, detected immunohistochemically, is restricted to the apical surface of the epithelium and is completely absent from the endothelium (Engelhardt et al., 2001). The level of ICAM-1 is also elevated when the inflammatory agent lipopolysaccharide is infused into the cerebral ventricles of rats (Endo et al., 1998). ICAM-1 is then readily demonstrable on the CP microvilli and on the fibroblasts constituting arachnoid trabeculae. The activated leukocytes migrate from the matrix vessels, thence the epithelium of the CP, into the ventricular CSF. At the same time, leukocytes migrate along the pia of the CP stroma to the pia-arachnoid extensions over the brain's surface. The choroid plexitis has expanded to a meningitis. Both the CP and leptomeninx play an important mutual role in the movement of leukocytes into the CSF compartment (Endo et al., 1998). A limited, periventricular migration of pathogen from the ventricles across the ependyma into the adjacent cerebral parenchyma may ensue.

Direct entry of activated leukocytes from blood across the barrier endothelium of brain parenchyma may also take place during inflammation. The path taken by migrating leukocytes across the brain's barrier endothelium may be paracellular, through tight junctions. Indirect evidence has implicated the junction associated molecule (JAM), which is 36–41 kDa and is a member of the immunoglobulin superfamily. The evidence was forthcoming when meningitis was induced in mice by the intraventricular infusion of the inflammatory cytokines tumor necrosis factor-α (TNF-α) and interleukin-1β (IL-1β). If the mice were first injected intravenously with an antibody to JAM, the migration of neutrophils and monocytes into the CSF was attenuated (Del Maschio et al., 1999). The question of whether JAM is associated with the tight junctions of the vessels and epithelium of the CP and whether it responds to the antibody with similar consequences warrants investigation in vivo and in vitro.

The conditions required for leukocytes to cross the endothelium, then the peri-endothelial basal lamina, into the perivascular space to aggregate as cellular cuffs are still incompletely defined. The intriguing observation has been made that, during experimental autoimmune encephalomyelitis (EAE) in mice, transendothelial migration takes place at sites where only laminin isoform 8 is expressed in the peri-endothelial basal lamina (Sixt et al., 2001). Activated leukocytes do not cross where other laminin isoforms are co-expressed with isoform 8. Does a similar, subtle change in the composition of basal lamina, in this case the basal lamina subtending CP epithelium, also determine where leukocytes cross the epithelium during inflammation? If so, a slight change in the constituents of the extracellular matrix might result in a profound change in the permeability of the epithelial junctions, whereby leukocytes can pass through them.

The CP initiates inflammation by secreting pro-inflammatory cytokines, which originate from cells in the CSF that are epithelial cell associates rather than from the epithelial cells themselves. These cells presumably include the resident epiplexus macrophages of Kolmer (Peters and Swan, 1979) and, perhaps, the leptomeningeal cells of the CP stroma. The secretion of cytokines into the CSF is an early inflammatory event that precedes the infiltration of circulating monocytes into the ventricles. The expression of mRNA for the pro-inflammatory cytokines TNF-α (Tarlow et al., 1993) and IL-1β (Erikson et al., 2000) is enhanced by the administration of bacterial endotoxin, i.e., lipopolysaccharide. The messenger RNA for both IL-1β and IL-1α is located in monocytes (Eriksson et al., 2000) including, presumably, the epiplexus, resident macrophages, rather than the epithelial cells of the CP. Similarly, the intraperitoneal injection of endotoxin in mice leads to a pronounced increase in the mRNA for IL-1β in both the CP and meninges (Garabedian et al., 2000). These authors concluded that, when whole brain is assayed for this cytokine, fully two thirds of the IL-1β is accounted for by the monocytes within the ventricles and in the meninges that still cling to the sampled brain tissue.

2.3.4 Lupus

Systemic lupus erythematosus (SLE) is a generalized, systemic disease that affects the CP and kidney. A prominent feature of SLE is the infiltration of the CP by lymphoid cells and the entrapment of circulating immune complexes in the perivascular basal lamina of these two organs (Atkins et al., 1972). Another autoimmune condition, serum sickness (Koss et al., 1974), has been produced in rats by the repeated, intravenous injection over time of bovine serum albumin in complete Freund's adjuvant and is manifested by proteinuria and disruption of the blood–brain barrier (Peress et al., 1977). A feature common to these autoimmune states is the embedment of immune complexes within the basal lamina of CP and renal glomerulus. The immune complexes appear as amorphous, linear, electron-dense deposits in the basal lamina. The functional significance of these intralaminar deposits is unclear. Although the blood–brain barrier to circulating ^{125}I-albumin (~67kDa) was disrupted (Peress et al., 1977), this barrier and the blood–CSF barrier to the smaller molecule, HRP (~43 kDa), was not opened in rats with chronic serum sickness (Peress and Tomkins, 1979). Peroxidase did not move, paracellularly, beyond the tight junctions of brain endothelium or CP epithelium. The authors recognized the limits of such cytological detection of probe molecules. It is also not apparent why the deposits infiltrated the basal lamina of vessels in certain circumventricular organs, such as the area postrema, subfornical organ, and pineal gland, but not the highly vascular neural lobe of the pituitary gland.

The renal involvement in these systemic, autoimmune states is not surprising because the fenestrated, permeable capillaries of the CP resemble those of the kidney rather than the continuous, barrier vessels of the nearby brain parenchyma. A feature common to the fenestrated capillary of various organs is the high density of anionic sites on their luminal and abluminal faces as well as their basal lamina. The fenestral diaphragm of CP vessels has a particularly high concentration of anionic sites (Dermietzel et al., 1983). The fenestrated capillaries of renal tubules (Charonis and Wissig, 1983) and of the CP (Taguchi et al., 1998) are surrounded by a basal lamina replete with fixed negative charges detected with the cation ruthenium red and by cationic colloidal iron, respectively. It is inferred that these anionic groups are contributed by sulfated proteoglycans, inasmuch as they are markedly diminished by exposing tissue sections of CP to enzymatic digestion with heparatinase and chondoitinase (Taguchi et al., 1998). Comparable treatment, in vivo, permeabilizes the basal lamina of the renal glomerulus to the large protein ferritin (~445 kDa), which in its native, anionic state is filtered out by the intact basal lamina (Kanwar et al., 1979). As in the kidney, the basal laminae of the CP appear to be charge-selective filters. In the lupus-like mouse, the associated nephritis is marked by proteinuria that may be, at least in part, the consequence of a focal loss of anionic sites in its perivascular basal lamina. A similar increase in permeability of CP basal laminae would be expected to follow the depletion of their fixed anionic loci.

The inflammation accompanying SLE is also expressed by the infiltration of monocytes into the CP and the perivascular spaces of parenchymal, barrier vessels. In the New Zealand strain of mice that serves as a model for lupus, leukocytes penetrate into the CP stroma, subarachnoid space, and perivascular spaces (Rudick and Eskin, 1983). Here, too, the pathway is the pia-arachnoid core of the choroid plexus.

2.4 CONCLUSIONS

The possibility that, as in other epithelia, the basal lamina and ECM of the CP are depots for the sequestration and release of factors mediating growth, differentiation, and migration of cells merits investigation. The alternative to this notion is the view that the choroidal matrix is not a depot but is still necessary for the anchoring of cell layers, thereby maintaining their polarity and paracellular permeability. A second inference is that the plexus's microvilli are labile in shape and length, sensitive to a variety of environmental conditions, and completely devoid of organelles involved in, e.g., protein synthesis. They are incapable of endocytosis and are easily damaged by a number of chemically unrelated agents. Yet they do present a large surface area bearing enzymes involved in CSF production, transporters, and adhesion molecules for leukocyte passage. Even if toxic damage were to be restricted to the apical microvilli, the plexus's formation of CSF would probably be appreciably curtailed. It would be of interest to see whether their plasma membranes are devoid of the proteins, clathrins, or caveolins required for endocytosis and transcytosis.

REFERENCES

Anders, J.J. and Brightman, M.W. (1986) Freeze-fracture studies of plasma membranes of astrocytes in freezing lesions. *Adv Neurol* 44: 765–774.

Atkins, C.J., Kondon, J.J., Quismorio, F.P. and Friou, G.J. (1972) The choroid plexus in systemic lupus erythematosis. *Ann Intern Med* 76: 65–72.

Aumailley, M., Pesch, M., Tunggal, L., Gaill, F. and Fassler, R. (2000) Altered synthesis of laminin 1 of basement membrane component deposition in (beta)1 integrin-deficient embryoid bodies. *J Cell Sci* 113: 259–268.

Barsky, S.H., Siegal, G.P., Jannotta, F. and Liotta, L.A. (1983) Loss of basement membrane components by invasive tumors but not by their benign counterparts. *Lab Invest* 49: 140–147.

Bashkin, P., Doctrow, S., Klagsbrun, M., Svahn, C.M., Folkman, J. and Vlodavsky, I. (1989) Basic fibroblast growth factor binds to subendothelial extracellular matrix and is released by heparatinase and heparin-like molecules. *Biochemistry* 28: 1737–1743.

Becker, N.H., Novikoff, A.B. and Zimmerman, H.M. (1967) Fine structure observations of the uptake of intravenously injected peroxidase by the rat choroid plexus. *J Histochem Cytochem* 15:160–165.

Berche, P. (1995) Bacteremia is required for invasion of the murine central nervous system by *Listeria monocytogenes. Microb Pathog* 18: 323–36.

Bernstein, J.J., Goldberg, W.J. and Laws, E.R. Jr. (1989) Human malignant astrocytoma xenografts migrate in rat brain: a model for central nervous system cancer research. *J Neurosci Res* 22: 134–143.

Bishop, N.E. (1997) An update on non-clathrin-coated endocytosis. *Rev Med Virol* 7: 199–209.

Bodenheimer, T.S. and Brightman, M.W. (1968) A blood-brain barrier to peroxidase in capillaries surrounded by perivascular spaces. *Am J Anat* 122: 249–267.

Brightman, M.W. (1967) Intracerebral movement of proteins injected into blood and cerebrospinal fluid. *Anat Rec* 157: 219.

Brightman, M.W. (1968) The intracerebral movement of proteins injected into blood and cerebrospinal fluid of mice. *Prog. Brain Res* 29: 19–37.

Brightman, M.W. (1975) Ultrastructural Characteristic of Adult Choroid Plexus: Relation to the Blood-Cerebrospinal Fluid Barriers. In Netsky, M. and Shuangshoti, S. (Eds.): *The Choroid Plexus in Health and Disease.* Charlottesville: University Press of Virginia, pp. 86–112.

Brightman, M.W. (1977) Morphology of blood-brain interfaces. *Exp Eye Res* 25: 1–25.

Brightman, M.W. (2002) The brain's interstitial clefts and their glial walls. *J. Neurocytol* 31: 595–603.

Brightman, M.W. and Kaya, M. (2000) Permeable endothelium and the interstitial space of brain. *Cell Mol Biol* 20: 111–130.

Brightman, M.W. and Reese, T.S. (1969) Junctions between intimately apposed cell membranes in the vertebrate brain. *J Cell Biol* 40: 648–677.

Broadwell, R.D., Baker-Cairns, B.J., Friden, P.M., Oliver, C., and Villegas, J.C. (1996) Transcytosis of protein through the mammalian cerebral epithelium and endothelium. III. Receptor-mediated transcytosis through the blood-brain barrier of blood-borne transferrin and antibody against the transferrin receptor. *Exp Neurol* 142: 47–65.

Carter, L.P., Beggs, J. and Waggener, J.D. (1972) Ultrastructure of three choroid plexus papillomas. *Cancer* 30: 1130–1136.

Castel, M., Sahar, A. and Erlij, D. (1974) The movement of lanthanum across diffusion barriers in the choroid plexus of the cat. *Brain Res* 67: 178–184.

Charonis, A.S. and Wissig, S.L. (1983) Anionic sites in basement membranes. Differences in their electrostatic properties in continuous and fenestrated capillaries. *Microvasc Res* 25: 265–285.

Courtoy, P.J., Timpl, R. and Farquhar, M.G. (1982) Comparative distribution of laminin, type IV collagen, and fibronectin in the rat glomerulus. *J Histochem Cytochem* 30: 874–886.

Danen, E.H.J., Sonneveld, P., Brakebusch, C., Fassler, R. and Sonnenberg, A. (2002) The fibronectin-binding integrins $\alpha 5\beta$–1 and $\alpha v\beta$–3 differentially modulate RhoA-GTP loading, organization of cell matrix adhesions, and fibronectin fibrillogenesis. *J Cell Biol* 159: 1071–1086.

Davet, J., Clavel, B., Datas, L., Mani-Ponset, L., Maurel, D., Herbute, S., Viso, M., Hinds, W., Jarvi, J. and Gsbrion, J. (1998) Choroidal readaptation to gravity in rats after spaceflight and head-down tilt. *J Appl Physiol* 64, 19–29.

Deguchi, Y., Okutsu, H., Okura, T., Yamada, S., Kimura, R., Yuge, T., Furukawa, A., Morimoto, K., Tachikawa, M., Otsuki, S., Hosoya, K. and Terasaki, T. (2002) Internalization of basic fibroblast growth factor at the mouse blood-brain barrier involves perlecan, a heparan sulfate proteoglycan. *J Neurochem* 83: 381–389.

Del Maschio, A., De Luigi, A., Martin-Padura, I., Brockhaus, M., Bartfai, T., Fruscella, P., Adorini, L., Martino, G., Furlan, R., De Simoni, M.G. and Dejana, E. (1999) Leukocyte recruitment in the cerebrospinal fluid of mice with experimental meningitis is inhibited by an antibody to junctional adhesion molecule (JAM). *J Exp Med* 190:1351–1356.

Dermietzel, R., Thurauf, N. and Kalweit, P. (1983) Surface charges associated with fenestrated brain capillaries. II. In vivo studies on the role of molecular charge in endothelial permeability. J *Ultrastruct Res* 84: 111–119.

DiPaolo, G. and DeCamilli, P. (2003) Does clathrin pull the fission trigger? *Proc Natl Acad Sci* 100: 4981–4983.

Duprez, T., Nzeusseu, A., Peeters, A. and Houssiau, F.A. (2001) Selective involvement of the choroid plexus on cerebral magnetic resonance images: a new radiological sign in patients with systemic lupus erythematosus with neurological symptoms. *J Rheumatol* 28: 387–391.

Elkin, M., Ilan, N., Ishai-Michaeli, R., Friedmann, Y., Papo, O., Pecker, I. and Vlodavsky, I. (2001) Heparanase as mediator of angiogenesis: mode of action. *FASEB J* 15: 1661–1663.

Endo, H., Sasaki, K., Tonosaki, A. and Kayama, T. (1998) Three-dimensional and ultrastructural ICAM-1 distribution in the choroid plexus, arachnoid membrane and dural sinus of inflammatory rats induced by LPS injection in the lateral ventricles. *Brain Res* 793: 297–301.

Engelhardt, B., Wolburg-Buchholz, K., and Wolburg, H. (2001) Involvement of the choroid plexus in central nervous system inflammation. *Micr Res Tech* 52: 112–129.

Eriksson, C., Nobel, S., Winblad, B. and Schultzberg, M. (1999) Expression of interleukin 1alpha and 1beta and interleukin 1 receptor antagonist mRNA in the rat central nervous system after peripheral administration of lipopolysaccharides. *Cytokine* 12: 423–431.

Flanc, R., Roberts, M., Strippoli, G., Chadban, S., Kerr, P. and Atkins, R. (2004) Treatment for lupus nephritis. *Cochrane Database Syst Rev* 1: CD002922.

Folkman, J., Klagsbrun, M., Sasse, J., Wadzinski, M., Ingber, D. and Vlodavsky, I. (1988) A heparin-binding angiogenic protein-basic fibroblast growth factor is stored within basement membrane. *Am J Pathol* 130: 393–400.

Ford-Perriss, M., Turner, K., Guimond, S., Apedaile, A., Haubeck, H.D., Turnbull, J. and Murphy, M. (2003) Localisation of specific heparan sulfate proteoglycans during the proliferative phase of brain development. *Dev Dyn* 227:170–84.

Furness, P.N., Lowe, J. and Tarrant, G.S. (1990) Subepithelial basement membrane deposition and intermediate filament expression in choroid plexus neoplasms and ependymomas. *Histopathology* 16: 251–255.

Garabedian, B.V., Lemaigre-Dubreuil, Y. and Mariani J. (2000) Central origin of IL-1beta produced during peripheral inflammation: role of meninges. *Mol Brain Res* 75: 259–263.

Gudeman, D., Brightman M.W., Merisko, E. and Merrill, C. (1989) Release from live choroid plexus of apical fragments and electrophoretic characterization of their synthetic products. *J Neurosci Res* 3:163–175.

Hammann, G.F., Okada, Y., Fitridge, R. and del Zeppo, G.J. (1995) Microvascular basal lamina antigens disappear during cerebral ischemia and reperfusion. *Stroke* 26: 2120–2126.

Handler, M., Yurchenco, P.D. and Iozzo, R.V. (1997) Developmental expression of perlecan during murine embryogenesis. *Dev Dyn* 210: 130–145.

Ishai-Michaeli, R., Eldor, A. and Vlodavsky, I. (1990) Heparanase activity expressed by platelets, neutrophils, and lymphoma cells release active fibroblast growth factor from extracellular matrix. *Cell Reg* 1: 833–842.

Johanson, C.E., Palm, D.E., Primiano, M.J., McMillan, P.N., Chan, P., Knuckey, N.W. and Stopa, E.G. (2001) Choroid plexus recovery after transient forebrain ischemia: role of growth factors and other repair mechanisms. *Cell Mol Neurobiol* 20: 197–216.

Kanwar, Y.S., Linker, A. and Farquhar, M.G. (1980) Increased permeability of the glomerular basement membrane to ferritin after removal of glycosaminoglycans (heparan sulfate) by enzyme digestion. *J Cell Biol* 86: 688–693.

Kaur, C. (1997) Effects of chloroquine on the ependyma, choroid plexus and epiplexus cells in the lateral ventricles of rats. *J Hirnforsch* 38: 99–106.

Keep, R.F. and Jones, H.C. (1990) A morphometric study on the development of the lateral ventricle choroid plexus, choroid plexus capillaries and ventricular ependyma in the rat. *Dev Brain Res* 56, 47–53.

Knox, S., Merry, C., Stringer, S., Melrose, J. and Whitelock, J. (2002) Not all perlecans are created equal: interactions with fibroblast growth factor (FGF) 2 and FGF receptors. *J Biol Chem* 277:14657–14665.

Kolb, S.A., Lahrtz, F., Paul, R., Leppert, D., Nadal, D., Pfister, H.-W. and Fontana, A. (1998) Matrix metalloproteinases and tissue inhibitors of metalloproteinases in viral meningitis: upregulation of MMP-9 and TIMP-1 in cerebrospinal fluid. *J Neuroimmunol* 84: 143–150.

Koshiba, K. (1987) Ultrastructure of the choroid plexus epithelium of pigeons treated with drugs: II. Effect of cytochalasin D and colchicines. *Am J Anat* 178: 133–43.

Koss, M.N., Chernack, W.J., Griswold, W.R. and McIntosh, R.M. (1973) The choroid plexus in acute serum sickness. Morpholgic, ultrastructural, and immunohistologic studies. *Arch Pathol* 96:331 and 334.

Levine, S. (1987) Choroid plexus: target for systemic disease and pathway to the brain. *Lab Invest* 56: 231–233.

Levine, S. and Sowinski, R. (1973) Choroid plexitis produced in rats by cyclophosphamide. *J Neuropathol Exp Neurol* 32: 365–70.

Lippoldt, A., Jansson, A., Kniesel, U., Andbjer, B., Andersson, A., Wolburg, H., Fuxe, K. and Haller, H. (2000) Phorbol ester induced changes in tight and adherens junctions in the choroid plexus epithelium and in the ependyma. *Brain Res* 854: 197–206.

Lo, C.-M., Keese, C.R. and Giaever, I. (1999) Cell-substrate contact: another factor may influence transepithelial electrical resistance of cell layers cultured on permeable filters. *Exp Cell Res* 250: 576–580.

Lum, H. and Malik, A.R. (1994) Regulation of vascular endothelial barrier function. *Am J Physiol* 11: L223-L241.

Masuzawa, T., Shimabukoro, H., Sato, F. and Saito, T. (1981) Ultrastructural localization of carbonic anhydrase activity in the rat choroid plexus epithelial cell. *Histochemistry* 73: 201–209.

Masseguin, C., Mani-Ponset, L., Herbute, S., Tixier-Vidal, A. and Gabrion, J. (2001) Persistence of tight junctions and changes in apical structures and protein expression in choroid plexus epithelium of rats after short-term head-down tilt. *J Neurocytol* 30: 365–377.

McCutcheon, I.A., Friend, K.E., Gerdes, T.M., Zhang, B.-M., Wildrick, D.M. and Fuller, G.N. (2000) Intracranial injection of human meniongioma cells in athymic mice: an orthopic model for meningioma growth. *J Neurosurg* 92: 306–314.

McIntosh, R.M., Koss, M.N., Chernack, W.B., Griswold, W.R., Copack, P.B. and Weil, III, R. (1974) Experimental pulmonary disease and autoimmune nephritis in the rabbit produced by homologous and heterologous choroid plexus (experimental Goodpasture's syndrome). *Proc Soc Exp Biol Med* 147: 216–223.

McNutt, N.S. (1978) A thin-section and freeze-fracture study of microfilament-membrane attachments in choroid plexus and intestinal microvilli. *J Cell Biol* 79: 774–787.

Mercier, F., Kitasako, J.T. and Hatton, G.I. (2002) Anatomy of the brain neurogenic zones revisited: fractones and the fibroblast/macrophage network. *J Comp Neurol* 451: 170–188. Erratum in: *J Comp Neurol* (2002) 454: 495.

Milhorat, T.H., Davis, D.A. and Hammock, M.K. (1976) Choroid plexus papilloma II. Ultrastructure and ultracytochemical localization of Na-K-ATPase. *Child's Brain* 2: 290–303.

Minshall, R.D., Sessa, W.C., Stan, R.V., Anderson, R.G.W. and Malik, A.B. (2003) Caveolin regulation of endothelial function. *Am J Physiol* 285: L1179-L1183.

Murdoch, A.D. and Iozzo, R.V. (1993) Perlecan: the multidomain heparan sulphate proteoglycan of basement membrane and extracellular matrix. *Virchows Arch A Pathol Anat Histopathol* 423: 237–242.

Nader, H.B., Dietrich, C.P., Buonassisi, V. and Colburn, P. (1987) Heparin sequences in the heparan sulfate chains of an endothelial cell proteoglycan. *Proc Natl Acad Sci USA* 84: 3565–3569.

Pagenstecher, A., Stalder, A.K., Kincaid, C.L., Shapiro, S.D. and Campbell, I.L. (1998) Differential expression of matrix metalloproteinase and tissue inhibitor of matrix metalloproteinase genes in the mouse central nervous system in normal and inflammatory states. *Am J Pathol* 152: 729–742.

Pardridge, W.M., Triguero, D., Buciak, J. and Yang, J. (1990) Evaluation of cationized rat albumin as a potential blood-brain barrier drug transport vector. *J Pharmacol Exp Ther* 255: 893–899.

Pascarlu, M., Bendayan, M. and Ghitescu, L. (2004) Correlated endothelial caveolin overexpression and increased transcytosis in experimental diabetes. *J Histochem Cytochem* 52: 65–76.

Peress, N.S., Miller, F. and Palu, W. (1977) The immunopathophysiological effects of chronic serum sickness on rat choroid plexus, ciliary process and renal glomerulus. *J Neuropathol Exp Neurol* 36: 726–733.

Peress, N.S. and Tompkins, D.T. (1979) Rat CNS in experimental chronic serum sickness: integrity of the zonulae occludentes of the choroid plexus epithelium and brain endothelium in experimental chronic serum sickness. *Neuropathol App Neurobiol* 5: 279–288.

Peters, A. (1974) The surface fine structure of the choroid plexus and ependymal lining of the rat lateral ventricle. *J Neurocytol* 3: 99–108.

Peters, A. and Swan, R.C. (1979) The choroid plexus of the mature and aging rat: the choroidal epithelium. *Anat Rec* 194: 325–354.

Pino, R.M. (1985) Restriction to endogenous plasma proteins by a fenestrated capillary endothelium. An ultrastructural immunocytochemical study of the choriocapillary endothelium. *Am J Anat* 172: 279–289.

Prats, N., Briones, V., Blanco, M.M., Altimira, J., Ramos, J.A., Dominguez, L. and Marco, A. (1992) Choroiditis and meningitis in experimental murine infection with Listeria monocytogenes. *Eur J Microbiol Infect Dis* 11: 744–747.

Quinton, P.M., Wright, E.M. and Tormey, J.M. (1973) Localization of sodium pumps in the choroid plexus epithelium. *J Cell Biol* 58: 724–730.

Raab, G. and Klagsbrun, M. (1997) Heparin-binding EGF-like growth factor. *Biochim Biophys Acta* 1333: F179–199.

Raats, C.J.I., Luca, M.E., Bakker, M.A.H., van der Wal, A., Heeringa, P., van Goor, H., van der Born, J., de Heer, E. and Berden, J.H.M. (1999) Reduction in glomerular heparan sulfate correlates with complement deposition and albuminuria in active Heymann nephritis. *J Am Soc Nephrol* 10: 1689–1699.

Rizzolo, L.J. (1999) Polarization of the Na^+,K^+-ATPase in epithelia derived from the neuroepithelium. *Int Rev Cytol* 185: 195–235.

Ross, H.J. and Wright, E.M. (1984) Neutral amino acid transport by plasma membrane vesicles of the rabbit choroid plexus. *Brain Res* 295: 155–160

Rothstein, R.P. and Levison, S.W. (2002) Damage to the choroid plexus, ependyma and subependyma as a consequence of perinatal hypoxia/ischemia. *Dev Neurosci* 24: 426–436.

Rudick, R.A. and Eskin, T.A. (1983) Neuropathological features of a lupus-like disorder in autoimmune mice. *Ann Neurol* 14: 325–332.

Saito, K., Naito, I., Seki, T., Oohashi, T., Kimura, E., Momota, R., Kishiro, Y., Sado, Y., Yoshioka, H. and Ninomiya, Y. (2000) Differential expression of mouse alpha5(IV) and alpha6(IV) collagen genes in epithelial basement membranes. *J Biochem* 128: 427–434.

Schmidley, J.W. and Wissig, S.L. (1986) Anionic sites on the luminal surface of fenestrated and continuous capillaries of the CNS. *Brain Res* 363: 265–271.

Sixt, M., Engelhardt, B., Pausch, F., Hallmann, R., Wendler, O. and Sorokin, L.M. (2001) Endothelial cell laminin isoforms, laminins 8 and 10, play decisive roles in T cell recruitment across the blood-brain barrier in experimental autoimmune encephalomyelitis. *J Cell Biol* 153: 933–946.

Smeland, S., Kolset, S.O., Lyon, M., Norum, K.R. and Blomhoff, R. (1997) Binding of perlecan to transthyretin *in vitro. Biochem J* 326: 829–836.

Taguchi, T., Ohtsuka, A. and Murakami, T. (1998) Light and electron microscopic detection of anionic sites in the rat choroid plexus. *Arch Hist Cyt* 61: 243–252.

Taipale, J. and Keski-Oja, J. (1997) Growth factors in the extracellular matrix. *FASEB J* 11: 51–59.

Tarlow, M.J., Jenkins, R., Comis, S.D., Osborne, M.P., Stephens, S., Stanley, P. and Crocker, J. (1993) Ependymal cells of the choroid plexus express tumor necrosis factor-alpha. *Neuropathol Appl Neurobiol* 19: 324–328.

Thomas, T., Stadler, E. and Dziadek, M. (1992) Effects of the extracellular matrix on fetal choroid plexus epithelial cells: changes in morphology and multicellular organization do not affect gene expression. *Exp Cell Res* 203: 198–213.

Tilling, T., Korte, D., Hoheisel, D. and Galla, H.-J. (1998) Basement membrane proteins influence brain capillary endothelial barrier and function *in vitro. J Neurochem* 71: 1151–1157.

Tuomanen, E. (1996) Entry of pathogens into the central nervous system. *FEMS Microbiol Rev* 18: 289–299.

Valois, A.A. and Webster, W.S. (1989) The choroid plexus as a target site for cadmium toxicity following chronic exposure in the adult mouse: an ultrastructural study. *Toxicology* 55: 193–205.

van der Flier, A. and Sonnenberg, A. (2001) Function and interactions of integrins. *Cell Tissue Res* 305: 285–298.

van Deurs, B., Von Bulow, F. and Moller, M. (1981) Vesicular transport of cationized ferritin by the epithelium of the rat choroid plexus. *J Cell Biol* 89: 131–139.

Villegas, J.C. and Broadwell, R.D. (1993) Transcytosis of protein through the mammalian cerebral epithelium and endothelium. II Adsorptive transcytosis of WGA–HRP and the blood-brain and brain-blood barriers. *J Neurocytol* 22: 67–80.

Vlodavsky, I., Fuks, Z., Ishai-Michaeli, R., Bashkin, P., Levi, E., Korner, G., Bar-Shavit, R. and Klagsbrun, M. (1991) Extracellular matrix-resident basic fibroblastic growth factor: implication for the control of angiogenesis. *J Cell Biochem* 45: 167–176.

Vorbrodt, A.W. (1989) Ultracytochemical characterization of anionic sites in the wall of brain capillaries. *J Neurocytol* 18: 359–368.

Whitelock, J.M., Murdoch, A.D., Iozzo, R.V. and Underwood, P.A. (1996) The degradation of human endothelial cell-derived perlecan and release of bound basic fibroblast growth factor by stromelysin, collagenase, plasmin and heparanases. *J Biol Chem* 271: 10079–10086.

Williams, A.E. and Blakemore, W.F. (1990) Pathogenesis of meningitis caused by Streptococcus suis type 2. *J Infect Dis* 162: 474–481.

Yebenes, de Garcia, E., Ho, A., Damani, T., Fillit, H. and Blum, M. (1999) Regulation of the heparan sulfate proteoglycan, perlecan, by injury and interleukin-1 alpha. *J Neurochem* 73: 812–820.

3 Junctional Proteins of the Blood–CSF Barrier

Susanne Angelow and Hans-Joachim Galla
Westfälische Wilhelms-Universität
Münster, Germany

CONTENTS

3.1 INTRODUCTION

The complex functions of the mammalian brain require controlled neuronal activity in a highly constant chemical environment. Two different anatomical structures, the so-called blood–brain barrier (BBB) and the blood–cerebrospinal fluid barrier

(BCB), serve to maintain brain homeostasis by separating the central nervous system (CNS) from the bloodstream. The BBB is located in the vessel wall of the cerebral capillaries that are lined by highly specialized endothelial cells.[1] This cell type is interconnected by strong intercellular junctions that are responsible for an effective closure of the paracellular shunt and that impede any uncontrolled diffusion of hydrophilic chemical species from the circulating blood flow into the interstitial fluid of the brain. Access to the CNS is thereby restricted to those compounds that are transported by corresponding channels or active transport systems in the plasma membrane of the endothelial cells.

The situation is somewhat different for hydrophobic molecules that are, in principle, capable of diffusing freely across the endothelial plasma membranes simply due to their physicochemical nature. In very limited regions of the brain, which amount approximately to 0.25% of the total weight of the adult brain, the endothelial cells in the vessel wall are not at all tight, but leaky, even for macromolecular substances and cells. These regions are located in the highly vascularized tissue of the choroid plexus (CP) that provide a direct but nevertheless regulated connection between the blood and the CSF within the CSF-filled ventricles of the brain. Since the CSF is separated from the interstitial fluid of the brain only by a permeable ependymal cell wall, the materials between these two fluid compartments can freely exchange. Thus, the passage across the BCB has to be controlled by a tightly connected choroidal epithelial layer located in the CP.

The CP also contributes by many vectorial transport processes to the nutrition of cerebral cells and to the secretion as well as the clearance of the CSF.[2,3] These complex functions are fulfilled by numerous transport proteins that are polarly distributed within the basolateral cell membrane, for example, nutrient transporters like the sodium-dependent vitamin C transporter SVCT2,[4] and the apical cell membrane like the organic anion transporters[5] that contribute to CSF detoxification. The choroidal epithelium also participates in controlling the composition of the CSF by producing and secreting key CSF proteins, e.g., transthyretin.[6]

The underlying morphological feature of the CP epithelium that effectively separates the CSF from the blood and exchanges molecules between these two fluid compartments by polarly distributed transporters is given by the heteropolymeric belt of integral proteins that encircles the apical end of the lateral membrane, called tight junctions (TJ). The TJ of adjacent cells are strongly interconnected within the intercellular space and act as a diffusion barrier by sealing the paracellular shunt. By this so-called gate function, the entry of water-soluble molecules into the CSF is restrictively controlled. But the TJ do not only constitute a simple diffusion barrier. Some members of the integral proteins of the claudin family, a major component of the TJ, have the ability to form selective pores in TJ strands and facilitate the passage of certain ions.[7] Besides the establishment of a diffusion restraint between two fluid compartments, TJ are also thought to limit lateral diffusion of lipids and proteins within the plasma membrane. This so-called fence function is, in addition to other mechanisms,[8] a key factor in maintenance of cell polarity by separation of the basolateral and the apical membrane compounds.

In this chapter we summarize the current knowledge about TJ models and proteins involved in the formation of junctional strands and focus on the specific barrier characteristics of the CP epithelium and factors manipulating this barrier.

3.2 MORPHOLOGICAL AND FUNCTIONAL ASPECTS OF BARRIER FORMING CELL-CELL CONTACTS

The major junctional complex of epithelial cells is located in well-separated regions at the most apical part of the lateral membrane (TJ), which mainly controls the paracellular pathway. Moving down to the basolateral side, the TJ are neighbored by the adherens junctions, the gap junctions, and the desmosomes (see Figure 3.1a). Here we mainly focus on the TJ, which are considered to form the barrier within this cell–cell contact. As already mentioned, the TJ exhibit a fence and a gate function. The fence function in terms of the polar distribution of proteins is still controversially discussed. Earlier studies describe both intermixture[9] as well as maintenance of polarly distributed proteins[10] after disruption of the junctional barrier. Undoubtedly, the junctional fence preserves lateral diffusion between the apical and the basolateral side of lipids, which are normally present in domain structures also called rafts. However, lipidic components that are not involved in direct biological functions and probably dispersed in a noninteracting manner may diffuse freely within the outer monolayer of the cellular plasma bilayer membrane from one cell to the other.[11] Diffusion across the TJ from the apical to the basolateral membrane part is completely restricted for probes inserted into the outer leaflet of the cell membrane.

The barrier function is built up by the closure of the intercellular cleft, which restricts the intercellular diffusion of solutes between the apical and the basolateral fluid compartment. The tightness of this barrier is an indispensable requirement for the physiological important maintenance of chemical and electrochemical gradients. The efficiency of the junctional barrier can be demonstrated by cultured epithelial cells derived from the porcine CP. After they are grown to confluence on permeable filter membranes, these cells are able to secrete liquor-like fluids into the apical compartment against a hydrostatic pressure corresponding to 40 mm water column, but only if the junctions possess a high tightness, which in this case was achieved by serum withdrawal (see below) (see Figure 3.2).

Thin-section electron micrographs clearly demonstrate the presence of close contacts between adjacent cells, which are also nicknamed "membrane kisses" to characterize the intimate contacts which resemble points of membrane fusion (see Figure 3.1c). Freeze fracture electron microscopy of epithelial cells exhibits a typical pattern of anastomosing strands of intramembraneous particles or fibrils (see Figure 3.1d). Most TJ particles appear at the protoplasmic fracture (P) face; accordingly, the exocytoplasic fracture (E) leaflet exhibits grooves containing only a few particles. However, although a variety of integral proteins specifically localized within these contact points are known (see Section 3), it is not clear whether those proteins necessary to form the contact are by themselves sufficient to build up a tight barrier as postulated in a pure protein model (see Figure 3.3a).

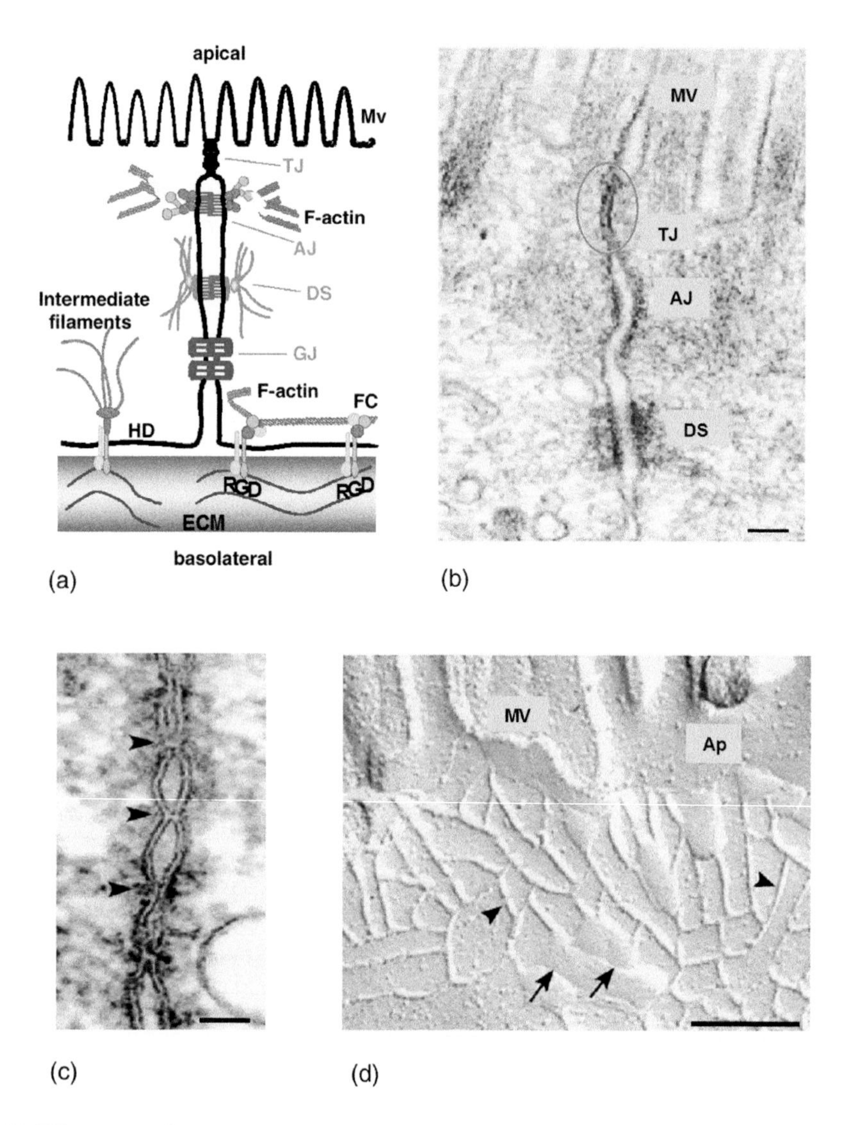

FIGURE 3.1 (a) Diagram and (b) electron micrograph of cell–cell junctions between two adjacent epithelial cells. TJ, tight junctions; AJ, adherens junctions; DS, desmosomes; GJ, gap junctions. On the cytoplasmic site of the membrane, adherens junctions are associated with the actin cytoskeleton, whereas desmosomes are linked to intermediate filaments, for instance, keratin filaments in epithelial cells. Anchorage of the cells to the extracellular matrix (ECM) is provided by focal contacts (FC) or hemidesmosomes (HD), which also differ with respect to their particular connections to the cytoskeleton. The apical membrane surface shows membrane protrusions (microvilli, Mv) that are typical of transporting epithelia. Bar, 200 nm. (c) High-resolution thin-section electron micrograph of the TJ area. The image is a magnification of the highlighted area shown in (b). Arrowheads indicate points of membrane kisses. Bar, 50 nm. (d) Freeze-fracture replicas of the TJ area show strands on the P-face (arrowheads) and corresponding grooves on the E-face (arrows). Ap, apical; Mv, microvilli. Bar, 200 nm. (b–d: From Tsukita et al.[16])

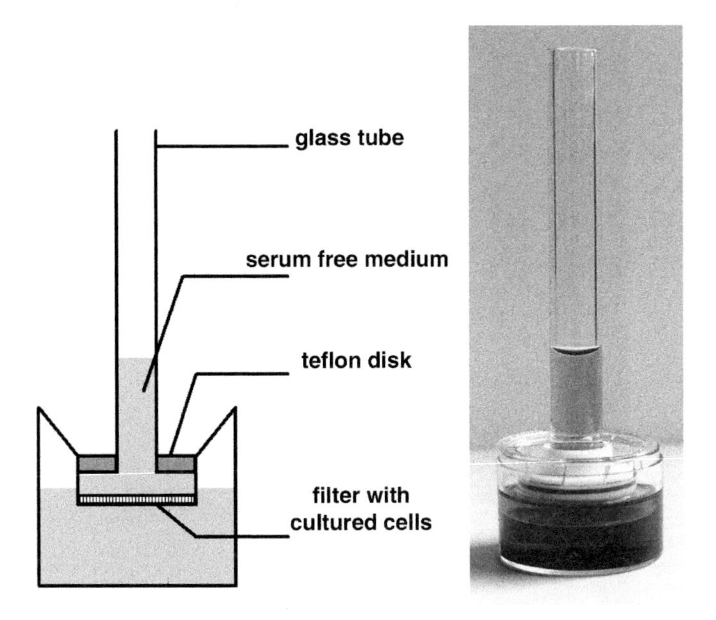

FIGURE 3.2 Cultured CP epithelial cells secret fluid into the apical chamber, a process that is comparable to the CSF secretion in vivo. In this experimental approach the monolayer resists a hydrostatic pressure corresponding to 40 mm water column (= 30% of in vivo pressure).

Lipids may also be involved in sealing the cleft by a special lipid-protein complex (lipid-protein model, Figure 3.3b), first hypothesized by Kachar and Reese in 1982.[12] Grebenkämper and Galla[11] presented experimental evidence that in fact lipid bridges are formed between adjacent cells, allowing externally incorporated fluorescent probes to diffuse within the outer leaflet of the cellular membrane to the neighboring cell. This was only possible under conditions where the TJ were closed. In another set of experiments,[13] our group demonstrated a good correlation with a model for the junctional tightness established by Claude,[14] who noted that the electrical resistance (R) of the monolayer increases logarithmically with the number of TJ strands (n) but also depends on a so-called open-close probability (p) of each fusion contact point ($R \sim p^{-n}$). These data fit well to a model that does not regard the TJ fibrils as fixed but dynamic barriers containing aqueous pores that fluctuate between an open and closed state. We explained this dependence by an equilibrium between inverted cylindrical lipid micelles (closed state) and a bilayer phase (open state).

From experiments using artificial lipid membrane systems, we were able to prove that in fact substances that open TJ shift this equilibrium to the bilayer phase and vice versa. This concept has been reviewed in detail by Wegener and Galla[15] and was taken up by Tsukita et al.[16] In summary, there is no doubt that membrane proteins are main constituents of the TJ. However, it seems to be evident that the extremely high electrical resistances in epithelial cells up to 5000·Ω cm^2 (MDCK I cells), which comes close to the electrical resistance of cell membrane lipid bilayers, is explainable only by involvement of lipids in the constitution of the TJ. We assume

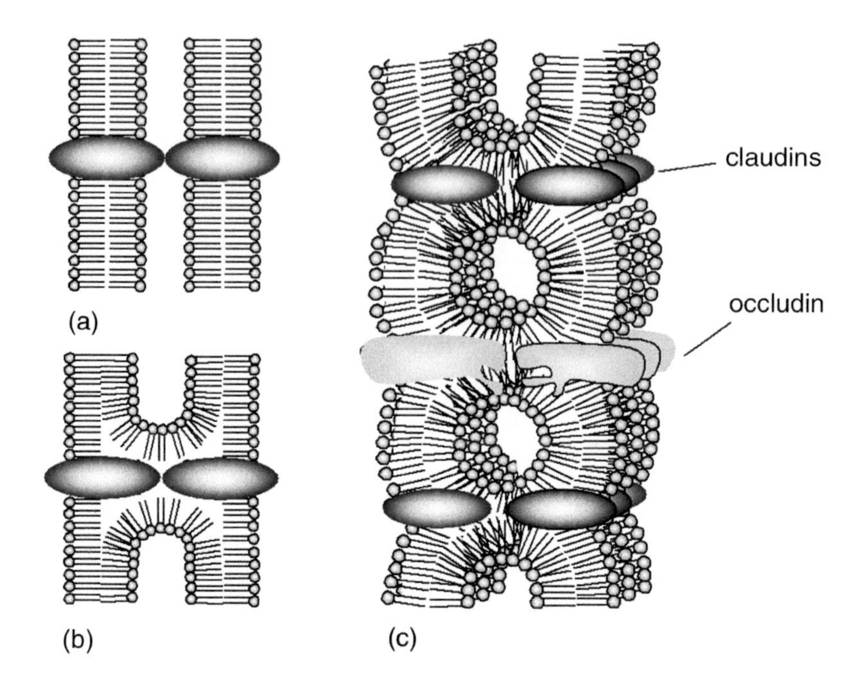

FIGURE 3.3 Models of TJ. (a) Protein model, (b) Lipid-protein model of the formation of lipid bridges by fusion of membranes of adjacent cells. (c) TJ fibrils consisting of polymerized integral proteins and overlaying lipids form cylindrical lipid-protein micelles in ZY-direction.

that hydrophobic proteins like occludin or the claudins or both form the contact between the cells, and the lipids may cover these hydrophobic proteins to form lipid bridges (see Figure 3.3c, XZ direction) or corresponding cylindrical lipid-protein micelles (see Figure 3.3c, ZY direction). This model can be viewed as the closed state as postulated by Claude. Loss of the lipid bridges leaving the pure protein-protein contact sketched in Figure 3.3a can be considered as the open state. A reduction in the insulating lipid layer as a result of weakened membrane-membrane interaction may lead to a higher electrical conductivity. This model may also explain the formation of aqueous pores postulated by Claude[14] and observed by Tsukita and Furuse,[7] which will be discussed in detail below.

3.3 JUNCTIONAL PROTEINS EXPRESSED BY THE CHOROID PLEXUS

3.3.1 Expression of Proteins Localized at Tight Junctional Strands

The previous section discusses anatomical aspects of TJ strands in general. This section will focus on TJ-associated proteins known to be expressed in the CP epithelium. For more general information, the reader is referred to the review articles

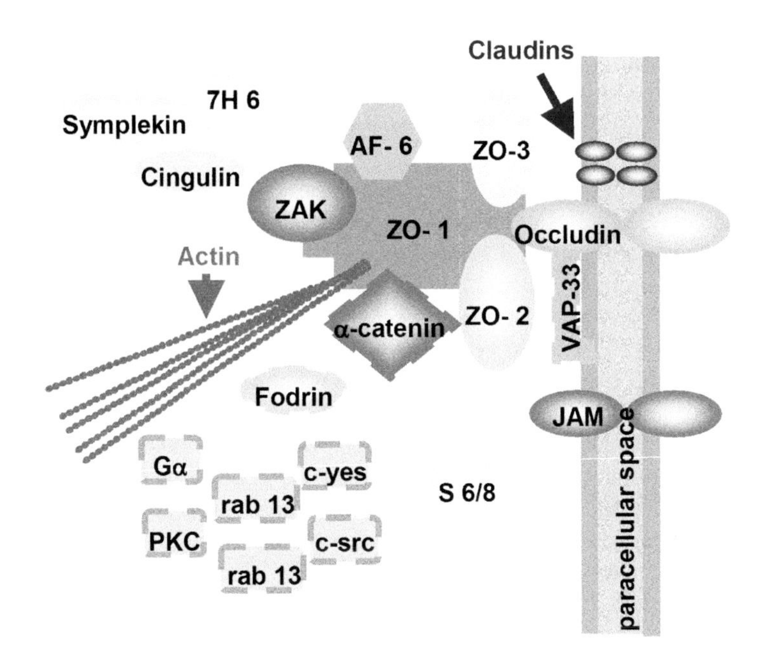

FIGURE 3.4 Schematic drawing of the molecular interaction within the TJ. Integral membrane proteins are associated with surface-bound proteins attached to the cytoskeleton. Regulatory proteins control the assembly and structural organization. (From Fanning et al.[17].)

in the literature.[16,17] The expression and distribution of proteins have been investigated by various methods involving immunofluorescence staining and fluorescence microscopy[18,19] or analysis at the ultrastructural level by electron microscopy using immunogold staining.[20] The junctional complex is illustrated in Figure 3.4.

3.3.1.1 Integral Membrane Proteins

Furuse et al.[21] discovered the first transmembrane component of TJ, called occludin. In this study immunoelectron microscopy revealed that this protein is exclusively localized at TJ of both epithelial and endothelial cells. Occludin, a hydrophobic protein of a molecular size of 60 to 65 kDa, possesses four transmembrane domains: a short cytoplasmatic N-terminal domain, a long cytoplasmatic C-terminal domain, and two extracellular loops that are considered to interact with occludin of an adjacent cell within the intercellular cleft. Two other isoforms of occludin originating by alternative splicing have been identified: the variant occludin 1B containing a 193 bp insertion encoding a longer form of occludin with a unique N-terminal sequence of 56 amino acids[22] and the occludin TM4− (found in primates) that exhibits a 162 bp depletion and lacks the fourth transmembrane domain.[23] Occludin can be phosphorylated on serine and threonine residues, which is considered a regulatory element for the protein's localization and barrier integrity.[24–27] The cytoplasmatic part of occludin interacts with other TJ-associated proteins like ZO-1 (see below), leading to an indirect linkage of occludin to the actin cytoskeleton. Interestingly, occludin does not seem to be an indispensable component of TJ strands since occludin-deficient mouse embryos are viable and do not show gross modifications of TJ strands,[28] although several histological abnormalities are observed. Furthermore, in some cell types TJ strands can also be formed without the involvement of occluding.[29,30]

Occludin has been shown to be expressed in the CP epithelial cells as well as less in CP endothelial cells by Lippoldt et al.[19] using immunofluorescence staining of thin frozen sections of rat brain. Whereas the epithelial cell borders of the CP are strongly immunoreactive to occludin, the interendothelial borders of the leaky CP capillaries show only moderate occludin immunoreactivity. To date the most detailed study of TJ expression in the CP was performed by Vorbrodt and Dobrogowska,[20] who used electron microscopy on immunogold-stained ultrathin fractions of mouse tissue. In this study, a high labeling density of occludin was found exclusively at the apical portion of TJ. In contrast, the immunogold reaction at the interendothelial junctions in the fenestrated CP capillaries was found to be negative or uncertain. The authors also studied the TJ composition of the endothelial cells at the BBB and found remarkable labeling of occludin at the TJ of these cells.

Immunofluorescence staining of cultured CP epithelial cells also showed a continuous pattern of occludin in in vitro models based on rat[31] and porcine[32] epithelial cells. These observations, combined with the results of freeze fracture and immunoelectron studies, are consistent with the establishments of TJ in cultured CP epithelial cells.

Besides the identification of occludin, Furuse et al.[33] also described the existence of two other proteins, claudin-1 and claudin-2, which colocalize with occludin at the TJ. Database search has identified other proteins with polypeptide sequence similar to claudin-1 and claudin-2. So far, more than 20 members of the claudin family are known,[34] which have in common the molecular weight of 22 to 24 kDa. The protein structure of claudins includes four transmembrane domains and two extracellular loops, resembling the structure of occludin, but they do not share any sequence homologies with occludin.

Claudins are widely expressed with varied expression patters depending on tissue. Tsukita et al.[16] outlined the growing evidence that claudins constitute the backbone of TJ. In contrast to occludin, claudin expressing mouse fibroblast transfectants showed $Ca^{2}+$-independent cell adhesion activity and many kissing points of TJ between adjacent cells formed by polymerized claudin molecules at the plasma membranes.[35] Only when transfected into fibroblasts expressing claudin fibrils is the occludin molecule recruited to these strands, suggesting that claudins are the major elements in driving TJ formation. For better understanding of claudin function and importance, three different knockout mice have been generated so far, which have different impacts on the phenotype such as neurological and reproductive defects in claudin-11 -/- ,[36] an increased BBB permeability for small molecules accompanied by death of the animals within 10 hours after birth in claudin-5 -/-,[37] and severe modification of the epidermal barrier with wrinkled skin and death within one day after birth in claudin-1 -/- mice.[38]

The expression of claudin-1, -2, and -11 in the CP epithelium has been shown by Wolburg et al.[39] by using immunofluorescence staining of mouse tissue. They also found claudin-1 and claudin-2 to be expressed at the TJ of the rat CP epithelial cells.[18] To a lesser extent a diffuse claudin-5 staining was observed in the vicinity of the junctional region. The situation seems to be reversed in the endothelial cells of the rat CP. In this cell type claudin-5 is highly expressed in contrast to a weak expression of claudin-1, whereas claudin-11 is not detectable. Claudin-1 and -2, but not claudin-5, expression was also observed in porcine CP epithelial cells in vitro (our unpublished data). In their electron microscopy studies, Vorbrodt and Dobrogowska[20] found low or questionable labeling at the CP epithelium and rather negative labelling at the CP endothelium of claudin-1 and -5 in the CP, but a significant labelling of claudin-5 in the endothelial cells of the brain capillaries.

The third type of integral membrane proteins of the TJ is represented by the junction adhesion molecules (JAM), which are assigned to the immunoglobulin superfamily. JAM are widely expressed at TJ of epithelial and endothelial cells where they span the intercellular cleft by their long extracellular domain anchored in the membrane only by one transmembrane domain. According to Martin-Padura et al.,[40] JAM-1 is expressed in CP vessels as well as in CP epithelial cells, proven by immunofluorescence staining. Vorbrodt and Dobrogowska[20] found a slightly positive or questionable JAM-1 labeling in the CP epithelium and no signal in the fenestrated endothelial cells of the capillaries. Analysis of murine and rat brain section shows JAM-1-positive immunostaining in brain capillaries.[20,40]

3.3.1.2 Peripheral Proteins

A peripheral membrane phosphoprotein with a mass of approximately 220 kDa, ZO-1, is the first protein identified as a TJ-associated component of various cell types.[41] It is not exclusively located at TJ, but found also to be expressed in cell types lacking TJ in colocalization with cadherins.[42] As a result of alternative splicing, this protein exists in two isoforms, ZO-1a$^+$ and ZO-1a$^-$, which differ by a 80-amino-acid sequence.[43] ZO-1 and its two homologues, ZO-2 and ZO-3, which have been identified by co-immunoprecipitation with ZO-1,[44,45] belong to a super family of MAGUKs (membrane-associated guanylate kinase-like homolog). ZO-1, -2, and -3 bind directly to the carboxy-termini of claudins by their PDZ1-domains[46] and to the carboxy-termini of occludin by their GUK-domain.[47,48,49] Further studies show that the sequence in occludin necessary for the ZO association is also indispensable for the TJ localization,[50] suggesting that ZOs form a scaffold for the organisation of TJ. All ZO proteins have also been demonstrated to bind directly to actin filaments[47,51] and are therefore considered to have a major function in linking the TJ to the cytoskeleton. Tsukita et al.[16] have suggested a regulatory function of this interaction on TJ. Besides their involvement in TJ assembly, ZOs are also thought to play a role in signaling cascades and transcriptional regulation.

ZO-1 is known to be a component of CP and brain microvascular TJ,[52] whereas the expression of ZO-2 and -3 in the CP has not yet been investigated. Regarding the blood–CSF barrier in vitro, Gath et al.[53] found ZO-1 immunofluorescence staining at cell borders in primary cultured porcine CP epithelial cells (PCPEC). Vorbrodt and Dobrogowska[20] detected a scattered cytoplasmatic ZO-1 staining along the entire length of the intercellular clefts of murine epithelial CP and microvascular endothelial cells using electron microscopy. The authors outline that such a localization supports the view that ZO-1 represents a peripheral membrane protein at both TJ and adherens junctions. A high signal density similar to that found in interepithelial junctions was also observed in the interendothelial junctions of the fenestrated CP vessels. The localization of other ZO molecules was unsuccessful. By summarizing their results, the authors conclude that ZO-1 is not a reliable marker of BBB TJ since it is also present in the interendothelial junction of nonbarrier vessels and it is not restricted to the TJ region. Furthermore, Krause et al.[54] detected ZO-1 expression in cultivated cerebral endothelial cells with depauperate TJ complexity and barrier functions.

Several other proteins have been identified that are thought to be peripheral components of the TJ. Cingulin, for instance, seems to act as a scaffolding protein since it has been shown by Cordenonsi et al.[55] to interact with ZO proteins, JAM, and myosin. Other proteins include 7H6[56] and symplecin,[57] but no data about their expression in the brain barriers are available at present.

3.3.2 Expression of Proteins Localized at Adherens Junctions

When taking into account junctional proteins besides TJ structures, the adherens junctions should be considered as well. The formation of adherens junctions involves a set of homologous integral proteins termed classical cadherins that mediate Ca^{2+}-dependent cell-cell adhesion at these circumferential belts located right underneath the TJ. Additionally, individual spots of cadherin-mediated cell adhesion are spread all over the contact area of opposing cells showing no ultrastructural organization. As reviewed by Tepass et al., cadherins interconnect adjacent cells by forming transdimers with varying lateral overlap in the extracellular region. Adherens junctions are considered to play a critical role in forming and maintaining the polarity of epithelial cells. In developing tissue and freshly seeded epithelial cells in culture, classical cadherins stabilize initial contacts.[59] Cadherins are tightly linked to the actin cytoskeleton by catenins, whose function resembles the role of ZO-1 in TJ. The mechanical bonding of filaments of individual cells gives epithelial cell layers a unique mechanical stability. Furthermore, the catenin-cadherin interaction is required as a key step for cadherin adhesion activity.[59]

The expression of adhesion molecules in rat CP epithelial cells has been investigated by Lippoldt et al.[19] Using a broadly reactive "pan-cadherin" antibody recognizing E, N, P and R cadherin, they found immunopositive fluorescence staining at the interfaces of the epithelial cells. Applying a comparable antibody Vorbrodt and Dobrogowska[20] detected in the above-mentioned study a weak immunostaining in rat and mouse CP epithelial cells along the lateral plasma membrane and occasionally in apical portions in close proximity to TJ. Altogether, the distribution pattern resembled the localization of ZO-1. The staining of the fenestrated endothelium of the CP capillaries was either very weak or questionable. All their attempts using specific antibodies for E, N, P, and VE cadherin gave negative or weak immunoreaction, whereas the latter protein was found in interendothelial junctions of BBB-type microvessels. Immunosignals for α-catenin and β-catenin were also found along the interepithelial and not interendothelial junctional cleft of CP cells, generally omitting the apical region representing the TJ.

3.4 FACTORS DETERMINING BARRIER TIGHTNESS: COMPARISON OF TIGHT JUNCTIONS OF ENDOTHELIAL AND EPITHELIAL CELLS OF BRAIN BARRIERS

3.4.1 Composition of Tight Junction Fibrils

Several factors may link TJ morphology and components to the permeability characteristics of the respective cell layer. One of the predominant aspects is the number of parallel arranged TJ fibrils, which by connecting to each other form a continuous network around the cell as demonstrated by freeze fracture images (see above).[60] When interpreting TJ strands as static paracellular permeability barriers, a linearly increased paracellular resistance is expected with increasing number of fibrils; however, these considerations have been refuted by Claude,[14] who found that the number of fibrils is proportional to the logarithm of the paracellular electrical resistance R (see above).

Exceptions also exist that do not follow the correlation found by Claude. A striking example is provided by Madin-Darby canine kidney (MDCK) I and II strains, which bear tight and leaky TJ, respectively, and can differ in the measured transepithelial electrical resistance (TER) by a factor of >30. Paradoxically, these two strains do not show any differences in the strand number and density of junction fibrils. A plot of Claude's function $R \sim p^{-n}$ assuming a constant fibril number of $n = 4$, experimentally determined for MDCK cells, is given in Figure 3.5.

Respecting the TER values for MDCK I (~6500 $\Omega \cdot cm^2$) and MDCK II (~300 $\Omega \cdot cm^2$), the open probability p can be calculated to be 0.2 and 0.6, respectively. Reagents, such as basic amino acids that weaken the junctional barrier and shift p to the higher values, for example $\Delta p = 0.1$, should decrease the TER values for MDCK I and MDCK II by different amounts. These hypothetical considerations have been experimentally proven by Hein et al.,[61] who investigated the modulation of the MDCK cell resistances by different factors. The idea that the MDCK strains are equipped with different channel characteristics has also been raised by Stevenson et al.[62] This group of investigators speculated that a different phosphorylation state of ZO-1 might cause higher junctional permeability in MDCK II cells.

About 13 years later, Furuse et al.[63] provided another explanation for the very different permeability characteristics of tight and leaky MDCK cells. They found that claudin-1 and claudin-4 were expressed in both MDCK I and MDCK II cells, whereas the expression of claudin-2 was restricted to MDCK II cells. Interestingly, when claudin-2 was introduced into MDCK I cells, the TJ were converted to leaky ones and the TER fell to the level of MDCK II cells. The increased permeability for ions caused by claudin-2 can be explained by the formation of aqueous pores with high conductance either by homotypic adhesion of claudin-2 or by the weak heterotypic adhesion between claudin-2 and claudin-1.

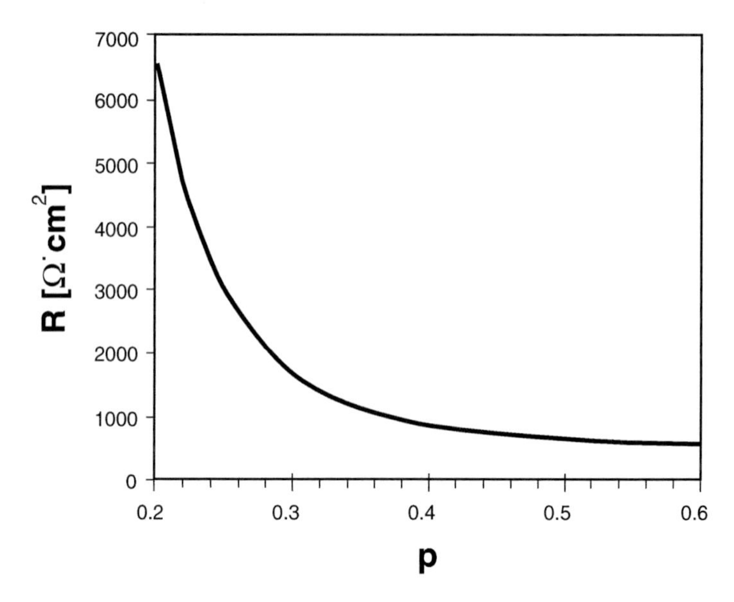

FIGURE 3.5 Calculated transepithelial electrical resistance R as a function of the open probability p and strand number n according to Claude.[14] The resistance was calculated following the dependence: $R_{tj} \cdot l_P = (R_{tj} \cdot l_P)_{min} \cdot p^{-n}$. The expression $(R_{tj} \cdot l_P)_{min}$ is the resistance of the intercellular cleft if no TJ are formed; l_P is the length of the cell contacts per area. To estimate the contact length, we assumed hexagonal cell shape and a cell density of 1106/cm^2. $R_{tj,min}$ is assumed to be $< 10\ \Omega \cdot cm^2$. The transcellular resistance, which is arranged in a circuit parallel to R_{tj}, is assumed to be 10 k$\Omega \cdot cm^2$. (From Hein et al.[61])

Referring to the lipid protein model presented above, the formation of aqueous pores might also be explained by the assumption that these proteins are not able to allow the formation of the lipid bridges shown in Figure 3.3, thus causing defects in the protein-adsorbed lipid monolayer enabling ion diffusion. Amasheh et al.[64] observed a 5.6-fold increase in paracellular conductance and the formation of cation-selective formation in an originally tight MDCK strain overexpressing claudin-2, whereas the fluxes for mannitol, lactulose, and 4 kDa dextran were not changed. Furthermore, Furuse et al.[63] found that in claudin-2–transfected MDCK I cells the freeze-fracture morphology of TJ was shifted from the P-face associated to the E-face-associated type.

The correlation of TJ tightness and the TJ association to the membranes' leaflets has been demonstrated by Wolburg et al.[65] They have shown that TJ in bovine brain capillaries are predominantly associated to the P-face. The E-face displays linear grooves occupied discontinuously with particles that most likely correspond to the spaces between the P-face strands. Loss of barrier function of cultured brain endothelial cells is accompanied by a switch to the E-face-associated type. This phenomenon could be reversed by coculturing the cells with astrocytes or an astrocyte-conditioned medium and by increased cAMP level, respectively. In cultures of endothelial cells of rat brain, the change of TJ particle association from the P-face to E-face was accompanied by a loss of claudin-1.[66] In contrast to the impact of claudin-2 in MDCK cells (see above), the presence of claudin-1 seems to be somehow consistent with P-face association of the TJ.

3.4.2 Particles and Continuity of Choroid Plexus Tight Junctions

Van Deurs and Koehler[67] studied freeze fracture images of the CP and found a surprisingly high number of junctional strands (n~7), which is atypical for a leaky epithelium according to Claude and Goodenough.[60] The TJ of the fixed CP of the rat appear on the P-face as a system of ridges formed by aligned particles. At the E-face, grooves with some scattered particles were observed. The authors distinguish different morphological types of junctions that consist of either parallel strands with very few interconnections (see Figure 3.6), relatively parallel and often anastomosing strands, or strands that appear as a wavy unstretched version of the other types. By analysis of complementary replicas, they tried to prove the existence of pores in the strands that might explain the relatively high conductance of this epithelium described by some authors.[68–70] By studying complementary particles in the E-face that might fill gaps in the P-face, they found that approximately 25% of the TJ were discontinuous. In contrast, the junctions of the "tight" small intestine epithelium showed only ~5% discontinuity. The data about the discontinuity of the TJ in the mammalian CP differs quite significantly in the literature. Mollgard et al.,[71] for example, investigated fetal sheep CP by the double replica technique and found that the few discontinuities of the P-face could be filled by particles in the complementary E-face, whereas Hirsch et al.[72] concluded from their double replica studies on rabbit CP that the TJ fibrils in this epithelium can be considered to be continuous.

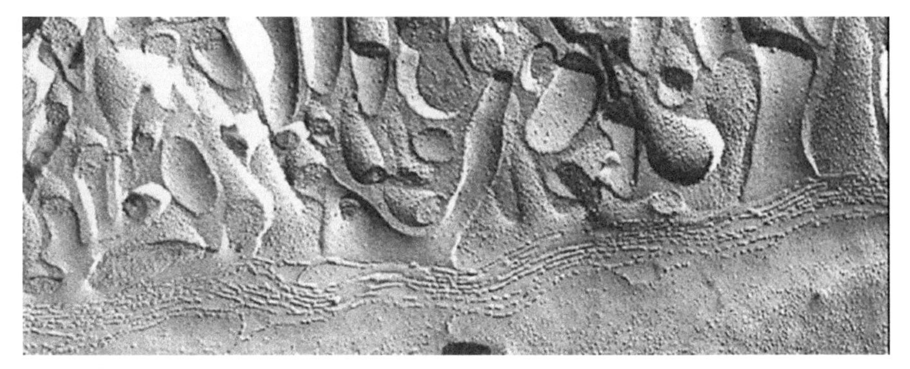

FIGURE 3.6 TJ of the CP epithelium of the rat visualized by freeze fracture (P-face). The junction includes 7 to 8 parallel strands (Type 1 geometry) composed of particles and short bars. (From van Deurs and Koehler.[67])

3.4.3 TER Values of Choroid Plexus Epithelial Cell Layers and the Blood–Brain Barrier Endothelium

The tightness of a cell layer is generally characterized by two methods: by measuring the flux of polar and uncharged radiolabeled tracer molecules that cannot pass the cell membranes, such as sucrose or mannitol, or by determining the transcellular electrical resistance.[17] The transepithelial or transendothelial electrical resistance is a combination of the paracellular and transcellular flux of small inorganic ions (predominantly Na^+ and Cl^-) across the monolayer. The latter contribution is commonly negligible due to the high lipid membrane resistance, but especially for monolayers with high paracellular tightness it may contribute to the measured TER value. In fact, the total resistance is dominated by elements with low resistance because it is a function of the inverse sums of individual resistances. Therefore, TER values are very sensitive to spots in the monolayer with low resistance like disrupted or dead cells, which can present an overall weak barrier.[17]

Comparing the permeability characteristics of the two brain barriers, i.e., the brain microvessels and the CP, the striking differences in their TER values attract attention. Unfortunately, only few data from in vivo measurements or experiments performed with intact tissue are available. The technique of in situ TER measurements was applied to determine the electrical resistance of the frog's brain microvascular endothelium, which was found to be 1870 $\Omega \cdot cm^2$, [73] consistent with typical values for "tight" endothelia. Butt et al.[74] investigated the permeability characteristics of brain pial vessels in rats of different developmental stages and found an increase in TER from 310 $\Omega \cdot cm^2$, measured in neonatal rats, to 1490 $\Omega \cdot cm^2$, measured in arterial vessels of adult rats. These data are in good correspondence to TER values obtained for an in vitro model of PBCEC that has been developed in our group.[75] Barrier properties of PBCEC are enormously improved under serum-free conditions and in the presence of hydrocortisone.

In contrast to the tight BBB endothelial cells, TER values obtained for the barrier constituted by the CP epithelial cells account for a rather leaky epithelium. Noticeably, however, these data were obtained exclusively from in vitro measurements. Saito et al.[76] measured an electrical resistance of 170 $\Omega \cdot cm^2$ across the isolated bullfrog CP mounted in an Ussing chamber, but their determination of the effective area of the folded epithelial surface seems to be problematic.

Several cell culture models predominantly based on primary cultured rat CP epithelium cells have shown TER values comparable to the data of Saito et al. Rat CP epithelial cells cultured on permeable filter supports are described to form confluent monolayers within 5 DIV with a TER of 80 $\Omega \cdot cm^2$ [77] to 100 $\Omega \cdot cm^2$.[78] Strazielle et al. measured TER values of 178 $\Omega \cdot cm^2$ and a sucrose permeability of $2 \cdot 10^{-4}$ cm/s using rat cells after 8 DIV.[31] CP epithelial cells derived from rabbit brain do not express such an efficient diffusion barrier, and TER values were found to be in the order of 43 $\Omega \cdot cm^2$.[79] Additionally, the electrical resistance of an immortalized cell line of murine CP was shown to be approximately 150 to 200 $\Omega \cdot cm^2$.[80] These model systems have in common that the culture medium is supplemented with serum to support cell growth.

In our group, a cell culture model system based on porcine CP epithelial cells was developed possessing comparable barrier tightness with TER values of 100 to 170 $\Omega \cdot cm^2$ in serum-containing medium.[53] Under serum-free conditions, the TER is dramatically increased up to 1700 $\Omega \cdot cm^2$ accompanied by a significant decrease of the monolayers' permeability for 4 kDa dextran (to $1 \cdot 10^{-7}$ cm/s) and sucrose ($2 \cdot 10^{-7}$ cm/s).[81] Consistent with the differences in the paracellular electrical properties, the sucrose permeability of PCPEC is very low compared to the rat cells. The effects of serum removal will be discussed in detail below.

By summarizing the results, two questions arise: what is the functional background for the leakiness of the CP epithelium compared to the BBB endothelium, and what are the factors responsible for these permeability differences? As to the latter question, Vorbrodt and Dobrogowska[20] suggest the differences in TJ composition to be responsible for differences in permeability between the CP epithelium and the endothelial cells of the BBB. Among the transmembrane proteins, only occludin expression is consistent between these cell types, whereas the immunolabeling of JAM-1 and VE-cadherin is restricted only to the BBB endothelium. As described above, striking differences also exist in the expression of claudins. The localization of the pore forming claudin-2 in the CP epithelium might have a connection with its increased permeability. As a result of TJ composition, morphological aspects like the P-face and E-face association might contribute to differences in barrier tightness.

3.5 REGULATION AND CHANGES OF PERMEABILITY OF THE CHOROID PLEXUS EPITHELIUM

3.5.1 Developmental Aspects

Early studies of the fetal brain revealed several characteristics that can misleadingly be explained by an immature and more permeable blood–CSF barrier. In the mature human brain, the CSF protein concentration is low and accounts for less than 1% of the plasma concentration.[82] In contrast, the CSF of the immature brain contains high concentration of proteins that are immunologically identical to those of the plasma. Furthermore, the entry of small lipid-insoluble molecules into the CSF seems to be increased in the immature brain according to higher steady-state CSF/plasma concentration ratios after application of tracer molecules. Interestingly, the TJ between CP epithelial cells of the mammalian brain appear at a very early stage of embryonic development and do not significantly differ from those in the adult CP as shown by freeze fracture studies.[83] As reviewed by Saunders et al.,[82] the TJ of the blood–brain and blood–CSF interfaces of the developing brain are able to prevent paracellular diffusion of proteins. The high CSF protein concentration in the fetal brain may be not a consequence of immature and leaky TJ, but rather of a transcellular mechanism that specifically transfers proteins across the CP epithelial cells, a pathway that also might account for low molecular weight lipid-insoluble molecules. Concerning the increased steady-state CSF/plasma concentration ratio of lipid-insoluble molecules, the interpretation of the data is complicated by the so-called sink effect described by Oldendorf and Davson.[84] Because of the continuous CSF flow due to permanent fluid secretion by the CP and drainage by the arachnoidvilli of the superior sagittal sinus, the steady state is decreased. At least in the fetal sheep where no significant change in the CSF turnover during maturation is observed, a decrease in permeability for small molecules in the adult CP may account for decreased steady-state CSF concentration (reviewed by Saunders et al.[85]).

Investigations addressing the regulation of the barrier tightness of the CP epithelium are rare because most studies that focus on basic aspects of paracellular epithelial permeability and TJ expression and modification are performed on cell lines like the human colon adenocarcinoma cell line (Caco-2) or MDCK cells. However, some regulatory elements found in other cells have been shown to be transferable to the epithelial cells of the CP.

3.5.2 Phorbol Esters

Numerous reports demonstrated the effect of tumor-promoting phorbol esters, such as the protein kinase C activator phorbol 12-myristate 13-acetate (PMA, also referred to as 12-*O*-tetradecanoylphorbol-13-acetate, TPA), on cell-cell contacts that generally leads to an increased transepithelial permeability, although barrier-improving effects are reported as well (reviewed by Karczewski and Groot[86]). TJ disruption after phorbol ester treatment has been shown to correlate with the dephosphorylation of occludin, e.g., in MDCK cells[26] or LLC-PK1 cells[25] after exposure to PMA, but the opposite effect on the occludin phosphorylation state with C-18 cells was observed as well.[27] Sjö et al.[87] observed occludin dephosphorylation in both MDCK I cells and HT-29 cells after exposure to PMA. Whereas the barrier in the “tight” MDCK I was impaired after PMA treatment, an improved barrier in “leaky” HT-29 cells was observed. The authors explain the different effects on barrier tightness with the different developmental stages of the cell line barriers.

Lippoldt et al.[18,19] investigated the effect of PMA on the CP in vivo by injection of PMA into the lateral ventricle of the rat brain and observed changes in TJ and adherens junction protein expression. At 24 hours after injection, the expression of cadherins and β-catenin was reduced in the CP epithelium and the ependymal cells of the lateral ventricle as shown by immunofluorescence staining. Furthermore, PMA induced a reduction of occludin and claudin-2 expression in the CP epithelium, whereas an increase of claudin-1 was observed accompanied by spreading of this protein from the junctions to the cytoplasm. The authors suggest an increased permeability of the ependymal border and the CP as a result of an impaired junctional phenotype.

3.5.3 cAMP

Throughout recent years it was found repeatedly that the barrier function of many epithelia is regulated by cAMP-dependent pathways.[88] However, tissue- and species-dependent differences have been observed such that an increase of intracellular cAMP levels brings about an enhancement or a partial loss of barrier integrity. Rubin et al.[89] and later Deli et al.[90] demonstrated that the barrier function of cerebral microvessel endothelial cells is strengthened when cAMP-dependent signal transduction cascades are released. Addressing the question as to whether this signal transduction pathway is also involved in the regulation of CP barrier function, we determined the impact of the membrane-permeable cAMP analogue CPT-cAMP and the adenylate cyclase activator forskolin on transepithelial electrical resistances and 4 kDa dextran permeabilities of porcine CP epithelial cell grown to confluence in the present of FCS.[91] In the presence of 10 μMM CPT-cAMP (see Figure 3.7) or 5 μM forskolin, the TER values almost doubled within 5 hours after application compared to the controls. Time-dependent experiments showed that barrier strengthe0ning became significant within 30 minutes after analog application. A corresponding decrease of dextran permeability showed that changes in TER are based on the decrease of the paracellular pathway. However, cautions must be taken in interpreting the data of TER measurements, because the current that traverses the epithelium has to flow through the filter pores and then through the narrow channels between the basal membrane and the substrate before it can escape through the paracellular shunt. This current underneath the cells provides an additional resistance

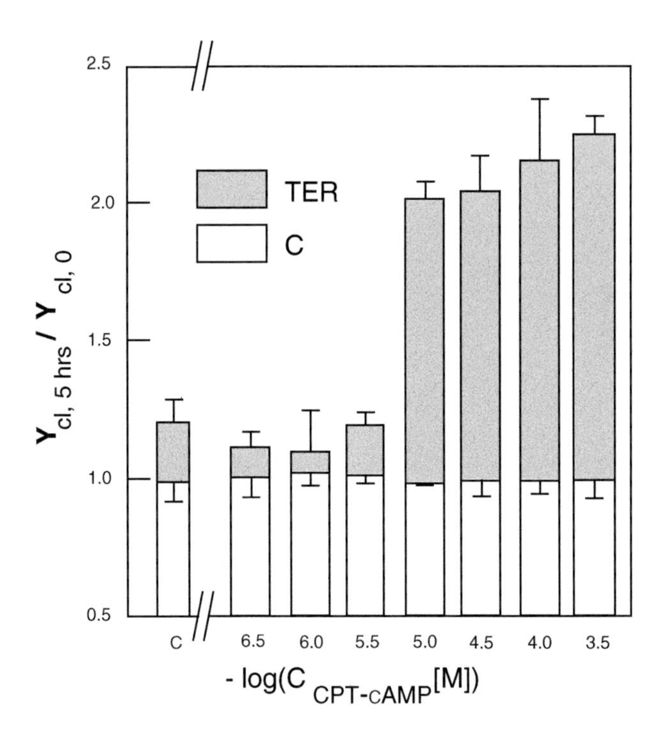

FIGURE 3.7 Dose-response relationships between relative changes of cell layer transepithelial electrical resistance (TER) (gray columns) or cell layer capacitance C (white columns) and CPT-cAMP concentration. Filter-grown PCPEC layers were exposed to the respective CPT-cAMP concentration in the basolateral fluid compartment for a period of 5 hours. C on the abscissa denotes the result of a control experiment in which only medium but no CPT-cAMP was added to the cell layer (±SD, n = 3). (From Wegener et al.[91])

that strongly depends on cell-substrate separation distance and has to be taken into consideration if the TJ electrical resistance is low.[92]

In order to elucidate cAMP-induced changes in this last-mentioned resistance barrier, our group has performed impedance analysis of CP cells grown on gold electrodes according to the electric cell-substrate impedance sensing (ECIS) technique, described by Giaever and Keese.[93] Analysis of the recorded data by a theoretical model, which takes into account the cell-substrate resistance contribution, confirmed a cAMP-induced increase in paracellular resistance. Additionally, the analysis revealed an increase in the resistance underneath the cell layer. These results indicate that not only the permeability of CP epithelial cells but also the adhesion of the cells to the substrate are regulated by cAMP-dependent pathways.[91] Thus, it is reasonable to assume that an appropriate hormonal stimulation, which is mediated by cAMP as a second messenger, may trigger a mutual tightening of both the BBB and the blood–CSF barrier.

The question remains as to the mechanism that translates an increase in intracellular cAMP levels into a strengthening of intercellular contacts. From our experimental data we cannot extract whether the observed enhancement of CP barrier function is due to changes in the TJ complex or due to a collapse of the entire intercellular cleft. In the last-mentioned mechanism, experimentally proven for the *Necturus* gallbladder epithelium,[94] increased cAMP levels induce an increased chloride conductance of the apical membrane; this, in turn, reduces the intracellular chloride concentration, leading to an influx of chloride from the intercellular space. Chloride influx is accompanied by an osmotically driven inward water flow that collapses the paracellular shunt. Since Deng and Johanson[95] have shown that cAMP analogs alter the apical chloride conductance in rat CP cells, the proposed mechanism may be applicable to our data as well.

3.5.4 Serum

The presence of serum in the culture medium is indispensable for the induction and promotion of cell proliferation and expansion of the culture. However, there are quite a few examples in the literature that serum components, most notably the growth factors and mitogens, can inhibit proper differentiation of cells in vitro.[96] In the light of this background we exposed CP epithelial cells that have been grown to confluence in the presence of 10% (v/v) fetal calf serum (FCS) to serum-free medium, and we followed the concomitant expression of the epithelial phenotype.[81] As mentioned previously, confluent cultures of porcine CP epithelial cells react with an enormous improvement of barrier tightness to removal of serum from the medium. Compared to TER values of roughly 200 $\Omega \cdot cm^2$ in serum-containing medium, one day of serum withdrawal is already sufficient to more than double the electrical tightness (see Figure 3.8). After seven to eight days in serum-free medium, the TER was enhanced to 1700 $\Omega \cdot cm^2$. Thus, it takes several days for the barrier function of CP epithelial cells to reach a new steady state. However, it is sufficient to reincubate the cells with serum-containing medium (10% FCS) for 4 hours to induce a complete reversal of this effect. The increase of TER in serum-free medium is accompanied by a significant decrease of the monolayer permeability for 4 kDa dextran (to $1 \cdot 10^{-7}$

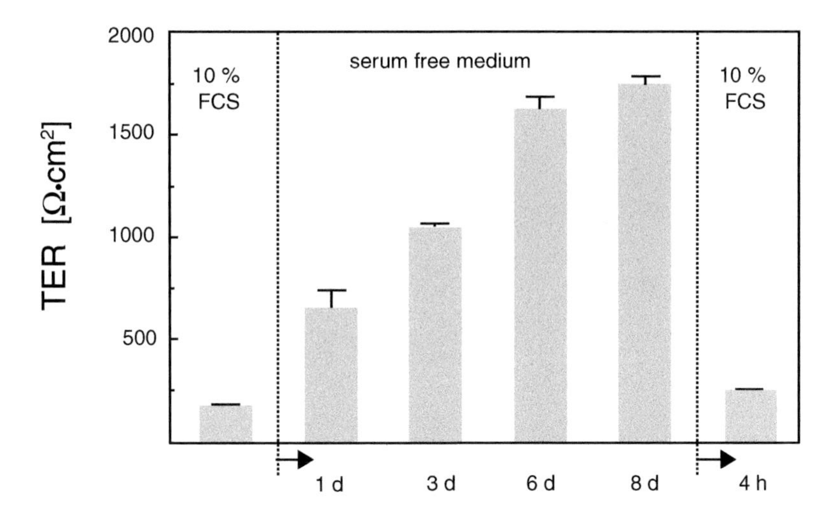

FIGURE 3.8 Transepithelial electrical resistance (TER) of PCPEC as measured by impedance analysis in the presence and absence of serum (±SD, $n = 3$). FCS: fetal calf serum. (From Hakvoort et al.[81])

cm/s) and sucrose ($2 \cdot 10^{-7}$ cm/s), suggesting that serum removal leads to an improvement of paracellular tightness.

Besides these functional alterations of the epithelial barrier, we also addressed the associated structural changes at the TJ that occur upon withdrawal of serum from the culture medium. Fluorescence staining of claudin-1 (A + B), ZO-1 (C + D), and occludin (E + F) shown in Figure 3.9 were performed when CP epithelial cells had been incubated in either medium supplemented with 10% FCS (A, C, and E) or serum-free medium (B, D, and F). Although under both conditions the proteins are exclusively located at the cell borders between adjacent cells, there are subtle differences. For cells that had been kept in serum-free medium (B, D, and F) the immunofluorescence signal for the proteins is a sharp line without any protrusions or invaginations along the cell perimeter. Those cells that were incubated in serum-containing medium do not show such a sharp and focused but fuzzy localization of the corresponding proteins (A, C, and E). Although these structural observations are not yet sufficient to explain the impairment of epithelial barrier function in the presence of serum, they clearly demonstrate that serum induces significant morphological alterations that are associated with barrier disintegration.

The question arises whether a serum-free in vitro culture system can be regarded as a physiological model of the blood–CSF barrier. Therefore, we investigated the influence of serum withdrawal on the expression of typical markers of the epithelial phenotype.[81] When the medium is replaced by serum-free medium after the cells have grown to confluence, the microvilli are trimmed and more dense and cover the whole apical cell membrane. As a further result of serum removal, Na^+K^+-ATPase is expressed more strongly and homogeneously along the apical cell surface. The amount of transthyretin secretion is increased under serum-free conditions, whereby the protein concentration is found to be much higher in the apical compartment.

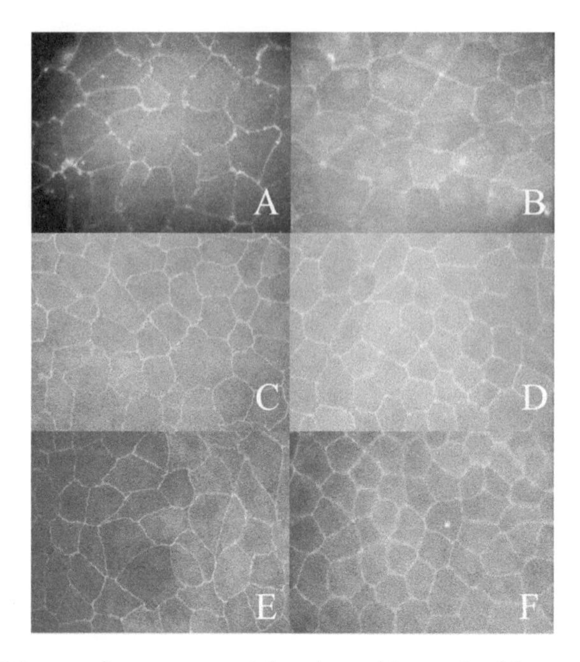

FIGURE 3.9 TJ immunofluorescence stainings imaged by confocal laser scanning microscopy. A/B: claudin-1, C + D: ZO-1, E + F: occludin. On the left side (A, C, E), cells under serum-containing conditions are shown (DIV 9); on the right side (B, D, F), serum-free conditions are illustrated (DIV 13). In serum-free medium, cell borders show a clear, well-defined circumferential pattern.

One can conclude that the differentiation state and the polarity of the cell layer in terms of morphology and protein targeting is improved under serum-free conditions. This differentiation process might be comparable to the postnatal morphological and physiological maturation of the CP cells in vivo.

The TJ opening effect of serum and serum factors on epithelial cells has been demonstrated by other investigators on MDCK cells[97,98] and on retinal pigmented epithelial (RPE) cells.[99] Further, serum not only prevents confluent porcine brain capillary endothelial cell (PBCEC) monolayers from the formation of high-resistance TJ, but also opens those already established TJ, as shown by a decreased TER.[100]

3.5.5 Epidermal Growth Factor and Other Serum Factors

For getting a more detailed insight into the serum compounds that might be responsible for decrease of TER, Hakvoort et al.[81] investigated the effect of the epidermal growth factor (EGF) on barrier properties of PCPEC. When EGF in physiological concentration of 1 ng/mL was added to the culture medium, the PCPEC failed to establish a barrier of comparable tightness that was measured under control conditions (see Figure 3.10). Additionally, filtration and heat inactivation of FCS showed that the TER decreasing compounds of FCS are contained in a low molecular weight fraction (≤10 kDa) and are not deactivated by heating, two characteristics that account for growth factors being involved in barrier weakening. Interestingly, in

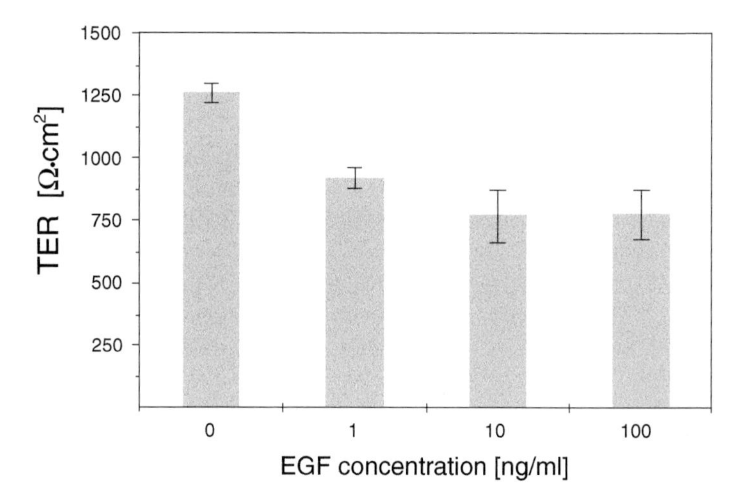

FIGURE 3.10 Effect of EGF on the TER of PCPEC. Cells were cultured in the presence of 10% FCS for 10 DIV and incubated either with serum-free medium (SFM) or SFM supplemented with different concentrations of EGF for 4 DIV. The TER was determined by impedance analysis (±SD, $n = 3$). (From Hakvoort et al.[81])

experiments performed with other epithelial cells like MDCK II cells[101] and primary cultured canine gastric mucosal cells,[102] EGF affected the barrier properties of the cell layers in a more positive way and led to an increase of TER. These striking differences in influence on barrier tightness compared to PCPEC might be explained by cell specificities. However, the experiments addressing the effect of EGF on PCPEC show that EGF cannot be the only serum component responsible for barrier weakening since the decrease of TER in the presence of only 1% FCS is much stronger than in the presence of high concentrations of EGF (up to 100 ng/mL EGF).

Recently, metallomatrix proteases (MMPs) present in the serum are reported to affect the permeability of endothelial cells of the BBB after tyrosine phosphatase inhibition.[103] Similar enzymatic activities might also account for the disintegration of the epithelial barrier. In future experiments additional factors that open TJ of PCPEC still have to be determined.

3.6 FUTURE PERSPECTIVES

The junctional protein complex formed by integral membrane proteins and peripheral proteins determine both the barrier and the fence function. The current lipid-protein model may explain the possibility to form cell monolayers with high electrical resistance/low permeability but to allow controlled opening of the barrier. In this model membrane lipids play an important role for the closure of the intercellular cleft. Factors determining the cellular barrier properties include the expression of proteins at the TJ that might form selective pores and the P-face and E-face association of the junctional particles. The permeability of the complex TJ structure is under strong regulatory control not only by second messenger pathways, but also by blood-derived effectors present in the serum of cell culture models. Developmental aspects, hormonal stimulation, phosphorylation events, as well as the effect of serum factors may all affect the integrity of the barrier. A molecular understanding of the assembly of the junctional complexes might further help to overcome the barrier to the pharmacological treatment of brain diseases.

The knowledge about characteristics and regulation of the BBB is important to understand pathological processes leading to the opening of these barriers. On the other hand, an artificial increase of the BBB permeability can be applied to facilitate the entry of drugs to the CNS, which is additionally complicated by numerous transporters, e.g., *P*-glycoprotein or MRP, that actively clear the CNS from many therapeutics.[104] This method has been reported to be applied for CNS chemotherapy in which osmotic opening of the BBB and widening of TJ is induced by infusing hypertonic solutions.[105]

Data about permeability regulation in the CP are rare. Some regulatory factors and pathways have been elucidated but should be investigated in greater detail, e.g., regulation of TJ expression, phosphorylation, and so on. Cell culture models of the CP will be a helpful tool in the investigation of barrier properties, because they allow the observation of a single cell type under defined conditions and easy determination of changes in the permeability and the electrophysical parameters such as the TER.

REFERENCES

1. Bradbury, M. W. The blood-brain barrier. *Exp Physiol* 78, 453–72 (1993).
2. Spector, R. Drug transport in the mammalian central nervous system: multiple complex systems. A critical analysis and commentary. *Pharmacology* 60, 58–73 (2000).
3. Speake, T., Whitwell, C., Kajita, H., Majid, A., and Brown, P. D. Mechanisms of CSF secretion by the choroid plexus. *Microsc Res Tech* 52, 49–59 (2001).
4. Angelow, S., Haselbach, M., and Galla, H. J. Functional characterisation of the active ascorbic acid transport into cerebrospinal fluid using primary cultured choroid plexus cells. *Brain Res* 988, 105–13 (2003).
5. Nagata, Y., Kusuhara, H., Endou, H., and Sugiyama, Y. Expression and functional characterization of rat organic anion transporter 3 (rOat3) in the choroid plexus. *Mol Pharmacol* 61, 982–8 (2002).
6. Schreiber, G., Richardson, S. J., and Prapunpoj, P. Structure and expression of the transthyretin gene in the choroid plexus: a model for the study of the mechanism of evolution. *Microsc Res Tech* 52, 21–30 (2001).
7. Tsukita, S. and Furuse, M. Pores in the wall: claudins constitute tight junction strands containing aqueous pores. *J Cell Biol* 149, 13–6 (2000).
8. Nelson, W. J. Regulation of cell surface polarity from bacteria to mammals. *Science* 258, 948–55 (1992).
9. Ziomek, C. A., Schulman, S., and Edidin, M. Redistribution of membrane proteins in isolated mouse intestinal epithelial cells. *J Cell Biol* 86, 849–57 (1980).
10. Jou, T. S., Schneeberger, E. E., and Nelson, W. J. Structural and functional regulation of tight junctions by RhoA and Rac1 small GTPases. *J Cell Biol* 142, 101–15 (1998).
11. Grebenkämper, K. and Galla, H. J. Translational diffusion measurements of a fluorescent phospholipid between MDCK-I cells support the lipid model of the tight junctions. *Chem Phys Lipids* 71, 133–43 (1994).

12. Kachar, B. and Reese, T. S. Evidence for the lipidic nature of tight junction strands. *Nature* 296, 464–6 (1982).
13. Hein, M., Post, A., and Galla, H. J. Implications of a non-lamellar lipid phase for the tight junction stability. Part I: Influence of basic amino acids, pH and protamine on the bilayer-hexagonal II phase behaviour of PS-containing PE membranes. *Chem Phys Lipids* 63, 213–21 (1992).
14. Claude, P. Morphological factors influencing transepithelial permeability: a model for the resistance of the zonula occludens. *J Membr Biol* 39, 219–32 (1978).
15. Wegener, J. and Galla, H. J. The role of non-lamellar lipid structures in the formation of tight junctions. *Chem Phys Lipids* 81, 229–255 (1996).
16. Tsukita, S., Furuse, M., and Itoh, M. Multifunctional strands in tight junctions. *Nat Rev Mol Cell Biol* 2, 285–93 (2001).
17. Fanning, A. S., Mitic, L. L., and Anderson, J. M. Transmembrane proteins in the tight junction barrier. *J Am Soc Nephrol* 10, 1337–45 (1999).
18. Lippoldt, A., Liebner, S., Andbjer, B., Kalbacher, H., Wolburg, H., Haller, H., and Fuxe, K. Organization of choroid plexus epithelial and endothelial cell tight junctions and regulation of claudin-1, -2 and -5 expression by protein kinase C. *Neuroreport* 11, 1427–31 (2000).
19. Lippoldt, A., Jansson, A., Kniesel, U., Andbjer, B., Andersson, A., Wolburg, H., Fuxe, K., and Haller, H. Phorbol ester induced changes in tight and adherens junctions in the choroid plexus epithelium and in the ependyma. *Brain Res* 854, 197–206 (2000).
20. Vorbrodt, A. W. and Dobrogowska, D. H. Molecular anatomy of intercellular junctions in brain endothelial and epithelial barriers: electron microscopist's view. *Brain Res Rev* 42, 221–42 (2003).
21. Furuse, M., Hirase, T., Itoh, M., Nagafuchi, A.,Yonemura, S., and Tsukita, S. Occludin: a novel integral membrane protein localizing at tight junctions. *J Cell Biol* 123, 1777–88 (1993).
22. Muresan, Z., Paul, D. L., and Goodenough, D. A. Occludin 1B, a variant of the tight junction protein occludin. *Mol Biol Cell* 11, 627–34 (2000).
23. Ghassemifar, M. R., Sheth, B., Papenbrock, T., Leese, H. J., Houghton, F. D., and Fleming, T. P. Occludin TM4(-): an isoform of the tight junction protein present in primates lacking the fourth transmembrane domain. *J Cell Sci* 115, 3171–80 (2002).
24. Sakakibara, A., Furuse, M., Saitou, M., Ando-Akatsuka, Y., and Tsukita, S. Possible involvement of phosphorylation of occludin in tight junction formation. *J Cell Biol* 137, 1393–401 (1997).
25. Clarke, H., Soler, A. P., and Mullin, J. M. Protein kinase C activation leads to dephosphorylation of occludin and tight junction permeability increase in LLC-PK1 epithelial cell sheets. *J Cell Sci* 113 (Pt 18), 3187–96 (2000).
26. Farshori, P. and Kachar, B. Redistribution and phosphorylation of occludin during opening and resealing of tight junctions in cultured epithelial cells. *J Membr Biol* 170, 147–56 (1999).
27. Marano, C. W., Garulacan, L. A., Ginanni, N., and Mullin, J. M. Phorbol ester treatment increases paracellular permeability across IEC-18 gastrointestinal epithelium *in vitro. Dig Dis Sci* 46, 1490–9 (2001).
28. Saitou, M., Furuse, M., Sasaki, H., Schulzke, J. D., Fromm, M., Takano, H., Noda, T., and Tsukita, S. Complex phenotype of mice lacking occludin, a component of tight junction strands. *Mol Biol Cell* 11, 4131–42 (2000).
29. Moroi, S., Saitou, M., Fujimoto, K., Sakakibara, A., Furuse, M., Yoshida, O., and Tsukita, S. Occludin is concentrated at tight junctions of mouse/rat but not human/guinea pig Sertoli cells in testes. *Am J Physiol* 274, C1708–17 (1998).

30. Saitou, M., Fujimoto, K., Doi, Y., Itoh, M., Fujimoto, T., Furuse, M., Takano, H., Noda, T., and Tsukita, S. Occludin-deficient embryonic stem cells can differentiate into polarized epithelial cells bearing tight junctions. *J Cell Biol* 141, 397–408 (1998).
31. Strazielle, N. and Ghersi-Egea, J. F. Demonstration of a coupled metabolism-efflux process at the choroid plexus as a mechanism of brain protection toward xenobiotics. *J Neurosci* 19, 6275–89 (1999).
32. Angelow, S., Wegener, J., and Galla, H.J. Transport and permeability characteristics of the blood-cerebrospinal fluid barrier in vitro. In *Blood-Spinal Cord and Brain Barriers in Health and Disease* 33–45 (Academic Press, San Diego, 2004).
33. Furuse, M., Fujita, K., Hiiragi, T., Fujimoto, K., and Tsukita, S. Claudin-1 and -2: novel integral membrane proteins localizing at tight junctions with no sequence similarity to occludin. *J Cell Biol* 141, 1539–50 (1998).
34. Gonzalez-Mariscal, L., Betanzos, A., Nava, P., and Jaramillo, B. E. Tight junction proteins. *Prog Biophys Mol Biol* 81, 1–44 (2003).
35. Kubota, K., Furuse, M., Sasaki, H., Sonoda, N., Fujita, K., Nagafuchi, A., and Tsukita, S. Ca(2+)-independent cell-adhesion activity of claudins, a family of integral membrane proteins localized at tight junctions. *Curr Biol* 9, 1035–8 (1999).
36. Gow, A., Southwood, C. M., Li, J. S., Pariali, M., Riordan, G. P., Brodie, S. E., Danias, J., Bronstein, J. M., Kachar, B., and Lazzarini, R. A. CNS myelin and sertoli cell tight junction strands are absent in Osp/claudin-11 null mice. *Cell* 99, 649–59 (1999).
37. Nitta, T., Hata, M., Gotoh, S., Seo, Y., Sasaki, H., Hashimoto, N., Furuse, M., and Tsukita, S. Size-selective loosening of the blood-brain barrier in claudin-5-deficient mice. *J Cell Biol* 161, 653–60 (2003).
38. Furuse, M., Hata, M., Furuse, K., Yoshida, Y., Haratake, A., Sugitani, Y., Noda, T., Kubo, A., and Tsukita, S. Claudin-based tight junctions are crucial for the mammalian epidermal barrier: a lesson from claudin-1-deficient mice. *J Cell Biol* 156, 1099–111 (2002).
39. Wolburg, H., Wolburg-Buchholz, K., Liebner, S., and Engelhardt, B. Claudin-1, claudin-2 and claudin-11 are present in tight junctions of choroid plexus epithelium of the mouse. *Neurosci Lett* 307, 77–80 (2001).
40. Martin-Padura, I., Lostaglio, S., Schneemann, M., Williams, L., Romano, M., Fruscella, P., Panzeri, C., Stoppacciaro, A., Ruco, L., Villa, A., Simmons, D., and Dejana, E. Junctional adhesion molecule, a novel member of the immunoglobulin superfamily that distributes at intercellular junctions and modulates monocyte transmigration. *J Cell Biol* 142, 117–27 (1998).
41. Stevenson, B. R., Siliciano, J. D., Mooseker, M. S., and Goodenough, D. A. Identification of ZO-1: a high molecular weight polypeptide associated with the tight junction (zonula occludens) in a variety of epithelia. *J Cell Biol* 103, 755–66 (1986).
42. Itoh, M., Nagafuchi, A., Yonemura, S., Kitani-Yasuda, T., and Tsukita, S. The 220-kD protein colocalizing with cadherins in non-epithelial cells is identical to ZO-1, a tight junction-associated protein in epithelial cells: cDNA cloning and immunoelectron microscopy. *J Cell Biol* 121, 491–502 (1993).
43. Willott, E., Balda, M. S., Heintzelman, M., Jameson, B., and Anderson, J. M. Localization and differential expression of two isoforms of the tight junction protein ZO-1. *Am J Physiol* 262, C1119–24 (1992).
44. Gumbiner, B., Lowenkopf, T., and Apatira, D. Identification of a 160-kDa polypeptide that binds to the tight junction protein ZO-1. *Proc Natl Acad Sci USA* 88, 3460–4 (1991).

45. Balda, M. S., Gonzalez-Mariscal, L., Matter, K., Cereijido, M., and Anderson, J. M. Assembly of the tight junction: the role of diacylglycerol. *J Cell Biol* 123, 293–302 (1993).
46. Itoh, M., Furuse, M., Morita, K., Kubota, K., Saitou, M., and Tsukita, S. Direct binding of three tight junction-associated MAGUKs, ZO-1, ZO-2, and ZO-3, with the COOH termini of claudins. *J Cell Biol* 147, 1351–63 (1999).
47. Fanning, A. S., Jameson, B. J., Jesaitis, L. A., and Anderson, J. M. The tight junction protein ZO-1 establishes a link between the transmembrane protein occludin and the actin cytoskeleton. *J Biol Chem* 273, 29745–53 (1998).
48. Itoh, M., Morita, K., and Tsukita, S. Characterization of ZO-2 as a MAGUK family member associated with tight as well as adherens junctions with a binding affinity to occludin and alpha catenin. *J Biol Chem* 274, 5981–6 (1999).
49. Haskins, J., Gu, L., Wittchen, E. S., Hibbard, J., and Stevenson, B. R. ZO-3, a novel member of the MAGUK protein family found at the tight junction, interacts with ZO-1 and occludin. *J Cell Biol* 141, 199–208 (1998).
50. Furuse, M., Itoh, M., Hirase, T., Nagafuchi, A., Yonemura, S., and Tsukita, S. Direct association of occludin with ZO-1 and its possible involvement in the localization of occludin at tight junctions. *J Cell Biol* 127, 1617–26 (1994).
51. Wittchen, E. S., Haskins, J., and Stevenson, B. R. Protein interactions at the tight junction. Actin has multiple binding partners, and ZO-1 forms independent complexes with ZO-2 and ZO-3. *J Biol Chem* 274, 35179–85 (1999).
52. Watson, P. M., Anderson, J. M., Vanltallie, C. M., and Doctrow, S. R. The tight-junction-specific protein ZO-1 is a component of the human and rat blood-brain barriers. *Neurosci Lett* 129, 6–10 (1991).
53. Gath, U., Hakvoort, A., Wegener, J., Decker, S., and Galla, H. J. Porcine choroid plexus cells in culture: expression of polarized phenotype, maintenance of barrier properties and apical secretion of CSF-components. *Eur J Cell Biol* 74, 68–78 (1997).
54. Krause, D., Mischeck, U., Galla, H. J., and Dermietzel, R. Correlation of zonula occludens ZO-1 antigen expression and transendothelial resistance in porcine and rat cultured cerebral endothelial cells. *Neurosci Lett* 128, 301–4 (1991).
55. Cordenonsi, M., D'Atri, F., Hammar, E., Parry, D. A., Kendrick-Jones, J., Shore, D., and Citi, S. Cingulin contains globular and coiled-coil domains and interacts with ZO-1, ZO-2, ZO-3, and myosin. *J Cell Biol* 147, 1569–82 (1999).
56. Zhong, Y., Saitoh, T., Minase, T., Sawada, N., Enomoto, K., and Mori, M. Monoclonal antibody 7H6 reacts with a novel tight junction-associated protein distinct from ZO-1, cingulin and ZO-2. *J Cell Biol* 120, 477–83 (1993).
57. Keon, B. H., Schafer, S., Kuhn, C., Grund, C., and Franke, W. W. Symplekin, a novel type of tight junction plaque protein. *J Cell Biol* 134, 1003–18 (1996).
58. Tepass, U., Truong, K., Godt, D., Ikura, M., and Peifer, M. Cadherins in embryonic and neural morphogenesis. *Nat Rev Mol Cell Biol* 1, 91–100 (2000).
59. Yeaman, C., Grindstaff, K. K., and Nelson, W. J. New perspectives on mechanisms involved in generating epithelial cell polarity. *Physiol Rev* 79, 73–98 (1999).
60. Claude, P., and Goodenough, D. A. Fracture faces of zonulae occludentes from "tight" and "leaky" epithelia. *J Cell Biol* 58, 390–400 (1973).
61. Hein, M., Madefessel, C., Haag, B., Teichmann, K., Post, A., and Galla, H. J. Implications of a non-lamellar lipid phase for the tight junction stability. Part II: Reversible modulation of transepithelial resistance in high and low resistance MDCK-cells by basic amino acids, Ca^{2+}, protamine and protons. *Chem Phys Lipids* 63, 223–33 (1992).

62. Stevenson, B. R., Anderson, J. M., Goodenough, D. A., and Mooseker, M. S. Tight junction structure and ZO-1 content are identical in two strains of Madin-Darby canine kidney cells which differ in transepithelial resistance. *J Cell Biol* 107, 2401–8 (1988).
63. Furuse, M., Furuse, K., Sasaki, H., and Tsukita, S. Conversion of zonulae occludentes from tight to leaky strand type by introducing claudin-2 into Madin-Darby canine kidney I cells. *J Cell Biol* 153, 263–72 (2001).
64. Amasheh, S., Meiri, N., Gitter, A. H., Schoneberg, T., Mankertz, J., Schulzke, J. D., and Fromm, M. Claudin-2 expression induces cation-selective channels in tight junctions of epithelial cells. *J Cell Sci* 115, 4969–76 (2002).
65. Wolburg, H., Neuhaus, J., Kniesel, U., Krauss, B., Schmid, E. M., Ocalan, M., Farrell, C., and Risau, W. Modulation of tight junction structure in blood-brain barrier endothelial cells. Effects of tissue culture, second messengers and cocultured astrocytes. *J Cell Sci* 107 (pt 5), 1347–57 (1994).
66. Liebner, S., Kniesel, U., Kalbacher, H., and Wolburg, H. Correlation of tight junction morphology with the expression of tight junction proteins in blood-brain barrier endothelial cells. *Eur J Cell Biol* 79, 707–17 (2000).
67. van Deurs, B., and Koehler, J. K. Tight junctions in the choroid plexus epithelium. A freeze-fracture study including complementary replicas. *J Cell Biol* 80, 662–73 (1979).
68. Wright, E. M. Solute transport across the frog choroid plexus. In *Fluid Environment of the Brain* (ed. Cserr, H. F., Fenstermacher, J. D) 139–156 (Academic Press Inc., New York, 1975).
69. Wright, E. M. Mechanisms of ion transport across the choroid plexus. *J Physiol* 226, 545–71 (1972).
70. Welch K. and Araki, H. Features of the choroid plexus of the cat, studied in vitro. In *Fluid Environment of the Brain* (ed. Cserr, H. F., Fenstermacher, J.D) 157–165 (Academic Press Inc., New York, 1975).
71. Mollgard, K., Lauritzen, B., and Saunders, N. R. Double replica technique applied to choroid plexus from early foetal sheep: completeness and complexity of tight junctions. *J Neurocytol* 8, 139–49 (1979).
72. Hirsch, M., Verge, D., Bouchaud, C., and Escaig, J. The tight junctions of rabbit choroid plexus and ciliary body epithelia. A comparative study using the double replica freeze-fracture technique. *Tissue Cell* 12, 437–47 (1980).
73. Crone, C. and Olesen, S. P. Electrical resistance of brain microvascular endothelium. *Brain Res* 241, 49–55 (1982).
74. Butt, A. M., Jones, H. C., and Abbott, N. J. Electrical resistance across the blood-brain barrier in anaesthetized rats: a developmental study. *J Physiol* 429, 47–62 (1990).
75. Hoheisel, D., Nitz, T., Franke, H., Wegener, J., Hakvoort, A., Tilling, T., and Galla, H. J. Hydrocortisone reinforces the blood-brain barrier properties in a serum free cell culture system. *Biochem Biophys Res Commun* 244, 312–6 (1998).
76. Saito, Y. and Wright, E. M. Bicarbonate transport across the frog choroid plexus and its control by cyclic nucleotides. *J Physiol* 336, 635–48 (1983).
77. Zheng, W., Zhao, Q., and Graziano, J. H. Primary culture of choroidal epithelial cells: characterization of an in vitro model of blood-CSF barrier. *In Vitro Cell Dev Biol Anim* 34, 40–5 (1998).
78. Southwell, B. R., Duan, W., Alcorn, D., Brack, C., Richardson, S. J., Kohrle, J., and Schreiber, G. Thyroxine transport to the brain: role of protein synthesis by the choroid plexus. *Endocrinology* 133, 2116–26 (1993).

79. Ramanathan, V. K., Hui, A. C., Brett, C. M., and Giacomini, K. M. Primary cell culture of the rabbit choroid plexus: an experimental system to investigate membrane transport. *Pharm Res* 13, 952–6 (1996).
80. Zheng, W. and Zhao, Q. Establishment and characterization of an immortalized Z310 choroidal epithelial cell line from murine choroid plexus. *Brain Res* 958, 371–80 (2002).
81. Hakvoort, A., Haselbach, M., Wegener, J., Hoheisel, D. and Galla, H. J. The polarity of choroid plexus epithelial cells *in vitro* is improved in serum-free medium. *J Neurochem* 71, 1141–50 (1998).
82. Saunders, N. R., Habgood, M. D., and Dziegielewska, K. M. Barrier mechanisms in the brain, II. Immature brain. *Clin Exp Pharmacol Physiol* 26, 85–91 (1999).
83. Mollgard, K., Milinowska, D. H., and Saunders, N. R. Lack of correlation between tight junction morphology and permeability properties in developing choroid plexus. *Nature* 264, 293–4 (1976).
84. Oldendorf, W. H., and Davson, H. Brain extracellular space and the sink action of cerebrospinal fluid. *Trans Am Neurol Assoc* 92, 123–7 (1967).
85. Saunders, N. R., Knott, G. W., and Dziegielewska, K. M. Barriers in the immature brain. *Cell Mol Neurobiol* 20, 29–40 (2000).
86. Karczewski, J. and Groot, J. Molecular physiology and pathophysiology of tight junctions III. Tight junction regulation by intracellular messengers: differences in response within and between epithelia. *Am J Physiol Gastrointest Liver Physiol* 279, G660–5 (2000).
87. Sjö, A., Magnusson, K. E., and Peterson, K. H. Distinct effects of protein kinase C on the barrier function at different developmental stages. *Biosci Rep* 23, 87–102 (2003).
88. Duffey, M. E., Hainau, B., Ho, S., and Bentzel, C. J. Regulation of epithelial tight junction permeability by cyclic AMP. *Nature* 294, 451–3 (1981).
89. Rubin, L. L., Hall, D. E., Porter, S., Barbu, K., Cannon, C., Horner, H. C., Janatpour, M., Liaw, C. W., Manning, K., and Morales, J. A cell culture model of the blood-brain barrier. *J Cell Biol* 115, 1725–35 (1991).
90. Deli, M. A., Dehouck, M. P., Abraham, C. S., Cecchelli, R., and Joo, F. Penetration of small molecular weight substances through cultured bovine brain capillary endothelial cell monolayers: the early effects of cyclic adenosine 3’, 5’-monophosphate. *Exp Physiol* 80, 675–8 (1995).
91. Wegener, J., Hakvoort, A., and Galla, H. J. Barrier function of porcine choroid plexus epithelial cells is modulated by cAMP-dependent pathways *in vitro*. *Brain Res* 853, 115–24 (2000).
92. Lo, C. M., Keese, C. R., and Giaever, I. Cell-substrate contact: another factor may influence transepithelial electrical resistance of cell layers cultured on permeable filters. *Exp Cell Res* 250, 576–80 (1999).
93. Giaever, I. and Keese, C. R. A morphological biosensor for mammalian cells. *Nature* 366, 591–2 (1993).
94. Kottra, G. and Fromter, E. Tight-junction tightness of Necturus gall bladder epithelium is not regulated by cAMP or intracellular Ca^{2+}. II. Impedance measurements. *Pflugers Arch* 425, 535–45 (1993).
95. Deng, Q. S. and Johanson, C. E. Cyclic AMP alteration of chloride transport into the choroid plexus-cerebrospinal fluid system. *Neurosci Lett* 143, 146–50 (1992).
96. Freshney, R. I. *Culture of Animal Cells: A Manual of Basic Techniques*. (Wiley-Liss, New York, 2000).

97. Conyers, G., Milks, L., Conklyn, M., Showell, H., and Cramer, E. A factor in serum lowers resistance and opens tight junctions of MDCK cells. *Am J Physiol* 259, C577–85 (1990).
98. Marmorstein, A. D., Mortell, K. H., Ratcliffe, D. R., and Cramer, E. B. Epithelial permeability factor: a serum protein that condenses actin and opens tight junctions. *Am J Physiol* 262, C1403–10 (1992).
99. Chang, C., Wang, X., and Caldwell, R. B. Serum opens tight junctions and reduces ZO-1 protein in retinal epithelial cells. *J Neurochem* 69, 859–67 (1997).
100. Nitz, T., Eisenblätter, T., Psathaki, K, and Galla, H. J. Serum-derived factors weaken the barrier properties of cultured porcine brain capillary endothelial cells *in vitro*. *Brain Res* 981, 30–40 (2003).
101. Singh, A.B. and Harris, R. C. EGF receptor activation differentially regulates claudin expression and enhances trans-epithelial resistance in MDCK cells. *J Biol Chem* 279, 3543–52 (2004).
102. Chen, M. C., Goliger, J., Bunnett, N., and Soll, A. H. Apical and basolateral EGF receptors regulate gastric mucosal paracellular permeability. *Am J Physiol Gastrointest Liver Physiol* 280, G264–72 (2001).
103. Lohmann, C., Krischke, M., Wegener, J., and Galla, H. J. Tyrosine phosphatase inhibition induces loss of blood-brain barrier integrity by matrix metalloproteinase-dependent and -independent pathways. *Brain Res* 995, 184–96 (2004).
104. Sun, H., Dai, H., Shaik, N., and Elmquist, W. F. Drug efflux transporters in the CNS. *Adv Drug Deliv Rev* 55, 83–105 (2003).
105. Rapoport, S. I. Advances in osmotic opening of the blood-brain barrier to enhance CNS chemotherapy. *Expert Opin Investig Drugs* 10, 1809–18 (2001).

Section Two

Physiology and Molecular Biology of the Blood–CSF Barrier

4 Fluid Compartments of the Central Nervous System

Malcolm B. Segal
King's College London,
London, U.K.

CONTENTS

4.1 FLUID COMPARTMENTS OF THE BRAIN

The brain has an interstitial fluid (ISF) volume of the same magnitude as that of other tissues of the body, i.e., about 15% of the body weight. This suggests that, if the brain weight is about 1500 g, the volume of ISF is quite large, in the order of 225 ml. Although this volume is considerable, it is difficult to measure because of the presence of the blood–brain barrier (BBB), which will restrict the distribution of most polar water-soluble marker molecules (Davson and Segal, 1995). The late Hugh Davson was one of the first to attempt to measure the volume of the brain ISF by studying the distribution of sodium and chloride isotopes in the rabbit. If these ions are used in vivo, the distribution "space" from the blood into the brain is very low. However, by comparing the values of distribution for the above ions measured in vitro, using brain slices, with those in strips of diaphragm muscle, Davson found that the values were very similar and much the same as in other tissues of the body (Davson and Spaziani, 1959). This in vitro method effectively removes the BBB, and later in vivo studies with Oldendorf confirmed these results by putting the marker ^{14}C-sucrose on both sides of the BBB in the rabbit and effectively removing the "sink action" of the cerebrospinal fluid (CSF). These experiments gave a value of about 18 to 20% for the volume of brain ISF (Oldendorf and Davson, 1967).

The source of the brain ISF has long intrigued those in the field, since there is little doubt that the brain interstitial fluid cannot be a simple ultrafiltrate of plasma since the resistance offered by the BBB would be too high (1500 Ω cm^2) to permit a simple coupling between the hydrostatic pressure and the colloid osmotic pressure (the Starling equilibrium) as is seen in most systemic tissues with low-resistance capillaries (Michel, 1984).

The source of the brain ISF has long been assumed to be the capillaries of BBB, which are unusual in that, as well as having the intercellular spaces between the cells sealed together with occluding tight junctions, the cells are encircled by astrocytic endfeet (Peters et al., 1991). The brain endothelial cells are claimed to have more mitochondria than other capillaries, reflecting their greater work capacity (Oldendorf and Brown, 1975).

Davson and Segal investigated the entry of sodium isotopes into the brain and CSF on the assumption that the active transport of sodium is responsible for the secretion of brain ISF, as it is in the gut and kidney. If this suggestion was true, it was thought that inhibition of the Na^+ pump with an Na^+K^+-ATPase "inhibitor," such as ouabain, should reduce the entry of the Na^+ into the brain. However, in spite of considerable effort, they were unable to demonstrate any effect of ouabain or of acetazolamide, another inhibitor of CSF secretion, on the entry of sodium into the brain of the intact rabbit. At the time it was thought that the actual amount of fluid being secreted by the BBB was in fact quite small in relation to the large surface area of the BBB interface and that the method used was not sensitive enough to detect such small changes (Davson and Segal, 1970). However, later studies by Keep et al. (1998) have demonstrated the active transport of sodium by the BBB of the rat, which supports the proposition that the brain ISF is secreted by the BBB. Recent calculations by Begley suggest that the volume of ISF secreted by the BBB is about 0.17 μl/g tissue per minute (Begley, 2004).

4.2 THE CHOROID PLEXUSES (CPs) AND CSF

The other major source of bulk fluid in the brain is the CSF, which is secreted into the ventricles by the choroid plexuses (CPs). These are highly vascular double-sided epithelial-like structures formed as an outpouching of the ependymal lining of the ventricles. The two largest CPs are found in the lateral ventricles with a small extension into the third ventricle, and there is also a single-sided choroid plexus forming the roof of the fourth ventricle (Davson and Segal, 1995).

The CPs of the lateral ventricle are leaf-like and double-sided and are covered with a layer of modified ependymal cells. The external covering of the CPs is continuous with the lining of the ventricles via a thin stalk at the base of the plexuses. This stalk enables the CPs to float in the CSF. Many of the ependymal cells have long cilia on the apical (CSF) side, which are actively motile, and they are thought to aid in the circulation of CSF and improve exchange by reducing unstirred layer effects between the CSF and the CP. The apical (CSF) side of the CPs is extensively expanded by microvilli, and on the baso-lateral blood side there are also complex intercellular clefts. These features greatly expand the interface between the blood and the CSF, and morphometric analysis of the size of the blood–CSF interface suggests that, if the area is added to the area of the dural blood vessels, the total blood–CSF interface would approach that of the BBB (Keep and Jones, 1990). The cells lining the outside of the CPs are sealed with occluding tight junctions (TJs) and are also rich in mitochondria, typical of an active fluid-secreting tissue. Although the CPs are the major source of CSF within the ventricles, a percentage of CSF may arise from the brain ISF and enter the ventricles across the ependymal lining. This extrachoroidal fluid is unlikely to be more than 5 to 10% of the total CSF, but the volume of this fluid entering from the brain is a matter of some debate (Curl and Pollay, 1968; Fenstermacher, 1984; Segal, 1999).

To enter the CSF from the brain, the ISF has to cross either the ependymal lining of the ventricles or the pia arachnoid on the external surfaces of the brain (see Figure 4.1). The ependyma is a single layer of cuboidal epithelium with many cell types (Del Bigio, 1995). In certain special regions, the circumventricular organs (CVOs), the cells of the ependyma, are sealed together on the CSF (ventricular) side by rings of occluding TJs. In contrast, the CVOs have an open BBB with fenestrated capillaries on the blood side and are sites of endocrine feedback (see Figure 4.1) (Weindl and Joynt, 1973; Gross et al., 1987). It has been argued that the neural and humoral pathways communicate from the blood via the CVOs, but how these molecules cross the TJs of the ependyma of the CVOs and enter the CSF is still a matter of some debate (Brightman, 1965; Dantzer et al., 2000). The fenestrated and leaky blood vessels of the CVOs might enable molecules to gain access to the brain around the

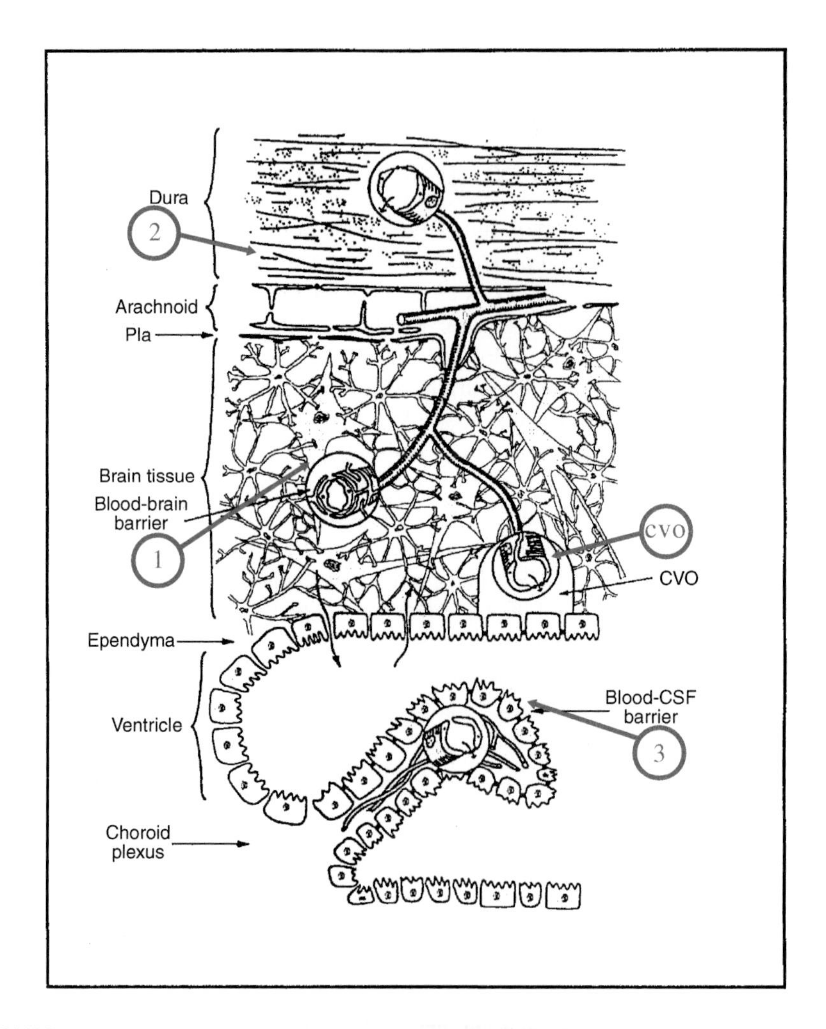

FIGURE 4.1 Location of barrier sites in the CNS. Barriers are present at three main sites: 1) the brain endothelium forming the blood–brain barrier (BBB), 2) the arachnoid epithelium forming the middle layer of the meninges, and 3) the choroid plexus epithelium, which secretes cerebrospinal fluid (CSF). At each site the physical barrier is caused by tight junctions that reduce the permeability of the paracellular (intercellular cleft) pathway. However, in the circumventricular organs (CVOs) that contain neurons specialized for neurosecretion or chemosensitivity or both, the endothelium is leaky. This allows tissue-blood exchange, but these sites are separated from the rest of the brain by an external glial barrier, the glia limitans, and from CSF by a barrier of tight junctions at the ependyma. Note that the endothelium of the choroid plexus is leaky; immune complexes can be deposited in the stroma of the choroid plexus and in CVOs, but are not generally found in brain parenchyma as a result of the BBB. (Reprinted with permission from Kluwer Academic Publishers, *The Blood–Brain Barrier, Amino Acids and Peptides,* by M.B. Segal and B.V. Zlokovic, 1990.)

BBB, but it is thought that the glia limitans around these areas and the rapid venous drainage might act as a functional barrier to prevent the entry of polar molecules into the brain from these regions (Segal, 1999).

4.3 DURA/ARACHNOID

The dura also has "leaky"-type fenestrated capillaries so molecules in theory could bypass the BBB by crossing into the dural ISF and gain access to the brain ISF. However, this is prevented by the double arachnoid layer, the cells of the outer layer of which are sealed together by TJs (see Figure 4.1) (Nabeshima et al., 1975).

4.4 COMPOSITION OF CSF

CSF has a composition similar to plasma but with some key differences, which was used by Davson to establish that CSF was a secreted fluid and not an ultrafiltrate of plasma. With modern analytical techniques, it is not easy to comprehend the difficulty and care needed by the early workers in the field using gravimetric and simple chemical techniques to accurately measure the composition of CSF and plasma. It took a whole day to measure a single value for the sodium concentration in these fluids (Davson, 1955).

As seen in Table 4.1, while the ionic composition differences between plasma and CSF apart from Ca^{2+}, Mg^{2+}, K^{+} are small, the $R_{Dialysis}$ ratios for ions are significantly different from a simple ultra-filtrate. From these studies it was clearly demonstrated that the CSF was a secreted fluid formed by active transport (Davson, 1955). The values for other mammalian species are reviewed in Davson and Segal (1995), and it can be observed that the composition for most species is very similar with a low value for proteins, sugars, and amino acids compared to plasma. Although, from these observations, the blood–CSF barrier restricts most proteins, studies by Felgenhauer (1980) have revealed that, although the concentration of protein in CSF is very low, the patterns of proteins in CSF and plasma are similar and are surprisingly largely independent of molecular weight. This implies that there must be a few large channels across the CP through which these macromolecules can diffuse into the CSF (Davson and Segal, 1995).

4.5 FUNCTIONS OF THE CPS

4.5.1 Secretion of CSF

The prime function of the CPs is to secrete CSF, and while there is wide variation in the overall rate of secretion in the various species, once these values are converted to a rate per mg of CP, the rate is very similar in all mammalian species (see Table 4.2) (Davson and Segal, 1995). The CPs are highly vascular structures with a blood flow of 3 to 4 ml/min/g, which is 10-fold higher than in most regions of the brain (Begley, 2004). While this high blood flow is needed to form CSF, it also has a function in

TABLE 4.1
Concentrations of Various Solutes in Plasma and Lumbar Cerebrospinal Fluid of Human Subjects

	Concentration (mEq/Kg H_2O)		
Substance	Plasma	CSF	R_{CSF}
Na	150.0	147.0	0.98
K	4.63	2.86	0.615
Mg	1.61	2.23	1.39
Ca	4.70	2.28	0.49
Cl	99.0	113.0	1.14
HCO_3	26.8	23.3	0.87
Inorganic P (mg/100 ml)	4.70	3.40	0.725
Amino acids	2.62	0.72	0.27
Osmolality	289.0	289.0	1.0
pH	7.397	7.307	—
PCO_2 (mmHg)	41.1	50.5	—

Reprinted with permission from CRC Press, *Physiology of the CSF and Blood–Brain Barriers,* by H. Davson and M.B. Segal, 1995.

TABLE 4.2
Rates of Secretion of CSF by Various Species Estimated by Ventriculo-Cisternal Perfusion

Species	μl/min	Rate (%/min)	μl/min/mg CP
Mouse	0.325	0.89	—
Rat	2.1–5.4	0.72–1.89	—
Guinea pig	3.5	0.875	—
Rabbit	10.0	0.43	0.43
Cat	20.0–22.0	0.45–0.50	0.50–0.55
Dog	47.0–66.0	0.40	0.625–0.77
Goat	164.0	0.65	0.36
Sheep	118.0	0.83	—
Monkey	28.6–41.0	—	—
Man	350.0–370.0	0.38	—

Reprinted with permission from CRC Press, *Physiology of the CSF and Blood–Brain Barriers,* by H. Davson and M.B. Segal, 1995.

removing molecules from the CSF back to the blood, and a variety of efflux systems have been identified for organic molecules and drugs (Begley, 2004). More information on the regulation of CSF formation and on choroidal blood flow as well as the molecular mechanisms of CSF secretion can be found in Chapters 5 and 6.

4.5.2 Barrier Functions

4.5.2.1 Passive Barrier

Being at the interface between the CSF and the blood, the CPs have a role as a simple physical barrier. This is composed of two aspects, the first of which is the lipid bilayer of the endothelial cell walls. Molecules can permeate across these cells related to their lipid solubility and their molecular weight. Lipid-soluble molecules will enter the lipid components of the cell wall easily but can have difficulty either entering the aqueous cytoplasm or returning to the plasma. An equilibrium will then be set up depending also on binding to protein and the limited solubility in the aqueous phases (Davson and Segal, 1995). Capillary depletion methods can be used to identify how much of a molecule has actually crossed the BBB and how much is trapped in the membrane lipids (Thomas and Segal, 1996).

4.5.2.2 TJ Permeability

The second aspect depends on the TJs of the CPs which differ in structure from those between the endothelial cells of the BBB. Wolberg and colleagues have shown differences in the claudin subtypes with the TJs of the CPs having claudin 11 as well as claudin 1 and 2, which may account for the slightly more leaky nature of these junctions compared to those of the BBB, which only have claudin 1 and 5. This may also account for the differences in morphology between these junctions, with those of the CPs being made up of parallel strands that are more sparsely interconnected, and there appears to be a tortuous open pathway through the strands (Wolberg et al., 2001). In spite of this potential pathway, the permeability of the CPs to small polar molecules such as mannitol is low (Deane and Segal, 1985).

4.5.3 Efflux Processes

It has long been known that the CPs have a carrier-mediated efflux pathway for a variety of molecules such as p-aminohippurate (PAH) and benzyl penicillin. These transport carriers have a broad specificity and have been reviewed elsewhere (Suzuki et al., 1997; Miller et al., 1999; Kusuhara and Sugyema, 2001; Ghersi-Egea and Strazielle, 2002; Begley, 2004). The enzyme activities of the CPs are analyzed in Chapter 7.

4.5.4 Enzyme Activity

The CPs contain high levels of detoxifying enzymes such as epoxide hydrolase, glutathione peroxidase, and the conjugating enzymes (Tayarani et al., 1989; Ghersi-Egea et al., 1995; Ghersi-Egea and Strazelle, 2002).

4.5.5 CSF and Brain Support

The brain "floats" in the CSF, which reduces the effective weight of the brain and cushions the brain from physical injury caused by striking the skull on hard objects. It also reduces the tension on the nerve roots and on the dura, which is surprisingly well supplied with pain fibers. Extensive neurosurgery can be performed using local anaesthesia as the brain has no pain fibers, but cutting the dura is quite painful (personal observation). This may account for the severe headaches caused if a patient stands up too soon after CSF has been removed by lumbar puncture. After this procedure the patient needs to be kept horizontal until this fluid has been replaced at about 0.5 ml/min by newly produced CSF and the ventricular system is refilled (Davson and Segal, 1995).

4.6 CSF TURNOVER AND DRAINAGE

The secretion of CSF within the ventricles and the resistance to drainage of CSF leads to the build-up of a CSF pressure of about 150 mm saline. This causes the circulation of CSF from the lateral ventricles through the foramen of Monro into the third ventricle (see Figure 4.2). Then CSF passes via the narrow aqueduct of Sylvius into the fourth ventricle and out into the subarachnoid space (SAS) via the foramina of Luschka and Magendie. From here the CSF flows over the surface of the brain and into the basal cisterns. Some will also pass down the outside of the spinal cord, and CSF can easily be sampled from the lumbar sac, via lumbar puncture. The flow of CSF over the surface of the brain allows fluid to reach the arachnoid granulations in the sagittal sinus, which have a valve-like nature, and if the pressure in the great vein of Galen is less than that of the CSF, the fluid drains into the venous system (see Figure 4.2). There are also arachnoid granulations in the spinal nerve routes, and here the fluid drains into the cervical lymphatics. From the observed rate of formation and the volume of CSF, the fluid is turned over some four times per 24 hours. This, however, depends on the size of the animal, with the rate of turnover being greater in animals with small brains (Begley and Brightman, 2003).

4.7 NASAL EPITHELIUM AND CERVICAL LYMPHATICS

There appears to be a route into and from the brain across the nasal epithelium, which is now a target of active research of the pharmaceutical industry seeking an inhaled form of insulin and its analogs. Earlier studies by Bradbury and others have shown that there is a route for the passage of large molecules from the CSF across the cribriform plate into the deep cervical lymphatics (Bradbury et al., 1983). The exact anatomical correlates of this route are still poorly defined, although it is affected by posture. Recent studies on gonadotropin-releasing hormone in sheep have shown that the neurons releasing this peptide migrate into the brain during development from the olfactory placode (Bruneau et al., 2003), which would suggest an embryological pathway in this region.

4.8 BULK FLOW OF BRAIN ISF

The elegant studies of the late Helen Cserr revealed a slow current of bulk flow of ISF through the brain toward the CSF. This work revealed that, when several molecules of widely differing sizes were carefully microinjected in a small volume

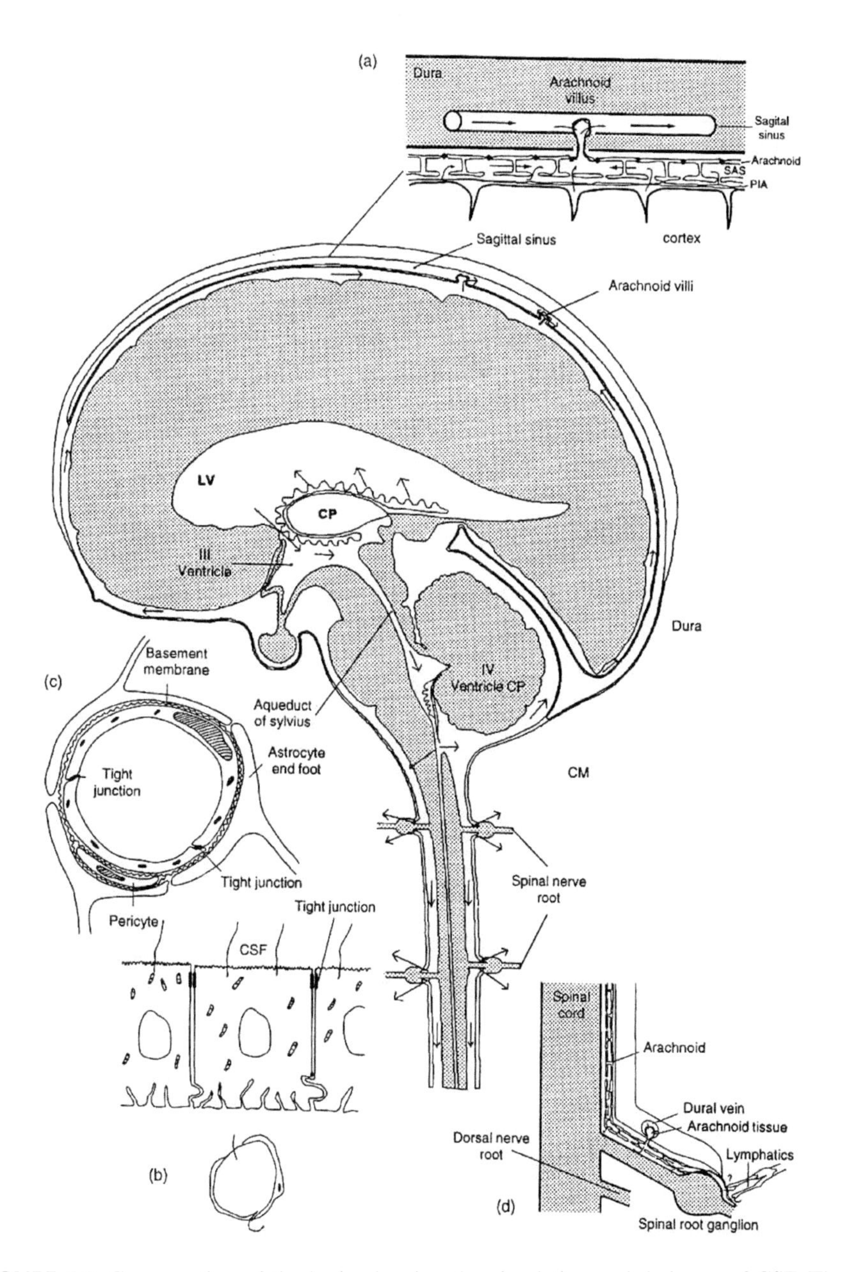

FIGURE 4.2 Cross section of the brain showing the circulation and drainage of CSF. The insets show the sites of the blood–CSF barrier at (a) the outer arachnoid layer and (b) the choroid plexus. The blood–brain barrier is shown in inset (c) at the level of the cerebral capillaries. The drainage routes for CSF are shown in inset (a) at the arachnoid villus and (d) the spinal nerve route. There is also a route via the olfactory bulb, which is not shown. CP = choroid plexus, CM = cisterna magna, LV = lateral ventricle, SAS = subarachnoid space. (Reprinted with permission from Whurr Publishers, *Intracranial and Inner Ear Physiology and Pathophysiology*, edited by A. Reid, R.J. Marchbanks, and A. Ernst, 1999.)

(to avoid a local rise in pressure) into the brain substance, they moved away from the injection site at the same rate, which is strong evidence for a bulk flow of brain ISF (Cserr et al., 1981). Rosenberg et al. (1980) investigated the effect of changing the arterial osmotic pressure and studied the movement of labeled mannitol from the ventriculo-cisternal perfusion into the white and gray matter. In the latter, diffusion seems to be the only process, whereas in white matter there is diffusion and bulk flow, which may contribute to the total production of CSF.

In gray matter there is a greater capillary density than in white matter, and around the blood vessels is the unusual feature of the Virchow-Robin spaces, which form a tortuous pathway between the SAS and the ventricles. It is thought that the fluid in these spaces is continuous with the CSF and the brain ISF, and it has been speculated that these paravascular pathways may serve as the equivalent of lymph channels. Cserr has shown that this is the route taken by lymphocytes entering the brain (Cserr et al., 1992; Cserr and Knopf, 1992).

4.9 BRAIN VOLUME REGULATION

The low hydraulic conductivity of the brain capillaries and strict autoregulation of cerebral blood flow limit the passage of water across the BBB. For example, an osmotic driving force that would produce a volume flow per unit area of membrane of 1 μl/min across the capillaries of the brain would produce a flow of 33 μl/min in skeletal muscle and 270 μl/min in the heart (Fenstermacher, 1984). The BBB acts as a "pure" lipid membrane with few, if any, water-filled channels. Water has therefore to cross directly through the luminal and abluminal membranes of the endothelial cells by dissolving in the lipids (Paulson et al., 1977; Bradbury, 1979). Although the water channel protein aquaporin 4 is found in the CPs, ependyma, and regions of the brain involved in osmoregulation and sodium intake, it was not reported in the endothelium of the CNS. In the brain, it is localized in the astrocyte endfeet and the glia limitans, but its role is still a matter of debate, although from its location at the brain fluid interfaces it could be a key component of CNS volume regulation (Venero et al., 1999; Vajda et al., 2000). Several groups have shown the rapid clearance of macromolecules from the brain ISF by special transport systems, which may osmotically remove excess water from the brain ISF (Lorenzo et al., 1972; Wagner et al., 1974; Blasberg, 1976; Fenstermacher, 1984).

Studies by Fenstermacher and his colleagues demonstrated a rapid movement of sucrose from the cisterna magna into many parts of the brain (Ghersi-Egea et al., 1996). This has led to the suggestion that there is a movement alongside the blood vessels within the brain along the paravascular spaces with some differences in direction between arteries and veins (Segal, 1999). The brain is unusual in that it lacks a lymphatic system found in other tissues, and this is possible since the high transcapillary resistance of the BBB prevents the leakage of protein into the ISF, which is normally recovered by the lymphatic pathway. Fenstermacher (1984) has suggested that this paravascular pathway may act as the so-called third circulation of the brain. The findings of Cserr suggested a slow bulk flow of fluid passing from the BBB through the brain interstitial space and draining into the CSF, which

Davson termed the "sink action" of the CSF, and it is thought to remove excess fluid and waste metabolic products from the brain (Fenstermacher, 1984; Davson and Segal, 1995).

4.10 THE EPENDYMA/CSF INTERFACE

As has been stated, the ependyma lines the ventricles and is permeable to all small molecules. Del Bigio (1995) has postulated in his excellent review on the ependyma that the tissue also has a number of functions, which may include axon guidance during development. The ependyma also possesses many enzymes, such as urokinase and plasminogen activator, involved in the proteolysis of extracellular matrix components. Glial fibrillary acidic protein is also present in the ependyma.

The cilia of the ependyma beat in a coordinated fashion, which may act to aid the circulation of the CSF. The ependyma also reacts during infection and upregulates adhesion molecules, and Cserr et al. (1992) considered these cells to have a role in immune activation. Another role is in the binding of copper, zinc, and some heavy metals. Zheng et al. (2001, 2003) suggest that the ependyma and the CPs may act to protect the brain from oxidative damage.

4.11 CSF VOLUMES

The volume of CSF in man is some 150 ml with some 23 ml (16%) in the ventricles and about 30 ml around the spinal cord, the majority of the fluid being found in the basal cisterns of the brain and the SAS. There is no real difference between males and females with regard to the volume of CSF, although the absolute volumes for men were higher (Luders et al., 2002). Some differences have been observed between ventricular and lumbar CSF as the fluid passes along the system (Davson and Segal, 1995).

4.12 EXCHANGES BETWEEN THE CSF AND BRAIN

While the largest interface between the blood and the brain is the capillary interface of the BBB, the CP/CSF interface has a considerable potential for exchange given the high flow through the CPs and the metabolic activity of this tissue.

4.13 VOLUME TRANSMISSION (VT)

Analysis of the fluid compartments of the brain should also include a discussion of volume transmission (VT). In the brain, non-synaptic VT via tortuous channels between the cells of the CNS is commonplace. VT can influence large regions of the CNS and is marked by the presence of multiple exocytoses in neurons which are often remote from the synaptic cleft. These molecules are released by this process into the extracellular fluid space at a considerable distance from their putative target receptors (see Figure 4.3). It is thought that the release of many transmitters, such as dopamine, substance P, calcitonin gene–related peptide, and vasopressin, into the extracellular spaces can influence the limbic system, neostriatum, hippocampus, and neocortex, which can play a role in the regulation of behavior. The VT explains the presence of many of these molecules in the CNS, which can operate as neuromodulators acting on the presynaptic side to control the amount of transmitter released or postsynaptically by regulating the sensitivity of receptors (Zoli et al., 1999; Descarries and Mechawar, 2000).

VT is now gaining acceptance as a key part of the control of information in the brain. Although at first sight, the lack of anatomically defined channels would appear a disadvantage, specialized routes such as the Virchow-Robin spaces, the paracellular spaces around the nerves, and the flow of CSF can mitigate these disadvantages (Agnati and Fuxe, 2000).

Charles Nicholson has been a pioneer in the field of diffusion and has worked with Helen Cserr developing her ideas on the movement of molecules through the brain. Nicholson examined the effect of diffusion in relation to molecular weight, and the effect of extracellular matrices on the flow through the brain extracellular space, which should impede the diffusion (Nicholson, 2000). However, the work of Cserr and her colleagues (Cserr et al., 1981) supported the concept of bulk flow through the extracellular space of the brain and showed that molecules of widely different molecular weight were cleared at the same rate.

As has been stated, much of this bulk flow does not occur freely through the brain but takes place in the Virchow-Robin spaces around the blood vessels. Rennels et al. (1985) using horseradish peroxidase provided evidence to support this special paravascular diffusion, as did Rosenberg et al. (1980), who studied the effect of diffusion and the influence of osmotic forces. These studies suggest a flow velocity of 10.5 μm/min toward the ventricles in white matter, but gray matter showed this movement only in respect to osmotic stress. Nicholson used fluorescent dextrans and showed a rapid long distance movement in the para-axonal spaces of the corpus callosum.

Bjelke et al. (1995) used tetraethyl ammonium (TMA^+), a small charged molecule, to show that the extracellular matrix places little restriction on the movement of small molecules through the brain. However, it does affect the distribution of large molecules, tortuosity does influence their distribution, and it has been shown that very large linear polymers of 10^6 molecular weight can move across as easily as TMA^+, which is a surprise (Nicholson, 2000). The effect of age and pathophysiology on VT has been studied by Sykova and her group (Sykova, 2000). This work has shown that glial cells can form diffusion barriers and modify VT. Dual probe microdialysis has been used to determine the diffusion of labeled molecules in vivo and can give data on clearance and metabolism of endogenous molecules such as dopamine (Kehr et al., 2000).

4.14 CONCLUSION

From this brief description, the importance of VT is expanding rapidly and will lead to a greater understanding of the mechanism by which the brain is controlled. The view that the CSF is a simple "sink" for the products of brain metabolism is now

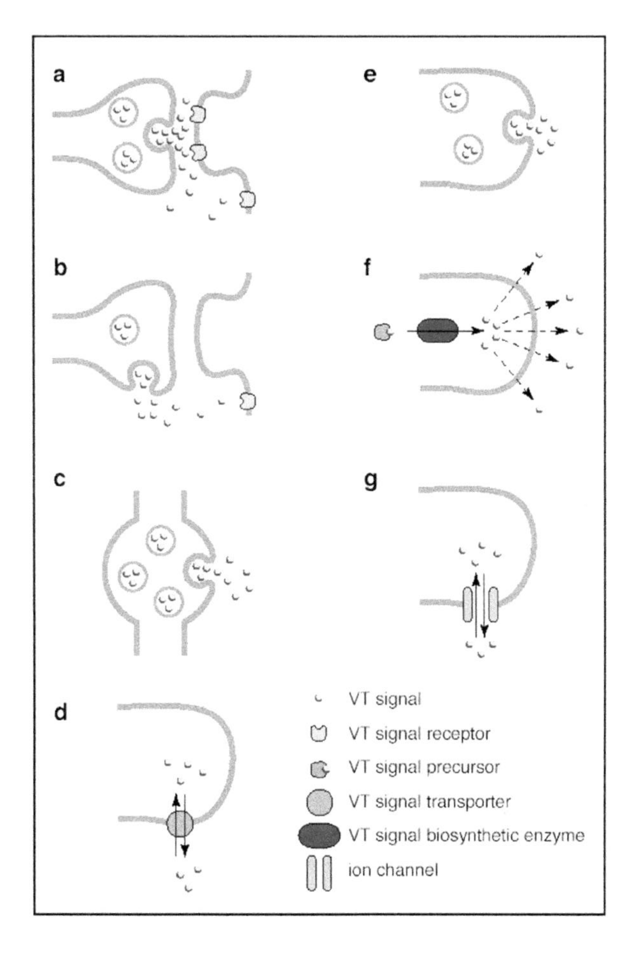

FIGURE 4.3 Schematic representation of the main sources of volume transmission (VT) signals in the CNS. (a) Open synaptic transmission (neuronal): intrasynaptic vesicular release followed by diffusion of the transmitter outside the synaptic cleft at effective concentration (synaptic spillover of, for example, amino acids). (b) Open synaptic transmission (neuronal): extrasynaptic vesicular release. The transmitter is released directly into the extracellular fluid outside the synaptic cleft (nonsynaptic release of, e.g., neuropeptides). (c) Paracrine transmission (neuronal): vesicular release from nonjunctional varicosities (e.g., diffuse catecholaminergic systems), that is, varicosities lacking presynaptic specializations and postsynaptic densities. (d) Paracrine transmission (neuronal or nonneuronal): reverse functioning of transmitter uptake carriers (e.g., release of glutamate, *D*-serine, and GABA from astroglia and glutamate and dopamine from neurons). (e) Paracrine transmission (nonneuronal): Ca^{2+}-dependent and constitutive vesicular release from nonneuronal cells (e.g., endothelin release from endothelial cells). (f) Paracrine transmission (neuronal and nonneuronal): release of gaseous transmitters (e.g., nitric oxide release from neurones and endothelial cells). (g) Local ion currents (neuronal and non-neuronal): changes in the extracellular fluid concentration of ions (e.g., K^+, H^+ and Ca^{2+}) induced by the activity of transmitter- or voltage-gated ion channels located in neurones or glia. (Reprinted with permission from Elsevier, *Trends Pharmacol. Sci.* 20, 1999, 142–150.)

too simplistic, and Davson's views must now be modified to include these exciting new findings. The distance involved over which VT operates has expanded from the "spillover" around synapse to global effects across the brain. While the signals for VT are beginning to be identified and range from inorganic molecules such as H^+, K^+, and Ca^{2+} as well as nitric oxide and CO_2 to the classical neurotransmitters, such as ATP, peptides, and macromolecules, all play a role both close to the synapse and now by propagation at considerable distances from the site of their release either in the Virchow-Robin spaces or with the circulation of CSF. VT has much to contribute to the understanding of the mechanism of brain function and can modify both the release of transmitters at the synapse and the receptor sensitivity at the postsynaptic membrane (Nicholson, 2000). Although the spaces between the neurons and glia are in theory complex and tortuous, the movement of very large molecules can occur easily, so more work must be done before a full understanding of VT can be achieved.

4.15 ACKNOWLEDGMENTS

The author wishes to thank Andreas Reichel for his help in the preparation of Figure 4.1.

REFERENCES

Agnati, L.F., Fuxe, K. (2000) Volume transmission as a key feature of information handling in the central nervous system: possible new interpretative value of the Turing's B-type machine. In: *Volume Transmission Revisited*, ed. by L.F. Agnati, K. Fuxe, C. Nicholson and E. Sykova, Elsevier, Amsterdam, pp. 3–20.

Begley, D.J. (2004) Efflux mechanisms in the central nervous system: a powerful influence on drug distribution within the brain. In: *Blood-Spinal Cord and Brain Barriers in Health and Disease*, ed. by H.S. Sharma and J. Westman, Elsevier, San Diego, CA, pp. 83–97.

Begley, D.J., Brightman, M.W. (2003) Structural and functional aspects of the blood-brain barrier. *Prog. Drug Res.* 61: 42–78.

Blasberg, R.G. (1976) Clearance of serum albumin from brain extracellular fluid. A possible role in cerebral edema. In: *Dynamics of Brain Edema*, ed. by H.M. Pappius and W. Feindel, Springer, Berlin, pp. 98–102.

Bradbury, M.W.B. (1979) *The Concept of a Blood-Brain Barrier*, John Wiley & Sons, Chichester.

Bradbury, M.W.B., Westrop, R.J. (1983) Factors influencing exit of substances from cerebrospinal fluid into deep cervical lymph of the rabbit. *J. Physiol.* 339: 519–534.

Brightman, M.W. (1965) The distribution within the brain of ferritin injected into cerebrospinal fluid compartments. I. Ependymal distribution. *J. Cell Biol.* 26: 99–123.

Bruneau, G., Izvolskaia, M., Ugrumov, M., Tillet, Y., Duittoz, A.H. (2003) Prolonged neurogenesis during early development of gonadotropin-releasing hormone neurones in sheep (*Ovis aries*): in vivo and in vitro studies. *Neuroendocrinology* 77: 177–186.

Cserr, H.F. (1975) Bulk flow of cerebral extracellular fluid as a possible mechanism of CSF-brain exchange. In: *Fluid Environment of the Brain*, ed. by H.F. Cserr, J.D. Fenstermacher and V. Fenci, Academic Press, New York, pp. 215–224.

Cserr, H.F., Cooper, D.N., Suri, P.K., Patlak, C.S. (1981) Efflux of radiolabeled polyethylene glycols and albumin from rat brain. *Am. J. Physiol.* 240: F319–F328.

Cserr, H.F., De Pasquale, M., Harling-Berg, C.J., Knopf, P.M. (1992) Afferent and efferent arms of the humoral immune response to CSF-administered albumin in a rat model with normal blood-brain barrier permeability. *J. Neuroimmunol.* 41: 195–202.

Cserr, H.F., Knopf, P.M. (1992) Cervical lymphatics, the blood-brain barrier and the immunoreactivity of the brain. *Immunol. Today* 13: 507–512.

Curl, F.D., Pollay, M. (1968) Transport of water and electrolytes between brain and ventricular fluid in the rabbit. *Exp. Neurol.* 20: 558–574.

Dantzer, R., Konsman, J.-P., Bluthe, R.-M., Kelley, K.W. (2000) Neural and humoral pathways of communication from the immune system to the brain: parallel or convergent? *Auton. Neurosci. Basic and Clin.* 85: 60–65.

Davson, H. (1955) A comparative study of the aqueous humour and cerebrospinal fluid in the rabbit. *J. Physiol.* 129: 111–133.

Davson, H., Segal, M.B. (1970) The effects of some inhibitors and accelerators of sodium transport on the turnover of ^{22}Na in the cerebrospinal fluid. *J. Physiol.* 209: 131–153.

Davson, H., Spaziani, E. (1959) The blood-brain barrier. *J. Physiol.* 149: 135–143.

Davson, H., Segal, M.B. (1995) *Physiology of the CSF and Blood-Brain Barriers*, CRC Press, Boca Raton, FL.

Deane, R., Segal, M.B. (1985) The transport of sugars across the perfused choroid plexus of the sheep. *J. Physiol.* 362: 245–260.

Del Bigio, M.R. (1995) The ependyma: A protective barrier between brain and cerebrospinal fluid. *Glia* 14: 1–13.

Descarries, L., Mechawar, N. (2000) Ultrastructural evidence for diffuse transmission by monoamine and acetylcholine neurons of the central nervous system. *Prog. Brain Res.* 125: 27–48.

Felgenhauer, K. (1980) Protein filtration and secretion at human body fluid barriers. *Pflugers Arch.* 384: 9–17.

Fenstermacher, J.D. (1984) Volume regulation of the central nervous system. In: *Edema*, ed. by N.C. Staub and A.E. Taylor, Raven Press, New York, pp. 383–404.

Ghersi-Egea, J.F., Finnegan, W., Chen, J.L., Fenstermacher, J.D. (1996) Rapid distribution of intraventricularly administered sucrose into cerebrospinal fluid cisterns via subarachnoid velae in rat. *Neuroscience* 75: 1271–1288.

Ghersi-Egea, J.F., Leininger-Muller, B., Cecchelli, R., Fenstermacher, J.D. (1995) Blood-brain interfaces: relevance to cerebral drug metabolism. *Toxicol. Lett.* 82/83: 645–653.

Ghersi-Egea, J.F., Strazielle, N. (2002) Choroid plexus transporters for drugs and other xenobiotics. *J. Drug Targeting* 11: 353–357.

Gross, P.M., Blasberg, R.G., Fenstermacher, J.D., Patlak, C.S. (1987) The microcirculation of rat cicumventricular organs and pituitary gland. *Brain Res. Bull.* 18: 73–85.

Keep, R.F., Jones, H.C. (1990) A morphometric study on the development of the lateral ventricle choroid plexus, choroid plexus capillaries and ventricular ependyma in the rat. *Dev. Brain Res.* 56: 47–53.

Keep, R.J., Ennis, S.R., Betz, A.L. (1998) Blood-brain barrier ion transport. In: *An Introduction to the Blood-Brain Barrier: Methodology, Biology and Pathology*, ed. by W.M. Pardridge, Cambridge University Press, Cambridge, pp. 207–213.

Kehr, J., Hoistad, M., Fuxe, K. (2000) Diffusion of radiolabeled dopamine, its metabolites and mannitol in the rat striatum studied by dual-probe microdialysis. *Prog. Brain Res.* 125: 179–192.

Kusuhara, H., Sugiyama, Y. (2001) Efflux transport systems for drugs at the blood-brain barrier and blood-cerebrospinal fluid barrier (part 2). *Drug Discov. Today.* 6: 206–212.

Lorenzo, A.V., Shiahage, I., Liang, M. Barlow, C.F. (1972) Temporary alteration of cerebrovascular permeability to plasma protein during drug-induced seizures. *Am. J. Physiol.* 223: 268–277.

Luders, E., Steinmetz, H. Jancke, L. (2002) Brain size and grey matter volume in the healthy human brain. *Cog. Neurosci. Neuropsychol.* 13: 2371–2374.

Michel, C.C. (1984) Fluid movement through capillary walls. In: *Handbook of Physiology, The Cardiovascular System*, Sect. 2, Vol. IV, ed. by E.M. Renkin, C.C. Michel and S.R. Geiger, American Physiological Society, Bethesda, MD, pp. 375–409.

Miller, D.S., Villalobos, A.R., Pritchard, J.B. (1999) Organic cation transport in rat choroid plexus cells studied by fluorescence microscopy. *Am. J. Physiol.* 276: C955–C968.

Nabeshima, S., Reese, T.S., Landis, D.M.D., Brightman, M.W. (1975) Junctions in the meninges and marginal glia. *J. Comp. Neurol.* 164: 127–170.

Nicholson, C., Chen, K.C., Hrabetova, S., Tao, L. (2000) Diffusion of molecules in brain extracellular space: theory and experiment. *Prog. Brain Res.* 125: 129–154.

Oldendorf, W. H., Davson, H. (1967) Brain extracellular space and the sink action of cerebrospinal fluid. *Arch. Neurol.* 17: 196–205.

Oldendorf, W.H., Brown, W.J. (1975) Greater number of capillary endothelial cell mitochondria in brain than in muscle. *Proc. Soc. Exp. Biol. Med.* 149: 736–738.

Paulson, O.B., Hertz, M.M., Bolwig, T.G., Lassen, N.A. (1977) Filtration and diffusion of water across the blood-brain barrier in man. *Microvasc. Res.* 13: 113–124.

Peters, A., Palay, S.I., Webster, H. deF. (1991) *The Fine Structure of the Nervous System, Neurons and Their Supporting Cells*, 3rd ed., Oxford University Press, New York, pp. 290–293.

Rennels, M., Gregory, T.F., Blaumanis, O.R., Fujimoto, K., Grady, P.A., (1985) Evidence for a "paravascular" fluid circulation in the mammalian central nervous system, provided by the rapid distribution of tracer protein throughout the brain from the subarachnoid space. *Brain Res.* 326: 47–53.

Rosenberg, G.A., Kyner, W.T., Estrada, E. (1980) Bulk flow of brain interstitial fluid under normal and hyperosmolar conditions. *Am. J. Physiol.* 238: F42–F49.

Segal, M.B. (1999) The anatomy and physiology of intracranial fluids. In: *Intracranial and Inner Ear Physiology and Pathophysiology*, ed. by A. Reid, R.J. Marchbanks and A. Ernst, Whurr Publishers Ltd., London, pp. 3–15.

Suzuki, H., Terasaki, T., Sugiyama, Y. (1997) Role of efflux transport across the blood–brain and blood–cerebrospinal fluid barrier on the disposition of xenobiotics in the central nervous system. *Adv. Drug Deliv. Rev.* 25: 257–285.

Sykova, E. (2001) Glial diffusion barriers during aging and pathological states. *Prog. Brain Res.* 132: 339–363.

Thomas, S.A., Segal, M.B. (1996) Identification of a saturable uptake system for deoxyribonucleosides at the blood-brain and blood cerebrospinal fluid barriers. *Brain Res.* 741: 230–239.

Vajda, Z., Promeneur, D., Doczi, T., Sulyok, E., Frokiaer, J., Ottersen, O.P., Nielsen, S. (2000) Increased aquaporin-4 immunoreactivity in rat brain in response to systemic hyponatremia. *Biochem. Biophys. Res. Comm.* 270: 495–503.

Venero, J.L., Vizuete, M.L., Ilundain, A.A., Machado, A., Echevarria, M., Cano, J. (1999) Detailed localization of aquaporin 4 messenger RNA in the CNS: preferential expression in periventricular organs. *Neuroscience* 94: 239–250.

Wagner, H.J., Pilgrim, C., Brandl, J. (1974) Penetration and removal of horseradish peroxidase injected into the cerebrospinal fluid: role of cerebral perivascular spaces, endothelium and microglia. *Acta. Neuropathol.* 27: 299–315.

Weindl, A., Joynt, R.J. (1973) Barrier properties of the subcommissural organ. *Arch. Neurol.* 29: 16–22.

Wolburg, K., Wolburg-Buchhoz, S., Liebner, S., Engelhardt, B. (2001) Claudin 1, Claudin 2 and Claudin 11 are present in tight junctions of choroid plexus epithelium of the mouse. *Neurosci. Lett.* 307: 72–80.

Zheng, W., Lu, T.M., Lu, G.Y., Zhao, Q., Cheung, O., Blaner, W.S. (2001) Transthyetin, thyroxine and retinol-binding protein in human cerebrospinal fluid: effect of lead exposure. *Toxicol. Sci.* 61: 107–114.

Zheng, W., Deane, R., Redzic, Z., Preston, J.E., Segal, M.B. (2003). Transport of [^{125}I]thyroxine by in situ perfused ovine choroid plexus: inhibition by lead exposure. *J. Toxicol. Environ. Health* 66: 435–451.

Zoli, M., Jansson, A., Sykova, E., Agnati, L.F., Fuxe, K. (1999) Volume transmission in the CNS and its relevance for neuropsychopharmacology. *Trends Pharmacol. Sci.* 20: 142–150.

5 Peptide-Mediated Regulation of CSF Formation and Blood Flow to the Choroid Plexus

Joanna Szmydynger-Chodobska and Adam Chodobski
Brown University School of Medicine,
Providence, Rhode Island, U.S.A.

CONTENTS

5.1 INTRODUCTION

The choroid plexus (CP) is the major source of cerebrospinal fluid (CSF). However, an estimated 10 to 30% of total CSF production is generated by the bulk flow of interstitial fluid from the brain parenchyma to the ventricles and subarachnoid space (Cserr, 1984). The CSF has a number of functions, including volume transmission and endocrine signaling (see Chapter 10), regulation of brain fluid homeostasis (Pullen et al., 1987), removal of CSF constituents ("sink" action) (Oldendorf and Davson, 1967), and communication between the brain and the immune system (Cserr et al., 1992; Knopf et al.,

1995). Blood flow to the CP is considerably higher compared to other brain regions (Szmydynger-Chodobska et al., 1994; Chodobski et al., 1995). High blood flow is necessary in order to maintain a continuous CSF production at a relatively high rate. Various factors have been identified that regulate choroidal blood flow. These include sympathomimetics, such as norepinephrine and phenylephrine, dopamine, serotonin, adenosine, and a number of vasoactive peptides (Townsend et al., 1984; Faraci et al., 1988a, 1988b, 1989; Schalk et al., 1989, 1992b; Kadel et al., 1990; Maktabi et al., 1990; Nilsson et al., 1991; Granstam et al., 1993; Chodobski et al., 1995, 1999). In rats and, possibly, other mammals, there is a steady increase in choroidal blood flow during the maturation of the central nervous system (CNS) (Szmydynger-Chodobska et al., 1994). Similar to choroidal blood flow, CSF production is also controlled by the sympathetic nervous system (Lindvall and Owman, 1981) and may be modulated by circulating and CSF-borne peptides (Steardo and Nathanson, 1987; Faraci et al., 1990; Nilsson et al., 1991; Chodobski et al., 1992b, 1994; Schalk et al., 1992a, 1992b; Maktabi et al., 1993). In this chapter we will focus on analyzing peptide-mediated regulation of choroidal blood flow and CSF formation and how the experimental data available relate to the normal as well as pathological situations.

5.2 REGULATION OF BLOOD FLOW TO THE CHOROID PLEXUS

Formation of CSF involves plasma ultrafiltration across the choroidal capillary walls and an active ion transport through the choroidal epithelial cells (see Chapter 6). The process of plasma ultrafiltration in the CP appears to be analogous to that seen in the kidney glomerulus, where the filtration rate is dependent upon blood flow (Baylis and Brenner, 1978). Therefore, choroidal blood flow is believed to control CSF production by limiting the amount of water and electrolytes available for vectorial transfer from the choroidal vascular compartment into the ventricular CSF (Cserr, 1971). However, the relationship between blood flow to the CP and the rate of CSF production has not yet been established.

In small rodents, such as rats, choroidal blood flow has been measured by the indicator fractionation method, using ^{123}I- or ^{125}I-isopropyliodoamphetamine (IMP) as a marker (Williams et al., 1991; Szmydynger-Chodobska et al., 1994; Chodobski et al., 1995, 1999). IMP behaves as a "chemical microsphere," i.e., it is almost entirely extracted by brain tissue during a single passage and sequestered thereafter (Kuhl et al., 1982; Obrenovitch et al., 1987; Bryan et al., 1988). Also, IMP metabolism while in the bloodstream is relatively slow. Accordingly, it is assumed that the amount of IMP accumulating in brain tissue is linearly related to blood flow. Tissue sequestration is critical for preventing the diffusional loss of a marker to CSF surrounding the CPs. For this reason, freely diffusible indicators, such as iodoantipyrine, are not suitable for measuring blood flow to the CP. Furthermore, the assumption of free diffusibility of iodoantipyrine is not valid for flow rates exceeding 1.8 $ml \cdot min^{-1} \cdot g^{-1}$ (Van Uitert et al., 1981), which, together with a diffusional loss of this marker to CSF, would result in significant underestimation of choroidal blood flow.

In adult rats, blood flow to the lateral ventricle CP is ~3 $ml{\cdot}min^{-1}{\cdot}g^{-1}$, whereas that to the 3rd and 4th ventricle CPs is ~4 $ml{\cdot}min^{-1}{\cdot}g^{-1}$. Therefore, choroidal blood flow is 5- to 7-fold higher than blood flow to the cerebral cortex (Szmydynger-Chodobska et al., 1994; Chodobski et al., 1995). Higher blood flow levels (~5 $ml{\cdot}min^{-1}{\cdot}g^{-1}$) that were similar in all CPs have also been reported (Williams et al., 1991). A gradual increase in choroidal blood flow has been observed in rats during postnatal development (Szmydynger-Chodobska et al., 1994). This could be considered as an adjustment of the choroidal vascular system to the steadily increasing ion transport capabilities of the maturing choroidal epithelium (Parmelee and Johanson, 1989; Preston et al., 1993). In larger animals, such as rabbits and dogs, radioactively labeled microspheres have been used to measure choroidal blood flow. In these animals, blood flow to the lateral ventricle CP ranged between 3 and 4 $ml{\cdot}min^{-1}{\cdot}g^{-1}$ (Faraci et al., 1988a, 1988b; Maktabi et al., 1990). It is presently unclear whether blood flow to the CP is autoregulated in a manner similar to that of cerebral blood flow (Chodobski et al., 1986). This issue is difficult to resolve experimentally because the sympathomimetic agents that are most commonly used to alter arterial blood pressure can by themselves affect choroidal blood flow (Faraci et al., 1988b).

Although a number of reports on peptide-mediated regulation of choroidal blood flow have been published, our understanding of how these bloodborne peptides control the choroidal blood supply is incomplete. To control blood flow, circulating peptides need to reach their cognate receptors on the smooth muscles of the choroidal arterioles or arteries or both. Available anatomical data does not indicate, however, that vasoactive peptides can readily penetrate the walls of the choroidal arterioles (van Deurs, 1979, 1980). On the other hand, the walls of the choroidal venules appear to be quite leaky (van Deurs, 1979, 1980), but these vessels are not able to control blood flow, as they lack a muscle coat (Bargmann et al., 1982). Scanning electron microscopic studies of corrosion casts of choroidal blood vessels have demonstrated that the choroidal arteries and veins are surrounded by a dense mesh of capillary vessels (see Figure 5.1A) (Motti et al., 1986; Weiger et al., 1986). It has been suggested that this feature of choroidal angioarchitecture is critical for the regulation of choroidal blood flow (Chodobski et al., 1999). According to this concept, the sheathing of arterioles and arteries by a network of capillaries possessing fenestrated endothelial walls allows circulating peptides to gain access to their receptors on the smooth muscles coating the large blood vessels (see Figure 5.1B). This provides a plausible explanation of how circulating peptides can control choroidal hemodynamics. However, further studies will be needed to test this hypothesis.

Among the vasoactive peptides, those whose effects on choroidal blood flow have been studied in detail are vasopressin (VP) and angiotensin II (Ang II). Limited information on the hemodynamic actions of endothelin-1 (ET-1), atrial natriuretic peptide (ANP), and vasoactive intestinal polypeptide (VIP) on the CP is also available. These data will be discussed below.

5.2.1 Vasopressin

Experiments performed in the perfused sheep CP have demonstrated that the circulating VP is a potent vasoconstrictor of choroidal arterioles with a resting diameter

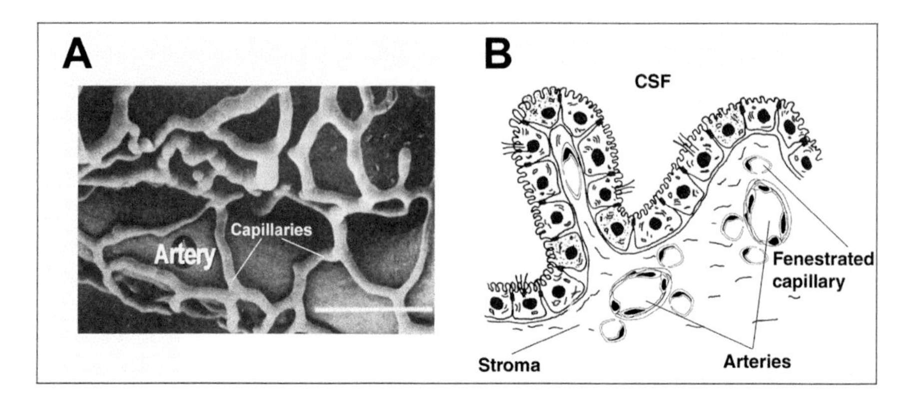

FIGURE 5.1 **(A)** Scanning electron microscopy image of a corrosion cast of the choroidal artery surrounded by a dense mesh of capillaries. (Reprinted with permission from Elsevier, *Brain Research* 378, 1986, 285–96.) **(B)** Schematic diagram of choroidal tissue illustrating the proposed mechanism of regulation of choroidal blood flow by circulating peptides. The sheathing of arteries by a network of capillaries that have fenestrated endothelial walls allows the circulating peptides to gain access to their receptors on smooth muscles coating the large blood vessels.

of 60 to 270 μm (Segal et al., 1992). By comparison, choroidal venules, with a resting diameter of 90 to 220 μm, did not respond to bloodborne VP. In rabbits, intravenous (iv) infusion of VP, producing plasma peptide concentrations that ranged between 60 and 440 pg/ml, lowered choroidal blood flow by 50 to 70% (Faraci et al., 1988a, 1990). The administration of a V_1 receptor antagonist, which by itself did not affect blood flow to the CP, abolished the hemodynamic VP actions on choroidal tissue. These observations indicate that circulating VP does not control the basal blood flow to the CP and that the vasoconstrictive VP actions on the CP require high plasma VP concentrations that are only observed during severe hemorrhage (Cameron et al., 1985; Wang et al., 1988; Shoji et al., 1993). Interestingly, the blockade of the V_{1a} receptors in rabbits subjected to hypoxia resulted in a ~2-fold increase in blood flow to the CP; however, in animals with normally functioning receptors, hypoxia did not change choroidal blood flow (Faraci et al., 1994). In these experiments, plasma VP levels were similar in normoxic and hypoxic animals. These findings suggest that, during hypoxia, the vasoconstrictive VP actions on the CP are balanced out by yet-to-be-identified vasodilators. However, the source(s) of this VP is unclear. Although VP is produced by the choroidal epithelium (Chodobski et al., 1997, 1998c), it is not known whether the choroidal synthesis of this neuropeptide is increased in hypoxia. In addition, VP appears to be released into the CSF across the apical plasma membrane and is therefore unlikely to affect choroidal blood flow. Further research into this area is needed to clarify this issue.

5.2.2 Angiotensin II

Ang II is a peptide hormone that plays an important role in the regulation of arterial blood pressure and the maintenance of body fluid homeostasis. Similar to VP, Ang

II is a potent vasoconstrictor (Osborn et al., 1987), and its plasma levels fluctuate considerably in various physiological and pathophysiological conditions (Huang et al., 1989; Mann et al., 1980, 1981). The effect of circulating Ang II on choroidal blood flow has been studied in rats. In these experiments, the synthetic hormone was infused iv to produce plasma Ang II concentrations estimated to range between 120 and 800 pg/ml (Chodobski et al., 1995). Such plasma Ang II levels have been observed in both normal and VP-deficient Brattleboro rats that were dehydrated for 12 to 48 hours, in animals fed a low-sodium diet, and in renal hypertensive rats (Mann et al., 1980; Huang et al., 1989). The highest plasma Ang II concentration produced in the above study (~2300 pg/ml) was close to that found in malignant renal hypertension (Mann et al., 1980) and was approximately two times lower than the plasma peptide levels observed in the rats that had their inferior vena cava ligated following the blockade of the Ang II receptors (Mann et al., 1981). Despite these broad changes in plasma Ang II concentrations, only a moderate decrease of 12 to 20% in blood flow to the lateral and 4th ventricle CPs was observed at estimated Ang II levels of ~300 to 400 pg/ml. Blood flow to the 3rd ventricle CP was not affected by bloodborne Ang II. By comparison, high Ang II doses significantly increased blood flow to the cerebral cortex, most likely functioning by way of endothelium-derived vasodilative mediators (Haberl et al., 1990).

Mechanistic insight into the hemodynamic actions of Ang II on choroidal tissue has been provided by the more recent pharmacological studies of Chodobski et al. (1999). Their analysis revealed a complex interplay of vasoconstrictive and vasodilative effects of Ang II on the choroidal vascular bed that maintains relatively stable levels of choroidal blood flow despite significant changes in plasma peptide concentrations. A model (see Figure 5.2) has been proposed in which Ang II controls choroidal blood flow not only by exerting a direct vasoconstrictive effect on choroidal vasculature, but also by modulating the activity of the sympathetic nervous system and regulating the synthesis of nitric oxide (NO) (Chodobski et al., 1999).

Choroidal blood vessels possess a dense sympathetic innervation originating from the superior cervical ganglia (Edvinsson et al., 1972, 1974). The density of these sympathetic nerve fibers varies among the CPs, with the 4th ventricle CP having fewer fibers than the lateral and 3rd ventricle CPs (Lindvall, 1979). Functional in vivo and in vitro studies have also shown that α- and β-adrenoceptors are expressed in the choroidal vasculature (Edvinsson and Lindvall, 1978; Wagerle and Delivoria-Papadopoulos, 1987; Faraci et al., 1988b). Previous studies have shown that Ang II potentiates sympathetic activity by promoting the release of noradrenaline from sympathetic nerve endings (Campbell et al., 1979; Szabo et al., 1990; Hilgers et al., 1993). The Ang II-dependent inhibition of noradrenaline reuptake by sympathetic nerve endings has also been demonstrated (Campbell et al., 1979). Since the selective blockade of α- and β-adrenoceptors affects the choroidal hemodynamic responses to Ang II (Chodobski et al., 1999), it is likely that Ang II controls choroidal vascular tone by interacting with the sympathetic nervous system. The differences in the density of the sympathetic innervation of CPs may contribute to the regional variability in the Ang II-mediated regulation of choroidal blood flow.

The role of NO in mediating the hemodynamic actions of Ang II on the CP has been revealed through the use of NO synthase (NOS) inhibitors (Chodobski et al.,

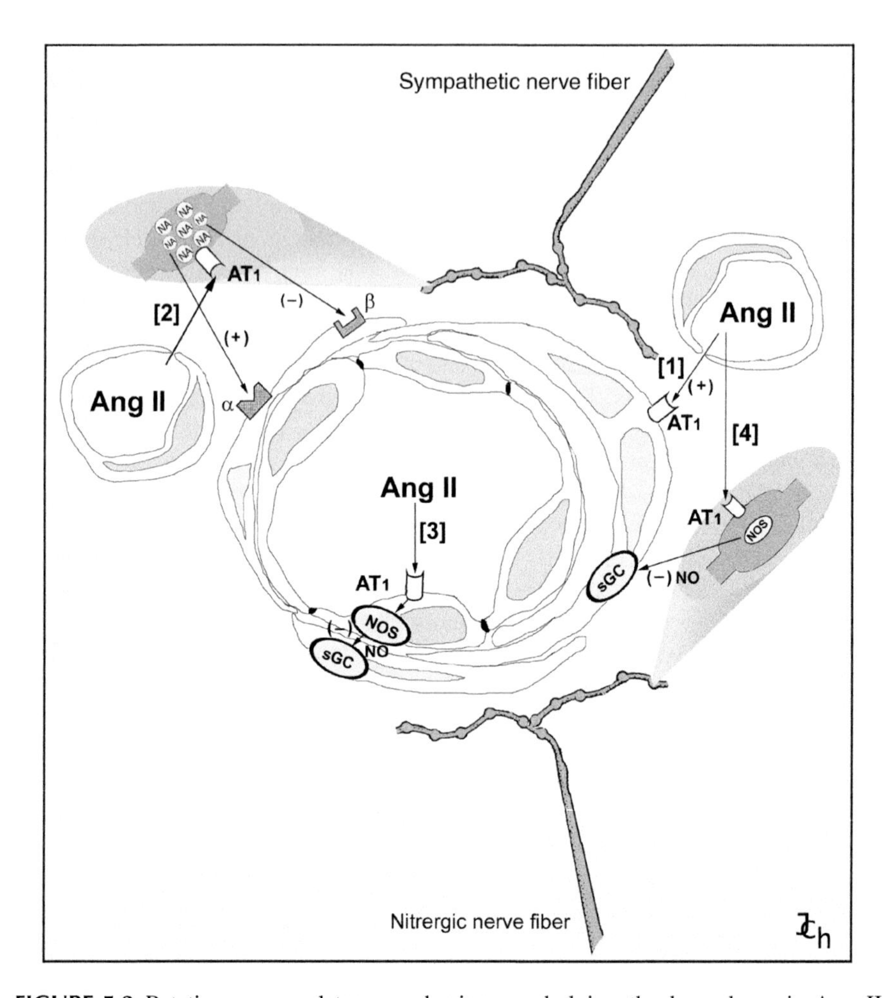

FIGURE 5.2 Putative vasoregulatory mechanisms underlying the hemodynamic Ang II actions on the rat CP. Some or all of these mechanisms may operate in various choroidal blood vessels. (+) and (−) denote the vasoconstrictive and vasodilative Ang II effects, respectively. [1] Direct vasoconstrictive Ang II effect mediated by the Ang II receptors (AT_1) that are expressed on the smooth muscle cells of choroidal arterioles and/or arteries. [2] Indirect vasoactive Ang II actions involving the potentiation of sympathetic activity, which results in the noradrenaline (NA)-mediated activation of α-adrenoceptors (α; vasoconstriction) or β-adrenoceptors (β; vasodilation). [3] Stimulation of nitric oxide synthase (NOS) activity in endothelial cells. [4] Interaction of Ang II with the nitrergic nerve terminals associated with the choroidal blood vessels. The endothelium-derived or nitrergic nerve-derived NO or both stimulates the soluble guanylyl cyclase (sGC) in smooth muscle cells. With increased activity of sGC, guanosine 3',5'-cyclic monophosphate accumulates in myocytes causing vasodilation.

1999). NO appears to play a major role in maintaining the basal blood flow to the CP (Faraci and Heistad, 1992; Chodobski et al., 1999). Nerve fibers containing NOS have been localized to the anterior choroidal artery and its branches (Lin et al., 1996). Also, the endothelial lining of choroidal blood vessels positively stain for NADPH-diaphorase, a histochemical marker for NOS (Szmydynger-Chodobska et al., 1996). These observations suggest that NO is produced by both the nitrergic nerve fibers and the vascular endothelium. It is, however, unclear which of these two NO sources is involved in maintaining the basal blood flow to the CP and mediating the vasodilative Ang II effect on the choroidal vasculature.

It is interesting to note that, in rabbits, the elevation of plasma Ang II levels by iv infusion of a synthetic Ang II has been found to increase choroidal blood flow by 12 to 60% at plasma peptide concentrations ranging between 190 and 2700 pg/ml (Maktabi et al., 1990). This suggests that the peptide-mediated regulation of choroidal hemodynamics may differ among mammalian species.

5.2.3 Other Peptides

The paucity of data does not allow for a detailed analysis of choroidal hemodynamic actions of other vasoactive peptides. Here, we will list the available, mostly phenomenological, observations on ET-1, ANP, and VIP. ET-1 is the most potent vasoconstrictor yet found (Levin, 1996; Schiffrin and Touyz, 1998), and, as expected, iv infusion of ET-1 has been shown to markedly decrease choroidal blood flow in rabbits and cats (Kadel et al., 1990; Granstam et al., 1993). Interestingly, the administration of a synthetic ET-1 into the ventricular CSF has also been found to significantly reduce choroidal blood flow (Granstam et al., 1993). This ET-1 action may be mediated in part by VP or the sympathetic nervous system, or both (Yamamoto et al., 1992; Rossi et al., 1997). However, the decrease in choroidal blood flow observed by Granstam et al. (1993) may have also resulted from the vasoconstrictive ET-1 effect on the choroidal vasculature following the clearance of ET-1 (via bulk CSF absorption) into the peripheral circulation. In contrast to ET-1, both iv and intracerebroventricular (icv) infusions of ANP (rabbits) and VIP (rats) increase blood flow to the CP (Schalk et al., 1989, 1992b; Nilsson et al., 1991). These findings demonstrate that the control of choroidal hemodynamics involves complex, multifactorial regulatory mechanisms.

5.3 REGULATION OF CSF FORMATION

As discussed earlier, the production of CSF is a two-step process involving the ultrafiltration of plasma across the walls of choroidal capillaries (first step), followed by active ion transport through the choroidal epithelium and the movement of fluid into the ventricular CSF space (second step). This suggests that CSF formation could be controlled by changing choroidal blood flow or modulating the activity of ion transporters in choroidal epithelium, or both. There is sufficient evidence for the peptide-mediated regulation of blood flow to the CP; however, vasoactive peptides, such as VP, may also affect ion transport in choroidal epithelial cells. Indeed, V_{1a} and V_{1b} receptors are expressed on the choroidal epithelium (Zemo and McCabe, 2001; Szmydynger-Chodobska et al., 2004). The activation of V_1 receptors stimulates protein kinase C (Thibonnier et al., 1998), which increases the phosphorylation and, consequently, inhibits the activity of choroidal Na^+, K^+-ATPase (Fisone et al., 1995, 1998), an ion pump that plays a key role in CSF formation (see Chapter 6).

Methodology associated with the measurement of CSF production is described in Chapter 24. Here, we will analyze the effects of various peptides on CSF formation.

5.3.1 Vasopressin

The elevation of plasma VP levels to 300 to 400 pg/ml has been shown to decrease CSF production by 35% (Faraci et al., 1990). At these plasma concentrations, VP also significantly lowers blood flow to the CP (Faraci et al., 1988a, 1990), suggesting that the inhibitory effect of circulating VP on CSF formation is associated with the hemodynamic action of this peptide. These VP levels are, however, quite high and are only observed in the setting of severe hemorrhage (Cameron et al., 1985; Wang et al., 1988; Shoji et al., 1993). In other pathophysiological situations, such as dehydration/hypernatremia, plasma VP concentrations are unlikely to attain the levels that would affect choroidal hemodynamics or CSF production (Szczepanska-Sadowska et al., 1983; Zerbe and Palkovits, 1984; Kadekaro et al., 1997). Indeed, CSF production does not change in moderately dehydrated sheep (Chodobski et al., 1998b). This indicates that bloodborne VP does not control CSF formation under normal physiological conditions.

5.3.2 Angiotensin II

Relatively small changes in choroidal blood flow have been observed despite considerable variation in plasma Ang II concentrations (Chodobski et al., 1995), suggesting that circulating Ang II has little effect on CSF formation. Indeed, in both rats and rabbits, iv infusion of Ang II has been shown not to have an effect on CSF production (Chodobski et al., 1992b, 1995). These findings are supported by the lack of change in CSF formation during dehydration (Chodobski et al., 1998b), a pathophysiological situation accompanied by moderately increased plasma Ang II levels (Mann et al., 1980). Interestingly, in another study performed on rabbits where comparable doses of Ang II were used, a 24% decrease in CSF formation was observed (Maktabi et al., 1993). It is important to note, however, that the baseline CSF production rate reported by these authors was lower by ~60% when compared to those observed by other investigators (Lindvall et al., 1979; Chodobski et al., 1992b). These differences may contribute to the discrepant results obtained.

Ang II does not only function as a hormone distributed by the peripheral circulation, but it is also synthesized within the CNS, where it may act as a neurotransmitter or neuromodulator, or both (Phillips, 1987; Wright and Harding, 1992; Andersson et al., 1995; Culman et al., 1995). Based on experiments mostly involving icv infusions of Ang II, this neuropeptide has been proposed to control various physiological functions, including the regulation of body fluid balance and arterial blood pressure. The existence of an intrinsic brain renin-angiotensin system (RAS) has been well documented. All components of RAS, such as angiotensinogen, renin, angiotensin I-converting enzyme, and Ang II, together with Ang II receptors, are widely distributed within the brain (Bunnemann et al., 1991; Phillips et al., 1993; Lenkei et al., 1998). In several brain regions, a mismatch in the distribution of various RAS components has been demonstrated (Bunnemann et al., 1991), suggesting that volume transmission plays a role in both parenchymal Ang II generation and signaling. The baseline Ang II levels in CSF are relatively low and do not exceed 10 to 30 pg/ml (Schelling et al., 1980; Cameron et al., 1985; Simon-Oppermann et al., 1986, 1987). However, they can increase 2- to 5-fold in response to dehydration and systemic hemorrhage (Cameron et al., 1985; Simon-Oppermann et al., 1986; Gray and Simon, 1987). These changes in Ang II levels in CSF occur independent of those in plasma (Simon-Oppermann et al., 1986), indicating that the Ang II detected in CSF is of central origin.

The effects of centrally released Ang II on CSF production have been investigated in rats and rabbits (Chodobski et al., 1992b, 1994). In these studies, a synthetic Ang II was infused icv to produce peptide levels in CSF estimated to range between 10^{-10} and 10^{-7} M. The actual CSF Ang II concentrations were most likely lower than the estimated Ang II levels, since this peptide is rapidly metabolized within the ventricular system (Harding et al., 1986). Low Ang II doses have been found to inhibit CSF formation by 16 to 36%, whereas at high Ang II infusion rates, CSF production was not altered. The use of receptor subtype-specific antagonists allowed these authors to determine that the inhibitory effect of Ang II on CSF formation is mediated by the AT_1 receptors. It is presently unclear why the high Ang II doses do not affect CSF production, but it is possible that, at high CSF Ang II levels, this neuropeptide exerts some additional effects that counter those produced by the low Ang II concentrations.

It has been demonstrated that the inhibitory effect of low Ang II doses on CSF formation is mediated by VP (Chodobski et al., 1998a). Although these findings appear to be consistent with the ability of Ang II to promote VP release into the bloodstream (Keil et al., 1975; Schölkens et al., 1982; Qadri et al., 1993), it is rather unlikely that circulating VP mediates the inhibitory effect of central Ang II on CSF production. Indeed, although high plasma VP concentrations, ranging between 300 and 400 pg/ml, have been found to reduce CSF production by 28 to 35% (Faraci et al., 1990; Chodobski et al., 1998a), only a moderate increase in plasma VP levels (up to 25 to 40 pg/ml) has been observed in rats in response to icv injection of 100 ng of Ang II. A 10^3-fold lower dose of Ang II was ineffective in stimulating VP secretion (Keil et al., 1975; Qadri et al., 1993). In this context it is important to note that, in rats, the Ang II-mediated decrease in CSF formation was observed when the peptide was infused icv at 0.5 to 5 pg/min (Chodobski et al., 1994). These findings suggest that the increase in plasma VP concentration following the icv infusion of low doses of Ang II is insufficient to inhibit CSF production. This raises the intriguing possibility that the inhibitory effect of central Ang II on CSF formation is mediated by VP that is synthesized or released or both within the CP. In this manner, sufficient VP levels to alter CSF production could be attained locally in choroidal tissue.

Neurophysin-containing axonal processes that originate from the paraventricular and supraoptic nuclei of the hypothalamus and innervate the CP have been identified (Brownfield and Kozlowski, 1977). Since the antibody used in this study recognized both VP- and oxytocin-associated neurophysins, it has been suggested that axonal processes releasing VP within the choroidal tissue exist. However, other investigators (Chodobski et al., 1997, 1998c) were unable to detect either the VP- or the VP-associated neurophysin-containing processes in the CP. Instead, VP has been found to be synthesized by the choroidal epithelium (Chodobski et al., 1997, 1998c). Since the AT_1 receptors have also been localized to the CP (Jöhren and Saavedra, 1996; Lenkei et al., 1998), it cannot be excluded that Ang II controls choroidal VP secretion. Further research will be needed to determine whether CP-derived VP plays a role in the regulation of CSF formation.

5.3.3 Other Peptides

Although circulating ET-1 is a potent vasoconstrictor of the choroidal vasculature (Kadel et al., 1990; Granstam et al., 1993), it only slightly decreases CSF formation (Schalk et al., 1992a). Interestingly, iv administration of the cyclooxygenase inhibitor, indomethacin, prior to iv infusion of ET-1 resulted in a larger reduction in CSF production (Schalk et al., 1992a), but the hemodynamic action of ET-1 on choroidal tissue is not affected by indomethacin (Granstam et al., 1993). These observations suggest that, although ET-1 curtails the water and electrolyte supply to the CP, the cyclooxygenase metabolites of arachidonic acid released in response to ET-1 act to maintain the fluid secretion. Consistent with this concept, the CP has been shown to synthesize a number of prostanoids, including prostaglandin E_2 (Goehlert et al., 1981; Yohai and Danon, 1987), which has the ability to stimulate ion transport in various epithelia (Bourke et al., 1995; Halm et al., 1995; Hogan et al., 1998).

CSF ET-1 levels are increased in subarachnoid hemorrhage (SAH) and in cerebral ischemia (Shirakami et al., 1994; Seifert et al., 1995; Lampl et al., 1997), which raises an important question about the possible effect of centrally released ET-1 on CSF formation. Studies in rabbits have demonstrated that icv administration of a synthetic ET-1 modestly inhibits CSF formation (Schalk et al., 1992a). In this context, it is important to note that the choroidal epithelium not only expresses ET-1 receptors (Angelova et al., 1996, 1997), but also synthesizes and secretes their ligand (Takahashi et al., 1998). Therefore, CP-derived ET-1 may regulate CSF formation through autocrine/juxtacrine signaling or paracrine signaling, or both.

Peripheral administration of ANP or VIP, two peptides having vasodilative effects on the choroidal vascular bed (Schalk et al., 1989; Nilsson et al., 1991), has not been found to alter CSF production (Nilsson et al., 1991; Chodobski et al., 1992a; Schalk et al., 1992b). This suggests that, with sufficient or superfluous blood supply to the CP, the rate of CSF formation depends mainly upon ion transport activity in the choroidal epithelium.

Elevated levels of ANP and VIP have been observed in CSF collected from patients with hydrocephalus (Sharpless et al., 1984; Diringer et al., 1990). In SAH, increased CSF ANP concentrations have been found (Dóczi et al., 1988; Rosenfeld et al., 1989; Diringer et al., 1990), whereas the CSF VIP levels were lower compared to control subjects (Juul et al., 1995). Although the central administration of ANP increases blood flow to the CP (Schalk et al., 1992b), it has been reported that there is either no change or a decrease in CSF production in response to icv infusion of ANP (Steardo and Nathanson, 1987; Chodobski et al., 1992a; Schalk et al., 1992b). In the isolated rat CP, ANP at a concentration of 10^{-7} M lowered the uptake of Na^+, but did not affect Cl^- transport (Johanson et al., 1990). These latter findings may explain, at least in part, the inhibitory effect of ANP on CSF formation observed in some studies. Similar to ANP, icv infusion of VIP increased choroidal blood flow, but curtailed CSF production (Nilsson et al., 1991). The mechanisms underlying these VIP actions are yet to be established.

5.4 CONCLUSIONS

Although various peptides have been shown to affect choroidal hemodynamics or CSF production, or both, the physiological significance of these findings remains incompletely understood. In moderate dehydration and hypernatremia, the pathophysiological conditions accompanied by increased synthesis and release of VP and Ang II, CSF formation appears to be unchanged, which may be important for maintaining brain fluid and electrolyte homeostasis. While a moderate hemorrhage does not seem to alter CSF formation, a severe hemorrhage most likely causes CSF production to fall because of the potent vasoconstrictive action of circulating VP on the choroidal vascular bed. Under such conditions, a decrease in blood flow to the highly perfused CP may help sustain an adequate blood supply to other brain regions. Unfortunately, the lack of sufficient mechanistic information does not allow for a thorough analysis of the available data. Further research is therefore needed to fill the gaps in our understanding of the physiological role of peptide-mediated regulation of choroidal blood flow and CSF formation.

5.5 ACKNOWLEDGMENTS

We are indebted to Dr. Gerald Silverberg for his comments on this review. The authors also wish to thank Dr. Arthur Messier for his critical reading and Andrew Pfeffer for his help in the preparation of this manuscript. This work was supported by grant from NIH NS-39921 and by research funds from the Neurosurgery Foundation and Lifespan, Rhode Island Hospital.

REFERENCES

Andersson, B., Eriksson, S. and Rundgren, M., Angiotensin and the brain, *Acta Physiol. Scand.* 155, 1995, 117–25.

Angelova, K., Fralish, G.B., Puett, D. and Narayan, P., Identification of conventional and novel endothelin receptors in sheep choroid plexus cells, *Mol. Cell. Biochem.* 159, 1996, 65–72.

Angelova, K., D. Puett, D. and Narayan, P., Identification of endothelin receptor subtypes in sheep choroid plexus, *Endocrine* 7, 1997, 287–93.

Bargmann, W., Oksche, A., Fix, J.D. and Haymaker, W., Meninges, choroid plexuses, ependyma, and their reactions. Part I. Histology and functional consideration, in W. Haymaker and R.D. Adams (eds.) *Histology and Histopathology of the Nervous System*, Springfild, IL, Charles C Thomas Publisher, 1982, pp. 560–641.

Baylis, C. and Brenner, B.M., The physiologic determinants of glomerular ultrafiltration, *Rev. Physiol., Biochem. Pharmacol.* 80, 1978, 1–46.

Bourke, J.R., Sand, O., Abel, K.C., Huxham, G.J. and Manley, S.W., Chloride channels in the apical membrane of thyroid epithelial cells are regulated by cyclic AMP, *J. Endocrinol.* 147, 1995, 441–8.

Brownfield, M.S. and Kozlowski, G.P., The hypothalamo-choroidal tract. I. Immunohistochemical demonstration of neurophysin pathways to telencephalic choroid plexuses and cerebrospinal fluid, *Cell Tissue Res.* 178, 1977, 111–27.

Bryan, Jr., R.M., Myers, C.L. and Page, R.B., Regional neurohypophysial and hypothalamic blood flow in rats during hypercapnia, *Am. J. Physiol.* 255, 1988, R295–302.

Bunnemann, B., Fuxe, K., Bjelke, B. and Ganten, D., The brain renin-angiotensin system and its possible involvement in volume transmission, in K. Fuxe and L.F. Agnati (eds.), *Volume Transmission in the Brain: Novel Mechanisms for Neural Transmission*, New York, Raven Press, 1991, pp. 131–58.

Cameron, V., Espiner, E.A., Nicholls, M.G., Donald, R.A. and MacFarlane, M.R., Stress hormones in blood and cerebrospinal fluid of conscious sheep: effect of hemorrhage, *Endocrinology* 116, 1985, 1460–5.

Campbell, W.B. and Jackson, E.K., Modulation of adrenergic transmission by angiotensins in the perfused rat mesentery, *Am. J. Physiol.* 236, 1979, H211–7.

Chodobski, A., Szmydynger-Chodobska, J. and Skolasinska, K., Effect of ammonia intoxication on cerebral blood flow, its autoregulation and responsiveness to carbon dioxide and papaverine, *J. Neurol., Neurosurg., Psychiatry* 49, 1986, 302–9.

Chodobski, A., Szmydynger-Chodobska, J., Cooper, E. and McKinley, M.J., Atrial natriuretic peptide does not alter cerebrospinal fluid formation in sheep, *Am. J. Physiol.* 262, 1992a, R860–4.

Chodobski, A., Szmydynger-Chodobska, J., Segal, M.B. and McPherson, I.A., The role of angiotensin II in regulation of cerebrospinal fluid formation in rabbits, *Brain Res.* 594, 1992b, 40–6.

Chodobski, A., Szmydynger-Chodobska, J., Vannorsdall, M.D., Epstein, M.H. and Johanson, C.E., AT_1 receptor subtype mediates the inhibitory effect of central angiotensin II on cerebrospinal fluid formation in the rat, *Regul. Peptides* 53, 1994, 123–9.

Chodobski, A., Szmydynger-Chodobska, J., Epstein, M.H. and Johanson, C.E., The role of angiotensin II in the regulation of blood flow to choroid plexuses and cerebrospinal fluid formation in the rat, *J. Cereb. Blood Flow and Metab.* 15, 1995, 143–51.

Chodobski, A., Loh, Y.P., Corsetti, S., Szmydynger-Chodobska, J., Johanson, C.E., Lim, Y.P. and Monfils, P.R., The presence of arginine vasopressin and its mRNA in rat choroid plexus epithelium, *Mol. Brain Res.* 48, 1997, 67–72.

Chodobski, A., Szmydynger-Chodobska, J. and Johanson, C.E., Vasopressin mediates the inhibitory effect of central angiotensin II on cerebrospinal fluid formation, *Eur. J. Pharmacol.* 347, 1998a, 205–9.

Chodobski, A., Szmydynger-Chodobska, J. and McKinley, M.J., Cerebrospinal fluid formation and absorption in dehydrated sheep, *Am. J. Physiol.* 275, 1998b, F235–8.

Chodobski, A., Wojcik, B.E., Loh, Y.P., Dodd, K.A., Szmydynger-Chodobska, J., Johanson, C.E., Demers, D.M., Chun, Z.G. and Limthong, N.P., Vasopressin gene expression in rat choroid plexus, *Adv. Exp. Med. Biol.* 449, 1998c, 59–65.

Chodobski, A., Szmydynger-Chodobska, J. and Johanson, C.E., Angiotensin II regulates choroid plexus blood flow by interacting with the sympathetic nervous system and nitric oxide, *Brain Res.* 816, 1999, 518–26.

Cserr, H.F., Physiology of the choroid plexus, *Physiol. Rev.* 51, 1971, 273–311.

Cserr, H.F., Convection of brain interstitial fluid, in K. Shapiro, A. Marmarou and H. Portnoy (eds.), *Hydrocephalus*, New York, Raven Press, 1984, pp. 59–68.

Cserr, H.F., Harling-Berg, C.J. and Knopf, P.M., Drainage of brain extracellular fluid into blood and deep cervical lymph and its immunological significance, *Brain Pathol.* 2, 1992, 269–76.

Culman, J., Höhle, S., Qadri, F., Edling, O., Blume, A., Lebrun, C. and Unger, T., Angiotensin as neuromodulator/neurotransmitter in central control of body fluid and electrolyte homeostasis, *Clin. Exp. Hypertension* 17, 1995, 281–93.

Diringer, M.N., Kirsch, J.R., Ladenson, P.W., Borel, C. and Hanley, D.F., Cerebrospinal fluid atrial natriuretic factor in intracranial disease, *Stroke* 21, 1990, 1550–4.

Dóczi, T., Joó, F., Vecsernyés, M. and Bodosi, M., Increased concentration of atrial natriuretic factor in the cerebrospinal fluid of patients with aneurysmal subarachnoid hemorrhage and raised intracranial pressure, *Neurosurgery* 23, 1988, 16–9.

Edvinsson, L., Owman, C., Rosengren, E. and West, K.A., Concentration of noradrenaline in pial vessels, choroid plexus, and iris during two weeks after sympathetic ganglionectomy or decentralization, *Acta Physiol. Scand.* 85, 1972, 201–6.

Edvinsson, L., Nielsen, K., Owman, CH. and West, K., Adrenergic innervation of the mammalian choroid plexus, *Am. J. Anat.*, 139, 1974, 299–307.

Edvinsson, L. and Lindvall, M., Autonomic vascular innervation and vasomotor reactivity in the choroid plexus, *Exp. Neurol.* 62, 1978, 394–404.

Faraci, F.M., Mayhan, W.G., Farrell, W.J. and Heistad, D.D., Humoral regulation of blood flow to choroid plexus: role of arginine vasopressin, *Circ. Res.* 63, 1988a, 373–9.

Faraci, F.M., Mayhan, W.G., Williams, J.K. and Heistad, D.D., Effects of vasoactive stimuli on blood flow to choroid plexus, *Am. J. Physiol.* 254, 1988b, H286–91.

Faraci, F.M., Mayhan, W.G. and Heistad, D.D., Effect of serotonin on blood flow to the choroid plexus, *Brain Res.* 478, 1989, 121–6.

Faraci, F.M., Mayhan, W.G. and Heistad, D.D., Effect of vasopressin on production of cerebrospinal fluid: possible role of vasopressin (V_1)-receptors, *Am. J. Physiol.* 258, 1990, R94–8.

Faraci, F.M. and Heistad, D.D., Does basal production of nitric oxide contribute to regulation of brain-fluid balance? *Am. J. Physiol.* 262, 1992, H340–4.

Faraci, F.M., Kinzenbaw, D. and Heistad, D.D., Effect of endogenous vasopressin on blood flow to choroid plexus during hypoxia and intracranial hypertension, *Am. J. Physiol.* 266, 1994, H393–8.

Fisone, G., Snyder, G.L., Fryckstedt, J., Caplan, M.J., Aperia, A. and Greengard, P., Na^+,K^+-ATPase in the choroid plexus. Regulation by serotonin/protein kinase C pathway, *J. Biol. Chemistry* 270, 1995, 2427–30.

Fisone, G., Snyder, G.L., Aperia, A. and Greengard, P., Na^+,K^+-ATPase phosphorylation in the choroid plexus: synergistic regulation by serotonin/protein kinase C and isoproterenol/cAMP-PK/PP-1 pathways, *Mol. Med.* 4, 1998, 258–65.

Goehlert, U.G., Ng Ying Kin, N.M. and Wolfe, L.S., Biosynthesis of prostacyclin in rat cerebral microvessels and the choroid plexus, *J. Neurochem.* 36, 1981, 1192–201.

Granstam, E., Wang, L. and Bill, A., Vascular effects of endothelin-1 in the cat; modification by indomethacin and L-NAME, *Acta Physiol. Scand.* 148, 1993, 165–76.

Gray, D.A. and Simon, E., Dehydration and arginine vasotocin and angiotensin II in CSF and plasma of Pekin ducks, *Am. J. Physiol.* 253, 1987, R285–91.

Haberl, R.L., Anneser, F., Villringer, A. and Einhäupl, K.M., Angiotensin II induces endothelium-dependent vasodilation of rat cerebral arterioles, *Am. J. Physiol.* 258, 1990, H1840–6.

Halm, D.R., Halm, S.T., DiBona, D.R., Frizzell, R.A. and Johnson, R.D., Selective stimulation of epithelial cells in colonic crypts: relation to active chloride secretion, *Am. J. Physiol.* 269, 1995, C929–42.

Harding, J.W., Yoshida, M.S., Dilts, R.P., Woods, T.M. and Wright, J.W., Cerebroventricular and intravascular metabolism of [^{125}I]angiotensins in rat, *J. Neurochem.* 46, 1986, 1292–7.

Hilgers, K.F., Veelken, R., Rupprecht, G., Reeh, P.W., Luft, F.C. and Mann, J.F., Angiotensin II facilitates sympathetic, transmission in rat hind limb circulation, *Hypertension* 21, 1993, 322–8.

Hogan, D.L., Yao, B. and Isenberg, J.I., Modulation of bicarbonate secretion in rabbit duodenum: the role of calcium, *Dig. Dis. Sci.* 43, 1998, 120–5.

Huang, H., Baussant, T., Reade, R., Michel, J.B. and Corvol, P., Measurement of angiotensin II concentration in rat plasma: pathophysiological applications, *Clin. Exp. Hypertension A* 11, 1989, 1535–48.

Johanson, C.E., Sweeney, S.M., Parmelee, J.T. and Epstein, M.H., Cotransport of sodium and chloride by the adult mammalian choroid plexus, *Am. J. Physiol.* 258, 1990, C211–6.

Jöhren, O. and Saavedra, J.M., Expression of AT_{1A} and AT_{1B} angiotensin II receptor messenger RNA in forebrain of 2-wk-old rats, *Am. J. Physiol.* 271, 1996, E104–12.

Juul, R., Hara, H., Gisvold, S.E., Brubakk, A.O., Fredriksen, T.A., Waldemar, G., Schmidt, J.F., Ekman, R. and Edvinsson, L., Alterations in perivascular dilatory neuropeptides (CGRP, SP, VIP) in the external jugular vein and in the cerebrospinal fluid following subarachnoid haemorrhage in man, *Acta Neurochirurg. (Wien)* 132, 1995, 32–41.

Kadekaro, M., Liu, H., Terrell, M.L., Gestl, S., Bui, V. and Summy-Long, J.Y., Role of NO on vasopressin and oxytocin release and blood pressure responses during osmotic stimulation in rats, *Am. J. Physiol.* 273, 1997, R1024–30.

Kadel, K.A., Heistad, D.D. and Faraci, F.M., Effects of endothelin on blood vessels of the brain and choroid plexus, *Brain Res.* 518, 1990, 78–82.

Keil, L.C., Summy-Long, J. and Severs, W.B., Release of vasopressin by angiotensin II, *Endocrinol.* 96, 1975, 1063–5.

Knopf, P.M., Cserr, H.F., Nolan, S.C., Wu, T.Y. and Harling-Berg, C.J., Physiology and immunology of lymphatic drainage of interstitial and cerebrospinal fluid from the brain, *Neuropathol. Appl. Neurobiol.* 21, 1995, 175–80.

Kuhl, D.E., Barrio, J.R., Huang, S.C., Selin, C., Ackermann, R.F., Lear, J.L., Wu, J.L., Lin, T.H. and Phelps, M.E., Quantifying local cerebral blood flow by *N*-isopropyl-*p*-[^{123}I]iodoamphetamine (IMP) tomography, *J. Nuclear Med.* 23, 1982, 196–203.

Lampl, Y., Fleminger, G., Gilad, R., Galron, R., Sarova-Pinhas, I. and Sokolovsky, M., Endothelin in cerebrospinal fluid and plasma of patients in the early stage of ischemic stroke, *Stroke* 28, 1997, 1951–5.

Lenkei, Z., Palkovits, M., Corvol, P. and Llorens-Cortes, C., Distribution of angiotensin type-1 receptor messenger RNA expression in the adult rat brain, *Neuroscience* 82, 1998, 827–41.

Levin, E.R., Endothelins as cardiovascular peptides, *Am. J. Nephrol.* 16, 1996, 246–51.

Lin, A.Y., Szmydynger-Chodobska, J., Rahman, M.P., Mayer, B., Monfils, P.R., Johanson, C.E., Lim, Y.P., Corsetti, S. and Chodobski, A., Immunohistochemical localization of nitric oxide synthase in rat anterior choroidal artery, stromal blood microvessels, and choroid plexus epithelial cells, *Cell Tissue Res.* 285, 1996, 411–8.

Lindvall, M., Fluorescence histochemical study on regional differences in the sympathetic nerve supply of the choroid plexus from various laboratory animals, *Cell Tissue Res.* 198, 1979, 261–7.

Lindvall, M., Edvinsson, L. and Owman, C., Effect of sympathomimetic drugs and corresponding receptor antagonists on the rate of cerebrospinal fluid production, *Exp. Neurol.* 64, 1979, 132–45.

Lindvall, M. and Owman, C., Autonomic nerves in the mammalian choroid plexus and their influence on the formation of cerebrospinal fluid, *J. Cereb. Blood Flow Metab.* 1, 1981, 245–66.

Maktabi, M.A., Heistad, D.D. and Faraci, F.M., Effects of angiotensin II on blood flow to choroid plexus, *Am. J. Physiol.* 258, 1990, H414–8.

Maktabi, M.A., Stachovic, G.C. and Faraci, F.M., Angiotensin II decreases the rate of production of cerebrospinal fluid, *Brain Res.* 606, 1993, 44–9.

Mann, J.F., Johnson, A.K. and Ganten, D., Plasma angiotensin II: dipsogenic levels and angiotensin-generating capacity of renin, *Am. J. Physiol.* 238, 1980, R372–7.

Mann, J.F., Johnson, A.K., Rascher, W., Genest, J. and Ganten, D., Thirst in the rat after ligation of the inferior vena cava: role of angiotensin II, *Pharmacol., Biochem. Behav.* 15, 1981, 337–41.

Motti, E.D., Imhof, H.G., Janzer, R.C., Marquardt, K. and Yasargil, G.M., The capillary bed in the choroid plexus of the lateral ventricles: a study of luminal casts, *Scanning Electron Microscopy* IV, 1986, 1501–13.

Nilsson, C., Lindvall-Axelsson, M. and Owman, C., Simultaneous and continuous measurement of choroid plexus blood flow and cerebrospinal fluid production: effects of vasoactive intestinal polypeptide, *J. Cereb. Blood Flow Metab.* 11, 1991, 861–7.

Obrenovitch, T.P., Clayton, C.B. and Strong, A.J., A double-radionuclide autoradiographic method using *N*-isopropyl-iodoamphetamine for sequential measurements of local cerebral blood flow, *J. Cereb. Blood Flow Metab.* 7, 1987, 356–65.

Oldendorf, W.H. and Davson, H., Brain extracellular space and the sink action of cerebrospinal fluid. Measurement of rabbit brain extracellular space using sucrose labeled with carbon 14, *Arch. Neurol.* 17, 1967, 196–205.

Osborn, Jr., J.W., Skelton, M.M. and Cowley, Jr., A.W., Hemodynamic effects of vasopressin compared with angiotensin II in conscious rats, *Am. J. Physiol.* 252, 1987, H628–37.

Parmelee, J.T. and Johanson, C.E., Development of potassium transport capability by choroid plexus of infant rats, *Am. J. Physiol.* 256, 1989, R786–91.

Phillips, M.I., Functions of angiotensin in the central nervous system, *Ann. Rev. Physiol.* 49, 1987, 413–35.

Phillips, M.I., Shen, L., Richards, E.M. and Raizada, M.K., Immunohistochemical mapping of angiotensin AT_1 receptors in the brain, *Regul. Peptides* 44, 1993, 95–107.

Preston, J.E., Dyas, M. and Johanson, C.E., Development of chloride transport by the rat choroid plexus, *in vitro*, *Brain Res.* 624, 1993, 181–7.

Pullen, R.G., DePasquale, M. and Cserr, H.F., Bulk flow of cerebrospinal fluid into brain in response to acute hyperosmolality, *Am. J. Physiol.* 253, 1987, F538–45.

Qadri, F., Culman, J., Veltmar, A., Maas, K., Rascher, W. and Unger, T., Angiotensin II-induced vasopressin release is mediated through *alpha*-1 adrenoceptors and angiotensin II AT1 receptors in the supraoptic nucleus, *J. Pharmacol. Exp. Ther.* 267, 1993, 567–74.

Rosenfeld, J.V., Barnett, G.H., Sila, C.A., Little, J.R., Bravo, E.L. and Beck, G.J., The effect of subarachnoid hemorrhage on blood and CSF atrial natriuretic factor, *J. Neurosurg.* 71, 1989, 32–7.

Rossi, N.F., O'Leary, D.S. and Chen, H., Mechanisms of centrally administered ET-1-induced increases in systemic arterial pressure and AVP secretion, *Am. J. Physiol.* 272, 1997, E126–32.

Schalk, K.A., Williams, J.L. and Heistad, D.D., Effect of atriopeptin on blood flow to cerebrum and choroid plexus, *Am. J. Physiol.* 257, 1989, R1365–9.

Schalk, K.A., Faraci, F.M. and Heistad, D.D., Effect of endothelin on production of cerebrospinal fluid in rabbits, *Stroke* 23, 1992a, 560–3.

Schalk, K.A., Faraci, F.M., Williams, J.L., VanOrden, D. and Heistad, D.D., Effect of atriopeptin on production of cerebrospinal fluid, *J. Cereb. Blood Flow Metab.* 12, 1992b, 691–6.

Schelling, P., Ganten, U., Sponer, G., Unger, T. and Ganten, D., Components of the renin-angiotensin system in the cerebrospinal fluid of rats and dogs with special consideration of the origin and the fate of angiotensin II, *Neuroendocrinol.* 31, 1980, 297–308.

Schiffrin, E.L. and Touyz, R.M., Vascular biology of endothelin, *J. Cardiovasc. Pharmacol.* 32, Suppl. 3, 1998, S2–13.

Schölkens, B.A., Jung, W., Rascher, W., Dietz, R. and Ganten, D., Intracerebroventricular angiotensin II increases arterial blood pressure in rhesus monkeys by stimulation of pituitary hormones and the sympathetic nervous system, *Experientia* 38, 1982, 469–70.

Segal, M.B., Chodobski, A., Szmydynger-Chodobska, J. and Cammish, H., Effect of arginine vasopressin on blood vessels of the perfused choroid plexus of the sheep, *Prog. Brain Res.* 91, 1992, 451–3.

Seifert, V., Loffler, B.M., Zimmermann, M., Roux, S. and Stolke, D., Endothelin concentrations in patients with aneurysmal subarachnoid hemorrhage. Correlation with cerebral vasospasm, delayed ischemic neurological deficits, and volume of hematoma, *J. Neurosurg.* 82, 1995, 55–62.

Sharpless, N.S., Thal, L.J., Perlow, M.J., Tabaddor, K., Waltz, J.M., Shapiro, K.N., Amin, I.M., Engel, Jr., J. and Crandall, P.H., Vasoactive intestinal peptide in cerebrospinal fluid, *Peptides* 5, 1984, 429–33.

Shirakami, G., Magaribuchi, T., Shingu, K., Kim, S., Saito, Y., Nakao, K. and Mori, K., Changes of endothelin concentration in cerebrospinal fluid and plasma of patients with aneurysmal subarachnoid hemorrhage, *Acta Anaesthesiol. Scand.* 38, 1994, 457–61.

Shoji, M., Kimura, T., Kawarabayasi, Y., Ota, K., Inoue, M., Yamamoto, T., Sato, K., Ohta, M., Funyu, T., Yamamoto, T. and Abe, K., Effects of acute hypotensive hemorrhage on arginine vasopressin gene transcription in the rat brain, *Neuroendocrinol.* 58, 1993, 630–6.

Simon-Oppermann, C., Gray, D.A. and Simon, E., Independent osmoregulatory control of central and systemic angiotensin II concentrations in dogs, *Am. J. Physiol.* 250, 1986, R918–25.

Simon-Oppermann, C., Eriksson, S., Simon, E. and Gray, D.A., Gradient of arginine vasopressin concentration but not angiotensin II concentration between cerebrospinal fluid of anterior 3rd ventricle and cisterna magna in dogs, *Brain Res.* 424, 1987, 163–8.

Steardo, L. and Nathanson, J.A., Brain barrier tissues: end organs for atriopeptins, *Science* 235, 1987, 470–3.

Szabo, B., Hedler, L., Schurr, C. and Starke, K., Peripheral presynaptic facilitatory effect of angiotensin II on noradrenaline release in anesthetized rabbits, *J. Cardiovasc. Pharmacol.* 15, 1990, 968–75.

Szczepanska-Sadowska, E., Gray, D. and Simon-Oppermann, C., Vasopressin in blood and third ventricle CSF during dehydration, thirst, and hemorrhage, *Am. J. Physiol.* 245, 1983, R549–55.

Szmydynger-Chodobska, J., Chodobski, A. and Johanson, C.E., Postnatal developmental changes in blood flow to choroid plexuses and cerebral cortex of the rat, *Am. J. Physiol.* 266, 1994, R1488–92.

Szmydynger-Chodobska, J., Monfils, P.R., Lin, A.Y., Rahman, M.P., Johanson, C.E. and Chodobski, A., NADPH-diaphorase histochemistry of rat choroid plexus blood vessels and epithelium, *Neurosci. Lett.* 208, 1996, 179–82.

Szmydynger-Chodobska, J., Chung, I., Kozniewska, E., Tran, B., Harrington, J.F., Duncan, J.A. and Chodobski, A., Increased expression of vasopressin V_{1a} receptors after traumatic brain injury, *J. Neurotrauma*, 2004 (in press).

Takahashi, K., Hara, E., Murakami, O., Totsune, K., Sone, M., Satoh, F., Kumabe, T., Tominaga, T., Kayama, T., Yoshimoto, T. and Shibahara, S., Production and secretion of endothelin-1 by cultured choroid plexus carcinoma cells, *J. Cardiovasc. Pharmacol.* 31, Suppl. 1, 1998, S367–9.

Thibonnier, M., Berti-Mattera, L.N., Dulin, N., Conarty, D.M. and Mattera, R., Signal transduction pathways of the human V_1-vascular, V_2-renal, V_3-pituitary vasopressin and oxytocin receptors, *Prog. Brain Res.* 119, 1998, 147–61.

Townsend, J.B., Ziedonis, D.M., Bryan, R.M., Brennan, R.W. and Page, R.B., Choroid plexus blood flow: evidence for dopaminergic influence, *Brain Res.* 290, 1984, 165–9.

van Deurs, B., Cell junctions in the endothelia and connective tissue of the rat choroid plexus, *Anat. Rec.* 195, 1979, 73–93.

van Deurs, B., Structural aspects of brain barriers, with special reference to the permeability of the cerebral endothelium and choroidal epithelium, *Int. Rev. Cytol.* 65, 1980, 117–91.

Van Uitert, R.L., Sage, J.I., Levy, D.E. and Duffy, T.E., Comparison of radio-labeled butanol and iodoantipyrine as cerebral blood flow markers, *Brain Res.* 222, 1981, 365–72.

Wagerle, L.C. and Delivoria-Papadopoulos, M., α-Adrenergic receptor subtypes in the cerebral circulation of newborn piglets, *Am. J. Physiol.* 252, 1987, R1092–8.

Wang, B.C., Flora-Ginter, G., Leadley, Jr., R.J. and Goetz, K.L., Ventricular receptors stimulate vasopressin release during hemorrhage, *Am. J. Physiol.* 254, 1988, R204–11.

Weiger, T., Lametschwandtner, A., Hodde, K.C. and Adam, H., The angioarchitecture of the choroid plexus of the lateral ventricle of the rabbit. A scanning electron microscopic study of vascular corrosion casts, *Brain Res.* 378, 1986, 285–96.

Williams, J.L., Jones, S.C., Page, R.B. and Bryan, Jr., R.M., Vascular responses of choroid plexus during hypercapnia in rats, *Am. J. Physiol.* 260, 1991, R1066–70.

Wright, J.W. and Harding, J.W., Regulatory role of brain angiotensins in the control of physiological and behavioral responses, *Brain Res. Rev.* 17, 1992, 227–62.

Yamamoto, T., Kimura, T., Ota, K., Shoji, M., Inoue, M., Sato, K., Ohta, M. and Yoshinaga, K., Central effects of endothelin-1 on vasopressin release, blood pressure, and renal solute excretion, *Am. J. Physiol.* 262, 1992, E856–62.

Yohai, D. and Danon, A., Effect of adrenergic agonists on eicosanoid output from isolated rabbit choroid plexus and iris-ciliary body, *Prostaglandins Leukotrienes Med.* 28, 1987, 227–35.

Zemo, D.A. and McCabe, J.T., Salt-loading increases vasopressin and vasopressin 1b receptor mRNA in the hypothalamus and choroid plexus, *Neuropeptides* 35, 2001, 181–8.

Zerbe, R.L. and Palkovits, M., Changes in the vasopressin content of discrete brain regions in response to stimuli for vasopressin secretion, *Neuroendocrinol.* 38, 1984, 285–9.

6 Ion Transporters and Channels Involved in CSF Formation

Peter D. Brown, Tracey Speake, Sarah L. Davies, and Ian D. Millar
University of Manchester,
Manchester, United Kingdom

CONTENTS

6.1 INTRODUCTION

The cerebrospinal fluid (CSF) fills the ventricles of the brain and the canals and spaces of the central nervous system. In humans the volume of the CSF is 140 ml, and this volume is totally replaced about four times each day. Thus, the total amount of CSF produced in 24 hours is approximately 600 ml (Wright, 1978). The majority of this CSF is produced by the epithelial cells of the choroid plexuses. In humans the choroid plexuses weigh about 2 g in total, thus the rate of CSF secretion is approximately 0.2 $ml.min^{-1}$ per g of tissue. This rate of secretion is significantly higher than in many other secretory epithelia, e.g., guinea pig pancreas = 0.06 $ml.min^{-1}$ per g tissue. The aim of this chapter is to review the mechanism by which CSF is produced by the choroid plexuses.

6.1.1 Epithelial Transport and Fluid Secretion

The choroid plexuses are branched structures made up of numerous villi that project into the ventricles of the brain (see Figure 6.1A). Each villus is composed of a single layer of epithelial cells overlying a core of connective tissue and blood capillaries (see Figure 6.1B). The capillaries in the choroid plexuses, unlike those in the majority of the cerebral blood circulation, are fenestrated and hence provide little resistance to the movement of small molecules, ions, and water (Segal, 1993). A barrier between the blood and CSF is formed (the blood–CSF barrier) by the epithelial cells, which are linked together by junctional complexes (see Figure 6.1B). The epithelium, however, is a selective barrier that facilitates the movement of nutrients, ions, and waste products into and out of the CSF.

The epithelium is superbly adapted for CSF secretion, a process that involves the unidirectional transport of ions and H_2O. Each cell contains many mitochondria and has a well-developed endoplasmic reticulum (ER), both of which are typical features of secretory cells. The single most impressive adaptation, however, is the large surface area of the epithelial cells. They have a very well-developed brush border (apical) membrane, which faces into the ventricle, while the basolateral membrane (blood side) displays many infoldings. Speake and Brown (2004) calculated from measurements of cell electrical capacitance* that the membrane area of

* Membrane capacitance is measured in electrophysiological experiments and is directly proportional to the membrane area of the cell (Kotera and Brown, 1994).

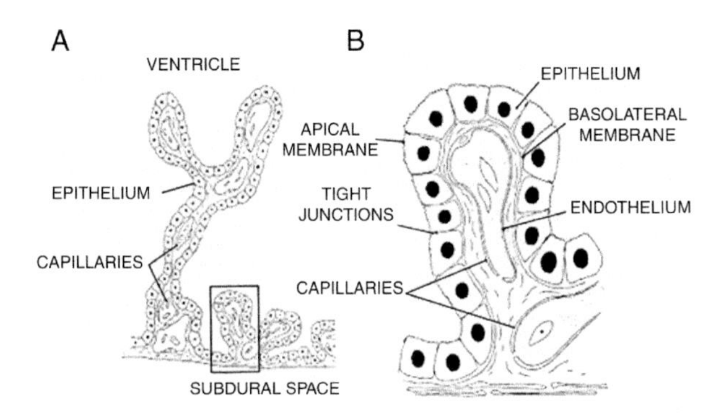

FIGURE 6.1 The structure of the choroid plexuses. (A) Branched structure of the choroid plexus, with villi projecting into the ventricle of the brain. **(B)** Each plexus consists of a network of capillaries covered by a single layer of cuboidal epithelial cells linked together by junctional complexes.

a cell from the rat fourth ventricle choroid plexus is as at least 10 times greater than that estimated from histological studies (Quay et al., 1966). This increase in surface area means that the total surface area of the blood–CSF barrier is of the same order of magnitude as that of the blood–brain barrier (Keep and Jones, 1990).

The large surface area of the epithelium helps explain the high rate of CSF secretion by the choroid plexus, but it does not explain how the CSF is secreted. To understand this process it is perhaps useful to consider fluid secretion by other epithelia. Fluid secretion in all epithelia is thought to be dependent on the unidirectional transport of ions. This creates an osmotic gradient, which drives the movement of water. Unidirectional transport of ions (either secretion or absorption) can be achieved due to the polarized distribution of ion transport proteins in the apical and basolateral membranes of the epithelial cells, i.e., different transport proteins and channels are expressed in these membranes (Steward and Case, 1989). Figure 6.2 shows the major pathways in a typical secretory epithelial cell. Cl^- is accumulated across the basolateral membrane by Na^+-K^+-$2Cl^-$ cotransporters, driven by the Na^+ gradient created by the Na^+,K^+-ATPases which are also expressed in the basolateral membrane. Cl^- then leaves the epithelial cell via anion channels in the apical membrane. This transcellular transport of Cl^- generates an electrical gradient for the transport of Na^+ across the epithelium via a paracellular route through the junctional complexes (see Figure 6.2). The resultant osmotic gradient drives H_2O transport via aquaporin water channels in the apical and basolateral membranes of the cell.

6.1.2 Overview of Ion Transport by the Choroid Plexus

Much of our understanding of the basic process of CSF secretion is based on experiments performed on amphibian choroid plexus. Tissue from the posterior ventricle

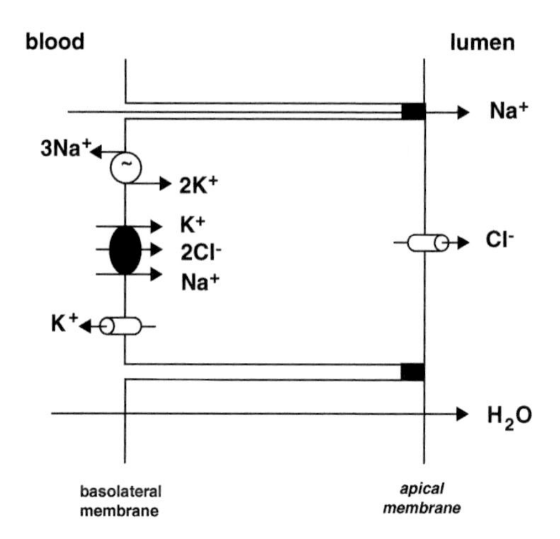

FIGURE 6.2 Model of fluid secretion by a secretory epithelium. The model was first proposed by Silva et al. (1977) to explain salt secretion by the shark rectal gland. It has subsequently been found to explain secretion in a wide range of epithelia. (From Steward et al., 1989.)

(fourth ventricle) of the bullfrog brain can be mounted in Ussing chambers*, enabling the measurement of unidirectional fluxes of ions across the epithelium. Wright (1972, 1978) showed that the epithelial cells of the choroid plexuses secrete Na^+, HCO_3^-, and Cl^- from the blood to the ventricles. Furthermore, secretion was found to be highly dependent on HCO_3^- (Saito and Wright, 1983, 1984). The absorption of a small amount of K^+ from the CSF to the blood was also observed (Wright, 1972). Wright (1978) proposed that, as in other secretory epithelia, the net movement of the ions from blood to CSF creates the osmotic gradient, which drives the movement of H_2O.

The technical difficulties of working with mammalian choroid plexus (i.e., the tissue cannot be mounted in Ussing chambers), mean that unidirectional transepithelial fluxes of ions have never actually been determined in the mammal. The general consensus, however, is that ion transport is very similar to that in the bullfrog. For instance, HCO_3^- appears to have a central role in secretion by the mammalian choroid plexus because carbonic anhydrase inhibitors reduce the rate of CSF secretion (Vogh et al., 1987; Cowan and Whitelaw, 1991; Maren et al., 1997). Furthermore, the composition of mammalian CSF (see Table 6.1) suggests that Na^+, Cl^-, and HCO_3^- are all secreted. However, a more significant observation is that the concentration of K^+ is lower in the CSF than in plasma (see Table 6.1), indicating that the mammalian choroid plexus absorbs K^+ from the CSF. Another interesting feature of CSF composition is that the concentrations of K^+ and HCO_3^- are carefully

* Ussing chambers are experimental apparatus in which sheets of epithelial cells can be mounted so that the epithelial cells form a barrier between the two compartments of the chamber. Movements of ions across the epithelium (i.e., between the compartments) can be measured using radioactive fluxes or electrophysiological methods (Steward and Case, 1989).

TABLE 6.1
Composition of the CSF and Plasma

	Plasma	CSF
Na^+ (mM)	155	151
K^+ (mM)	4.6	3.0
Mg^{2+}(mM)	0.7	1.0
Ca^{2+}(mM)	**2.9**	**1.4**
Cl^- (mM)	121	133
HCO_3^- (mM)	**26.2**	**25.8**
Glucose (mM)	6.3	4.2
Amino acids (mM)	**2.3**	**0.8**
pH	7.4	7.4
Osmolality (mosmol.Kg H_2O^{-1})	300	305
Protein (mg.100g^{-1})*	6500	25

Values are for dog CSF (except * rabbit) and are taken from Davson et al. (1987). The concentrations of the components in **bold type** are regulated by the activity of transporters in the choroid plexuses.

regulated (see Table 6.1)*. Thus, when plasma concentrations of these ions are experimentally varied, the concentrations in nascent CSF remain relatively constant (Husted and Reed, 1976, 1977; Jones and Keep, 1987). These data suggest that the epithelial cells of the choroid plexus express specific mechanisms for the transport K^+ and HCO_3^-, in addition to the transporters for Na^+ and Cl^- found in most secretory epithelia.

6.2 MOLECULAR AND CELLULAR IDENTIFICATION OF ION TRANSPORT PROTEINS

Until the late 1980s our knowledge of the expression of ion transporters and channels in the mammalian choroid plexus was based largely on the results of in vitro radioisotope flux studies in intact tissue preparations, and of experiments examining the effects of transport inhibitors on CSF production. These studies produced much significant data (for reviews, see Segal, 1983; Brown and Garner, 1993; Johanson, 1999), but the results were often confusing and sometimes even contradictory. Problems arose in the interpretation of data from such experiments because of the lack of specificity of the transport inhibitors employed and because the polarity of transporter expression was often difficult to determine using these methods.

In the last 15 years the application of molecular biological methods such as RT-PCR, in situ hybridization, and immunocytochemistry has enabled the precise

* The concentration of Ca^{2+} in the CSF is also regulated (Murphy et al., 1986), but the transport of Ca^{2+} contributes little to the osmotic gradient across the epithelium. It is therefore not discussed in this chapter.

determination of transporter expression. Furthermore, patch clamp electrophysiological methods have been used to characterize the major ion conductances in mammalian choroid plexus cells. The following sections provide details on the molecular identity, localization, and properties of the ion transport proteins identified to date in mammalian choroid plexus. In general, transporters or channels are referred to by their common, descriptive names and by their Human Genome Organization (HUGO)–designated names.

6.2.1 The Na^+,K^+-ATPase Pump (ATP1A1, ATP1B1, ATP1B2)

The Na^+,K^+-ATPase is a heterodimer normally of one α and one β subunit (Blanco and Mercer, 1998; Scheiner-Bobis, 2002). The α subunit is a multispanning membrane protein, which contains the ATP and cation-binding sites. It is this subunit that mediates the transport of 3 Na^+ out of the cell and 2 K^+ into the cell at the expense of a single molecule of ATP. The glycoprotein β subunit crosses the plasma membrane only once, but is required for the correct functioning of the pump (Blanco and Mercer, 1998). Four different α subunits (1 to 4) and three β subunits (1 to 3) have been identified in mammalian cells (Blanco and Mercer, 1998; Scheiner-Bobis, 2002). A γ subunit, which is also thought to modulate pump activity, has been observed in some tissues (Scheiner-Bobis, 2002).

The activity of the Na^+,K^+-ATPase pump is closely associated with the secretion of CSF. Inhibitors of the pump, such as the cardiac glycoside ouabain, have been shown to reduce CSF secretion and the movement of Na^+ into the CSF (Davson and Segal, 1970; Wright, 1972; Pollay et al., 1986). Wright (1972) observed that ouabain was most effective at inhibiting Na^+ transport when applied to the apical membrane of bullfrog choroid plexus, suggesting that Na^+-K^+ ATPase was in this membrane. This was in marked contrast to most epithelial cells where the Na^+-K^+ ATPase pump is expressed only in the basolateral membrane (Rizzolo, 1999). Quinton et al. (1973) provide further evidence for the apical localization by demonstrating that ^3H-oubain bind to the apical membrane of bullfrog choroid plexus. However, the apical localization was not established in mammalian choroid plexus until immunocytochemical methods were employed with rat and mouse choroid plexus (Masuzawa et al., 1984; Ernst et al., 1986).

More recent studies of Na^+-K^+ATPase expression have employed RT-PCR and Western blot analysis to determine that α_1, β_1, and β_2 subunits are all expressed in rat ventricle choroid plexus (Watts et al., 1991; Zlokovic et al., 1993; Klarr et al., 1997). The expression of two β subunits is somewhat unusual, and the reason for this is not known. However, the different β subunits do modify the Na^+ and K^+ affinities of the α subunits (Blanco and Mercer, 1998), so their presence in the choroid plexus may help to increase the range of concentrations over which the pumps are optimally active. Alternatively, Rizzolo (1999) speculated that the presence of the β_2 subunit may be significant in determining the apical localization of Na^+,K^+-ATPase. Expression of γ subunits of Na^+,K^+-ATPase has not been observed in choroid plexus, but phospholemman (a homologue of the γ subunit) appears to be associated with Na^+,K^+-ATPase in choroid plexus (Freschenko et al., 2003). Once again, the significance of this observation remains to be determined.

The apical localization of Na^+,K^+-ATPase is important in the choroid plexus because it is the main route by which Na^+ is transported into the CSF. It may also be important, because in this membrane the Na^+,K^+-ATPase can directly contribute to K^+ absorption from the CSF. This hypothetical role is supported by the fact that the expression of all three Na^+,K^+-ATPase subunits is increased in rats with experimentally induced hyperkalemia (Klarr et al., 1997). This presumably increases the capacity of the choroid plexuses to transport K^+ out of the CSF in these animals.

The central role of Na^+,K^+-ATPases in secretion means that they are potential targets for the regulation of CSF production. Indeed, there is evidence for the posttranslational modification of Na^+,K^+-ATPase activity. Serotonin acting at 5-HT_{2C} reduces Na^+,K^+-ATPase activity by stimulating PKC-mediated phosphorylation of the $\alpha_1$subunit (Fisone et al., 1995). Activation of the muscarinic acetylcholine receptors in the choroid plexus epithelium also inhibits Na^+,K^+-ATPase. This effect, however, is thought to be due to PKG-mediated phosphorylation, which is the result of the generation of NO and cGMP (Ellis et al., 2000). Both serotonin and acetylcholine have previously been reported to inhibit CSF secretion (Nilsson et al., 1992).

6.2.2 Cation-Chloride Cotransporters (SLC12 Family)

This is a family of seven proteins that transport Cl^- and cations (Na^+ or K^+ or both) in the same direction across the cell membrane (Hebert et al., 2004). The Na^+-Cl^- cotransporter (NCC) and Na^+-K^{+-}2Cl cotransporters (NKCC1 and NKCC2) mediate ion influx, driven by the chemical gradient for Na^+. The K^+-Cl^- cotransporters (KCC1–4), on the other hand, mediate ion efflux driven by the K^+ chemical gradient, which exists across the cell membrane. The possible expression of cation-chloride cotransporters in the choroid plexus and their potential role in CSF secretion has until recently been the subject of some controversy, i.e., there was evidence for and against the expression of Na^+- and K^+-coupled cotransporters (Javaheri, 1991). It is now apparent that much confusion arose because the choroid plexus epithelium actually expresses several different cation-chloride cotransporters in both the apical and basolateral membrane.

6.2.2.1 Na^+-K^+-$2Cl^-$ Cotransporter (SLC12A2)

There are good functional data for the expression of the Na^+-K^+-$2Cl^-$ cotransporter in choroid plexus, based on flux studies using the specific inhibitor bumetanide (Bairamian et al., 1991; Keep et al., 1994). Expression of the NKCC1 isoform was demonstrated in the rat choroid plexus by in situ hybridization and Western blot analysis (Plotkin et al., 1997). By contrast, there is no evidence for the expression of NKCC2 or NCC in rat choroid plexus by RT-PCR (S.L. Davies and P.D. Brown, unpublished data).

Immunocytochemical studies demonstrate that the NKCC1 protein is expressed in the apical membrane of epithelial cells from the rat choroid plexus (Plotkin et al., 1997; Wu et al., 1998). This observation, while contrary to some expectations that NKCC1 would be in the basolateral membrane (see Javaheri, 1991), is consistent with the functional data of Keep et al. (1994). The role of the NKCC1 in the apical membrane, however, is unclear. Keep et al. (1994), using estimates of intracellular ion activities, calculated that the cotransporter should export ions from the epithelial cells into the CSF (i.e., contribute to Na^+ and Cl^- secretion). On the other hand, Wu et al. (1998) found that inhibiting the NKCC1 with bumetanide reduced choroid plexus epithelial cell volume, suggesting that the cotransporter mediates ion transport into the epithelial cell. More recently Angelow et al. (2003) found that Na^+-dependent transport of vitamin C into choroid plexus cells was enhanced in the presence of bumetanide. This observation supports the hypothesis that NKCC1 mediates Na^+ (K^+ and Cl^-) influx to the cell from the CSF. The significance of this transport, however, remains to be determined.

6.2.2.2 K^+-Cl^- Cotransporters (SLC12A4, SLC12A6, SLC12A7)

KCCs were first identified in red blood cells, where they have a role in regulatory volume decrease (RVD), which occurs in response to cell swelling (Hebert et al., 2004). They have subsequently been found to play a role in transepithelial ion transport, e.g., in the kidney and colon (Hebert et al., 2004). Most recently they have been found to maintain intracellular [Cl^-] below equilibrium in neurons (Hubner et al., 2001). Molecular biological methods have demonstrated the expression of several different KCCs in the choroid plexus epithelium, in both the apical and the basolateral membranes.

In situ hybridization studies have shown that mRNA encoding KCC1 (SLC12A4) is expressed in the choroid plexus from rats and mice (Kanaka et al., 2001; Li et al., 2002; LeRouzic et al., 2004). Protein expression, however, has not been investigated. The expression of mRNA for KCC4 (SLC12A7) has been determined in choroid plexus by in situ hybridization (Li et al., 2002; LeRouzic et al., 2004). A recent immunocytochemical study has shown that the KCC4 protein is expressed in the apical membrane of mouse choroid plexus (Karadsheh et al., 2004). There is no evidence for the expression KCC2 (SLCA5) in choroid plexus either by in situ hybridization (Kanaka et al., 2001) or by RT-PCR (S.L. Davies and P.D. Brown, unpublished data).

There are two isoforms of KCC3 (SLC12A6). KCC3b is a truncated form of KCC3a (a 50-amino-acid deletion at the N-terminal which contains several putative phosphorylation sites; Mount et al., 1999). DiFluvio et al. (2003) demonstrated differences in the regulation of KCC3a and KCC3b in vascular smooth muscle, but whether there are any functional differences between the two isoforms remains to be determined. Expression of KCC3a but not KCC3b expression was determined in mouse choroid plexus by Northern analysis (Pearson et al., 2001). By contrast we have found evidence of both KCC3a and KCC3b expression by RT-PCR (S.L. Davies and P.D. Brown, unpublished data). Pearson et al. (2001) determined that KCC3 protein is expressed in the basolateral membrane of rat choroid plexus. The antibody used in these experiments did not discriminate between the KCC3a and KCC3b isoforms. Thus, if the two isoforms are expressed, they must both be located in the basolateral membrane.

There is only very limited data on the function of the K^+-Cl^- cotransporters in the choroid plexus. Zeuthen (1994) found evidence for the cotransport of K^+, Cl^- and H_2O in the apical membrane of amphibian (*Necturus maculosus*) choroid plexus, but whether this is mediated by an amphibian KCC isoform remains to be determined. Keep et al. (1994), on the other hand, was unable to identify DIOA-sensitive components to K^+ influx or efflux in rat choroid plexus epithelial cells (DIOA being a KCC inhibitor) (Culliford et al., 2003). Thus at present the roles of the KCCs in the apical and basolateral membranes remain hypothetical. The KCC3 transporters in the basolateral membrane may have a significant role in the absorption of K^+ from the CSF to the blood, as they are to date the only transporter or channel identified in the basolateral membrane capable of mediating K^+ efflux. The role of KCC4 in the apical membrane is less obvious, but they may contribute to Cl^- transport into the CSF and to the recycling of K^+ that occurs at the apical membrane (Zeuthen and Wright, 1981; Speake et al., 2004). A role for KCC1 is impossible to predict without knowledge of the membrane in which they are expressed, but they presumably must contribute to K^+ (and Cl^-) efflux at one or other of the epithelial membranes.

6.2.3 HCO_3^- Transporters (SCL4 Family)

A diverse range of transport proteins has been identified, which mediate both the influx and efflux of HCO_3^-, e.g., Cl^--HCO_3^- exchangers (AE1–3), Na^+-HCO_3^- cotransporters, and Na^+-dependent Cl^--HCO_3^- exchangers (Romero et al., 2004)*. Several of these transporters are expressed in the choroid plexus, and they may have a role in CSF secretion.

6.2.3.1 Cl^--HCO_3^- Exchanger (SLC4A2)

Cl^--HCO_3^- exchangers are expressed in many different types of cell. Their main function is in the regulation of intracellular pH (i.e., they mediate HCO_3^- efflux helping to acidify the cytoplasm), but they can also play a role in cell volume regulation and in epithelial transport (Alper et al., 2002). The first indication that the Cl^--HCO_3^- exchangers may be involved in CSF secretion came from experiments in which DIDS (an inhibitor of Cl^--HCO_3^- exchange, but also other anion transporters) was found to inhibit Cl^- transport into the CSF (Frankel and Kazemi, 1983; Deng and Johanson, 1989). Additional evidence for the expression of the Cl^--HCO_3^- exchanger was provided by studies of intracellular pH regulation in rat choroid plexus cells (Johanson et al., 1985). As a result of these studies, it was

* The molecular identity of many of these transporters has only recently been determined, and several remain incompletely characterized. An accepted nomenclature of the transporters has yet to be established, but we have adopted the names suggested by Romero et al. (2004).

suggested that Cl^--HCO_3^- exchangers are expressed in the apical membrane of the choroid plexus and directly contribute to the secretion of HCO_3^- into the CSF.

Three isoforms of functional exchanger (AE1–3) have been isolated: AE1 is expressed in red blood cells, AE2 is the epithelial isoform, and AE3 is expressed in neurons (Romero et al., 2004). Lindsey et al. (1990) identified the expression of the AE2 isoform in the choroid plexus epithelium. Immunolocalization showed that the AE2 protein is expressed exclusively in the basolateral membrane of mouse choroid plexus cells (Lindsey et al., 1990), an observation that has since been confirmed by Alper et al. (1994) and Wu et al. (1998) in rat choroid plexus. This observation ended the speculation that the exchanger provides a route for HCO_3^- entry into the CSF. Lindsey et al. (1990) suggested, instead, that the main role for AE2 in the choroid plexus may be in the regulation of intracellular pH. However, AE2 may also provide an important route for Cl^- influx across the basolateral membrane of the choroid plexus. Indirect evidence for this hypothesis was provided by data from single channel, patch clamp studies of rat choroid plexus (Brown and Garner, 1993).

6.2.3.2 Electrogenic Na^+-HCO_3^- Cotransporters (SLC4A6)

Na^+-HCO_3^- cotransporters (NBC) can be described as either electrogenic or electroneutral (Romero et al., 2004). An electrogenic transporter is one that generates an electric current by mediating the net movement of ionic charge across a membrane. In the case of an NBC this occurs because more than one HCO_3^- is transported with each Na^+. By contrast, the electroneutral transporters mediate the movement of an equal number of positive and negative charges.

NBCs were first identified in the kidney proximal tubule (Yoshitomi et al., 1985; Soleimani and Aronson, 1989). The kidney transporter (NBCe1 or SLC4A4) mediates the electrogenic efflux HCO_3^- from the cell by transporting 3 HCO_3^- for 1 Na^+. At this stoichiometry, the outward gradient for HCO_3^- efflux is thermodynamically more favorable than that for the inward movement of Na^+. The molecular identity of the kidney NBCe1 was determined by Romero et al. in 1997. Since then several other NBCe1 isoforms have been identified in pancreas, heart, and brain (Romero et al., 2004). These all exhibit a 2:1 stoichiometry and hence mediate HCO_3^- influx driven by the Na^+ electrochemical gradient. A second family of electrogenic NBC (NBCe2 or SLC4A5) has now also been cloned from the cardiac muscle (Pushkin et al., 2000). NBCe2 has a wide tissue distribution (Pushkin et al., 2000) and displays a 2:1 stoichiometry (Sassini et al., 2002; Virkki et al., 2002).

Electrogenic NBCs play a critical role in secretion by other HCO_3^- secreting epithelia, e.g., pancreatic duct cells (Case et al., 2004). However, the only functional evidence supporting the expression of these transporters in choroid plexus is from a study of intracellular pH (Mayer and Sanders-Bush, 1993). This study demonstrated that a Na^+-dependent, DIDS-sensitive transporter, possibly an NBC, contributes to pH regulation in choroid plexus cells. Molecular biological studies of electrogenic NBC expression have been inconclusive. Schmidt et al. (2001) used in situ hybridization to demonstrate that mRNA for NBCe1 is expressed in rat choroid plexus. Speake et al. (2001) also reported preliminary RT-PCR data suggesting NBCe1 expression in the rat. More recent RT-PCR experiments have identified mRNA for NBCe2, but not NBCe1 in rat and mouse choroid plexus (Praetorius et al., 2004)*. The expression of NBCe1 and NBCe2 protein in choroid plexus has not been investigated. Thus, the question as to whether electrogenic NBCs are expressed in the choroid plexus remains to be confirmed.

6.2.3.3 Electroneutral Na^+-HCO_3^- Cotransporter (SLC4A7)

The electroneutral NBCn-1 transports 1 HCO_3^- for 1 Na^+ (Choi et al., 2000). These transporters were first cloned from human skeletal muscle (Pushkin et al., 1999) and have subsequently been identified in many other tissues (Alamal et al., 1999; Choi et al., 2000). The 1:1 stoichiometry means that they are driven by the inwardly directed Na^+ gradient. Praetorius et al. (2004) identified the expression of mRNA for NBCn-1 in rat and mouse choroid plexus. Furthermore, NBCn-1 protein was found in the basolateral membrane of rat and mouse choroid plexus by immunocytochemical methods (Praetorius et al., 2004). Thus, NBCn-1 may work in tandem with AE2 to mediate the accumulation of Na^+, HCO_3^-, and Cl^- at the basolateral membrane. These events are thought to be an important step in the process of CSF secretion (Brown and Garner, 1993; Segal, 1993).

6.2.3.4 Na^+-Dependent Cl^--HCO_3^- Exchange (SLC4A10)

The molecular structure of two isoforms of Na-dependent Cl^--HCO_3^- exchangers have been determined, i.e., NDCBE (SLC4A8) (Gritchenko et al., 2001) and NBCE (SLC4A10) (Wang et al., 2000). These transporters mediate the efflux of one Cl^- in exchange for the influx of one Na^+ and 2 HCO_3^-. Praetorius et al. (2004) have demonstrated using RT-PCR and Western blotting that NCBE is expressed in rat and mouse choroid plexus. Immunocytochemistry also shows that the protein is expressed in the basolateral membrane of the choroid plexus epithelium. Na^+-dependent Cl^--HCO_3^- exchangers are generally thought to have a role in the regulation of intracellular pH (Romero et al., 2004), but they could potentially also have a role in epithelial anion transport. Thus they may act with NBCn1 to mediate Na^+ and HCO_3^- influx.

6.2.4 Na^+-H^+ Exchange (SLC9 Family)

The Na^+-H^+ exchangers transport H^+ out of the cell in exchange for Na^+ (Orlowski and Grinstein, 2004). Na^+-H^+ exchangers are widely expressed and mainly have a housekeeping role within the cell, e.g., regulation of intracellular pH and cell volume. However, they also contribute to transepithelial ion transport (Orlowski and Grinstein, 2004). Evidence of Na^+-H^+ exchanger expression in the choroid plexus comes mainly from the work of Murphy and Johanson (1989 and 1990), who showed that Na^+ transport into the CSF is reduced by amiloride (an inhibitor of the Na^+-H^+

* There is a greater than 50% identity between the amino acid sequences of NBCe1 and NBCe2, and thus it is possible that Schmidt et al. (2001) and Speake et al. (2001) identified mRNA for NBCe2 (or NBCn1 = 39% identical; or NCBE = 41% identical), rather than NBCe1 in their experiments.

exchanger). Furthermore, because of the sidedness of the effect of amiloride, it was postulated that the exchanger is expressed in the basolateral membrane of rat lateral and fourth ventricle choroid plexus. Additional evidence for the presence of the exchanger in the rabbit choroid plexus was provided by studies of intracellular pH regulation in epithelial cells maintained in primary culture (Mayer and Sanders-Bush, 1993). The presence of the exchanger in the basolateral membrane would provide a mechanism for Na^+ transport into the choroid plexus epithelial cell and help remove H^+ produced as a by-product of carbonic anhydrase activity.

A total of eight isoforms of the Na^+-H^+ exchangers (NHE1 to NHE8) have been identified (Orlowski and Grinstein, 2004). The expression of NHE1 in the pig lateral ventricle choroid plexus was indicated from the results of amiloride-binding studies and by RT-PCR methods (Kalaria et al., 1998). Alper et al. (1994), however, found no evidence for the expression of NHE1 in the rat lateral ventricle choroid plexus using immunocytochemical methods. The molecular identity and membrane localization of the exchanger in the choroid plexus therefore remain to be determined.

6.2.5 Aquaporins (AQP1, AQP4)

The aquaporins are a family of 10 proteins (AQP0-AQP9), all of which facilitate the movement of H_2O across cell membranes. Some aquaporins (AQP3, AQP7, AQP8, and AQP9) are also permeable to other small electrolytes, e.g., urea and glycerol (Baudaut et al., 2002). The expression of the aquaporins has not been studied systematically in the choroid plexus. There is, however, evidence that both AQP1 and AQP4 are expressed in the epithelium.

6.2.5.1 AQP1

The expression AQP1 has been determined in the apical membrane of the rat choroid plexus epithelium by using immunocytochemical methods (Nielsen et al., 1993; Wu et al., 1998; Speake et al., 2003). Furthermore, Mobasheri and Marple (2004) demonstrated that the choroid plexus exhibits the highest expression of AQP1 in any human tissue using microarray methods. AQP1 is therefore likely to have a major role in mediating water transport across the apical membrane of the choroid plexus during CSF secretion. Indeed, it has recently been shown that CSF production is significantly reduced in AQP1 knock-out mice (Oshio et al., 2003).

6.2.5.2 AQP4

This aquaporin is widely expressed in the brain (Baudaut et al., 2002). It is also expressed in the basolateral membrane of many epithelial cells (Frigeri et al., 1995). It is therefore a prime candidate as the route for water transport across the basolateral membrane of the choroid plexus. Data on AQP4 expression in the choroid plexus, however, is equivocal. In situ hybridization studies have provided evidence for (Venero et al., 1999) and against (Hasegawa et al., 1994) the expression of AQP4. Studies of protein expression are unclear. Frigeri et al. (1995) and Nielsen et al.

(1997) reported no expression in choroid plexus by immunocytochemistry, although it is expressed in basolateral membrane of the ependymal cells lining each ventricle of the brain (Nielsen et al., 1997). By contrast, Speake et al. (2003) found evidence of AQP4 protein expression in choroid plexus by Western blot analysis. Immunocytochemical methods also demonstrated expression of AQP4 in rat choroid plexus cells. Expression, however, was confined to the cytoplasm (possibly in organellar membranes), and AQP4 did not appear to be expressed in either the apical or basolateral membrane (Speake et al., 2003). The route for water transport across the basolateral membrane of the choroid plexus therefore remains to be determined.

6.2.6 K^+ Channels

K^+ channels expressed in the choroid plexus have a number of important roles in CSF secretion (Speake et al., 2001): (1) they help generate the membrane potential of the epithelial cell (Vm) which makes important contributions to the electrochemical gradients favoring anion efflux via channels in the apical membrane; (2) they act as a leak pathway for K^+ accumulated in the cell through the actions of the Na^+,K^+-ATPase; (3) they participate in the absorption of K^+ from the CSF. Whole-cell patch clamp experiments have identified two major components of the K^+ conductance in cells from rat choroid plexus (Kotera and Brown, 1994; Speake and Brown, 2004). Figure 6.3A shows that a time-independent, inward-rectifying conductance (Kir) is observed at hyperpolarizing Vm, while at depolarizing Vm a time-dependent outward-rectifying conductance (Kv) is observed. Significant progress has been made over the last 5 years in determining which channel proteins carry these conductances.

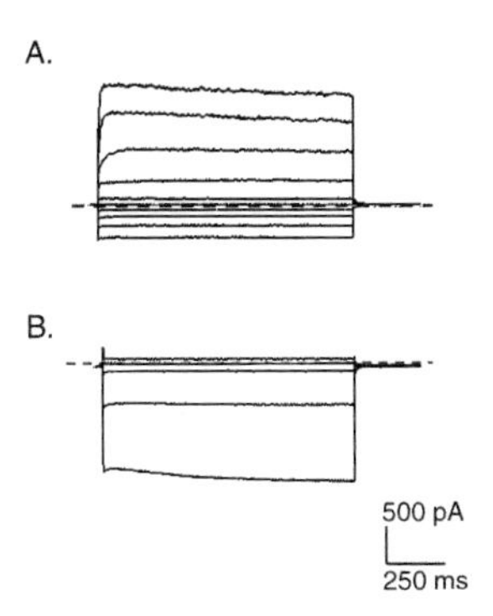

FIGURE 6.3 Whole-cell conductances in rat choroid plexus epithelial cells. (A) Current profiles for the two components of the K^+ conductance (Kv and Kir). The profiles were produced by applying voltage steps from -120 to 60 mV. The time-dependent Kv currents are observed at potentials greater than -20 mV. The time-independent Kir currents are observed at potentials less than -80 mV. A K^+-rich electrode solution was used, and the bath solution was an artificial CSF. **(B)** The inward-rectifying anion conductance. Currents were recorded in K^+-free solutions at Vm $= 0$ to -140 mV.

6.2.6.1 Kir 7.1 Channels (KCNJ13)

The time-independent, inward-rectifying conductance in choroid plexus shares many properties with Kir channels (Kotera and Brown, 1994). Kir channels are expressed in many mammalian cells and are mainly involved in maintaining Vm. The Kir channel family is divided into seven subfamilies (Kir1-Kir7). All Kir channels share a similar molecular structure, i.e., each Kir protein has two transmembrane-spanning domains, with four proteins assembling to produce a functional channel (Coetzee et al., 1999).

The first indication that the Kir conductance in choroid plexus was carried by the Kir7.1 was produced by in situ hybridization studies, which demonstrated that mRNA for Kir7.1 is highly expressed in the choroid plexus epithelium (Döring et al., 1998). Furthermore, Döring et al. (1998) showed that when Kir7.1 channels are expressed in *Xenopus* oocytes, they produce an inward-rectifying conductance with identical properties to those of the Kir in the choroid plexus. This strongly suggests that Kir7.1 makes a major contribution to the Kir conductance in the choroid plexus. Nakamura et al. (1999) demonstrated that the Kir7.1 channels are expressed in the apical membrane of the choroid plexus using immunocytochemical methods. These channels must therefore contribute to K^+ efflux at this membrane.

6.2.6.2 Kv1 Family of Channels (KCNA1, KCNA3, KCNA6)

Kv channels are members of the six transmembrane-spanning domain family of voltage-gated and Ca^{2+}-activated K^+ channels (Coetzee et al., 1999), which are expressed in many excitable and nonexcitable cells. All Kv channels are activated by a depolarization of Vm and exhibit outward rectification. In the choroid plexus, Kv channels contribute to the time-dependent, outward-rectifying K^+ conductance (see Figure 6.3A). This conductance is activated at depolarizing Vm, but is Ca^{2+}-independent (Kotera and Brown, 1994).

The properties of the Kv conductance in choroid plexus are similar to those of the Kv1 family of channels (Coetzee et al., 1999). Western analysis was therefore used to investigate Kv1 protein expression. Kv1.1 (KCNA1) (Speake et al., 2004), Kv1.3 (KCNA3) (Speake et al., 2004), and Kv1.6 (KCNA6) (T. Speake and P.D. Brown, unpublished data) were all identified by this method in rat choroid plexus. Neither Kv1.4 nor Kv1.5 was observed, and Kv1.2 expression has not been determined (T. Speake and P.D. Brown, unpublished data). The whole-cell K^+ currents in choroid plexus were partially inhibited by dendrotoxin-K or margatoxin (highly specific blockers of Kv1.1 and Kv1.3 channels, respectively) (Speake et al., 2004). These observations suggest that Kv1.1, Kv1.3, and possibly Kv1.6 contribute to the whole-cell K^+ conductance. Immunocytochemical studies have demonstrated that the expression of Kv1.1 and Kv1.3 is confined to the apical membrane of rat choroid plexus (Speake et al., 2004). Thus, Kv1.1 and Kv1.3 channels are thought to contribute to K^+ efflux at the apical membrane of the choroid plexus.

The magnitudes of the Kv currents in rat choroid plexus can be significantly reduced by serotonin (Speake et al., 2004). Hung et al. (1993) also observed that serotonin reduces apical K^+ channel activity in single channel recording experiments on mouse choroid plexus. The effect of serotonin is via 5-HT_{2C} receptors and is mediated intracellularly by protein kinase C (Speake et al., 2004). Serotonin is known to inhibit CSF secretion (Nilsson et al., 1992), and the effect on K^+ channel activity may be significant in this inhibition. This is because reducing K^+ channel activity may depolarize Vm and hence reduce the driving force for anion secretion.

6.2.7 Anion Channels

Anion channels are vital components of the secretory process in most epithelia (Jentsch et al., 2002). This is vividly demonstrated in the disease cystic fibrosis, where a defect in the expression of apical membrane anion channels results in a dramatic reduction in fluid secretion. In the epithelia affected by cystic fibrosis, the anion channel is called the cystic fibrosis transmembrane conductance regulator (CFTR), which is activated by cAMP (Jentsch et al., 2002). In many other epithelia Ca^{2+}-activated Cl^- channels are involved in secretion (Jentsch et al., 2002). None of these types of channel are expressed in the choroid plexus (Kibble et al., 1996, 1997). Whole cell patch clamp experiments have, however, identified two anion channels in the choroid plexus: an inward-rectifying anion channel and a volume-sensitive anion channel.

6.2.7.1 Inward-Rectifying Anion Channels

Anion channels with inward-rectifying current-voltage relationships have been observed in whole-cell recordings from the choroid plexus of rat (Kibble et al., 1996), mouse (Kibble et al., 1997), pig (Kajita et al., 2000a), and the amphibian *Necturus maculosus* (Birnir et al., 1987). The channels (1) exhibit time-dependent activation at hyperpolarizing Vm (see Figure 6.3B), (2) have a uniquely high permeability to HCO_3^- ($P_{HCO3}:P_{Cl} = 1.5$), and are more permeant to I^- than Cl^- or Br^-, (3) are blocked by Cl^- channel blockers (DIDS or the anthracene derivative NPPB), divalent cations (Ba^{2+}, Cd^{2+}, and Zn^{2+}) and H^+, and (4) are activated by cAMP and protein kinase A, but inhibited by protein kinase C (Kibble et al., 1996; Kajita and Brown, 1997; Kajita et al., 2000b). Some of these properties are similar to those of the ClC-2 channel (Speake et al., 2001; Jentsch et al., 2002), but the conductance is unchanged in ClC-2 knock-out mice (Speake et al., 2002). This indicates that ClC2 channels do not contribute to the conductance (Speake et al., 2002), and to date the molecular identity of the channel remains unknown.

Inward-rectifying anion channels make a significant contribution to the whole-cell conductance at the resting Vm (Kibble et al., 1996), suggesting that they have a role in CSF secretion. Furthermore, the regulation of these channels by cAMP and the high permeability to HCO_3^- suggest that they are similar to the apical HCO_3^- channels, which have a major role in CSF secretion by bullfrog choroid plexus (Saito and Wright, 1984). The properties are also consistent with the observation that Cl^- efflux from the rat choroid plexus is stimulated by cAMP (Deng and Johanson, 1992) and inhibited by agonists such as vasopressin, which activate protein kinase C (Johanson et al., 1999). To participate in CSF secretion, the inward-rectifying channels must be located in the apical membrane; however, the site of their expression has yet to be determined.

6.2.7.2 Volume-Sensitive Anion Channels

Volume-sensitive anion channels are expressed in choroid plexus from rats and mice (Kibble et al., 1996, 1997). These channels are activated by cell swelling and have the following properties: they are dependent on intracellular ATP, and they are blocked by DIDS and NPPB (Kibble et al., 1997). The channels are therefore very similar to volume-sensitive anion channels described in many other cells (Okada et al., 1997; Jentsch et al., 2002). The molecular structure of these channels remains unknown, although several candidate proteins have been identified and then discarded (Okada et al., 1997; Jentsch et al., 2002).

Immunocytochemical methods cannot be directly used to determine the location of either the volume-sensitive channel or inward-rectifying anion channel in choroid plexus, because their molecular identity is unknown. However, single channel patch clamp methods may help indirectly in determining where these channels are expressed. Garner and Brown (1992) reported the expression of anion channels with conductances of 27 pS in the apical membranes of rat choroid plexus. The activity (open probability) of these channels can be increased by serotonin (Garner et al., 1993). This observation suggests that the 27 pS channels do not contribute to the inward-rectifying whole-cell conductance, which is inhibited by PKC (Kajita et al., 2000b). Furthermore, the other properties of the 27 pS channel, e.g., linear I-V and opening with depolarization, are consistent with those of the volume-sensitive conductance. Thus it is possible that these channels carry the volume-sensitive anion conductance.*

The volume-sensitive anion channels in the choroid plexus epithelium make little, if any, contribution to the whole-cell conductance at normal cell volumes (Kibble et al., 1997). Thus, they are unlikely to have a significant role in CSF secretion. They are more likely to be involved in cell volume regulation, as has been observed in many other cells (Okada, 1997; Jentsch et al., 2002). The mechanisms of volume regulation in choroid plexus cells, however, have not yet been determined.

6.3 COMPLICATIONS OF STUDYING CHOROID PLEXUS: SPECIES DIFFERENCES AND INTERVENTRICLE VARIATIONS

Much of the data on ion transporter and ion channel expression are obtained from studies on rat choroid plexus. The question as to how representative these tissues are of the choroid plexuses from other species is unclear. The data from rats are generally in a good agreement with those from other mammalian species, e.g., mouse, rabbit, pig, and even human. However, there are some differences in ion channel expression between amphibian and mammalian choroid plexus. For exam-

* The 26pS channel could be activated by serotonin as a result of cell swelling, caused as a consequence of the inhibition of the Kv and inward-rectifying anion conductances.

ple, large conductance Ca^{2+}-activated K^+ channels are the major K^+ channel in amphibia (Loo et al., 1988), but are not expressed in mammalian choroid plexus, where Kv and Kir channels predominate (Kotera and Brown, 1994). Thus, the differences may indeed exist among mammalian species. Systematic comparisons of choroid plexus functions in different species thus become necessary.

Another complication in studying the choroid plexuses is that there may be variations in ion transport in plexus tissues from the different ventricles of the brain (see Pershing and Johanson, 1982; Harbut and Johanson, 1986). Most of the experimental data on the lateral ventricle and fourth ventricle choroid plexus agree reasonably well. For example, Speake and Brown (2004) demonstrated that the ion channels expressed in choroid plexus from rat fourth and lateral ventricles are qualitatively and quantitatively very similar. Less, however, is known about the physiology of the third ventricle choroid plexus. Immunocytochemical and patch clamp methods, however, should enable the properties of this plexus to be investigated in the near future.

A final potential problem in studying the choroid plexus is that there may be different subpopulations of cells within the choroid plexus epithelium. The choroid plexus epithelium is normally thought to contain a single type of cuboidal cell. This hypothesis is supported by electrophysiological data, which shows that the capacitances of choroid plexus epithelial cells exhibit a normal distribution with a single mean (Speake et al., 2004). Morphological studies have, however, identified "dark" and "light" cells within the choroid plexus epithelium (Dohrmann, 1970; Liszczak et al., 1986; Johanson et al., 1999), raising the possibility that there may be more than one type of cell in the epithelium. Lisczak et al. (1986) and Johanson et al. (1999) both observed that the number of dark cells increased in tissue exposed to arginine vasopressin, which also inhibits Cl^- efflux from the choroid plexus tissue. This suggests that the dark cells may develop as of changes in transport activity (i.e., posttranslational modification of transporters expressed), rather than an inherent difference between the cells. The future use of immunocytochemical methods should help assert this conclusion.

6.4 MODELS OF CSF SECRETION

Figure 6.4 illustrates various models of ion transport by the choroid plexus, which incorporate all of the recently identified ion transport proteins. The major movements of ions across the choroid plexus epithelium are summarized in Figure 6.4A. Figure 6.4B is a model to explain the unidirectional transport of Na^+, HCO_3^-, and Cl^- from the blood to CSF side of the epithelium. The model in Figure 6.4B is actually based on another, which was first proposed to explain electrophysiological data on bullfrog choroid plexus (Saito and Wright, 1983). This model has gradually evolved over the last 20 years to include nearly all of what has subsequently been determined about ion transporter and channel expression in the mammalian choroid plexus. Central to the model are the Na^+,K^+-ATPases in the apical membrane. The Na^+ pumps are the only route for Na^+ transport into the CSF; the pumps also provide the energy, in the form of the electrochemical gradient for Na^+, which either directly or indirectly drives events at the basolateral membrane of the epithelium.

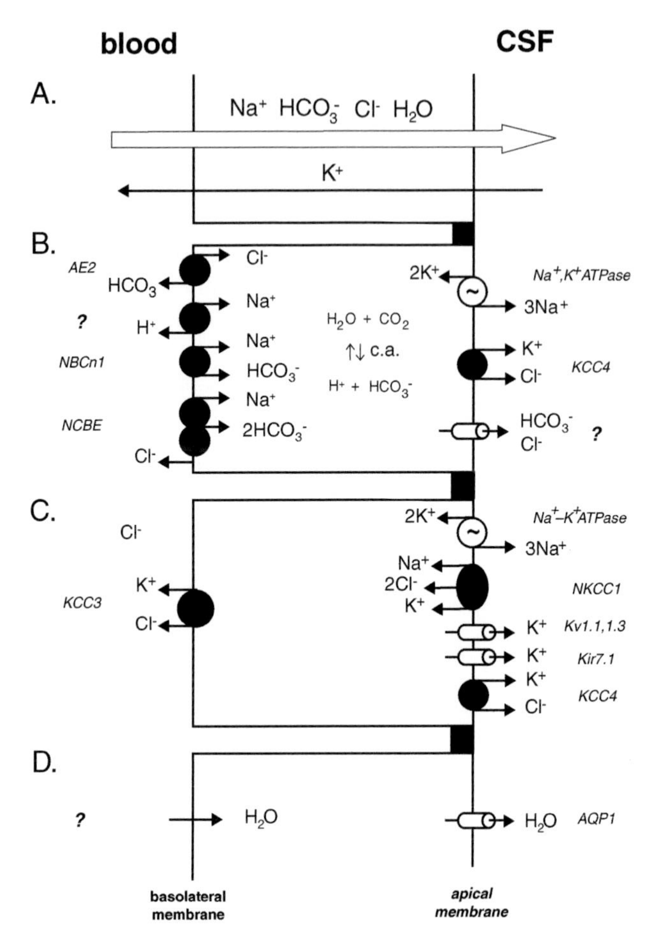

FIGURE 6.4 Ion transport by the choroid plexus. (**A**) Major fluxes of ions across the chroroid plexus epithelium. (**B**) Ion transporters involved in Na^+, HCO_3^-, and Cl^- secretion by the choroid plexus. (**C**) Mechanism of K^+ absorption. (**D**) H_2O transport in the choroid plexus epithelium.

6.4.1 Ion Transport at the Basolateral Membrane

Figure 6.4B shows the wide range of transporters that are expressed in the basolateral membrane. The net action of these transporters is to mediate the accumulation of Na^+, HCO_3^-, and Cl^- in the cytoplasm of epithelial cells. The NBCn1 and NBCE use the Na^+ gradient to drive HCO_3^- accumulation into the cell. HCO_3^- is also produced within the epithelial cells through the actions of the carbonic anhydrases (Speake et al., 2001). Protons, a by-product carbonic anhydrase activity, are exchanged across the basolateral membrane on the Na^+-H^+ exchangers. An outwardly directed HCO_3^- gradient is created, and this drives Cl^- accumulation on the AE2 in the basolateral membrane. Thus, in essence, Cl^- is accumulated into the cytoplasm by a Na^+-dependent and HCO_3^--dependent process.

6.4.2 Ion Transport at the Apical Membrane

The Na^+,K^+-ATPase in this membrane pumps Na^+ out of the cell and subsequently into the CSF. HCO_3^- and Cl^- can leave the cell down their respective electrochemical gradients via anion channels (probably the inward-rectifying channel), which are expressed in the apical membrane of the epithelium. The electrochemical gradients for anion efflux are generated by the anion transporters in the basolateral membrane along with a negative Vm. Cl^- efflux may also take place via the KCC4, which is also expressed in the apical membrane.

6.4.3 Transepithelial Transport of Water

The routes by which H_2O crosses the choroid plexus epithelium are illustrated in Figure 6.4D. AQP1 is likely to have a major role in transport at the apical membrane. The pathway across the basolateral membrane, however, remains unknown. Neither AQP1 nor AQP4 is expressed in the basolateral membrane, but other members of the aquaporin family may be involved. Furthermore, recent research has shown that a variety of solute transporters may mediate H_2O transport (Loo et al., 2002). Many of the original observations in support of this theory were made by Zeuthen (1994), who, using amphibian choroid plexus as a model, reported that H_2O was transported across the apical membrane via K^+-Cl^- cotransporters. These data have yet to be confirmed in mammalian choroid plexus. However, it is possible that KCCs expressed in the basolateral membrane may also play a role in H_2O transport.

6.4.4 Model of K^+ Absorption from the CSF by Choroid Plexus

The movement of K^+ across the choroid plexus epithelium can also be explained using our knowledge of transporter and channel expression (see Figure 6.4C). Once again the activity of Na^+,K^+-ATPase is central to this model. The Na^+ pump is the one efficient mechanism by which K^+ can be transported into the epithelial cells from the CSF against a large electrochemical gradient. The vast majority of this K^+ is then simply recycled across the apical membrane. Zeuthen and Wright (1981) suggested that more than 90% of the K^+ leaves bullfrog epithelial cells via K^+ channels in the apical membrane. In the rat the Kv1 channels and Kir7.1 probably provide the major route for K^+ efflux. However, KCC4 may also contribute to K^+ transport. The recycling of K^+ is necessary to limit the transport of K^+ across the epithelium, so that the CSF does not become denuded of K^+.

To explain K^+ absorption, a small fraction of the K^+ pumped into the cells via the Na^+,K^+-ATPase must leave the cell across the basolateral membrane. Zeuthen and Wright (1981) estimated that 10% of the K^+ conductance in bullfrog choroid plexus resides in the basolateral membrane. K^+ channels, however, have yet to be identified in the basolateral membrane of mammalian choroid plexus. To date the only transporter mediating K^+ efflux at the basolateral membrane is KCC3. It is also possible that KCC1 and other, as yet unidentified, channels may also contribute to K^+ transport at this membrane.

6.4.5 Testing the Models of Ion Transport by the Choroid Plexus

The use of molecular biological methods has generated a wealth of information about ion transport protein expression in the mammalian choroid plexus. Figure 6.4 shows that this information can be incorporated into existing models of ion transport. Furthermore, it is possible to envisage how additional information on transporter expression could improve these models. For example, KCC1 may have a role in K^+ absorption or K^+ recycling, depending upon whether it is expressed in either the basolateral or apical membrane. Electrogenic NBCs, on the other hand, could play a role in anion accumulation at the basolateral membrane, or even HCO_3^- efflux at the apical membrane. These observations suggest that the models in Figure 6.4 may be accurate representations of ion transport by the choroid plexus. It is important to remember, however, that these are proposed models, and any model comes with limitations.

The first limitation stems from the use of the molecular methods, which are at best only semi-quantitative. While it is possible to identify the expression of mRNAs or proteins of a particular transporter, the lack of quantitative and functional information makes it difficult to estimate the relative contributions of transporters or channels to ion transport. An example of this is the model of K^+ absorption (see Figure 6.4C), where KCC3 is the only pathway for K^+ transport at the basolateral membrane. However, to date there are no functional data to indicate that KCC3 makes a significant contribution to K^+ transport in the basolateral membrane. Without such data the model in Figure 6.4C can be regarded only as hypothetical.

A second limitation is that many of the transporters identified either by function or by molecular methods can have roles in maintaining cell homeostasis. The cation-chloride cotransporters, for example, have major roles in volume regulation in many cells, whereas many of the HCO_3^- transporters play a role in the regulation of intracellular pH. Thus, even observing the functional activity of these transporters in the choroid plexus does not necessarily mean that they have a direct role in the process of CSF secretion.

The third limitation pertains to the availability of in vitro models for ion transport study. To determine the precise role of particular transporters in CSF secretion, there is a need to measure unidirectional fluxes of ions across the choroid plexus epithelium and to determine the effects of transport inhibitors on these fluxes. Until very recently this would have been difficult due to a lack of good in vitro preparations of mammalian choroid plexus. However, there are now primary cultures of both rat (detailed in Chapter 23) and pig choroid plexus epithelial cells (Haselbach et al., 2001), which form monolayers when grown on permeable filters. These epithelial monolayers can be mounted in Ussing chambers allowing transepithelial transport to be measured (Strazielle and Gersi-Egea, 1999; Haselbach et al., 2001). The pig cells, in particular, offer an exciting possibility to study ion transport processes, since they form a very tight epithelial barrier in culture (Haselbach et al., 2001) and appear to express all the ion channels of the intact choroid plexus (Kajita et al., 2000a). Data on ion transport across these monolayer cultures are eagerly awaited.

The value of measuring unidirectional fluxes cannot be underestimated, because these experiments will determine the basic functions of the mammalian choroid plexus, i.e., which ions are secreted and which are absorbed. However, the usefulness of the method in determining the specific roles of individual transporters may be limited. This is because of the lack of specific transporter inhibitors, particularly for anion transporters, e.g., the same compounds inhibit both cation-chloride cotransporters and the HCO_3^- transporters (Culliford et al., 2003). Thus, determining the precise mechanisms involved in CSF secretion will require the development of improved transport inhibitors.

The use of knock-out mice may provide an alternative means for the study of choroid plexus function. To date, however, only a few studies of ion transport have employed these methods. Two studies of anion channel expression show that neither CFTR (Kibble et al., 1997) nor ClC-2 (Speake et al., 2002) contribute to the whole cell anion conductance of choroid plexus cells. Flagella et al. (1999) have also observed abnormalities in the choroid plexus tissue isolated from mice in which NKCC1 had been deleted, although the significance of this observation is unclear. Finally, a recent study has found that the rate of CSF secretion greatly reduced in mice lacking AQP1 (Oshio et al., 2003). These experiments demonstrate not only the importance of AQP1 in the transport of H_2O across the choroid plexus, but also the potential of using genetically modified animals in studies of CSF secretion.

6.5 CONCLUSIONS

Choroid plexus epithelial cells secrete cerebrospinal fluid by a process that involves the movement of Na^+, Cl^-, and HCO_3^- from the blood to the ventricles of the brain. This creates an osmotic gradient, which drives the secretion of H_2O. The unidirectional movement of the ions is achieved due to the polarity of the epithelium, i.e., the ion transport proteins in the blood-facing (basolateral) membranes are different from those in the ventricular (apical) membranes. A variety of methods has been used to determine the expression of ion transporters and channels in the choroid plexus epithelium. The location of most of these transporters has now been identified, e.g., Na^+,K^+-ATPase, K^+ channels, Na^+-$2Cl^-$$K^+$ cotransporters, and aquaporin 1 are expressed in the apical membrane. By contrast, the basolateral membrane contains Cl^--HCO_3^- exchangers, a variety of Na^+-coupled HCO_3^- transporters, and K^+-Cl^- cotransporters. A model of CSF secretion by the mammalian choroid plexus is proposed, which accommodates these transporter/channel proteins. The model also explains the mechanisms by which K^+ is transported from the CSF to the blood. However, the model needs to be further tested in order to determine exactly how these proteins contribute to CSF secretion.

6.6 ACKNOWLEDGMENTS

This work was supported by the Wellcome Trust (Grant 070139/Z/02).

REFERENCES

Alper SL, Stuart-Tilley A, Simmons CF, Brown D, Drenckhahn D. (1994) The fodrin-ankyrin cytoskeleton of choroid plexus preferentially colocalizes with apical Na^+K^+-ATPase rather than with basolateral anion exchanger AE2. J Clin Invest 93:1430–1438.

Alper SL, Darman RB, Chernova MN, Dahl NK. (2002) The AE gene family of Cl^-/HCO_3^- exchangers. J Nephrol 15: S41–S53.

Amlal H, Burnham CE, Soleimani M. (1999) Characterization of Na^+/HCO^-_3 cotransporter isoform NBC-3. Am J Physiol 276: F903–F913.

Angelow S, Haselbach M, Galla HJ. (2003) Functional characterisation of the active ascorbic acid transport into cerebrospinal fluid using primary cultured choroid plexus cells. Brain Res 988: 105–113.

Badaut J, Lasbennes F, Magistretti PJ, Regli L. (2002). Aquaporins in brain: distribution, physiology and pathophysiology. J Cereb Blood Flow Metab 22: 367–378.

Bairamian D, Johanson CE, Parmalee JT, Epstein MH. (1991) Potassium cotransport with sodium and chloride in the choroid plexus. J Neurochem 56: 1632–1629.

Birnir B, Loo DDF, Brown PD, Wright EM. (1989) Whole cell currents in *Necturus* choroid plexus. FASEB J 2: A1722.

Blanco G, Mercer RW. (1998) Isozymes of the Na-K-ATPase: heterogeneity in structure, diversity in function. Am J Physiol 275: F633–F650.

Brown PD, Garner C. (1993) Cerebrospinal fluid secretion: the transport of fluid and electrolytes by the choroid plexus. In: The Subcommissural Organ, Oksche A, Rodriguez EM, Fernandez-Llebrez P., eds. Springer International, Heidelberg, pp 233–242.

Case RM, Ishiguro H, Steward MC. (2004) Pancreatic bicarbonate secretion. In: Encyclopedia of Gastroeneterology, Johnson LR, ed. Academic Press, New York pp 38–40.

Choi I, Aalkjaer C, Boulpaep EL, Boron WF. (2000) An electroneutral sodium/bicarbonate cotransporter NBCn1 and associated sodium channel. Nature 405: 571–575.

Coetzee WA, Amarillo Y, Chiu J, Chow A, Lau D, McCormack T, Moreno H, Nadal MS, Ozaita A, Pountney D, Saganich M, Vega-Saenz De Miera E, Rudy B. (1999) Molecular diversity of K^+ channels. Ann NY Acad Sci 868: 233–285.

Cowan F, Whitelaw A. (1991) Acute effects of acetazolamide on cerebral blood flow velocity and pCO_2 in the newborn infant. Acta Pediatr Scand 80: 22–27.

Culliford S, Ellory C, Lang HJ, Englert H, Staines H, Wilkins R. (2003) Specificity of classical and putative Cl^- transport inhibitors on membrane transport pathways in human erythrocytes. Cell Physiol Biochem 13: 181–188.

Davson H, Segal MB. (1970) The effects of some inhibitors and accelerators of sodium transport on the turnover of ^{22}Na in the cerebrospinal fluid and the brain. J Physiol 209: 139–153.

Davson H, Welch K, Segal MB. (1987) The physiology and pathophysiology of the cerebrospinal fluid. Churchill Livingstone, Edinburgh.

Deng QS, Johanson CE. (1989) Stilbenes inhibit exchange of chloride between blood, choroid plexus and the cerebrospinal fluid. Brain Res 510: 183–187.

Deng QS, Johanson CE. (1992) Cyclic AMP alteration of chloride transport into the choroid plexus–cerebrospinal fluid system. Neurosci Lett 143: 146–150.

DiFulvio M, Lauf PK, Adragna NC. (2003) The NO signaling pathway differentially regulates KCC3a and KCC3b mRNA expression. Nitric Oxide 9: 165–171.

Dohrmann GJ. (1970) Dark and light epithelial cells in the choroid plexus of mammals. J Ultrastruct. Res 32: 268–273.

Döring F, Derst C, Wischmeyer E, Karschin C, Schneggenburger R, Daut J, Karschin A. (1998) The epithelial inward rectifier channel Kir 7.1 displays unusual K^+ permeation properties. J Neurosci 18: 8625–8636.

Ellis DZ, Nathanson JA, Sweadner KJ. (2000) Carbachol inhibits Na^+,K^+-ATPase activity in choroid plexus via stimulation of the NO/cGMP pathway. Am J Physiol 279: C1685–C1693.

Ernst SA Palacios JR, Siegel GJ. (1986) Immunocytochemical localization of Na^+,K^+-ATPase catalytic polypeptide in mouse choroid plexus. J Histochem Cytochem 34: 189–195.

Fisone G, Snyder GL, Fryckstedt J, Caplan MJ, Aperia A, Greengard P. (1995) Na^+,K^+-ATPase in the choroid plexus. Regulation by serotonin/protein kinase C pathway. J Biol Chem 270: 2427–2430.

Flagella M, Clarke LL, Miller ML, Erway LC, Giannella RA, Andringa A, Gawenis LR, Kramer J, Duffy JJ, Doetschman T, Lorenz JN, Yamoah EN, Cardell EL, Shull GE. (1999) Mice lacking the basolateral Na-K-2Cl cotransporter have impaired epithelial chloride secretion and are profoundly deaf. J Biol Chem 274: 26946–26955.

Frankel H, Kazemi H. (1983) Regulation of CSF composition-blocking chloride-bicarbonate exchange. J Appl Physiol 55: 177–182.

Freschenko MS, Donnet C, Wetzel RK, Asinovski NK, Jones LR, Sweadner KJ. (2003) Phospholemman, a single-span membrane protein, is an accessory protein of Na,K-ATPase in cerebellum and choroid plexus. J Neurosci 23: 2161–2169.

Frigeri A, Gropper MA, Umenishi F, Kawashima M, Brown D, Verkman AS. (1995) Localization of MIWC and GLIP water channel homologs in neuromuscular, epithelial and glandular tissues. J Cell Sci 108: 2993–3002.

Garner C, Brown PD. (1992) Two types of chloride channel in the apical membrane of rat choroid plexus epithelial cells. Brain Res 591: 137–145.

Garner C, Feniuk W, Brown PD. (1993) Serotonin activates chloride channels in the apical membranes of rat choroid plexus epithelial cells. Eur J Pharmacol 239, 31–37.

Grichtchenko II, Choi I, Zhong X, Bray-Ward P, Russell JM, Boron WF. (2001) Cloning, characterization, and chromosomal mapping of a human electroneutral Na^+-driven Cl-HCO_3 exchanger. J Biol Chem 276: 8358–8363.

Harbut RE, Johanson CE. (1986) Third ventricle choroid plexus function and its response to acute perturbations in plasma chemistry. Brain Res 374: 137–146.

Hasegawa H, Ma T, Skach W, Matthay MA, Verkman AS. (1994) Molecular cloning of a mercurial-insensitive water channel expressed in selected water-transporting tissues J Biol Chem 269: 5497–5500.

Haselbach M, Wegener J, Decker S, Engelbertz C, Galla HJ. (2001) Porcine choroid plexus epithelial cells in culture: regulation of barrier properties and transport processes. Microsc Res Tech 52:137–152.

Hebert SC, Mount DB, Gamba G. (2004) Molecular physiology of cation-coupled Cl^- cotransport: the SLC12 family. Plügers Archiv 447: 580–593.

Hubner CA, Stein V, Hermans-Borgmeyer I, Meyer T, Ballanyi K, Jentsch TJ. (2001) Disruption of KCC2 reveals an essential; role of K-Cl cotransport already in early synaptic inhibition. Neuron 30: 515–524.

Hung BCP, Loo DDF, Wright EM. (1993) Regulation of mouse choroid plexus apical Cl^- and K^+ channels by serotonin. Brain Res 617: 285–295.

Husted RF, Reed DJ. (1976) Regulation of cerebrospinal fluid potassium by the cat choroid plexus. J Physiol 259: 213–221.

Husted RF, Reed DJ. (1977) Regulation of cerebrospinal fluid bicarbonate by the cat choroid plexus. J Physiol 267: 411–428.

Javaheri S. (1991) Role of NaCl cotransport in cerebrospinal fluid production: effects of loop diuretics. J Appl Physiol 71: 795–800.

Jentsch TJ, Stein V, Weinreich F, Zdebik AA. (2002) Molecular structure and physiological function of chloride channels. Physiol Rev 82: 503–568.

Johanson CE (1999) Choroid plexus. In: Encyclopedia of Neuroscience, Adelman G, Smith BH, eds. Elsevier Science, New York, pp 384–387.

Johanson CE, Parandoosh Z, Smith QR. (1985). Cl-HCO_3 exchange in choroid plexus: analysis by the DMO method for cell pH. Am J Physiol 249: F478–F484.

Johanson CE, Preston JE, Chodobski A, Stopa E.G., Szmydynger-Chodobski J, McMillan PN. (1999) AVP V_1 receptor-mediated decrease in Cl^- efflux and increase in dark cell number in choroid plexus epithelium. Am J Physiol 276: C82–C90.

Jones HC, Keep F. (1987) The control of potassium concentration in the cerebrospinal fluid and brain interstitial fluid of developing rats. J Physiol 383: 441–453.

Kajita H, Brown PD. (1997) Inhibition of the inward-rectifying Cl^- channel in the rat choroid plexus by a decrease in extracellular pH. J Physiol 498: 703–707.

Kajita H, Omori K, Matsuda H. (2000a) The chloride channel ClC-2 contributes to the inwardly rectifying Cl^- conductance in cultured porcine choroid plexus epithelial cells. J Physiol 523: 313–324.

Kajita H, Whitwell C, Brown PD. (2000b) Properties of the inward-rectifying Cl^- channel in rat choroid plexus: regulation by intracellular messengers and inhibition by divalent cations. Pflügers Arch 440, 933–940.

Kalaria RN, Premkumar DR, Lin CW, Kroon SN, Bae JY, Sayre LM, LaManna JC. (1998) Identification and expression of the Na^+/H^+ exchanger in mammalian cerebrovascular and choroidal tissues: characterisation by amiloride-sensitive [^{3}H]MIA binding and RT-PCR analysis. Brain Res Mol Brain Res 58: 178–187.

Kanaka C, Ohno K, Okabe A, Kuriyama K, Itoh T, Fukuda A, Sato K. (2001). The differential expression patterns of messenger RNAs encoding K-Cl cotransporters (KCC1, 2) and Na-K-2Cl cotransporter (NKCC1) in the rat nervous system. Neuroscience 104: 933–946.

Karadsheh MF, Byun N, Mount DB, Delpire E. (2004) Localization of the KCC4 potassium-chloride cotransporter in the nervous system. Neuroscience 123: 381–391.

Keep RF, Jones HC. (1990) A morphometric study on the development of the lateral ventricle choroid plexus, choroid plexus capillaries and ventricular ependyma in the rat. Brain Res Dev Brain Res 56: 47–53.

Keep RF, Xiang J, Betz AL. (1994) Potassium cotransport at the rat choroid plexus. Am J Physiol 267: C1616–C1622.

Kibble JD, Tresize AO, Brown PD. (1996) Properties of the cAMP-activated Cl^- conductance in choroid plexus epithelial cells isolated from the rat. J Physiol 496: 69–80.

Kibble JD, Garner C, Kajita H, Colledge WH, Evans MJ, Radcliff R, Brown PD. (1997) Whole-cell Cl^- conductances in mouse choroid plexus epithelial cells do not require CFTR expression. Am J Physiol 272: C1899–C1907.

Klarr SA, Ulanski LJ, Stummer W, Xiang J, Betz AL, Keep RF. (1997) The effects of hypo- and hyperkalemia on choroid plexus potassium transport. Brain Res 758: 39–44.

Kotera T, Brown PD. (1994) Two types of potassium current in rat choroid plexus epithelial cells. Pflügers Archiv 237: 317–324.

Le Rouzic P, Ivanov TR, Stanley PJ, Chan F, Pinteaux E, Baudoin FM, Best L, Brown PD, Luckman SM. (2004) The expression of cation-chloride cotransporters in the rat brain: regulation of KCC2 in glucose-sensing regions of the hypothalamus. Neurosci. (Submitted)

Li H, Tornberg J, Kaila K, Airaksinen MS, Rivera C. (2002) Patterns of cation-chloride cotransporter expression during embryonic rodent CNS development. Eur J Neurosci 16: 2358–2370.

Lindsey AE, Schneider K, Simmons DM, Baron R, Lee BS, Kopito RR. (1990) Functional expression and subcellular localization of an anion exchanger cloned from choroid plexus. Proc Natl Acad Sci USA 87: 5278–5282.

Liszczak TM, Black PM, Foley L. (1986) Arginine vasopression causes morphological changes suggestive of fluid transport in rat choroid plexus epithelium. Cell Tissue Res 246: 379–385.

Loo DDF, Brown PD, Wright EM. (1988) Ca^{2+} activated K^+ currents in *Necturus* choroid plexus. J Membrane Biol 105: 221–231.

Loo DDF, Wright EM, Zeuthen T. (2002) Water pumps. J Physiol 542: 53–60.

Maren TH, Conroy CW, Wynns GC, Godman DR. (1997) Renal and cerebrospinal fluid formation pharmacology of a high molecular weight carbonic anhydrase inhibitor. J Pharmacol Exp Ther 280: 98–104.

Masuzawa T, Ohta T, Kawamura M, Nakahara N, Sato F. (1984) Immunohistochemical localization of Na^+, K^+-ATPase in the choroid plexus. Brain Res 302: 357–362.

Mayer SE, Sanders-Bush E. (1993) Sodium-dependent antiporters in choroid plexus epithelial cultures from rabbit. J Neurochem 60: 1308–1316.

Mobasheri A, Marples D. (2004) Expression of the AQP-1 water channel in normal human tissues: a semi-quantitative study using tissue mircroarray technology. Am J Physiol 286: C529–C537.

Mount DB, Mercado A, Song L, Xu J, George AL Jr, Delpire E, Gamba G. (1999) Cloning and characterization of KCC3 and KCC4, new members of the cation-chloride cotransporter gene family. J Biol Chem 274: 16355–16362.

Murphy VA, Smith QR, Rapoport SI. (1986) Homeostasis of brain and cerebrospinal fluid calcium concentrations during chronic hypo- and hypercalcemia. J Neurochem 47: 1735–1741.

Murphy VA, Johanson CE. (1989) Alteration of sodium transport by the choroid plexus with amiloride. Biochim Biophys Acta 979: 187–92.

Murphy VA, Johanson CE. (1990) Na^+-H^+ exchange in choroid plexus and CSF in acute metabolic acidosis or alkalosis. Am J Physiol 258: F1528–F1537.

Nakamura N, Suzuki Y, Sakuta H, Ookata K, Kawahara K, Hirose S. (1999) Inwardly rectifying K^+ channel Kir7.1 is highly expressed in thyroid follicular cells, intestinal epithelial cells and choroid plexus epithelial cells: implication for a functional coupling with Na^+,K^+-ATPase. Biochem J 342: 329–336.

Nielsen S, Smith BL, Christensen EI, Agre P. (1993) Distribution of the aquaporin CHIP in secretory and resorptive epithelia and capillary endothelia. Proc Natl Acad Sci USA 90: 7275–7279.

Nielsen S, Nagelhus EA, Amiry-Moghaddam M, Bourque C, Agre P, Ottersen OP. (1997) Specialized membrane domains for water transport in glial cells: high-resolution immunogold cytochemistry of aquaporin-4 in rat brain. J Neurosci 17: 171–180.

Nilsson C, Lindvall-Axelsson M, Owman C. (1992) Neuroendocrine regulatory mechanisms in the choroid plexus–cerebrospinal fluid system. Brain Res Rev 17: 109–138.

Okada Y. (1997) Volume expansion-sensing outward-rectifier Cl^- channel: fresh start to the molecular identity and volume sensor. Am J Physiol 273: C755–C789.

Orlowski J, Grinstein S. (2004) Diversity of the mammalian sodium/proton exchanger SLC9 gene family. Pfügers Archiv 447: 549–565.

Oshio K, Song Y, Verkman AS, Manley GT. (2003) Aquaporin-1 deletion reduces osmotic water permeability and cerebrospinal fluid production. Acta Neurchir Suppl 86: 525–528.

Pearson MM, Lu J, Mount DB, Delpire E. (2001) Localization of the K^+-Cl^- cotransporter, KCC3, in the central and peripheral nervous systems: expression in the choroid plexus, large neurons and white matter tracts. Neuroscience 103: 481–491.

Pershing LK, Johanson CE. (1982) Acidosis-induced enhanced activity of the Na-K exchange pump in the *in vivo* choroid plexus: an ontogenetic analysis of possible role in cerebrospinal fluid pH homeostasis. J Neurochem 38: 322–332.

Plotkin MD, Kaplan MR, Peterson LN, Gullans SR, Hebert SC, Delpire E. (1997) Expression of the Na^+-K^+-$2Cl^-$ cotransporter BSC2 in the nervous system. Am J Physiol 272: C173–C183.

Pollay M, Hisey B, Reynolds E, Tomkins P, Stevens A, Smith R. (1985) Choroid plexus Na^+/K^+-activated adenosine triphophatase and cerebrospinal fluid secretion. Neurosurgery 17: 768–772.

Praetorius J, Nejsum LN, Nielsen S. (2004) A SCL4A10 gene product maps selectively to the basolateral membrane of choroid plexus epithelial cells. Am J Physiol 286: C601–C610.

Pushkin A, Abuladze N, Lee I, Newman D, Hwang J, Kurtz I. (1999) Cloning, tissue distribution, genomic organization, and functional characterization of NBC3, a new member of the sodium bicarbonate cotransporter family. J Biol Chem 274: 16569–16575.

Pushkin A, Abuladze N, Newman D, Lee I, Xu G, Kurtz I. (2000) Cloning, characterization and chromosomal assignment of NBC4, a new member of the sodium bicarbonate cotransporter family. Biochim Biophys Acta 1493: 215–218.

Quay WB. (1966) Regional differences in metabolism and composition of choroid plexuses. Brain Res 2: 378–389.

Quinton PM, Wright EM, Tormey JM. (1973) Localization of sodium pumps in the choroid plexus epithelium. J Cell Biol 58: 724–730.

Rizzolo LJ. (1999) Polarization of the Na^+,K^+-ATPase in epithelia derived from the neuroepithelium. Int Rev Cytol 185: 195–235.

Romero MF, Hediger MA, Boulpaep EL, Boron WF. (1997) Expresion cloning and characterization of a renal electrogenic Na^+/HCO_3 cotransporter. Nature 387: 409–413.

Romero MF, Fulton CM, Boron WF. (2004) The SLC4 family of HCO_3^- transporters. Pflügers Archiv 447: 495–509

Saito Y, Wright EM. (1983) Bicarbonate transport across the frog choroid plexus and its control by cyclic nucleotides. J Physiol 336: 635–648.

Saito Y, Wright EM. (1984) Regulation of bicarbonate transport across the brush border membrane of the bull-frog choroid plexus. J Physiol 350: 327–342.

Sassani P, Pushkin A, Gross E, Gomer A, Abuladze N, Dukkipati R, Carpenito G. (2002) Functional characterization of NBC4: a new electrogenic sodium-bicarbonate cotransporter. Am J Physiol 282: C408–C416.

Scheiner-Bobis G. (2002) The sodium pump. Its molecular properties and mechanics of ion transport. Eur J Biochem 269: 2424–2433.

Schmitt BM, Berger UV, Douglas RM, Bevensee MO, Hediger MA, Haddad GG, Boron WF. (2000) Na/HCO_3 cotransporters in rat brain: expression in glia, neurons, and choroid plexus. J Neurosci 20: 6839–6848.

Segal MB. (1993) Extracellular and cerebrospinal fluid. J Inherit Metabol Dis 16: 617–638.

Silva P, Stoff J, Field M, Fine L, Forrest JN, Epstein FH. (1977). Mechanism of active chloride secretion by shark rectal gland: role of Na-K-ATPase in chloride transport. Am J Physiol 233: F298–F306.

Soleimani M, Aronson PS. (1989) Ionic mechanism of Na^+-HCO_3^- cotransport in rabbit renal basolateral membrane vesicles. J Biol Chem 264: 18302–18308.

Speake T, Brown PD. (2004) Ion channels in epithelial cells of the choroid plexus isolated from the lateral ventricle of rat brain. Brain Res 1005: 60–66.

Speake T, Freeman LJ, Brown PD. (2003) Expression of aquaporin-1 and aquaporin-4 water channels in rat choroid plexus. Biochim Biophys Acta 1609: 80–86.

Speake T, Kajita H, Smith CP, Brown PD. (2002) Inward-rectifying anion channels are expressed in the epithelial cells of choroid plexus isolated from ClC-2 'knock-out' mice. J Physiol 539: 385–390.

Speake T, Kibble JD, Brown PD. (2004) Kv1.1 and Kv1.3 channels contribute to the delayed-rectifying K^+ conductance in rat choroid plexus epithelial cells. Am J Physiol 286: C611–C620.

Speake T, Whitwell C, Kajita H, Brown PD. (2001). Mechanism of CSF secretion by the choroid plexus. Microsc Res Tech 52: 49–59.

Steward MC, Case RM. (1989) Principles of ion and water transport across epithelia. In: Gastrointestinal Secretion, Davison JS, ed. Wright, London, pp 1–31.

Strazielle N, Gersi-Egea JF. (1999) Demonstration of a coupled mechanism–efflux process at the choroid plexus as a mechanism of brain protection toward xenobiotics. J Neurosci 19: 6275–6289.

Venero JL, Vizuete ML, Ilundain AA, Machado A, Echevarria M, Cano J. (1999) Detailed localization of aquaporin-4 messenger RNA in the CNS: preferential expression in periventricular organs. Neuroscience 94: 239–250.

Virkki LV, Wilson DA, Vaughan-Jones RD, Boron WF. (2002) Functional characterization of human NBC4 as an electrogenic Na^+-HCO_3^- cotransporter (NBCe2). Am J Physiol 282: C1278–1289.

Vogh BP, Godman DR, Maren TH. (1987) The effect of $AlCl_3$ and other acids on cerebrospinal fluid production: a correction. J Pharmacol Exp Ther 243: 35–39.

Wang CZ, Yano H, Nagashima K, Seino S. (2000). The Na^+-driven Cl^-/HCO_3^- exchanger. Cloning, tissue distribution, and functional characterization. J Biol Chem 275: 35486–35490.

Watts AG, Sanchez-Watts G, Emanuel JR, Levenson R. (1991) Cell-specific expression of mRNAs encoding Na^+,K^+-ATPase alpha- and beta-subunit isoforms within the rat central nervous system. Proc Natl Acad Sci USA 88: 7425–7429.

Wright EM. (1972) Mechanisms of ion transport across the choroid plexus. J. Physiol 226: 545–571.

Wright EM. (1978) Transport processes in the formation of the cerebrospinal fluid. Rev Physiol Pharmacol 83: 1–34

Wu Q, Delpire E, Hebert SC, Strange K. (1998) Functional demonstration of Na^+-K^+-$2Cl^-$ cotransporter activity in isolated, polarized choroid plexus cells. Am J Physiol 275: C1565–C1572.

Yoshitomi K, Burckhardt BC, Fromter E. (1985) Rheogenic sodium-bicarbonate cotransport in the peritubular cell membrane of rat renal proximal tubule. Pflugers Arch 405: 360–366.

Zeuthen T. (1994) Cotransport of K^+, Cl^- and H_2O by membrane proteins from choroid plexus epithelium of *Necturus maculosus.* J Physiol 478: 203–219.

Zeuthen T, Wright EM. (1981) Epithelial potassium transport: tracer and electrophysiological studies in choroid plexus. J Membrane Biol 60: 105–128.

Zlokovic BV, Makic JB, Wang L, McComb JG, McDonough A. (1993) Differential expression of Na,K-ATPase α and β subunit isoforms at the blood-brain barrier and the choroid plexus. J Biol Chem 268: 8019–8025.

7 The Molecular Basis of Xenobiotic Transport and Metabolism in Choroid Plexus

David S. Miller
National Institute of Environmental Health Sciences
National Institutes of Health
Research Triangle Park, North Carolina, U.S.A.

Simon Lowes
AstraZeneca Pharmaceuticals
Mereside, Alderley Park,
Cheshire, United Kingdom

John B. Pritchard
National Institute of Environmental Health Sciences
National Institutes of Health
Research Triangle Park, North Carolina, U.S.A.

CONTENTS

The cells of the central nervous system (CNS) are particularly sensitive to chemical injury and, thus, require a protected and highly regulated extracellular environment. This is accomplished largely by two selectively permeable barrier tissues: (1) the choroid plexus, an epithelium that separates blood and cerebrospinal fluid (CSF), and (2) the brain vasculature, which contains a large proportion of nonfenestrated capillaries that limit exchange of solutes between blood and brain (blood–brain barrier, BBB). Through restricted entry, specific efflux, and metabolic conversion, these tissues protect the CNS from potentially toxic waste products of normal metabolism and foreign chemicals (xenobiotics). However, these mechanisms do not distinguish between potentially toxic chemicals and therapeutic drugs. As a consequence, it has been extremely difficult to treat certain diseases of the CNS, e.g., cancer and bacterial and viral infections, since peripherally effective drugs penetrate poorly into the CNS. In addition, competition for transport and metabolism may underlie drug-drug and drug-metabolite interactions that alter therapeutic effectiveness. Therefore, understanding the mechanisms that underlie the blood–brain and blood–CSF barriers is an important step in devising strategies to target therapeutic drugs to the CNS while at the same time limiting entry of toxic chemicals and preserving an optimal extracellular environment.

The present chapter is focused on the cellular and molecular mechanisms that underlie the ability of the choroid plexus to (1) transport potentially toxic xenobiotics from the cerebrospinal fluid (CSF) into blood for eventual elimination through urine and bile and (2) metabolize xenobiotics to modify their pharmacological and toxicological properties. Our intent is to review the functional properties of transporters and drug-metabolizing enzymes known to be expressed in the tissue, attempt to reconcile transport phenomenology with the molecular machinery, propose a model for coordinated regulation of transport and metabolism, and suggest problems toward which to channel future research efforts.

7.1 XENOBIOTIC TRANSPORT ACROSS THE CHOROIDAL EPITHELIUM

The capacity of the choroid plexus to transport xenobiotics is evident from in vivo experiments in which xenobiotics injected into the ventricles of test animals are rapidly cleared to the blood and in vitro experiments in which isolated tissue rapidly accumulates and concentrates radiolabeled xenobiotics added to artificial CSF (aCSF) bathing the ventricular surface (Pritchard and Miller, 1993; Lee et al., 2001; Ghersi-Egea and Strazielle, 2002; Graff and Pollack, 2004). To traverse the choroid plexus epithelium, xenobiotics must cross the plasma membrane at one side of the cell, move through the cytoplasm, and exit across the plasma membrane at the other side. The first and last steps, transport across functionally and morphologically distinct regions of the plasma membrane, may involve simple diffusion, but solutes often receive help from one or more specialized carrier proteins (transporters). Equilibrative transport across the membrane, whether by simple or facilitated diffusion, is driven solely by the solute's electrochemical gradient (concentration gradient plus electrical potential difference, PD). When transport across the membrane is energetically uphill, it must be coupled to a source of potential energy, e.g., ATP splitting or ion or electrical gradients. Solute movement through the cell interior may involve simple diffusion through an aqueous cytoplasm, but binding to cytoplasmic proteins or organelle surfaces, sequestration within organelles and biotransformation to metabolites with altered transport properties may occur. Thus, both plasma membrane and intracellular elements influence the path and time of transit through the epithelium.

The three steps responsible for transepithelial transport can be visualized in intact, living choroid plexus using fluorescent xenobiotics and confocal microscopy. Figure 7.1 shows bright-field and corresponding confocal images of tissue incubated to steady state in medium containing the organic anion fluorescein (FL). At low magnification, two step-wise changes in fluorescence intensity are seen: one at the (CSF side) apical membrane of the epithelial cells and the other at the basolateral membrane (blood side) (see Figures 7.1A and 7.1B). Both steps represent carrier-mediated processes, since at each membrane, transport is saturable and inhibited by other organic anions (Breen et al., 2002). Both steps are also concentrative. That is, the FL concentration in the subepithelial/vascular spaces is about 3-fold greater than in the epithelial cells, which in turn is 3- to 5-fold greater than in the medium on the CSF side of the tissue. Thus, for FL, the epithelium generates at least a 10-fold concentration gradient from aCSF to blood. Breen et al. (2002) have demonstrated that both uphill steps are energetically coupled to cellular metabolism. Apical uptake is indirectly coupled to the Na gradient, whereas basolateral efflux is driven by the PD across the membrane. At higher magnification, it becomes clear that FL is not uniformly distributed in the epithelial cells (see Figures 7.1C and 7.1D). Rather, the compound appears to be both diffusely distributed in the cytoplasm and to accumulate in vesicular structures. It is not, at present, certain how vesicular accumulation influences transcellular transport of organic anions in the choroid plexus (Breen et al., 2002), although a case has been made for the involvement of vesicular transcytosis in the transport of organic cations (Miller et al., 1999) (see below).

7.2 TRANSPORTERS OF THE CHOROID PLEXUS

Two main features determine the efficacy of drug fluxes across the choroidal epithelium, the transporters present (discussed below), and the polarized distribution

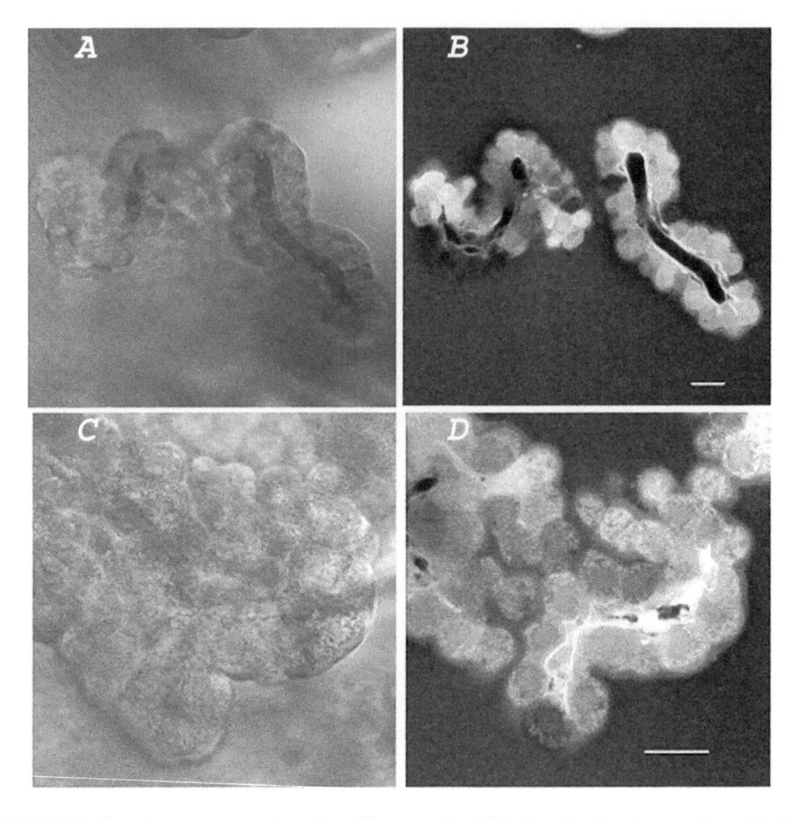

FIGURE 7.1 Steady-state organic anion (fluorescein, FL) distribution in rat choroid plexus. Shown are transmitted light micrographs (A and C) and corresponding confocal sections (B and D). The pattern of fluorescence indicates a two-step mechanism of transepithelial transport. Both steps, uptake at the apical membrane and efflux at the basolateral membrane, are concentrative. Note the high fluorescence levels in the subepithelial and vascular spaces and the lack of fluorescence in the trapped red blood cells, which exclude FL. At high magnification (D), punctate sites of FL accumulation are visible in the cytoplasm of the epithelial cells, suggesting vesicular sequestration. The white bars indicate 20 μm.

of those transporters. The choroid plexus is atypical in its polarization. For example, in kidney or liver, excretory transport moves solutes from the basolateral, or blood, side of the epithelium to the apical, or luminal, side. Consistent with this functional polarization, transporters including Na,K-ATPase as well as drug excretory transporters, e.g., OAT1 and 3 and OCT2 (the transporters mediating the entry step in organic anion and cation secretion), are present in the basolateral membrane (Sweet et al., 1999a, 1999b). Efflux transporters, e.g., p-glycoprotein and the multidrug resistance-associated protein 2 (MRP2), are expressed in the apical membrane (Kruh and Belinsky, 2003; Haimeur et al., 2004). The same basolateral to apical pattern is seen in the capillary endothelium that makes up the BBB (Sun et al., 2003). However, consistent with the direction of transport in the choroid plexus—CSF to blood rather than blood to urine or bile—this pattern is reversed for many transporters. Thus, Na,K-ATPase is an apical protein (CSF side), and many of the drug transporters expressed basolaterally in kidney and liver are apical in choroid plexus (Ernst et al., 1986; Pritchard et al., 1999; Sweet et al., 2001). As a result, we must consider not only the molecular properties of the transporters (e.g., substrate specificity, driving forces), but also their pattern of distribution of this polarized epithelium before we can begin to draw firm conclusions about the potential of a given transporter to mediate elimination or uptake of a drug from the brain (see Figure 7.2).

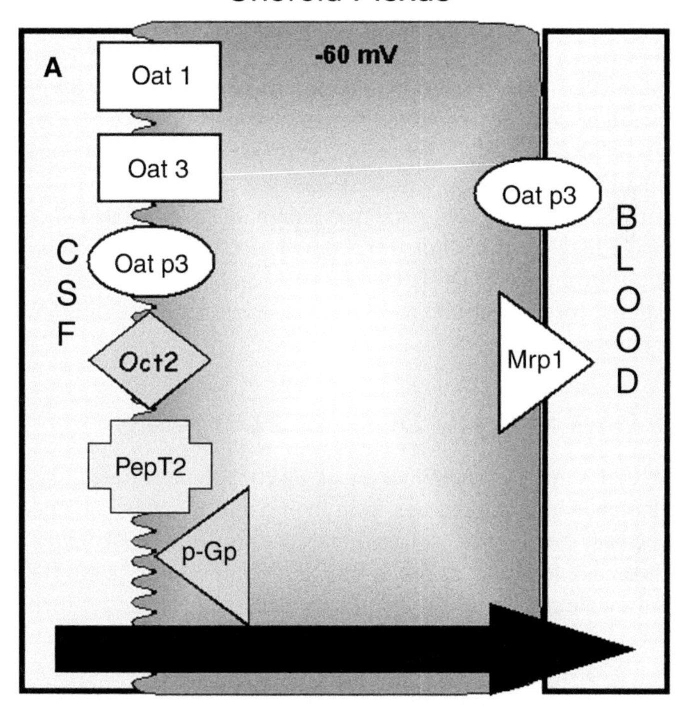

B Other Transporters Expressed (mRNA):

High/Moderate Levels: Oat2, OctN1, OctN2, Cnt1, Cnt2, Ent1, Mrp4, Mrp4, Mrp5,

Low Levels: Oct1, Oct2, Oatp1, Oatp4, Oatp5, Oatp12, Oat-K1/2, Pept1, Ent2, Mrp2, Mrp3

FIGURE 7.2 Drug transporters in the choroid plexus. (A) Transporters with known subcellular locations, based on specific immunostaining for transporter protein (see text). P-Glycoprotein, although localized to the apical side of choroid plexus epithelial cells, seems to be found mainly in an intracellular compartment (Rao et al., 1999; D. S. Miller et al., unpublished observations). It is not clear whether this is a storage form of the transporter, waiting for the proper signal for recruitment to the plasma membrane; alternatively, this transporter may contribute to active pumping of xenobiotics into vesicles. (B) Additional transporters for which mRNA has been detected in rat choroid plexus (Choudhuri et al., 2003).

In choroid plexus, a variety of drug transport proteins is poised to modulate the entrance into or efflux from the CNS. Chief among these are the members of the solute carrier 22 family, which includes the organic cation transporters (OCTs; SLC22A1-5) and the organic anion transporters (OATs; SLC22A6-12). Members of the organic anion transporting polypeptide family (OATPs; SLC21) also play prominent roles. The nucleoside and nucleotide transporters (SLC28 and SLC29 families) and the polypeptide transporters (SLC15 family) may also contribute to both drug entry and elimination from the brain/CSF. Finally, the ATP-driven drug pumps of the ABC families (the ABC-Bs, the p-glycoprotein group [p-gp] for cationic and neutral compounds, and the ABC-Cs, the multidrug resistance-associated protein group [MRPs] for anionic drugs and metabolites) certainly contribute to the effectiveness of the BBB at the cerebral capillary endothelium, but as discussed in more detail below, their contribution at the blood–CSF barrier (BCSFB) is less clear. In the sections below, each of these groups will be examined individually, with the central focus on where transporters are expressed and how they work. In addition, the substrates handled by each group are summarized in Figures 7.3 and 7.4; for more extensive lists see Russel et al. (2002). These lists emphasize both the multispecific nature of the OAT, OCT, OATP, MRP, and MDR families of transporters and the potential for competition for transport as one basis for drug-drug and drug-metabolite interactions.

7.3 THE OATS AND THE OCTS (SLC22A FAMILY)

The early studies of Pappenheimer et al. (1961) demonstrated the probenecid-sensitive clearance of anionic drugs from the CSF of the intact animal. Subsequent studies indicated that anionic solutes ranging from neurotransmitter metabolites, e.g., 5-hydroxyindole acteic acid (HIAA) and homovanillic acid (HVA), to drugs including salicylate and penicillin and to anionic herbicides like 2,4-dichlorophenoxy acetic acid (2,4-D) were all actively taken up from the CSF (Cserr and Van Dyke, 1971; Forn, 1972; Lorenzo and Spector, 1973; Spector and Lorenzo, 1974; Pritchard, 1980). Similarly, cationic molecules like the radiocontrast agent, iodipamide, were

ABC Transporters: Highly multispecific, ATP-driven, drug export pumps.

p-Glycoprotein transports organic cations, weak organic bases, some uncharged drugs and some polypeptide derivatives – chemotherapeutics, Ca channel blockers, digoxins, cyclosporins, HIV protease inhibitors

Mrps transport organic anions, weak organic acids, some uncharged drugs, some polypeptide derivatives and products of Phase II drug metabolism – MTX, statins, endothelin, cyclic nucleotides (AZT, PMEA; Mrp4, Mrp5), cephalosporins, GSH, glucuronide and sulfate conjugates

FIGURE 7.3 Specificity characteristics of ABC family transporters expressed in choroid plexus.

Organic Anion Transporters (OATs; SLC22A): Multispecific, organic anion transporters/exchangers; Oat1 and Oat3 are coupled to Na

NSAIDS, fluoroquinalones, β-lactam antibiotics, nucleoside phosphonates (Oat1), PAH (Oat1 and Oat3), AZT, prostaglandins, eicosanoids, statins, cimetidine (Oat3), MTX, thyroxine, GSH, glucuronide and sulfate conjugates

Organic Anion Transporting Polypeptides (OATPs; SLC21): Multispecific, organic anion transporters/exchangers; not Na-coupled

MTX, thyroxines, digoxin (Oatp2), bile salts, peptides, eicosanoids, statins, GSH, glucuronide and sulfate conjugates of hormones

Organic Cation Transporters (OCTs; SLC22A): Multispecific, organic cation transporters/exchangers.

Oct1,2,3 - neurotransmitter derivatives, cimetidine, amantidine, Ca channel blockers, quinacrine, quinine

OctN1,N2 - carnitine and some tetraalkylammoniums

Nucleoside transporters (SLC28A, SLC29A): ENT1-2 exchangers; CNT1-3 Na coupled cotransporters

5dFU, fluorouridine, gemcitabine, floxidine, didanosine, AZT

Peptide transporters (Pept; SLC15A): H^+ coupled cotransporters

di- and tripeptides, β-lactam antibiotics, valacyclovir, ACE inhibitors, bestatin

FIGURE 7.4 Specificity characteristics of SLC family transporters expressed in choroid plexus.

also shown to be transported by the choroid plexus (Barany, 1972, 1973). Nevertheless, decades would pass before specific transport proteins capable of mediating these functional properties were identified and cloned (reviewed in Sweet and Pritchard, 1999). Soon thereafter, messages encoding four OATs (1–4) and two OCTs (2 and 3) were detected for the first time in the rat choroid plexus by PCR (Pritchard et al., 1999; Sweet et al., 2001). The following sections examine the properties and functional roles of these transporters.

7.3.1 Organic Anion Transporters (OATs)

Functional studies using intact rat plexus in vitro confirmed the presence of organic anion/dicarboxylate exchange in the apical membrane of this tissue (Pritchard et al., 1999). Since both OAT1 (SLC22A6) and OAT3 (SLC22A8) are now known to exchange anionic substrates for intracellular dicarboxylates and to be capable energetically of coupling organic anion uptake to the Na gradient (Sekine et al., 1997; Sweet et al., 1997, 2003), one or both are positioned to mediate concentrative uptake of organic anions from the CSF in exchange for intracellular α-ketoglutarate. These observations have been confirmed in separate studies of rat and human choroid plexus in both immuno-histochemical and functional studies (Nagata et al., 2002; Alebouyeh et al., 2003). However, recent studies of the choroid plexus from OAT3 knockout (KO) mice indicate that additional complexities remain to be unraveled (Sykes et al., 2004). On the one hand, the KO studies show that mOAT3 is responsible for essentially all of the choroid plexus transport of certain small organic anions, e.g., FL and *p*-aminohippurate (PAH), with little if any contribution by mOAT1. As one would expect, based on experiments with the cloned transporter, mediated transport in the intact tissue was blocked when bath Na was removed or when tissue was exposed to estrone sulfate (ES) or PAH, conditions that functionally define OAT3 action. This result is consistent with the recent quantitative PCR studies of Choudhuri et al. (2003), which indicate far greater absolute expression of OAT3 message than of OAT1 in rat plexus. They are also consistent with data from rat choroid plexus showing that essentially all transport of both PAH and FL is mediated by OAT3 (Breen et al., 2002; S. Lowes et al., unpublished data).

On the other hand, for ES, a well-documented OAT3 substrate that is not transported by OAT1, only about 30% of uptake was lost in choroid plexus tissue from OAT3 KO mice (Sykes et al., 2004). At least two additional pathways for apical uptake were revealed: Na-dependent and Na-independent. The identities of these additional transporters are as yet unresolved, although as discussed below, one or more of the OATPs, probably OATP3 (Ohtsuki et al., 2003) is most likely responsible for the Na-independent component. Note that in rat choroid plexus the kinetics of ES uptake is described by a single saturable component (Kitazawa et al., 2000). In spite of this, our recent experiments indicate that there are at least two mediated components of ES uptake in rat choroid plexus. As in the mouse, less than half of mediated uptake is both Na-dependent and inhibited by saturating concentrations of PAH, indicating uptake on OAT3 (S. Lowes et al., unpublished data). It is likely that at least a portion of residual ES transport in rat choroid plexus may be mediated by OATP3.

ES is not the only substrate for which uptake at the apical membrane involves a Na-dependent component that is not mediated by OAT3 or OAT1. First, uptake of taurocholate (a substrate for OAT3 and OATPs) (see Figure 7.3) is not reduced in choroid plexus from OAT3-null mice, nor is taurocholate uptake inhibited by PAH (Sykes et al., 2004). Uptake is, however, partially Na-dependent. Recent experiments in rat choroid plexus also show no evidence that taurocholate uptake is inhibited by PAH even though it is partly Na-dependent (S. Lowes et al., unpublished data). Second, apical uptake of a fluorescent methotrexate derivative (FL-MTX) appears to be Na-dependent in choroid plexus from rat and mouse, but FL-MTX transport is not reduced in tissue from the OAT3-null mouse (Sweet et al., 2002; Breen et al., 2004). No candidate for the residual Na-dependent component of organic anion uptake has, thus far, been identified in choroid plexus. However, SOAT, a recently cloned member of the SCL10 family of bile salt transporters, was shown to be expressed in brain and to mediate Na-dependent ES transport with modest affinity (K_m ~ 30 μM) (Geyer et al., 2004). If expressed in choroid plexus, it could account for the Na-dependent ES transport. Of the remaining transporters shown by PCR to be present in choroid plexus (Pritchard et al., 1999), it is reasonable to hypothesize that OAT2 (SLC22A7), which is basolateral in liver and in human kidney (Sekine et al., 1998; Enomoto et al., 2002), might, by analogy with OATs 1 and 3, also be apical in the plexus and account for a portion of this transport. At present, the functional properties of this transporter are poorly understood, and there has been no demonstration of its expression in choroid plexus.

The nature of the efflux step for small organic anions from choroid plexus cell to the blood is not well understood, and the molecular identity of the transporter(s) involved is uncertain. The imaging studies of Breen et al. (2002) demonstrate that the efflux step for the small organic anion FL is potential sensitive and carrier mediated, but at present none of the cloned OATs appear to match with this functional fingerprint. This may simply reflect the lack of choroid plexus data on subcellular distribution and functional roles, if any, of the remaining OATs expressed in choroid plexus, OAT2 and OAT4 (SLC22A11) (Choudhuri et al., 2003). In fact, given the reversed polarity of transporter expression in the choroid plexus relative to the kidney, OAT4, which is apical in kidney (Babu et al., 2002; Ekaratanawong et al., 2004), and possibly OAT2, which is apical in the distal segments of the rat tubule (Kojima et al., 2002) (although it is basolateral in the proximal tubule of the human [Enomoto et al., 2002], as discussed above) might well be present in the basolateral membrane of the plexus. Another possibility is that the basolateral transporter is the rodent homolog of OATv1, a novel organic anion transporter from pig renal cortex recently cloned by Jutabha et al. (2003). This PD-driven transporter handles PAH and a number of other organic anions. It has been localized to the apical membrane of renal proximal tubule cells where it could mediate PD-driven organic anion efflux into the tubular lumen. Although the message for this transporter could not be detected by Northern blot in whole brain, expression in specific regions of the brain, e.g., choroid plexus, was not examined. If OATv1 or a similar transporter were expressed in the basolateral membrane of the choroid plexus epithelial cells, it could mediate PD-driven organic anion efflux. Clearly, until the appropriate expression and localization studies are performed, firm conclusions on the identity of a basolateral OAT in choroid plexus are not possible.

7.3.2 Organic Cation Transporters (OCTs)

Far less is known about the molecular basis of organic cation transport in choroid plexus. It is likely that uptake at the apical membrane is mediated by a PD-driven transporter and that organic cations are sequestered in a vesicular compartment within the cells (Miller et al., 1999; Villalobos et al., 1999). At present we know nothing about how basolateral efflux is accomplished, although evidence has been presented that specific vesicular transport could contribute to transepithelial flux (Miller et al., 1999) (see Figure 7.5). All five of the organic cation-transporting SCL22A members, OCT1, OCT2, OCT3, OCTN1, and OCTN2, are expressed in rat choroid plexus at some level, but quantitative PCR indicates that both OCT1 and

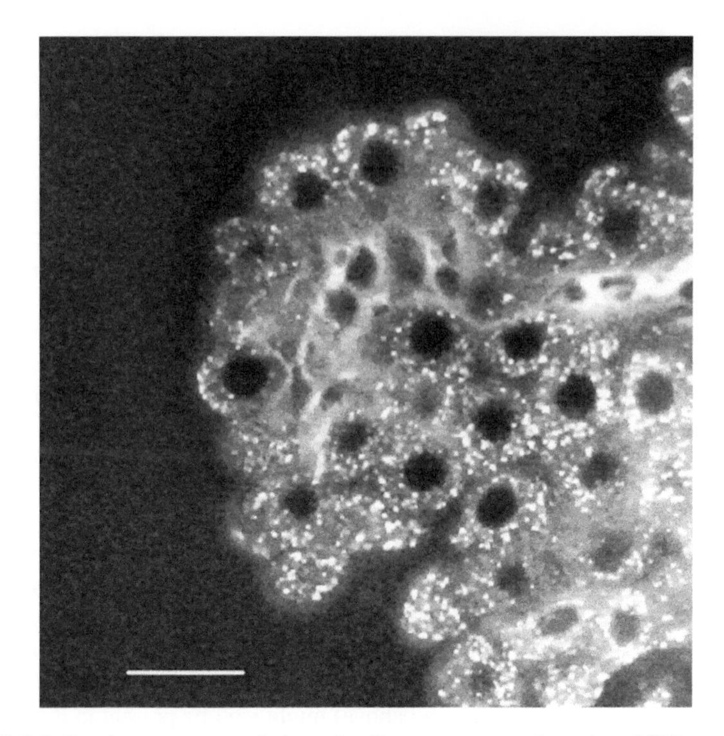

FIGURE 7.5 Steady-state accumulation of a fluorescent organic cation (NBD-tetramethylammonium) in rat choroid plexus. The image shows concentrative transepithelial transport. Measured blood vessel-to-medium fluorescence ratios are as high as 50:1. Cellular accumulation is dominated by intensely fluorescent vesicular structures in the cytoplasm; nuclei are devoid of these structures and exhibit low fluorescence. Red blood cells within blood vessels also exhibit low fluorescence. The white bar indicates 20 μm.

OCT2 are expressed at low levels (Sweet et al., 2001; Choudhuri et al., 2003). The subcellular distribution of the potential driven OCTs (1, 2, and 3) have only been assessed for OCT2. Based on expression of an OCT2-GFP construct, OCT2 appears to be apically expressed in choroid plexus (Sweet et al., 2001). Thus, it could mediate uptake of cationic drugs and metabolites from the CSF. Since this transporter is basolateral in kidney, like OATs 1 and 3, it is quite likely, but unproven, that the polarity of the choroid plexus is reversed for the potential-driven OCTs (1–3) and that all three are apical there. Thus, they would be poised to mediate clearance of positively charged drugs and metabolites from the CSF. Functional studies have only been done for OCT2. This data indicates that OCT2 is able to mediate the clearance of choline from the CFS (Sweet et al., 2001). However, the contribution of other, as yet unidentified, apical organic cation transporters to the transport of choline and of cationic drugs remains likely. OCTN1 is proposed to be apical in kidney based primarily on its function as a proton/organic cation exchanger (Koepsell and Endou, 2004), and OCTN2 is apical in renal proximal tubule and BBB endothelium (Kido et al., 2001; Tamai et al., 2001; Inano et al., 2003). However, there are no data on their subcellular distribution or their functional role(s), if any, in the choroid plexus.

Nevertheless, expression levels for both OCTNs are high in choroid plexus, much greater than either OCT1 or OCT2 (Choudhuri et al., 2003).

Two aspects of organic cation transport in choroid plexus deserve further discussion. First, the weak organic base, cimetidine, is a substrate for cloned OCT2 and OAT3 (Grundemann et al., 1999; Kusuhara et al., 1999). In renal proximal tubule, which expresses both transporters, it appears to be transported by an OCT (likely OCT2) (Boom and Russel, 1993). However, in choroid plexus cimetidine uptake is inhibited by organic anions, not by organic cations, indicating that its transport is mediated by an organic anion transporter (likely OAT3) (Suzuki et al., 1986; Whittico et al., 1990; Nagata et al., 2002). Second, confocal imaging of the transport of a fluorescent organic cation in rat choroid plexus shows the expected concentrative transport from a CSF to blood, but much of the cellular accumulation of the compound appears to be within a vesicular compartment (see Figure 7.4). This image graphically highlights the possibility that vesicular trafficking could contribute to the transcellular movement of organic cations across the epithelium, as suggested previously by Miller et al. (1999). These authors showed that quinacrine accumulated in vesicles within choroid plexus epithelial cells and that those vesicles moved predominantly in the apical-to-basal direction by a microtubule-dependent mechanism. In addition, they provided evidence for fusion of quinacrine-loaded vesicles with the basolateral plasma membrane, dumping their contents into the subepithelial space.

7.4 THE ORGANIC ANION TRANSPORTING POLYPEPTIDES (OATPS, SLC21 FAMILY)

The second major group of drug-transporting proteins is the OATPs. These transporters can be roughly divided into two groups: those that appear to have an excretory or barrier function, i.e., show broad substrate specificity and are expressed principally in liver, kidney and/or barrier tissues, and those with a more limited set of substrates that may have a more specific function, e.g., the prostaglandin transporter hPGT (SCL21A2, or OATP2A1) (Hagenbuch and Meier, 2003, 2004). More than 50 OATPs have now been cloned (Hagenbuch and Meier, 2004). Of these, none has been identified in human choroid plexus, and only a few are known to be expressed in choroid plexus of other species. An additional complication is that there is no direct one-to-one correspondence between the human and rodent forms, making extrapolation of results from model species to humans most uncertain. This noncongruence has also led to an extremely convoluted nomenclature, and in current literature, common names, solute carrier family names (SLC21), and the new HUGO nomenclature (SLCO), which is based on evolutionary relationships and is divided into families and subfamilies, all appear (Hagenbuch and Meier, 2003). Since only a few of the OATPs have been definitively demonstrated in choroid plexus, all three of the current nomenclatures will be used to identify each OATP transporter discussed below. Of course, because of limited data, few firm conclusions may be drawn about the contributions of the OATPs to drug transport by the choroid plexus, particularly in the human.

Choudhuri et al. (2003) used quantitative PCR to assess the expression of 8 OATPs in rat choroid plexus. They showed that messages for four of these were virtually absent in the plexus—OATP1 (SLC21a1, or OATP1a1), OATP4 (SLC21a10, or OATP1b2), OATP5 (SLC21a13, or OATP1a6), and OAT-K (SLC21a4, or OATP1a4). Of the remaining OATPs, OATP3 (SLC21a7, or OATP1a5) was highly expressed at levels far exceeding those in liver or kidney. OATP3 has been shown by immunohistochemistry to be expressed in the apical membrane of the rat choroid plexus, where it is thought to participate in the efflux of anionic substrates like ES from the CSF (Kusuhara et al., 2003; Ohtsuki et al., 2003; Hagenbuch and Meier, 2004). OATP2 has also been detected in choroid plexus by quantitative PCR (Choudhuri et al., 2003), and immunohistochemistry indicates that it is basolateral in expression (Gao et al., 1999). OATP2 is unique among organic anion transporters because of its high affinity for the cardiac glycoside digoxin, and recent experiments in rat choroid plexus show that basolateral efflux of FL-MTX is blocked by low concentrations of digoxin, implicating OATP2 (Breen et al., 2004). The subcellular distributions and, thus, the functional roles of the other two OATPs detected by quantitative PCR, OATP9 (SLC21a9, or OATP2b1) and OATP12 (SLC21a12, or OATP4a1) (Choudhuri et al., 2003), are unknown.

7.5 THE NUCLEOSIDE AND NUCLEOTIDE TRANSPORTERS (SCL28 AND SLC29 FAMILIES)

There are three members of the concentrative nucleoside transporter family, CNT1-3 (SCL28A1-3) (Gray et al., 2004) and four members of the equilibrative nucleoside transporter family, ENT1-4 (SLC29A1-4) (Baldwin et al., 2004). As implied by their names, the CNTs are capable of sodium-driven, concentrative cellular uptake of purine and pyrimidine nucleosides and their analogs. CNT1 prefers pyrimidines, CNT2 transports purines, and CNT3 handles both. In polarized epithelia, the CNTs are apically expressed. On the other hand, the ENTs mediate facilitative diffusion and will not support energetically uphill transport (Baldwin et al., 2004). All four ENTs are widely distributed in various cell types, and ENT1 and 2 are known to be basolateral in the polarized epithelium of the kidney. Thus, it appears that in combination the apical CNTs and the basolateral ENTs are organized to mediate absorptive transport (toward the blood) of nucleobases, nucleosides, and their therapeutic analogs.

Of the members of these two families, expression of four (CNTs 1 and 2 and ENTs 1 and 2) have been demonstrated in choroid plexus by quantitative PCR (Choudhuri et al., 2003). CNTs 1 and 2 and ENT1 were found to be highly expressed, i.e., at or above the highest levels in the reference organs (liver, kidney, and ileum) tested concurrently. ENT2 was detected at more modest levels. Unfortunately, the cellular localization of these transporters in choroid plexus is unknown, so their possible roles in entry or efflux of nucleoside analog drugs must remain speculative. However, their distribution in other epithelia does suggest some possibilities. A polar distribution similar to kidney (apical CNTs and basolateral ENTs) should mediate efflux of nucleosides from the CSF and contribute to the barrier function of the choroid plexus, whereas a reversal of polarity, as seen for Na,K-ATPase and a number of the drug transporters discussed above, could make the plexus a portal for facilitated entry of nucleoside drugs into the CNS.

7.6 THE PEPTIDE TRANSPORTERS (SLC15 FAMILY)

This family consists of four members. All four are proton-coupled cotransporters that transport di- and tri-peptides in addition to peptidomimetric drugs and prodrugs (Rubio-Aliaga and Daniel, 2002; Daniel and Kottra, 2004). PEPT1 and PEPT2 (SLC15A1 and A2) are predominantly expressed in polarized epithelia, where they mediate the apical uptake step in peptide absorption or reabsorption. For most substrates PEPT1 is a high-capacity, low-affinity system, whereas PEPT2 generally has a lower capacity and higher affinity. In intestine and kidney, they mediate the effective removal of peptide substrates from the lumen to the blood. Real-time PCR indicates that both PEPT1 and PEPT2 are expressed by the choroid plexus, with PEPT2 levels being much greater. Two PEPT2 knockout mice have been generated (Rubio-Aliaga et al., 2003; Ocheltree et al., 2004). Both mice are generally healthy, but do demonstrate reduced transport of model peptides. No compensatory increase in PEPT1 expression was seen in the KO animals. Ocheltree et al. (2004) have focused on the impact of PEPT2 loss on choroid plexus function. Their data indicates that PEPT2 is expressed in the apical membrane of the choroid plexus in the wild type, where it mediates uptake of peptides and peptidomimetric drugs from the CSF. Upon loss of PEPT2, uptake of the cephalosporin antibiotic cefadroxil was greatly reduced. Further reduction of cefadroxil uptake in choroid plexus was produced by the organic anion PAH, indicating that OAT family transporters (probably OAT3, see above) accounted for the remaining mediated cefadroxil uptake seen in the PEPT2 knockout mice. This study suggests that 80 to 85% of the cefadroxil uptake by the wild-type mouse is mediated by PEPT2, 10 to 15% via OATs, and 5% by nonspecific mechanisms. These data illustrate the important point that a given drug may be transported by more than one of the transporters expressed in the choroid plexus and that the relative proportions mediated by a given system may vary depending on transporter expression levels, the kinetics of transport, and other drugs (competitors) present during administration of the drug of interest.

There is much less information available on the PHT transporters (PHT1 [SLC15A4] and PHT2 [SLC15A3]). Both transport histidine in addition to peptides in a proton-dependent manner (Yamashita et al., 1997; Sakata et al., 2001; Rubio-Aliaga and Daniel, 2002). PHT1 is known to be expressed in many regions of the brain, and its message is expressed at high levels in choroid plexus.

7.7 THE ATP-DRIVEN DRUG PUMPS (ABC-B AND ABC-C FAMILIES)

7.7.1 THE P-GLYCOPROTEINS (ABC-B FAMILY)

Although originally discovered as a result of their ability to confer resistance to anticancer drugs in mammalian tumor cells, it is now apparent that members of this group, MDR1 and MDR3 in human and Mdr1a/1b and Mdr2 in rodents, play physiological roles as well (Schinkel, 1999; Hagenbuch et al., 2002). In both rodents and humans, Mdr1/MDR1 transport drugs, xenobiotics, and metabolites at barrier and excretory sites (Schinkel, 1997; Jonker and Schinkel, 2004), whereas Mdr2/MDR2/3 are primarily responsible for lipid transport (Borst et al., 2000; Fuchs and Stange, 2001). Thus, in brain capillaries, MDR1/Mdr1a/1b are important in both speeding the elimination of cationic drugs and metabolites from the brain and in preventing their entry from the blood into the brain (see Figure 7.4). In keeping with this function, they are expressed on the luminal, or blood, side of the capillary endothelium (Schinkel, 1999; Hagenbuch et al., 2002). In contrast to the BBB, as first demonstrated by Rao et al. (1999) and discussed recently by Sun et al. (2003), Mdr1 is expressed apically or subapically in the choroid plexus. Thus, it is positioned to mediate ATP-driven entry into the CSF and the brain, rather than efflux from them. Understanding of this apparent anomaly awaits further study.

7.7.2 The Multidrug Resistance-Associated Proteins (MRPs, ABC-C Family)

Whereas the MDR transporters mediate the ATP-driven efflux of lipophilic cationic and neutral compounds from the cells expressing them, the MRPs handle predominantly anionic substrates (see Figure 7.4). Evidence from knockout mice indicates that MRP1 is expressed in the basolateral membrane of the choroid plexus, where it greatly enhances cell-to-blood flux of its substrates (Wijnholds et al., 2000; Breen et al., 2004) and is thus poised to reduce CSF and brain concentration of these agents. This is the only MRP with known subcellular distribution in the choroid plexus. MRP2 is well established as an efflux transporter at the apical membrane of the proximal tubule, the canalicular membrane of the hepatocyte, and the capillary endothelium of the brain (Miller et al., 2000; Hagenbuch et al., 2002). However, examination of MRP2 message expression in the plexus of rat (Choudhuri et al., 2003) and mouse (Lee et al., 2004) indicate that it is relatively low to nonexistent in the choroid plexus. On the other hand, these same studies indicate that mRNAs for MRPs 4, 5, and 6 are all present in the choroid plexus. The functional roles and subcellular localizations of these transporters in the plexus have yet to be established.

7.8 XENOBIOTIC METABOLISM

The principal site of xenobiotic metabolism in the body is the liver, which is well suited to the task, expressing a rich complement of enzymes and efflux transporters, which together detoxify compounds and eliminate the metabolites into bile. The liver is by far the best characterized organ with respect to its metabolizing capacity, but other tissues, such as the intestine and kidney, are recognized as important secondary sites of xenobiotic metabolism. In contrast, we know very little about xenobiotic metabolism in the barrier tissues of the brain. It is clear that extracts from choroid plexus exhibit many of the same enzymatic activities as the primary drug-metabolizing tissues, i.e., liver, kidney, and intestine, and that genes coding for phase I and phase II drug-metabolizing enzymes are expressed in choroid plexus (below).

However, at the molecular level, the list of enzymes expressed and their cellular locations is incomplete, and, at the tissue level, the contribution of the choroid plexus to overall xenobiotic metabolism in brain has not been assessed.

7.8.1 Phase I Metabolism

The overall process of xenobiotic detoxification comprises three broad phases, the first two involving enzymatic conversion to more hydrophilic metabolites and the last being excretory transport. Phase I metabolism is functionalization, a process that exposes an existing functional group within a molecule or introduces a new functional group, e.g., -OH, -NH_2, or -SH. This prepares the molecule for subsequent phase II metabolism (see below). Functionalization is achieved through several different types of enzyme-driven reactions, including oxidation, reduction, hydration, and hydrolysis, which in turn are carried out by a variety of metabolizing enzymes. Quantitatively, the most important set of reactions is oxidation. The oxidation reactions carried out by the microsomal mixed function oxidase (MFO) system are of particular importance in xenobiotic metabolism and are responsible for the modification of an extremely broad range of compounds. This is due in part to the extensive range of enzymes involved in this system and the broad, overlapping substrate specificities of these enzymes. The principal components of the MFO system are a cytochrome P450 isoform and a NADPH-cytochrome P450 reductase, together with molecular oxygen and NADPH. The cytochrome P450 (CYP) superfamily of enzymes is an extensive and continually growing group of hemoproteins expressed in bacteria to mammals (Nelson, 1999). There are at least 481 CYP genes and 22 pseudogenes recognized across species (Nelson et al., 1996). CYP enzymes show a diverse range of functions, from the generation of steroids and cholesterol to the metabolism of xenobiotics. CYP families 1 to 4 are the major subclasses involved with xenobiotic metabolism, whereas the higher CYPs are typically involved with endogenous biochemical reactions (Smith et al., 1998); CYP7A1 (cholesterol 7 α-hydroxylase), for example, is of central importance in bile acid metabolism and cholesterol homeostasis (Schwarz et al., 1998). The major CYP isozymes responsible for xenobiotic metabolism in humans include CYP1A2, CYP2B6, CYP2C9, CYP2C19, CYP2D6, and CYP3A4 (Lewis et al., 2001). Although the CYP isoforms expressed in humans and rodents often overlap, there are also several species-specific enzymes. In addition to the CYPs and MFOs, other enzymes that catalyze phase I metabolism include alcohol dehydrogenase (ADH), xanthine oxidase, and monoamine oxidase.

The most thoroughly studied CYP isoforms in choroid plexus are members of the CYP1A subfamily. CYP1 enzymes are involved in the metabolic activation of many mutagens and carcinogens (Gonzalez and Kimura, 2003). CYP1A1 is expressed mainly at extrahepatic sites, its substrates include caffeine and phenacetin, and it also activates carcinogenic polycyclic aromatic hydrocarbons such as benzo(*a*)pyrene (Smith et al., 1998). CYP1A2 on the other hand is expressed predominantly in hepatic tissue and accounts for between 10 and 15% of the CYP component of human liver. It is the major CYP responsible for the metabolism of caffeine, theophylline, imipramine, clozapine, phenacetin, and propranolol (Brosen, 1995; Smith et al., 1998). It is also involved in the conversion of heterocyclic amines to the carcinogenic and mutagenic forms, as well as the metabolism of endogenous compounds, such as the steroid hormone estradiol (Brosen, 1995; Lewis, 2003). Both rat and human isoforms of CYP1B1 can activate carcinogenic benzo(*a*)pyrene (Kim et al., 1998).

Morse et al. (1998) probed whole-brain microsomes for CYP1A1 protein expression by Western blotting and function as measured by ethoxyresorufin *O*-deethylase (EROD) activity. They found no CYP1A1 expression or EROD activity in control animals, but after treatment with β-naphthoflavone (BNF), a known CYP1A1 inducer (Zhang et al., 1997), low levels of CYP1A1 protein and EROD activity were detectable. However, prior removal of choroid plexus and arachnoid membranes abolished detectable protein and EROD activity despite BNF treatment, suggesting that one or both of these tissues were the source of CYP1A1. Further investigation confirmed inducible CYP1A1 expression and EROD activity in both the choroid plexus and arachnoid membrane, suggesting regulation by an aryl hydrocarbon receptor (Granberg et al., 2003). In contrast to CYP1A1, Morse et al. (1998) found that CYP1A2 protein was detectable in all areas of brain tested, irrespective of prior removal of choroid plexus or arachnoid membrane, though expression levels were considerably lower than those detected in liver (Morse et al., 1998). BNF treatment did not affect levels of CYP1A2. It is still unclear where precisely in the choroid plexus and arachnoid tissue CYP1A1 and CYP1A2 are expressed, though it appears that they are certainly located within the endothelial cells of the blood capillaries (Brittebo, 1994; Dey et al., 1999; Granberg et al., 2003). CYP1B1 has also been localized to the smooth muscle cells of the arterioles in choroid plexus tissue, but not to the capillary endothelial cells (Granberg et al., 2003).

There are two members of the CYP2B subfamily in humans, CYP2B6 and CYP2B7, the former being expressed in liver. CYP2B6 substrates include cyclophosphamide and ifosphamide (Chang et al., 1993), as well as 6-aminochrysene and nicotine (Mimura et al., 1993). The CYP3A family of enzymes plays a crucial role in xenobiotic metabolism. In particular, CYP3A4 is the most abundant CYP expressed in human liver and intestine, and it is estimated that 36% of all drugs undergo CYP3A4-mediated metabolism (Lewis, 2003). CYP3A4 substrates comprise a broad, structurally dissimilar range of compounds, including verapamil, erythromycin, etoposide, midazolam, and certain steroid hormones (Lewis, 2003). In choroid plexus, protein expression of CYP2B1/CYP2B2 in rats and CYP2C29 and CYP3A in mice is also reported (Volk et al., 1995; Rosenbrock et al., 2001).

Finally, other phase I enzyme components showing significant functional activities in choroid plexus are NADPH cytochrome P450 reductase and mEH, with levels comparable to those in the liver (Ghersi-Egea et al., 1994; Strazielle and Ghersi-Egea, 1999). Expression of alcohol deyhdrogenase (ADH) mRNA (class III but not classes I or IV) has also been reported (Galter et al., 2003).

7.8.2 Phase II Metabolism

Phase II enzymes conjugate Phase I metabolites and some parent xenobiotics to, for example, glucuronic acid, sulfate and glutathione, producing even less toxic, more polar compounds that are readily excreted in urine and bile. The UDP-glucuronosyl transferases catalyze covalent addition of glucuronic acid to an electron-rich atom (N, O, or S) by means of an amide, ester, or thiol bond. This pathway is quantitatively the most important conjugation route for xenobiotic and endogenous compounds including phenols, steroids, bile acids, and bilirubin (Bock et al., 1987). Glutathione *S*-transferases play numerous roles within cells, including protecting against oxidative stress, xenobiotic conjugation, and others not related directly to detoxification (Sheehan et al., 2001). They conjugate phase I metabolites and other electrophiles to glutathione. The resulting glutathione *S*-conjugates may also be excreted or further metabolized by γ-glutamyltranfserase and peptidases, which cleave glutamate and glycine, respectively, to yield cysteine *S*-conjugates.

Works on phase II xenobiotic-metabolizing enzymes in choroid plexus have focused primarily on UDP-glucuronosyl transferase and GST. UDP-glucuronosyl transferase activity was detected in choroid plexus (Ghersi-Egea et al., 1994; Strazielle and Ghersi-Egea, 1999), and the isoform responsible for conjugating 1-naphthol was found to be inducible, suggesting dynamic regulation of phase II enzymes in this tissue. Leininger-Muller et al. (1994) pre-treated rats with phenobarbital and 3-methylcholanthrene (3MC), both of which are known to induce different UDP-glucuronosyl transferase isoforms in liver, and measured the ability of tissue UDP-glucuronosyl transferase to conjugate 1-naphthol. 3MC treatment increased 1-naphthol conjugation in choroid plexus and liver, but not in cerebral cortex. Phenobarbital, on the other hand, had no effect on 1-napthol conjugation in any of these tissues (Leininger-Muller et al., 1994).

For GST, immunohistochemistry in rat choroid plexus confirmed the presence of various α-, μ-, and π-class isoforms (Johnson et al., 1993). There is also experimental evidence for functional activity of GSTs, γ-glutamyltransferase, and epoxide hydrolase in rat choroid plexus (Anderson et al., 1989; Ghersi-Egea et al.; Strazielle and Ghersi-Egea, 1999).

7.8.3 Metabolism Coupled to Excretory Transport

The possibility that phase II xenobiotic metabolism is functionally coupled to excretory transport was investigated by Strazielle and Ghersi-Egea (1999). They found that transport of the cytotoxic compound 1-naphthol across monolayers of choroidal epithelial cells was solely by simple diffusion. However, its glucuronide-conjugate, 1-naphthyl-β-*D*-glucuronide, which is produced by intracellular UDP-glucuronosyl transferase, is transported preferentially in the apical-to-basal (CSF-to-blood) direction, supporting active efflux into the blood. Efflux was temperature-sensitive, reversibly blocked by probenecid, an inhibitor of organic anion transport, and competitively inhibited by 2,4-dinitrophenyl *S*-glutathione (DNP-SG), an MRP substrate. Finally, when choroidal epithelial cells were incubated with 4-methylumbelliferone, another UDP-glucuronosyl transferase substrate, the glucuronide metabolite that was produced was preferentially transported into the basal compartment (Strazielle and Ghersi-Egea, 1999). At a minimum it is likely that metabolite efflux was mediated by basolateral MRP1. However, other isoforms (MRP4, 5, and 6) are expressed in the tissue (Choudhuri et al., 2003). Without information on their subcellular location, one cannot say whether they also could have mediated metabolite efflux.

7.8.4 Coordinate Regulation of Xenobiotic Metabolism and Transport

The central role of the liver and intestine in xenobiotic detoxification and excretion is reflected not only in their rich complement of transporters and enzymes, but also in their ability to regulate these components so that hepatocytes and enterocytes are able to adapt to changes in physiologic balance. To coordinate xenobiotic metabolism and transport with load, these cells are endowed with a regulatory network through which ligand-activated nuclear receptors transcriptionally regulate xenobiotic metabolizing enzymes and xenobiotic transporters. The principal xenobiotic receptors involved in such regulation are the farnesoid X receptor (FXR), pregnane X receptor (PXR; also known in humans as the steroid and xenobiotic receptor, SXR), liver X receptor (LXR), peroxisome proliferator-activated receptor (PPAR), and constitutive androstane receptor (CAR). Once activated by ligand binding, these receptors couple with their obligate partner, RXR. The heterodimer formed then binds to xenoboitic response elements on target genes, activating transcription (Lehmann et al., 1998).

PXR/SXR and CAR play a central role in defense against xenobiotics. PXR is activated by a variety of xenobiotics, including calcium channel blockers, HIV protease inhibitors, statins, and anticancer drugs, among many others (Handschin and Meyer, 2003). These include the classical CYP inducers rifampin in humans and pregnenolone 16α-carbonitrile (PCN) in rodents. CAR is activated by, for example, phenobarbital, chlorpromazine, and DDT. Both nuclear transcription factors are also activated by physiological ligands such as bile acids, steroid hormones, and androstane metabolites, so xenobiotic metabolism and transport are clearly altered pathophysiologically, e.g., during obstructive or drug-induced cholestasis (Francis et al., 2003; Trauner and Boyer, 2003).

Target genes for each of these ligand-activated receptors include numerous components of Phase I, Phase II, and Phase III detoxification. For example, in liver, PXR and CAR together regulate CYP isoforms from the 1A, 3A, and 2B families, GSTs, sulfotransferases, UDP-glucuronosyl transferases, p-glycoprotein, MRP1-3, and OATP2 (Handschin and Meyer, 2003; Honkakoski et al., 2003) As discussed above, several of these target enzymes and transporters are expressed in choroid plexus. The extent to which ligand-activated nuclear receptors modulate xenobiotic metabolism and transport in the tissue is unknown. However, as a first step in understanding this aspect of regulation, using RT-PCR, we have detected mRNA for PXR, LXR, RXR, and FXR in rat choroid plexus (D.S. Miller et al., unpublished results). The structure of the gene network regulated by these receptors in the choroid plexus and extent to which the expression of specific xenobiotic metabolizing enzymes and transporters are affected by receptor ligands remain to be determined.

7.9 PERSPECTIVES

It is evident from the above discussion that research on xenobiotic metabolism and transport in choroid plexus has entered the molecular age. We are indeed well along in identifying the pertinent genes expressed in the tissue and in understanding how the resulting gene products function in model systems and, to some extent, in native tissues. It is clear, however, that some important aspects of enzyme and transport function remain to be addressed. In closing, we would like to highlight two areas of research that appear critical for advancing the understanding of metabolism and transport in this tissue.

7.9.1 Matching Substrates to Transporters

One challenge of modern physiology is to match genes and their encoded proteins with events at the molecular, cellular, tissue, and organism levels, and thus provide a multilevel understanding of gene function and dysfunction. How well this can be done for xenobiotic transport and metabolism depends on a knowledge of the genes expressed in the tissue, the cellular locations of the gene products, and our ability to match substrates with transporters. Figure 7.2 summarizes current knowledge of the distribution of xenobiotic transporters in rat choroid plexus (we have no such inventory for other species). Note that of the nearly 30 transporters known to be expressed in the tissue (13 at high to moderate mRNA levels [Choudhuri et al., 2003]), only 8 can be placed with certainty on one side or the other of the epithelial cells. Moreover, as discussed above, it is not known whether several newly discovered transporters are expressed in choroid plexus. This lack of information about the players and their locations within the cells makes it difficult to assign substrates to specific metabolic and transport pathways. Uncertainty as to which proteins actually handle specific substrates in the intact tissue limits our ability to predict drug-drug and drug-metabolite interactions as well as the consequences of altered transporter or enzyme function due to regulation or mutations. What is needed is a full inventory of xenobiotic metabolizing enzymes and xenobiotic transporters expressed, some measure of relative levels of each protein, and, certainly for transporters, firm data on where the protein is located within the cells.

Improved tools to match substrates to their transporters in native tissues and in intact animals are also needed. At present, how well we can do this depends on the ability to implement two strategies, generally used in concert. One is to correlate xenobiotic transport and metabolism in cells, tissues, and organisms with the properties of cloned genes overexpressed in heterologous systems, e.g., *Xenopus laevis* oocytes or cell lines. This approach assumes that transporter and enzyme kinetics, specificity, and energetics in native tissues will be the same as in expression systems and that they will be sufficiently distinct so that the contribution of the transporter or enzyme to tissue function can be dissected out. This strategy is used widely, but it suffers from lack of certainty when applied to tissues that express multiple, polyspecific transporters and enzymes. The second strategy is to determine the consequences of naturally occurring or engineered mutations that cause loss or gain of function. Barring toxicity, lethality, and the complications of specific compensation, knockouts can provide simple answers to questions of gene function, as we have seen for OAT3, Pept2, and MRP1 (see above). However, generation of a knockout animal is expensive and time-consuming and at present very difficult in species other than mice. Newer approaches that may avoid many of these pitfalls include tissue-specific and conditional knockouts as well as knockouts and knockdowns using RNA-silencing technology.

7.9.2 Regulation of Transport and Metabolism

Xenobiotic transport and metabolism in excretory and barrier tissues appears to be dynamic, changing with altered gene expression, altered tissue environment, and disease. These processes are regulated at both the transcriptional and functional levels. Consider, for example, the OAT family of transporters. In kidney, expression of multiple family members is transcriptionally modulated by estrogen and androgen (Ljubojevic et al., 2004) and functionally modulated by hormones that activate the mitogen-activated protein kinase, protein kinase A, and protein kinase C intracellular signaling pathways (Dantzler and Wright, 2003). In contrast, we know little about how transport and metabolism in choroid plexus are regulated. The potential for transcriptional regulation is indicated by substrate-level induction of CYP1A isoforms presumably acting through an aryl hydrocarbon receptor (Granberg et al., 2003) and the detection in rat choroid plexus of mRNA coding for a number of other ligand-activated nuclear receptors (see above). Functional regulation may be the basis for reduced transport of organic anions found when tissue is exposed to lipopolysaccharide (Han et al., 2002) or inflammatory cytokines (Strazielle et al., 2003). Thus, the roadmap that describes possible routes available for xenobiotics to cross the tissue may in fact be dynamic. A dynamic map implies that routes taken by specific chemicals will not necessarily be fixed, but will be contingent on tissue state. If this turns out to be the case, one consequence may be that not only will individual xenobiotic fluxes change, but so will the nature of competitive interactions between xenobiotics. Clearly, if we wish to be able to manipulate the system to facilitate CNS therapy, it will be important to identify regulatory mechanisms that alter xenobiotic transport and metabolism in choroid plexus.

REFERENCES

Alebouyeh, M., Takeda, M., Onozato, M. L., Tojo, A., Noshiro, R., Hasannejad, H., Inatomi, J., Narikawa, S., Huang, X. L., Khamdang, S., Anzai, N., and Endou, H. 2003, Expression of human organic anion transporters in the choroid plexus and their interactions with neurotransmitter metabolites, *J. Pharmacol. Sci.,* vol. 93, pp. 430–436.

Anderson, M. E., Underwood, M., Bridges, R. J., and Meister, A. 1989, Glutathione metabolism at the blood-cerebrospinal fluid barrier, *FASEB J.,* vol. 3, pp. 2527–2531.

Babu, E., Takada, M., Narikawa, S., Kobayashi, H., Enomoto, A., Tojo, A., Cha, S. H., Sekine, T., Sakthisekaran, D., and Endou, H. 2002, Role of human organic anion transporter 4 in the transport of ochratoxin A, *Biochim. et Biophys. Acta,* vol. 1590, pp. 64–75.

Baldwin, S. A., Beal, P. R., Yao, S. Y., King, A. E., Cass, C. E., and Young, J. D. 2004, The equilibrative nucleoside transporter family, SLC29, *Pflügers Arch.,* vol. 447, pp. 735–743.

Barany, E. H. 1972, Inhibition by hippurate and probenecid of in vitro uptake of iodipamide and *o*-iodohippurate. A composite uptake system for iodipamide in choroid plexus, kidney cortex and anterior uvea of several species, *Acta Physiol. Scand.,* vol. 86, pp. 12–27.

Barany, E. H. 1973, The liver-like anion transport system in rabbit kidney, uvea, and choroid plexus. I. Selectivity of some inhibitors, direction of transport, possible physiological substrates, *Acta Physiol. Scand.,* vol. 88, pp. 412–429.

Bock, K. W., Lilienblum, W., Fischer, G., Schirmer, G., and Bock-Henning, B. S. 1987, The role of conjugation reactions in detoxication, *Arch. Toxicol.*, vol. 60, pp. 22–29.

Boom, S. P. and Russel, F. G. 1993, Cimetidine uptake and interactions with cationic drugs in freshly isolated proximal tubular cells of the rat, *J. Pharmacol. Exp. Ther.*, vol. 267, pp. 1039–1044.

Borst, P., Zelcer, N., and van Helvoort, A. 2000, ABC transporters in lipid transport, *Biochim. Biophys. Acta*, vol. 1486, pp. 128–144.

Breen, C. M., Sykes, D., Baehr, C., Fricker, G., and Miller, D. S. 2004, Fluorescein-methotrexate transport in rat choroid plexus analyzed using confocal microscopy, *Am. J. Physiol.*, vol. 287, pp. F562–F569.

Breen, C. M., Sykes, D. B., Fricker, G., and Miller, D. S. 2002, Confocal imaging of organic anion transport in intact rat choroid plexus, *Am. J. Physiol. Renal Physiol.*, vol. 282, pp. F877-F885.

Brittebo, E. B. 1994, Metabolism-dependent binding of the heterocyclic amine Trp-P-1 in endothelial cells of choroid plexus and in large cerebral veins of cytochrome P450-induced mice, *Brain Res.*, vol. 659, pp. 91–98.

Brosen, K. 1995, Drug interactions and the cytochrome P450 system. The role of cytochrome P450 1A2, *Clin. Pharmacokinet.,* vol. 29 suppl 1, pp. 20–25.

Chang, T. K., Weber, G. F., Crespi, C. L., and Waxman, D. J. 1993, Differential activation of cyclophosphamide and ifosphamide by cytochromes P-450 2B and 3A in human liver microsomes, *Cancer Res.,* vol. 53, pp. 5629–5637.

Choudhuri, S., Cherrington, N. J., Li, N., and Klaassen, C. D. 2003, Constitutive expression of various xenobiotic and endobiotic transporter mRNAs in the choroid plexus of rats, *Drug Metab. Dispos.,* vol. 31, pp. 1337–1345.

Cserr, H. F. and Van Dyke, D. H. 1971, 5-Hydroxyindoleacetic acid accumulation by isolated choroid plexus, *Am. J. Physiol.*, vol. 220, pp. 718–723.

Daniel, H. and Kottra, G. 2004, The proton oligopeptide cotransporter family SLC15 in physiology and pharmacology, *Pflügers Arch.,* vol. 447, pp. 610–618.

Dantzler, W. H. and Wright, S. H. 2003, The molecular and cellular physiology of basolateral organic anion transport in mammalian renal tubules, *Biochim. Biophys. Acta*, vol. 1618, pp. 185–193.

Dey, A., Jones, J. E., and Nebert, D. W. 1999, Tissue- and cell type-specific expression of cytochrome P450 1A1 and cytochrome P450 1A2 mRNA in the mouse localized in situ hybridization, *Biochem. Pharmacol.*, vol. 58, pp. 525–537.

Ekaratanawong, S., Anzai, N., Jutabha, P., Miyazaki, H., Noshiro, R., Takeda, M., Kanai, Y., Sophasan, S., and Endou, H. 2004, Human organic anion transporter 4 is a renal apical organic anion/dicarboxylate exchanger in the proximal tubules, *J. Pharmacol. Sci.,* vol. 94, pp. 297–304.

Enomoto, A., Takada, M., Shimoda, M., Narikawa, S., Kobayashi, Y., Yamamoto, M., Sekine, T., Cha, S. H., Niwa, T., and Endou, H. 2002, Interaction of human organic anion transporters 2 and 4 with organic anion transport inhibitors, *J. Pharmacol. Exp. Ther.*, vol. 301, pp. 797–802.

Ernst, S. A., Palacios, J. R., Aas, A. T., and Siegel, G. J. 1986, Immunocytochemical localization of Na^+,K^+-ATPase catalytic polypeptide in mouse choroid plexus, *J. Histochem. Cytochem.,* vol. 34, pp. 189–195.

Forn, J. 1972, Active transport of 5-hydroxyacetic acid by rabbit choroid plexus in vitro: blockade by probenecid and metabolic inhibitors, *Biochem. Pharmacol.*, vol. 21, pp. 619–624.

Francis, G. A., Fayard, E., Picard, F., and Auwerx, J. 2003, Nuclear receptors and the control of metabolism, *Annu. Rev. Physiol.*, vol. 65: 261–311.

Fuchs, M. and Stange, E. F. 2001, Cholesterol and cholestasis: a lesson from the Mdr2 (-/-) mouse, *J. Hepatol.*, vol. 34, pp. 339–341.

Galter, D., Carmine, A., Buervenich, S., Duester, G., and Olson, L. 2003, Distribution of class I, III and IV alcohol dehydrogenase mRNAs in the adult rat, mouse and human brain, *Eur. J. Biochem.,* vol. 270, pp. 1316–1326.

Gao, B., Stieger, B., Noe, B., Fritschy, J. M., and Meier, P. J. 1999, Localization of the organic anion transporting polypeptide 2 (Oatp2) in capillary endothelium and choroid plexus epithelium of rat brain, *J. Histochem. Cytochem.,* vol. 47, pp. 1255–1264.

Geyer, J., Godoy, J. R., and Petzinger, E. 2004, Identification of a sodium-dependent organic anion transporter from rat adrenal gland, *Biochem. Biophys. Res. Commun.*, vol. 316, pp. 300–306.

Ghersi-Egea, J. F., Leninger-Muller, B., Suleman, G., Siest, G., and Minn, A. 1994, Localization of drug-metabolizing enzyme activities to blood-brain interfaces and circumventricular organs, *J. Neurochem.*, vol. 62, pp. 1089–1096.

Ghersi-Egea, J. F. and Strazielle, N. 2002, Choroid plexus transporters for drugs and other xenobiotics, *J. Drug Target*, vol. 10, pp. 353–357.

Gonzalez, F. J. and Kimura, S. 2003, Study of P450 function using gene knockout and transgenic mice, *Arch. Biochem. Biophys.,* vol. 409, pp. 153–158.

Graff, C. L. and Pollack, G. M. 2004, Drug transport at the blood-brain barrier and the choroid plexus, *Curr. Drug Metab.*, vol. 5, pp. 95–108.

Granberg, L., Ostergren, A., Brandt, I., and Brittebo, E. B. 2003, CYP1A1 and CYP1B1 in blood-brain interfaces: CYP1A1-dependent bioactivation of 7,12-dimethylbenz(a)anthracene in endothelial cells, *Drug Metab. Dispos.,* vol. 31, pp. 259–265.

Gray, J. H., Owen, R. P., and Giacomini, K. M. 2004, The concentrative nucleoside transporter family, SLC28, *Pflügers Arch.,* vol. 447, pp. 728–734.

Grundemann, D., Liebich, G., Kiefer, N., Koster, S., and Schomig, E. 1999, Selective substrates for non-neuronal monoamine transporters, *Mol. Pharmacol.,* vol. 56, pp. 1–10.

Hagenbuch, B., Gao, B., and Meier, P. J. 2002, Transport of xenobiotics across the blood-brain barrier, *News Physiol. Sci.,* vol. 17, pp. 231–234.

Hagenbuch, B. and Meier, P. J. 2003, The superfamily of organic anion transporting polypeptides, *Biochim. Biophys. Acta*, vol. 1609, pp. 1–18.

Hagenbuch, B. and Meier, P. J. 2004, Organic anion transporting polypeptides of the OATP/SLC21 family: phylogenetic classification as OATP/SLCO superfamily, new nomenclature and molecular/functional properties, *Pflügers Arch.,* vol. 447, pp. 653–665.

Haimeur, A., Conseil, G., Deeley, R. G., and Cole, S. P. 2004, The MRP-related and BCRP/ABCG2 multidrug resistance proteins: biology, substrate specificity and regulation, *Curr. Drug Metab*, vol. 5, pp. 21–53.

Han, H., Kim, S. G., Lee, M. G., Shim, C. K., and Chung, S. J. 2002, Mechanism of the reduced elimination clearance of benzylpenicillin from cerebrospinal fluid in rats with intracisternal administration of lipopolysaccharide, *Drug Metab. Dispos.,* vol. 30, pp. 1214–1220.

Handschin, C. and Meyer, U. A. 2003, Induction of drug metabolism: the role of nuclear receptors, *Pharmacol. Rev.*, vol. 55, pp. 649–673.

Honkakoski, P., Sueyoshi, T., and Negishi, M. 2003, Drug-activated nuclear receptors CAR and PXR, *Ann. Med.*, vol. 35, pp. 172–182.

Inano, A., Sai, Y., Nikaido, H., Hasimoto, N., Asano, M., Tsuji, A., and Tamai, I. 2003, Acetyl-*L*-carnitine permeability across the blood-brain barrier and involvement of carnitine transporter OCTN2, *Biopharm. Drug Dispos.*, vol. 24, pp. 357–365.

Johnson, J. A., el Barbary, A., Kornguth, S. E., Brugge, J. F., and Siegel, F. L. 1993, Glutathione S-transferase isoenzymes in rat brain neurons and glia, *J. Neurosci.*, vol. 13, pp. 2013–2023.

Jonker, J. W. and Schinkel, A. H. 2004, Pharmacological and physiological functions of the polyspecific organic cation transporters: OCT1, 2, and 3 (SLC22A1-3), *J. Pharmacol. Exp. Ther.*, vol. 308, pp. 2–9.

Jutabha, P., Kanai, Y., Hosoyamada, M., Chairoungdua, A., Kim, d. K., Iribe, Y., Babu, E., Kim, J. Y., Anzai, N., Chatsudthipong, V., and Endou, H. 2003, Identification of a novel voltage-driven organic anion transporter present at apical membrane of renal proximal tubule, *J. Biol. Chem.*, vol. 278, pp. 27930–27938.

Kido, Y., Tamai, I., Ohnari, A., Sai, Y., Kagami, T., Nezu, J., Nikaido, H., Hashimoto, N., Asano, M., and Tsuji, A. 2001, Functional relevance of carnitine transporter OCTN2 to brain distribution of *L*-carnitine and acetyl-*L*-carnitine across the blood-brain barrier, *J. Neurochem.*, vol. 79, pp. 959–969.

Kim, J. H., Stansbury, K. H., Walker, N. J., Trush, M. A., Strickland, P. T., and Sutter, T. R. 1998, Metabolism of benzo[a]pyrene and benzo[a]pyrene-7,8-diol by human cytochrome P450 1B1, *Carcinogenesis*, vol. 19, pp. 1847–1853.

Kitazawa, T., Hosoya, K., Takahashi, T., Sugiyama, Y., and Terasaki, T. 2000, In-vivo and in-vitro evidence of a carrier-mediated efflux transport system for oestrone-3-sulphate across the blood-cerebrospinal fluid barrier, *J. Pharmacy Pharmacol.*, vol. 52, pp. 281–288.

Koepsell, H. and Endou, H. 2004, The SLC22 drug transporter family, *Pflügers Arch.*, vol. 447, pp. 666–676.

Kojima, R., Sekine, T., Kawachi, M., Cha, S. H., Suzuki, Y., and Endou, H. 2002, Immunolocalization of multispecific organic anion transporters, OAT1, OAT2, and OAT3, in rat kidney, *J. Am. Soc. Nephrol.*, vol. 13, pp. 848–857.

Kruh, G. D. and Belinsky, M. G. 2003, The MRP family of drug efflux pumps, *Oncogene*, vol. 22, pp. 7537–7552.

Kusuhara, H., He, Z., Nagata, Y., Nozaki, Y., Ito, T., Masuda, H., Meier, P. J., Abe, T., and Sugiyama, Y. 2003, Expression and functional involvement of organic anion transporting polypeptide subtype 3 (Slc21a7) in rat choroid plexus, *Pharm. Res.*, vol. 20, pp. 720–727.

Kusuhara, H., Sekine, T., Utsunomiya-Tate, N., Tsuda, M., Kojima, R., Cha, S. H., Sugiyama, Y., Kanai, Y., and Endou, H. 1999, Molecular cloning and characterization of a new multispecific organic anion transporter from rat brain, *J. Biol. Chem.*, vol. 274, pp. 13675–13680.

Lee, G., Dallas, S., Hong, M., and Bendayan, R. 2001, Drug transporters in the central nervous system: brain barriers and brain parenchyma considerations, *Pharmacol. Rev.*, vol. 53, pp. 569–596.

Lee, Y. J., Kusuhara, H., and Sugiyama, Y. 2004, Do multidrug resistance-associated protein-1 and -2 play any role in the elimination of estradiol-17beta-glucuronide and 2,4-dinitrophenyl-S-glutathione across the blood-cerebrospinal fluid barrier?, *J. Pharm. Sci.*, vol. 93, pp. 99–107.

Lehmann, J. M., McKee, D. D., Watson, M. A., Willson, T. M., Moore, J. T., and Kliewer, S. A. 1998, The human orphan nuclear receptor PXR is activated by compounds that regulate CYP3A4 gene expression and cause drug interactions, *J. Clin. Invest*, vol. 102, pp. 1016–1023.

Leininger-Muller, B., Ghersi-Egea, J. F., Siest, G., and Minn, A. 1994, Induction and immunological characterization of the uridine diphosphate-glucuronosyltransferase conjugating 1-naphthol in the rat choroid plexus, *Neurosci. Lett.*, vol. 175, pp. 37–40.

Lewis, D. F. 2003, Human cytochromes P450 associated with the phase 1 metabolism of drugs and other xenobiotics: a compilation of substrates and inhibitors of the CYP1, CYP2 and CYP3 families, *Curr. Med. Chem.*, vol. 10, pp. 1955–1972.

Lewis, D. F., Modi, S., and Dickins, M. 2001, Quantitative structure-activity relationships (QSARs) within substrates of human cytochromes P450 involved in drug metabolism, *Drug Metab. Drug Interact.*, vol. 18, pp. 221–242.

Ljubojevic, M., Herak-Kramberger, C. M., Hagos, Y., Bahn, A., Endou, H., Burckhardt, G., and Sabolic, I. 2004, Rat renal OAT1 and OAT3 exhibit gender differences determined by both androgen stimulation and estrogen inhibition, *Am. J. Physiol. Renal Physiol.*, vol. 287, pp. F124–F138.

Lorenzo, A. V. and Spector, R. 1973, Transport of salicylic acid by choroid plexus in vitro, *J. Pharmacol. Exp. Ther.*, vol. 184, pp. 465–471.

Miller, D. S., Nobmann, S. N., Gutmann, H., Toeroek, M., Drewe, J., and Fricker, G. 2000, Xenobiotic transport across isolated brain microvessels studied by confocal microscopy, *Mol. Pharmacol.*, vol. 58, pp. 1357–1367.

Miller, D. S., Villalobos, A. R., and Pritchard, J. B. 1999, Organic cation transport in rat choroid plexus cells studied by fluorescence microscopy, *Am. J. Physiol.*, vol. 276, p. C955-C968.

Mimura, M., Baba, T., Yamazaki, H., Ohmori, S., Inui, Y., Gonzalez, F. J., Guengerich, F. P., and Shimada, T. 1993, Characterization of cytochrome P-450 2B6 in human liver microsomes, *Drug Metab. Dispos.*, vol. 21, pp. 1048–1056.

Morse, D. C., Stein, A. P., Thomas, P. E., and Lowndes, H. E. 1998, Distribution and induction of cytochrome P450 1A1 and 1A2 in rat brain, *Toxicol. Appl. Pharmacol.*, vol. 152, pp. 232–239.

Nagata, Y., Kusuhara, H., Endou, H., and Sugiyama, Y. 2002, Expression and functional characterization of rat organic anion transporter 3 (rOat3) in the choroid plexus, *Mol. Pharmacol.*, vol. 61, pp. 982–988.

Nelson, D. R. 1999, Cytochrome P450 and the individuality of species, *Arch. Biochem. Biophys.*, vol. 369, pp. 1–10.

Nelson, D. R., Koymans, L., Kamataki, T., Stegeman, J. J., Feyereisen, R., Waxman, D. J., Waterman, M. R., Gotoh, O., Coon, M. J., Estabrook, R. W., Gunsalus, I. C., and Nebert, D. W. 1996, P450 superfamily: update on new sequences, gene mapping, accession numbers and nomenclature, *Pharmacogenetics*, vol. 6, pp. 1–42.

Ocheltree, S. M., Shen, H., Hu, Y., Xiang, J., Keep, R. F., and Smith, D. E. 2004, Mechanisms of cefadroxil uptake in the choroid plexus: studies in wild-type and PEPT2 knockout mice, *J. Pharmacol. Exp. Ther.*, vol. 308, pp. 462–467.

Ohtsuki, S., Takizawa, T., Takanaga, H., Terasaki, N., Kitazawa, T., Sasaki, M., Abe, T., Hosoya, K., and Terasaki, T. 2003, In vitro study of the functional expression of organic anion transporting polypeptide 3 at rat choroid plexus epithelial cells and its involvement in the cerebrospinal fluid-to-blood transport of estrone-3-sulfate, *Mol. Pharmacol.*, vol. 63, pp. 532–537.

Pappenheimer, J. R., Heisy, S. R., and Jordan, E. F. 1961, Active transport of Diodrast and phenolsulfonphthalein from cerebrospinal fluid to blood., *Am. J. Physiol.*, vol. 200, pp. 1–10.

Pritchard, J. B. 1980, Accumulation of anionic pesticides by rabbit choroid plexus in vitro, *J. Pharmacol. Exp. Ther.*, vol. 212, pp. 354–359.

Pritchard, J. B. and Miller, D. S. 1993, Mechanisms mediating renal secretion of organic anions and cations, *Physiol. Rev.,* vol. 73, pp. 765–796.

Pritchard, J. B., Sweet, D. H., Miller, D. S., and Walden, R. 1999, Mechanism of organic anion transport across the apical membrane of choroid plexus, *J. Biol. Chem.*, vol. 274, pp. 33382–33387.

Rao, V. V., Dahlheimer, J. L., Bardgett, M. E., Snyder, A. Z., Finch, R. A., Sartorelli, A. C., and Piwnica-Worms, D. 1999, Choroid plexus epithelial expression of MDR1 P glycoprotein and multidrug resistance-associated protein contribute to the blood-cerebrospinal-fluid drug-permeability barrier, *Proc. Natl. Acad. Sci. USA*, vol. 96, pp. 3900–3905.

Rosenbrock, H., Hagemeyer, C. E., Ditter, M., Knoth, R., and Volk, B. 2001, Expression and localization of the CYP2B subfamily predominantly in neurones of rat brain, *J. Neurochem.*, vol. 76, pp. 332–340.

Rubio-Aliaga, I. and Daniel, H. 2002, Mammalian peptide transporters as targets for drug delivery, *Trends Pharmacol. Sci.,* vol. 23, pp. 434–440.

Rubio-Aliaga, I., Frey, I., Boll, M., Groneberg, D. A., Eichinger, H. M., Balling, R., and Daniel, H. 2003, Targeted disruption of the peptide transporter Pept2 gene in mice defines its physiological role in the kidney, *Mol. Cell Biol.,* vol. 23, pp. 3247–3252.

Russel, F. G., Masereeuw, R., and van Aubel, R. A. 2002, Molecular aspects of renal anionic drug transport, *Annu. Rev. Physiol.*, vol. 64, pp. 563–594.

Sakata, K., Yamashita, T., Maeda, M., Moriyama, Y., Shimada, S., and Tohyama, M. 2001, Cloning of a lymphatic peptide/histidine transporter, *Biochem. J.,* vol. 356, pp. 53–60.

Schinkel, A. H. 1997, The physiological function of drug-transporting P-glycoproteins, *Semin. Cancer Biol.,* vol. 8, pp. 161–170.

Schinkel, A. H. 1999, P-Glycoprotein, a gatekeeper in the blood-brain barrier, *Adv. Drug Deliv. Rev.,* vol. 36, pp. 179–194.

Schwarz, M., Lund, E. G., and Russell, D. W. 1998, Two 7 alpha-hydroxylase enzymes in bile acid biosynthesis, *Curr. Opin. Lipidol.,* vol. 9, pp. 113–118.

Sekine, T., Cha, S. H., Tsuda, M., Apiwattankul, N., Nakajima, N., Kanai, Y., and Endou, H. 1998, Identification of multispecific organic anion transporter 2 expressed predominantly in liver, *FEBS Lett.,* vol. 429, pp. 179–182.

Sekine, T., Watanabe, N., Hosoyamada, M., Kanai, Y., and Endou, H. 1997, Expression cloning and characterization of a novel organic anion transporter, *J. Biol. Chem.*, vol. 272, pp. 18526–18529.

Sheehan, D., Meade, G., Foley, V. M., and Dowd, C. A. 2001, Structure, function and evolution of glutathione transferases: implications for classification of non-mammalian members of an ancient enzyme superfamily, *Biochem. J.,* vol. 360, pp. 1–16.

Smith, G., Stubbins, M. J., Harries, L. W., and Wolf, C. R. 1998, Molecular genetics of the human cytochrome P450 monooxygenase superfamily, *Xenobiotica*, vol. 28, pp. 1129–1165.

Spector, R. and Lorenzo, A. V. 1974, Inhibition of penicillin transport from the cerebrospinal fluid after intracisternal inoculation of bacteria, *J. Clin. Invest.,* vol. 54, pp. 316–325.

Strazielle, N. and Ghersi-Egea, J. F. 1999, Demonstration of a coupled metabolism-efflux process at the choroid plexus as a mechanism of brain protection toward xenobiotics, *J. Neurosci.,* vol. 19, pp. 6275–6289.

Strazielle, N., Khuth, S. T., Murat, A., Chalon, A., Giraudon, P., Belin, M. F., and Ghersi-Egea, J. F. 2003, Pro-inflammatory cytokines modulate matrix metalloproteinase secretion and organic anion transport at the blood-cerebrospinal fluid barrier, *J. Neuropathol. Exp. Neurol.,* vol. 62, pp. 1254–1264.

Sun, H., Dai, H., Shaik, N., and Elmquist, W. F. 2003, Drug efflux transporters in the CNS, *Adv. Drug Deliv. Rev.,* vol. 55, pp. 83–105.

Suzuki, H., Sawada, Y., Sugiyama, Y., Iga, T., and Manabu, H. 1986, Comparative uptake of cimetidine by rat choroid plexus between the lateral and the 4th ventricles, *J. Pharmacobio-Dyn.,* vol. 9, pp. 327–329.

Sweet, D. H., Chan, L. M., Walden, R., Yang, X. -P., Miller, D. S., and Pritchard, J. B. 2003, Organic anion transporter 3 (Slc22a8) is a dicarboxylate exchanger indirectly coupled to the Na^+ gradient, *Am. J. Physiol.*, vol. 284, p. F763-F769.

Sweet, D. H., Miller, D. S., and Pritchard, J. B. 1999a, Cellular localization of an organic anion transporter (rROAT/GFP) fusion construct in intact proximal tubules, *Am. J. Physiol.*, vol. 276, p. F864-F873.

Sweet, D. H., Miller, D. S., and Pritchard, J. B. 1999b, Localization of an organic anion transporter-GFP fusion construct (rROAT1-GFP) in intact proximal tubules, *Am. J. Physiol.*, vol. 276, p. F864-F873.

Sweet, D. H., Miller, D. S., and Pritchard, J. B. 2001, Ventricular choline transport: a role for organic cation transporter 2 expressed in choroid plexus, *J. Biol. Chem.*, vol. 276, pp. 41611–41619.

Sweet, D. H., Miller, D. S., Pritchard, J. B., Fujiwara, Y., Beier, D. R., and Nigam, S. K. 2002, Impaired organic anion transport in kidney and choroid plexus of organic anion transporter 3 (Oat3 (Slc22a8)) knockout mice, *J. Biol. Chem.*, vol. 277, pp. 26934–26943.

Sweet, D. H. and Pritchard, J. B. 1999, The molecular biology of renal organic anion and organic cation transporters, *Cell Biochem. Biophysics.,* vol. 31, pp. 89–118.

Sweet, D. H., Wolff, N. A., and Pritchard, J. B. 1997, Expression cloning and characterization of ROAT1: The basolateral organic anion transporter in rat kidney, *J. Biol. Chem.*, vol. 272, pp. 30088–30095.

Sykes, D., Sweet, D. H., Lowes, S., Nigam, S. K., Pritchard, J. B., and Miller, D. S. 2004, Organic anion transport in choroid plexus from wild-type and organic anion transporter 3 (Slc22a8)-null mice, *Am. J. Physiol. Renal Physiol.*, vol. 286, p. F972-F978.

Tamai, I., China, K., Sai, Y., Kobayashi, D., Nezu, J., Kawahara, E., and Tsuji, A. 2001, $Na(^+)$-coupled transport of L-carnitine via high-affinity carnitine transporter OCTN2 and its subcellular localization in kidney, *Biochim. Biophys. Acta,* vol. 1512, pp. 273–284.

Trauner, M. and Boyer, J. L. 2003, Bile salt transporters: molecular characterization, function, and regulation, *Physiol. Rev.,* vol. 83, pp. 633–671.

Villalobos, A. R., Parmelee, J. T., and Renfro, J. L. 1999, Choline uptake across the ventricular membrane of neonate rat choroid plexus, *Am. J. Physiol.*, vol. 276, pp. C1288–C1296.

Volk, B., Meyer, R. P., von Lintig, F., Ibach, B., and Knoth, R. 1995, Localization and characterization of cytochrome P450 in the brain. In vivo and in vitro investigations on phenytoin- and phenobarbital-inducible isoforms, *Toxicol. Lett.,* vol. 82–83: 655–62., pp. 655–662.

Whittico, M. T., Yuan, G., and Giacomini, K. M. 1990, Cimetidine transport in isolated brush border membrane vesicles from bovine choroid plexus, *J. Pharmacol. Exp. Ther.*, vol. 255, pp. 615–623.

Wijnholds, J., deLange, E. C., Scheffer, G. L., van den Berg, D. J., Mol, C. A., van der Valk, M., Schinkel, A. H., Scheper, R. J., Breimer, D. D., and Borst, P. 2000, Multidrug resistance protein 1 protects the choroid plexus epithelium and contributes to the blood-cerebrospinal fluid barrier, *J. Clin. Invest*, vol. 105, pp. 279–285.

Yamashita, T., Shimada, S., Guo, W., Sato, K., Kohmura, E., Hayakawa, T., Takagi, T., and Tohyama, M. 1997, Cloning and functional expression of a brain peptide/histidine transporter, *J. Biol. Chem.*, vol. 272, pp. 10205–10211.

Zhang, Q. Y., Wikoff, J., Dunbar, D., Fasco, M., and Kaminsky, L. 1997, Regulation of cytochrome P4501A1 expression in rat small intestine, *Drug Metab. Dispos.*, vol. 25, pp. 21–26.

8 Homeostasis of Nucleosides and Nucleobases in the Brain: The Role of Flux between the CSF and the Brain ISF, Transport across the Choroid Plexus and the Blood–Brain Barrier, and Cellular Uptake

Zoran B. Redzic
University of Cambridge,
Cambridge, United Kingdom

CONTENTS

8.1 METABOLIC FUNCTIONS OF NUCLEOSIDES AND NUCLEOBASES IN MAMMALIAN TISSUES

Nucleosides consist of a nucleobase ring and a ribose sugar moiety attached to N1 of this ring. The nucleobase ring can be either purine or pyrimidine. The following nucleobases are found in mammalian tissues: (1) the aminopurines adenine and guanine, (2) products of aminopurine oxidation—oxypurines hypoxanthine and xanthine, and (3) the pyrimidines cytosine, thymine, uracil, and orotate. The nomenclature of purines and pyrimidines is shown in Table 8.1. The formation of ester bonds between two terminal phosphoric acid molecules require >30 kJ/M of energy, so nucleotide diphosphates (NDPs) and nucleotide triphosphates (NTPs) have been selected during evolution as molecules that store the energy released in various intracellular catabolic processes in these high-energy bonds. Therefore, NDP and NTP play an important role in the metabolism of all tissues.

8.2 NUCLEOTIDE SYNTHESIS

The major source of nucleotides in a number of tissues is de novo synthesis (see Figure 8.1), which in the case of the purines consists of 10 sequential reactions, producing inosinic acid (IMP) (see Figure 8.1A). Finally, IMP is converted to either adenosine monophosphate (AMP) or guanosine monophosphate (GMP) through a well-balanced two-branched pathway. This anabolic pathway is extremely energy consuming, since it requires six high-energy bonds for the production of a single AMP or GMP molecule. The de novo synthesis of pyrimydines is a less complex process and requires less energy than the de novo synthesis of purines (see Figure 8.1B).

Brain cells differ from most cells in other tissues in their profile of nucleotide synthesis. That difference is most likely caused by the brain's great requirement for

TABLE 8.1
Nomenclature of Purine and Pyrimidine Bases and Their Corresponding Nucleosides and Nucleotides

Base	Nucleoside (base-sugar)[a]	Nucleotide[b] (base-sugar phosphate)
	Purines	
Adenine (6-aminopurine)	Adenosine	Adenosine monophosphate (AMP) or adenylic acid
	Deoxyadenosine	Deoxyadenosine monophosphate (dAMP)
Guanine (2-amino-6-oxypurine)	Guanosine	Guanosine monophosphate (GMP) or guanylic acid
	Deoxyguanosine	Deoxyguanosine monophosphate (dGMP)
Hypoxanthine	Inosine	Inosine monophosphate (IMP) or inosinic acid
	Deoxyinosine	Deoxyinosine monophosphate (dIMP)
Xanthine	Xanthosine[c]	Xanthosine monophosphate (XMP)[c] or xanthosynic acid
	Pyrimidines	
Cytosine (2-oxy-4-aminopyrimidine)	Cytidine	Cytidine monophosphate (CPM)
	Deoxycytidine	Deoxycytidine monophosphate (dCMP)
Thymine (2,4-dioxy-5-methylpyrimidine)	Thymidine (thymine deoxyriboside)	Thymidine monophosphate (TMP) (thymidine deoxyribotide)
Uracil	Uridine	Uridine monophosphate (UMP)

Purines and pyrimidines are amino, methyl, or oxy derivates of nitrogen base rings.

[a]The sugar residue in nucleosides can be ribose or 2-deoxyribose. If it is deoxyribose, it is identified as deoxynucleoside/nucleotide; otherwise it is assumed to be ribose, with the exception of thymidine, which is deoxyriboside.

[b]A nucleotide is nucleoside monophosphate; monophosphates are acids since the phosphate group has an excessive tendency to release protons.

[c]Xanthine is, in general, only the product of metabolic degradation of purines, and the rate of further oxidation of this purine into uric acid is very high, so the nucleoside xanthosine and nucleotide xanthosine monophosphate (XMP) are present in tracer concentrations in mammalian tissues.

energy, which is mainly used to generate and maintain membrane potential (Clarke and Sokoloff, 1999). For this reason brain cells have a very limited potential for a number of anabolic pathways, including de novo synthesis of nucleotides (Linden, 1999). Therefore, this tissue is performing synthesis of nucleotides through so-called

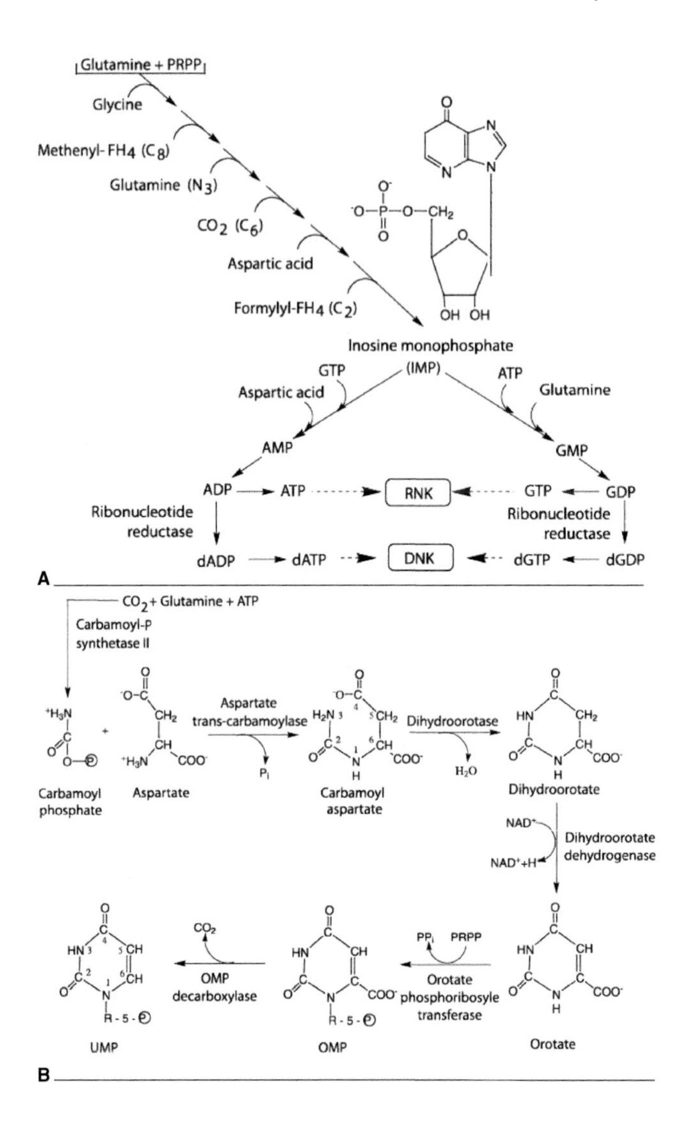

FIGURE 8.1 (**A**) De novo synthesis of purine nucleotides. Synthesis starts from an "activated" ribose ring, 5'-phosphoribosyl-1'-pyrophosphate (PRPP), which is synthesized from ribose-5'-phosphate through the action of PRPP synthetase, a reaction that expends two high-energy bonds. In the six following reactions the purine ring is built on this ribose moiety, expanding two more high-energy bonds and giving IMP as a product. Since inosylates are not of particular importance for either RNK/DNK synthesis or metabolism, IMP is converted to either AMP or GMP, and this expands two high-energy bonds in both cases, giving the total of six high-energy bonds for the de novo synthesis of single AMP/GMP. These molecules could then be further phosphorylated into corresponding nucleotide diphosphates, molecules that serve as a source of nucleotides for both RNA and DNA synthesis, through the action of ribonucleotide reductase. (**B**) Synthesis of pyrimidine nucleotides. This process is much less complex than the synthesis of purines, and the energy cost is lower (four vs. six high-energy bonds). The final product is UMP, which can be converted into other pyrimidines.

salvage pathways. In these pathways, nucleobases and nucleosides, which are waste products of nucleotide metabolic degradation, are reconverted to their corresponding nucleotides through the action of various transferases/kinases:

$$\text{Purine ring} + \text{P-ribose-P-P} \rightarrow \text{purine-ribose-P} + \text{PPi} \quad (8.1)$$

$$\text{Purine-base} + \text{ATP} \rightarrow \text{base-ribose-phosphate} + \text{ADP} \quad (8.2)$$

There are two main salvage pathways of nucleobases in brain cells: salvage of hypoxanthine and guanine into IMP and GMP, respectively, through the action of hypoxanthine-guanine phosphoribosyl transferase (HGPRT) and the salvage of adenine into AMP through the action of adenine phosphoribosyl transferase (APRT). It is estimated that the production of nucleotides through these two pathways exceeds that of all other sources of purine nucleotides in the brain (Linden, 1999). Nucleosides can be also salvaged into nucleotides through the action of various nucleoside kinases. There are at least three nucleoside kinases in the brain—inosine kinase, guanosine kinase, and adenosine kinase—the third one being the most abundant. Overall, salvage of adenosine is copious in the brain, which is an important part of the homeostatic mechanism that keeps a low concentration of this molecule in the interstitial fluid (ISF) of this tissue (see below). The majority of nucleosides, with the exception of adenosine, are produced intracellularly and leave the cell via various nucleoside transporters to enter the brain ISF. Adenosine is produced abundantly in the extracellular space of the brain through the action of ecto-ATP hydrolases on ATP (released from neurons as a neurotransmitter). This reaction produces AMP, which is then a substrate for ecto-5'-nucleotidases, giving adenosine as the final product of ATP hydrolysis (Zimmermann, 1996).

8.3 NUCLEOTIDE CATABOLISM

Degradation of purine nucleotides in the brain occurs during the turnover of either endogenous nucleic acids or NTPs (see Figure 8.2). Through the action of various 5'-nucleotidases, NMPs hydrolyze into free nucleoside and phosphate (see Figure 8.2). This is the major source of intracellular nucleosides in the brain, except for adenosine, and the further breakdown of these nucleosides leads to the production of hypoxanthine. Vertebrates do not express the enzyme capable of opening the purine ring, so in most tissues this nucleobase is oxidized to xanthine and then further to insoluble uric acid through the action of xanthine oxidase (Corry, 1997). However, the activity of xanthine oxidase in brain homogenates is very low (Betz, 1985), which is in accordance with the low levels of xanthine and uric acid detected in this tissue. It is believed that the majority of purines in the brain are metabolized into hypoxanthine, which is then a substrate for salvage pathways.

Pyrimidine catabolism takes place mainly in the liver, and there is no direct evidence that pyrimidine-metabolizing enzymes are active in the brain. In mammals, including humans, the pyrimidine ring opens in the liver, producing soluble β-alanine and β-aminoisobutyric acid (see Figure 8.3), which are partially excreted through the kidneys and partially reused in the liver.

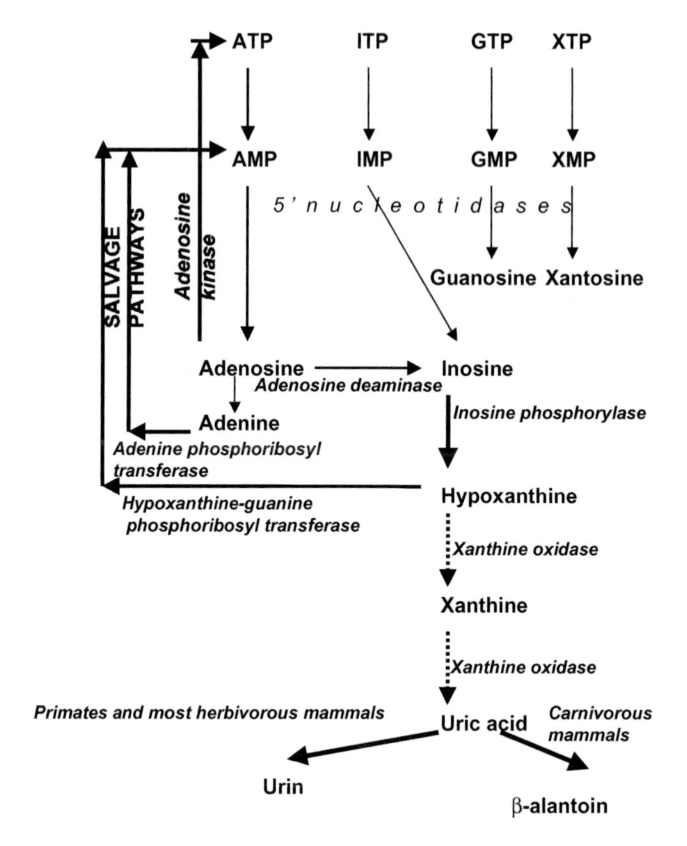

FIGURE 8.2 Catabolism of purine nucleotides. This is one of the most complicated catabolic pathways in mammals, and it is of particular importance for the brain. During various metabolic processes/RNA recycling, NTPs are hydrolyzed into corresponding NMPs, which can then be either rephosphorylated into NDPs and NTPs or further hydrolyzed into nucleosides through the action of 5'-nucleotidases. Nucleosides can then be further catabolized into the corresponding nucleobases inside the cells. Contrary to all other tissues, in which all purine nucleobases are finally oxidized into the insoluble uric acid, the final product of purine oxidation in the brain appears to be hypoxanthine, which can then be salvaged into IMP through the action of HGPRT-ase. Adenine can also be salvaged into AMP through the action of APRT-ase. These two salvage pathways are more abundant in the brain than in other tissues.

8.4 TRANSPORT OF NUCLEOSIDES AND NUCLEOBASES ACROSS CELLULAR MEMBRANES

Naturally occurring nucleosides and nucleobases cannot diffuse through the phospholipid layers of cellular membranes, so specific carrier proteins are required to mediate their transport. They can either facilitate the diffusion of these molecules across the membranes (equilibrative transport) or mediate concentrative, secondary active co-transport with Na^+ (Simon and Jarvis, 1996) (see Figure 8.4).

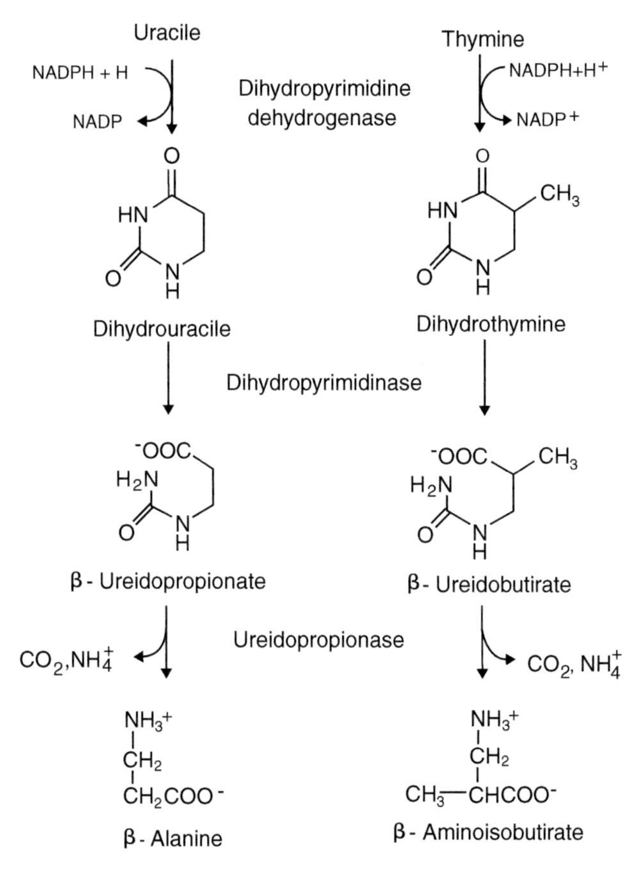

FIGURE 8.3 Catabolism of pyrimidines. This process is less complex than the catabolism of purines: all vertebrates possess enzymes that can open pyrimidine ring, yielding the soluble molecules β-alanine and β-aminoisobutyrate as the final products of pyrimidine catabolism.

8.4.1 Equilibrative Nucleoside Transport

Two distinct forms of equilibrative transport systems exist, which are classified on the basis on their sensitivity to inhibition by the synthetic thio-purine nitrobenzylthioinosine (NBTI or NBMPR) (see Figure 8.4) (Plagemman et al., 1988; Griffith and Jarvis, 1996). The best characterized is the equilibrative system, which is sensitive to nanomolar concentrations of NBTI (known as "equilibrative sensitive," or *es*) and to a number of other synthetic nucleoside analogues (e.g., dipyridamole and dilazep) (Plagemman et al., 1988; Simon and Jarvis, 1996). The purification of the protein responsible for the typical *es*-type transport activity of human erythrocytes led to the cloning of the corresponding cDNA from placenta (Griffits et al., 1997b). This cDNA encoded a 456-residue protein, the human equilibrative nucleoside transporter 1 (hENT1). This protein was then expressed in oocytes from the African clawed toad (*Xenopus laevis*) which is a good experimental model since these oocytes do not normally express nucleoside transporters (Crawford et al., 1998). Kinetic

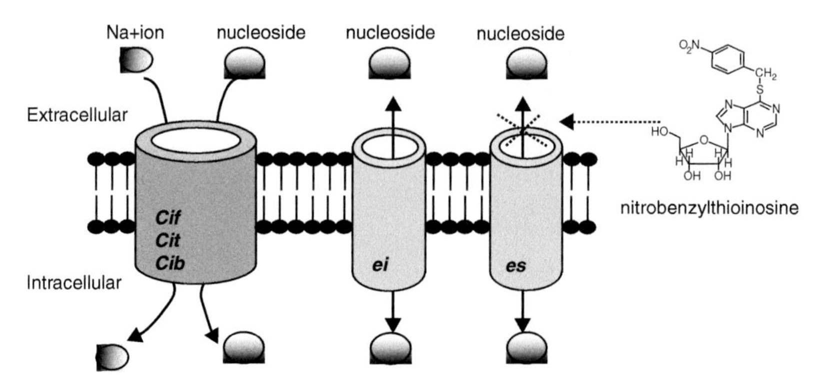

FIGURE 8.4 A schematic representation of nucleoside transporters. These membrane proteins mediate either facilitated bi-directional diffusion of nucleosides (equilibrative nucleoside transporters) or unidirectional cotransport of nucleosides with Na^+, driven by the concentration or electrochemical gradient of this ion. Equilibrative transporters are further divided by functional studies into two subclasses: equilibrative transporter sensitive to nanomolar concentrations of synthetic analogue NBTI-*es* or an equilibrative transporter insensitive to nanomolar concentrations of NBTI-*ei*. Although not shown in this figure, concentrative transporters can also be divided by functional studies into several subclasses according to the affinity for particular nucleosides or nucleobases. [Reprinted with permission from Elsevier (*Mol. Med. Today*, 5, 1999, 216–224).]

characterization of hENT1 following this expression showed it to be a typical *es*-type transporter, with a broad selectivity for purine and pyrimidine nucleosides but not nucleobases (see Figure 8.5) (Griffiths et al., 1997b; Ward et al., 2000). In addition to natural substrates, the *es* transport system has also been shown to transport a number of synthetic nucleoside analogues used in cancer chemotherapy, but it does not show any affinity for the antiviral drugs zidovudine, zalcitabine, and didanosine (Mackey et al., 1998, 1999; Young et al., 2000). The rat homologue of hENT1, rENT1, a 457-residue protein, is about 80% identical in sequence to hENT1 and has similar kinetic properties (Yao et al., 1997). This transporter is widely expressed, with the highest expression reported in erythrocytes, placenta, liver, and brain (Griffits et al., 1997b; Choi et al., 2000).

The second equilibrative system is much less sensitive to NBTI (0.5% of inhibition with 1 μM of NBTI) and therefore is known as "equilibrative insensitive," or *ei*. Following the cloning of cDNA encoding a human *ei*-type transporter, the identification of the encoded protein, hENT2, was accomplished rapidly by PCR amplification of cDNA from placenta (Griffits et al., 1997a). The rat homologue, rENT2, has been cloned from jejunum (Yao et al., 1997). The 456-residue proteins encoded by ENT2 cDNAs are 50% identical in sequence to ENT1s. When expressed in *Xenopus* oocytes, hENT2 was found to exhibit the functional properties of mammalian *ei*-type transport (Crowford et al., 1998; Ward et al., 2000). It has a broad range of natural purine substrates (Griffiths et al., 1997a; Crawford et al., 1998; Ward et al., 2000) and also shows affinity to nucleobases, particularly for hypoxanthine (see Figure 8.5), so it is believed that the majority of equilibrative nucleobase transport is mediated via the ENT2 transport system. This is of particular importance for the

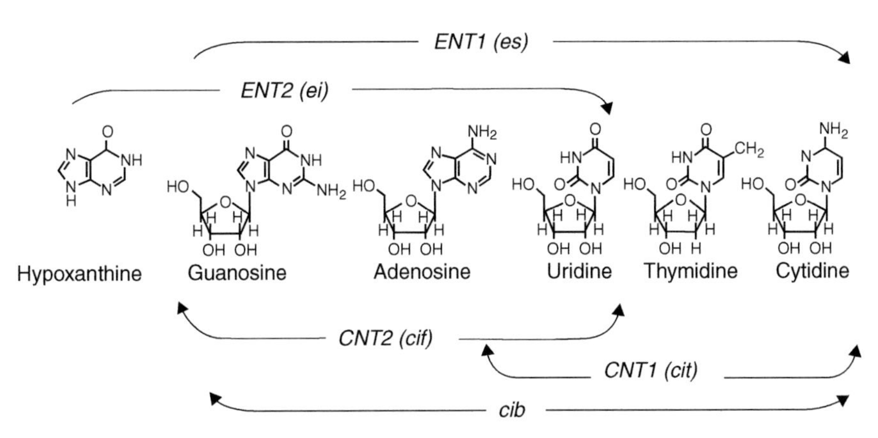

FIGURE 8.5 The affinity of cloned nucleoside transporters for endogenous nucleosides/nucleobases. In general, equilibrative transporters ENT1 and ENT2 show affinity for most of the purine and pyrimidine nucleosides, with the major difference that ENT2 shows a much higher affinity for nucleobases than ENT1. Concentrative transporters have a more limited range of substrates: CNT1 mediates primarily the transport of pyrimidines and has very weak affinity for purines, while CNT2 shows affinity for purines and the pyrimidine uridine.

brain, since, as mentioned above, it appears that hypoxanthine is the major end product of purine catabolism in this tissue. This transport system also mediates equilibrative transport of synthetic antiviral nucleosides (Young et al., 2000).

As predicted by computer modeling, both ENT proteins possess 11 transmembrane helices, with a cytoplasmic N-terminus and central hydrophobic loop region and an extracellular C-terminus (see Figure 8.6). The loops from the transmembrane domain (TD) 3–6 are responsible for the binding of dilazep, dipyridamole, and NBTI, while loops from TD 3–5, with a contribution from several residues in the region 1–99 are responsible for the binding of natural substrates (Sundaram et al., 1998).

8.4.2 Concentrative Nucleoside Transport

This type of transport is less widely distributed in mammalian tissues than equilibrative transport and mediates the transport of nucleosides against a concentration gradient, using the concentration or electrochemical gradient of Na^+ (see Figure 8.4). In a number of cells concentrative nucleoside transport exists alongside equilibrative transport, although not necessarily in the same membrane domain/side of the cell. At least five different concentrative transport systems were identified in functional studies (Griffith and Jarvis, 1996), but only two are widely distributed in tissues.

The first to be identified by functional studies was a transporter that primarily shows an affinity for purines and uridine (see Figure 8.5), with the synthetic molecule formycin B as the substrate model (Griffith and Jarvis, 1996), called *cif*. The other concentrative transport process was identified soon after *cif*, and functional studies revealed it to show affinity for pyrimidines, especially thymidine, and very weak affinity for purines; this system was named *cit* (Griffith and Jarvis, 1996). Following

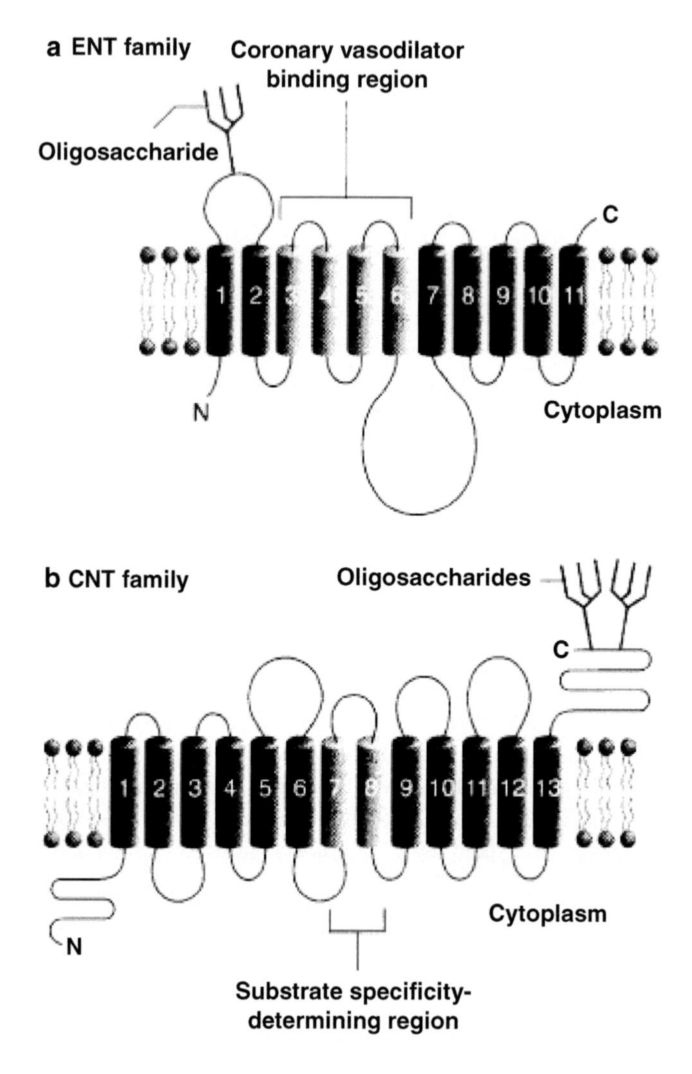

FIGURE 8.6 The structure of ENT and CNT protein families as proposed by a computer modeling. Among the proteins in one family there is a great homology in sequence (e.g., ENT1 *vs.* ENT2), usually more than 65%. However, there is no homology between ENT and CNT families or between the CNT family and any other mammalian protein that mediates co-transport of substrate with Na^+. It is believed that all members of the ENT family have 11 transmembrane domains, with the section between loops 3 and 6 being recognized as most important for binding of substrates and synthetic analogues. CNT proteins appear to have 13 transcellular domains, with heavily glucosylated C-terminal and substrate-binding regions and extracellular loops between 7 and 8. This strong glucosylation of CNT proteins might be one reason that they show weak affinity for nucleobases. [Reprinted with permission from Elsevier (*Mol. Med. Today,* 5, 1999, 216–224).]

the cloning of equilibrative transporters, the first concentrative nucleoside transporter was cloned from rat jejunum and designated rat concentrative nucleoside transporter 1 (rCNT1) (Huang et al., 1994). Soon afterwards, another concentrative nucleoside transporter was cloned from rat intestine and called rat concentrative nucleoside transporter 2 (rCNT2) (Ritzel et al., 1997, 1998; Wang et al., 1997). Human analogues of both rCNTs, designated hCNT1 and hCNT2, have subsequently been identified (Ritzel et al., 1997, 1998; Wang et al., 1997). When expressed in *Xenopus* oocytes, CNT1 proteins showed them to be typical *cit*-type transporters and sodium dependent and transporting pyrimidines to be preferable substrates. CNT2 proteins exhibited sodium-dependent nucleoside transport characteristics typical of *cif*-type transporters, with adenosine and other purine nucleosides as the main substrates (Ritzel et al., 1997, 1998; Wang et al., 1997). rCNT1 is expressed in intestine, kidney (Huang et al., 1994), liver (Felipe et al., 1998), and brain (Anderson et al., 1996), while rCNT2 is more widely distributed. Mammalian CNT1 proteins exhibit >60% sequence identity to CNT2 proteins. However, they show no similarity either to the equilibrative transporters or to other mammalian Na^+-dependent transporters. Therefore, they form a novel nucleoside transporter family called the concentrative nucleoside transporter (CNT) family (Cass et al., 1999). Although not completely confirmed by experimental evidence, it appears that CNT proteins span the membrane 13 times in the form of α-helices (see Figure 8.6), with helices 7 and 8 determining substrate selectivity (Hamilton et al., 1997).

8.5 NUCLEOTIDES AND NUCLEOSIDES AS INTERCELLULAR MESSENGERS

The important role of nucleotides in cellular metabolism and their consequent ubiquitous nature in the brain were probably the basis for the evolutionary selection of NTPs, NDPs, and nucleosides for the complex processes of intercellular signaling and signal modulation in the central nervous system (CNS). NTPs and nucleosides with a role in neurotransission/neuromodulation are classified as the purinergic system of the brain, which includes ATP, diadenosine polyphosphates, and adenosine (Linden, 1999). However, some evidence suggests a role for the pyrimidines UTP and UDP as well as the nucleoside uridine in intercellular signaling within the brain (Connolly and Duleu, 1999; Kimura et al., 2001).

ATP and diadenosine polyphosphates (DAPP) are neurotransmitters, released not only from purinergic neurons, but also from cholinergic and adrenergic neurons, with an estimated ratio for ATP:Ach/noradrenaline of 1:10 in the nerve terminals (Linden, 1999). This fact suggests that release of ATP from neurons is relatively constant. Once in the ISF of the brain, it interacts with several classes of receptors, the ionotropic P2X receptors and the G-protein–coupled P2Y receptors (Ralevic and Burnstock, 1998). However, ATP in the brain ISF is undetectable due to constant and rapid enzymatic degradation first into AMP and then into adenosine (Linden, 1999; Zimmermann, 1996). It is estimated that complete degradation of ATP into adenosine under normal resting conditions occurs within 1 second of release, so the constant release of ATP from neurons results in the constant production of adenosine

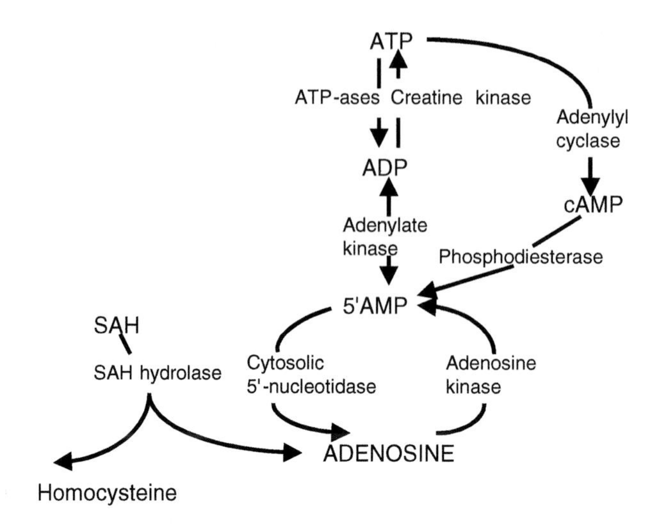

FIGURE 8.7 Main intracellular pathways of adenosine formation and utilization. The energy charge of the brain is usually ~0.95, which means that the intracellular ATP/AMP ratio is very high due to a rapid phosphorylation of 5'AMP into ATP, so cytosolic 5'-nucleotidase presence is far from saturation. However, in the case of any energy deprivation (e.g., brain hypoxia/ischemia), the rate of ATP hydrolysis exceeds the rate of formation, so 5'AMP concentration increases promptly, finally leading to the saturation of 5'-nucleotidase and an increase in the ISF adenosine concentration. Besides this main pathway of adenosine production in the cell, some amount of this nucleoside could be produced by the hydrolysis of S-adenosyl homocysteine, but this production does not depend directly on the energy charge of the cell. Some intracellular adenosine can then leave the cell via nucleoside transporters, the rest being salvaged into AMP through the action of adenosine kinase (AK). AK isoenzymes in the brain have a very low K_m value, which is believed to be important to maintain the low concentration of free adenosine in the brain ISF.

in the brain ISF. In addition, a certain amount of adenosine is produced intracellularly from ATP and S-adenosyl homocysteine through complex regulated pathways (see Figure 8.7) and leaves the cells via certain nucleoside transporters in the membrane. This is a minor source under normal resting conditions, but may become predominant when a mismatch between energy supply and demand occurs (Hagberg et al., 1987; Melani et al., 1999). Another minor pathway of extracellular adenosine production is the hydrolysis of extracellular cAMP through the action of ecto-cAMP phosphodiesterase (Rosenberg and Li, 1995).

In a number of tissues, extracellular adenosine acts as a metabolic messenger, but in the CNS it also acts as a neuromodulator; it is very likely that this neuromodulatory effect evolved during the evolution of its original role as an intercellular metabolic messenger. Its neuromodulatory effects in the brain are mediated through the activation of four classes of P1 purinergic receptors. Under normal resting conditions the effects of adenosine are mediated mainly through high affinity A_1 and A_{2A} receptors, which implies that these receptors are responsible for the basal "tone" of adenosine in the CNS, producing inhibition of adenylate cyclase, inhibition of

Ca^{2+} channels, and activation of phospholipase C. These effects cause hyperpolarization of neurons, inhibition of neuronal release of glutamate and of other excitatory neurotransmitters, and attenuation of glutamate receptor activity (von Lubitz, 1999; Dunwiddie and Masino, 2001). Therefore, it is believed that the accumulation of adenosine in the brain ISF within physiological limits (which is likely to be 220 nM) (Melani et al., 1999) causes sedation of the CNS, could induce sleep and has a general neuroprotective effect (von Lubitz, 1999; Parkinson et al., 2000). However, when adenosine in the ISF increases excessively above physiological limits (e.g., during hypoxia, ischemia, or hypoglycemia), it can also produce neurotoxic effects, possibly through the activation of low-affinity A_{2B} and A_3 receptors (Latini and Pedata, 2001). Therefore, although highly regulated, the net effect of adenosine in the brain also depends on its concentration in the brain ISF, which indicates the importance of adenosine homeostasis in the ISF for the normal CNS function.

There are also experimental data that another nucleoside, uridine, may also act as a neuromodulator with effects similar to those of adenosine (Kimura et al., 2001). Some recent studies produced pharmacokinetic evidence that uridine receptors, distinct from adenosine receptors, exist in the CNS. Uridine in the ISF is mainly produced by extracellular degradation of UTP and UDP released from neurons as neurotransmitters (Linden, 1999). It binds to its receptor, which is particularly abundant in the regions of the brain that control sleep (Kimura et al., 2001). There is also some evidence that neuromodulation by uridine is important for normal cognitive function in humans (Fornai et al., 2002). However, there are no available data for the structure of the uridine receptor or the sequence that encodes it, which limits further research on this issue.

Some recent studies also indicated that other pyrimidine nucleosides, such as cytidine, might play a role in modulation and could have some importance in cognitive function in primates and humans (Fornai et al., 2002).

8.6 CONCENTRATION GRADIENTS BETWEEN ISF AND CSF

Measurements of nucleoside concentration in various brain regions have been performed using several methods attempting to minimize postmortem breakdown of NTP into NMP and nucleosides. These methods include the immersion of animals/decapitated heads into liquid nitrogen, in situ freezing of the brain with liquid nitrogen (Ponten et al., 1973; Isakovic et al., 2004), and 3–10 kW microwave irradiation of the head (Delaney and Geiger, 1996). Although these methods are suitable for various metabolic studies, the data they produce concerns the average concentrations of these molecules in the tissue and not in the ISF, so most of the available data on nucleoside concentrations in the brain ISF has been obtained by sampling of the ISF in vivo using microdialysis probes. These samples were then analyzed by high-performance liquid chromatography (HPLC) with the detection of ultraviolet (UV) light absorption or, alternatively, by HPLC-fluorimetric analysis (Zhang et al., 1991; Saito et al., 1999; Isakovic et al., 2004).

Most of the data on nucleoside/nucleobase concentrations in the CSF in experimental animals has been obtained by the analysis of samples collected by the puncture of the cisterna magna. Therefore, it represents mainly ventricular CSF, and there is a complete lack of data regarding concentrations of these molecules in the subarachnoid CSF of experimental animals, which could be of importance in elucidating the flux of these molecules between the two fluid compartments, CSF and ISF. On the other hand, all available data in humans has been obtained from CSF collected by lumbar puncture (see Eells and Spector, 1983).

Figure 8.8 depicts typical chromatograms obtained by HPLC analysis of the CSF, brain ISF (microdialysate), and deproteinized whole brain homogenate samples (Dobolyi et al., 1998, 2000). The estimated concentrations of these molecules are presented in Table 8.2. From these values it is obvious that the concentrations of purine nucleosides in the CSF is very low, within the nanomolar range, while the estimated concentrations of these molecules in the brain ISF, especially in the areas remote from the ventricular system (Hagberg et al., 1987), are much higher, which indicates a net ISF-to-CSF concentration gradient. The concentrations of purine nucleobases in the CSF are one order of magnitude higher than the concentrations of nucleosides, within the micromolar range. However, concentrations of purine nucleobases in the brain ISF are also very high when compared to the concentrations of nucleosides, probably due to constant production via catabolism of nucleotides. Therefore, it appears that no concentration gradient for these molecules exists between CSF and ISF. In the case of pyrimidines, the concentration of both nucleosides and nucleobases in the CSF is higher than the concentration of purines and also higher than the concentration of pyrimidines in the ISF, indicating a net CSF-to-ISF concentration gradient for these molecules. However, as shown in Table 8.2, the estimated concentrations of purines in the ISF sampled by microdialysis from the area near the cerebral ventricles are closer to the concentrations of these molecules in the CSF, indicating that the distance from the cerebral ventricles/subarachnoid space and not the presence of ependymal layer appears to be the limiting factor for the free diffusion of these molecules between brain ISF and CSF.

Due to its importance in neuromodulation, adenosine concentration in the brain ISF, CSF, and plasma has been extensively studied. However, a number of studies over the last decade have reported a wide range of adenosine concentrations in the ISF, from 100 nM to 5 μM (see Table 8.3). Such a wide range of concentrations might be confusing, but after careful examination of the experimental conditions used to obtain ISF samples, it is clear that these differences were probably the consequence of various degrees of tissue damage/hypoxia during and after insertion of the microdialysis probe: all the studies in which samples had been collected immediately or within 24 hours of probe insertion reported very high, micromolar concentrations, while studies that allowed >24 hours between probe insertion and ISF sampling reported concentrations within the narrow limit of 120 and 250 nM (see Table 8.3). The concentration of this molecule in rat CSF appears to be within the same range, about 180 nM (Isakovic et al., 2004).

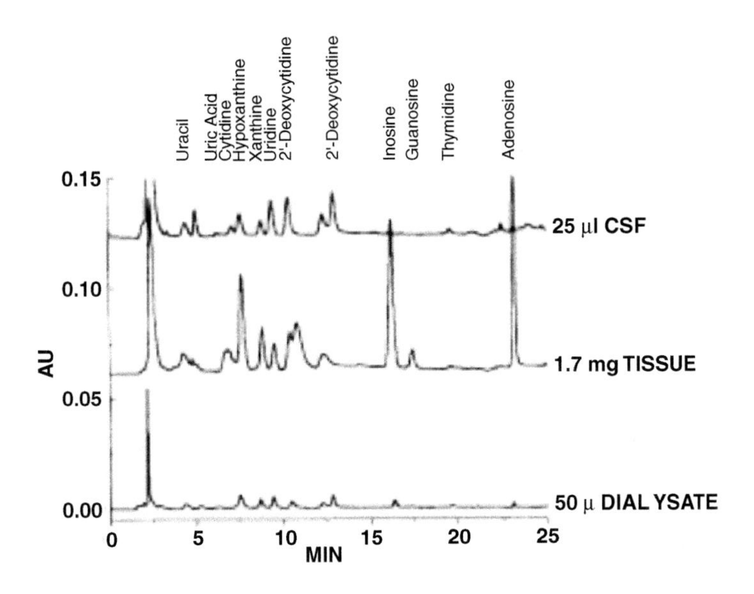

FIGURE 8.8 Typical HPLC chromatogram of the CSF, brain tissue homogenate, and brain ISF samples. Brain ISF samples were obtained by microdialysis and the curves in the chromatograms represent UV light absorption over the time of elution. The concentration of nucleosides/nucleobases in these samples was then estimated (see Table 8.2). [Reprinted with permission from Elsevier (*Neurochem. Int.* 32, 1998, 247–256).]

8.7 CONCENTRATION GRADIENTS BETWEEN THE BRAIN FLUIDS AND PLASMA FOR ADENOSINE, URIDINE, AND HYPOXANTHINE

The reported values for nucleoside and nucleobase concentrations in plasma indicate a wide range of concentrations, especially for adenosine and inosine; moreover, these concentrations seem to depend largely on the conditions of the subject at the time of blood collection. For example, the plasma concentration of adenosine increases in moderate hypoxia, exercise, fever, respiratory stimulation, or allergic reactions (Reid et al., 1991; Saito et al., 1999; Obata et al., 2001; Lo et al., 2001; Vizi et al., 2002). In contrast, the concentration of this molecule in the brain ISF and CSF remains constant, which emphasizes the importance of the blood–brain barrier (BBB) and the blood–cerebrospinal fluid barrier (BCSFB) in the homeostasis of adenosine in CNS.

Inosine or uridine or both are the principal nucleosides in plasma of most mammals and are present in micromolar concentrations (Cory, 1997). The reported values for nucleobase concentrations in plasma differ from 10^2 nM in humans

TABLE 8.2
Concentrations of Purine and Pyrimidine Nucleobases, Nucleosides, and Deoxynucleosides in Rat Brain ISF and Rat CSF

		Brain Interstitial Fluid (μM)	
	Cerebrospinal Fluid (μM)	Close to Ventricles	Remote from Ventricles
Purines			
Adenosine	0.09 ± 0.01	0.97 ± 0.04[a]	1.92 ± 0.35[c]
Inosine	0.14 ± 0.005	0.72 ± 0.06[a]	1.50 ± 0.17[c]
Hypoxanthine	3.78 ± 0.27	4.62 ± 0.24[a]	4.82 ± 0.43[c]
Xanthine	1.31 ± 0.08	1.26 ± 0.11[a]	3.95 ± 0.47[c]
Uric acid	0.96 ± 0.06	1.15 ± 0.13[b]	
Guanosine	0.02 ± 0.004	0.24 ± 0.02[a]	
2'-deoxyadenosine	0.008 ± 0.002	0.17 ± 0.04[b]	
Pyrimidines			
Uracil	2.66 ± 0.027	1.18 ± 0.21[a]	
Uridine	3.83 ± 0.40	0.79 ± 0.07[a]	
Cytidine	0.51 ± 0.05	0.11 ± 0.01[b]	
2'-deoxycytidine	7.35 ± 0.43	1.10 ± 0.07[a]	
2'-deoxyuridine	4.82 ± 0.19	0.96 ± 0.04[a]	
Thymidine	0.76 ± 0.02	0.22 ± 0.02[a]	

Concentrations in brain ISF were estimated by HPLC analysis of an aliquots obtained by sampling fluid from the rat brain by microdialysis. Concentrations in the CSF were estimated by HPLC analysis of fluid sampled by the puncture of cisterna magna.

Source:
[a,b] Concentrations of these molecules sampled by microdialysis from the rat thalamus, relatively close to cerebral ventricles (Dobolyi et al., 2000).
[a,b] Concentrations of these molecules sampled by microdialysis from the regions remote from cerebral ventricles (Hagberg et al., 1987).

(Eells and Spector, 1983) to micromolar values in sheep (Redzic et al., 2002). The concentration of uric acid in human plasma appeared to be extremely high, ~0.26 mM, almost 10^3-fold higher than the concentration of nucleobases (Eells and Spector, 1983), while the concentration of this molecule in sheep plasma is 150-fold lower than in humans, which might also be a consequence of the different fate of urate in primates versus other mammals (Bhaghavan, 2001).

Table 8.4 shows the concentrations of nucleosides and nucleobases estimated in sheep CSF and plasma by HPLC with UV detection and HPLC-fluorimetric analysis (the latter was applied to estimate adenosine concentration). If these values are compared, it is clear that a concentration gradient from CSF toward plasma exists for adenosine, guanosine, and hypoxanthine. This suggests that these molecules probably originate from the catabolism of nucleotides in the brain, so choroid plexus (CP) epithelium might play a role in providing additional efflux of these molecules from the brain. On the other hand, a high gradient toward CSF exists for xanthine

TABLE 8.3
Adenosine Concentrations in the Brain ISF, CSF, and Plasma

Brain ISF (μM)	CSF (μM)	Plasma (μM)
1.9[1] (1.5 h)	0.09[7]	0.016–0.24[9]
1.8[2] (2 h)	0.18[8]	0.09[8]
10[3] (0 h)	0.16[16]	0.07[11]
0.12[4] (>24 h)	0.016[12] (humans)	0.059[17] (humans)
0.072[5] (>24 h)	0.018[14] (humans)	0.02[18] (humans)
0.21[6] (>24 h)	0.014[15] (humans)	
0.012–0.025[13]		
0.95[7]		

All of the reported values for ISF and most of the values for CSF and plasma were obtained using samples from rat; some data for CSF and plasma were obtained from human samples. The concentrations were estimated by HPLC or HPLC-fluorometric analysis of the samples; in the case of ISF they were collected by microdialysis from various brain regions. CSF was sampled by the puncture of cisterna magna (rat) or by lumbar puncture (humans). Plasma samples were obtained from blood, which was collected using a syringe with the stopping solution (for the composition of most commonly used stopping solutions, see Zhang et al., 1991 or Saito et al., 1999).

Source: [1]Hagberg et al., 1987; [2]Zettestorm et al., 1982; [3]Ballarin et al., 1987; [4]Pazzagli et al., 1994; [5]Pazzagli et al., 1993; [6]Melani et al., 1999; [7]Dobolyi et al., 1998; [8]Isakovic et al., 2004; [9]Yada et al., 1999; [11]Lo et al., 2001; [12]Ohisalo et al., 1983; [13]Bashera et al., 1999; [14]Shore et al., 2004; [15]Clark et al., 1997; [16]Meno et al., 1991; [17]Vizi et al., 2002; [18]Saito et al., 1999.

and uric acid, meaning that CP epithelium constitutes an important barrier that prevents influx of these molecules into the CSF.

In the case of brain ISF, a concentration gradient from ISF toward plasma exists for hypoxanthine and probably the nucleosides adenosine and uridine. Such a gradient might be the driving force for facilitated diffusion of these molecules from the ISF to the blood. Such an efflux process would have strong physiological logic, since accumulation of these two nucleosides could affect brain function due to their effects as neuromodulators, and hypoxanthine appears to be the final product of purines catabolism in the brain. On the other hand, a concentration gradient in the opposite direction appears to exist for inosine, thymidine, xanthine, uric acid, and uracil, which indicates that the BCSFB disables to some degree the entry of these molecules into the ISF.

The fact that a concentration gradient exists between CSF/ISF and plasma for a number of nucleosides/nucleobases emphasizes the importance of transport and

TABLE 8.4
Concentrations of Nucleosides and Nucleobases in Sheep Plasma and CSF Determined by HPLC Analysis[a]

	Inosine	Adenosine	Guanosine	Hypoxanthine	Xanthine	Uric acid	Adenine	Uracil
Plasma	3.12 ± 0.41	0.11 ± 0.02	0.06 ± 0.02	7.23 ± 0.77	2.77 ± 0.13	1.81 ± 0.14	19.74± 3.32	3.68 ± 0.35
CSF	2.08 ± 0.31	0.19 ± 0.04	0.09 ± 0.02	15.13± 0.84	0.81 ± 0.21	0.43 ± 0.17	7.87 ± 2.03	3.61 ± 0.34

[a]All values in μM.
CSF and blood samples were collected by puncture of cisterna magna and jugular vein of anesthetized sheep, respectively. Each syringe contained stopping solution designed to prevent any exchange of molecules between plasma and blood cells (for details, see Saito et al., 1999, and Isakovic et al., 2004). Deproteinized CSF and plasma samples were then injected into the C-18 HPLC column and analysis performed as explained in Redzic et al., 1998, and Isakovic et al., 2004, with the UV light absorbance detection. The limit of detection under these conditions was 0.4 to 0.5 μM. Adenosine concentration was well below that limit, so it was estimated using the HPLC-fluorometric analysis (for details, see Zhang et al., 1991; Saito et al., 1999; Isakovic et al., 2004). Values are presented as mean ± SEM; number of samples for each value was 4–6.

metabolic processes at the cerebral capillaries, which form the BBB, and at the CP epithelium, which forms the BCSFB in vivo.

8.8 NET FLUX OF NUCLEOSIDES AND NUCLEOBASES FROM THE CP ISF TO CSF

Although various aspects of nucleoside and nucleobase homeostasis in the brain have been extensively studied, there is a surprising lack of data regarding the net flux of nucleosides and nucleobases from CP ISF into the CSF and vice versa. The main reason for this could be that no technique presently available can measure net transepithelial flux of test molecules in vivo.

8.8.1 Studies on Isolated CP

Pioneering studies in the early 1980s investigated nucleoside transport in isolated rabbit CP. These studies provided evidence that rabbit CP contains both concentrative systems and an equilibrative efflux system for nucleosides (Spector and Eells, 1984, Spector and Huntoon, 1984, Spector 1985, 1985a, 1986). Using the same technique, Wu demonstrated that a *cib* Na^+-nucleoside transport system is operable in rabbit CP (Wu et al., 1992, 1994). Using ATP-depleted rabbit CP slices, they showed that uridine and thymidine accumulated in the slices against a concentration gradient in the presence of an inwardly directed Na^+ gradient, and this Na^+-driven uptake was saturable with K_m values of 18.1 and 13.0 μM, respectively. Na^+-driven uridine uptake was inhibited by naturally occurring ribo- and deoxyribo-nucleosides, while both purine and pyrimidine nucleosides were potent inhibitors of Na^+-dependent thymidine transport with IC_{50} values ranging between 5 and 23 μM. Data from these studies suggest that both purine and pyrimidine nucleosides are substrates of a single Na^+-nucleoside transport system, later designated as the *cib* system. These studies also suggested the presence of *es* and *ei* transporters at the CSF-facing side of the rabbit CP.

Using isolated rat CP incubated in medium with radiolabeled nucleosides, Wu et al. (1993) demonstrated the presence of both Na^+-dependent and Na^+-independent nucleoside transport mechanisms at the CSF-facing side. Formycin B accumulated in the presence of an inwardly directed Na^+ gradient, suggesting a Na^+-dependent *cif* transport system at the apical side of CP epithelium. Additional studies were carried out in the presence of NBTI. In the absence of Na^+, the volume of distribution of formycin B decreased significantly with the addition of NBTI, suggesting that an *es* transport system is also present in the apical membrane.

Although these studies revealed that both Na^+-dependent and Na^+-independent nucleoside transport was present in the CP tissue, the fact that they used isolated CP tissue incubated in the medium suggests that the results represent mainly transport processes across the apical (CSF-facing) side of the tissue and do not provide any data regarding the real flux of these molecules across the CP epithelium.

8.8.2 Intraventricular Injections of Radiolabeled Test Molecules: Preliminary Data on Transport from the CSF

In order to study elimination from the CSF via the *cif* system in rat, [^{3}H]formycin B (used as a model substrate) was directly injected into the lateral ventricle with [^{14}C]inulin as a marker of bulk flow of the CSF. Then the radioactivity of CSF samples in [^{3}H] and [^{14}C] channels was determined over time. Formycin B showed a clearance value higher than that of inulin, suggesting that it is eliminated by pathways other than bulk flow, and this process appeared to be saturable. When NBTI was injected simultaneously, the clearance of [^{3}H]formycin B was significantly reduced, confirming the presence of an equilibrative transport process at the apical side, which was also demonstrated in the in vitro studies, as explained above.

It has been shown that, after intraventricular injection of [^{14}C]hypoxanthine, the [^{14}C]radioactivity was rapidly cleared from the CSF into the blood as intact [^{14}C]hypoxanthine or accumulated in the brain mainly as [^{14}C] NTPs and NDPs (Spector, 1988). Negligible amounts of [^{14}C]xanthine and no [^{14}C]uric acid or allantoin were formed (Spector, 1988). Contrary to this rapid clearance from the CSF, it has been shown that, during intravenous infusion of [^{3}H]hypoxanthine in rabbit, [^{3}H]radioactivity entered CSF slowly and was converted in the brain to nucleotides. This suggests that efflux of hypoxanthine from the CSF and not influx into CSF across the CP epithelium might be the physiological role of CPs in nucleobase/nucleoside homeostasis in the brain and also confirms other evidences that hypoxanthine is the main end product of purine catabolism in the brain.

8.8.3 Studies Using Isolated Sheep CP Perfused In Situ

A number of studies have been performed using continuous in situ perfusion of isolated sheep choroid plexus. This technique is basically designed to measure the "loss" of radioactive tracer (test molecule) from perfusate during a single pass through the choroidal circulation; therefore, it enables the determination of transport properties of the basolateral side of CP epithelium, facing the ISF of the CP. Using this technique, it was revealed that the basolateral membrane is very permeable to purine nucleosides and less permeable to pyrimidines (see Figure 8.9) (Redzic et al., 1997). The values for the uptake of purine nucleobases were very close to those for purine nucleosides, while the uptake of pyrimidine nucleobases was significantly lower (Redzic et al., 2001). This uptake was Na^+-independent and sensitive to NBTI, indicating the presence of the *es* transport system in the basolateral membrane of the sheep CP. Kinetic properties of hypoxanthine uptake indicated that the physiological role of this transport system might be to remove (from the cells) hypoxanthine that has entered these cells from the apical (CSF) side (Redzic et al., 2002). When radiolabeled adenosine was infused through the CP capillaries, only radiolabeled hypoxanthine appeared in the newly formed CSF, collected from the apical side of the isolated CP, indicating that CP epithelium represents an enzymatic barrier for circulating adenosine (Redzic et al., 1997).

All studies mentioned above provided important data on the transport of nucleosides and nucleobases at the CP. However, none of them provided accurate data about the net transport (flux) of nucleosides and nucleobases across the CP epithelium in vivo.

8.8.4 Primary Culture of CP Epithelium: A Tool to Study Transepithelial Flux of Nucleosides and Nucleobases

In order to overcome this problem, the primary culture of sheep CP epithelium as a monolayer on permeable supports (plastic filters in a multiwell plate) was developed (see Figure 8.10). The studies performed so far indicate that, for all the tested nucleosides and nucleobases, flux from the apical chamber (which corresponds to the CSF side in vivo) to the basolateral chamber (which corresponds to the CP ISF side in vivo) exceeds flux in the opposite direction. This supports the hypothesis that the efflux of these molecules from the CSF into the blood could be an important physiological function of CPs. Only the net flux of the nucleobase adenine was in the opposite direction, from the basolateral to the apical side. RT-PCR studies revealed the presence of ENT1 and ENT2 equilibrative transporters as well as CNT2 concentrative transporter in CP epithelial cells in primary culture (Z.B. Redzic et al., unpublished). Functional studies that measured the cellular uptake across the membrane facing one of these two sides revealed that the Na^+-dependent uptake is present only on the apical side. Equilibrative transport, not sensitive to the presence

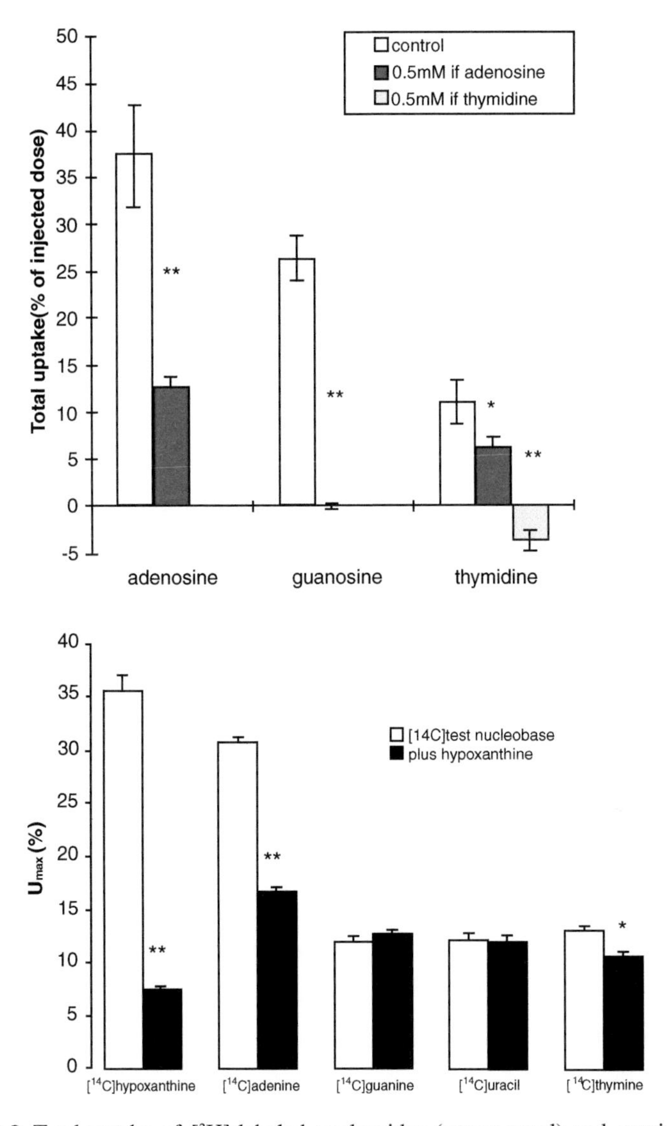

FIGURE 8.9 Total uptake of [^{3}H]-labeled nucleosides (upper panel) and maximal uptake (U_{max}) of [^{14}C]-labeled nucleobases (lower panel) after single pass through the capillaries of the isolated sheep CP perfused in situ in the control group and after addition of unlabeled molecules. Values are mean ± S.D. $n = 4$–6. $^{**}p < 0.05$; $^{*}p < 0.01$. [Reprinted with permission from Elsevier (*Brain Res.* 767, 1997, 26–33; upper panel) and (*Brain Res.* 888, 2001, 66–74; lower panel).]

of 1μM of NBTI, was also detected on this side, while the uptake of adenosine across the other side (basolateral) was >40% sensitive to NBTI and independent of Na^+. The proposed model of transport across these cells is shown in Figure 8.10.

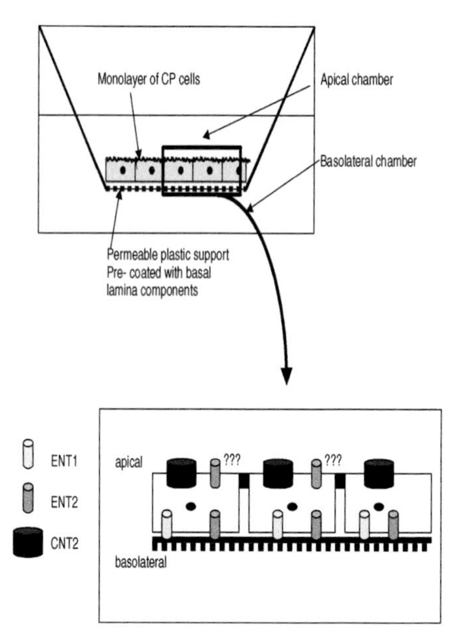

FIGURE 8.10 A schematic representation of the model of sheep CP epithelial cells in primary culture on plastic permeable supports (filters). The enlarged rectangle represents a proposed model for nucleoside transporter distribution in these cells based on PCR data and functional uptake studies. The CNT2 transporter appears to be present only at the apical side, so it could mediate concentrative transport of purines and hypoxanthine from the apical chamber (which corresponds to the CSF in vivo) into the cell. Uptake studies suggest that the ENT2 transporter might also be present on this side, but due to the lack of specific synthetic inhibitor of this transporter, this conclusion is not completely convincing. Functional studies suggest that only equilibrative transport mediates the uptake of nucleosides across the basolateral side, which means, together with data from PCR analysis, that both ENT1 and ENT2 transporters are operable in the basolateral membrane.

8.8.5 Studies on Human CP

Studies on nucleoside transport between blood and brain in humans or other primates are quite limited. However, the presence of Na^+-dependent nucleoside transport in isolated choroid plexus from both human and rhesus monkey has been demonstrated (Washington et al., 1998). In the absence of Na^+, both thymidine and guanosine accumulated in ATP-depleted CP slices, while in the presence of an inwardly directed Na^+ gradient, both thymidine and guanosine accumulated in the tissue slices against the nucleoside concentration gradient. Inhibition of Na^+-dependent thymidine uptake by various nucleosides was also studied using choroid plexus from rhesus monkey. When 0.1 mM of cytidine, formycin B, and guanosine were present in the medium, the uptake of [^{3}H]thymidine was significantly inhibited, suggesting the presence of a broadly selective nucleoside transporter (Washington et al., 1998).

Overall, most of the results mentioned above suggest that transport of nucleosides and nucleobases across the apical (CSF) side of the CP epithelium is Na^+-dependent, mediated by some of the concentrative transporters, while transport across the basolateral membrane is equilibrative. This suggests that efflux transport from the CSF into the blood across the CP epithelium might be an important physiological role of CPs in vivo.

8.9 ROLE OF THE BBB IN FACILITATING THE EFFLUX OF EXCESS NUCLEOSIDES AND NUCLEOBASES FROM THE BRAIN

The homeostasis of nucleosides and nucleobases in the brain, as well as the relative importance of transport of these molecules across the CPs, cannot be elucidated without a comprehension of the role of the BBB, since this represents the largest surface of contact between the blood and the brain ISF. Therefore, a number of studies have been performed using various techniques, from whole animal studies to three-dimensional reconstitution of the BBB in vitro, to elucidate transport of nucleosides across the BBB.

8.9.1 BUI Studies: Uptake of Nucleosides and Nucleobases from the Blood into the Brain

Initial data on this uptake were obtained using the brain uptake index (BUI) technique (Oldendorf, 1970). This technique has a great advantage when compared to brain perfusion techniques, developed later, in that the cerebral circulation is intact during the experiment. The injected bolus does not mix significantly with the circulating blood (Pardridge et al., 1985), and cerebral blood flow (CBF) under anesthesia is reduced only 30 to 40% when compared to that of a conscious rat (Pardridge and Ferier, 1985; Isakovic et al., 2004).

These early studies indicated that the BUI values for nucleosides and nucleobases were significantly higher than those for the vascular space markers, but lower than those for the essential amino acids or hexoses. It was found that the uptake of adenine, hypoxanthine, adenosine, guanosine, inosine, and uridine was saturable (Cornford and Oldendorf, 1975). These studies also indicated that the brain uptake of nucleosides was mediated by a separate transport system from that mediating the brain uptake of nucleobases. They estimated K_m values of 27 μM and 18 μM for the nucleobase and nucleoside transporter, respectively (Cornford and Oldendorf, 1975). The kinetics of adenosine brain uptake were also examined using the BUI technique, and this study revealed that the K_m of adenosine uptake was 23 μM (Pardridge, 1983), close to the values mentioned above.

No analytical technique at that time was sensitive enough to accurately measure the concentration of nucleosides and nucleobases, especially adenosine, in rat plasma. Therefore, based purely on the findings of BUI studies, it was believed that the BBB represents an important surface for the entry of nucleosides and nucleobases into the brain ISF. However, keeping in mind the concentration of adenosine in rat plasma estimated by HPLC-fluorometric analysis (~0.1 μM, see above), it is clear that this transport system at the blood side of the BBB is less than 0.5% saturated under normal resting conditions, so the real influx of adenosine from the blood into the brain endothelial cells in vivo is likely to be negligible.

Specific nucleobase transport has been postulated at the BBB in a study in which [^{3}H] hypoxanthine was continuously infused intravenously (Spector, 1988) and also using BUI technique (Betz, 1985). Data from these studies suggests that transport of hypoxanthine from plasma into the brain is mediated by a saturable transport system at the BBB (Spector, 1988). The results from both studies also suggest that the brain, contrary to other tissues, does not catabolize hypoxanthine to xanthine and uric acid, but uses most of the injected radiolabeled hypoxanthine for salvage pathways (which are shown in the Figure 8.2), to produce purine nucleotides (Spector, 1988). However, after the intracarotid injection of radiolabeled hypoxanthine, it was revealed that brain endothelial cells convert hypoxanthine into uric acid (Betz, 1985).

8.9.2 Brain Perfusion Studies

A brain perfusion study in the rat has revealed rapid unidirectional uptake of adenosine into the brain (Pardridge et al., 1994), with a K_m of only ~1 μM and V_{max} of ~200 pmol/min/g. This study also indicated that the *ei* transport system was not operable at the blood-facing side of cerebral capillaries, since NBTI did not inhibit adenosine uptake. This study revealed for the first time that adenosine taken up by the endothelial cells was rapidly metabolized inside these cells, so only ~10% of radioactivity still resided in the free adenosine pool after 15 seconds of perfusion. This was the first study that showed that the BBB might represent an enzymatic barrier for circulating nucleosides.

Brain perfusion in the guinea pig has been used to clarify the transport of other nucleosides and deoxyribonucleosides from the blood into the brain. These studies indicated presence of a NBTI-sensitive (*es*) facilitative transport system for thymidine and the likely presence of a NBTI-insensitive and/or sodium-dependent transport system at the BBB of the guinea pig (Thomas and Segal, 1996, 1997). These studies also indicated that the transport system for thymidine at the BBB has a low affinity (K_m ~200 μM), but a relatively high capacity (V_{max} = 1.06 ± 0.08 nmol min-1 g-1).

Overall, BUI and brain perfusion studies have provided useful data on the brain uptake of nucleosides and nucleobases. However, these techniques measured only the brain uptake of radiolabeled molecules and provided no information about the reverse process: efflux of these molecules from the brain over the same period of time or any kind of confirmation of nucleoside transporters present at the BBB. Another disadvantage of brain perfusion techniques was that the perfusion medium in the case of nucleoside transport studies did not contain erythrocytes, which might severely change delivery of oxygen to the brain and significantly affect the energy charge of the brain cells and the concentration of free nucleosides. Based purely on the results of these techniques, it was believed that the BBB might represent the route of supply of the brain with nucleosides and nucleobases.

8.9.3 Nucleoside Transporters at the BBB

The techniques mentioned above provided no information about the fate of test molecules once they had entered brain endothelial cells, which means that it could not be clarified whether they passed into the brain as intact molecules, were metabolized and then passed into the brain ISF as metabolites, or were just trapped inside endothelial cells by phosphorylation. Another limitation of whole-animal techniques is that the possibility of isolating fresh endothelial cells after the perfusion is quite limited, since all available techniques for the rapid isolation of these cells from brain homogenates give a cell fraction that is heavily contaminated with other cell types. Therefore, most of the accurate data about the expression and presence of nucleoside transporters has been acquired using primary culture of rat brain endothelial cells or culture of the immortalized rat brain endothelial cell line RBE4 (Abbott et al., 1995).

Functional studies on RBE4 cells grown on plastic permeable supports revealed that both concentrative and equilibrative nucleoside transporters exist in these cells. Equlibrative transport was inhibited by NBTI, and the uptake of radiolabeled adenosine exhibited a biphasic sigmoid curve in the presence of this inhibitor, with IC_{50} values of 20 nM and 5.6 μM (see Figure 8.11A) (Chishty et al., 2003). This figure indicates that both *ei* and *es* transport systems were operable in these cells. Adenosine uptake was also inhibited in the presence of formycin-B and thymidine in the medium, but not in the presence of tubercidin (see Figure 8.11B), indicating that *cit* and *cif* but not *cib* concentrative nucleoside transport systems are operational in these cells (Chishty, 2003). When total RNA was isolated from these cells and subjected to RT-PCR, the presence of mRNA for rENT1, rENT2, rCNT1, rCNT2, and rCNT3 nucleoside transporters was revealed after 35 cycles, although the expression of these mRNA (relative to the expression of mRNA for the housekeeper gene GAPDH) differed among each other (Dolman et al., unpublished). Keeping in mind the results of the functional studies mentioned above (Chishty et al., 2003), it seems that mRNA for CNT3 (which corresponds to the transporter classified as *cib* in functional studies) exists in these cells, but it is likely that it is not being translated into protein or that the protein is not functional in the membrane.

Primary culture of rat brain endothelial cells was also used to elucidate expression of various transporters recently. Real-time PCR studies have revealed that mRNA for rENT1 and rENT2 equilibrative transporters are present in these cells and the amount [relative to the amount of mRNA for housekeeper gene β-micro globulin (BMG)] did not differ with the number of passages (Z.B. Redzic et al., unpublished); mRNA for rCNT2 was also present in all samples (<1% of the level of mRNA for BMG), while the level of mRNA for rCNT1 and rCNT3 differs: these mRNA were present in the cells after the 1st and 2nd passage at the levels of <0.5% than the level of mRNA for BMG, but the levels increased after the 6th passage, indicating that the presence of these transporters might be a sign of dedifferentiation of brain endothelial cells.

Preliminary functional studies on primary culture of rat brain endothelial cells (RBECs) on plastic permeable supports indicate that uptake of radiolabeled adenosine from the apical chamber (which corresponds to blood side in vivo) is

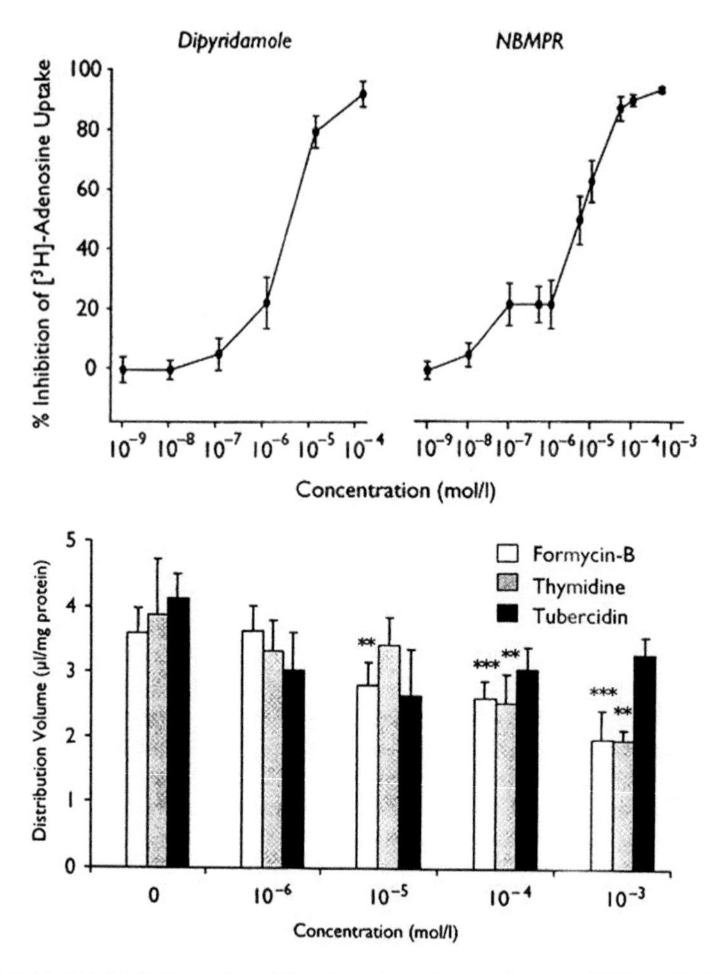

FIGURE 8.11 (**A**) Inhibition of equilibrative adenosine uptake into RBE4 cells by dipyridamole and nitrobenzylmercaptopurine riboside (NBMPR). Logarithmic plot of the percent inhibition of [^{3}H]-adenosine uptake by RBE4 cells in Na$^+$-free medium as a function of dipyridamole (left) and NBMPR (right) concentration. Values are mean ± SEM, n = 6 wells. (**B**) Inhibition of concentrative adenosine uptake by increasing concentrations of thymidine (substrate model for *cit* transport, gray bars), formycin-B (substrate-model for *cif* transport, white bars,) and tubercidin (substrate model for *cib* transport, black bars). Incubation time was 2 min; values are mean ± SEM, *n* = 6 wells. [Reprinted with permission from Lippincott, Williams, and Wilkins (*Neuroreport* 14, 2003, 1087–1090).]

Na$^+$-independent and sensitive to dypiridamole, while the uptake from the lower chamber (corresponding to the abluminal side in vivo) is >50% Na$^+$-sensitive and also partially sensitive to dipyridamole (Z.B. Redzic et al., unpublished). These data could indicate a polarity of nucleoside transporters at the BBB, with the rCNTs present exclusively at the side of the cells facing brain ISF, while equilibrative transporters are likely to be present at both sides. Such a distribution of transporters could indicate that the role of the BBB in the homeostasis of nucleosides and nucleobases in the brain appears to be the removal of an excess of these molecules from the ISF.

8.9.4 BBB Efflux and Influx Clearance of Purines

Although numerous experimental data have been gathered using cell cultures, these results cannot be directly extrapolated to conditions in vivo. Therefore, in addition to in vitro studies, a series of experiments was performed using the brain efflux index technique (BEI) (Kakee et al., 1996) and the BUI technique, which measure BBB efflux transport and BBB influx transport, respectively. These two techniques are available only for BBB transport studies in vivo on the brain with intact vascular circulation and CBF values of >50% of the CBF values in the conscious rat.

The BEI technique revealed that the efflux of [^{14}C]purines from brain ISF to blood after intracerebral microinjection is initially very rapid, especially for nucleobases (see Table 8.5). The estimated K_{eff} values for nucleosides were one order lower than that for nucleobases (see Table 8.5) (Isakovic et al., 2002). The volumes of distribution of these molecules in the brain slices were then calculated and the BBB efflux clearances estimated. These values are also shown in Table 8.5. The BUI values for the same molecules were determined using carotid microinjections with [^{3}H]water as a reference (Oldendorf, 1970), then CBF under thiopentone anesthesia estimated using [^{14}C]N-isopropyl-p-iodoamphetamine hydrochloride (Pardridge and Fierer, 1985). The BBB influx clearance for these molecules was calculated using these BUI and CBF values (see Table 8.5). From the values in this table, it is obvious that efflux clearances are at least 2.85- to 30.25-fold greater than the corresponding influx clearances, which further supports our findings in RBECs in primary culture that the distribution of nucleoside transporters is not equal between the blood facing and the membrane facing the brain ISF.

TABLE 8.5
Estimated Values of BBB Efflux Clearances and BBB Influx Clearances for Four [^{14}C]-Labeled Purines

Test Molecule	K_{eff} (min^{-1})	Cl_{eff} (µl/min/g)	BUI (%)	Cl_{inf} (µl/min/g)	Cl_{eff}/Cl_{inf}
Hypoxanthine	210.0 ± 60.0[b]	223.3 ± 79.4	1.8 ± 0.33[c]	7.37 ± 1.27	30.25
Adenine	133.3 ± 6.0[b]	141.7 ± 8.22	2.0 ± 0.12[c]	8.61 ± 0.34	16.45
Inosine	69.3 ± 32.0[b]	72.1 ± 34.2	2.8 ± 0.41[c]	11.52 ± 1.46	6.26
Adenosine	25.1 ± 2.61[a]	27.6 ± 5.22[a]	2.4 ± 0.3[a]	9.69 ± 1.26[a]	2.85

Values are presented as mean ± SEM, n = 3–5.

[a]From Isakovic et al., 2004.
[b]From Isakovic et al., 2002.
[c]From Redzic et al., 1998.

The concentration of adenosine in plasma was almost the half the concentration reported for the brain ISF. BEI studies have revealed that both this gradient and kinetic properties of transport across endothelial cells are the driving forces for a constant net efflux of adenosine from the rat brain under normal resting conditions (Isakovic et al., 2004). However, when the kinetic parameters of efflux were compared to the kinetic parameters of adenosine uptake by brain cells (Phillips and Newsholme, 1979; Bender et al., 1980), it appeared that BBB transport makes a negligible contribution to adenosine removal from the brain ISF. The greater effectiveness of the glial/neuronal population in removing adenosine from the ISF reflects the much larger effective membrane area of glia and neurons compared to the endothelium. However, under pathological conditions such as hypoxia/ischemia, when the adenosine concentration in the ISF can increase to 15 to 40 μM (Hagberg et al., 1987; Pedata et al., 2001), tending to saturate the glial/neuronal transport system, removal by the brain endothelium may play a more significant role, accounting for up to 25% of clearance (Isakovic et al., 2004).

Another important finding from the BEI studies was that the BBB represents not only a physical but also an enzymatic barrier for adenosine, since after the intracerebral injection of [^{14}C]adenosine only [^{14}C]hypoxanthine and [^{14}C]adenine appeared in the blood collected from the jugular vein (Isakovic et al., 2004) (see Figure 8.12). This means that [^{14}C]adenosine had been metabolized almost completely into nucleobases inside endothelial cells, so only nucleobases appeared in the circulation, indicating a powerful enzymatic barrier to adenosine inside endothelial cells. Although adenosine in plasma can increase two- to threefold under certain physiological conditions, such as exercise (Vizi et al., 2002) or hypoxia (Saito et al., 1999), rapid metabolic sequestration of adenosine into nucleobases in the brain endothelium, as demonstrated in this study as well as polarised distribution of nucleoside transporters, could be important in protecting the brain from circulating adenosine. Rapid sequestration of adenosine by the brain endothelial cells in vitro was also reported for the cells grown as a cell culture cartridge, with the aim of representing a three-dimensional model of the BBB in vitro (Parkinson et al., 2003).

Overall, it could be concluded that the BBB provides an important surface for exchange of nucleosides and nucleobases between the blood and the brain ISF. Brain endothelial cells express a number of nucleoside transporters, and it appears that the distribution of these transporters between luminal and abluminal membranes is unequal, indicating polarity of the nucleoside transport at the BBB. BEI studies indicated that the possible role of the BBB in the brain homeostasis of these molecules is to provide the pathway for their efflux transport. In addition, it appears that the BBB represents a powerful enzymatic barrier to circulating adenosine, which further prevents influx of this neuroactive nucleoside from plasma.

8.10 CONCLUSIONS

Results so far suggest that the brain constantly produces nucleosides and nucleobases as products of nucleotide catabolism. Most of these molecules are then rapidly taken up by neurons and glia, which use them as a substrate for the salvage into nucleotides,

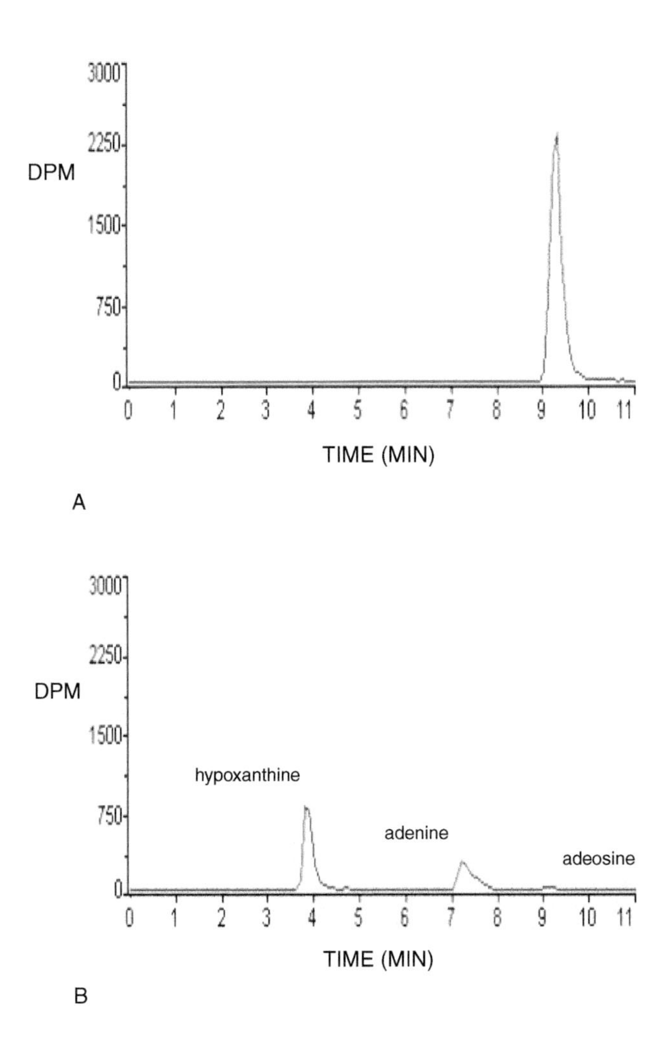

FIGURE 8.12 HPLC-radiodetector analysis of (**A**) [^{14}C]adenosine standard and of (**B**) deproteinized plasma obtained from blood collected from the jugular vein after intracerebral microinjection of bolus that contained [^{14}C]adenosine. Under these conditions of analysis, peak elution of radioactivity in the standard occurred at 9.28 min. However, a negligible amount of radioactivity was eluted at this time in the plasma sample, where about half of the radioactivity appeared in the hypoxanthine peak and the rest in the adenine peak. These peaks were identified by retention times and spectral analysis of HPLC chromatograms (not shown in the figure). [Reprinted with permission from Blackwell Publishing (*J. Neurochem.* 90, 2004, 272–286).]

a process that saves a large amount of energy for the brain. However, BBB and CP epithelium may play additional roles to remove an excess of these molecules from the brain. Both barriers (BBB and BCSFB) appear to be enzymatic barriers for circulating nucleosides, which further prevents influx of these molecules from blood into the CNS.

8.11 ACKNOWLEDGMENTS

I gratefully acknowledge Hiran Cooray for his constructive and helpful suggestions on this manuscript and Dr. Sonja Misirlic Dencic for her excellent assistance in creating the figures.

REFERENCES

N.J. Abbott, F. Roux, P.O. Couraud, et al. Studies on an immortalised brain endothelial cell line: characterisation, permeability and transport. In: J. Grenwood, D.J. Begley and M.B. Segal (eds.), New Concept of a Blood-Brain Barrier. Plenum Press, New York, 1995, 239–249.

R. Alcolado, R.O. Weller, E.P. Parish, D. Garrod. The cranial arachnoid and pia mater in man: anatomical and ultrastructural observations. *Neuropathol. Appl. Neurobiol.* 14, 1988, 1–17.

C.M. Anderson, et al. Demonstration of the existence of mRNAs encoding N1/*cif* and N2/*cit* sodium/nucleoside cotransporters in rat brain. *Mol. Brain Res.* 42, 1996, 358–361.

M. Ballarin, M. Herrera-Marschitz, M. Casas, U. Ungerstedt. Striatal adenosine levels measured '*in vivo*' by microdialysis in rats with unilateral denervation, *Neurosci. Lett.* 83, 1987, 338–344.

R. Basheera, T. Porkka-Heiskanena, D. Stenbergb, W.B. McCarleya. Adenosine and behavioral state control: adenosine increases c-Fos protein and AP1 binding in basal forebrain of rats. *Mol. Brain Res.* 73 (1–2), 1999, 1–10.

A.S. Bender, P.H. Wu, J.W. Phillis. The characterization of [^{3}H]adenosine uptake into rat cerebral cortical synaptosomes., *J. Neurochem.,* 35, 1980, 629–640.

A.L. Betz. Identification of hypoxanthine transport and xanthine oxidase activity in brain capillaries. *J. Neurochem.* 44, 1985, 574–579.

N.V. Bhagavan. *Medical Biochemistry.* Academic Press, New York, 2001, pp. 615–644.

C.E. Cass, J.D. Young, S.A. Baldwin, M.A. Cabrita, K.A. Graham, M. Griffiths, M.L. Jennings, J.R. Mackey, A. Ng, M.W.L. Ritzel, M.F. Vickers, S.Y.M. Yao. Nucleoside transporters of mammalian cells. In G.L. Amidon and W. Sadée (eds.), *Membrane Transporters as Drug Targets.* Plenum Publishing Corporation, New York, 1999, pp. 313–352.

M. Chishty, D. Begley, N.J. Abbott, A. Reichel. Functional characterisation of nucleoside transport in rat brain endothelial cells. *NeuroReport* 14, 2003, 1087–1090.

D. Choi, M. Handa, H. Young, A.S. Gordon, I. Diamond, R.O. Messing. Genomic organization and expression of mouse equlibrative, NBTI sensitive nucleoside transporter 1 (ENT1) gene. *Biochem. Biophys. Res. Commun.* 277, 2000, 200–208.

R.S. Clark, J.A. Carcillo, P.M. Kochanek, W.D. Obrist, E.K. Jackson, Z. Mi, S.R. Wisneiwski, M.J. Bell, D.W. Marion. Cerebrospinal fluid adenosine concentration and uncoupling of cerebral blood flow and oxidative metabolism after severe head injury in humans. *Neurosurgery* 41(6), 1997, 1284–1292.

D.D. Clarke, L. Sokoloff. Circulation and energy metabolism of the brain. In: Siegel G.J., Agranoff B.W., Albers R.W., Fisher S.K., Uhler M.D. (eds.), *Basic Neurochemistry.* 6th ed. Lippincot-Raven Press, New York, 1999, pp 637–671.

G.P. Connolly, J.A. Duley. Uridine and its nucleotides: biological actions, therapeutic potentials. *TiPS* 20, 1999, 218–225.

E.M. Cornford, W.H. Oldendorf. Independent blood-brain barrier transport systems for nucleic acid precursors. *Biochim. Biophys. Acta* 394, 1975, 211–219.

J.G. Cory. Purine and pyrimidine nucleotide metabolism. In: T.M. Devlin (ed.) *Textbook of Biochemistry with Clinical Correlations.* John Wiley & Sons, New York, 1997, pp. 489–524.

C.R. Crawford, D.K. Patel, C. Naeve, J.A. Belt. Cloning of the human equilibrative, nhrobenzylmercaptopurine riboside (NBMPR)-insensitive nucleoside transporter ei by functional expression in a transport-deficient cell line. *J. Biol. Chem.* 273, 1998, 5288–5293

M.P. Dacevic, J.S. Tasic, V.M. Pejanovic, D.D. Ugljesic-Kilibarda, A.J. Isakovic, D.J. Begley, M.B. Segal, L.M. Rakic, ZX.B. Redzic. The linkage of glucose to tiazofurin decreases the *in vitro* uptake into the rat glioma C6 cells. *J. Drug Target.* 10(8), 2002, 633–636.

S.M. Delaney, J.D. Geiger. Brain regional levels of adenosine and adenosine nucleotides in rats killed by high energy focused microwave irradiation. *J. Neurosci. Methods* 64, 1996, 151–156.

A. Dobolyi, A. Reichart., T. Szikra, N. Szilagyi, A.K. Kékesi, T. Karancsi, P. Slegel, M. Palkovits, G. Juhasz. Analysis of purine and pyrimidine bases, nucleosides and deoxynucleosides in brain microsamples (microdyalisates and micropunches) and cerebrospinal fluid. *Neurochem. Int.* 32, 1998, 247–256.

A. Dobolyi, A. Reichart, T. Szikra, G. Nyitrai, K.A. Kékesi, G. Juhaasz. Sustained depolarisation induces changes in the extracellular concentrations of purine and pyrimidine nucleosides in the rat thalamus. *Neurochem. Int.* 37, 2000, 71–79.

T.V. Dunwiddie, S.A. Masino. The role and regulation of adenosine in the central nervous system. *Ann. Rev. in Neurosci.* 24, 2001, 31–55.

J.T. Eells, R. Spector. Purine and pyrimidine base and nucleoside concentrations in human cerebrospinal fluid and plasma. *Neurochem. Res.* 8, 1983, 1451–1457.

A. Felipe et al. Na^+-dependent nucleoside transport in liver: two different isoforms from the same gene family are expressed in liver cells. *Biochem. J.*, 330, 1998, 997–1001.

F. Fornai, C.L. Busceti, M. Ferrucci, G. Lazzeri, S. Ruggieri. Is there a role for uridine and pyrimidine nucleosides in the treatment of vascular dementia? *Funct. Neurol.* 17, 2002, 93–99.

J.F. Ghersi-Egea, P.D. Gorevic, J. Ghiso, B. Frangione, C.S. Patlak, J.D. Fenstermacher. Fate of cerebrospinal fluid-borne amyloid β-peptide: rapid clearance into blood and appreciable accumulation by cerebral arteries. *J. Neurochem.* 67, 1996, 880–883.

D.A. Griffith, S.M. Jarvis. Nucleoside and nucleobase transport systems of mammalian cells. *Biochim. Biophys. Acta* 1286, 1996, 153–181.

M. Griffiths, S.Y.M. Yao, F. Abidi, S.E.V. Phillips, C.E. Cass, J.D. Young, S.A. Baldwin. Molecular cloning and characterization of a NBTI-insensitive (*ei*) equilibrative nucleoside transporter from human placenta. *Biochem. J.* 328, 1997a, 739–743.

M. Griffiths, N. Beaumont, S.Y.M. Yao, M. Sundaram, C.E. Bounah, A. Davies, F.Y.P. Kwong, I. Coe, C.E. Cass, J.D. Young, S.A. Baldwin. Cloning of a human nucleoside transporter implicated in the cellular uptake of adenosine and chemotherapeutic drugs. *Nat. Med.*, 3, 1997b 89–93.

H. Hagberg, P. Andersson, P. Lacarewic, I. Jacobson, S. Butcher, M. Sanberg. Extracellular adenosine, inosine, hypoxanthine and xanthine in relation to tissue nucleotides and purines in rat striatum during transient ischemia. *J. Neurochem.* 49, 1987, 227–231.

S.R. Hamilton et al. Anti-peptide antibodies as probes of the structure and subcellular distribution of Na^+-dependent nucleoside transporter rCNT1. *J. Physiol.* 499P, 1997, P50–P51.

Q.Q. Huang et al. Cloning and functional expression of a complementary DNA encoding a mammalian nucleoside transport protein. *J. Biol. Chem.* 269, 1994, 17757–17760.

A.J. Isakovic, M. Segal, B. Milojkovic, M. Dacevic, S. Misirlic, M. Rakic, Z. Redzic. The efflux of purine nucleobases and nucleosides from the rat brain. *Neurosci. Lett.* (2002), 318, 65–68.

A.J. Isakovic, J.N. Abbott, Z.B. Redzic. Brain to blood efflux transport of adenosine: blood-brain barrier studies in the rat. *J. Neurochem.* (2004), 90, 272–277.

Kakee, T. Terasaki, Y. Sugiyama. Brain efflux index as a novel method of analysing efflux transport at the blood brain barrier. *J. Pharmacol. Exp. Ther.* 277, 1996, 1550–1559.

T. Kimura, I.K. Ho, I. Yamamoto. Uridine receptor: discovery and its involvement in sleep mechanism. *Sleep* 24(3), 2001, 261–60.

S. Latini, F. Pedata. Adenosine in the central nervous system: release mechanisms and extracellular concentrations. *J. Neurochem.* 79, 2001, 463–484.

J.Y. Li, R.J. Broado, W.M. Pardridge. Differential kinetics of transport of 2', 3'-dideoxyinosine and adenosine via concentrative Na^+ nucleoside transporter CNT2 cloned from rat blood-brain barrier. *J. Pharmacol. Exp. Ther.* 299, 2001, 735–740.

J.M. Linden. Purinergic system. In: Siegel G.J., Agranoff B.W., Albers R.W., Fisher S.K., Uhler M.D. (eds.), Basic Neurochemistry, Lippincot-Raven Press, New York, 6th edition, 1999, pp 347–363.

S.M. Lo, F.M. Mo, H.J. Ballard. Interstitial adenosine concentration in rat red or white skeletal muscle during systemic hypoxia or contractions. *Exp. Physiol.* 86, 2001, 593–598.

J.R. Mackey, S.A. Baldwin, J.D. Young, C.E. Cass. Nucleoside transport and its significance for anticancer drug resistance. *Drug Resist. Updates* 1, 1998, 310–324.

J.R. Mackey, S.M. Yao, K.M. Smith, E. Karpinski, S.A. Baldwin, C.E. Cass, J.D. Young. Gemcitabine transport in *Xenopus* oocytes expressing recombinant plasma membrane mammalian nucleoside transporters. *J. Nat. Cancer Inst.* 91, 1999, 1876–1881.

A. Melani, L. Pantoni, C. Corsi, L. Bianchi, A. Monopoli, R. Bertorelli, G. Pepeu, F. Pedata. Striatal outflow of adenosine, excitatory amino acids, gamma-aminobutyric acid and taurine in awake freely moving rats after middle cerebral artery occlusion. Correlation with neurological deficit and histopathological damage. *Stroke* 30, 1999, 2448–2455.

J.R. Meno, A.C. Ngai, S. Ibayashi, H.R. Winn. Adenosine release and changes in pial arteriolar diameter during transient cerebral ischemia and reperfusion. *J. Cereb. Blood Flow Metab.* 11(6), 1991, 986–93.

T. Obata, S. Kubota, Y. Yamanaka. Histamine increases interstitial adenosine concentration via activation of ecto-5'-nucleotidase in rat hearts *in vivo*. *J. Pharmacol. Exp. Ther.* 298, 2001, 71–76.

J.J. Ohisalo, K. Murros, B.B. Fredholm, T.A. Hare. Concentrations of gamma-aminobutyric acid and adenosine in the CSF in progressive myoclonus epilepsy without Lafora's bodies. *Arch. Neurol.* 40 (10), 1983, 553–558.

W.H. Oldendorf. Measurement of brain uptake of radiolabelled substances using a tritiated water internal standard. *Brain Res.* 24, 1970, 372–376.

W.M. Pardridge. Brain metabolism: a perspective from the blood-brain barrier. Physiological Rev. 63, 1983, 1481–1535.

W.M. Pardridge, G. Fierer. Blood brain barrier transport of butanol and water relative to N-isopropyl-p-iodoamphetamine as the internal reference. *J. Cereb. Blood Flow Metab.* 5, 1985, 275–281.

W.M. Pardridge, E.M. Landaw, L.P. Miller, L.D. Braun, W.H. Oldendorf. Carotid artery injection technique: bounds for bolus mixing by plasma and by brain. *J. Cereb. Blood Flow Metab.* 253, 1985, 576–583.

W.M. Pardridge, T. Yoshikawa, Y.S. Kang, L.P. Miller. Blood-brain barrier transport and brain metabolism of adenosine and adenosine analogues. *J. Pharmacol. and Exp. Therapy* 268, 1994, 14–18.

F.E. Parkinson, J. Friesen, L. Krizanac-Bengez, D. Janigro. Use of a three-dimensional *in vitro* model of the rat blood-brain barrier to assay nucleoside efflux from brain. *Brain Res.* 980, 2003, 233–241.

F.E. Parkinson, Y.W. Zhang, P.N. Shepel, S.C. Greenway, J. Peeling, J.D. Geiger. Effects of nitrobenzylthioinosine on neuronal injury, adenosine levels and adenosine receptor activity in rat forebrain ischemia. *J. Neurochem.* 75, 2000, 795–802.

M. Pazzagli, F. Pedata, G. Pepeu. Effect of K^+ depolarization, tetrodotoxin and NMDA receptor inhibition on extracellular adenosine levels in rat striatum. *Eur. J. Pharmacol.* 254, 1993, 277–282.

M. Pazzagli, C. Corsi, S. Latini, F. Pedata, G. Pepeu. In vivo regulation of extracellular adenosine levels in the cerebral cortex by NMDA and muscarinic receptors. *Eur. J. Pharmacol.* 254, 1994, 277–282.

F. Pedata, C. Corsi, A. Melani, F. Bordoni, S. Latini. Adenosine extracellular brain concentrations and role of A2A receptors in ischemia. *An. NY Acad. Sci.* 939, 2001, 79–84.

E. Phillips, E.A. Newsholme. Maximum activities, properties and distribution of 5'-nucleotidase, adenosine kinase and adenosine deaminase in rat and human brain. *J. Neurochem.* 33, 1979, 217–225.

P.G.W. Plagemann, R.M. Wohlhueter, C. Woffendin. Nucleoside and nucleobase transport in animal cells. *Biochim. Biophys. Acta* 947 (1988) 405–443.

U. Ponten, R.A. Ratcheson, L.G. Salford, B.K. Seisjo. Optimal freezing conditions for cerebral metabolites in rat. *J. Neurochem.* 21, 1973, 1127–1138.

V. Ralevic, G. Burnstock. Receptors for purines and pyrimidines. *Pharmacol. Rev.* 50, 1998, 413–490.

J. Ralph, K. Hyde, C.E. Cass, J.D. Young, S.A. Baldwin. The ENT family of eukaryote nucleoside and nucleobase transporters: recent advances in the investigation of structure/function relationships and the identification of novel isoforms. *Mol. Membrane Biol.* 18, 2001, 53–63.

Z.B. Redzic, M.B. Segal, I.D. Markovic, J.M. Gasic, V.P. Vidovic, L.M. Rakic. The characteristics of nucleoside transport on the blood "side" of the sheep choroid plexus and the effect of NO inhibition on these processes. *Brain Res.* 767, 1997, 26–33.

Z.B. Redzic, A.J. Isakovic, M.B. Segal, S.A. Thomas, L.M. Rakic. The kinetics of hypoxanthine efflux from the rat brain. *Brain Res.* 899, 2001, 248–250.

Z.B. Redzic, I.D. Markovic, V.P. Vidovic, V.P. Vranic, J.M. Gasic, B.M. Djuricic, M. Pokrajac, J.B. Djordjevic, M.B. Segal, L.M. Rakic. Endogenous nucleosides in the guinea pig eye: analysis of transport and metabolites. *Exp. Eye Res.* 66, 1998, 315–325.

Z.B. Redzic, J.M. Gasic, I.D. Markovic, V.P. Vojvodic, V.P. Vranic, S.S. Jovanovic, L.M. Rakic. The effects of NO synthesis inhibition on the uptake of endogenous nucleosides into the rat brain. *Neurosci. Res. Commun.* 22 (1), 1998, 11–20.

Z.B. Redzic, M.B. Segal, J.M. Gasic, I.D. Markovic, A.J. Isakovic, L.M. Rakic. The kinetics of tiazofurin uptake by the isolated perfused choroid plexus of the sheep. *Methods Findings Exp. Clin. Pharmacol.* 22 (3), 2000, 149–154.

Z.B. Redzic, M.B. Segal, J.M. Gasic, I.D. Markovic, V.P. Vojvodic, A.J. Isakovic, S.A. Thomas, L.M. Rakic. The characteristics of nucleobase transport and metabolism by the perfused sheep choroids plexus. *Brain Res.* 888, 2001, 66–74.

Z.B. Redzic, M.B. Segal. The structure of the choroid plexus and the physiology of the choroid plexus epithelium. *Advanced Drug Del. Rev.* 56, 2004, *in press*.

P.G. Reid, A.H. Watt, W.J. Penny, A.C. Newby, A.P. Smith, P.A. Routledge. Plasma adenosine concentrations during adenosine-induced respiratory stimulation in man. *Eur. J. Clin. Pharmacol.* 40, 1991, 175–180.

M.W.L. Ritzel et al.. Molecular cloning and functional expression of cDNAs encoding a human Na^+-nucleoside co-transporter (hCNT1). *Am. J. Physiol.* 41, 1997, C707–C714.

M.W.L. Ritzel et al. Molecular cloning, functional expression and chromosomal localization of a cDNA encoding a human Na^+-nucleoside co-transporter (hCNT2) selective for purine nucleosides and uridine. *Mol. Membrane Biol.* 15, 1998, 203–211.

P.A. Rosenberg, Y. Li. Adenylyl cyclase activation underlines intracellular cyclic AMP accumulation, cyclic AMP transport and extracellular adenosine concentration evoked by beta-adrenergic receptor stimulation in mixed cultures of neurons and astrocytes derived from rat cerebral cortex. *Brain Res.* 692, 1995, 227–232.

H. Saito, M. Nishimura, H. Shinano, H. Makita, I. Tsujino, I. Shibuya, F. Sato, K. Miyamoto, Y. Kawakami. Plasma concentration of adenosine during normoxia and moderate hypoxia in humans. *Am. J. Respir. Crit. Care Med.* 159, 1999, 1014–1018.

P.M. Shore, E.K. Jackson, S.R. Wisniewski, R.S. Clark, P.D. Adelson, P.M. Kochanek. Vascular endothelial growth factor is increased in cerebrospinal fluid after traumatic brain injury in infants and children. *Neurosurgery* 54(3), 2004, 605–612.

R. Spector. Thymidine transport and metabolism in choroid plexus: effect of diazepam and thiopental. *J. Pharmacol. Exp. Ther.* 235, 1985, pp. 16–19.

R. Spector. Nucleoside and vitamin homeostasis in the mammalian central nervous system. *Ann. NY Acad. Sci.* 481, 1986, 221–230.

R. Spector. Uridine transport and metabolism in the central nervous system. *J. Neurochem.* 45, 1985a, 1411–1418.

R. Spector, J. Eells. Deoxynucleoside and vitamin transport into the central nervous system. *Fed. Proc.* 43, 1984, 196–200.

R. Spector, S. Huntoon. Specificity and sodium dependence of the active nucleoside transport system in choroid plexus. *J. Neurochem.* 42, 1984, 1048–1052

R. Spector. Hypoxanthine transport and metabolism in the central nervous system. *J. Neurochem.* 50, 1988, 969–978.

M. Sundaram, S.Y.M. Yao, A.M.L. Ng, M. Griffiths, C.E. Cass, S.A. Baldwin, J.D. Young. Chimeric constructs between human and rat equilibrative nucleoside transporters (hENT1 and rENT1) reveal hENT1 structural domains interacting with coronary vasoactive drugs. *J. Biol. Chem.* 273, 1998, 21519–21525.

S.A. Thomas, M.B. Segal. Identification of a saturable uptake system for deoxyribonucleosides at the blood-brain and blood-cerebrospinal fluid barrier. *Brain Res.* 741, 1996, 230–239.

S.A. Thomas, M.B. Segal. Saturation kinetics, specificity and NBMPR sensitivity of thymidine entry into the central nervous system. *Brain Res.* 760, 1997, 59–67.

E. Vizi, F. Huzsar, Z. Czoma, G. Boszormenyi-Nagy, E. Barat, I. Horvath, I. Herjavetz, M. Kollai. Plasma adenosine concentration increases during exercise: a possible contributing factor in exercise-induced bronchoconstriction in asthma. *J. Allergy Clin. Immunol.* 109, 2002, 446–448.

D.K. Von Lubitz. Adenosine and cerebral ishemia: therapeutic future or death of a brave concept? *Eur. J. Pharmacol.* 365, 1999, 9–25.

J. Wang et al. Na^+-dependent purine nucleoside transporter human kidney: cloning and functional characterization. *Am. J. Physiol.* 42, 1997, F1058–F1065.

J.L. Ward, A. Sherali, Z.P. Ma, C.M. Tse. Kinetic and pharmacological properties of cloned human equiibrative nucleoside transporters, ENT1 and ENT2, stably expressed in nucleoside transporter-deficient PK15 cells. ENT2 exhibits a low affinity for guanosine and cytidine but a high affinity for inosine. *J. Biol. Chem.* 275, 2000, 8375–8381.

C.B. Washington, K.M. Giacomini, C.M. Brett. Nucleoside transport in isolated human and rhesus choroid plexus tissue slices. *Pharmacol. Res.* 15, 1998, 1145–1147.

X. Wu, M.M. Gutierrez, K.M. Giacomini. Further characterization of the sodium-dependent nucleoside transporter (N3) in choroid plexus from rabbit. *Biochim. Biophys. Acta* 1191, 1994, 190–196.

X. Wu, A.C. Hui, K.M. Giacomini. Formycin B elimination from the cerebrospinal fluid of the rat. *Pharmacol. Res.* 10, 1993, 611–615.

X. Wu, G. Yuan, C.M. Brett, A.C. Hui, K.M. Giacomini. Sodium-dependent nucleoside transport in choroid plexus from rabbit. Evidence for a single transporter for purine and pyrimidine nucleosides. *J. Biol. Chem.* 267, 1992, 8813–8818.

T. Yada, K.N. Richmond, R. van Bibber, K. Kroll, E.O. Feigl. Role of adenosine in local metabolic coronary vasodilation. *Am. J. Physiol.* 276, 1999, H1425–H1433.

S.Y.M. Yao, A.L. Ng, W.R. Muzyka, M. Griffiths, C.E. Cass, S.A. Baldwin, J.D. Young. Molecular cloning and functional characterization of nitrobenzylthioinosine (NBMPR)-sensitive (es) and NBMPR-insensitive (ei) equilibrative nucleoside transporter praters (rENT1 and rENT2) from rat tissues. *J. Biol. Chem.* 272, 1997, 28423–28430.

J.D. Young, C.I. Cheeseman, J.R. Mackey, C.E. Cass, S.A. Baldwin. Molecular mechanisms of nucleoside and nucleoside drug transport. In: K.E. Barren and M. Donowitz (eds.), Gastrointestinal Transport. Academic Press, San Diego, CA, 2000, pp. 329–378.

T. Zetterstom, L. Vernet, U. Tossman, B. Jonzon, B.B. Fredholm. Purine levels in the intact brain, studies with an implanted perfused hollow fibre. *Neurosci. Lett.* 29, 1982, 111–115.

Y. Zhang, J.D. Geiger, W.W. Lautt. Improved high pressure liquid chromatographic-fluorometric assay for measurement of adenosine in plasma. *Am. J. Physiol.* 260, 1991, G658–664.

H. Zimmermann. Biochemistry, localization and functional roles of ecto-nucleotidases in the nervous system. *Prog. Neurobiol.* 49, 1996, 589–618.

9 Regulation of Neuroactive Metals by the Choroid Plexus

G. Jane Li and Wei Zheng
Purdue University,
West Lafayette, Indiana, U.S.A.

CONTENTS

9.1 INTRODUCTION

The choroid plexus plays a pivotal role in maintaining the homeostasis of essential metal ions in the central nervous system (CNS), which embraces the cerebrospinal fluid (CSF) compartment, the interstitial fluid (ISF) or extracellular fluid compartment, and the intracellular compartment. The choroid plexus, where the blood–CSF barrier is located, separates the CSF compartment from the systemic blood compartment. Research in the last several decades has revealed that at least 11 metals—lead (Pb), mercury (Hg), cadmium (Cd), manganese (Mn), arsenic (As), iron (Fe), copper (Cu), zinc (Zn), silver (Ag), gold (Au), and tellurium (Te)—accumulate in the choroid plexus (Zheng, 2001a, 2002), making the tissue a major target in brain for toxicities associated with environmental exposure to heavy metals.

Metals acting on the choroid plexus can be categorized into three major groups. The first group of metals directly damages the choroid plexus structure, such as Pb, Hg, and Cd—the name-direct choroid plexus toxicants. Metals in the second group can impair specific plexus regulatory pathways that are critical to brain development and function, but they do not necessarily induce massive pathological alteration as do the metals in the first group. Thus, metals in the second group are called selective choroid plexus toxicants. Typical examples include Mn, Cu, and aluminum (Al). The third group of metals can be sequestered in the choroid plexus. Sequestration may represent an essential defense mechanism for the barrier system to protect against insults from the blood circulation. Metals in this group, including Zn, Fe, Ag, and Au, are thus called sequestered choroid plexus toxicants.

Trace metals in the CNS compartment are essential to brain development and function. For example, Zn, Cu, and Mn are required for optimal CNS function. They play important roles as catalysts, second messengers, and gene expression regulators. Being essential cofactors for functional expressions of many proteins, these elements are needed to activate and stabilize enzymes, such as superoxide dismutase (SOD), metalloproteases, protein kinases, and transcriptional factors containing zinc finger proteins. Consequently, the concentrations of these metal ions in the CNS compartment must be maintained at an optimal level, for both deficiency and excess can result in aberrant CNS function.

Other metals, such as Pb, Hg, and Al, have no known beneficial utility to the brain. The presence of these metals at even a low level causes defects in brain development and function or even degeneration. Thus, preventing their entry into the CNS compartment at the brain barriers is fundamental to the chemical stability of the CNS.

This chapter will review the current understanding of the role of choroid plexus in metal-associated neurotoxicity. The anatomical location in the brain and the structural characteristics of the choroid plexus, which render the tissue vulnerable to metal insults, will be briefly discussed. Since extensive reviews on the topic of metals in the choroid plexus have been done previously (Zheng, 1996, 2001a,b; Zheng et al., 2003), this chapter will focus exclusively on three neuroactive metals (i.e., Cu, Zn, and Al), their CNS homeostasis, their neurotoxicities, and the possible relationship to the choroid plexus. Finally, the implications of the choroid plexus in neurotoxicology and future research needs concerning the toxicology of the choroid plexus are discussed.

9.2 VULNERABILITY OF THE CHOROID PLEXUS TO METAL INSULTS

As a barrier between the blood and CSF, several unique anatomical and physiological features of the choroid plexus render the tissue more vulnerable to insults from the circulating bloodstream or from the CSF. As discussed earlier in this volume, the choroid plexus occupies all the brain ventricles. The large surface area of the choroidal microvilli increases the chance of this tissue being exposed to toxic metals from either side of the barrier.

The leaky endothelial layer in the choroid plexus allows metals to readily gain access to the choroidal epithelial cells. A faster blood flow at the choroid plexus than elsewhere in the brain brings toxic metals in the blood circulation to the choroid plexus.

The tight junctions between the epithelial cells in the blood–CSF barrier are less tightly connected than those between endothelial cells of the blood–brain barrier. The looseness of the choroidal barrier not only may permit metals to leak into the CSF, but it may also provide a pathway for metals present in the CSF to enter the choroid plexus, serving as a cleansing mechanism in the CNS.

It is unclear if the location of the choroid plexus within the cerebral ventricles is an important factor in metal-induced neurotoxicities. For example, the choroid plexus in the lateral ventricle is adjacent to hippocampal formation and other neuronal structures. Toxic Pb is known to accumulate in the choroid plexus and have access to the hippocampus. Whether or not there is a connection between these two seemingly separate events is unknown.

Because of these anatomical and physiological characteristics, the choroid plexus is a vital functional compartment that readily accumulates toxic metals in the brain.

Cu and Zn, under normal physiological conditions, are present in the serum and CSF. The concentrations of Cu and Zn in the CSF are about 30- and 80-fold less than those in the serum, respectively (see Table 9.1). Both metals are known to accumulate in the choroid plexus. Normal human brains contain Al at a concentration

TABLE 9.1
Concentration of Copper, Zinc, and Aluminum in Serum, Brain Tissue, CSF, and Choroid Plexus of Normal Human Subjects

Metals	Serum	Brain	CSF	Choroid Plexus
Cu	1060 μg/L (1)	0.1–0.5 mM (2)	33.9 ± 2.88 μg/L (3)	0.70 μg/g (1)
Zn	897 ± 110.5 μg/L (4)	0.1–0.5 mM (2)	10.4 ± 2.60 μg/L (4)	39.0 μg/g (1)
	975 μg/L (&rat) (5)		7 ± 5.9 μg/L (6)	
	920 μg/L (1)			
	850 ~1100 μg/L (7)			
Free Zn:	$6.5 \times 10^{-6} \sim 10^{-5}$ μg/L (5)			
Al	<10 μg/L) (8)	0.1–0.5 mM (2)	< 270 μg/L (8)	Unknown
		0.017 mM (8)		

Source: (1): Reviewed in Zheng, 1996; (2): Lovell et al., 1998; Atwood et al., 1999; (3): Burhanoglu et al., 1996.; (4): Palm and Hallmans, 1982.; (5): Bradbury, 1992; (6): el-Yazigi et al., 1986; (7): Goyer, 1998; (8): Markesbery et al., 1984; Yokel, 2002.

of 2 μg/g of tissue weight. In one Al intoxication case, the Al concentration in the cortex and subcortex reached as high as 9.3 μg/g along with a significant accumulation in the choroid plexus (Reusche et al., 2001). While it is postulated that Al may enter CSF to gain access to the brain parenchyma, the exact concentration of Al in the CSF remains uncertain.

Table 9.2 summarizes the binding species for Cu, Zn, and Al in the blood and in the CSF compartment. All three metals are bound to large molecular weight proteins in plasma, which represent the predominant forms of three metals in blood circulation. In the CSF, however, the metals are bound to ligands of much smaller molecular weight, such as histidine, citrate, or metallothionein.

9.3 COPPER (CU)

9.3.1 Cu in Brain Function and Dysfunction

Cu is an essential trace metal element in all living organisms. Inadequate or excessive intake of Cu can be pathogenic and life-threatening. Cu serves as a cofactor for a variety of proteins, including more than 30 enzymes involved in biological reactions, such as photosynthesis and respiration, free radical eradication, connective tissue formation, iron metabolism, and so on. In mammals the balance between Cu supply and consumption is maintained at both the cellular and tissue levels. Genetic factors that affect systemic levels of Cu, either excess or deficiency, cause clinically well-defined syndromes.

TABLE 9.2
Binding Species/Ligands for Copper, Zinc, and Aluminum in Plasma, Brain ISF, and CSF

Metals	Plasma	Brain ISF/CSF
Cu	Exchangeable:	Cu-histidine
	Cu-albumin (88%)	
	Cu-transferrin (11%)	
	Cu-histidine and cysteine (1%)	
	Nonexchangeable:	
	Cu-ceruloplasmin (15 ~ 20%) (1)	
Zn	Exchangeable:	
	Zn-albumin (70%)	Zn proteins/metallothioneins (2)
	Zn-histidine and cysteine (3)	
	Unknown:	
	Zn-α_2-macroglobulin (50%) (1)	
	Zn-transferrin	
Al (4)	Al-transferrin (81 ~ 91%)	Al-citrate (90%)
	Al-citrate (7 ~ 20%)	Al-transferrin (4%)
		Hydroxide (free) (5%)

The numbers represent the percentage of metals predicted to be associated with that ligand.

Source: (1): Bradbury, 1992; (2): Frederikson, 1989; St Croix et al., 2002; (3): Harris and Keen, 1989; (4): Öhman and Martin, 1994; Yokel, 2002.

9.3.1.1 Wilson's Disease

Wilson's disease is a genetic disorder that affects one in 50,000 to 100,000 people worldwide. A mutation in a Cu-transport ATPase gene, which is located on the long arm (q) of chromosome 13 (13q14.3), causes an overload of Cu in the body, initially mainly in the liver. When the capacity of the liver to handle Cu is exceeded, the metal is then released from the liver and begins to accumulate in other organs of the body, particularly the brain, eyes, and kidneys. In essence, Wilson's disease is the dysfunction of Cu transport in the body.

The build-up of Cu in the brain causes neurological symptoms, including tremor of the head, arms, or legs; generalized, impaired muscle tone and sustained muscle contractions that produce abnormal postures, twisting, and repetitive movements (dystonia); and slowness of movements (bradykinesia), particularly those of the tongue, lips, and jaw. Patients may also experience clumsiness, difficulty with balance, and impaired coordination of voluntary movements, such as walking (ataxia), or slowness of finger movements and loss of fine motor skills (Kodama, 1996).

Wilson's disease may be associated with a damaged blood–brain barrier. In one clinical study with four patients who had cerebral manifestation of Wilson's disease, the ratio of albumin concentrations in the CSF and serum was used as the indicator of barrier intactness in addition to other parameters such as Cu concentrations in the CSF and serum. The investigators reported that in all cases there was an increase in the albumin ratio, suggesting a disturbed blood–brain barrier. All patients showed an initial worsening of the neurological condition, which corresponded to a rise in the albumin ratio. The maximal rise in albumin ratio was reached after about seven months. The ratio declined and finally returned to the normal level during and after therapeutic treatment (Stuerenburg, 2000). These authors conclude that blood–brain barrier permeability may play a role in the early stages of disease development. The normalization of the CSF Cu concentration in patients is a slow process, even if the therapy is sufficient (Stuerenburg, 2000).

9.3.1.2 Menkes' Disease

Menkes' disease and a related disorder, occipital horn syndrome (OHS), result from a significant Cu deficiency due to a failure to transport Cu across membranes of intestinal enterocytes and across cerebrovascular cells of the blood–brain barrier. Deprivation of Cu directly affects the activity of intracellular cuproenzymes, such as cytochrome c oxidase, CuZnSOD, lysyl oxidase, tyrosinase, ascorbic acid oxidase, ceruloplasmin, and dopamine β-hydroxylase (Harris, 2003). The Menkes gene is located on the long arm of the X chromosome at Xq13.3; the gene product is a 1500-amino-acid P-type adenosine triphosphatase (ATPase), which has 17 domains: six Cu binding, eight transmembrane, one phosphatase, one phosphorylation, and one ATP binding. The predominant sites of Menkes gene expression are the placenta, gastrointestinal tract, and blood–brain barrier. As the gene is an X-linked gene, the disease primarily affects male infants.

Patients with Menkes' disease, primarily infants, have abnormally low levels of Cu in the liver and brain, but higher than normal levels in the kidney and intestinal lining. Affected infants may be born prematurely. Symptoms appear during infancy, including seizures, psychomotor deterioration, failure to thrive, temperature instability (hypothermia), and strikingly peculiar hair. There can be extensive neurodegeneration in the gray matter of the brain. Arteries in the brain can also be twisted with frayed and split inner walls. This can lead to rupture or blockage of the arteries (Dexter et al., 1991).

The clinical manifestations of Menkes' disease are due to a lack of Cu in certain regions; this could, in turn, be due to a deficient transport of Cu by the blood–brain barrier. By using reverse transcription-polymerase chain reaction (RT-PCR), it has been shown that cerebrovascular endothelial cells that comprise the blood–brain barrier express the gene for the Cu-ATPase. Functional analysis using an ATP7A inhibitor also reveals that Cu efflux can be blocked by p-chloromercuribenzoate (p-CMB), a potent inhibitor of ATP7A (Qian et al., 1998). Thus, these results provide strong evidence that a Cu-ATPase is present at the blood–brain barrier, and the transport of Cu at the barrier may control the entrance of Cu into brain parenchyma. Understandably, a genetic disorder in the expression of the Cu-ATPase transporter in cerebrovascular endothelial cells would lead to a low brain Cu level in Menkes' disease. However, whether the mutation of the Menkes gene in the choroid plexus contributes to the progress of the disease is unknown.

9.3.1.3 Alzheimer's Disease

Alzheimer's disease (AD) is characterized by the chronic deposition of β-amyloid peptides (Aβ) in senile plaques and hyperphosphorylated Tau protein in neurofibrillary tangles. Essential metals such as Cu, Zn, and Fe accumulate in Aβ deposits in the cortex along with sugar-derived glycation end products (Waggoner et al., 1999). Aβ peptide is generated from amyloid precursor protein (APP) by the proteolytic activity of Aβ- and γ-secretase (Checler, 1995). The presence of binding sites for Cu and Zn on the precursor Aβ partly explains the enrichment of these metals in the plaques (Atwood et al., 2000).

The levels of Cu in the CSF and serum, as well as in brain tissues in Alzheimer's patients are rather inconsistent and sometimes controversial (Basun et al., 1991; Cuajungco and Lees, 1997; Cuajungco and Fagét, 2003). Some investigators report a significant decrease of brain Cu concentrations in AD patients (Deibel et al., 1996), while others present a 2-fold increase of Cu levels in the CSF (Basun et al., 1991), serum (Gonzalez et al., 1999), and amyloid plaque rim (Lovell et al., 1998), as well as an increase in brain and CSF ceruloplasmin levels. The latter is a known Cu-binding/transporting protein (Loeffler et al., 1996) that is synthesized by the choroid plexus. This discrepancy may result from the differences in analytical approaches, technical variations during tissue sampling and processing, and the limited sample sizes.

$A\beta_{1-40}$ has two binding sites for Cu: the higher affinity site of log Kapp 10 and lower affinity site of log Kapp 7.0. But the binding affinity of Cu to $A\beta_{1-42}$ is much greater than that to $A\beta_{1-40}$ (Atwood et al., 2000). Binding of Cu to Aβ can lead to an augmented oxidation of Aβ (Huang et al., 1999) and accelerate the formation of covalently crosslinked glycation end products (Loske et al., 2000).

Cu can also directly bind to amyloid precursor protein (APP). Specific and saturable binding sites for copper (APP 135–155; K_D = 10 nM) have been identified within the cysteine-rich region of the APP-695 ectodomain (Hesse et al., 1994). While Zn binding to APP is believed to perform a structural role, binding of Cu to APP reduces Cu(II) to Cu(I), which, in turn, results in an oxidation of Cys144 and Cys155. The formation of ensuing intramolecular disulfide bridge renders the APP-Cu(I) complex more prone to redox reactions, leading to a random APP fragmentation (Multhaup et al., 1996).

The choroid plexus contains a high amount of Cu. The extensive precipitation of amyloid plaques in the AD choroid plexus (Miklossy et al., 1999) may be associated with Cu accumulation in this tissue.

9.3.1.4 Amyotrophic Lateral Sclerosis

Amyotrophic lateral sclerosis (ALS) is a neurodegenerative disorder of motor neurons in the CNS. While the cause remains unsolved, a connection of the ALS to Cu has been established through CuZnSOD. A clinical study has observed a dominant mutation in the copper/zinc superoxide dismutase (SOD1) gene on chromosome 21 in 15 to 20% of familial ALS (FALS) cases (Reinholz et al., 1999). In brains of ALS patients, most neurons stained weakly, or not at all, with both anti-Cu/Zn- and

Mn-SOD antibodies, whereas the pia mater and the epithelial cells of choroid plexus stained intensely (Uchino et al., 1994).

In another study with a transgenic ALS mouse model having a point mutation in the gene encoding Cu/Zn SOD, continuous subcutaneous administration of polyamine-modified catalase, which increases the permeability of the blood–brain barrier, delays the onset of symptoms, and increases animal survival rate, suggested a possible role of brain barriers in ALS disease (Poduslo et al., 2000).

9.3.1.5 Prion Disease and Other Brain Disorders

Prion disease has been linked to an abnormal form of the prion protein (PrP) in neurons. PrP molecules are believed to bind Cu at multiple sites along the peptide chain (Hornshaw et al., 1995), serving to stabilize the PrP. Excess brain Mn, due to rising environmental levels of Mn, may replace Cu and causes malfunction of PrP, which clinically manifests itself as Creutzfeldt-Jakob disease. Current evidence has shown that both the blood–brain barrier and the blood–CSF barrier are target sites for Mn toxicity, at least in cases of Fe-associated neurotoxicities. However, whether or not overexposure to Mn can lead to an altered Cu transport at brain barriers is unknown and deserves further investigation.

Cu may be involved in congenital hydrocephalus. Mori and colleagues (1993) report that the amount of Cu/Zn-SOD in the brain of a congenitally hydrocephalic rat model is less than in the control, especially in the choroid plexus. The authors postulate that a congenitally reduced SOD activity may impair the function of choroid plexus as a result of increased oxygen species in the plexus tissue. A combination of reduced SOD in the choroid plexus, hippocampus, ependymal cells of ventricles, and aqueduct may promote the development of hydrocephalus.

9.3.2 Transport of Cu by Brain Barriers

Under normal physiological conditions, brain barriers are impermeable to Cu. Movement of Cu across brain barriers between two fluid compartments requires specific Cu transport system(s). However, under certain pathological conditions, where the barrier's permeability is increased, Cu may enter the CSF via passive diffusion. For example, in children with meningitis, the elevation of CSF Cu levels is associated with an elevated protein level in the CSF, suggesting that the high CSF Cu is a result of the breakdown of the blood–brain barrier and subsequent leakage of the trace element along with proteins from serum to the CSF (Burhanoglu et al., 1996).

The levels of CSF-Tau protein also correlate significantly with the serum levels of Cu in Alzheimer's disease patients treated with clioquinol; the latter is a chelator that crosses the blood–brain barrier and has greater affinity for Zn and Cu ions than for Ca and Mg ions (Regland et al., 2001). Again, the damaged blood–brain barrier may permit the passage of the trace elements, including Cu, into the subarachnoid space (Kapaki et al., 1989).

Cu transport at brain barriers is achieved by a coordinate series of interactions between passive and active transport proteins. There are several membrane Cu

TABLE 9.3
Possible Transporters Involved in the Influx and Efflux of Copper, Zinc, and Aluminum at Brain Barriers

Metal	Fluxes	Metal Transporter	Present at BBB	Present at BCB
Cu	Influx	Ctrl (1)	Unknown	Yes (2)
		Nramp2//DMT1/DCT1(3) (energy-independent)	Yes (4)	Yes (4)
		ATP7A	Yes (5)	Yes (2)
		ATP7B (6)	Yes	Unknown
	Efflux	MTP1 (2)	Yes	Yes (7)
Zn		PHT1[a]	Unknown	Yes (8)
		DMT1[b]	Yes (4)	Yes (9)
		ZIP (ZRT1. IRT1-like protein) (10)	Unknown[e]	
		ZnT-1 through ZnT-4 (11)	Unknown[e]	
Al	Influx (15)	TfR-ME (3)[c]	Yes (12)	Yes (13)
		MCT1[d]	Yes	Yes (14)
		MCT7	Unknown	Unknown
		MCT8	Unknown	Unknown
	Efflux (16)	MCT	Yes	Yes (14)
		Oatp	Yes	Yes (17)

[a]peptide/histidine transporter; [b]divalent metal transporter; [c]TfR-ME: transferring-receptor-mediated endocytosis; [d]MCT: monocarboxylate transporter—Na independent, pH independent, and energy dependent, but not dependent on transporter Na/K-ATPase; [e]hZIP-1 mRNA and hZnT-1 mRNA present in prostate epithelial cell line LnCap, PC3, and CRL2220; hZnT-4 present in LnCap (Beck et al., 2004).

Source: (1): Dancis et al., 1994; (2): Nishihara et al., 1998; (3): Rolfs and Hediger, 1999; (4): Burdo et al., 2001; (5): Qian et al., 1998; (6): Hamza, 1999; (7): Wu et al., 2004; (8): Yamashita et al., 1997; (9): Burdo et al., 2001; Siddappa et al., 2002 (10): Grotz et al., 1998; Gaither and Eide, 2000; (11): Ebadi et al., 1995; (12): Roskams and Connor, 1990; (13): Zheng, 2002; (14): Leino et al., 1999; (15): Yokel et al., 2002; (16): Ackley and Yokel, 1997, 1998; Koehler-Stec et al., 1998; (17): Choudhuri et al., 2003.

transporters present in the brain barriers, including copper transporter 1 (Ctr1), divalent metal transporter 1 (DMT1), and Cu transport ATPases (see Table 9.3). In addition, Cu is transported by certain vesicles and soluble peptides such as chaperones (Harris, 2001).

9.3.2.1 Copper Transporter 1

The yeast *Ctr1* gene was the first eukaryotic gene discovered that codes for a Cu transport protein (Dancis et al., 1994). A Cu transport gene in human cells (*hCtr1*) can transport Cu as Cu(I) in a reaction that does not require the energy of ATP, but is stimulated by K ions and an acidic pH (Lee et al., 2002). Human fibroblasts transfected with *hCtrl* cDNA favorably take up Cu (Moller et al., 2000). By inactivating the *Ctr1* gene in mice via targeted mutagenesis, Kuo and his colleagues (2001) also provide the evidence that *Ctr1* is essential for embryonic growth and development and is required for Cu transport into the brain. *Ctr1* expression is abundant in a variety of epithelial-derived tissues, including the choroid plexus (Nishihara et al., 1998).

9.3.2.2 Divalent Metal Transporter 1

DMT1 (Slc11a2), also known as Nramp2 (natural resistance-associated microphage protein) and DCT1 (divalent cation transporter), was originally identified as the transporter responsible for intestinal nonheme iron apical uptake and trafficking and also believed to mediate influx of Cu via an energy-independent mechanism (Rolfs and Hediger,1999; Arredondo et al., 2003). Yet its role in Cu uptake is probably less specific (Harris, 2003; Garrick et al., 2003).

Using the branched DNA signal amplification method, Choudhuri et al. (2003) demonstrated that DMT1 mRNA was expressed in the choroid plexus at a higher level than in liver, kidney, and ileum of rats. During the perinatal period, DMT1 is expressed in the choroid plexus epithelial cells, the cerebral blood vessels, and ependyma in developing rat brain (Burdo et al., 2001; Siddappa et al., 2002). However, Moos and Morgan (2004) recently failed to detect DMT1 in brain capillary endothelial cells, although they identified DMT1 in neurons and choroid plexus.

9.3.2.3 ATPases Involved in Cu Transport

Two diseases, Menkes' syndrome and Wilson's disease, have led to the discovery of the Cu-transport ATPases in humans. ATP7A (coded by 8.5 kb mRNA) and ATP7B (coded by 7.5 kb mRNA), belonging to a subclass of ATPases, conduct ATP-dependent transport of Cu across brain barriers in mammals (Solioz and Vulpe, 1996). ATP7A is a component of Cu efflux from brain endothelial cells. A shutdown of ATP7a gene expression can result in Cu accumulation in brain capillaries of brindled and macular mutant mice (Yoshimura et al., 1995), suggesting that a dysfunction in Cu transport from blood to brain via the blood–brain barrier may contribute to the etiology of Menkes' disease. Qian et al. (1998) also provide further evidence to support the view that a lesion to the Cu-ATPase at brain barriers may be the primary cause of low brain Cu levels in Menkes' disease. In Wilson's disease, the overload of Cu in brain may be associated with the mutated ATP7b in the blood–brain barrier (Stuerenburg, 2000).

ATP7A is highly expressed in the choroid plexus (Nishihara et al., 1998); whether ATP7B is also present in the choroid plexus remains unknown.

9.3.2.4 Export via Vesicles

The movement of Cu within the cells through the cytosol is controlled by Golgi-derived vesicles and soluble polypeptides called chaperones. Vesicles containing embedded Cu-ATPase protein are believed to pinch off from the trans-Golgi membrane and move to the outer membranes (Voskoboinik et al., 1998). The movement is seen as an emergency response to a toxic threat and is consistent with cells in the act of releasing Cu. The intracellular vesicle movement is also viewed as a necessary event in transporting Cu across the blood–brain barrier, but the event has not been identified in the blood–CSF barrier. The mobility of vesicles depends upon the availability of heavy metal binding (Hmb) domains in the protein (Harris, 2003).

9.3.2.5 Chaperones

The Cu chaperones are small Cu-binding peptides. As part of a larger family of metallochaperones, they are structurally adapted to bind Cu, recognize recipients, and conduct facile exchange by ligand exchange reaction with target proteins. Thus, chaperones typically have the Cu-binding motif that is found in the N-terminus of ATP7A and ATP7B (Larin et al., 1999). To date, at least four chaperones are known to perform Cu transport functions in eukaryotes. The chaperone ATOX1 (or HAH1/ATX1) is capable of transporting Cu to membrane-bound CuATPases. The transfer of cytosolic Cu to the mitochondria is a function of a small peptide called COX (cytochrome oxidase enzyme complex) (Glerum et al., 1996). CCS (copper chaperone for superoxide dismutase)/LYS7 is the chaperone required to insert Cu into the Cu_2Zn_2 superoxide dismutase (CuZnSOD) in mammals (Culotta et al., 1997; Harris, 2003). Besides these target-specific chaperones, glutathione (GSH) is a general Cu mobilizer and a nonspecific Cu transport factor.

By in situ hybridization, Ctr1, ATX1, and ATP7A have been found to have a high expression highly in the choroid plexus (Nishihara et al., 1998). Immunostaining for COX shows a remarkable signal in choroidal epithelial cells (Taskinalp et al., 2000). In addition, staining for cytochrome oxidase activity displays an intense staining in the mitochondria of choroidal epithelium (Kim et al., 1990). Thus, it seems reasonable that the choroid plexus, by adapting the Cu intracellular trafficking from uptake to export, may participate in the regulation of Cu in the CSF.

9.3.2.6 Metallothionein in the Choroid Plexus

Metallothionein (MT) is abundantly expressed in the choroid plexus, much more so than in the rest of brain (Nishimura et al., 1992). MT is a low molecular weight protein inducible by metals such as Cu, Cd, and Zn. MT functions to regulate Cu and Zn homeostasis and to sequester metals so as to reduce the cytotoxicity induced by metals. A study on a Cu-poisoned sheep showed that elevated brain Cu is associated with increased MT immunoreactivity in astrocytes, pia mater, choroid plexus, and ependyma. The authors suggest that these sites may sequester Cu and possibly modulate CNS Cu homeostasis (Dincer et al., 1999).

Cu as Cu(I) may enter the cells through the Ctr1 transporter and distribute intracellularly to enzymes (CuZnSOD), storage proteins (metallothionein), or organelles (mitochondria) via GSH and Cu chaperones (CCS, COX, and ATOX1). Efflux from nonhepatic cells is via the trans-Golgi network, which uses Cu-ATPase ATP7A and vesicles that cycle between the trans-Golgi and the plasma membrane (Harris, 2003). Since the genes for many Cu regulatory proteins are intensely expressed in the choroid plexus (Iwase et al., 1996; Murata et al., 1997), this tissue may serve as an important Cu port for the brain (Murata et al., 1997; Nishihara et al., 1998).

9.3.3 Toxicity of Cu at Brain Barriers

Based on the study of Cu and angiogenesis, some researchers have proposed that Cu may selectively target endothelial cells during the developmental period (Harris, 2003). While the mechanism is not completely understood, Cu appears to have the capacity to mobilize endothelial cells during the development of brain blood vessels. A 48-hour exposure to 500 μM Cu in serum-free medium nearly doubled the number of human endothelial cells derived from umbilical artery and vein, but it had no effect on the growth of dermal fibroblasts or arterial smooth muscle cells (Hu, 1998). In addition, Cu appears to induce the synthesis of vascular endothelial growth factor, thereby promoting wound healing through an angiogenesis process in the wound area (Sen et al., 2002). Damage by Cu to the brain barrier system, particularly the blood–CSF barrier, is not well documented.

Cu itself is not a barrier destroyer, but the presence of Cu may alter the transport of Fe. This was seen in rats fed a Cu-containing diet, where the influx of Fe into brain was significantly decreased compared to that of rats fed a control diet (Crowe and Morgan, 1996).

Exposure to other metals may interfere with Cu transport by brain barriers. Qian et al. (1999) demonstrated that Pb accumulation in C6 rat glioma cells altered the membrane transport properties for Cu, leading to an increased uptake and a decreased efflux of Cu. However, in another study on rats exposed to Pb, Cd, or combination of these two metals, no changes in brain Cu levels were observed (Skoczynska et al., 1994). The interaction of Cu and Pb in the choroid plexus has not been explored.

9.4 ZINC (Zn)

Zn, second only to Fe, is one of the most needed essential elements in the human body (Vallee and Falchuk, 1993; Choi and Koh, 1998). The normal plasma Zn level ranges from 85 to 110 μg/dL (Goyer, 1998). Zn in blood binds primarily to albumin, constituting the largest component of exchangeable Zn pool (see Table 9.2). Zn also binds to other plasma proteins, such as transferrin and α_2-macroglobulin. Albumin-bound Zn is not essential for Zn transport into brain (Takeda et al., 1997), while the function of α_2-macroglobulin-bound Zn remains unknown.

In addition to protein binding, Zn also forms complexes with amino acids such as histidine and cysteine, which makes up the second largest pool of exchangeable Zn (Harris and Keen, 1989). Evidence has also shown that L-histidine may play a role in the transport of Zn into the brain at brain barrier systems. The transfer of Zn from plasma proteins to histidine determines the brain permeability to Zn (Buxani-Rice et al., 1994; Keller et al., 2000).

9.4.1 Zn in Brain Function and Dysfunction

Zinc is essential for the normal growth, development, and function of the CNS. Involvement of Zn in enzymatic reactions has been known for more than half a century. Zn-requiring/containing enzymes whose three-dimensional (3D) structures are well defined number more than 200 (Maret, 2001). Zn performs three major functions in Zn enzymes: catalysis, coactivity, and structure (Vallee and Falchuk, 1993).

9.4.1.1 Zn Deficiency

Deficiency in the dietary supply of Zn causes alteration of Zn homeostasis in the brain, leading to brain dysfunctions, e.g., mental disorders (Golub et al., 1995). Zn deficiency during brain development in experimental animals causes permanent malformations; the activities of Zn metalloenzymes (e.g., glutamate dehydrogenase and 2'-3'-cyclic nucleotide 3'-phosphohydrolase) are reduced in brains of Zn-deficient rats (Dreosti et al., 1981); animals with a decreased hippocampal Zn develop convulsive seizures (Fukahori et al., 1988). In humans, Zn deficiency in young or mature individuals can lead to abnormal neuromotor and cognitive function (Keen et al., 1993).

Brain dysfunction due to zinc deprivation may be indirectly related to a generalized decrease in zinc-dependent processes, such as protein synthesis, DNA and RNA synthesis, or cell membrane stability (Keen et al., 1993). More recently, Zn was found to be involved in the release of synaptic neurotransmitters (Frederickson et al., 2000; Huntington et al., 2001).

Systemic Zn deficiency can also affect the permeability of brain barriers. Noseworthy and Bray (2000) reported that Zn deficiency significantly increases the permeability of the blood–brain barrier and further suggested that the diminished brain barrier integrity may be a free radical-mediated process, as the ratio of oxidized to reduced glutathione (GSSG:GSH) is significantly elevated.

9.4.1.2 Zn in Alzheimer's Disease

Involvement of Zn in Alzheimer's disease has been recognized since the early 1990s. Both APP and Aβ can be characterized as Cu/Zn metalloproteins, with APP in the CSF being more easily precipitated than Aβ peptides (Brown et al., 1997). Zn can bind to a specific, cysteine-rich region of the APP-695 ectodomain (Bush et al., 1994b; Cuajungco and Fagét, 2003). Noticeably, Zn(II) is the only physiologically available metal ion capable of precipitating Aβ at pH 7.4. The consequence of Zn binding to APP or Aβ remains controversial. Some suggest that binding results in Zn-induced oxidative stress or Aβ-mediated oxidative stress or both in affected brain regions (Moreira et al., 2000), while others indicate that Zn may inhibit Aβ-mediated cytotoxicity, particularly at low concentrations, playing a protective role against Aβ cytotoxicity (Yoshiike et al., 2001).

Reports on Zn concentrations in the systemic circulation and brain among Alzheimer's patients have been contradictory (Basun et al., 1991; Cuajungco and Lees, 1997; Gonzalez et al., 1999; Rulon et al., 2000; Tully et al., 1995). Earlier studies suggested no observable changes in CSF concentrations of Zn in patients (Basun et al., 1991). A more recent study indicated that CSF Zn level is significantly decreased in AD as compared to the age-matched controls (Molina et al., 1998). When Zn was supplemented at 200 to 400 mg/day for 1 year, Alzheimer's patients showed a significant slowing of cognitive decline (Van Rhijn et al., 1990; Potocnik et al., 1997).

Thus, an altered brain homeostasis of Zn appears to contribute to the etiology of Alzheimer's disease (Bush et al., 1994a,b). Brain turnover of Zn is rather slow, with a half-life ($T_{1/2}$) of 7 to 42 days. The half-life lasts even longer in brain regions that contain vesicular Zn (Kasarskis, 1984). Zn concentrations above 300 nmol/L can rapidly precipitate synthetic human $A\beta_{1-40}$ (Bush et al., 1994a). Since Zn in the brain is regulated by the transport mechanisms located at the blood–brain barrier, any disruption of the barrier's function and permeability to Zn could conceivably lead to an increased influx of Zn into the brain. With regard to Zn regulation by the blood–CSF barrier in the choroid plexus, it is known that certain Zn transporters are located in the choroid plexus, e.g., PHT1 and DMT1; whether Zn is transported via the choroid plexus and in which direction the choroid plexus modulates Zn fluxes remain unknown.

9.4.1.3 Protective Effect of Zn-Containing Enzymes at Brain Barriers

Following trauma, ischemia, and reperfusion injury, the permeability of the blood–brain barrier can be altered, resulting in pathogenesis of brain edema. The oxygen-derived radicals, particularly superoxide, may contribute to barrier damage. The cytosolic antioxidant Cu/Zn superoxide dismutase (CuZnSOD) may be an important factor in protecting against oxidative damage to barriers. In transgenic mice with overexpression of CuZnSOD, thromboembolic cortical ischemia causes much less oxidative DNA damage, less DNA fragmentation, and much less ensuing blood–brain barrier disruption compared to wild-type mice (Kim et al., 2001). Studies on a reperfusion-ischemia model using transgenic mice deficient in CuZn-SOD provide the opposite results and thus reach the similar conclusion (Gasche et al., 2001). Chan and colleagues (1991) have also proved that an increased level of superoxide dismutase activity in the brain reduces the development of vasogenic brain edema and infarction.

9.4.2 Transport of Zn by Brain Barriers

9.4.2.1 Brain Regional Distribution

The concentration of Zn in the CSF of normal subjects is about 1% that of plasma, i.e., 0.15 μM (Franklin et al., 1992). Zn concentration in brain extracellular fluid is nearly the same as that in the CSF (Palm et al., 1986). The half-life of ^{65}Zn in rat brain tissue ranged between 16 and 43 days (Takeda et al., 1995). A considerable difference exists in elimination of ^{65}Zn from various brain regions. Zn may be eliminated from the arachnoid villi or arachnoid granulations, or both via the CSF.

The levels of Zn in the brain vary during the developmental stage. The brain concentration of Zn increases with growth after birth and is maintained constant in the adult brain (Sawashita et al., 1997). Concentrations of Zn along with Cu and Fe in gray matter are in the same order of magnitude as that of magnesium (0.1 to 0.5

mM) (Lovell et al., 1998). Because of its relatively high concentration in the brain, Zn is hardly regarded as a "trace" element in the brain.

In the brain, Zn exhibits a regional specific distribution, with the highest level found in the limbic system, including the hippocampus and amygdala. Autoradiographic study using in vivo ^{65}Zn shows a low level of radioactivity in the white matter and a high level in choroid plexus and cerebral cortex, particularly in the dentate gyrus (Franklin et al., 1992). Approximately 90% of the total brain Zn is bound in Zn proteins in neurons and glial cells. The rest appears to be present in presynaptic vesicles as a neuro-modulator in synaptic neurotransmission (Howell and Frederickson, 1990).

9.4.2.2 Transport by Blood–Brain Barrier

Zn in the blood circulation is transported to the brain via the blood–brain barrier. With prolonged perfusion (>30 min) in rats, the uptake of Zn is unidirectional with an influx rate constant (K_{in}) of approximately 5×10^{-4} mL/min/g. If the perfusion time is more than 30 minutes, ^{65}Zn fluxes between blood and brain become bidirectional, with an influx K_{in} greater than 5×10^{-4} mL/min/g (Pullen et al., 1991).

In a Transwell model using cultured brain capillary endothelial cells (BCEC), treatment with low concentrations of Zn in culture media (3 μmol/L) results in an increased rate of Zn uptake into BCEC and an increased rate of Zn transport across the BCEC. The same experiment with a high Zn concentration (50 μmol Zn/L), however, decreases the rate of Zn transport across BCEC, although the uptake of Zn into BCEC is indeed increased (Lehmann et al., 2002). Apparently, at high concentration, Zn ions are sequestered by intracellular binding proteins, which prevent the metal from further entering the other side of the compartment. It is well known that Zn induces metallothionein.

9.4.2.3 Transport by Blood–CSF Barrier

Zn can also be transported to the CSF compartment via the blood–CSF barrier in the choroid plexus. One hour after intravenous injection of $^{65}ZnCl_2$ to rats, ^{65}Zn is highly concentrated in the choroid plexus. It subsequently accumulates in brain parenchyma with an apparent decrease in choroidal ^{65}Zn. The time sequence of radioactive Zn in brain suggests that Zn ion gradually reaches the brain mainly via the CSF secreted by the choroid plexus (Takeda et al., 1997). The permeability of choroid plexus to Zn, with a K_{in} of 61×10^{-4} mL/min/g, is about 12 times higher than that of the cerebral capillaries ($K_{in} = 5 \times 10^{4}$ mL/min/g). However, overall influx of Zn to brain via the choroid plexus is less than 5% that across cerebral capillaries. Thus, some authors suggest that the primary supply of Zn to the brain is perhaps via the blood–brain barrier, while the choroid plexus may contribute to the slow, prolonged flux to the brain (Franklin et al., 1992).

The main contribution of the choroid plexus to brain Zn regulation may come from mediation of efflux, i.e., by transporting Zn out of the CSF compartment (Kasarskis, 1984). Brain Zn is maintained at a fairly stable level; it does not change significantly even if plasma Zn concentrations rise 10 times above the normal level

(Blair-West et al., 1990). In Zn-deficient animals, brain Zn levels are also scarcely affected (O'Dell et al., 1989). The controlled influx at both barriers and efflux by the choroid plexus may explain the tight regulation in the brain homeostasis of Zn.

9.4.2.4 Mechanism of Zn Transport

The mechanisms by which Zn is transported from brain capillary endothelial cells or choroidal epithelial cells to the brain extracellular fluid and the CSF remain unclear. Several transport systems are suggested to be responsible for Zn transport at the cell membrane or inside of the cells (see Table 9.3). ZnT-1, 2, 3, and 4 (originally known as MT-I through MT-IV) are Zn-binding proteins whose sequences have been cloned (McMahon and Cousins, 1998). ZnT-1/MT-I and ZnT-2/MT-II can be induced by Zn and Cu in astrocytes (Hidalgo and Carrasco, 1998). Functionally speaking, ZnT-1/MT-I may be associated with Zn efflux. ZnT-3/MT-III is expressed in Zn-containing glutamatergic neurons and is supposedly involved in Zn transport from the cytosol to synaptic vesicles in neurons (Palmiter et al., 1996; Wenzel et al., 1997; Tsuda et al., 1997; McMahon and Cousins, 1998; Cole et al., 1999). These proteins are involved in intracellular Zn homeostasis (Ebadi et al., 1995). It is unclear if all these Zn-binding proteins exist at both barriers. The other transporter that may be involved in Zn transport across the brain barrier is ZIP (zipper) (ZRT1, IRT1-like protein, also named ZIRTL, zinc-iron regulated transporter-like gene) (Grotz et al., 1998; Gaither and Eide, 2000).

9.5 ALUMINUM (Al)

9.5.1 Al in Brain Dysfunction

Al is found throughout the environment (Flaten et al., 1996). It apparently serves no essential biological function and is a known neurotoxicant potentially associated with a number of neurodegenerative disorders (Yokel, 2000; Spencer, 2000). Potential sources of Al exposure include food, vaccines, antiperspirants, drinking water, antacids, and intravenous solutions. A direct Al neurotoxicity seen in clinics has been dialysis encephalopathy, where Al is a major source of toxicity (Flaten et al., 1997). During the disease development, patients display Al toxicity-associated symptoms from neurological, skeletal, and hematological systems and in most cases with advanced renal failure (Alfrey, 1991).

One of the major Al-induced neurotoxicities that remains debatable is the role of Al in Alzheimer's disease (Flaten et al., 1997; Savory et al., 1997; Spencer, 2000). In particular, the evidence from human studies of AD patients has been highly controversial. Reusche (1997) reported an increased level of Al in neurons of an AD patient who suffered from a dialysis-associated encephalopathy (DAE). Zatta et al. (2003) also found that Al was more often associated with the neurofibrillary tangles (NFT) than with the plaques. Using a laser microprobe mass analysis technique, Good et al. (1992) demonstrated a significant increase of Al and Fe in NFT among AD patients. However, Landsberg et al. (1992) failed to reveal the presence of Al in senile plaques from AD brains by autopsy. Lovell et al. (1993) found no significant increase of Al in NFT in AD brain; Murray et al. (1992) further reported a diminished Al concentration in NFT. Several epidemiological studies also failed to establish the associations between AD occurrence and Al contamination in drinking water (Forster et al., 1995; Martyn et al., 1997), although some studies suggested a small, yet not statistically significant, increase in the risk of dementia illness for people living in areas with higher Al concentrations in drinking water (Neri and Hewitt, 1991; McLachlan et al., 1996).

Deposition of amyloid β polypeptide in cerebral vessels is a common phenomenon in β-amyloid diseases, including Alzheimer's disease. Al is known to alter the structure and function of Aβ by inhibiting the enzymes (metalloproteases) associated with the processing and degradation of Aβ. For example, Banks et al. (1996) found that Al altered the enzymatic degradation of Aβ through a calcium-sensitive process. A reduced degradation of Aβ by Al treatment as demonstrated in vitro could alter the entry of Aβ to brain. Al can also change the permeability of the blood–brain barrier to peptides with the molecular weight similar to Aβ. Kaya et al. (2003) found that exposure to a high level of Al results in an additional increase in blood–brain barrier permeability to Evans blue (EB) dye in chemical-induced chronic hypertensive rats. These studies indicate that Al is able to alter the access of bloodborne Aβ to the CNS either by increasing the permeability of the BBB or by affecting its enzymatic degradation.

Al also induces other neurotoxicities. Reusche et al. (2001) reported a case of Al intoxication. The patient underwent a bone reconstruction with an Al-containing cement, which had about 30 mg of Al. Autopsy data revealed the presence of Al-containing intracytoplasmic argyrophilic inclusions in choroid plexus epithelia, neurons, and cortical neuroglia. Atomic absorption spectrometry to analyze Al showed an increase of Al concentrations in the cortex and subcortex up to 9.3 μg/g (normal range < 2 μg/g). Apparently Al in blood passes across the brain barriers to gain access to brain parenchyma via a CSF leakage.

9.5.2 Transport of Al by Brain Barriers

Even though Al is the third most ubiquitous element on earth, very little Al is absorbed into the human body. Two major barrier systems may be designed to rigorously exclude Al from entering the systemic circulation and targeted tissues such as brain. The barrier in the gastrointestinal tract forms the first defense line for Al toxicity; the blood–brain barrier serves as the second line to protect the brain from Al insults (Banks et al., 1996).

9.5.2.1 Toxicokinetics of Al

Approximately 80% of plasma Al is bound to transferrin. Another 11% of plasma Al forms Al-citrate complex, which is the predominant small molecular weight Al species in plasma (Öhman and Martin, 1994). In contrast, about 90% of Al in brain extracellular fluid may be in the form of Al-citrate complex and only 4% of Al bound to transfer (Yokel, 2002).

Oral ingestion of Al leads to an increased amount of Al in brain tissue. Rats receiving oral doses of ^{26}Al for 1 to 2 weeks showed a significant accumulation of ^{26}Al in brain tissues (Walton et al., 1995). Determination of ^{26}Al in rat brain at various times after intravenous ^{26}Al suggested a prolonged brain ^{26}Al half-life. The half-life of brain ^{26}Al was estimated to be about 150 days (Yokel, 2002). Microdialysis allows one to sample brain extracellular fluid in different brain compartments and to detect unbound Al in brain extracellular fluid. Using this method, Allen and Yokel (1992) reported that systemically administered Al-citrate can be quickly detected in brain extracellular fluid, indicating that Al in the blood circulation is capable of penetrating the blood–brain barrier to enter the brain and accumulate there. It also suggests that the blood–brain barrier is the primary site for Al-citrate transport.

9.5.2.2 Influx of Al

Like many chemicals, Al in the blood circulation may enter brain parenchyma through either the blood–brain barrier or the blood–CSF barrier in the choroid plexuses. Yokel et al. (1999) suggested that Al enters the brain primarily through brain microvessels, rather than through choroid plexuses, because Al has a much more rapid appearance in the frontal cortex than in other brain regions following injection. In addition, the Al concentration is always lower in the lateral ventricle than in the frontal cortex, suggesting a lack of concentration gradient for Al to distribute from the choroid plexus in the ventricles to other brain regions.

There appear to be at least two mechanisms mediating Al distribution across the blood–brain barrier (see Table 9.3). One involves receptor-mediated brain influx of Al, which is bound to transferrin; the other involves transport via small molecular weight species, such as Al-citrate complex.

Roskams and Connor (1990) have shown that transferrin-bound Al can be detected in brain less than one hour following dose administration. These authors suggest that a TfR-mediated endocytosis may be responsible for fast influx of Al into the brain, and that the mechanism is similar to Fe transport via TfR at the blood–brain barrier. The assumption is reasonable, because Al binds extensively to plasma transferrin and because there are many similarities between the chemistry of Fe and Al. In vitro uptake studies further support the TfR-mediated mechanism, since the presence of transferrin in the culture media increases in vitro Al uptake into neuroblastoma cells and oligodendroglia but not into astrocytes (Golub et al., 1999).

Transport of Al into brain by Al-citrate conjugates appears to be much faster than TfR-mediated endocytosis or other diffusion processes (Yokel et al., 1999). Akeson and Munns (1989) suggest that Al uptake in the presence of citrate may be carrier mediated, because the process is energy dependent and proton driven.

The monocarboxylate transporter (MCT) may be a candidate carrier for brain Al-citrate influx. The MCT is located at both the luminal and abluminal surfaces of the blood–brain barrier (Gerhart et al., 1997) and is believed to mediate bidirectional transport of lactate, pyruvate, and other monocarboxylates. It has the second highest V_{max} (60 nmol g^{-1} brain min^{-1}) of all known carriers at the blood–brain barrier

(Laterra et al., 1999). Al-citrate conjugates are formed by coordinative binding of Al with the hydroxyl group and the two terminal carboxylates of citrate (Gregor and Powell, 1986). This leaves a free carboxylate, which is dissociated at physiological pH, perhaps serving as the binding site for the MCT. This hypothesis, however, has not been tested.

Other candidate carriers for Al-citrate include the anion-exchanger family (SCL4A), the organic anion transporter (OAT), also known as multidrug resistance-associated protein (MRP) family (ABC subfamily C), or the organic anion transporting polypeptide (Oatp) family (SCL21). The question as to whether and how these carriers mediate Al influx at the blood–brain barrier remains a topic for further investigation.

The other potential route for Al to enter brain may be via olfactory absorption from the nasal cavity. In Alzheimer's disease, it has been recognized that there is a tendency for the development of NFTs among neurons of cortical regions associated with the olfactory system. Perl and Good (1991) found that neurofibrillary tangle-bearing neurons in cortical regions contained highly elevated levels of Al. This not only indicates that Al may enter the central nervous system via the olfactory pathways, it also implies an association between inhaled Al and the risk of AD.

Although Al is known to accumulate in the choroid plexus (Reusche et al., 2001), whether and to what extent the blood–CSF barrier may contribute to the homeostasis of Al in the brain is unclear.

9.5.2.3 Efflux of Al

Clearance of Al out of the brain is a rather slow process. In human, brain Al concentration increases with age (Markesbery et al., 1984; Ehmann et al., 1986). The concentration of Al in normal adult brains is 17 μM, which is 42 times higher than the normal blood Al concentration (<0.4 μM) (Markesbery et al., 1984). Results from animal studies also support prolonged retention of Al in brain. Following systemic ^{26}Al injection to rats, brain ^{26}Al concentrations decreased only slightly even 35 days after dose administration. These results suggest that Al may be intracellularly distributed in the brain, possibly due to tight tissue binding. Alternatively, Al may be very slowly cleared or removed from the brain compartment.

The efflux of Al from the brain likely takes place at the blood–brain barrier rather than at the blood–CSF barrier (Allen et al., 1992). The efflux of Al occurs shortly after the Al enters the brain in a small quantity. The efflux is energy-dependent (Allen et al., 1995; Ackley and Yokel, 1997); it is unlikely to be TfR mediated (Yokel et al., 1999). Other studies suggest that the MCT is more efficient in efflux than influx of Al at the blood–brain barrier (Deguchi et al., 1997). It has been demonstrated that at least one MCT isoform, MCT1, is expressed at the blood–brain barrier (Yokel, 2002). Ackley and Yokel (1997, 1998) found that Al has been transported out of brain extracellular fluid by a proton-dependent MCT at the blood–brain barrier, probably MCT1.

The choroid plexus occupies the entire brain ventricles. The large surface area of the choroidal epithelia makes the tissue an ideal place to remove unwanted materials from the CSF. Many transporter proteins for Al have been identified in the choroid plexus. It is surprising, however, that the role of the choroid plexus in clearance of CSF Al has never been addressed.

9.5.3 Toxicity of Al at Brain Barriers

As discussed earlier, Al exposure increases the permeability of the blood–brain barrier. The effect is rapid in onset, quickly reversible, and largely dependent on the metal species expressed by the different physicochemical properties of the various metal ligands (Banks et al., 1996). Despite a reported increase of Al in the choroid plexus of an Al-intoxicated patient (Reusche et al., 2001), no study so far has been conducted to identify and characterize the specific damage cause by Al.

9.6 SUMMARY AND PERSPECTIVES

The choroid plexus, as a barrier between the blood and CSF, possesses numerous transporters for metals, metal–amino acid conjugates, and metal–protein complexes. It is clear that dysregulation of metal transport by brain barriers will lead to a significant change in metal homeostasis in the cerebral compartment. This could further contribute to the etiology of metal-associated neurodegenerative disorders. While an effort has been made to understand the underlying mechanisms and therefore to structure the strategy of prevention and therapy, many questions remain. For example, how do these metals get into the brain? Are these metals removed from the CNS by the choroid plexus, or by other mechanisms? Does the disease status become significant only after the barrier loses its protective effect? Our knowledge about the molecular and cellular mechanisms associated with the transport of these metals across brain barriers, as well as the properties and kinetics of molecular movement at the barriers, remains incomplete.

In addition, there are many channels, transporters, carriers, and receptors at the brain barriers. Too little has been learned about how Cu, Zn, and Al interact with endothelial and epithelial cells and to what degree these interactions contribute to neurological disorders. The mechanism whereby the choroid plexus interacts with toxic metals remains largely unknown. The significance of these metals acting on the brain barrier system for the etiology and pathology of CNS disease also needs to be clarified.

9.7 ACKNOWLEDGMENT

Dr. Zheng's research has been supported by NIH Grants RO1 ES-07042, RO1 ES-08146, PO1 ES-09089, Calderone Foundation, Johnson & Johnson Focused Giving Funds, Lily Research Funds, and Burroughs Wellcome Foundation.

REFERENCES

Ackley DC, Yokel RA. (1997). Aluminum citrate is transported from brain into blood via the monocarboxylic acid transporter located at the blood-brain barrier. *Toxicology* 120, 89–97.

Ackley DC, Yokel RA. (1998). Aluminum transport out of brain extracellular fluid is proton dependent and inhibited by mersalyl acid, suggesting mediation by the monocarboxylate transporter (MCT1). *Toxicology* 127, 59–67.

Akeson MA, Munns DN. (1989). Lipid bilayer permeation by neutral aluminum citrate and by three alpha-hydroxy carboxylic acids. *Biochim Biophys Acta* 984, 200–206.

Alfrey AC. (1991). Aluminum intoxication recognition and treatment, In: *Aluminum in Chemistry, Biology and Medicine*, Nicolini M, Zatta P, Corain B, ed. Raven Press, New York, pp. 73–84.

Allen DD and Yokel RA, (1992). Dissimilar aluminum and gallium permeation of the blood-brain barrier demonstrated by in-vivo microdialysis. *J Neurochem* 58, 903–908.

Allen DD, Crooks PA, Yokel RA. (1992). 4-Trimethylammonium antipyrine: a quaternary ammonium nonradionuclide marker for blood-brain barrier integrity during *in vivo* microdialysis. *J Pharmacol Toxicol Meth* 28, 129–135.

Allen DD, Orvig C, Yokel RA. (1995). Evidence for energy-dependent transport of aluminum out of brain extracellular fluid. *Toxicology* 98, 31–39.

Arredondo M, Munoz P, Mura CV, Nunez MT. (2003). DMT1, a physiologically relevant apical Cu1+ transporter of intestinal cells. *Am J Physiol Cell Physiol* 284, C1525–1530.

Atwood CS, Huang X, Moir RD, Tanzi RE, Bush AI. (1999). Role of free radicals and metal ions in the pathogenesis of Alzheimer's disease. *Met Ions Biol Syst* 36, 309–364.

Atwood CS, Scarpa RC, Huang X, Moir RD, Jones WD, Fairlie DP, Tanzi RE, Bush AI. (2000). Copper interactions with Alzheimer amyloid beta peptides: identification of an attomolar-affinity copper binding site on amyloid beta1–42. *J Neurochem* 75, 1219–1233.

Banks WA, Maness LM, Banks MF, Kastin AJ. (1996). Aluminum-sensitive degradation of amyloid beta-protein 1–40 by murine and human intracellular enzymes. *Neurotoxicol Teratol* 18, 671–677.

Basun H, Forssell LG, Wetterberg LG, Winblad B. (1991). Metals and trace elements in plasma and cerebrospinal fluid in normal aging and Alzheimer's disease. *J Neural Transm Park Dis Dement Sect* 3, 231–258.

Beck FW, Prasad AS, Butler CE, Sakr WA, Kucuk O, Sarkar FH. (2004). Differential expression of hZnT-4 in human prostate tissues. *Prostate* 58, 374–381.

Blair-West JR, Denton DA, Gibson AP, McKinely MJ. (1990). Opening the blood-brain barrier to zinc. *Brain Res* 507, 6–10.

Bradbury MW. (1992). An approach to study of transport of trace metals at the blood-brain barrier. *Progr in Brain Res* 91, 133–138.

Brown AM, Tummolo DM, Thodes KJ, Hofmann JR, Jacobsen JS, Sonnenberg-Reines J. (1997). Selective aggregation of endogenousβ-amyloid peptide and soluble amyloid precursor protein in cerebrospinal fluid by zinc. *J Neurochem* 69, 1204–1212.

Burdo JR, Menzies SL, Simpson IA, Garrick LM, Garrick MD, Dolan KG, Haile DJ, Beard JL, Connor JR. (2001). Distribution of divalent metal transporter 1 and metal transport protein 1 in the normal and Belgrade rat. *J Neurosci Res* 66, 1198–1207.

Burhanoglu M, Tutuncuoglu S, Coker C, Tekgul H, Ozgur T. (1996). Hypozincaemia in febrile convulsion. *Eur J Pediatr* 155, 498–501.

Bush AI, Pettingell WH, Multhaup G, Paradis MD, Vonsattel JP, Gusella JF, Beyreuther K, Masters CL,Tanzi RE. (1994a). Rapid induction of Alzheimer Aβ amyloid formation by zinc. *Science* 265, 1464–1467.

Bush AI, Pettingell WH, Paradis MD Tanzi RE, (1994b). The amyloid-β protein precursor and its mammalian homologues. Evidence for a zinc-modulated heparin-binding superfamily. *J Biol Chem* 269, 26618–26621.

Buxani-Rice S, Ueda F, Bradbury MW. (1994). Transport of zinc-65 at the blood-brain barrier during short cerebrovascular perfusion in the rat: its enhancement by histidine. *J Neurochem* 62, 665–672.

Chan PH, Yang GY, Chen SF, Carlson E, Epstein CJ. (1991). Cold-induced brain edema and infarction are reduced in transgenic mice overexpressing CuZn-superoxide dismutase. *Ann Neurol* 29, 482–486.

Checler F. (1995). Processing of the β-amyloid precursor protein and its regulation in Alzheimer's disease. *J Neurochem* 65, 1431–1444.

Choi DW, Koh JY. (1998). Zinc and brain injury. *Annu Rev Neurosci* 21, 347–375.

Choudhuri S, Cherrington NJ, Li N, Klaassen CD. (2003). Constitutive expression of various xenobiotic and endobiotic transporter mRNAs in the choroid plexus of rats. *Drug Metab Dispos* 31, 1337–1345.

Cole TB, Wenzel HJ, Kafer KE, Schwartzkroin PA, Palmiter RD. (1999). Elimination of zinc from synaptic vesicles in the intact mouse brain by disruption of the ZnT3 gene. *Proc Natl Acad Sci USA* 96, 1716–1721.

Crowe A, Morgan EH. (1996). Iron and copper interact during their uptake and deposition in the brain and other organs of developing rats exposed to dietary excess of the two metals. *J Nutr* 126, 183–194.

Cuajungco MP, Lees GJ. (1997). Zinc and Alzheimer's disease: Is there a direct link? *Brain Res Brain Res Rev* 23, 219–236.

Cuajungco, MP and Fagét KY. (2003). Zinc takes the center stage: its paradoxical role in Alzheimer's disease. *Brain Res Rev* 41, 44–56.

Culotta VC, Klomp LW, Strain J, Casareno RL, Krems B, Gitlin JD. (1997). The copper chaperone for superoxide dismutase. *J Biol Chem* 272, 23469–23472.

Dancis A, Yuan DS, Haile D, Askwith C, Eide D, Moehle C, Kaplan J, Klausner RD. (1994). Molecular characterization of a copper transport protein in *S. cerevisiae*: an unexpected role for copper in iron transport. *Cell* 76, 393–402.

Deguchi Y, Nozawa K, Yamada S, Yokoyama Y, Kimura R. (1997). Quantitative evaluation of brain distribution and blood-brain barrier efflux transport of probenecid in rats by microdialysis: possible involvement of the monocarboxylic acid transport system. *J Pharmacol Exp Ther* 280, 551–560.

Deibel MA, Ehmann WD, Markesbery WR. (1996). Copper, iron, and zinc imbalances in severely degenerated brain regions in Alzheimer's disease: possible relation to oxidative stress. *J Neurol Sci* 143, 137–142.

Dexter DT, Carayon A, Javoy-Agid F, Agid Y, Wells FR, Daniel SE, Lees AJ, Jenner P, Marsden CD. (1991). Alterations in the levels of iron, ferritin and other trace metals in Parkinson's disease and other neurodegenerative diseases affecting the basal ganglia. *Brain* 114, 1953–1975.

Dincer Z, Haywood S, Jasani B. (1999). Immunocytochemical detection of metallothionein (MT1 and MT2) in copper-enhanced sheep brains. *J Comp Pathol* 120, 29–37.

Dreosti IE, Manuel SJ, Buckley RA, Fraser FJ, Record IR. (1981) The effect of late prenatal and/or early postnatal zinc deficiency on the development and biochemical aspects of the cerebellum and hippocampus in rats. *Life Sci* 28, 2133–2141.

Ebadi M, Iversen PL, Hao R, Cerutis DR, Rojas P, Happe HK, Murrin LC, Pfeiffer RF. (1995). Expression and regulation of brain metallothionein. *Neurochem Int* 27, 1–22.

Ehmann WD, Markesbery WR, Alauddin M, Hossain TI, Brubaker EH. (1986). Brain trace elements in Alzheimer's disease. *Neurotoxicology* 7, 195–206.

el-Yazigi A, Al-Saleh I, Al-Mefty O. (1986). Concentrations of zinc, iron, molybdenum, arsenic, and lithium in cerebrospinal fluid of patients with brain tumors. *Clin Chem* 32, 2187–2190.

Flaten TP, Alfrey AC, Birchall JD, Savory J Yokel RA. (1996). The status and future concerns of clinical and environmental aluminum toxicology. *J Toxicol Environ Health* 48, 527–542

Flaten TP, Alfrey AC, Birchall JD, Savory J, Yokel RA. (1997). In: *Research Issues in Aluminum Toxicity*, Yokel RA, Golub MS, ed. Taylor & Francis, Washington, DC, pp. 1–15.

Forster DP, Newens AJ, Kay DW, Edwardson JA. (1995). Risk factors in clinically diagnosed presenile dementia of the Alzheimer type: a case-control study in northern England. *J Epidemiol Community Health* 49, 253–258.

Franklin PA, Pullen RG, Hall GH. (1992). Blood-brain exchange routes and distribution of 65Zn in rat brain. *Neurochem Res* 17, 767–771.

Frederickson CJ. 1989. Neurobiology of zinc and zinc-containing neurons. *Int Rev Neurobiol* 31, 145–238.

Frederickson CJ, Suh SW, Silva D, Frederickson CJ, Thompson RB. (2000). Importance of zinc in the central nervous system: the zinc-containing neuron. *J Nutr* 130, 1471S–1483S.

Fukahori M, Itoh M, Oomagari K, Kawasaki H. (1988). Zinc content in discrete hippocampal and amygdaloid areas of the epilepsy (El) mouse and normal mice. *Brain Res* 455, 381–384.

Gaither LA, Eide DJ. (2000). Functional expression of the human hZIP2 zinc transporter. *J Biol Chem* 275, 5560–5564.

Garrick MD, Nunez MT, Olivares M, Harris ED. (2003). Parallels and contrasts between iron and copper metabolism. *Biometals* 16, 1–8.

Gasche Y, Copin JC, Sugawara T, Fujimura M, Chan PH. (2001). Matrix metalloproteinase inhibition prevents oxidative stress-associated blood-brain barrier disruption after transient focal cerebral ischemia. *J Cereb Blood Flow Metab* 21, 1393–1400.

Gerhart DZ, Emerson BE, Zhdankina OY and Leino RL, (1997). Expression of monocarboxylate transporter MCT1 by brain endothelium and glia in adult and suckling rats. *Am J Physiol* 273, E207–E213.

Glerum DM, Shtanko A, Tzagoloff A. (1996). Characterization of COX17, a yeast gene involved in copper metabolism and assembly of cytochrome oxidase. *J Biol Chem* 271, 14504–14509.

Golub MS, Han B, Keen CL. (1999). Aluminum uptake and effects on transferrin mediated iron uptake in primary cultures of rat neurons, astrocytes and oligodendrocytes. *Neurotoxicology* 20, 961–970.

Golub MS, Keen CL, Gershwin ME, Hendrickx AG. (1995). Developmental zinc deficiency and behavior. *J Nutr* 125, 2263S–2271S.

Gonzalez C, Martin T, Cacho J, Brenas MT, Arroyo T, Garcia-Berrocal B, Navajo JA, Gonzalez-Buitrago JM. (1999). Serum zinc, copper, insulin and lipids in Alzheimer's disease epsilon 4 apolipoprotein E allele carriers. *Eur J Clin Invest* 29, 637–642.

Good PF, Perl DP, Bierer LM, Schmeidler J. (1992). Selective accumulation of aluminum and iron in the neurofibrillary tangles of Alzheimer's disease: a laser microprobe (LAMMA) study. *Ann Neurol* 31, 286–292.

Goyer RA. (1998). Toxic effects of metals. In: *Casarett & Doull's Toxicology: The Basic Science of Poisons*, 5th ed. Klaassen CD, ed., McGraw-Hill, New York, 1998, p. 721.

Gregor JE and Powell HKJ. (1986). Aluminum(III)-citrate complexes: a potentiometric and 13C-NMR study. *Aust J Chem* 39, 1851–1864.

Grotz N, Fox T, Connolly E, Park W, Guerinot ML, Eide D. (1998). Identification of a family of zinc transporter genes from *Arabidopsis* that respond to zinc deficiency. *Proc Natl Acad Sci USA* 95, 7220–7224.

Hamza I, Schaefer M, Klomp LW, Gitlin JD. (1999). Interaction of the copper chaperone HAH1 with the Wilson disease protein is essential for copper homeostasis. *Proc Natl Acad Sci USA* 96, 13363–13368.

Harris WR, Keen C. (1989). Calculations of the distribution of zinc in a computer model of human serum. *J Nutr* 119, 1677–1682.

Harris ED. (2001). Copper homeostasis: the role of cellular transporters. *Nutr Rev* 59, 281–285.

Harris ED. (2003). Basic and clinical aspects of copper. *Crit Rev Clin Lab Sci* 40, 547–586.

Hesse L, Beher D, Masters CL, Multhaup G. (1994). The βA4 amyloid precursor protein binding to copper. *FEBS Lett* 349, 109–116.

Hidalgo J, Carrasco J. (1998). Regulation of the synthesis of brain metallothioneins. *Neurotoxicology* 19, 661–666.

Hornshaw MP, McDermott JR, Candy JM (1995). Copper binding to the N-terminal tandem repeat regions of mammalian and avian prion protein. *Biochem Biophys Res Commun* 207, 621–629.

Howell GA, Frederickson CJ. (1990). A retrograde transport method for mapping zinc-containing fiber systems in the brain. *Brain Res* 515, 277–286.

Hu GF. (1998). Copper stimulates proliferation of human endothelial cells under culture. *J Cell Biochem* 69, 326–335.

Huang X, Cuajungco MP, Atwood CS, Hartshorn MA, Tyndall J, Hanson GR, Stokes KC, Leopold M, Multhaup G, Goldstein LE, Scarpa RC, Saunders AJ, Lim J, Moir RD, Glabe C, Bowden EF, Masters CL, Fairlie DP, Tanzi RE, Bush AI. (1999). Cu(II) potentiation of Alzheimer Abeta neurotoxicity. Correlation with cell-free hydrogen peroxide production and metal reduction. *J Biol Chem* 274, 37111–37116.

Huntington C, Shay N, Grouzmann E, Arseneau L, Beverly J. (2001) Zinc status affects neurotransmitter activity in the paraventricular nucleus of rats. *J Nutr* 132, 270–275.

Iwase T, Nishimura M, Sugimura H, Igarashi H, Ozawa F, Shinmura K, Suzuki M, Tanaka M, Kino I. (1996). Localization of Menkes gene expression in the mouse brain; its association with neurological manifestations in Menkes model mice. *Acta Neuropathol* (Berl) 91, 482–488.

Kapaki E, Segditsa J, Papageorgiou C. (1989). Zinc, copper and magnesium concentration in serum and CSF of patients with neurological disorders. *Acta Neurol Scand* 79, 373–378.

Kasarskis EJ. (1984). Zinc metabolism in normal and zinc-deficient rat brain. *Exp Neurol* 85, 114–127.

Kaya M, Kalayci R, Arican N, Kucuk M, Elmas I. (2003). Effect of aluminum on the blood-brain barrier permeability during nitric oxide-blockade-induced chronic hypertension in rats. *Biol Trace Elem Res* 92, 221–230.

Keen CL, Taubeneck MW, Daston GP, Rogers JM, Gershwin ME. (1993). Primary and secondary zinc deficiency as factors underlying abnormal CNS development. *Ann NY Acad Sci* 678, 37–47.

Keller KA, Chu Y, Grider A, Coffield JA. (2000). Supplementation with L-histidine during dietary zinc repletion improves short-term memory in zinc-restricted young adult male rats. *J Nutr* 130, 1633–1640.

Kim CS, Roe CR, Ambrose WW. (1990). L-Carnitine prevents mitochondrial damage induced by octanoic acid in the rat choroid plexus. *Brain Res* 536, 335–338.

Kim GW, Lewen A, Copin J, Watson BD, Chan PH. (2001). The cytosolic antioxidant, copper/zinc superoxide dismutase, attenuates blood-brain barrier disruption and oxidative cellular injury after photothrombotic cortical ischemia in mice. *Neuroscience* 105, 1007–1018.

Kodama H. (1996). Genetic disorders of copper metabolism. In: *Toxicology of Metals*, Chang LW, ed. CRC Press, Boca Raton, FL, pp. 371–386.

Koehler-Stec EM, Simpson IA, Vannucci SJ, Landschulz KT, Landschulz WH. (1998). Monocarboxylate transporter expression in mouse brain. *Am J Physiol* 275, E516–524.

Kuo YM, Zhou B, Cosco D, Gitschier J. (2001). The copper transporter CTR1 provides an essential function in mammalian embryonic development. *Proc Natl Acad Sci USA* 98, 6836–6841.

Landsberg JP, McDonald B, Watt F. (1992). Absence of aluminium in neuritic plaque cores in Alzheimer's disease. *Nature* 360, 65–68.

Larin D, Mekios C, Das K, Ross B, Yang AS, Gilliam TC. (1999). Characterization of the interaction between the Wilson and Menkes disease proteins and the cytoplasmic copper chaperone, HAH1p. *J Biol Chem* 274, 28497–28504.

Laterra J, Keep R, Betz AL, Goldstein GW, (1999). In: *Basic Neurochemistry: Molecular, Cellular and Medical Aspects*, 6th ed., Siegel GJ, et al., ed. Lippincott-Raven Publishers, Philadelphia, pp. 671–689.

Lee JY, Cole TB, Palmiter RD, Suh SW, Koh JY. (2002). Contribution by synaptic zinc to the gender-disparate plaque formation in human Swedish mutant APP transgenic mice. *Proc Natl Acad Sci USA* 99, 7705–7710.

Lehmann HM, Brothwell BB, Volak LP, Bobilya DJ. (2002). Zinc status influences zinc transport by porcine brain capillary endothelial cells. *J Nutr* 132, 2763–2768.

Leino RL, Gerhart DZ, Drewes LR. (1999). Monocarboxylate transporter (MCT1) abundance in brains of suckling and adult rats: a quantitative electron microscopic immunogold study. *Brain Res Dev Brain Res* 113, 47–54.

Loeffler DA., Lewitt PA, Juneau PL, Sima AA, Nguyen HU, Demaggio AJ, Brickman CM, Brewer GJ, Dick RD, Troyer MD, Kanaley L. (1996). Increased regional brain concentrations of ceruloplasmin in neurodegenerative disorders. *Brain Res* 738, 265–274.

Loske C, Gerdemann A, Schepl W, Wycislo M, Schinzel R, Palm D, Riederer P, Munch G. (2000). Transition metal-mediated glycoxidation accelerates crosslinking of β-amyloid peptide. *Eur J Biochem* 267, 4171–4178.

Lovell MA, Robertson JD, Teesdale WJ, Campbell JL, Markesbery WR. (1998). Copper, iron and zinc in Alzheimer's disease senile plaques. *J Neurol Sci* 158, 47–52.

Lovell MA, Ehmann WD, Markesbery WR. (1993). Laser microprobe analysis of brain aluminum in Alzheimer's disease. *Ann Neurol* 33, 36–42.

Maret W. (2001). Zinc biochemistry, physiology, and homeostasis—recent insights and current trends. *Biometals* 14, 187–190.

Markesbery WR, Ehmann WD, Alauddin M, Hossain TI. (1984). Brain trace element concentrations in aging. *Neurobiol Aging* 5, 19–28.

Martyn CN, Coggon DN, Inskip H, Lacey RF, Young WF. (1997). Aluminum concentrations in drinking water and risk of Alzheimer's disease. *Epidemiology* 8, 281–286.

McLachlan DR, Bergeron C, Smith JE, Boomer D, Rifat SL. (1996). Risk for neuropathologically confirmed Alzheimer's disease and residual aluminum in municipal drinking water employing weighted residential histories. *Neurology* 46, 401–405.

McMahon RJ, Cousins RJ. (1998). Mammalian zinc transporters. *J Nutr* 128, 667–670.

Miklossy J, Taddei K, Martins R, Escher G, Kraftsik R, Pillevuit O, Lepori D, Campiche M. (1999). Alzheimer disease: curly fibers and tangles in organs other than brain. *J Neuropathol Exp Neurol* 58, 803–814.

Molina JA, Jimenez-Jimenez FJ, Aguilar MV, Meseguer I, Mateos-Vega CJ, Gonzalez-Munoz MJ, de Bustos F, Porta J, Orti-Pareja M, Zurdo M, Barrios E, Martinez-Para MC. (1998). Cerebrospinal fluid levels of transition metals in patients with Alzheimer's disease. *J Neural Transm* 105, 479–488.

Moller LB, Petersen C, Lund C, Horn N. (2000). Characterization of the hCTR1 gene: genomic organization, functional expression, and identification of a highly homologous processed gene. *Gene* 257, 13–22.

Moos T, Morgan EH. (2004). The significance of the mutated divalent metal transporter (DMT1) on iron transport into the Belgrade rat brain. *J Neurochem* 88, 233–245.

Moreira P, Pereira C, Santos MS, Oliveira C. (2000). Effect of zinc ions on the cytotoxicity induced by the amyloid beta-peptide. *Antioxid Redox Signal* 2, 317–325.

Mori K, Miyake H, Kurisaka M, Sakamoto T. (1993). Immunohistochemical localization of superoxide dismutase in congenital hydrocephalic rat brain. *Childs Nerv Syst* 9, 136–141.

Multhaup G, Schlicksupp A, Hesse L, Beher D, Ruppert T, Masters CL, Beyreuther K. (1996). The amyloid precursor protein of Alzheimer's disease in the reduction of copper(II) to copper(I). *Science* 271, 1406–1409.

Murata Y, Kodama H, Abe T, Ishida N, Nishimura M, Levinson B, Gitschier J, Packman S. (1997). Mutation analysis and expression of the mottled gene in the macular mouse model of Menkes disease. *Pediatr Res* 42, 436–442.

Murray FE, Landsberg JP, Williams RJ, Esiri MM, Watt F. (1992). Elemental analysis of neurofibrillary tangles in Alzheimer's disease using proton-induced x-ray analysis. *Ciba Found Symp* 169, 201–216.

Neri LC, Hewitt D. (1991). Aluminium, Alzheimer's disease, and drinking water. *Lancet* 338, 390.

Nishihara E, Furuyama T, Yamashita S, Mori N. (1998). Expression of copper trafficking genes in the mouse brain. *Neuroreport* 9, 3259–3263.

Nishimura N, Nishimura H, Ghaffar A, Tohyama C. (1992). Localization of metallothionein in the brain of rat and mouse. *J Histochem Cytochem* 40, 309–315.

Noseworthy MD, Bray TM. (2000). Zinc deficiency exacerbates loss in blood-brain barrier integrity induced by hyperoxia measured by dynamic MRI. Proc *Soc Exp Biol Med* 223, 175–182.

O'Dell BL, Becker JK, Emery MP, Browning JD. (1989). Production and reversal of the neuromuscular pathology and related signs of zinc deficiency in guinea pigs. *J Nutr* 119, 196–201.

Öhman LO, Martin RB. (1994). Citrate as the main small molecule binding $Al3+$ in serum. *Clin Chem* 40, 598–601.

Palm R, Hallmans G. (1982). Zinc concentrations in the cerebrospinal fluid of normal adults and patients with neurological diseases. *J Neurol Neurosurg Psychiatry* 45, 685–690.

Palm R, Strand T, Hallmans G. (1986). Zinc, total protein, and albumin in CSF of patients with cerebrovascular diseases. *Acta Neurol Scand* 74, 308–313.

Palmiter RD, Cole TB, Quaife CJ, Findley SD. (1996). ZnT-3, a putative transporter of zinc into synaptic vesicles. *Proc Natl Acad Sci USA* 93, 14934–14939.

Perl DP, Good PF. (1991). Aluminum, Alzheimer's disease, and the olfactory system. *Ann NY Acad Sci* 1991, 640: 8–13.

Poduslo JF, Whelan SL, Curran GL, Wengenack TM. (2000). Therapeutic benefit of polyamine-modified catalase as a scavenger of hydrogen peroxide and nitric oxide in familial amyotrophic lateral sclerosis transgenics. *Ann Neurol* 48, 943–947.

Potocnik FC, van Rensburg SJ, Park C, Taljaard JJF Emsley RA. (1997). Zinc and platelet membrane microviscosity in Alzheimer's disease: the *in vivo* effect of zinc on platelet membranes and cognition. *S Afr Med J* 87, 1116–1119.

Pullen RG, Franklin PA, Hall GH. (1991). 65Zn uptake from blood into brain in the rat. *J Neurochem* 56, 485–489.

Qian Y, Mikeska G, Harris ED, Bratton GR, Tiffany-Castiglioni E (1999). Effect of lead exposure and accumulation on copper homeostasis in cultured C6 rat glioma cells. *Toxicol Appl Pharmacol* 158, 41–49.

Qian Y, Tiffany-Castiglioni E, Welsh J, Harris ED. (1998). Copper efflux from murine microvascular cells requires expression of the menkes disease Cu-ATPase. *J Nutr* 128, 1276–1282.

Regland B, Lehmann W, Abedini I, Blennow K, Jonsson M, Karlsson I, Sjogren M, Wallin A, Xilinas M, Gottfries CG. (2001). Treatment of Alzheimer's disease with clioquinol. *Dement Geriatr Cogn Disord* 12, 408–414.

Reinholz MM, Merkle CM, Poduslo JF. (1999). Therapeutic benefits of putrescine-modified catalase in a transgenic mouse model of familial amyotrophic lateral sclerosis. *Exp Neurol* 159, 204–216.

Reusche E. (1997). Argyrophilic inclusions distinct from Alzheimer neurofibrillary changes in one case of dialysis-associated encephalopathy. *Acta Neuropathol* (Berl). 94, 612–616.

Reusche E, Pilz P, Oberascher G, Lindner B, Egensperger R, Gloeckner K, Trinka E, Iglseder B. (2001). Subacute fatal aluminum encephalopathy after reconstructive otoneurosurgery: a case report. *Human Pathol* 32, 1136–1140.

Rolfs A, Hediger MA. (1999). Metal ion transporters in mammals: structure, function and pathological implications. *J Physiol* 518, 1–12.

Roskams AJ, Connor JR. (1990). Aluminum access to the brain: a role for transferrin and its receptor. *Proc Natl Acad Sci* 87, 9024–9027.

Rulon LL, Robertson JD, Lovell MA, Deibel MA, Ehmann WD Markesbery WR. (2000). Serum zinc levels and Alzheimer's disease. *Biol Trace Elem Res* 75, 79–85.

Savory J, Exley C, Forbes WF, Huang Y, Joshi JG, Kruck T, McLachlan DRC, Wakayama I. (1997). In: *Research Issues in Aluminum Toxicity*, Yokel RA and Golub MS, ed. Taylor & Francis, Washington, DC, pp. 185–205.

Sawashita J, Takeda A, Okada S. (1997). Change of zinc distribution in rat brain with increasing age. *Brain Res Dev Brain Res* 102, 295–298.

Sen CK, Khanna S, Venojarvi M, Trikha P, Ellison EC, Hunt TK, Roy S. (2002). Copper-induced vascular endothelial growth factor expression and wound healing. *Am J Physiol Heart Circ Physiol* 282, H1821–1827.

Siddappa AJ, Rao RB, Wobken JD, Leibold EA, Connor JR, Georgieff MK. (2002). Developmental changes in the expression of iron regulatory proteins and iron transport proteins in the perinatal rat brain. *J Neurosci Res* 68, 761–775.

Skoczynska A, Smolik R, Milian A. (1994). The effect of combined exposure to lead and cadmium on the concentration of zinc and copper in rat tissues. *Int J Occup Med Environ Health* 7, 41–49.

Solioz M, Vulpe C. (1996). Cpx-type ATPases: a class of P type ATPases that pump heavy metals. *Trends Biochem Sci* 21, 237–241.

Spencer PS. (2000). Aluminum and its compounds. In: *Experimental and Clinical Neurotoxicology*, Spencer PS, Schaumburg HH, ed. Oxford University Press, New York, pp. 142–151.

St Croix CM, Wasserloos KJ, Dineley KE, Reynolds IJ, Levitan ES, Pitt BR, (2002). Nitric oxide-induced changes in intracellular zinc homeostasis are mediated by metallothionein/thionein. *Am J Physiol Lung Cell Mol Physiol* 282, L185–L192.

Stuerenburg HJ. (2000). CSF copper concentrations, blood-brain barrier function, and coeruloplasmin synthesis during the treatment of Wilson's disease. *J Neural Transm* 107, 321–329.

Takeda A, Sawashita J, Okada S. (1995). Biological half-lives of zinc and manganese in rat brain. *Brain Res* 695, 53–58.

Takeda A, Kawai M, Okada S. (1997). Zinc distribution in the brain of Nagase analbuminemic rat and enlargement of the ventricular system. *Brain Res* 769, 193–195.

Taskinalp O, Aktas RG, Cigali B, Kutlu AK. (2000). Immunohistochemical demonstration of cytochrome oxidase in different parts of the central nervous system: a comparative experimental study. *Anat Histol Embryol* 29, 345–349.

Tsuda M, Imaizumi K, Katayama T, Kitagawa K, Wanaka A, Tohyama M, Takagi T. (1997). Expression of zinc transporter gene, ZnT-1, is induced after transient forebrain ischemia in the gerbil. *J Neurosci* 17, 6678–6684.

Tully CL, Snowdon DA, Markesbery WR. (1995). Serum zinc, senile plaques, and neurofibrillary tangles: findings from the Nun study. *NeuroReport* 6, 2105–2108.

Uchino M, Ando Y, Tanaka Y, Nakamura T, Uyama E, Mita S, Murakami T, Ando M. (1994). Decrease in Cu/Zn- and Mn-superoxide dismutase activities in brain and spinal cord of patients with amyotrophic lateral sclerosis. *J Neurol Sci* 127, 61–67.

Vallee BL, Falchuk KH. (1993). The biochemical basis of zinc physiology. *Physiol Rev* 73, 79–118.

Van Rhijn AG, Prior CA, Corrigan FM. (1990). Dietary supplementation with zinc sulphate, sodium selenite and fatty acids in early dementia of Alzheimer's type. *J Nutr Med* 1, 259–266.

Voskoboinik I, Brooks H, Smith S, Camakaris J. (1998). ATP dependent copper transport by the Menkes protein in membrane vesicles isolated from cultured Chinese hamster ovary cells. *FEBS Lett* 435, 178–182.

Waggoner DJ, Bartnikas TB, Gitlin JD. (1999). The role of copper in neurodegenerative disease. *Neurobiol Dis* 6, 221–230.

Walton J, Tuniz C, Fink D, Jacobsen G, Wilcox D. (1995).Uptake of trace amounts of aluminum into the brain from drinking water. *Neurotoxicology* 16, 187–190.

Wenzel HJ, Cole TB, Born DE, Schwartzkroin PA, Palmiter RD. (1997). Ultrastructural localization of zinc transporter-3 (ZnT-3) to synaptic vesicle membranes within mossy fiber boutons in the hippocampus of mouse and monkey. *Proc Natl Acad Sci USA* 94, 12676–12681.

Wu LJ, Leenders AG, Cooperman S, Meyron-Holtz E, Smith S, Land W, Tsai RY, Berger UV, Sheng ZH, Rouault TA. (2004). Expression of the iron transporter ferroportin in synaptic vesicles and the blood-brain barrier. *Brain Res* 1001, 108–117.

Yamashita T, Shimada S, Guo W, Sato K, Kohmura E, Hayakawa T, Takagi T, Tohyama M. (1997). Cloning and functional expression of a brain peptide/histidine transporter. *J Biol Chem* 272, 10205–10211.

Yokel RA. (2000). The toxicology of aluminum in the brain: a review. *Neurotoxicology* 21, 813–828.

Yokel RA, Allen DD, Ackley DC. (1999). The distribution of aluminum into and out of the brain. *J Inorg Biochem* 76, 127–132.

Yokel RA. (2002). Brain uptake, retention, and efflux of aluminum and manganese. *Environ Health Perspect* 110, 699–704.

Yokel RA, Wilson M, Harris WR, Halestrap AP. (2002).Aluminum citrate uptake by immortalized brain endothelial cells: implications for its blood-brain barrier transport. *Brain Res* 930, 101–110.

Yoshiike Y, Tanemura K, Murayama O, Akagi T, Murayama M, Sato S, Sun X, Tanaka N, Takashima A, (2001). New insights on how metals disrupt amyloid β-aggregation and their effects on amyloid-β cytotoxicity. *J Biol Chem* 276, 32293–32299.

Yoshimura N, Kida K, Usutani S, Nishimura M. (1995). Histochemical localization of copper in various organs of brindled mice after copper therapy. *Pathol Int* 45, 10–18.

Zatta P, Lucchini R, Susan Rensburg SJ, Taylor A. (2003). The role of metals in neurodegenerative processes: aluminum, manganese, and zinc. *Brain Res Bull* 62, 15–28.

Zheng W. (1996). The choroid plexus and metal toxicities. In: *Toxicology of Metals*, Chang LW, ed. CRC Press, New York, pp 609–626.

Zheng W. (2001a). Toxicology of choroid plexus: special reference to metal-induced neurotoxicities. *Microsc Res Techn* 52, 89–103.

Zheng W. (2001b). Neurotoxicology of the brain barrier system: new implications. *J Toxicol Clin Toxicol* 39, 711–719.

Zheng W. (2002). Blood-brain barrier and blood-CSF barrier in metal-induced neurotoxicities. In: Handbook of *Neurotoxicology*, Vol. 1 Massaro EJ, ed. Humana Press, Totowa, NJ, pp. 161–193.

Zheng W, Aschner M, Ghersi-Egea JF. (2003). Brain barrier systems: a new frontier in metal neurotoxicological research. *Toxicol Appl Pharmacol* 192, 1–11.

10 The Role of the Choroid Plexus in the Transport and Production of Polypeptides

Adam Chodobski
Brown University School of Medicine,
Providence, Rhode Island, U.S.A.

Gerald D. Silverberg
Stanford University School of Medicine,
Stanford, California, U.S.A.

Joanna Szmydynger-Chodobska
Brown University School of Medicine,
Providence, Rhode Island, U.S.A.

CONTENTS

10.1 THE CONCEPT OF VOLUME TRANSMISSION IN THE BRAIN

Conventionally, the intercellular communication within the central nervous system (CNS) is thought of as synaptic and gap junction-mediated signaling. This traditional view has been recently expanded to incorporate the concept of volume transmission into central signaling (Agnati et al., 1995; Zoli and Agnati, 1996; Zoli et al., 1998). Volume transmission may be defined as both the short-distance (diffusional) and long-range (convective) movement of molecules within the extracellular fluid space of the brain. The convective traffic of molecules within neural tissue is mediated by the bulk flow of interstitial fluid (Rosenberg et al., 1978, 1980; Cserr, 1984). The continual formation and flow of cerebrospinal fluid (CSF), on the other hand, facilitates the distribution of molecules within the CSF space, i.e., the cerebral ventricles, the perivascular, Virchow-Robin, spaces that penetrate the neuropil, and the subarachnoid space of the brain and spinal cord. In the mid-1970s Rodríguez proposed that the CSF plays an important role in neuroendocrine integration within the CNS (Rodríguez, 1976). More recent studies have supported this hypothesis by providing several lines of evidence for CSF-mediated signaling, such as the propagation of immune signals and the widespread activation of astroglia (Proescholdt et al., 2002) (see also Chapter 18). In an elegant behavioral study, Silver et al. (1996) have also shown that recovery of circadian rhythms in hamsters following lesions of the hypothalamic suprachiasmatic nuclei (SCN) occurs after grafting encapsulated SCN tissue onto the wall of the 3rd ventricle. While these observations indicate that the CSF can effectively convey a "message" within the CNS, it remains incompletely understood how CSF-borne molecules, such as polypeptides, reach their target cells within the brain parenchyma. Indeed, although the ependyma and the pial-glial lining do not represent a major barrier to polypeptide penetration (Brightman and Rees, 1969), the diffusional movement of these molecules within the extracellular space of the neuropil is limited (Nicholson and Syková, 1998). It is possible that the biological actions of some CSF-borne polypeptides are mediated by other central neuroendocrine systems (see Sec. 10.2.1). Various CSF-borne polypeptides may also exert their biological effects by way of receptor-mediated retrograde transport in neurons whose axonal processes are located near the ependymal or the pial-glial lining (Ferguson and Johnson, 1991; Ferguson et al., 1991; Mufson et al., 1999).

In this review we will focus on two aspects of choroid plexus (CP) functions that are intimately associated with volume transmission. The first is hormonal signaling across the choroidal blood–CSF barrier (BCSFB). This signaling involves a vectorial, receptor-mediated transport of peptide hormones through the choroidal epithelium. The second CP function that we will discuss is related to the ability of the CP to synthesize and secrete polypeptides. Increasing evidence indicates that the choroidal epithelium is highly active in polypeptide synthesis. For some proteins, such as transthyretin (TTR) and insulin-like growth factor II (IGF-II), the CP is their major source of supply to the CNS. Therefore, it is likely that the bulk flow of CSF distributes these CP-derived polypeptides to their distal target cells in the brain. Evidence supporting this hypothesis will be critically examined in this chapter.

10.2 TRANSPORT OF PEPTIDE HORMONES ACROSS THE BLOOD–CSF BARRIER

Our knowledge of polypeptide transport across the choroidal BCSFB and its role in hormonal signaling is limited. In recent years, leptin transport across the BCSFB and the blood–brain barrier (BBB) has attracted considerable interest. The transport of prolactin (PRL) across the BCSFB has also been demonstrated, and evidence for its physiological regulation has been provided. Earlier studies by Zlokovic et al. (1990, 1991) have demonstrated choroidal uptake and transport across the BBB of circulating vasopressin (VP). Although choroidal VP uptake and peptide influx into the brain were V_1 receptor-dependent and exhibited saturable kinetics, their Michaelis-Menten constants (K_ms) were high (~30 nM for uptake and 2 to 3 μM for transport) when compared to plasma VP levels observed in various physiological or pathophysiological conditions (Szczepanska-Sadowska et al., 1983; Cameron et al., 1985; Wang et al., 1988). Therefore, the physiological significance of VP transport across the BCSFB or the BBB is uncertain. More recently, the transport of secretin from the blood into the brain has attracted attention because of the possible use of this peptide for the treatment of autism (McQueen and Heck, 2002). Studies performed in mice have shown that a saturable transport of circulating secretin across the BCSFB exists (Banks et al., 2002). However, the functional importance of this transport remains unclear. Further discussion of hormonal signaling across the choroidal BCSFB will focus on the two peptide hormones, leptin and PRL.

10.2.1 LEPTIN

Leptin, a 16-kDa circulating protein, plays an important role in maintaining energy homeostasis (Ahima and Flier, 2000). This hormone is produced primarily, but not exclusively, by white adipose tissue (Ahima and Flier, 2000), and its levels in serum positively correlate with the percentage of body fat (Maffei et al., 1995b; Considine et al., 1996). Gender is an important factor in determining serum leptin levels. Women have considerably higher hormone concentrations than men for any given body fat content (Saad et al., 1997). The major sites of leptin actions, related to the regulation of energy balance, are located in the hypothalamus, and, more specifically, include the neurons in the arcuate (ARC), paraventricular (PVN), and ventromedial (VMN) nuclei (Choi et al., 1999). Selective lesions of these nuclei cause obesity and render rodents insensitive to leptin (Choi et al., 1999). The regulatory role of hypothalamic neurons has recently been demonstrated in mice with a neuron-specific disruption of leptin receptors (Ob-Rs) (Cohen et al., 2001). In these animals, the extent of obesity negatively correlated with the levels of hypothalamic Ob-R expression, and the obese mice expressing low levels of Ob-R showed increased plasma leptin concentrations.

Physiological actions of leptin have largely been investigated in rodents. In these studies, peripheral or intracerebroventricular (icv) administration of the recombinant leptin to lean mice or rats was found to reduce food intake and body weight (Campfield et al., 1995; Cusin et al., 1996; Seeley et al., 1996; Al-Barazanji et al.,

1997; Halaas et al., 1997; Flynn et al., 1998; Harris et al., 1998; Choi et al., 1999). In obese *ob/ob* mice lacking functional leptin, the effect of the recombinant hormone on food intake and body weight appeared to be amplified when compared to lean animals (Harris et al., 1998), which, at least in part, might have resulted from increased hypothalamic expression of Ob-Rs (Fei et al., 1997; Baskin et al., 1998). On the other hand, obese *db/db* mice and *fa/fa* Zucker rats, in which the obesity was a consequence of an abnormal splicing of (Lee et al., 1996), and a point mutation in, Ob-R (Chua et al., 1996), either did not respond or had reduced sensitivity to leptin (Campfield et al., 1995; Cusin et al., 1996; Seeley et al., 1996; Al-Barazanji et al., 1997).

Ob-R has been identified through expression cloning by Tartaglia et al. (1995). Further studies have demonstrated that there are at least six isoforms of murine Ob-Rs generated from one gene by alternative RNA splicing (Lee et al., 1996; Takaya et al., 1996; Wang et al., 1996; Chua et al., 1997). Four short isoforms, Ob-Ra, Ob-Rc, Ob-Rd, and Ob-Rf, have been identified. Ob-Ra, a dominant short isoform of the leptin receptor, highly expressed in the CP (Guan et al., 1997; Hileman et al., 2002), has rather limited signaling capabilities (Ghilardi et al., 1996; Bjørbæk et al., 1997), and it was suggested, therefore, that this isoform had a role to play in leptin transport (see below). The long isoform of the leptin receptor, Ob-Rb, is abundantly expressed in the hypothalamus (Fei et al., 1997; Guan et al., 1997) and is believed to mediate the biological effects of leptin (Ahima and Flier, 2000). The levels of hypothalamic expression of this receptor appear to be negatively regulated by circulating leptin, since they are elevated in leptin-deficient *ob/ob* mice and in normal fasted rats (Fei et al., 1997; Baskin et al., 1998). In comparison, systemic administration of the recombinant leptin to *ob/ob* mice significantly reduces hypothalamic levels of Ob-Rb mRNA (Baskin et al., 1998). In addition to plasma membrane-associated isoforms of Ob-R discussed above, a soluble isoform, Ob-Re, has also been identified in rodents (Takaya et al., 1996; Chua et al., 1997). While the transcripts for Ob-Re have not been found in humans (Chua et al., 1997), the metalloprotease-mediated cleavage of ectodomains of human Ob-Ra and Ob-Rb, producing a soluble leptin-binding protein, has been reported (Maamra et al., 2001). These soluble receptors likely modulate the biological activity of leptin.

The cell surface Ob-Rs are single membrane-spanning receptors belonging to the class I cytokine receptor family (Zabeau et al., 2003). They are most related to the granulocyte-colony stimulating factor receptor and the glycoprotein 130 (gp130) family of receptors, including gp130, the leukemia inhibitory factor, and the oncostatin M receptors. The binding of leptin to Ob-Rb results in the activation of several signal transduction pathways, including Janus kinase-2 (JAK2), signal transducer and activator of transcription (STAT)-3, phosphatidylinositol 3-kinase (PI3K), and extracellular signal-regulated kinase (ERK) (Sweeney, 2002; Zabeau et al., 2003). Suppressor of cytokine signaling-3, which is rapidly activated following leptin binding, negatively regulates the Ob-Rb signaling. The selective disruption of Ob-Rb–mediated activation of STAT-3 and the use of highly specific PI3K inhibitors have provided evidence that the signaling via STAT-3 and PI3K is critical for the leptin-dependent regulation of body energy homeostasis (Niswender et al., 2001; Bates et al., 2003). The signaling capability of the short isoforms of the leptin receptor is a matter of debate. Ob-Ra is unable to activate STAT-3 (Ghilardi et al., 1996) and can only weakly stimulate the ERK signaling cascade (Bjørbæk et al., 1997). However, like Ob-Rb, the short isoform has been shown to mediate the induction of immediate early genes, such as c-*fos*, c-*jun*, and *jun-B* (Murakami et al., 1997).

The limited signaling capability of Ob-Ra suggests that the short receptor isoforms play only a minor role in mediating the biological leptin effects and that these isoforms are instead involved in hormone transport via a receptor-dependent transcytosis. While this concept is supported by the observation that leptin internalization (Uotani et al., 1999) and transcytosis (Hileman et al., 2000) are Ob-Ra-mediated, the efficiency of transcellular transport of the intact hormone is rather low, because significant amounts of internalized leptin are degraded by lysosomes. Based on the observations of strong leptin binding to the CP, it has been proposed that the BCSFB plays an important role in transporting the hormone into the brain (Tartaglia et al., 1995). Three short receptor isoforms, Ob-Ra, Ob-Rc, and Ob-Rf, have been identified in the CP (Guan et al., 1997; Hileman et al., 2002); however, discrepant results regarding the relative abundance of these three isoforms have been obtained in these studies. The transport of leptin into the CSF has recently been studied using two experimental models: in situ rat brain perfusion (Zlokovic et al., 2000) and perfused sheep CP (Thomas et al., 2001). Although these two studies provided evidence for the transport of leptin across the BCSFB, they came up with quite different K_m estimates for this transport: ~1 ng/ml by Zlokovic et al., and ~260 ng/ml by Thomas et al. Since other reports have suggested that leptin transport into the CSF is saturated at plasma hormone levels of ~10 ng/ml (Caro et al., 1996; Wiedenhöft et al., 1999), the latter K_m estimate is probably too high. Future studies are likely to provide a more precise characterization of leptin transport across the BCSFB.

In addition to the CP, cerebral microvessels have also been shown to express short isoforms of the leptin receptor (Hileman et al., 2002). The Ob-Rs identified in brain microvessels appear to be as abundant as those found in the CP (Hileman et al., 2002). Consistent with these findings, leptin has been demonstrated to enter the brain by crossing not only the BCSFB, but also the BBB (Banks et al., 1996; Banks et al., 2000; Zlokovic et al., 2000). Variable estimates of the K_m for this transport have been provided, ranging between ~90 and ~350 ng/ml for various brain regions, e.g., the cerebral cortex, caudate nucleus, and hippocampus (Zlokovic et al., 2000). Whole brain estimates of the transport K_m average ~16 ng/ml (Banks et al., 2000). The latter study suggests that the transport of leptin across the BBB is partially saturated even at plasma hormone levels observed in nonobese subjects (Maffei et al., 1995b; Considine et al., 1996).

Some authors have emphasized the importance of the BBB for central leptin signaling and questioned the physiological significance of leptin transport across the choroidal BCSFB and subsequent hormone delivery to its hypothalamic targets via the CSF pathway (Maness et al., 1998). Based on the observations of limited penetration of the subependymal parenchyma by CSF-borne leptin, these authors argued that, especially in large brains, this polypeptide might not reach sufficient numbers of its neuronal receptors to elicit an adequate biological response. This reasoning, however, does not take into consideration the possible interactions of

CSF-borne leptin with other neuroendocrine systems, which could play a mediatory role in leptin-dependent regulation of energy balance. Indeed, it has recently been reported that the anorexic effects of leptin are mediated by interleukin-1 (IL-1) (Luheshi et al., 1999). In these studies, hypothalamic IL-1β synthesis was found to be upregulated in response to the icv infusion of leptin. Furthermore, the inhibition of food intake in response to icv or peripheral administration of leptin was attenuated by the IL-1 receptor antagonist, and the inhibition of food intake by this hormone was completely abolished in the IL-1 receptor type I-deficient mice. These observations are consistent with the presence of IL-1β-immunopositive neuronal fibers that are located close to the wall of the 3rd ventricle and innervate ARC, PVN, and VMN (Breder et al., 1988). The contact surface between the ventricular lining and the neural tissue is enhanced by the presence of 3rd ventricle invaginations that extend laterally into the mediobasal hypothalamus (Amat et al., 1999) and likely facilitate the hypothalamic actions of CSF-borne leptin.

Any discussion of leptin signaling should also include the possible role of leaky capillaries within the median eminence (ME), a circumventricular organ. Previous studies have shown that the peripheral administration of gold thioglucose, a toxin producing lesions in brain areas adjacent to the circumventricular organs (Levine and Sowinski, 1983), causes obesity (Maffei et al., 1995a) and significantly decreases Ob-Rb mRNA in the hypothalamus (Fei et al., 1997). These observations suggest that some of the neuronal leptin targets in the hypothalamus are also reached by the hormone via permeable cerebrovascular endothelium within the ME. Further research is clearly needed to increase our understanding of leptin signaling. At present, we may tentatively assume that both the BBB and the choroidal BCSFB are involved in leptin delivery to its central target sites.

The obesity seen so commonly in humans rarely results from mutations in the *LEP* or the *OBR* gene (O'Rahilly et al., 2003). Whereas single nucleotide polymorphisms in the *OBR* gene, such as Gln223Arg, have been reported to affect the adiposity in some populations (Quinton et al., 2001), the common observation in the majority of obese individuals is elevation in their serum leptin level (Maffei et al., 1995b; Considine et al., 1996). The latter finding suggests that obesity in humans is associated with leptin resistance (receptor insensitivity or decreased hypothalamic leptin levels or both). Clinical studies by Caro et al. (1996), in which the CSF/serum ratio for leptin was analyzed in both lean and obese individuals, suggest that leptin transport into the brain is saturable and this transport capacity is reduced in obesity (see Figure 10.1). Impaired transport of leptin across the BBB has been found in obese CD-1 and New Zealand obese (NZO) mice that do not have any known defects in the *Obr* gene and in mice fed a high-fat diet (Banks et al., 1999; Hileman et al., 2002; Banks and Farell, 2003). These studies indicate that the reduction in the transport rate of leptin associated with obesity is not simply related to the saturation of the transporter because of high serum leptin levels, but is due to the decreased capacity of the BBB to transport the hormone. Leptin resistance may therefore be associated with defective hormone transport into the brain. The mechanisms underlying this defect in leptin signaling remain unclear. Indeed, the levels of Ob-Ra mRNA in the cerebral microvessels of NZO and high-fat diet-fed mice were found to be similar to those observed in lean animals (El-Haschimi et al., 2000; Hileman

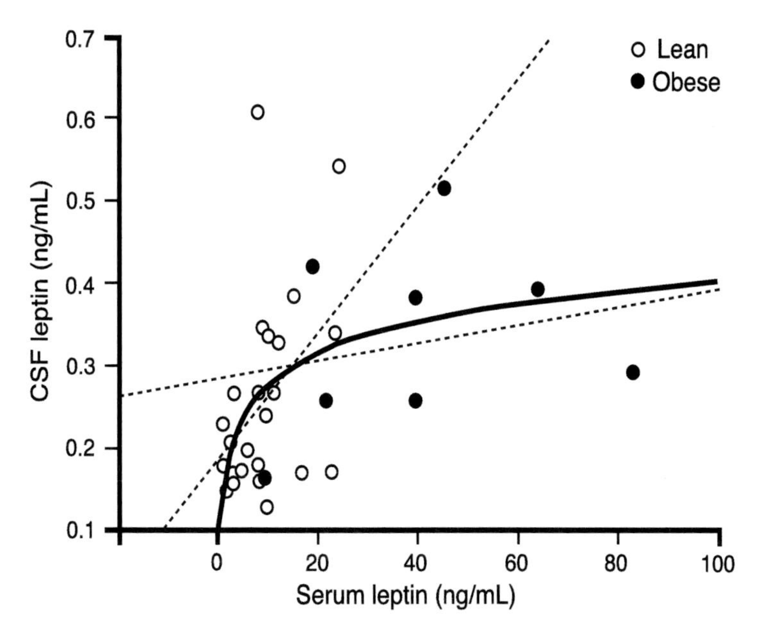

FIGURE 10.1 Relationship between CSF and serum leptin levels in lean and obese humans. These data suggest that leptin transport into the brain is saturable and its capacity is reduced in obesity. Solid line = logarithmic function; broken lines = linear regression. [Reprinted with permission from Elsevier (*Lancet* 348, 1996, 159–61).]

et al., 2002). Also, the rats appeared to respond to a high-fat diet with increased cerebrovascular Ob-Ra expression at both the messenger and protein level (Boado et al., 1998). It is possible that, in obesity, there is increased lysosomal degradation or decreased efficiency of leptin transcytosis or both across the cerebrovascular endothelium that may contribute to the impairment of hormone transport across the BBB. Interestingly, leptin resistance may result not only from inadequate leptin delivery to the brain, but may also be related to defective signal transduction in hypothalamic Ob-Rb (El-Haschimi et al., 2000). Further studies, including an analysis of leptin transport across the BCSFB, are required to shed more light on leptin resistance in obesity.

10.2.2 Prolactin

PRL is a peptide hormone originally found to promote mammary gland development and to stimulate lactation in pregnancy (Bole-Feysot et al., 1998; Freeman et al., 2000). This hormone also regulates a number of brain functions, including maternal behavior, suppression of fertility, food intake, and the response to stress. The PRL receptor (PRL-R) belongs to the class I cytokine receptor family (Goffin and Kelly, 1997; Clevenger and Kline, 2001). Several isoforms of PRL-R have been identified. These include short, intermediate, and long isoforms. These isoforms have identical extracellular and transmembrane domains, but differ with regard to the length of their intracellular parts. Recently, a new PRL-R isoform, with a long intracellular tail, but truncated extracellular domain, has been described (Kline et al., 2002). Furthermore, a PRL-binding protein representing the extracellular domain of PRL-R has been found (Kline and Clevenger, 2001). This protein can modulate the biological activity of PRL. Similar to Ob-Rb, the binding of PRL to its receptor activates the JAK2 and STAT-5 signaling cascades (Goffin and Kelly, 1997; Clevenger and Kline, 2001). Interestingly, unlike the Ob-Rs, stimulation of both long and short isoforms of PRL-R results in the activation of JAK2 and STAT-5 (Bignon et al., 1999). The families of Src and Tec tyrosine kinases are also involved in PRL-R signaling (Goffin and Kelly, 1997; Clevenger and Kline, 2001).

Intravenous injection of ^{125}I-PRL was found to heavily label the CP, whereas the cerebral vasculature did not appear to bind the radioiodinated PRL (Walsh et al., 1978). This suggests that circulating PRL is transported into the brain across the choroidal BCSFB. Subsequent studies by the same group (Walsh et al., 1987) have corroborated this hypothesis, demonstrating the presence of a saturable transport of bloodborne PRL into the CSF. Both short and long isoforms of PRL-R are expressed in the CP, with the short isoform being expressed at higher levels than the long isoform (Bakowska and Morrell, 1997, 2003; Pi and Grattan, 1998). The message for PRL-R is increased during pregnancy and lactation (Grattan et al., 2001), which correlates well with increased PRL uptake by the CP and the augmented transport of the hormone from the blood into the CSF observed in hyperprolactinemic rats (Mangurian et al., 1992). While these findings provide solid evidence for receptor-mediated transport of PRL across the choroidal epithelial barrier, it is uncertain which isoform of PRL-R is involved in this process. Further studies will be needed to clarify this issue and to enhance our understanding of the physiological importance of PRL transport across the BCSFB.

10.3 PEPTIDE AND PROTEIN SYNTHESIS IN THE CHOROIDAL EPITHELIUM

The number of polypeptides known to be synthesized in the CP is growing steadily. Although the functional data on these CP-derived polypeptides are still limited, we are beginning to appreciate the physiological significance of the ability of the CP to manufacture and secrete peptides and proteins into the CSF. For many CP-derived polypeptides, the cognate receptors have been found to be expressed on choroidal epithelium, suggesting that these polypeptides act in an autocrine manner or juxtacrine/paracrine manner or both. However, as proposed earlier in this chapter, many CP-derived polypeptides may also exert their biological effects on distant brain targets after being distributed through the CSF pathways. In the following paragraphs, we will analyze the putative actions of selected peptides and proteins produced by the choroidal epithelium.

10.3.1 Vasopressin

The levels of VP in the CSF collected from the 3rd ventricle are 7- to 10-fold higher than those found in plasma (Simon-Oppermann et al., 1983; Szczepanska-Sadowska et al., 1983), indicating that this neuropeptide is not only released into the peripheral circulation, but also centrally secreted. Various physiological stimuli, such as hypernatremia, dehydration, and systemic hemorrhage, have been demonstrated to increase CSF concentrations of VP (Szczepanska-Sadowska et al., 1983). There is a dramatic increase in VP levels in the CSF in response to general anesthesia and brain surgery (Simon-Oppermann et al., 1983). The physiological significance of centrally released VP is supported by data obtained in animal experiments involving icv administration of the synthetic peptide. In these studies, a number of physiological and behavioral responses to VP have been observed, including changes in the diffusional permeability of the BBB to water (Raichle and Grubb, 1978), hormonal secretion (Matthews and Parrott, 1994), peripheral hemodynamics (Harland et al., 1989), thermoregulation (Terlouw et al., 1996), the sleep–wake cycle (Arnauld et al., 1989), as well as learning and memory consolidation (Le Moal et al., 1987). Animal studies have also demonstrated that, after ischemic brain injury, central VP production promotes the formation of edema (Dickinson and Betz, 1992), which is consistent with the clinical observation that increased concentrations of VP in the CSF occur in ischemic stroke (Sørensen et al., 1985).

One possible source of CSF VP is the VP-containing axonal processes originating in the PVN that are localized close to the lateral ventricles (Buijs et al., 1978). The presence of high VP concentrations in the 3rd ventricle CSF suggests that similar VP-containing neuronal processes are located near the 3rd ventricle as well. A prominent daily rhythm of CSF VP levels has been noted in various mammalian species (Reppert et al., 1982; Schwartz et al., 1983; Günther et al., 1984). This rhythm appears to reflect the circadian variations in the activity of the VP-ergic neurons in the SCN. However, unlike the VP-containing neurons in PVN and the supraoptic hypothalamic nuclei, the VP-ergic neurons located in the SCN do not respond to hyperosmotic stress by increasing VP synthesis (Sherman et al., 1986). The choroidal epithelium is also a likely source of CSF VP (Chodobski et al., 1997, 1998c). VP and other polypeptide components of the prohormone for VP, such as the VP-associated neurophysin and glycopeptide, were identified in the CP, indicating that, in this tissue, the biosynthetic processing of the prohormone for VP is similar to that in the hypothalamus (see Figure 10.2). Consistent with these findings, hypernatremia has recently been shown to increase choroidal VP mRNA in a manner similar to that seen in the hypothalamus (Zemo and McCabe, 2001).

The VP-immunoreactive product is predominantly localized close to the apical membrane domain of choroidal epithelial cells (see Figure 10.2), suggesting that VP is secreted into the CSF. Since the V_{1a} receptors are highly expressed on the apical (CSF facing) membrane of choroidal epithelium (Szmydynger-Chodobska et al., 2004), VP synthesized in, and released from, epithelial cells may exert feedback on these cells as an autocrine regulator or juxtacrine/paracrine regulator or both. Indeed, CP-derived VP appears to mediate the inhibitory effect of central angiotensin II on CSF formation (Chodobski et al., 1998b), possibly by decreasing choroidal extrusion of Cl^- (Johanson et al., 1999a). The transport of Cl^- and the movement of water are tightly coupled in the choroidal epithelium (Zeuthen, 1991). Therefore, the VP-dependent decrease in choroidal Cl^- efflux is likely to reduce CSF production. The underlying mechanisms for this VP action are presently unknown, but may involve the VP-mediated stimulation of the activity of the ubiquitously expressed

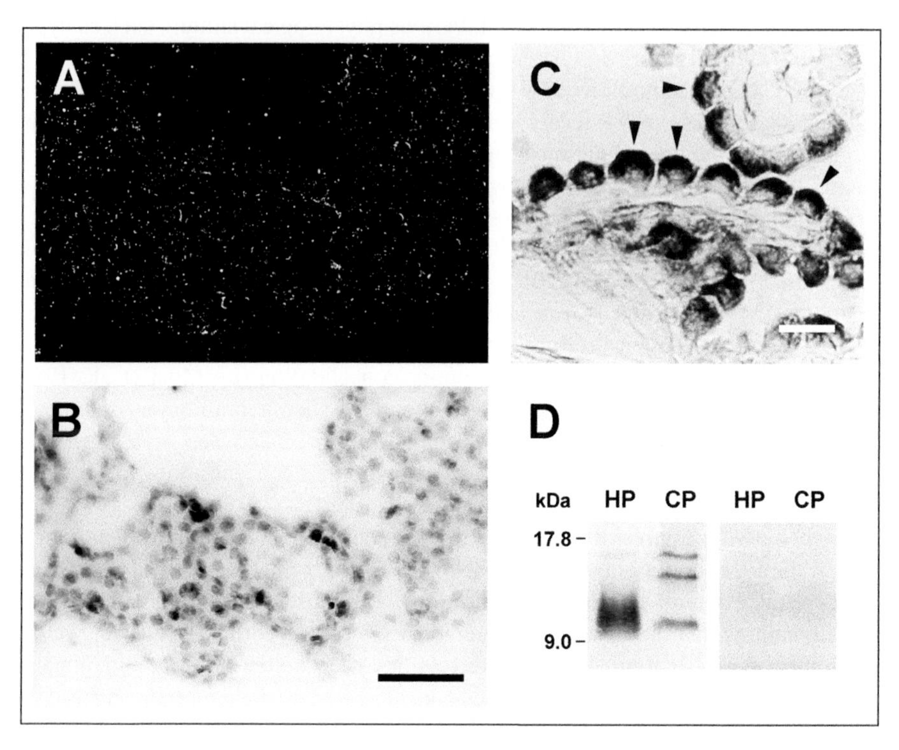

FIGURE 10.2 (A and B) Dark- and bright-field (methyl green staining) microphotographs showing the localization of VP mRNA in the rat CP, detected by in situ hybridization histochemistry. In situ hybridization was performed on cryostat sections of choroidal tissue using the ^{35}S-labeled VP riboprobe. (C) Immunohistochemical localization of the VP-associated neurophysin (VP-NP) in the rat choroidal epithelium. The immunohistochemical procedures were performed on free-floating choroidal tissue. A polyclonal rabbit anti-VP-NP antibody, recognizing the 14-amino-acid C-terminus sequence of rat VP-NP, was used. Note the subapical localization of the VP-NP-immunoreactive product (arrowheads). (D) Immunoprecipitation and immunoblot analysis of rat choroidal protein extracts. Protein extracts were immunoprecipitated with VP-NP antiserum, and recovered immune complexes were immunoblotted with the same antibody following separation via a polyacrylamide gel electrophoresis (left panel). As expected, VP-NP was detected as a ~10-kDa polypeptide in the hypothalamic (HP) protein extracts. A similar band, with lower signal intensity, was found in CP. Two additional bands of ~14 and ~16 kDa may represent the VP-NP intermediates produced during the processing of the prohormone for VP. In a control experiment (right panel), VP-NP antiserum was preabsorbed with the antigenic peptide. Scale bars = 50 μm for A and B; 10 μm for C. [Reprinted with permission from Kluwer/Plenum Publishers (*Adv. Exp. Med. Biol.* 449, 1998, 59–65).]

isoform 1 of the Na^+-K^+-$2Cl^-$ cotransporter (NKCC1) (O'Donnell, 1991; O'Donnell et al., 1995). NKCC1 is involved in transepithelial ion transport (Russell, 2000) and has been localized to the apical membrane of the choroidal epithelium (Plotkin et al., 1997; Wu et al., 1998). Studies in dissociated choroidal epithelial cell preparations from the rat indicate that NKCC1 mediates the influx of C^- into the choroidal epithelium (Wu et al., 1998). These observations suggest that, by stimulating the activity of NKCC1, VP may reduce the net Cl^- flux across the apical membrane of choroidal epithelium and, consequently, inhibit CSF formation. While this hypothesis is plausible, it remains unclear as to how the epithelial cells would then maintain their intracellular volume. Indeed, NKCC1 is not only involved in transepithelial ion transport, it also plays a key role in cell volume regulation (Russell, 2000). The inhibition of NKCC1 activity with bumetanide has been shown to cause a rapid shrinkage of choroidal epithelial cells (Wu et al., 1998), suggesting that VP, by controlling NKCC1 activity, may also regulate the volume of choroidal epithelial cells. Indirect evidence indicates that VP also modulates the movement of water through aquaporin-4 (Niermann et al., 2001), a water channel that is expressed in the CP (Venero et al., 1999). This latter VP action may facilitate the regulation of epithelial cell volume.

The above VP actions are somewhat inconsistent with the VP-dependent increase in the number of dark, clearly shrunken, epithelial cells in choroidal tissue (Schultz et al., 1977; Liszczak et al., 1986; Johanson et al., 1999a). The physiological importance of this V_1 receptor-mediated induction of dark cells is unclear. It has been speculated that the increase in the number of dark cells is indicative of fluid absorption across the choroidal epithelium back into the bloodstream (Schultz et al., 1977). This possibility is rather unlikely, as it would require the hydrostatic pressure gradient to be reversed in order to promote the movement of ventricular CSF into the vascular CP compartment. Since the hydrostatic pressure in choroidal microvessels is high due to the low resistance of the choroidal vascular bed (Szmydynger-Chodobska et al., 1994), an excessively high intraventricular CSF pressure, seldom seen in clinical practice, would be required to maintain fluid absorption across the choroidal epithelium.

In addition to its local actions, VP produced by choroidal epithelium and released into the CSF may exert distal, endocrine-like effects on its target cells reached through CSF pathways. This raises intriguing questions about the possible mediatory role of CP-derived VP in the regulation of fluid balance in both the normal and injured brain. Future studies are likely to provide some answers to these important questions.

10.3.2 Adrenomedullin

Adrenomedullin (ADM) is a multifunctional peptide that was initially isolated from human pheochromocytoma cells (Kitamura et al., 2002). Human ADM has 52 amino acid residues and is a member of the calcitonin family of peptides (Kitamura et al., 2002). The deletion of the *Adm* gene in mice is lethal to the embryo, with extreme hydrops foetalis and cardiovascular abnormalities (Caron and Smithies, 2001; Shindo et al., 2001). Heterozygote $Adm^{+/-}$ animals survive to adulthood, but have elevated arterial blood pressure and diminished nitric oxide production (Shindo et al., 2001). This indicates that ADM is not only critical for normal mammalian development, but also plays an important role in maintaining cardiovascular homeostasis in the adult organism. High-affinity ADM receptors are made up of two molecules: a calcitonin receptor-like receptor (CRLC) and a single transmembrane domain protein, receptor activity-modifying protein 2 (RAMP2) or RAMP3 (McLatchie et al., 1998; Poyner et al., 2002). ADM can also bind, albeit with lower affinity, to receptors for other members of the calcitonin family of peptides (Hinson et al., 2000; Poyner et al., 2002). Signal transduction in the ADM receptors predominantly involves the activation of adenylyl cyclase, causing an increased production of adenosine 3',5'-cyclic monophosphate (cAMP) (Hinson et al., 2000).

Both ADM and its cognate receptors are expressed in the CP (Takahashi et al. 1997; Kobayashi et al., 2001); and ADM has been shown to be secreted by the choroidal epithelium (Takahashi et al., 1997). ADM increases the choroidal synthesis of cAMP (Kobayashi et al., 2001). It is thus possible that this peptide regulates CSF formation by affecting the activity of the Na^+, K^+-ATPase and Cl^- channels in choroidal epithelial cells (Kotera and Brown, 1994; Kibble et al., 1996; Fisone et al., 1998). Whereas the biological effects of CP-derived ADM are largely unclear, it is evident that this peptide plays an important role in the regulation of BBB function and cerebral blood flow (Kis et al., 2001a, 2001b; 2003). ADM is expressed by the cerebrovascular endothelium at a very high level and is predominantly released across the luminal membrane of endothelial cells (Kis et al., 2002). CRLC, as well as RAMP2 and RAMP3, have also been found in the cerebrovascular endothelium (Kis et al., 2001a), suggesting some autocrine and/or juxtacrine/paracrine ADM actions on the BBB. Short- and long-term exposure of cerebral endothelial cell cultures to ADM increases transendothelial electrical resistance and reduces the permeability of endothelial monolayers to low molecular weight markers (Kis et al. 2001b; 2003). However, ADM does not appear to change the expression of tight junction proteins, such as occludin, claudin-1, and zonula occludens-1 (Kis et al., 2003). Hypoxia upregulates endothelial ADM expression (Ladoux and Frelin, 2000). It is thus possible that this peptide acts to diminish the increased permeability of the BBB caused by hypoxia (Schoch et al., 2002). The underlying mechanisms of these ADM actions have yet to be characterized.

Increased levels of ADM mRNA and the augmented release of immunoreactive ADM into conditioned media have been observed in cultures of human choroidal epithelial cells exposed to various cytokines, including IL-1β and tumor necrosis factor-α (Takahashi et al., 1997). These cytokines are released into the CSF in response to brain injury (Marion et al., 1997; Csuka et al., 1999; Stover et al., 2000; 2001; Singhal et al., 2002), and they may promote the synthesis and secretion of choroidal ADM. This idea is supported by the observation that increased CSF ADM levels are seen in patients with traumatic brain injury (Robertson et al., 2001). It is tempting to speculate that this CP-derived ADM protects the BCSFB in a manner similar to that in the BBB. However, the beneficial effects of CSF-borne ADM are uncertain. Indeed, icv administration of ADM in the rat model of focal cerebral ischemia has been found to exacerbate the ischemic brain damage (Wang et al., 1995). Further studies are needed to clarify the physiological significance of CP-derived ADM.

10.3.3 Insulin-Like Growth Factor II

IGF-II is a growth factor structurally related to insulin (Daughaday and Rotwein; 1989; Sara and Hall, 1990; LeRoith and Roberts, 1993). The transcripts for IGF-II are abundant in the CP from early development to maturity (Hynes et al., 1988; Bondy et al., 1992). Studies performed in the primary cultures of sheep and rat choroidal epithelial cells suggest that IGF-II is secreted into the CSF (Holm et al., 1994; Nilsson et al., 1996). The *Igf2* gene is parentally imprinted in rodents with the monoallelic expression of *Igf2* from the paternal allele observed in all tissues except the CP and leptomeninges, in which both the paternal and maternal alleles are expressed (DeChiara et al., 1991; Overall et al., 1997). In humans, there is also biallelic *IGF2* expression in the CP (Ohlsson et al., 1994). It is possible that the biallelic expression of the *IGF2* gene in the choroidal epithelium results in a more abundant synthesis and secretion of this growth factor. In addition to IGF-II, the CP synthesizes insulin-like growth factor-binding proteins (Holm et al., 1994; Stenvers et al., 1994; Walter et al., 1997). These proteins play an important role in controlling the interactions of IGF-II with its receptors and therefore modulating its biological effects (Sara and Hall, 1990; Clemmons et al., 1993).

Two types of IGF-II receptors, IGF-1R and IGF-2R, have been identified, with the type 2 receptor having a higher IGF-II binding affinity than type 1 (Sara and Hall, 1990; LeRoith et al., 1993). Interestingly, the mouse embryo produces transcripts of *Igf2r* only from the maternal chromosome (Barlow et al., 1991), whereas the human *IGF2R* gene is expressed from both alleles (Kalscheuer et al., 1993). The IGF-1R shows both structural and functional similarities to the insulin receptor. It consists of two extracellular α subunits and two transmembrane β subunits containing a tyrosine kinase domain in the intracellular segment. The IGF-2R is structurally unrelated to the IGF-1R, or the insulin receptor, and does not exhibit tyrosine kinase activity. The physiological significance of IGF-II binding to IGF-2R is unclear. It has been hypothesized that IGF-2R plays a role in the sequestration and subsequent degradation of IGF-II, which would limit embryonic growth (Haig and Graham, 1991). Consistent with this idea, mouse mutants maternally inheriting a targeted disruption of the imprinted *Igf2r* gene had increased serum and tissue levels of IGF-II and exhibited overgrowth (Ludwig et al., 1996).

Since the choroidal epithelium expresses both receptors for IGF-II (Bondy et al., 1992; Nilsson et al., 1992; Kar et al., 1993), it has been proposed that CP-derived IGF-II is an autocrine regulator of epithelial growth (Nilsson et al., 1996). Whereas IGF-II may promote choroidal development in an autocrine manner or juxtacrine/paracrine manner, or both, it is rather unlikely that, under normal conditions, IGF-II is a mitogen for the mature choroidal epithelium (McDonald and Green, 1988). Instead, due to volume transmission involving CSF pathways, CP-derived IGF-II may exert important endocrine-like effects in the adult brain. Indeed, augmented secretion of IGF-II into the CSF, presumably originating from the choroidal epithelium, has been observed in response to a localized brain injury (Walter et al., 1999). The CSF IGF-II concentration increased transiently, reaching maximal levels at

7 days postinjury, which coincided with the increased levels of immunoreactive IGF-II in the injured parenchyma. During this acute postinjury period, message for IGF-II was not upregulated in any cells of the affected parenchyma, suggesting that the IGF-II protein detected in the injured parenchyma is of CP origin. While these observations support the putative endocrine-like actions of CP-derived IGF-II, its local effects on choroidal function in adult animals remain unclear.

10.3.4 Vascular Endothelial Growth Factor

Vascular endothelial growth factor (VEGF), a potent mitogen for vascular endothelial cells, is not only indispensable to vasculogenesis, but is also critical for the angiogenesis associated with reproduction, wound healing, ischemia, and tumor growth (Ferrara et al., 1992; Dvorak et al., 1995; Ferrara, 2000; Robinson and Stringer, 2001). Five major isoforms of human VEGF exist, with the number of amino acid residues ranging between 121 and 206. Although $VEGF_{165}$ is the predominant isoform, the transcripts for $VEGF_{121}$ and $VEGF_{189}$ are also present in the majority of cells and tissues expressing the *VEGF* gene. VEGF isoforms differ considerably in regard to their bioavailability, which is related to the varying heparin-binding properties of these isoforms. The different biochemical properties of various VEGF isoforms may play an important role in the finely tuned regulation of the biological activity of this growth factor (Carmeliet et al., 1999; Ruhrberg et al., 2002).

VEGF is frequently referred to as a vascular permeability factor due to its ability to increase microvascular permeability to low molecular weight markers and macromolecules. This important property of VEGF has been well documented through the observation that increased permeability of the BBB is induced in rats receiving intracerebral infusions of the recombinant protein (Harrigan et al., 2002; Krum et al., 2002). Brain tumors expressing VEGF are accompanied by prominent vasogenic edema (Berkman et al., 1993; Strugar et al., 1995), and this growth factor has also been shown to cause a breakdown of the blood-retinal barrier in diabetes (Murata et al., 1996; Qaum et al., 2001). Available data suggest that the VEGF-mediated increase in vascular permeability is related to a rapid phosphorylation of several components of the endothelial tight and adherens junction complexes (Kevil et al., 1998; Esser et al., 1998a; Antonetti et al., 1999; Pedram et al., 2002). In addition, VEGF has been demonstrated to induce endothelial fenestrations in both in vitro and in vivo experimental settings (Roberts and Palade, 1995; Esser et al., 1998b).

In the CP, VEGF is constitutively expressed by the epithelial cells, whereas in the leaky choroidal microvessels, it has not been detected (Breier et al., 1992; Naito et al., 1995; Chodobski et al., 2003). Based on these findings and the observation that VEGF is expressed in the epithelial cells of the kidney glomerulus that, similar to choroidal epithelial cells, are associated with fenestrated endothelium, it has been proposed that, in these tissues, VEGF induces the fenestrated endothelial phenotype in a paracrine regulatory manner (Breier et al., 1992). This hypothesis has been challenged by the observations of Naito et al. (1995) made in grafting experiments in which skeletal muscle tissue was transplanted to the CP. The authors noticed that, in the E16 fetal muscle grafts, the majority of invading choroidal blood vessels maintained their fenestrated phenotype, even though the muscle produced no VEGF and the invading vessels did not express VEGF receptors. These results indicate that the fenestrated phenotype of choroidal microvasculature is unlikely to result from the paracrine actions of epithelium-derived VEGF.

VEGF has been found to bind to heparan sulfate proteoglycans (HSPGs) associated with the apical membrane of the choroidal epithelium (Chodobski et al., 2003), suggesting that this growth factor is predominantly secreted into the CSF. Two types of VEGF receptors, VEGFR-1 (Flt-1) and VEGFR-2 (Flk-1/KDR), are expressed in choroidal tissue (Marti and Risau, 1998), and VEGFR-2 appears to be localized to the apical domain of epithelial cells (Naito et al., 1995). This suggests that, similar to VP, VEGF may act as an autocrine and/or juxtacrine/paracrine regulator of choroidal epithelial function. However, the biological significance of these presumed VEGF actions is presently unclear. Interestingly, even though VEGF has been shown to increase the permeability of epithelial monolayers (Hartnett et al., 2003), CP-derived VEGF does not appear to do so in the normal choroidal epithelium. The latter phenomenon may be explained by recent studies in which the tissue inhibitor of metalloproteinase-3, an inhibitor of matrix metalloproteinases that is highly expressed in the CP (Pagenstecher et al., 1998), has been demonstrated to block the binding of VEGF to its type 2 receptor (Qi et al., 2003).

It is interesting to note that, in the normal brain, VEGF is produced not only by the choroidal epithelium, but is also expressed in astrocytic processes located close to the CSF space, i.e., under the ependyma of the lateral and 3rd ventricles, within and below the pial-glial lining, and around the velum interpositum (Chodobski et al., 2003). This raises the intriguing possibility that VEGF is constitutively released into the CSF and distributed within the brain via CSF pathways. The physiological role of this putative endocrine-like action of VEGF awaits future investigation.

10.3.5 Fibroblast Growth Factor 2

Fibroblast growth factor 2 (FGF2), also known as basic fibroblast growth factor, is a heparin-binding protein originally demonstrated to be a mitogen for fibroblasts and a potent promoter of angiogenesis (Gospodarowicz et al., 1986). It is now recognized that, similar to other growth factors, such as IGF-II, VEGF, and transforming growth factor-β (TGF-β), FGF2 is a multifunctional peptide with significant neurotrophic and neuroprotective properties (Abe and Saito, 2001). The exocytosis of FGF2 has been a matter of debate because FGF2 lacks a hydrophobic signal sequence that could direct its secretion through the classical endoplasmic reticulum-Golgi system. However, several laboratories, using specific anti-FGF2 antibodies in various cell culture systems, have provided unequivocal evidence for the secretion of this growth factor (see, e.g., Mignatti et al., 1991). Although FGF2 is produced by the CP (Gonzalez et al., 1995; Fuxe et al., 1996), normal CSF FGF2 levels are low (Malek et al., 1997; Huang et al., 1999), possibly because, upon its release, FGF2 binds to the cell surface- and/or extracellular matrix-associated HSPGs (Vlodavsky et al., 1991). Intracisternal or icv administration of recombinant FGF2 has been shown to be neuroprotective in cerebral ischemia and other forms of brain injury (Yamada et al., 1991; Cummings et al., 1992; Koketsu et al., 1994; Ma et al., 2001). However, it is unclear how FGF2 can reach its target cells, given the strong heparin-binding properties of this growth factor and, consequently, its limited penetration of the brain parenchyma. It is possible that the neuroprotective actions of CSF-borne FGF2 involve, at least in part, the receptor-mediated retrograde transport of this growth factor in neurons whose axonal processes are located near the ependyma or the pial-glial lining (Ferguson and Johnson, 1991; Mufson et al., 1999).

The icv infusion of recombinant FGF2 in adult rats has been found to have a profound effect on neurogenesis in the subventricular zone (SVZ) of the lateral ventricles (Kuhn et al., 1997). SVZ is located close to the ventricular CSF space and contains a rapidly dividing population of stem cells (Alvarez-Buylla and Garcia-Verdugo, 2002; Galli et al., 2003). FGF2 was noted to expand the SVZ progenitor population and increase the number of newborn neurons in the olfactory bulb (Kuhn et al., 1997), an area of the brain into which the SVZ neuronal progenitors migrate along the rostral migratory stream (Alvarez-Buylla and Garcia-Verdugo, 2002; Galli et al., 2003). Because of the anatomical location of the SVZ, it is possible that CP-derived FGF2 plays a role in promoting both neurogenesis in the SVZ and the migration of the SVZ neuronal progenitors to the olfactory bulb. In this context it is important to note that the CP also produces Slit2, a chemorepellent that controls cell migration along the rostral migratory stream (Hu, 1999; Nguyen-Ba-Charvet et al., 2004).

Chronic icv administration of recombinant FGF2 has been found to produce hydrocephalus in various mammalian species, such as monkey, rat, and mouse (Denton et al., 1995; Pearce et al., 1996; Johanson et al., 1999b). Detailed analysis of the changes in CSF dynamics in hydrocephalic rats revealed that, compared to control animals, the FGF2-infused rats had decreased CSF production, increased resistance to CSF absorption, and unchanged intraventricular CSF pressure (Johanson et al., 1999b). The presence of severe ventriculomegaly, with a patent cerebral aqueduct and normal CSF pressure, has been interpreted to be indicative of an ex vacuo hydrocephalus. The increased resistance to CSF absorption has been attributed to FGF2-mediated deposition of collagen fibrils in the arachnoid granulations, a major site of CSF absorption. Further studies have suggested that the FGF2-induced ventricular enlargement is, at least in part, associated with neuronal apoptosis (Chodobski et al., 1998a). Consistent with these findings are in vitro observations of a proapoptotic FGF2 action in several cell types, including the neural retina, myofibroblasts, and breast cancer cells (Funato et al., 1997; Yokoyama et al., 1997; Wang et al., 1998). Other growth factors, such as nerve growth factor and brain-derived neurotrophic factor, have also been shown to promote neuronal apoptosis (Frade et al., 1996; Bamji et al., 1998; Frade and Barde, 1999). While FGF2-dependent neuronal death may contribute to ventricular enlargement, it is important to note that the doses of FGF2 used to produce hydrocephalus were relatively high. Therefore, the relevance of the above animal model to the clinical setting of hydrocephalus has yet to be established.

As discussed above, icv infusion of FGF2 has been found to decrease CSF production (Johanson et al., 1999b). The inhibitory effect of FGF2 on CSF formation has also been demonstrated in primary cultures of porcine choroidal epithelial cells (Hakvoort and Johanson, 2000). It has been proposed that these FGF2 actions are mediated by CP-derived VP (Szmydynger-Chodobska et al., 2002). This idea is supported by the observation that similar choroidal responses to FGF2 and VP occur, such as the induction of dark epithelial cells (Schultz et al., 1977; Liszczak et al., 1986; Johanson et al., 1999a, 1999b) and a decrease in CSF formation (Chodobski et al., 1998b; Johanson et al., 1999b). Furthermore, fibroblast growth factor receptors have been found to be expressed apically on choroidal epithelial cells and to colocalize with VP (Szmydynger-Chodobska et al., 2002). Although it is possible that the CP-derived VP mediates the inhibitory effect of FGF2 on CSF formation, further testing of the proposed model will be required. Evaluation of choroidal VP release in response to FGF2 together with the use of V_{1a} receptor antagonists would provide the necessary information on whether these FGF2 actions are mediated by VP.

10.3.6 Transforming Growth Factor-β

TGF-β is a multipotent cytokine that plays an important regulatory role in brain development, neuronal survival, and neurodegeneration (Krieglstein et al., 1995, 2002; Pratt and McPherson, 1997; Flanders et al., 1998). The biological effects of TGF-β are mediated by a heteromeric transmembrane receptor consisting of two subunits designated TβRI and TβRII (Massagué et al., 1994). Three isoforms of TGF-β (β1, β2, and β3) together with their cognate receptors are expressed in the CP (Knuckey et al., 1996; Morita et al., 1996). Choroidal TGF-β expression is upregulated in response to transient forebrain ischemia in adult rats (Knuckey et al., 1996) and combined hypoxia/ischemia (Klempt et al., 1992) in infant rats. The enhanced CP expression of TGF-β, after a localized injury to the cerebral cortex, has also been reported (Logan et al., 1992). This augmented synthesis of choroidal TGF-β, presumably followed by an increased release of this growth factor into the CSF, may play an important role in promoting neuronal survival in brain regions that are in close contact with the CSF space. Indeed, icv infusion of a recombinant TGF-β in rats subjected to transient forebrain ischemia has been shown to be protective for hippocampal neurons (Henrich-Noack et al., 1996). In other rodent studies where permanent middle cerebral artery occlusion was produced or where the animals were subjected to a hypoxic/ischemic injury, icv administration of recombinant TGF-β was found to decrease the area of infarction (Henrich-Noack et al., 1994; McNeill et al., 1994). TGF-β appears to promote neuronal survival by interfering with the signaling pathways mediating programmed cell death. The molecular mechanisms underlying this TGF-β action likely involve the upregulation of Bcl-2 expression, the phosphorylation of Bad, the inhibition of caspase-3, and the maintenance of Ca^{2+} homeostasis (Prehn et al., 1994; Zhu et al., 2001, 2002).

In addition to its neuroprotective actions following brain injury, TGF-β has been demonstrated to promote the survival of neuronal cultures exposed to amyloid β-peptides (Aβs) (Prehn et al., 1996; Ren and Flanders, 1996; Kim et al., 1998), thought to be involved in the pathogenesis of Alzheimer's disease (AD) (Cappai and White, 1999; Small and McLean, 1999; Storey and Cappai, 1999; Wilson et al., 1999). CSF TGF-β levels are elevated in patients with AD (Chao et al., 1994), suggesting that this growth factor may be beneficial in slowing down the progression of neurodegeneration. This concept is consistent with the observation that there is a decreased amyloid load in the hippocampus and cerebral cortex, and the reduction in the number of dystrophic neurites in mice carrying the transgenes for the human amyloid precursor protein and porcine TGF-β1 (Wyss-Coray et al., 2001). These findings are, however, at variance with the observation that there is increased amyloid deposition in organotypic hippocampal slice cultures exposed to various isoforms of TGF-β (Harris-White et al., 1998). Moreover, although in the above-mentioned double transgenic mice the amyloid burden in the brain parenchyma was decreased, these animals demonstrated accelerated accumulation of β-amyloid in the cerebral blood vessels and meninges (Wyss-Coray et al., 1997, 2001). Further research into this area is likely to provide an explanation for these conflicting results.

Excessive TGF-β synthesis in the CNS may result in hydrocephalus. Studies in rodents have shown that the intracerebral infusion of the recombinant TGF-β1 (Tada et al., 1994) or the transgene-driven overexpression of this cytokine in the CNS (Galbreath et al., 1995; Wyss-Coray et al., 1995) induces communicating hydrocephalus. In transgenic mice, in which the cerebral expression of porcine TGF-β1 was driven by a promoter of the glial fibrillary acidic protein gene, high levels of TGF-β production either resulted in severe ventriculomegaly with seizures and motor deficits, or these animals died between birth and 3 weeks of age. These transgenic mice express high levels of fibronectin and laminin (Wyss-Coray et al., 1995), two components of the extracellular matrix, and the ultrastructural studies of leptomeninges from mice receiving intracerebral injection of the recombinant TGF-β1 have demonstrated an abundant deposition of collagen fibers (Nitta and Tada, 1998). These latter observations are consistent with the well-know profibrotic activity of TGF-β (Ignotz and Massagué, 1986; Roberts et al., 1986; Montesano and Orci, 1988) and suggest that, in these mice, hydrocephalus resulted from fibrotic obstruction of the CSF outflow pathways.

Recently, a new member of the TGF-β superfamily, named growth/differentiation factor-15 or macrophage-inhibiting cytokine-1 (GDF-15/MIC-1), has been identified by several groups (Bootcov et al., 1997; Böttner et al., 1999). This growth factor is highly expressed in both neonatal and adult rat CP (Böttner et al., 1999; Schober et al., 2001). GDF-15/MIC-1 has been demonstrated to be a potent survival-promoting factor for cultured dopaminergic neurons (Strelau et al., 2000). In a rat model of Parkinson's disease in which the nigrostriatal dopaminergic neurons were unilaterally lesioned with 6-hydroxydopamine (6-OHDA), icv administration of recombinant GDF-15/MIC-1 prior to the 6-OHDA injections reduced the loss of dopaminergic neurons and prevented the behavioral changes induced by the toxin (Strelau et al., 2000). These observations suggest that, due to volume transmission, CP-derived GDF-15/MIC-1 may exert distal neurotrophic effects or neuroprotective effects, or both on nigrostriatal dopaminergic neurons.

10.3.7 TRANSTHYRETIN

TTR, also known as prealbumin, is highly expressed in the choroidal epithelium (Aleshire et al., 1983; Herbert et al., 1986) and is secreted into the CSF (Schreiber et al., 1990; Southwell et al., 1993). The role of TTR in thyroxine transport into the brain is analyzed in Chapter 11. TTR has also been demonstrated to affect the deposition of β-amyloid (Schwarzman et al., 1994), and this putative TTR function will now be discussed briefly.

Using a battery of in vitro assays, Schwarzman et al. (1994) found that TTR has the ability to sequester soluble Aβ (sAβ) and prevent amyloid fibril formation. The authors have also shown that TTR-sAβ complexes are formed in the CSF. The concentration of TTR in the CSF is two orders of magnitude higher than the concentration of sAβ and is also higher than the CSF levels of other known sAβ-binding proteins (see discussion in Schwarzman et al., 1994). Although isoforms 3 and 4 of apolipoprotein E (apoE) are considerably more effective than TTR in inhibiting the aggregation of Aβ, apoE-sAβ complexes have not been found in the CSF (Schwarzman et al., 1994). Since the CSF TTR levels in patients with AD are reduced, when compared to normally aging subjects (Riisøen, 1988; Serot et al., 1997; Merched et al., 1998), the bulk flow-mediated clearance of sAβ is likely to be compromised in these patients. TTR-dependent inhibition of β-amyloid formation is supported by the observations of reduced amyloid deposition in the nematode *Caenorhabditis elegans* that, in addition to expressing human Aβ, carried the human *TTR* transgene (Link, 1995). Delayed amyloid deposition and a lack of neurodegeneration in mice overexpressing a mutant form of human amyloid precursor protein have also been attributed to increased hippocampal expression of TTR observed in these transgenic animals (Stein and Johnson, 2002). Collectively, these findings suggest that TTR plays an important role in controlling the biological activity of sAβ and in its clearance from the CNS.

A considerable number of mutations in the *TTR* gene have been reported (Gambetti and Russo, 1998). Some of these mutations were found to affect the binding of TTR to thyroxine (Refetoff et al., 1986; Rosen et al., 1993), but it is unclear whether they could also influence the sequestration of sAβ by TTR. Indeed, no correlation has been observed between the presence of mutations in the *TTR* gene and AD (Palha et al., 1996). The most common phenotype associated with mutations of the *TTR* gene is a condition identified as familial amyloidotic polyneuropathy. However, it is important to note that, in individuals having wild-type TTR, a senile systemic amyloidosis resulting from the aggregation of TTR may develop as well (Cornwell et al., 1988). Interestingly, the involvement of the CNS in patients with TTR amyloidosis is very rare (Gambetti and Russo, 1998), suggesting that circulating factors or tissue-specific factors, or both play a role in amyloid formation in peripheral organs.

10.3.8 Apolipoprotein J

Apolipoprotein J (apoJ)/clusterin is a heterodimeric, secreted glycoprotein expressed by a wide variety of tissues and present in various body fluids (Jones and Jomary, 2002; Trougakos and Gonos, 2002). ApoJ interacts with high affinity (K_d = 2 nM) with sAβ, and under in vitro conditions, the apoJ-sAβ complexes cannot be dissociated by other competing proteins, such as apoE3, apoE4, and TTR (Matsubara et al., 1995). Similar to TTR, apoJ has been shown to inhibit the aggregation of Aβ (Oda et al., 1994; Matsubara et al., 1996; Hammad et al., 1997), suggesting that this protein plays an important role in the regulation of amyloid metabolism in the CNS. Since apoJ is produced by the CP (Aronow et al., 1993; Page et al., 1998) and apoJ-sAβ complexes are present in the CSF (Ghiso et al., 1993; Golabek et al., 1995), it is possible that CP-derived apoJ also facilitates the clearance of CSF-borne sAβ. The clearance of sAβ may not only be associated with the bulk absorption of CSF, but may also involve receptor-mediated uptake by the choroidal epithelium.

One possible receptor involved in the choroidal uptake of sAβ is the low-density lipoprotein receptor-related protein 2 (LRP2), also known as megalin or glycoprotein 330, which is a member of the low-density lipoprotein receptor gene family (Gliemann, 1998; Hussain et al., 1999). LRP2 is an endocytic receptor for apoJ (Kounnas et al., 1995) and has been found by immunohistochemistry to be expressed at high levels on the apical surface of choroidal epithelial cells of mouse embryos (Kounnas et al., 1994). Based on these observations, it has been proposed by Hammad et al. (1997) that LRP2 mediates the clearance and enzymatic degradation of CSF-borne sAβ by the choroidal epithelium. The authors used the mouse teratocarcinoma F9 cells, which highly express LRP2, to demonstrate the endocytosis and subsequent lysosomal degradation of apoJ-sAβ complexes. sAβ by its own was not internalized by these cells, indicating that the formation of apoJ-sAβ complexes is critical for LRP2-mediated clearance of sAβ. While choroidal LRP2 may be involved in sAβ clearance during embryonic development, it is unclear whether this endocytic receptor plays a similar role in the adult CP. Indeed, using immunohistochemistry, Zheng et al. (1994) were unable to detect LRP2 in the CP of young adult rats. Although LRP2 has been found by Western blotting to be expressed in adult rat CP (Chun et al., 1999), the message for LRP2 could not be detected in CP either by in situ hybridization or reverse-transcriptase polymerase chain reaction (RT-PCR) (Page et al., 1998; Chun et al., 1999). Using RT-PCR, we were able to detect LRP2 mRNA in adult rat CP, albeit at much lower levels than those seen in the kidney (unpublished observations). Further research will therefore be needed to determine whether LRP2-mediated clearance of sAβ by an adult choroidal epithelium exists.

10.3.9 Gelsolin

Using suppression subtractive hybridization, Matsumoto et al. (2003) have found that gelsolin (GSN) is abundantly expressed in the CP. GSN is a member of the large superfamily of actin-binding proteins (McGough et al., 2003). However, GSN is also a secreted protein, the transcripts of which are generated by alternative RNA splicing (Kwiatkowski et al., 1988). GSN has recently been demonstrated also to bind sAβ (Chauhan et al., 1999), which indicates that GSN may be involved in controlling the levels of sAβ in the CSF. Consistent with its ability to bind sAβ, further studies have shown that GSN inhibits the fibrillization of sAβ and promotes the disaggregation of preformed amyloid fibrils (Ray et al., 2000). These observations suggest that CP-derived GSN plays an important role in limiting the deposition of β-amyloid in brain tissue.

10.4 FUTURE DIRECTIONS

Our intention in this chapter was not to catalog all of the polypeptides known to be synthesized in the choroidal epithelium or transported across the choroidal BCSFB, but rather to provide the reader with a selective analysis of pertinent functional data that are available in the literature. This analysis indicates that the CP plays an important role in hormonal signaling and provides a central source of biologically active polypeptides. However, we are still far from being able to grasp all of the intricacies of these CP functions. Several important issues need to be addressed. First, significant biological effects have been demonstrated for various CSF-borne polypeptides; however, it is largely unknown where the target cells for these polypeptides are located and how these targets are reached. Answering these questions will allow us to better understand the role of volume transmission signaling in the CNS. Second, volume transmission is a key element in hormonal signaling across the BCSFB and the BBB; however, for peptide hormones, such as leptin, the physiological importance of receptor-mediated transport across the choroidal epithelium versus transport across the cerebrovascular endothelium remains to be established. Third, choroidal epithelium produces a number of secreted peptides and proteins whose receptors are not only located on the parenchymal and non-parenchymal cells in the brain, but are also expressed in the CP. Accordingly, it will be important to determine the physiological significance of local (autocrine or juxtacrine/paracrine or both) versus distal (mediated by bulk flow of CSF) actions of CP-derived polypeptides. Fourth, it remains to be determined what role the CP-derived polypeptides play in brain development and maintenance of normal brain function, and whether the choroidal production of polypeptides is affected by aging or neurodegenerative diseases or both, such as AD or both. Furthermore, although reduced CSF levels of proteins that maintain sAβ in solution are associated with AD, it is unclear as to whether these reduced levels are causative or reactive in the progression of the disease. Fifth, the increased choroidal production of several polypeptides has been observed in brain injury. It is unclear, however, whether the changes in this synthetic activity of the choroidal epithelium contribute to the progression of the injury or play a role in the recovery from and repair after the injury, or both.

10.5 ACKNOWLEDGMENTS

The authors wish to thank Andrew Pfeffer for his help in the preparation of this manuscript. This work was supported by grant from NIH NS-39921 and by research funds from the Neurosurgery Foundation and Lifespan, Rhode Island Hospital.

REFERENCES

K. Abe and H. Saito, Effects of basic fibroblast growth factor on central nervous system functions, Pharmacological Researchfunctions, *Pharmacol. Res.* 43, 2001, 307–12.

L.F. Agnati, M. Zoli, I. Stromberg and K. Fuxe, Intercellular communication in the brain: wiring versus volume transmission, *Neuroscience* 69, 1995, 711–26.

R.S. Ahima and J.S. Flier, Leptin, *Ann. Rev. Physiol.* 62, 2000, 413–37.

K.A. Al-Barazanji, R.E. Buckingham, J.R. Arch, A. Haynes, D.E. Mossakowska, D.L. McBay, S.D. Holmes, M.T. McHale, X.M. Wang and I.S. Gloger, Effects of intracerebroventricular infusion of leptin in obese Zucker rats, *Obesity Res.* 5, 1997, 387–94.

S.L. Aleshire, C.A. Bradley, L.D. Richardson and F.F. Parl, Localization of human prealbumin in choroid plexus epithelium, *J. Histochem. Cytochem.* 31, 1983, 608–12.

A. Alvarez-Buylla and J.M. Garcia-Verdugo, Neurogenesis in adult subventricular zone, *J. Neurosci.* 22, 2002, 629–34.

P. Amat, F.E. Pastor, J.L. Blazquez, B. Pelaez, A. Sanchez, A.J. Alvarez-Morujo, D. Toranzo and G. Amat-Peral, Lateral evaginations from the third ventricle into the rat mediobasal hypothalamus: an amplification of the ventricular route, *Neurosci.* 88, 1999, 673–7.

D.A. Antonetti, A.J. Barber, L.A. Hollinger, E.B. Wolpert and T.W. Gardner, Vascular endothelial growth factor induces rapid phosphorylation of tight junction proteins occludin and zonula occludens 1. A potential mechanism for vascular permeability in diabetic retinopathy and tumors, *J. Biol. Chem.* 274, 1999, 23463–7.

E. Arnauld, V. Bibene, J. Meynard, F. Rodriguez and J.D. Vincent, Effects of chronic icv infusion of vasopressin on sleep-waking cycle of rats, *Am. J. Physiol.* 256, 1989, R674–84.

B.J. Aronow, S.D. Lund, T.L. Brown, J.A. Harmony and D.P. Witte, Apolipoprotein J expression at fluid-tissue interfaces: potential role in barrier cytoprotection, *Proc. Natl. Acad. Sci. USA* 90, 1993, 725–9.

J.C. Bakowska and J.I. Morrell, Atlas of the neurons that express mRNA for the long form of the prolactin receptor in the forebrain of the female rat, *J. Compar. Neurol.* 386, 1997, 161–77.

J.C. Bakowska and J.I. Morrell, The distribution of mRNA for the short form of the prolactin receptor in the forebrain of the female rat, *Mol. Brain Res.* 116, 2003, 50–8.

S.X. Bamji, M. Majdan, C.D. Pozniak, D.J. Belliveau, R. Aloyz, J. Kohn, C.G. Causing and F.D. Miller, The p75 neurotrophin receptor mediates neuronal apoptosis and is essential for naturally occurring sympathetic neuron death, *J. Cell Biol.* 140, 1998, 911–23.

W.A. Banks, C.M. Clever and C.L. Farrell, Partial saturation and regional variation in the blood-to-brain transport of leptin in normal weight mice, *Am. J. Physiol.* 278, 2000, E1158–65.

W.A. Banks, C.R. DiPalma and C.L. Farrell, Impaired transport of leptin across the blood-brain barrier in obesity, *Peptides* 20, 1999, 1341–5.

W.A. Banks and C.L. Farrell, Impaired transport of leptin across the blood-brain barrier in obesity is acquired and reversible, *Am. J. Physiol.* 285, 2003, E10–5.

W.A. Banks, M. Goulet, J.R. Rusche, M.L. Niehoff and R. Boismenu, Differential transport of a secretin analog across the blood-brain and blood-cerebrospinal fluid barriers of the mouse, *J. Pharmacol. Exp. Thera.* 302, 2002, 1062–9.

W.A. Banks, A.J. Kastin, W. Huang, J.B. Jaspan and L.M. Maness, Leptin enters the brain by a saturable system independent of insulin, *Peptides* 17, 1996, 305–11.

D.P. Barlow, R. Stöger, B.G. Herrmann, K. Saito and N. Schweifer, The mouse insulin-like growth factor type-2 receptor is imprinted and closely linked to the *Tme* locus, *Nature* 349, 1991, 84–7.

D.G. Baskin, R.J. Seeley, J.L. Kuijper, S. Lok, D.S. Weigle, J.C. Erickson, R.D. Palmiter and M.W. Schwartz, Increased expression of mRNA for the long form of the leptin receptor in the hypothalamus is associated with leptin hypersensitivity and fasting, *Diabetes* 47, 1998, 538–43.

S.H. Bates, W.H. Stearns, T.A. Dundon, M. Schubert, A.W. Tso, Y. Wang, A.S. Banks, H.J. Lavery, A.K. Haq, E. Maratos-Flier, B.G. Neel, M.W. Schwartz and M.G. Myers, Jr., STAT3 signalling is required for leptin regulation of energy balance but not reproduction, *Nature* 421, 2003, 856–9.

R.A. Berkman, M.J. Merrill, W.C. Reinhold, W.T. Monacci, A. Saxena, W.C. Clark, J.T. Robertson, I.U. Ali and E.H. Oldfield, Expression of the vascular permeability factor/vascular endothelial growth factor gene in central nervous system neoplasms, *J. Clin. Invest.* 91, 1993, 153–9.

C. Bignon, N. Daniel, L. Belair and J. Djiane, In vitro expression of long and short ovine prolactin receptors: activation of Jak2/STAT5 pathway is not sufficient to account for prolactin signal transduction to the ovine beta-lactoglobulin gene promoter, *J. Mol. Endocrinol.* 23, 1999, 125–36.

C. Bjørbæk, S. Uotani, B. Da Silva and J.S. Flier, Divergent signaling capacities of the long and short isoforms of the leptin receptor, *J. Biol. Chem.* 272, 1997, 32686–95.

R.J. Boado, P.L. Golden, N. Levin and W.M. Pardridge, Up-regulation of blood-brain barrier short-form leptin receptor gene products in rats fed a high fat diet, *J. Neurochem.* 71, 1998, 1761–4.

C. Bole-Feysot, V. Goffin, M. Edery, N. Binart and P.A. Kelly, Prolactin (PRL) and its receptor: actions, signal transduction pathways and phenotypes observed in PRL receptor knockout mice, *Endocr. Rev.* 19, 1998, 225–68.

C. Bondy, H. Werner, C.T. Roberts, Jr. and D. LeRoith, Cellular pattern of type-I insulin-like growth factor receptor gene expression during maturation of the rat brain: comparison with insulin-like growth factors I and II, *Neurosci.* 46, 1992, 909–23.

M.R. Bootcov, A.R. Bauskin, S.M. Valenzuela, A.G. Moore, M. Bansal, X.Y. He, H.P. Zhang, M. Donnellan, S. Mahler, K. Pryor, B.J. Walsh, R.C. Nicholson, W.D. Fairlie, S.B. Por, J.M. Robbins and S.N. Breit, MIC-1, a novel macrophage inhibitory cytokine, is a divergent member of the TGF-β superfamily, *Proc. Natl. Acad. Sci. USA* 94, 1997, 11514–9.

M. Böttner, C. Suter-Crazzolara, A. Schober and K. Unsicker, Expression of a novel member of the TGF-β superfamily, growth/differentiation factor-15/macrophage-inhibiting cytokine-1 (GDF-15/MIC-1) in adult rat tissues, *Cell Tissue Res.* 297, 1999, 103–10.

C.D. Breder, C.A. Dinarello and C.B. Saper, Interleukin-1 immunoreactive innervation of the human hypothalamus, *Science* 240, 1988, 321–4.

G. Breier, U. Albrecht, S. Sterrer and W. Risau, Expression of vascular endothelial growth factor during embryonic angiogenesis and endothelial cell differentiation, *Development* 114, 1992, 521–32.

M.W. Brightman and T.S. Reese, Junctions between intimately apposed cell membranes in the vertebrate brain, *J. Cell Biol.* 40, 1969, 648–77.

R.M. Buijs, D.F. Swaab, J. Dogterom and F.W. van Leeuwen, Intra- and extrahypothalamic vasopressin and oxytocin pathways in the rat, *Cell Tissue Res.* 186, 1978, 423–33.

V. Cameron, E.A. Espiner, M.G. Nicholls, R.A. Donald and M.R. MacFarlane, Stress hormones in blood and cerebrospinal fluid of conscious sheep: effect of hemorrhage, *Endocrinol.* 116, 1985, 1460–5.

L.A. Campfield, F.J. Smith, Y. Guisez, R. Devos and P. Burn, Recombinant mouse OB protein: evidence for a peripheral signal linking adiposity and central neural networks, *Science* 269, 1995, 546–9.

R. Cappai and A.R. White, Amyloid β, *Int. J. Biochemical Cell Biol.* 31, 1999, 885–9.

P. Carmeliet, Y.S. Ng, D. Nuyens, G. Theilmeier, K. Brusselmans, I. Cornelissen, E. Ehler, V.V. Kakkar, I. Stalmans, V. Mattot, J.C. Perriard, M. Dewerchin, W. Flameng, A. Nagy, F. Lupu, L. Moons, D. Collen, P.A. Damore and D.T. Shima, Impaired myocardial angiogenesis and ischemic cardiomyopathy in mice lacking the vascular endothelial growth factor isoforms $VEGF_{164}$ and $VEGF_{188}$, *Nat. Med.* 5, 1999, 495–502.

J.F. Caro, J.W. Kolaczynski, M.R. Nyce, J.P. Ohannesian, I. Opentanova, W.H. Goldman, R.B. Lynn, P.L. Zhang, M.K. Sinha and R.V. Considine, Decreased cerebrospinal-fluid/serum leptin ratio in obesity: a possible mechanism for leptin resistance, *Lancet* 348, 1996, 159–61.

K.M. Caron and O. Smithies, Extreme hydrops fetalis and cardiovascular abnormalities in mice lacking a functional adrenomedullin gene, *Proc. Natl. Acad. Sci. USA* 98, 2001, 615–9.

C.C. Chao, S. Hu, W.H. Frey, 2nd, T.A. Ala, W.W. Tourtellotte and P.K. Peterson, Transforming growth factor β in Alzheimers disease, *Clin. Diagn. Lab. Immunol.* 1, 1994, 109–10.

V.P. Chauhan, I. Ray, A. Chauhan and H.M. Wisniewski, Binding of gelsolin, a secretory protein, to amyloid β-protein, *Biochemical Biophys. Res. Commun.* 258, 1999, 241–6.

A. Chodobski, I. Chung, E. Kozniewska, T. Ivanenko, W. Chang, J.F. Harrington, J.A. Duncan and J. Szmydynger-Chodobska, Early neutrophilic expression of vascular endothelial growth factor after traumatic brain injury, *Neurosci.* 122, 2003, 853–67.

A. Chodobski, Y.P. Loh, S. Corsetti, J. Szmydynger-Chodobska, C.E. Johanson, Y.P. Lim and P.R. Monfils, The presence of arginine vasopressin and its mRNA in rat choroid plexus epithelium, *Mol. Brain Res.* 48, 1997, 67–72.

A. Chodobski, J. Szmydynger-Chodobska, K.A. Dodd, E.H. Kill, E.G. Stopa, A. Baird and C.E. Johanson, Central administration of fibroblast growth factor-2 causes neuronal death and cerebroventricular enlargement, *Soc. Neurosci. Abstr.* 24, 1998a, 1807.

A. Chodobski, J. Szmydynger-Chodobska and C.E. Johanson, Vasopressin mediates the inhibitory effect of central angiotensin II on cerebrospinal fluid formation, *Eur. J. Pharmacol.* 347, 1998b, 205–9.

A. Chodobski, B.E. Wojcik, Y.P. Loh, K.A. Dodd, J. Szmydynger-Chodobska, C.E. Johanson, D.M. Demers, Z.G. Chun and N.P. Limthong, Vasopressin gene expression in rat choroid plexus, *Advances Exp. Med. Biol.* 449, 1998c, 59–65.

S. Choi, R. Sparks, M. Clay and M.F. Dallman, Rats with hypothalamic obesity are insensitive to central leptin injections, *Endocrinol.* 140, 1999, 4426–33.

S.C. Chua, Jr., I.K. Koutras, L. Han, S.M. Liu, J. Kay, S.J. Young, W.K. Chung and R.L. Leibel, Fine structure of the murine leptin receptor gene: splice site suppression is required to form two alternatively spliced transcripts, *Genomics* 45, 1997, 264–70.

S.C. Chua, Jr., D.W. White, X.S. Wu-Peng, S.M. Liu, N. Okada, E.E. Kershaw, W.K. Chung, L. Power-Kehoe, M. Chua, L.A. Tartaglia and R.L. Leibel, Phenotype of *fatty* due to Gln269Pro mutation in the leptin receptor (*Lepr*), *Diabetes* 45, 1996, 1141–3.

J.T. Chun, L. Wang, G.M. Pasinetti, C.E. Finch and B.V. Zlokovic, Glycoprotein 330/megalin (LRP-2) has low prevalence as mRNA and protein in brain microvessels and choroid plexus, *Exp. Neurol.* 157, 1999, 194–201.

D.R. Clemmons, J.I. Jones, W.H. Busby and G. Wright, Role of insulin-like growth factor binding proteins in modifying IGF actions, *Ann. NY Acad. Sci.* 692, 1993, 10–21.

C.V. Clevenger and J.B. Kline, Prolactin receptor signal transduction, *Lupus* 10, 2001, 706–18.

P. Cohen, C. Zhao, X. Cai, J.M. Montez, S.C. Rohani, P. Feinstein, P. Mombaerts and J.M. Friedman, Selective deletion of leptin receptor in neurons leads to obesity, *J. Clin. Invest.* 108, 2001, 1113–21.

R.V. Considine, M.K. Sinha, M.L. Heiman, A. Kriauciunas, T.W. Stephens, M.R. Nyce, J.P. Ohannesian, C.C. Marco, L.J. McKee, T.L. Bauer and J.F. Caro, Serum immunoreactive-leptin concentrations in normal-weight and obese humans, *N. Engl. J. Med.* 334, 1996, 292–5.

G.G. Cornwell, 3rd, K. Sletten, B. Johansson and P. Westermark, Evidence that the amyloid fibril protein in senile systemic amyloidosis is derived from normal prealbumin, *Biochem. Biophys. Res. Commun.* 154, 1988, 648–53.

B.J. Cummings, G.J. Yee and C.W. Cotman, bFGF promotes the survival of entorhinal layer II neurons after perforant path axotomy, *Brain Res.* 591, 1992, 271–6.

I. Cusin, F. Rohner-Jeanrenaud, A. Stricker-Krongrad and B. Jeanrenaud, The weight-reducing effect of an intracerebroventricular bolus injection of leptin in genetically obese *fa/fa* rats. Reduced sensitivity compared with lean animals, *Diabetes* 45, 1996, 1446–50.

H.F. Cserr, Convection of brain interstitial fluid, in K. Shapiro, A. Marmarou and H. Portnoy (eds.), *Hydrocephalus*, New York, Raven Press, 1984, pp. 59–68.

E. Csuka, M.C. Morganti-Kossmann, P.M. Lenzlinger, H. Joller, O. Trentz and T. Kossmann, IL-10 levels in cerebrospinal fluid and serum of patients with severe traumatic brain injury: relationship to IL-6, TNF-α, TGF-β1 and blood-brain barrier function, *J. Neuroimmunol.* 101, 1999, 211–21.

W.H. Daughaday and P. Rotwein, Insulin-like growth factors I and II. Peptide, messenger ribonucleic acid and gene structures, serum, and tissue concentrations, *Endocr. Rev.* 10, 1989, 68–91.

T.M. DeChiara, E.J. Robertson and A. Efstratiadis, Parental imprinting of the mouse insulin-like growth factor II gene, *Cell* 64, 1991, 849–59.

D.A. Denton, J.R. Blair-West, M. McBurnie, R.S. Weisinger, A. Logan, A.M. Gonzalez and A. Baird, Central action of basic fibroblast growth factor on ingestive behaviour in mice, *Physiol. Behav.* 57, 1995, 747–52.

L.D. Dickinson and A.L. Betz, Attenuated development of ischemic brain edema in vasopressin-deficient rats, *J. Cereb. Blood Flow Metab.* 12, 1992, 681–90.

H.F. Dvorak, L.F. Brown, M. Detmar and A.M. Dvorak, Vascular permeability factor/vascular endothelial growth factor, microvascular hyperpermeability, and angiogenesis, *Am. J. Pathol.* 146, 1995, 1029–39.

K. El-Haschimi, D.D. Pierroz, S.M. Hileman, C. Bjørbæk and J.S. Flier, Two defects contribute to hypothalamic leptin resistance in mice with diet-induced obesity, *J. Clin. Invest.* 105, 2000, 1827–32.

S. Esser, M.G. Lampugnani, M. Corada, E. Dejana and W. Risau, Vascular endothelial growth factor induces VE-cadherin tyrosine phosphorylation in endothelial cells, *J. Cell Sci.* 111, 1998a, 1853–65.

S. Esser, K. Wolburg, H. Wolburg, G. Breier, T. Kurzchalia and W. Risau, Vascular endothelial growth factor induces endothelial fenestrations *in vitro*, *J. Cell Biol.* 140, 1998b, 947–59.

H. Fei, H.J. Okano, C. Li, G.H. Lee, C. Zhao, R. Darnell and J.M. Friedman, Anatomic localization of alternatively spliced leptin receptors (Ob-R) in mouse brain and other tissues, *Proc. Natl. Acad. Sci. USA* 94, 1997, 7001–5.

I.A. Ferguson and E.M. Johnson, Jr., Fibroblast growth factor receptor-bearing neurons in the CNS: identification by receptor-mediated retrograde transport, *J. Compar. Neurol.* 313, 1991, 693–706.

I.A. Ferguson, J.B. Schweitzer, P.F. Bartlett and E.M. Johnson, Jr., Receptor-mediated retrograde transport in CNS neurons after intraventricular administration of NGF and growth factors, *J. Compar. Neurol.* 313, 1991, 680–92.

N. Ferrara, Vascular endothelial growth factor and the regulation of angiogenesis, *Recent Prog. Hormone Res.* 55, 2000, 15–35.

N. Ferrara, K. Houck, L. Jakeman and D.W. Leung, Molecular and biological properties of the vascular endothelial growth factor family of proteins, *Endocr. Rev.* 13, 1992, 18–32.

G. Fisone, G.L. Snyder, A. Aperia and P. Greengard, Na^+,K^+-ATPase phosphorylation in the choroid plexus: synergistic regulation by serotonin/protein kinase C and isoproterenol/cAMP-PK/PP-1 pathways, *Mol. Med.* 4, 1998, 258–65.

K.C. Flanders, R.F. Ren and C.F. Lippa, Transforming growth factor-βs in neurodegenerative disease, *Prog. Neurobiol.* 54, 1998, 71–85.

M.C. Flynn, T.R. Scott, T.C. Pritchard and C.R. Plata-Salaman, Mode of action of OB protein (leptin) on feeding, *Am. J. Physiol.* 275, 1998, R174–9.

J.M. Frade, A. Rodríguez-Tébar and Y.A. Barde, Induction of cell death by endogenous nerve growth factor through its p75 receptor, *Nature* 383, 1996, 166–8.

J.M. Frade and Y.A. Barde, Genetic evidence for cell death mediated by nerve growth factor and the neurotrophin receptor p75 in the developing mouse retina and spinal cord, *Development* 126, 1999, 683–90.

M.E. Freeman, B. Kanyicska, A. Lerant and G. Nagy, Prolactin: structure, function, and regulation of secretion, *Physiol. Rev.* 80, 2000, 1523–631.

N. Funato, K. Moriyama, H. Shimokawa and T. Kuroda, Basic fibroblast growth factor induces apoptosis in myofibroblastic cells isolated from rat palatal mucosa, *Biochem. Biophys. Res. Commun.* 240, 1997, 21–6.

K. Fuxe, B. Tinner, M. Zoli, R.F. Pettersson, A. Baird, G. Biagini, G. Chadi and L.F. Agnati, Computer-assisted mapping of basic fibroblast growth factor immunoreactive nerve cell populations in the rat brain, *J. Chem. Neuroanat.* 11, 1996, 13–35.

E. Galbreath, S.J. Kim, K. Park, M. Brenner and A. Messing, Overexpression of TGF-β1 in the central nervous system of transgenic mice results in hydrocephalus, *J. Neuropath. Exp. Neurol.* 54, 1995, 339–49.

R. Galli, A. Gritti, L. Bonfanti and A.L. Vescovi, Neural stem cells: an overview, *Circ. Res.* 92, 2003, 598–608.

P. Gambetti and C. Russo, Human brain amyloidoses, *Nephrol., Dialysis, Transplant.* 13, Suppl. 7, 1998, 33–40.

N. Ghilardi, S. Ziegler, A. Wiestner, R. Stoffel, M.H. Heim and R.C. Skoda, Defective STAT signaling by the leptin receptor in *diabetic* mice, *Proc. Natl. Acad. Sci. USA* 93, 1996, 6231–5.

J. Ghiso, E. Matsubara, A. Koudinov, N.H. Choi-Miura, M. Tomita, T. Wisniewski and B. Frangione, The cerebrospinal-fluid soluble form of Alzheimer's amyloid beta is complexed to SP-40,40 (apolipoprotein J), an inhibitor of the complement membrane-attack complex, *Biochem. J.* 293, 1993, 27–30.

J. Gliemann, Receptors of the low density lipoprotein (LDL) receptor family in man. Multiple functions of the large family members via interaction with complex ligands, *Biol. Chem.* 379, 1998, 951–64.

V. Goffin and P.A. Kelly, The prolactin/growth hormone receptor family: structure/function relationships, *J. Mamm. Gland Biol. Neoplasia* 2, 1997, 7–17.

A. Golabek, M.A. Marques, M. Lalowski and T. Wisniewski, Amyloid β binding proteins *in vitro* and in normal human cerebrospinal fluid, *Neurosci. Lett.* 191, 1995, 79–82.

A.M. Gonzalez, M. Berry, P.A. Maher, A. Logan and A. Baird, A comprehensive analysis of the distribution of FGF2 and FGFR1 in the rat brain, *Brain Res.* 701, 1995, 201–26.

D. Gospodarowicz, G. Neufeld and L. Schweigerer, Fibroblast growth factor, *Mol. Cell. Endocrinol.* 46, 1986, 187–204.

D.R. Grattan, X.J. Pi, Z.B. Andrews, R.A. Augustine, I.C. Kokay, M.R. Summerfield, B. Todd and S.J. Bunn, Prolactin receptors in the brain during pregnancy and lactation: implications for behavior, *Hormones Behav.* 40, 2001, 115–24.

X.M. Guan, J.F. Hess, H. Yu, P.J. Hey and L.H. van der Ploe.g., Differential expression of mRNA for leptin receptor isoforms in the rat brain, *Mol. Cell. Endocrinol.* 133, 1997, 1–7.

O. Günther, R. Landgraf, J. Schuart and H. Unger, Vasopressin in cerebrospinal fluid (CSF) and plasma of conscious rabbits—circadian variations, *Exp. Clin. Endocrinol*ogy 83, 1984, 367–9.

D. Haig and C. Graham, Genomic imprinting and the strange case of the insulin-like growth factor II receptor, *Cell* 64, 1991, 1045–6.

A. Hakvoort and C.E. Johanson, Growth factor modulation of CSF formation by isolated choroid plexus: FGF2 vs. TGF-β1, *Eur. J. Pediatr. Surg.* 10, Suppl. I, 2000, 44–6.

J.L. Halaas, C. Boozer, J. Blair-West, N. Fidahusein, D.A. Denton and J.M. Friedman, Physiological response to long-term peripheral and central leptin infusion in lean and obese mice, *Proc. Natl. Acad. Sci. USA* 94, 1997, 8878–83.

S.M. Hammad, S. Ranganathan, E. Loukinova, W.O. Twal and W.S. Argraves, Interaction of apolipoprotein J-amyloid β-peptide complex with low density lipoprotein receptor-related protein-2/megalin. A mechanism to prevent pathological accumulation of amyloid β-peptide, *J. Biol. Chem.* 272, 1997, 18644–9.

D. Harland, S.M. Gardiner and T. Bennett, Differential cardiovascular effects of centrally administered vasopressin in conscious Long Evans and Brattleboro rats, *Circ. Res.* 65, 1989, 925–33.

M.R. Harrigan, S.R. Ennis, T. Masada and R.F. Keep, Intraventricular infusion of vascular endothelial growth factor promotes cerebral angiogenesis with minimal brain edema, *Neurosurg.* 50, 2002, 589–98.

R.B. Harris, J. Zhou, S.M. Redmann, Jr., G.N. Smagin, S.R. Smith, E. Rodgers and J.J. Zachwieja, A leptin dose-response study in obese (*ob/ob*) and lean (+/?) mice, *Endocrinology* 139, 1998, 8–19.

M.E. Harris-White, T. Chu, Z. Balverde, J.J. Sigel, K.C. Flanders and S.A. Frautschy, Effects of transforming growth factor-β (isoforms 1–3) on amyloid-β deposition, inflammation, and cell targeting in organotypic hippocampal slice cultures, *J. Neurosci.* 18, 1998, 10366–74.

M.E. Hartnett, A. Lappas, D. Darland, J.R. McColm, S. Lovejoy and P.A. D'amore, Retinal pigment epithelium and endothelial cell interaction causes retinal pigment epithelial barrier dysfunction via a soluble VEGF-dependent mechanism, *Exp. Eye Res.* 77, 2003, 593–9.

P. Henrich-Noack, J.H. Prehn and J. Krieglstein, Neuroprotective effects of TGF-β1, *J. Neural Transmission* 43, Suppl., 1994, 33–45.

P. Henrich-Noack, J.H. Prehn and J. Krieglstein, TGF-β1 protects hippocampal neurons against degeneration caused by transient global ischemia. Dose-response relationship and potential neuroprotective mechanisms, *Stroke* 27, 1996, 1609–14.

J. Herbert, J.N. Wilcox, K.T. Pham, R.T. Fremeau, Jr., M. Zeviani, A. Dwork, D.R. Soprano, A. Makover, D.S. Goodman, E.A. Zimmerman, J.L. Roberts and E.A. Schon, Transthyretin: a choroid plexus-specific transport protein in human brain, *Neurol.* 36, 1986, 900–11.

S.M. Hileman, D.D. Pierroz, H. Masuzaki, C. Bjørbæk, K. El-Haschimi, W.A. Banks and J.S. Flier, Characterization of short isoforms of the leptin receptor in rat cerebral microvessels and of brain uptake of leptin in mouse models of obesity, *Endocrinology* 143, 2002, 775–83.

S.M. Hileman, J. Tornøe, J.S. Flier and C. Bjørbæk, Transcellular transport of leptin by the short leptin receptor isoform ObRa in Madin-Darby canine kidney cells, *Endocrinology* 141, 2000, 1955–61.

J.P. Hinson, S. Kapas and D.M. Smith, Adrenomedullin, a multifunctional regulatory peptide, *Endocr. Rev.* 21, 2000, 138–67.

N.R. Holm, L.B. Hansen, C. Nilsson and S. Gammeltoft, Gene expression and secretion of insulin-like growth factor-II and insulin-like growth factor binding protein-2 from cultured sheep choroid plexus epithelial cells, *Mol. Brain Res.* 21, 1994, 67–74.

H. Hu, Chemorepulsion of neuronal migration by Slit2 in the developing mammalian forebrain, *Neuron* 23, 1999, 703–11.

C.C. Huang, C.C. Liu, S.T. Wang, Y.C. Chang, H.B. Yang and T.F. Yeh, Basic fibroblast growth factor in experimental and clinical bacterial meningitis, *Pediatr. Res.* 45, 1999, 120–7.

M.M. Hussain, D.K. Strickland and A. Bakillah, The mammalian low-density lipoprotein receptor family, *Ann. Rev. Nutr.* 19, 1999, 141–72.

M.A. Hynes, P.J. Brooks, J.J. Van Wyk and P.K. Lund, Insulin-like growth factor II messenger ribonucleic acids are synthesized in the choroid plexus of the rat brain, *Mol. Endocrinol.* 2, 1988, 47–54.

R.A. Ignotz and J. Massagué, Transforming growth factor-β stimulates the expression of fibronectin and collagen and their incorporation into the extracellular matrix, *J. Biol. Chem.* 261, 1986, 4337–45.

C.E. Johanson, J.E. Preston, A. Chodobski, E.G., Stopa, J. Szmydynger-Chodobska and P.N. McMillan, AVP V_1 receptor-mediated decrease in Cl^- efflux and increase in dark cell number in choroid plexus epithelium, *Am. J. Physiol.* 276, 1999a, C82–90.

C.E. Johanson, J. Szmydynger-Chodobska, A. Chodobski, A. Baird, P. McMillan and E.G. Stopa, Altered formation and bulk absorption of cerebrospinal fluid in FGF2-induced hydrocephalus, *Am. J. Physiol.* 277, 1999b, R263–71.

S.E. Jones and C. Jomary, Clusterin, *Int. J. Biochem. Cell Biol.* 34, 2002, 427–31.

V.M. Kalscheuer, E.C. Mariman, M.T. Schepens, H. Rehder and H.H. Ropers, The insulin-like growth factor type-2 receptor gene is imprinted in the mouse but not in humans, *Nat. Genet.* 5, 1993, 74–8.

S. Kar, J.G. Chabot and R. Quirion, Quantitative autoradiographic localization of [^{125}I]insulin-like growth factor I, [^{125}I]insulin-like growth factor II, and [^{125}I]insulin receptor binding sites in developing and adult rat brain, *J. Compar. Neurol.* 333, 1993, 375–97.

C.G. Kevil, D.K. Payne, E. Mire and J.S. Alexander, Vascular permeability factor/vascular endothelial cell growth factor-mediated permeability occurs through disorganization of endothelial junctional proteins, *J. Biol. Chem.* 273, 1998, 15099–103.

J.D. Kibble, A.E. Trezise and P.D. Brown, Properties of the cAMP-activated $C1^-$ current in choroid plexus epithelial cells isolated from the rat, *J. Physiol. (Lond.)* 496, 1996, 69–80.

E.S. Kim, R.S. Kim, R.F. Ren, D.B. Hawver and K.C. Flanders, Transforming growth factor-β inhibits apoptosis induced by β-amyloid peptide fragment 25–35 in cultured neuronal cells, *Mol. Brain Res.* 62, 1998, 122–30.

B. Kis, C.S. Ábrahám, M.A. Deli, H. Kobayashi, A. Wada, M. Niwa, H. Yamashita and Y. Ueta, Adrenomedullin in the cerebral circulation, *Peptides* 22, 2001a, 1825–34.

B. Kis, M.A. Deli, H. Kobayashi, C.S. Ábrahám, T. Yanagita, H. Kaiya, T. Isse, R. Nishi, S. Gotoh, K. Kangawa, A. Wada, J. Greenwood, M. Niwa, H. Yamashita and Y. Ueta, Adrenomedullin regulates blood-brain barrier functions *in vitro*, *Neuroreport* 12, 2001b, 4139–42.

B. Kis, H. Kaiya, R. Nishi, M.A. Deli, C.S. Ábrahám, T. Yanagita, T. Isse, S. Gotoh, H. Kobayashi, A. Wada, M. Niwa, K. Kangawa, J. Greenwood, H. Yamashita and Y. Ueta, Cerebral endothelial cells are a major source of adrenomedullin, *J. Neuroendocrinol.* 14, 2002, 283–93.

B. Kis, J.A. Snipes, M.A. Deli, C.S. Ábrahám, H. Yamashita, Y. Ueta and D.W. Busija, Chronic adrenomedullin treatment improves blood-brain barrier function but has no effects on expression of tight junction proteins, *Acta Neurochir.* 86, Suppl., 2003, 565–8.

K. Kitamura, K. Kangawa and T. Eto, Adrenomedullin and PAMP: discovery, structures, and cardiovascular functions, *Microsc. Res. Techn.* 57, 2002, 3–13.

N.D. Klempt, E. Sirimanne, A.J. Gunn, M. Klempt, K. Singh, C. Williams and P.D. Gluckman, Hypoxia-ischemia induces transforming growth factor β_1 mRNA in the infant rat brain, *Mol. Brain Res.* 13, 1992, 93–101.

J.B. Kline and C.V. Clevenger, Identification and characterization of the prolactin-binding protein in human serum and milk, *J. Biol. Chem.* 276, 2001, 24760–6.

J.B. Kline, M.A. Rycyzyn and C.V. Clevenger, Characterization of a novel and functional human prolactin receptor isoform (ΔS1PRLr) containing only one extracellular fibronectin-like domain, *Mol. Endocrinol.* 16, 2002, 2310–22.

N.W. Knuckey, P. Finch, D.E. Palm, M.J. Primiano, C.E. Johanson, K.C. Flanders and N.L. Thompson, Differential neuronal and astrocytic expression of transforming growth factor beta isoforms in rat hippocampus following transient forebrain ischemia, *Mol. Brain Res.* 40, 1996, 1–14.

H. Kobayashi, S. Shiraishi, S. Minami, H. Yokoo, T. Yanagita, T. Saitoh, M. Mohri and A. Wada, Adrenomedullin receptors in rat choroid plexus, *Neurosci. Lett.* 297, 2001, 167–70.

N. Koketsu, D.J. Berlove, M.A. Moskowitz, N.W. Kowall, C.G. Caday and S.P. Finklestein, Pretreatment with intraventricular basic fibroblast growth factor decreases infarct size following focal cerebral ischemia in rats, *Ann. Neurol.* 35, 1994, 451–7.

T. Kotera and P.D. Brown, Cl^- current activation in choroid plexus epithelial cells involves a G protein and protein kinase A, *Am. J. Physiol.* 266, 1994, C536–40.

M.Z. Kounnas, C.C. Haudenschild, D.K. Strickland and W.S. Argraves, Immunological localization of glycoprotein 330, low density lipoprotein receptor related protein and 39 kDa receptor associated protein in embryonic mouse tissues, *In Vivo* 8, 1994, 343–51.

M.Z. Kounnas, E.B. Loukinova, S. Stefansson, J.A. Harmony, B.H. Brewer, D.K. Strickland and W.S. Argraves, Identification of glycoprotein 330 as an endocytic receptor for apolipoprotein J/clusterin, *J. Biol. Chem.* 270, 1995, 13070–5.

K. Krieglstein, M. Rufer, C. Suter-Crazzolara and K. Unsicker, Neural functions of the transforming growth factors β, *Int. J. Develpomental Neurosci.*13, 1995, 301–15.

K. Krieglstein, J. Strelau, A. Schober, A. Sullivan and K. Unsicker, TGF-β and the regulation of neuron survival and death, *J. Physiol. (Paris)* 96, 2002, 25–30.

J.M. Krum, N. Mani and J.M. Rosenstein, Angiogenic and astroglial responses to vascular endothelial growth factor administration in adult rat brain, *Neurosci.* 110, 2002, 589–604.

H.G. Kuhn, J. Winkler, G. Kempermann, L.J. Thal and F.H. Gage, Epidermal growth factor and fibroblast growth factor-2 have different effects on neural progenitors in the adult rat brain, *J. Neurosci.* 17, 1997, 5820–9.

Kwiatkowski, R. Mehl and H.L. Yin, Genomic organization and biosynthesis of secreted and cytoplasmic forms of gelsolin, *J. Cell Biol.* 106, 1988, 375–84.

A. Ladoux and C. Frelin, Coordinated up-regulation by hypoxia of adrenomedullin and one of its putative receptors (RDC-1) in cells of the rat blood-brain barrier, *J. Biol. Chem.* 275, 2000, 39914–9.

G.H. Lee, R. Proenca, J.M. Montez, K.M. Carroll, J.G. Darvishzadeh, J.I. Lee and J.M. Friedman, Abnormal splicing of the leptin receptor in diabetic mice, *Nature* 379, 1996, 632–5.

M. Le Moal, R. Dantzer, B. Michaud and G.F. Koob, Centrally injected arginine vasopressin (AVP) facilitates social memory in rats, *Neurosci. Lett.* 77, 1987, 353–9.

D. LeRoith and C.T. Roberts, Jr., Insulin-like growth factors, *Ann. NY Acad. Sci.* 692, 1993, 1–9.

D. LeRoith, H. Werner, T.N. Faria, H. Kato, M. Adamo and C.T. Roberts, Jr., Insulin-like growth factor receptors. Implications for nervous system function, *Ann. NY Acad. Sci.* 692, 1993, 22–32.

S. Levine and R. Sowinski, Localization of gold thioglucose and bipiperidyl mustard lesions near artificial disruptions of the blood-brain barrier, *Exp. Neurol.* 79, 1983, 462–71.

C.D. Link, Expression of human β-amyloid peptide in transgenic *Caenorhabditis elegans*, *Proc. Natl. Acad. Sci. USA* 92, 1995, 9368–72.

T.M. Liszczak, P.M. Black and L. Foley, Arginine vasopressin causes morphological changes suggestive of fluid transport in rat choroid plexus epithelium, *Cell Tissue Res.* 246, 1986, 379–85.

A. Logan, S.A. Frautschy, A.M. Gonzalez, M.B. Sporn and A. Baird, Enhanced expression of transforming growth factor β1 in the rat brain after a localized cerebral injury, *Brain Res.* 587, 1992, 216–25.

T. Ludwig, J. Eggenschwiler, P. Fisher, A.J. D'Ercole, M.L. Davenport and A. Efstratiadis, Mouse mutants lacking the type 2 IGF receptor (IGF2R) are rescued from perinatal lethality in *Igf2* and *Igf1r* null backgrounds, *Dev. Biol.* 177, 1996, 517–35.

G.N. Luheshi, J.D. Gardner, D.A. Rushforth, A.S. Loudon and N.J. Rothwell, Leptin actions on food intake and body temperature are mediated by IL-1, *Proc. Natl. Acad. Sci. USA* 96, 1999, 7047–52.

J. Ma, J. Qiu, L. Hirt, T. Dalkara and M.A. Moskowitz, Synergistic protective effect of caspase inhibitors and bFGF against brain injury induced by transient focal ischaemia, *Br. J. Pharmacol.* 133, 2001, 345–50.

M. Maamra, M. Bidlingmaier, M.C. Postel-Vinay, Z. Wu, C.J. Strasburger and R.J. Ross, Generation of human soluble leptin receptor by proteolytic cleavage of membrane-anchored receptors, *Endocrinology* 142, 2001, 4389–93.

M. Maffei, H. Fei, G.H. Lee, C. Dani, P. Leroy, Y. Zhang, R. Proenca, R. Negrel, G. Ailhaud and J.M. Friedman, Increased expression in adipocytes of ob RNA in mice with lesions of the hypothalamus and with mutations at the *db* locus, *Proc. Natl. Acad. Sci. USA* 92, 1995a, 6957–60.

M. Maffei, J. Halaas, E. Ravussin, R.E. Pratley, G.H. Lee, Y. Zhang, H. Fei, S. Kim, R. Lallone, S. Ranganathan, et al., Leptin levels in human and rodent: measurement of plasma leptin and *ob* RNA in obese and weight-reduced subjects, *Nat. Med.* 1, 1995b, 1155–61.

A.M. Malek, S. Connors, R.L. Robertson, J. Folkman and R.M. Scott, Elevation of cerebrospinal fluid levels of basic fibroblast growth factor in moyamoya and central nervous system disorders, *Pediatr. Neurosurg.* 27, 1997, 182–9.

L.M. Maness, A.J. Kastin, C.L. Farrell and W.A. Banks, Fate of leptin after intracerebroventricular injection into the mouse brain, *Endocrinology* 139, 1998, 4556–62.

L.P. Mangurian, R.J. Walsh and B.I. Posner, Prolactin enhancement of its own uptake at the choroid plexus, *Endocrinology* 131, 1992, 698–702.

D.W. Marion, L.E. Penrod, S.F. Kelsey, W.D. Obrist, P.M. Kochanek, A.M. Palmer, S.R. Wisniewski and S.T. DeKosky, Treatment of traumatic brain injury with moderate hypothermia, *N. Engl. J. Med.* 336, 1997, 540–6.

H.H. Marti and W. Risau, Systemic hypoxia changes the organ-specific distribution of vascular endothelial growth factor and its receptors, *Proc. Natl. Acad. Sci. USA* 95, 1998, 15809–14.

J. Massagué, L. Attisano and J.L. Wrana, The TGF-β family and its composite receptors, *Trends Cell Biol.* 4, 1994, 172–8.

E. Matsubara, B. Frangione and J. Ghiso, Characterization of apolipoprotein J-Alzheimer's Aβ interaction, *J. Biol. Chem.* 270, 1995, 7563–7.

E. Matsubara, C. Soto, S. Governale, B. Frangione and J. Ghiso, Apolipoprotein J and Alzheimer's amyloid β solubility, *Biochem. J.* 316, 1996, 671–9.

N. Matsumoto, H. Kitayama, M. Kitada, K. Kimura, M. Noda and C. Ide, Isolation of a set of genes expressed in the choroid plexus of the mouse using suppression subtractive hybridization, *Neurosci.* 117, 2003, 405–15.

S.G. Matthews and R.F. Parrott, Centrally administered vasopressin modifies stress hormone (cortisol, prolactin) secretion in sheep under basal conditions, during restraint and following intravenous corticotrophin-releasing hormone, *Eur. J. Endocrinol.* 130, 1994, 297–301.

T.F. McDonald and K. Green, Cell turnover in ciliary epithelium compared to other slow renewing epithelia in the adult mouse, *Curr. Eye Res.* 7, 1988, 247–52.

A.M. McGough, C.J. Staiger, J.K. Min and K.D. Simonetti, The gelsolin family of actin regulatory proteins: modular structures, versatile functions, *FEBS Lett.* 552, 2003, 75–81.

L.M. McLatchie, N.J. Fraser, M.J. Main, A. Wise, J. Brown, N. Thompson, R. Solari, M.G. Lee and S.M. Foord, RAMPs regulate the transport and ligand specificity of the calcitonin-receptor-like receptor, *Nature* 393, 1998, 333–9.

H. McNeill, C. Williams, J. Guan, M. Dragunow, P. Lawlor, E. Sirimanne, K. Nikolics and P. Gluckman, Neuronal rescue with transforming growth factor-β1 after hypoxic-ischaemic brain injury, *Neuroreport* 5, 1994, 901–4.

J.M. McQueen and A.M. Heck, Secretin for the treatment of autism, *Ann. Pharmacother.* 36, 2002, 305–11.

A. Merched, J.M. Serot, S. Visvikis, D. Aguillon, G. Faure and G. Siest, Apolipoprotein E, transthyretin and actin in the CSF of Alzheimer's patients: relation with the senile plaques and cytoskeleton biochemistry, *FEBS Lett.* 425, 1998, 225–8.

P. Mignatti, T. Morimoto and D.B. Rifkin, Basic fibroblast growth factor released by single, isolated cells stimulates their migration in an autocrine manner, *Proc. Natl. Acad. Sci. USA* 88, 1991, 11007–11.

R. Montesano and L. Orci, Transforming growth factor β stimulates collagen-matrix contraction by fibroblasts: implications for wound healing, *Proc. Natl. Acad. Sci. USA* 85, 1988, 4894–7.

N. Morita, T. Takumi and H. Kiyama, Distinct localization of two serine-threonine kinase receptors for activin and TGF-β in the rat brain and down-regulation of type I activin receptor during peripheral nerve regeneration, *Mol. Brain Res.* 42, 1996, 263–71.

E.J. Mufson, J.S. Kroin, T.J. Sendera and T. Sobreviela, Distribution and retrograde transport of trophic factors in the central nervous system: functional implications for the treatment of neurodegenerative diseases, *Prog. Neurobiol.* 57, 1999, 451–84.

T. Murakami, T. Yamashita, M. Iida, M. Kuwajima and K. Shima, A short form of leptin receptor performs signal transduction, *Biochem. Biophys. Res. Commun.* 231, 1997, 26–9.

T. Murata, K. Nakagawa, A. Khalil, T. Ishibashi, H. Inomata and K. Sueishi, The relation between expression of vascular endothelial growth factor and breakdown of the blood-retinal barrier in diabetic rat retinas, *Lab. Invest.* 74, 1996, 819–25.

S. Naito, L. Chang, K. Pettigrew, S. Ishihara and M. Brightman, Conditions that may determine blood vessel phenotype in tissues grafted to brain, *Exp. Neurol.* 134, 1995, 230–43.

K.T. Nguyen-Ba-Charvet, N. Picard-Riera, M. Tessier-Lavigne, A. Baron-Van Evercooren, C. Sotelo and A. Chédotal, Multiple roles for slits in the control of cell migration in the rostral migratory stream, *J. Neurosci.* 24, 2004, 1497–506.

C. Nicholson and E. Syková, Extracellular space structure revealed by diffusion analysis, *Trends Neurosci.* 21, 1998, 207–15.

H. Niermann, M. Amiry-Moghaddam, K. Holthoff, O.W. Witte and O.P. Ottersen, A novel role of vasopressin in the brain: modulation of activity-dependent water flux in the neocortex, *J. Neurosci.* 21, 2001, 3045–51.

C. Nilsson, P. Blay, F.C. Nielsen and S. Gammeltoft, Gene expression and receptor binding of insulin-like growth factor-II in pig choroid plexus epithelial cells, *J. Neurochem.* 58, 1992, 923–30.

C. Nilsson, B.M. Hultberg and S. Gammeltoft, Autocrine role of insulin-like growth factor II secretion by the rat choroid plexus, *Eur. J. Neurosci.* 8, 1996, 629–35.

K.D. Niswender, G.J. Morton, W.H. Stearns, C.J. Rhodes, M.G. Myers, Jr. and M.W. Schwartz, Key enzyme in leptin-induced anorexia, *Nature* 413, 2001, 794–5.

J. Nitta and T. Tada, Ultramicroscopic structures of the leptomeninx of mice with communicating hydrocephalus induced by human recombinant transforming growth factor-β1, *Neurol. Medico-Chir. (Tokyo)* 38, 1998, 819–24.

T. Oda, G.M. Pasinetti, H.H. Osterburg, C. Anderson, S.A. Johnson and C.E. Finch, Purification and characterization of brain clusterin, *Biochem. Biophys. Res. Commun.* 204, 1994, 1131–6.

M.E. O'Donnell, Endothelial cell sodium-potassium-chloride cotransport. Evidence of regulation by Ca^{2+} and protein kinase C, *J. Biol. Chem.* 266, 1991, 11559–66.

M.E. O'Donnell, A. Martinez and D. Sun, Cerebral microvascular endothelial cell Na-K-Cl cotransport: regulation by astrocyte-conditioned medium, *Am. J. Physiol.* 268, 1995, C747–54.

R. Ohlsson, F. Hedborg, L. Holmgren, C. Walsh and T.J. Ekström, Overlapping patterns of *IGF2* and *H19* expression during human development: biallelic *IGF2* expression correlates with a lack of *H19* expression, *Development* 120, 1994, 361–8.

S. O'Rahilly, I.S. Farooqi, G.S. Yeo and B.G. Challis, Minireview: human obesity—lessons from monogenic disorders, *Endocrinology* 144, 2003, 3757–64.

M. Overall, M. Bakker, J. Spencer, N. Parker, P. Smith and M. Dziadek, Genomic imprinting in the rat: linkage of *Igf2* and *H19* genes and opposite parental allele-specific expression during embryogenesis, *Genomics* 45, 1997, 416–20.

K.J. Page, R.D. Hollister and B.T. Hyman, Dissociation of apolipoprotein and apolipoprotein receptor response to lesion in the rat brain: an *in situ* hybridization study, *Neurosci.* 85, 1998, 1161–71.

A. Pagenstecher, A.K. Stalder, C.L. Kincaid, S.D. Shapiro and I.L. Campbell, Differential expression of matrix metalloproteinase and tissue inhibitor of matrix metalloproteinase genes in the mouse central nervous system in normal and inflammatory states, *Am. J. Pathol.* 152, 1998, 729–41.

J.A. Palha, P. Moreira, T. Wisniewski, B. Frangione and M.J. Saraiva, Transthyretin gene in Alzheimer's disease patients, *Neurosci. Lett.* 204, 1996, 212–4.

R.K. Pearce, P. Collins, P. Jenner, C. Emmett and C.D. Marsden, Intraventricular infusion of basic fibroblast growth factor (bFGF) in the MPTP-treated common marmoset, *Synapse* 23, 1996, 192–200.

A. Pedram, M. Razandi and E.R. Levin, Deciphering vascular endothelial cell growth factor/vascular permeability factor signaling to vascular permeability. Inhibition by atrial natriuretic peptide, *J. Biol. Chem.* 277, 2002, 44385–98.

X.J. Pi and D.R. Grattan, Differential expression of the two forms of prolactin receptor mRNA within microdissected hypothalamic nuclei of the rat, *Mol. Brain Res.* 59, 1998, 1–12.

M.D. Plotkin, M.R. Kaplan, L.N. Peterson, S.R. Gullans, S.C. Hebert and E. Delpire, Expression of the Na^+-K^+-$2Cl^-$ cotransporter BSC2 in the nervous system, *Am. J. Physiol.* 272, 1997, C173–83.

D.R. Poyner, P.M. Sexton, I. Marshall, D.M. Smith, R. Quirion, W. Born, R. Muff, J.A. Fischer and S.M. Foord, International Union of Pharmacology. XXXII. The mammalian calcitonin gene-related peptides, adrenomedullin, amylin, and calcitonin receptors *Pharmacol. Rev.* 54, 2002, 233–46.

B.M. Pratt and J.M. McPherson, TGF-β in the central nervous system: potential roles in ischemic injury and neurodegenerative diseases, *Cytokine Growth Factor Rev.* 8, 1997, 267–92.

J.H. Prehn, V.P. Bindokas, J. Jordan, M.F. Galindo, G.D. Ghadge, R.P. Roos, L.H. Boise, C.B. Thompson, S. Krajewski, J.C. Reed and R.J. Miller, Protective effect of transforming growth factor-β1 on β-amyloid neurotoxicity in rat hippocampal neurons, *Mol. Pharmacol.* 49, 1996, 319–28.

J.H. Prehn, V.P. Bindokas, C.J. Marcuccilli, S. Krajewski, J.C. Reed and R.J. Miller, Regulation of neuronal Bcl2 protein expression and calcium homeostasis by transforming growth factor type β confers wide-ranging protection on rat hippocampal neurons, *Proc. Natl. Acad. Sci. USA* 91, 1994, 12599–603.

M.G. Proescholdt, S. Chakravarty, J.A. Foster, S.B. Foti, E.M. Briley and M. Herkenham, Intracerebroventricular but not intravenous interleukin-1β induces widespread vascular-mediated leukocyte infiltration and immune signal mRNA expression followed by brain-wide glial activation, *Neurosci.* 112, 2002, 731–49.

T. Qaum, Q. Xu, A.M. Joussen, M.W. Clemens, W. Qin, K. Miyamoto, H. Hassessian, S.J. Wiegand, J. Rudge, G.D. Yancopoulos and A.P. Adamis, VEGF-initiated blood-retinal barrier breakdown in early diabetes, *Invest. Ophthalmol. Visual Sci.* 42, 2001, 2408–13.

J.H. Qi, Q. Ebrahem, N. Moore, G. Murphy, L. Claesson-Welsh, M. Bond, A. Baker and B. Anand-Apte, A novel function for tissue inhibitor of metalloproteinases-3 (TIMP3): inhibition of angiogenesis by blockage of VEGF binding to VEGF receptor-2, *Nat. Med.* 9, 2003, 407–15.

N.D. Quinton, A.J. Lee, R.J. Ross, R. Eastell and A.I. Blakemore, A single nucleotide polymorphism (SNP) in the leptin receptor is associated with BMI, fat mass and leptin levels in postmenopausal Caucasian women, *Human Genet.* 108, 2001, 233–6.

M.E. Raichle and R.L. Grubb, Jr., Regulation of brain water permeability by centrally-released vasopressin, *Brain Res.* 143, 1978, 191–4.

I. Ray, A. Chauhan, J. Wegiel and V.P. Chauhan, Gelsolin inhibits the fibrillization of amyloid beta-protein, and also defibrillizes its preformed fibrils, *Brain Res.* 853, 2000, 344–51.

S. Refetoff, F.E. Dwulet and M.D. Benson, Reduced affinity for thyroxine in two of three structural thyroxine-binding prealbumin variants associated with familial amyloidotic polyneuropathy, *J. Clin. Endocrinol. Metab.* 63, 1986, 1432–7.

R.F. Ren and K.C. Flanders, Transforming growth factors-β protect primary rat hippocampal neuronal cultures from degeneration induced by β-amyloid peptide, *Brain Res.* 732, 1996, 16–24.

S.M. Reppert, R.J. Coleman, H.W. Heath and H.T. Keutmann, Circadian properties of vasopressin and melatonin rhythms in cat cerebrospinal fluid, *Am. J. Physiol.* 243, 1982, E489–98.

H. Riisøen, Reduced prealbumin (transthyretin) in CSF of severely demented patients with Alzheimer's disease, *Acta Neurol. Scand.* 78, 1988, 455–9.

W.G. Roberts and G.E. Palade, Increased microvascular permeability and endothelial fenestration induced by vascular endothelial growth factor, *J. Cell Sci.* 108, 1995, 2369–79.

A.B. Roberts, M.B. Sporn, R.K. Assoian, J.M. Smith, N.S. Roche, L.M. Wakefield, U.I. Heine, L.A. Liotta, V. Falanga, J.H. Kehrl and A.S. Fauci, Transforming growth factor type β: rapid induction of fibrosis and angiogenesis *in vivo* and stimulation of collagen formation *in vitro*, *Proc. Natl. Acad. Sci. USA* 83, 1986, 4167–71.

C.L. Robertson, N. Minamino, R.A. Ruppel, K. Kangawa, S.R. Wisniewski, T. Tsuji, K.L. Janesko, H. Ohta, P.D. Adelson, D.W. Marion and P.M. Kochanek, Increased adrenomedullin in cerebrospinal fluid after traumatic brain injury in infants and children, *J. Neurotrauma* 18, 2001, 861–8.

C.J. Robinson and S.E. Stringer, The splice variants of vascular endothelial growth factor (VEGF) and their receptors, *J. Cell Sci.* 114, 2001, 853–65.

E.M. Rodríguez, The cerebrospinal fluid as a pathway in neuroendocrine integration, *J. Endocrinol.* 71, 1976, 407–43.

H.N. Rosen, A.C. Moses, J.R. Murrell, J.J. Liepnieks and M.D. Benson, Thyroxine interactions with transthyretin: a comparison of 10 different naturally occurring human transthyretin variants, *J. Clin. Endocrinol. Metab.* 77, 1993, 370–4.

G.A. Rosenberg, W.T. Kyner and E. Estrada, Bulk flow of brain interstitial fluid under normal and hyperosmolar conditions, *Am. J. Physiol.* 238, 1980, F42–9.

G.A. Rosenberg, L.I. Wolfson and R. Katzman, Pressure-dependent bulk flow of cerebrospinal fluid into brain, *Exp. Neurol.* 60, 1978, 267–76.

C. Ruhrberg, H. Gerhardt, M. Golding, R. Watson, S. Ioannidou, H. Fujisawa, C. Betsholtz and D.T. Shima, Spatially restricted patterning cues provided by heparin-binding VEGF-A control blood vessel branching morphogenesis, *Genes Development* 16, 2002, 2684–98.

J.M. Russell, Sodium-potassium-chloride cotransport, *Physiol. Rev.* 80, 2000, 211–76.

M.F. Saad, S. Damani, R.L. Gingerich, M.G. Riad-Gabriel, A. Khan, R. Boyadjian, S.D. Jinagouda, K. El-Tawil, R.K. Rude and V. Kamdar, Sexual dimorphism in plasma leptin concentration, *J. Clinincal Endocrinol. Metab.* 82, 1997, 579–84.

V.R. Sara and K. Hall, Insulin-like growth factors and their binding proteins, *Physiol. Rev.* 70, 1990, 591–614.

A. Schober, M. Böttner, J. Strelau, R. Kinscherf, G.A. Bonaterra, M. Barth, L. Schilling, W.D. Fairlie, S.N. Breit and K. Unsicker, Expression of growth differentiation factor-15/macrophage inhibitory cytokine-1 (GDF-15/MIC-1) in the perinatal, adult, and injured rat brain, *J. Compar. Neurol.* 439, 2001, 32–45.

H.J. Schoch, S. Fischer and H.H. Marti, Hypoxia-induced vascular endothelial growth factor expression causes vascular leakage in the brain, *Brain* 125, 2002, 2549–57.

G. Schreiber, A.R. Aldred, A. Jaworowski, C. Nilsson, M.G. Achen and M.B. Segal, Thyroxine transport from blood to brain via transthyretin synthesis in choroid plexus, *Am. J. Physiol.* 258, 1990, R338–45.

W.J. Schultz, M.S. Brownfield and G.P. Kozlowski, The hypothalamo-choridal tract. II. Ultrastructural response of the choroid plexus to vasopressin, *Cell Tissue Res.* 178, 1977, 129–41.

W.J. Schwartz, R.J. Coleman and S.M. Reppert, A daily vasopressin rhythm in rat cerebrospinal fluid, *Brain Res.* 263, 1983, 105–12.

A.L. Schwarzman, L. Gregori, M.P. Vitek, S. Lyubski, W.J. Strittmatter, J.J. Enghilde, R. Bhasin, J. Silverman, K.H. Weisgraber, P.K. Coyle, M.G. Zagorski, J. Talafous, M. Eisenberg, A.M. Saunders, A.D. Roses and D. Goldgaber, Transthyretin sequesters amyloid β protein and prevents amyloid formation, *Proc. Natl. Acad. Sci. USA* 91, 1994, 8368–72.

R.J. Seeley, G. van Dijk, L.A. Campfield, F.J. Smith, P. Burn, J.A. Nelligan, S.M. Bell, D.G. Baskin, S.C. Woods and M.W. Schwartz, Intraventricular leptin reduces food intake and body weight of lean rats but not obese Zucker rats, *Hormone Metab. Res.* 28, 1996, 664–8.

J.M. Serot, D. Christmann, T. Dubost and M. Couturier, Cerebrospinal fluid transthyretin: aging and late onset Alzheimer's disease, *J. Neurol., Neurosurg. Psychiatry* 63, 1997, 506–8.

T.G. Sherman, J.F. McKelvy and S.J. Watson, Vasopressin mRNA regulation in individual hypothalamic nuclei: a northern and *in situ* hybridization analysis, *J. Neurosci.* 6, 1986, 1685–94.

T. Shindo, Y. Kurihara, H. Nishimatsu, N. Moriyama, M. Kakoki, Y. Wang, Y. Imai, A. Ebihara, T. Kuwaki, K.H. Ju, N. Minamino, K. Kangawa, T. Ishikawa, M. Fukuda, Y. Akimoto, H. Kawakami, T. Imai, H. Morita, Y. Yazaki, R. Nagai, Y. Hirata and H. Kurihara, Vascular abnormalities and elevated blood pressure in mice lacking adrenomedullin gene, *Circ.* 104, 2001, 1964–71.

R. Silver, J. LeSauter, P.A. Tresco and M.N. Lehman, A diffusible coupling signal from the transplanted suprachiasmatic nucleus controlling circadian locomotor rhythms, *Nature* 382, 1996, 810–3.

C. Simon-Oppermann, D. Gray, E. Szczepanska-Sadowska and E. Simon, Vasopressin in blood and third ventricle CSF of dogs in chronic experiments, *Am. J. Physiol.* 245, 1983, R541–8.

A. Singhal, A.J. Baker, G.M. Hare, F.X. Reinders, L.C. Schlichter and R.J. Moulton, Association between cerebrospinal fluid interleukin-6 concentrations and outcome after severe human traumatic brain injury, *J. Neurotrauma* 19, 2002, 929–37.

D.H. Small and C.A. McLean, Alzheimer's disease and the amyloid β protein: What is the role of amyloid? *J. Neurochem.* 73, 1999, 443–9.

P.S. Sørensen, A. Gjerris and M. Hammer, Cerebrospinal fluid vasopressin in neurological and psychiatric disorders, *J. Neurol., Neurosurg. Psychiatry* 48, 1985, 50–7.

B.R. Southwell, W. Duan, D. Alcorn, C. Brack, S.J. Richardson, J. Köhrle and G. Schreiber, Thyroxine transport to the brain: role of protein synthesis by the choroid plexus, *Endocrinology* 133, 1993, 2116–26.

T.D. Stein and J.A. Johnson, Lack of neurodegeneration in transgenic mice overexpressing mutant amyloid precursor protein is associated with increased levels of transthyretin and the activation of cell survival pathways, *J. Neurosci.* 22, 2002, 7380–8.

K.L. Stenvers, E.M. Zimmermann, M. Gallagher and P.K. Lund, Expression of insulin-like growth factor binding protein-4 and -5 mRNAs in adult rat forebrain, *J. Compar. Neurol.* 339, 1994, 91–105.

E. Storey and R. Cappai, The amyloid precursor protein of Alzheimer's disease and the Aβ peptide, *Neuropath. Appl. Neurobiol.* 25, 1999, 81–97.

J.F. Stover, B. Schöning, T.F. Beyer, C. Woiciechowsky and A.W. Unterberg, Temporal profile of cerebrospinal fluid glutamate, interleukin-6, and tumor necrosis factor-α in relation to brain edema and contusion following controlled cortical impact injury in rats, *Neurosci. Lett.* 288, 2000, 25–8.

J.F. Stover, B. Schöning, O.W. Sakowitz, C. Woiciechowsky and A.W. Unterberg, Effects of tacrolimus on hemispheric water content and cerebrospinal fluid levels of glutamate, hypoxanthine, interleukin-6, and tumor necrosis factor-α following controlled cortical impact injury in rats, *J. Neurosurg.* 94, 2001, 782–7.

J. Strelau, A. Sullivan, M. Böttner, P. Lingor, E. Falkenstein, C. Suter-Crazzolara, D. Galter, J. Jaszai, K. Krieglstein and K. Unsicker, Growth/differentiation factor-15/macrophage inhibitory cytokine-1 is a novel trophic factor for midbrain dopaminergic neurons *in vivo*, *J. Neurosci.* 20, 2000, 8597–603.

J.G. Strugar, G.R. Criscuolo, D. Rothbart and W.N. Harrington, Vascular endothelial growth/permeability factor expression in human glioma specimens: correlation with vasogenic brain edema and tumor-associated cysts, *J. Neurosurg.* 83, 1995, 682–9.

G. Sweeney, Leptin signalling, *Cell. Signall.*14, 2002, 655–63.

E. Szczepanska-Sadowska, D. Gray and C. Simon-Oppermann, Vasopressin in blood and third ventricle CSF during dehydration, thirst, and hemorrhage, *Am. J. Physiol.* 245, 1983, R549–55.

J. Szmydynger-Chodobska, A. Chodobski and C.E. Johanson, Postnatal developmental changes in blood flow to choroid plexuses and cerebral cortex of the rat, *Am. J. Physiol.* 266, 1994, R1488–92.

J. Szmydynger-Chodobska, Z.G. Chun, C.E. Johanson and A. Chodobski, Distribution of fibroblast growth factor receptors and their co-localization with vasopressin in the choroid plexus epithelium, *Neuroreport* 13, 2002, 257–9.

J. Szmydynger-Chodobska, I. Chung, E. Kozniewska, B. Tran, J.F. Harrington, J.A. Duncan and A. Chodobski, Increased expression of vasopressin V_{1a} receptors after traumatic brain injury, *J. Neurotrauma*, 21, 2004, 1090–1102.

T. Tada, M. Kanaji and S. Kobayashi, Induction of communicating hydrocephalus in mice by intrathecal injection of human recombinant transforming growth factor-β1, *J. Neuroimmunol.* 50, 1994, 153–8.

K. Takahashi, F. Satoh, E. Hara, O. Murakami, T. Kumabe, T. Tominaga, T. Kayama, T. Yoshimoto and S. Shibahara, Production and secretion of adrenomedullin by cultured choroid plexus carcinoma cells, *J. Neurochem.* 68, 1997, 726–31.

K. Takaya, Y. Ogawa, N. Isse, T. Okazaki, N. Satoh, H. Masuzaki, K. Mori, N. Tamura, K. Hosoda and K. Nakao, Molecular cloning of rat leptin receptor isoform complementary DNAs—identification of a missense mutation in Zucker fatty (*fa/fa*) rats, *Biochem. Biophys. Res. Commun.* 225, 1996, 75–83.

L.A. Tartaglia, M. Dembski, X. Weng, N. Deng, J. Culpepper, R. Devos, G.J. Richards, L.A. Campfield, F.T. Clark, J. Deeds, C. Muir, S. Sanker, A. Moriarty, K.J. Moore, J.S. Smutko, G.G. Mays, E.A. Woolf, C.A. Monroe and R.I. Tepper, Identification and expression cloning of a leptin receptor, OB-R, *Cell* 83, 1995, 1263–71.

E.M. Terlouw, S. Kent, S. Cremona and R. Dantzer, Effect of intracerebroventricular administration of vasopressin on stress-induced hyperthermia in rats, *Physiol. Behav.* 60, 1996, 417–24.

S.A. Thomas, J.E. Preston, M.R. Wilson, C.L. Farrell and M.B. Segal, Leptin transport at the blood-cerebrospinal fluid barrier using the perfused sheep choroid plexus model, *Brain Res.* 895, 2001, 283–90.

I.P. Trougakos and E.S. Gonos, Clusterin/apolipoprotein J in human aging and cancer, *Int. J. Biochem. Cell Biol.* 34, 2002, 1430–48.

S. Uotani, C. Bjørbæk, J. Tornøe and J.S. Flier, Functional properties of leptin receptor isoforms: internalization and degradation of leptin and ligand-induced receptor downregulation, *Diabetes* 48, 1999, 279–86.

J.L. Venero, M.L. Vizuete, A.A. Ilundáin, A. Machado, M. Echevarria and J. Cano, Detailed localization of aquaporin-4 messenger RNA in the CNS: preferential expression in periventricular organs, *Neurosci.* 94, 1999, 239–50.

I. Vlodavsky, R. Bar-Shavit, R. Ishai-Michaeli, P. Bashkin and Z. Fuks, Extracellular sequestration and release of fibroblast growth factor: a regulatory mechanism? *Trends Biochem. Sci.* 16, 1991, 268–71.

R.J. Walsh, B.I. Posner, B.M. Kopriwa and J.R. Brawer, Prolactin binding sites in the rat brain, *Science* 201, 1978, 1041–3.

R.J. Walsh, F.J. Slaby and B.I. Posner, A receptor-mediated mechanism for the transport of prolactin from blood to cerebrospinal fluid, *Endocrinology* 120, 1987, 1846–50.

H.J. Walter, M. Berry, D.J. Hill, S. Cwyfan-Hughes, J.M. Holly and A. Logan, Distinct sites of insulin-like growth factor (IGF)-II expression and localization in lesioned rat brain: possible roles of IGF binding proteins (IGFBPs) in the mediation of IGF-II activity, *Endocrinology* 140, 1999, 520–32.

H.J. Walter, M. Berry, D.J. Hill and A. Logan, Spatial and temporal changes in the insulin-like growth factor (IGF) axis indicate autocrine/paracrine actions of IGF-I within wounds of the rat brain, *Endocrinology* 138, 1997, 3024–34.

B.C. Wang, G. Flora-Ginter, R.J. Leadley, Jr. and K.L. Goetz, Ventricular receptors stimulate vasopressin release during hemorrhage, *Am. J. Physiol.* 254, 1988, R204–11.

Q. Wang, P. Maloof, H. Wang, E. Fenig, D. Stein, G. Nichols, T.N. Denny, J. Yahalom and R. Wieder, Basic fibroblast growth factor downregulates Bcl-2 and promotes apoptosis in MCF-7 human breast cancer cells, *Exp. Cell Res.* 238, 1998, 177–87.

X. Wang, T.L. Yue, F.C. Barone, R.F. White, R.K. Clark, R.N. Willette, A.C. Sulpizio, N.V. Aiyar, R.R. Ruffolo, Jr. and G.Z. Feuerstein, Discovery of adrenomedullin in rat ischemic cortex and evidence for its role in exacerbating focal brain ischemic damage, *Proc. Natl. Acad. Sci. USA* 92, 1995, 11480–4.

M.Y. Wang, Y.T. Zhou, C.B. Newgard and R.H. Unger, A novel leptin receptor isoform in rat, *FEBS Lett.* 392, 1996, 87–90.

A. Wiedenhöft, C. Müller, R. Stenger, W.F. Blum and C. Fusch, Lack of sex difference in cerebrospinal fluid (CSF) leptin levels and contribution of CSF/plasma ratios to variations in body mass index in children, *J. Clin. Endocrinol. Metab.* 84, 1999, 3021–4.

C.A. Wilson, R.W. Doms and V.M. Lee, Intracellular APP processing and Aβ production in Alzheimer disease, *J. Neuropath. Exp. Neurol.* 58, 1999, 787–94.

Q. Wu, E. Delpire, S.C. Hebert and K. Strange, Functional demonstration of Na^+-K^+-$2Cl^-$ cotransporter activity in isolated, polarized choroid plexus cells, *Am. J. Physiol.* 275, 1998, C1565–72.

T. Wyss-Coray, L. Feng, E. Masliah, M.D. Ruppe, H.S. Lee, S.M. Toggas, E.M. Rockenstein and L. Mucke, Increased central nervous system production of extracellular matrix components and development of hydrocephalus in transgenic mice overexpressing transforming growth factor-β1, *Am. J. Pathol.* 147, 1995, 53–67.

T. Wyss-Coray, C. Lin, F. Yan, G.Q. Yu, M. Rohde, L. McConlogue, E. Masliah and L. Mucke, TGF-β1 promotes microglial amyloid-β clearance and reduces plaque burden in transgenic mice, *Nat. Med.* 7, 2001, 612–8.

T. Wyss-Coray, E. Masliah, M. Mallory, L. McConlogue, K. Johnson-Wood, C. Lin and L. Mucke, Amyloidogenic role of cytokine TGF-β1 in transgenic mice and in Alzheimer's disease, *Nature* 389, 1997, 603–6.

K. Yamada, A. Kinoshita, E. Kohmura, T. Sakaguchi, J. Taguchi, K. Kataoka and T. Hayakawa, Basic fibroblast growth factor prevents thalamic degeneration after cortical infarction, *J. Cereb. Blood Flow Metab.* 11, 1991, 472–8.

Y. Yokoyama, S. Ozawa, Y. Seyama, H. Namiki, Y. Hayashi, K. Kaji, K. Shirama, M. Shioda and K. Kano, Enhancement of apoptosis in developing chick neural retina cells by basic fibroblast growth factor, *J. Neurochem.* 68, 1997, 2212–5.

L. Zabeau, D. Lavens, F. Peelman, S. Eyckerman, J. Vandekerckhove and J. Tavernier, The ins and outs of leptin receptor activation, *FEBS Lett.* 546, 2003, 45–50.

D.A. Zemo and J.T. McCabe, Salt-loading increases vasopressin and vasopressin 1b receptor mRNA in the hypothalamus and choroid plexus, *Neuropeptides* 35, 2001, 181–8.

T. Zeuthen, Secondary active transport of water across ventricular cell membrane of choroid plexus epithelium of *Necturus maculosus*, *J. Physiol. (Lond.)* 444, 1991, 153–73.

G. Zheng, D.R. Bachinsky, I. Stamenkovic, D.K. Strickland, D. Brown, G. Andres and R.T. McCluskey, Organ distribution in rats of two members of the low-density lipoprotein receptor gene family, gp330 and LRP/α2MR, and the receptor-associated protein (RAP), *J. Histochem. Cytochem.* 42, 1994, 531–42.

Y. Zhu, B. Ahlemeyer, E. Bauerbach and J. Krieglstein, TGF-β1 inhibits caspase-3 activation and neuronal apoptosis in rat hippocampal cultures, *Neurochem. Int.* 38, 2001, 227–35.

Y. Zhu, G.Y. Yang, B. Ahlemeyer, L. Pang, X.M. Che, C. Culmsee, S. Klumpp and J. Krieglstein, Transforming growth factor-β1 increases Bad phosphorylation and protects neurons against damage, *J. Neurosci.* 22, 2002, 3898–909.

B.V. Zlokovic, S. Hyman, J.G. McComb, M.N. Lipovac, G. Tang and H. Davson, Kinetics of arginine-vasopressin uptake at the blood-brain barrier, *Biochim. et Biophys. Acta* 1025, 1990, 191–8.

B.V. Zlokovic, S. Jovanovic, W. Miao, S. Samara, S. Verma and C.L. Farrell, Differential regulation of leptin transport by the choroid plexus and blood-brain barrier and high affinity transport systems for entry into hypothalamus and across the blood-cerebrospinal fluid barrier, *Endocrinology* 141, 2000, 1434–41.

B.V. Zlokovic, M.B. Segal, J.G. McComb, S. Hyman, M.H. Weiss and H. Davson, Kinetics of circulating vasopressin uptake by choroid plexus, *Am. J. Physiol.* 260, 1991, F216–24.

M. Zoli and L.F. Agnati, Wiring and volume transmission in the central nervous system: the concept of closed and open synapses, *Prog. Neurobiol.* 49, 1996, 363–80.

M. Zoli, C. Torri, R. Ferrari, A. Jansson, I. Zini, K. Fuxe and L.F. Agnati, The emergence of the volume transmission concept, *Brain Res. Rev.* 26, 1998, 136–47.

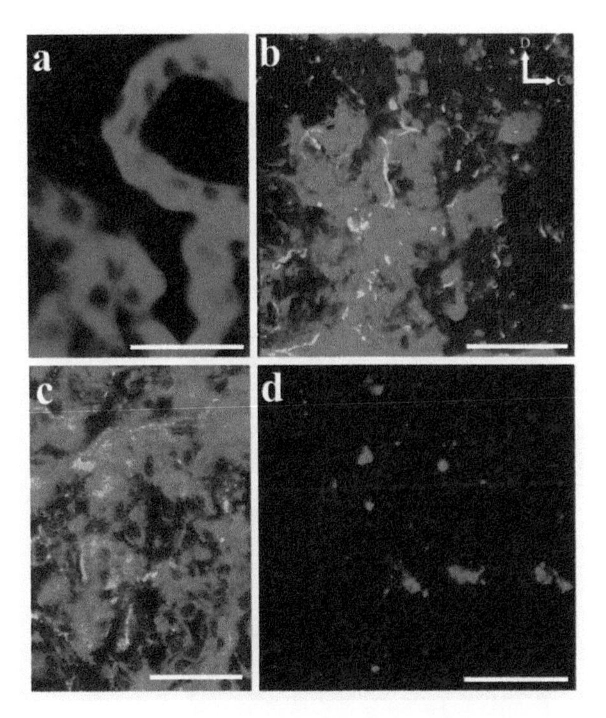

COLOR FIGURE 21.4 Grafting of DiI-labeled CPECs into the rat spinal cord. (a) Only CPECs were exclusively stained by DiI, which had been injected into the 4th ventricle before grafting. (b) Regenerating axons immunostained for neurofilaments (green) are in close contact with grafted CPECs (red). D and C: dorsal and caudal direction, respectively. (c) CGRP fibers (green) intimately interact with CPECs (red). (d) DiI-labeled CPECs were treated repetitively by freezing to kill cellular elements before grafting. No CPECs remained after 7 days in the spinal cord. Some macrophages containing DiI-positive debris in the cytoplasm are seen. Bar, 50 μm. (From Ide et al., 2001.)

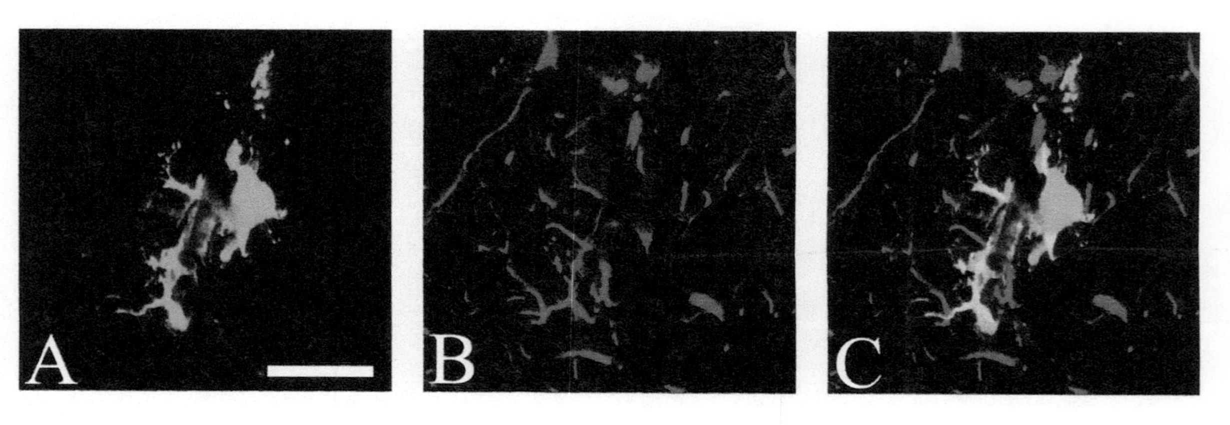

COLOR FIGURE 21.6 CPECs derived from GFP-positive mice were cultured and grafted 1 week before fixation: (A) grafted GFP-positive CPECs; (B) GFAP-immunostaining; (C) CPECs double-positive for GFP and GFAP are yellow-colored in this merged image. Bar, 20 μm. (From Kitada et al., 2001.)

11 Expression of Transthyretin in the Choroid Plexus: Relationship to Brain Homeostasis of Thyroid Hormones

Samantha J. Richardson
The University of Melbourne,
Victoria, Australia

CONTENTS

11.1 TRANSTHYRETIN

This chapter provides an overview of transthyretin (TTR). Most aspects will be addressed in greater detail in relation to the choroid plexus in subsequent sections within this chapter.

TTR was first discovered in human cerebrospinal fluid (CSF) (Kabat et al., 1942a,b), and then later the same year in serum (Siebert and Nielson, 1942). As it was the only protein that migrated ahead of albumin during electrophoresis, it was named "prealbumin," or PA, and found to have a molecular mass of 55 kDa. In those days, electrophoresis was conducted using barbital buffers. It was not known then that barbital inhibits the binding of thyroid hormones to PA, so it was not until 1958 when Ingbar used a Tris-malate buffer to analyze thyroid hormone-binding proteins in serum by electrophoresis that PA was found to be a thyroid hormone distributor protein (Ingbar, 1958). To reflect its function, PA's name was changed to "thyroxine-binding prealbumin," or TBPA. A decade later, the laboratory of Dewitt Goodman discovered that TBPA bound retinol-binding protein (RBP) (Raz and Goodman 1969), and in 1981 its name was changed to "transthyretin" to indicate its roles in the transport of thyroid hormones and retinol-binding protein (Nomenclature Committee of the International Union of Biochemistry, 1981). In 1974 Colin Blake's group published the first x-ray crystal structure of (human) TTR. TTR was found to be a tetramer composed of four identical subunits that come together to form a central channel where thyroid hormones are bound (Blake et al., 1974). There are two potential thyroid hormone-binding sites, but due to negative cooperativity, only one site is occupied under physiological conditions (Wojczak et al., 2001). A high-resolution human TTR x-ray crystal structure was the first x-ray crystal structure of a protein from which its amino acid sequence could be determined (Blake et al., 1978). The x-ray crystal structures for TTRs from rat (Wojczak, 1997), chicken (Sunde et al., 1996) (see Figure 11.1) and fish (sea bream, *Sparus aurata*) (Folli, 2003) have also been determined. The conservation of the 3D structure of TTR from fish to human is extremely high, as is the amino acid sequence (see Sec. 11.3). The laboratory of Hugo Monaco published the structure of the TTR-RBP complex in 1995 (Monaco et al., 1995) and clarified that up to two molecules of RBP can bind TTR's external surface, i.e., sites independent of the thyroid hormone-binding sites.

In humans, the major sites of TTR synthesis are the liver, the choroid plexus, and the retinal pigment epithelium. TTR synthesized by the liver is secreted into the blood (see Schreiber, 1987) where it acts as a thyroid hormone distributor protein and binds RBP, thereby also being involved in retinol distribution. TTR synthesized by the choroid plexus is secreted into the cerebrospinal fluid (CSF) (Schreiber et al., 1990) and is involved with the transport of thyroid hormone from the blood into the brain (Dickson et al., 1987; Schreiber et al., 1990; Southwell et al., 1993). TTR also acts as a thyroid hormone distributor protein in the cerebrospinal fluid (see Schreiber, 2002). Studies of the TTR null mouse have raised questions about the function of TTR synthesized and secreted by the choroid plexus (Palha, 2002). Data on TTR and RBP synthesized by the retinal pigment epithelium has come mainly from the laboratory of Jo Herbert (e.g., Dwork et al., 1990; Herbert et al., 1991; Mizuno et al., 1992). It has been suggested that the synthesis of TTR and RBP by the retinal pigment epithelium of the eye is involved in the delivery of retinol to amacrine and Muller cells (Ong et al., 1994). Retinol is then converted to retinal, which is required for the normal function of photoreceptors (Bridges et al., 1984). However, studies of TTR null mice suggest that additional routes of retinol delivery

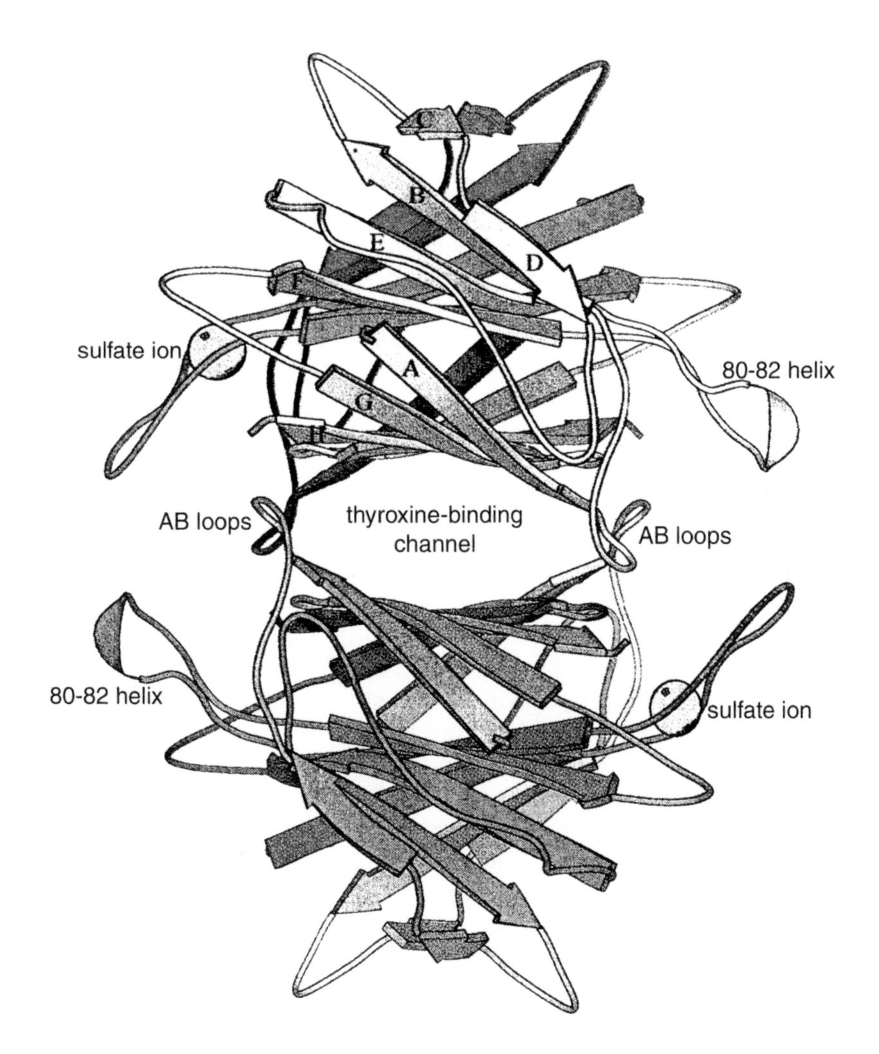

FIGURE 11.1 X-ray crystal structure of chicken TTR. (From Sunde et al., 1996.)

to the eye other than the classical retinol/RBP/TTR complex must exist (Wei et al., 1995; Bui et al., 2001).

In humans, TTR is a negative acute phase plasma protein, i.e., during inflammation, surgery, or trauma, the rate of TTR mRNA synthesis is decreased in the liver (but not in the choroid plexus) (Dickson et al., 1986). For this reason, TTR is used as an indicator of postoperative recovery and of nutritional status (Ingenbleek and Young, 1994).

TTR is usually a very stable protein, both in vitro and in vivo. However, it can form amyloid fibrils naturally in vivo, and can be induced to form amyloid in vitro (Colon and Kelly, 1992). There are two forms of amyloid, and TTR can form both types in humans. Familial amyloidotic polyneuropathy (FAP) is a specific form of autosomal dominant hereditary polyneuropathy, which initially manifests as systemic deposition of amyloid involving the peripheral nerves, but later affects many visceral organs. At least 80 of the 127 amino acids in the TTR subunit have point mutations resulting in FAP (Connors et al., 2000). Senile systemic amyloidosis (SSA) is age-dependent, and the TTR fibrils are formed from wild-type protein. Studies have shown that 64% of people over 70 years of age have TTR SSA (see Calkins, 1974).

TTR also binds a variety of drugs and pollutants, including the nonsteroidal anti-inflammatory drugs (NSAIDs), lead, plant flavonoids, and industrial pollutants (Köhrle et al., 1988; Munro et al., 1989; Craik et al., 1996; Zheng et al., 2003). This has major implications for unnatural compounds competing with thyroid hormones for binding to TTR, crossing the blood–CSF barrier, and subsequent toxicity and disease. By contrast, Kelly and colleagues have published a paper demonstrating that binding of NSAIDs to TTR confers increased stability and protection against TTR amyloid formation in vitro (Miller et al., 2004).

11.2 THYROID HORMONES AND THEIR EFFECTS ON THE BRAIN

Thyroid hormones play an important role in brain development in vertebrates. Iodine is essential for the synthesis of thyroid hormones and is derived solely from the diet. In human fetuses, insufficient thyroid hormones during development can lead to irreversible brain damage resulting from a multitude of abnormalities including stunting of dendritic growth, reduced deposition of myelin, a reduction in the number of synapses, decrease in the number of dendritic spines of pyramidal cells in the hippocampus and the neocortex, and death of granule cells (Jacobsen, 1978; Dussault and Ruel, 1987; Morreale de Escobar et al., 1987; Bernal et al., 2003). Endemic cretinism is when an abnormally high number of people in a population have irreversible physical and intellectual developmental abnormalities due to hypothyroidism. This type of hypothyroidism is usually due to insufficient iodine in the diet.

The two main forms of thyroid hormones are 3′,5′,3,5-tetraiodo-L-thyronine (T4) and 3′,3,5-triiodo-L-thyronine (T3) (see Figure 11.2). They are lipophilic compounds that partition between the lipid phase and the aqueous phase at 20,000:1 (Dickson et al., 1987). Therefore, they are bound to specific plasma proteins to ensure an even distribution throughout tissues and that a pool of adequate size circulates in the blood and is kept within the CSF (Mendel et al., 1987; Schreiber and Richardson, 1997). In human blood, the concentration of T4 is 100 nM, and of this, 26 pM is free in solution and 99.97% is bound to protein. The concentration of T3 is 2.2 nM, with 9 nM being free in solution and 99.70% bound to plasma proteins (Mendel, 1989).

In humans, the plasma proteins that bind and distribute thyroid hormones in the blood are albumin, TTR, and thyroxine-binding globulin (TBG). As their function is not simply the binding of thyroid hormones, but to ensure appropriate distribution of thyroid hormones, they should be referred to as "thyroid hormone distributor proteins." Of these, TBG has the highest affinity for T4 and T3 (1.0×10^{10} M^{-1} and

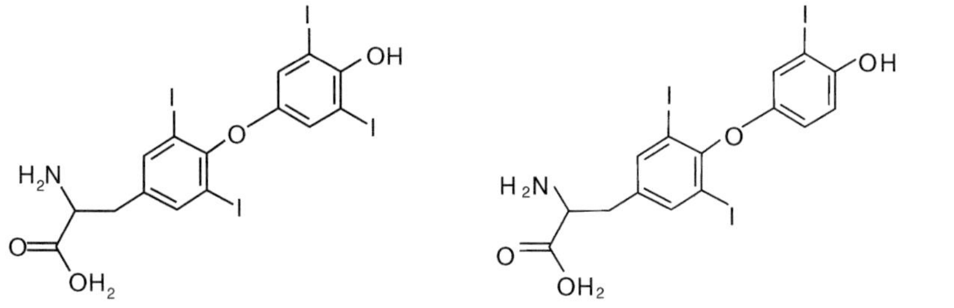

FIGURE 11.2 The structure of 3′,5′,3,5-tetraiodo-L-thyronine (T4) and 3′,3,5-triiodo-L-thyronine (T3).

4.6×10^8 M^{-1}, respectively), TTR has intermediate affinity (7.0×10^7 M^{-1} and 1.4×10^7 M^{-1}, respectively), and albumin has the lowest affinity (7.0×10^5 M^{-1} and 1.0×10^5 M^{-1}, respectively). TBG is present in the lowest concentration in blood (0.015 g/L), TTR has an intermediate concentration (0.25 g/L), and albumin has the highest concentration (42 g/L). Overall, TBG distributes about 75% of thyroid hormones, TTR 15%, and albumin transports about 10% of vascular thyroid hormones (Robbins, 2000). Because TBG binds about 75%, it is sometimes referred to as the most important thyroid hormone distributor. However, this is too simplistic, as the biological importance is related to the delivery of thyroid hormones to cells, which is dependent on the dissociation rates of T3 and T4 from the distributor proteins.

The Free Hormone Transport Hypothesis states that the hormone enters tissue from the bloodstream via the pool of free hormone, following spontaneous dissociation from binding proteins within the capillaries of the tissue. Therefore, uptake of a hormone into a tissue cannot exceed its dissociation rate from the distributor proteins. The Free Hormone Hypothesis states that it is the concentration of free hormone in the blood, not of the protein-bound hormone, that is important for biological activity (see Robbins and Rall, 1957). Both hypotheses hold true for thyroid hormones (Mendel, 1989).

To determine which of the three thyroid hormone distributor proteins contributes most effectively to hormone delivery to tissues, we need to examine the dissociation rates and the capillary transit times. In brief, the dissociation rates for T4 and T3 from TBG are 0.018 sec^{-1} and 0.16 sec^{-1}, respectively, from TTR are 0.094 sec^{-1} and 0.69 sec^{-1}, respectively, and from albumin are 1.3 sec^{-1} and 2.2 sec^{-1}, respectively (Mendel and Weisiger, 1990). Thus, given the capillary transit times for various tissues (see Mendel et al., 1989), TTR is responsible for much of the immediate delivery of thyroid hormones to tissues (Robbins, 2000). According to Ingbar (1958): "TBG is the savings account for thyroxine and TBPA is the checking account."

The main form of thyroid hormone in mammalian blood is T4, and the thyroid hormone distributor proteins in mammals have higher affinity for T4 than T3 (see above). Consequently, the idiom has arisen that T4 is the "transport" form of the hormone, and that T3 is the "active" form of the hormone, since the thyroid hormone receptors have higher affinity for T3 than T4. However, this does not hold true for nonmammalian species (see Richardson, 2002) (see also Sec. 11.6).

In mammals, T4 is the principal form of the hormone, which dissociates from the thyroid hormone distributor proteins and partitions into cell membranes. There, a family of enzymes, deiodinases, can either activate T4 by removing an iodine atom from the outer ring to produce T3 or can inactivate T4 by removing an iodine atom from the inner ring to produce rT3. For a review of deiodinases and their tissue specificity, see Köhrle (2002). Within the nucleus, T3 binds a thyroid hormone receptor, which then dimerizes. Dimers can then accumulate co-activator or co-repressor proteins, bind to specific thyroid hormone response elements, and positively or negatively regulate transcription of specific genes. For normal development to progress, it is crucial that the timing and level of expression of genes are tightly regulated.

Many genes in the brain are regulated by thyroid hormones. Some of these genes are under the control of only thyroid hormones during brain development. Examples are calbindin, myo-inositol-1,4,5-triphosphate (IP3) receptor, and Purkinje cell protein 2 (PCP-2), all of which are involved in Purkinje cell development and differentiation in the cerebellum (Strait et al., 1992). Another example is myelin basic protein, which is responsible for the myelination of oligodendrocytes throughout the brain (Farsetti et al., 1991). Under conditions of hypothyroidism, these and many other proteins have their respective mRNA levels reduced, and brain development does not proceed normally. Strait and co-workers quantitated the mRNA levels for calbindin, IP3 receptor, PCP-2, and myelin basic protein for the first few months after birth of hypothyroid mice. By about day 60, the levels of mRNA for these proteins had reached euthyroid levels (Strait et al., 1992).

There are several studies linking depression in adults with restricted central nervous system (CNS) hypothyroidism associated with reduced levels of TTR in CSF, although in the periphery the patients are euthyroid (Hatterer et al., 1993; Sullivan et al., 1999). The molecular basis for thyroid hormone-related depression is not yet understood, but it is thought to be related to regulation of expression of thyroid hormone-responsive genes in the brain. There is also growing evidence for behavioral responses due to the presence of unliganded receptors (Bernal et al., 2003).

Hypothyroidism in adult humans can manifest in several ways, including development of a goiter, fatigue, mental impairment, depression, and weight gain (for review, see Braverman and Utiger, 2000). While administration of thyroxine can restore thyroid hormone, TRH, and TSH concentrations to euthyroid levels, the interactions of molecules and changes in regulations of pathways (particularly in the brain) from decreased free thyroid hormone levels in the blood to the wide range of hypothyroid symptoms are not yet fully understood.

11.3 STRUCTURE AND REGULATION OF THE TRANSTHYRETIN GENE IN THE CHOROID PLEXUS

The human TTR gene was first cloned by Tsuzuki et al. in 1985 and found to consist of four exons with three intervening introns. Today, TTR amino acid sequences have been determined directly or derived from cDNA sequences from 20 species including fish [sea bream, *Sparus aurata* (Santos and Power, 1999)], amphibians [bullfrog, *Rana catesbeiana* (Yamauchi et al., 1998); clawed African toad, *Xenopus laevis* (Prapunopj et al., 2000b)], reptiles [stumpy-tailed lizard, *Tiliqua rugosa* (Achen et al., 1993); salt water crocodile, *Crocodylus porosus* (Prapunpoj et al., 2002)], a bird [chicken, *Gallus gallus* (Duan et al., 1991)], marsupials [stripe-faced dunnart, *Sminthopsis macroura*; American short-tailed gray opossum, *Monodelphis domestica*; sugar glider, *Petaurus breviceps* (Duan et al., 1995a); tammar wallaby, *Macropus eugenii* (Brack et al., 1995); Eastern gray kangaroo, *Macropus giganteus* (Aldred et al., 1997)], and eutherians [human, *Homo sapiens* (Tsuzuki et al., 1985); rat, *Rattus norvegicus* (Dickson et al., 1985; Duan et al., 1989); mouse, *Mus musculus* (Wakasugi et al., 1985); hedgehog, *Erinaceus europaeus*; shrew, *Sorex ornatus* (Prapunopj et al., 2000a); pig, *Sus scrofa* (Duan et al., 1995b); cow, *Bos taurus* (Irikara et al., 1997); sheep, *Ovis aries* (Tu et al., 1989); rabbit, *Oryctolagus cuniculus* (Sundelin et al., 1985)]. The alignment of these TTR amino acid sequences demonstrates the high level of conservation throughout the evolution of vertebrates (see Figure 11.3).

The sequence of rat TTR cDNA from a choroid plexus library was found to be identical to that from a rat liver library (Duan et al., 1989). Together with data from Southern analyses (Fung et al., 1988), it was concluded that there was only one TTR gene per haploid genome, which was expressed in both liver and choroid plexus (Duan et al., 1989). This has been subsequently confirmed for human and mouse TTR genes by scanning the respective genome databases.

The exact positions of all three introns (by comparison of genomic with cDNA sequences) are known only for TTR genes from human (Mita et al., 1984; Tsuzuki et al., 1985) and mouse (Wakasugi et al., 1985, 1986), and are in identical positions for both species, i.e., intron 1 is between amino acids 3 and 4, intron 2 in between amino acids 41 and 42, and intron 3 is between amino acids 92 and 93. (Amino acid numbering from N-terminus, as for human TTR.)

During a series of studies on the evolution of transthyretin, in order to confirm the position of the N-terminus of the TTR subunit suggested by the sequencing of the cDNAs, several TTRs were purified from serum, and 8 to 10 amino acids from the N-termini were sequenced by Edman degradation (Duan et al., 1995a). It quickly became apparent that TTRs from eutherian species had shorter and more hydrophilic N-terminal regions, whereas TTRs from reptiles and birds had longer N-terminal regions which were more hydrophobic in character (Duan et al., 1995a). This was found to be due to the stepwise shift in the intron 1/exon 2 border in the 3′ direction, effectively moving sequences from the 5′ end of exon 2 into the 3′ end of exon 1

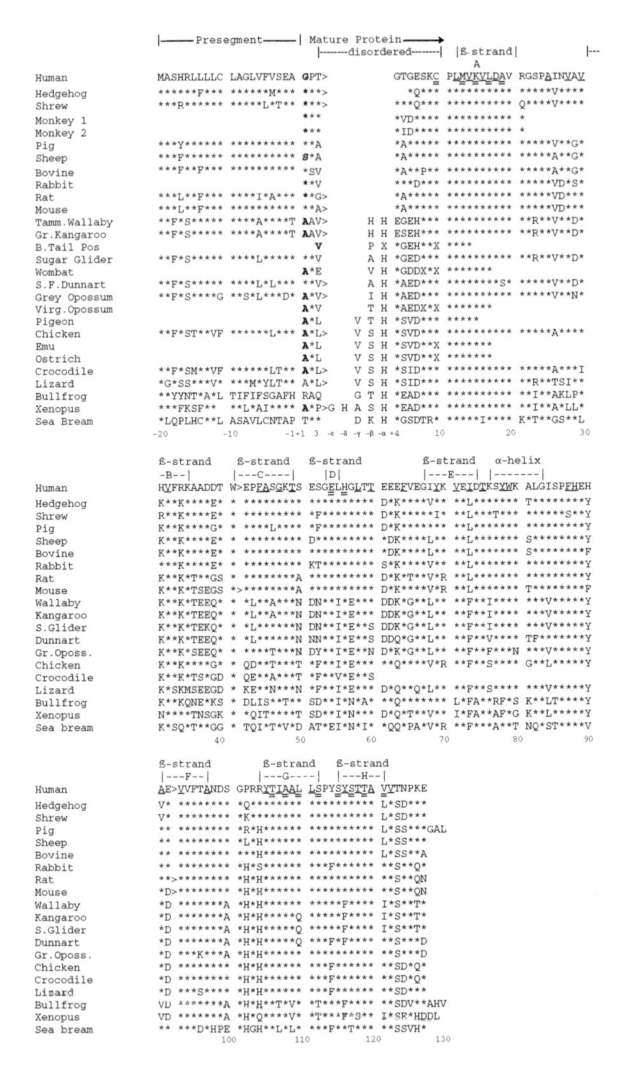

FIGURE 11.3 Alignment of 20 complete and 8 N-terminal TTR sequences. Structural features for human TTR are indicated above the sequences; single underlined residues are in the core of the subunit; double underlined residues are in the binding site; numbering is as for human TTR, and -α to -ϵ are introduced for additional residues in the N-terminal regions of marsupial, avian, reptilian, amphibian, and piscine sequences; asterisks mean the amino acid is identical to that in human; gaps were introduced to aid alignment; The N-terminal amino acid is highlighted in bold, for sequences where this is known; > indicates exon/exon boundary; X indicates amino acid not resolved by Edman degradation. Sources of complete sequences are stated in the text. Partial sequences: rhesus monkey, *Macaca mulatto* (van Jaarsveld et al., 1973); brushtail possum, *Trichosurus caninus*; Southern hairy-nosed wombat, *Lasiorhinus latifrons* (Richardson et al., 1994); Virginia opossum, *Didelphis virginiana* (Richardson et al., 1996); pigeon, *Columba livia*; emu, *Dromaius novaehollandiae*; ostrich, *Struthio camelus*. (From Chang et al., 1999.)

(Aldred et al., 1997). At the protein level, this has the effect of shortening the N-terminal region of each of the four TTR subunits. This is illustrated in Figure 11.4. The biological consequences for the choroid plexus of the shortening of the N-termini of TTRs during evolution will be discussed below in section 6.

TTR has long been known to be a "negative acute phase plasma protein," meaning that following trauma, surgery, inflammation, or malnutrition the TTR gene in the liver is downregulated and protein levels in blood become dramatically reduced (see Schreiber and Howlett, 1983). As there is only one TTR gene per haploid genome in rats, it was investigated whether the TTR gene in the choroid plexus was also under negative acute phase control. Surprisingly, the TTR gene in the choroid plexus was found not to be under negative acute phase regulation, but to be regulated independently from the TTR gene in the liver (Dickson et al., 1986). It was speculated that if TTR synthesis was involved with transporting thyroid hormones into the brain, given the dependence of brain development on thyroid hormones and the sensitivity of the adult brain to the effects of thyroid hormones, it would make biological sense for the brain to be protected and to continue to receive the usual supply of thyroid hormones even when the body is experiencing trauma or inflammation (Dickson et al., 1986).

Costa and co-workers discovered the vast majority of what is known about the regulation of TTR gene transcription in liver, choroid plexus, and visceral yolk sac of rats. In 1990 they found that the start site for mRNA synthesis was identical for the TTR genes in the liver and the choroid plexus and that both genes had a distal enhancer sequence at −1.86 to −1.96 kbp and a promoter-proximal region at −70 to −200 bp from the mRNA cap site. Most intriguingly, they found that, for choroid plexus specific TTR gene expression, positive element(s) within the region 3 kbp upstream of the sequence were required (but not required for TTR gene expression in the liver) (Tan et al., 1990). In the same year, they reported identification of four nuclear transcription factors (HNF-1, HNF-3, HNF-4, and C/EBP) involved in the regulation of the TTR gene in the liver, but that the same four did not regulate the TTR gene in the choroid plexus or the visceral yolk sac. This clearly demonstrated that the differential cellular distribution of positively acting transcription factors was responsible for tissue-specific expression of the TTR gene (Costa et al., 1990).

Costa and Grayson showed that the enhancer sequence and the HNF-3 strong site were essential for TTR gene expression and that participation by other factors was also necessary (Costa and Grayson, 1991). It was then discovered that the decrease in hepatic TTR gene expression during the acute phase response was due to the downregulation of HNF-3α (Qian et al., 1995) and HNF-3β (Samadani et al., 1996). They also demonstrated that adenovirus-mediated increase of HNF-3α stimulated TTR gene expression in cell culture (Tan et al., 2001). This explains the findings of Dickson et al. (1986), who found the TTR gene in the liver, but not in the choroid plexus, to be under negative acute phase regulation.

The identification of the element(s) in the 3 kbp region upstream of the TTR gene, which is required for expression of the gene in the choroid plexus, remains to be elucidated.

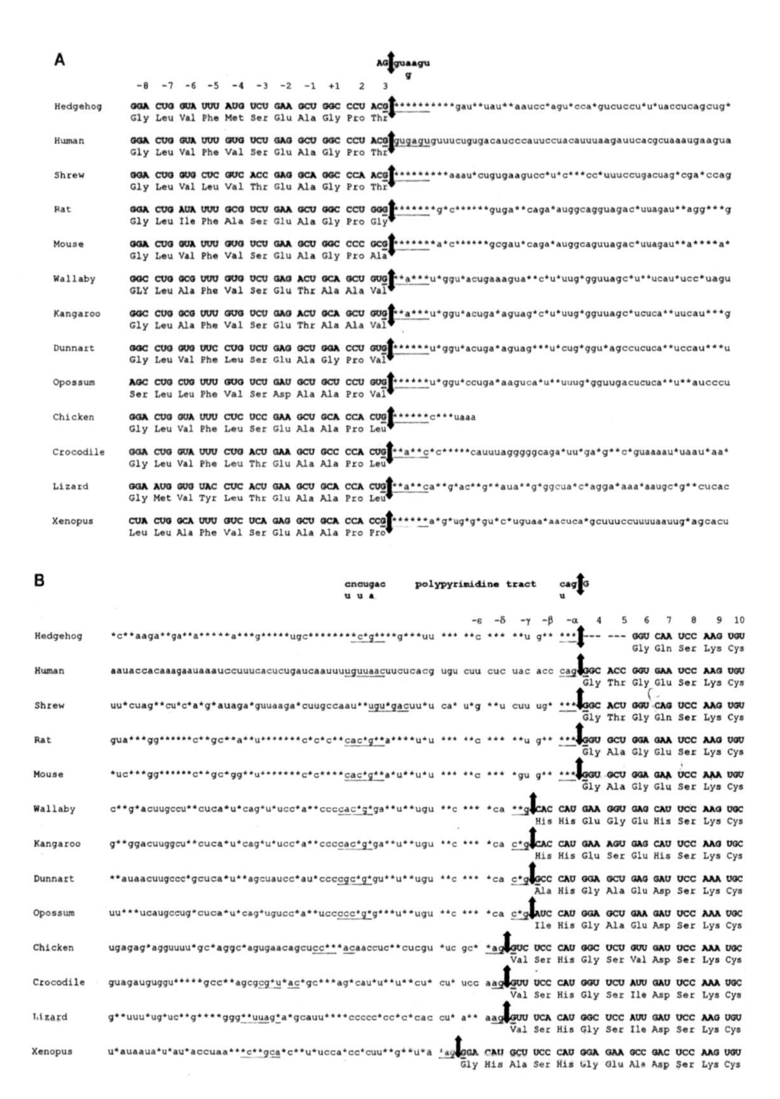

FIGURE 11.4 Alignment of (A) exon 1 / intron 1 splice sites and of (B) intron 1 / exon 2 splice sites, demonstrating the shift of the latter in the 3' direction during evolution of the TTR gene in vertebrates. Birds and reptiles consequently have longer, more hydrophobic TTR N-terminal regions, whereas eutherian species have shorter, more hydrophilic TTR N-terminal regions. Nucleotide sequences on either side of the intron 1 precursor mRNA are shown. Consensus splice site recognition sequences are given above the sequences, and identical sequences in the introns are underlined. The splice sites are indicated by double-headed arrows. The nucleotide sequences in the exons, which are translated, are indicated in bold. The corresponding amino acids are written below them. Asterisks indicate that the base is identical to that in the human sequence. (From Prapunpoj et al., 2002.)

11.4 TRANSTHYRETIN GENE EXPRESSION IN THE CHOROID PLEXUS DURING DEVELOPMENT

The choroid plexus has the highest concentration of TTR mRNA in the body. The adult rat choroid plexus has 4.4 μg TTR mRNA per gram wet weight tissue compared to only 0.39 μg TTR mRNA per gram wet weight liver, i.e., an 11-fold difference (Schreiber et al., 1990). In chickens, the adult choroid plexus has 7.2 μg TTR mRNA per gram wet weight, whereas the liver has only 0.33 μg TTR mRNA per gram wet weight, i.e., a 22-fold difference (Duan et al., 1991). In reptiles, birds, and mammals, TTR is the major protein secreted by the choroid plexus (Harms et al., 1991; Richardson et al., 1994).

TTR is synthesized specifically by the cuboidal epithelial cells of the choroid plexus (Stauder et al., 1986; Thomas et al., 1988). Cells immediately adjacent to the choroid plexus epithelial cells do not have TTR mRNA transcripts detected.

In rats, TTR mRNA was detected in the choroid plexus primordia initially in the fourth ventricle at E12.5, then in the lateral ventricles at E13.5, and finally in the third ventricle by E17.5 (in this strain of rats, gestation is 24 days). In situ hybridization revealed that TTR mRNA was only present in the choroid plexus epithelial cells, and not in the adjacent mesenchymal tissue or the neural tissue, which is continuous with the choroid plexus epithelium. Although cells expressing the TTR gene can be readily identified by in situ hybridization, these cells cannot be distinguished morphologically within the neuroepithelium (Thomas et al., 1988). In the fetal brain it is only a small population of cells that synthesize TTR (Fung et al., 1988).

In rats, the proportion of TTR mRNA in total RNA increased 40-fold from E12.5 until birth (Fung et al., 1988) (see Figure 11.5, upper panel). Because the choroid plexus develops faster than many other parts of the brain (Sturrock, 1979), part of this 40-fold increase in TTR mRNA could have been due to the increase in size of the choroid plexus with respect to the size of the brain. The maximum of TTR mRNA as a proportion of total brain mRNA (about 140% level in adults) occurred 2 days prior to birth, just before the period of fastest brain growth. By 8 days after birth, TTR mRNA levels decreased to adult levels (Fung et al., 1988).

Most animals can be described as either precocial (e.g., chickens, sheep) or altricial (e.g., rats, mice). In precocial animals, the brain is significantly further developed at birth compared to the brains of altricial animals at birth. Therefore, it is interesting to compare TTR gene expression in the choroid plexus of altricial animals with that in precocial animals.

In sheep (precocial), TTR mRNA was detected in choroid plexus from embryos with a crown-rump length of 1 cm, corresponding to just a few days gestation (Tu et al., 1990). Similarly to the situation in rats, the maximum increase in size of the choroid plexus (E70) occurred before that of the brain (E105) (total gestation is 155 days). At E40 the proportion of TTR mRNA in total mRNA was 34% of that found for adult choroid plexus. By E90, this had increased to 70% of the adult value and remained constant throughout the rest of gestation (Tu et al., 1990) (see Figure 11.5, lower panel).

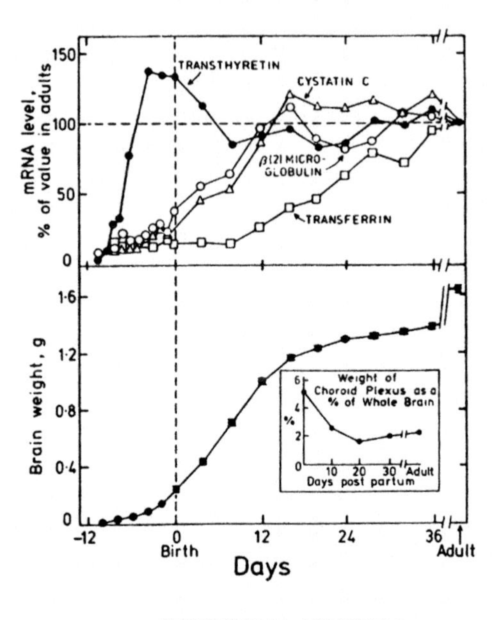

PRECOCIAL (SHEEP)

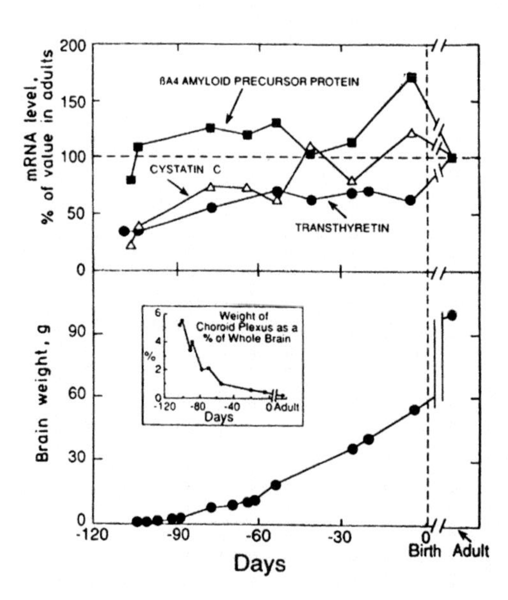

FIGURE 11.5 Comparison of growth of brains and patterns of mRNA levels for TTR, transferrin, cystatin C, β2-microtubulin and βA4-amyloid precursor protein during development in altricial (rat) and precocial (sheep) animals. mRNA levels are based on quantitation of Northern analyses from Fung et al., 1988, and Tu et al., 1990. (From Schreiber and Aldred, 1993.)

In rats, which are altricial, the maximum increase in brain weight was about 9 days after birth (Fung et al., 1988), whereas that for precocial sheep was 50 days before birth (Tu et al., 1990). The ratio of choroid plexus weight to total brain weight in rats decreased to a stable value about 15 days after birth (Fung et al., 1988), whereas that for sheep occurred about 20 days before birth (Tu et al., 1990). For both species, the choroid plexus had its maximal growth rate prior to that of the rest of the brain.

A study of TTR mRNA levels in choroid plexus of chickens (precocial) during development gave similar results to those obtained for sheep (precocial). Furthermore, low levels of TTR mRNA were found to be present in the choroid plexus Anlage cells on the day that thyroid hormones first appeared in the serum, and the levels of TTR mRNA in the choroid plexus increased rapidly at the same time as the thyroid hormone levels in serum increased (Southwell et al., 1991).

The critical period of brain development that is dependent on thyroid hormones is later for rats (altricial) than for sheep (precocial) (Fisher and Polk, 1989). The maximum TTR mRNA level in the choroid plexus of rats occurs 2 days prior to birth, whereas in sheep it occurs during the first half of gestation. If TTR is involved with transporting thyroid hormone across the blood–CSF barrier into the brain and acts as a thyroid hormone distributor protein in the CSF, in each of these species the timing of the maximal expression of the TTR gene is highly appropriate.

Given that (1) thyroid hormones have a profound effect on brain development, (2) the blood–brain barrier starts to develop as early as the first blood vessel grows into the brain (Saunders et al., 1999), (3) the choroid plexus develops more rapidly than other parts of the brain, (4) the choroid plexus presumably has an important function in regulating the composition of the CSF, and (5) the timing of maximal TTR mRNA in the choroid plexus of altricial and precocial animals is just prior to the maximal growth rate of the brain, it would follow that the synthesis of TTR from the stage of the choroid plexus primordia would appear to have an important regulatory role in determining the access of thyroid hormones to the brain.

11.5 EVOLUTION OF TRANSTHYRETIN SYNTHESIS BY THE CHOROID PLEXUS

To analyze the proteins synthesized and secreted by the choroid plexus, the choroid plexus was dissected from the lateral, third, and fourth ventricles of freshly killed animals and incubated in culture medium that contained ^{14}C-leucine. Thus, proteins that were newly synthesized by the choroid plexus in culture incorporated the radioactive label, and those that were secreted by the choroid plexus could be analyzed by sampling the culture medium. Proteins originating from the blood were not radioactively labeled and therefore not detected by autoradiography following separation of the proteins in the medium by SDS-PAGE.

This was done for a range of species, including representatives of fish, amphibians, reptiles, birds, monotremes, marsupials, and eutherians. (Monotremes are mammals that lay eggs, e.g., the Australian platypus and echidnas. Marsupials are

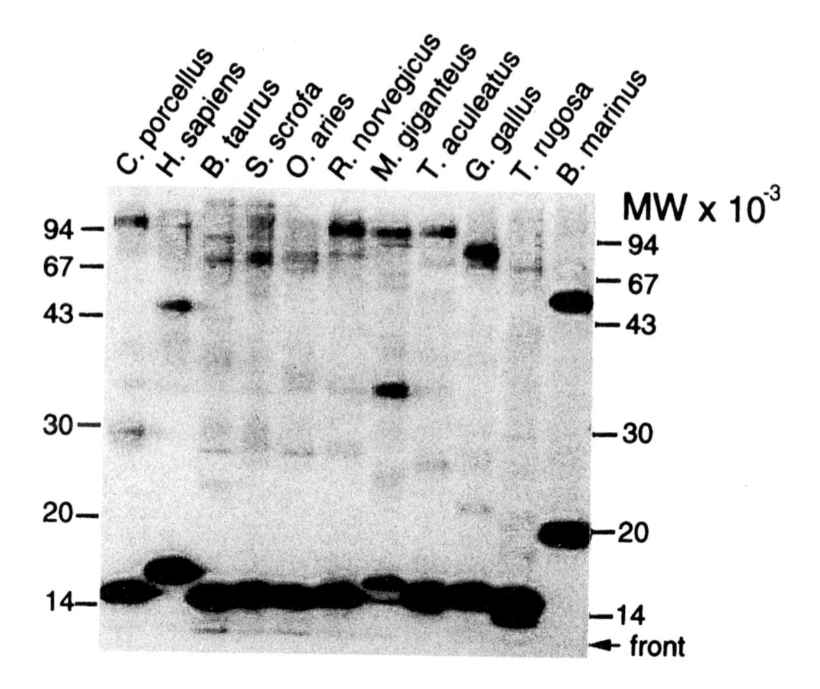

FIGURE 11.6 Autoradiograph following SDS-PAGE analysis of media (containing ^{14}C-leucine) in which choroid plexus were incubated. The major protein synthesized and secreted by choroid plexus from eutherians, marsupials, monotremes, birds, and reptiles is TTR. The major protein synthesized and secreted by amphibian choroid plexus is the lipocalin prostaglandin D synthetase. Positions of molecular weight markers are indicated. The human sample was from a choroid plexus papilloma, which had been removed from an 18-month-old child. *Cavia porcellus*: guinea pig; *Bos taurus*: cow; *Sus scrofa*: pig; *Ovis aries*: sheep; *Rattus norvegicus*: rat; *Macropus giganteus*: Eastern gray kangaroo; *Tachyglossus aculeatus*: short-beaked echidna; *Gallus gallus*: domestic chicken; *Tiliqua rugosa*: stumpy-tailed lizard; *Bufo marinus*: cane toad. (From Richardson et al., 1994.)

mammals that give birth to very altricial young after a relatively short period of gestation. These young spend a period permanently attached to their mothers' teats and undergo a relatively long period of lactation, e.g., American opossums, Australian kangaroos, koalas, and Tasmanian devils. Eutherians are sometimes referred to as "placental mammals," even though this is technically wrong, as all mammals have placentas. Examples of eutherians are humans, whales, and bats.)

TTR was found to be the major protein synthesized and secreted by the choroid plexus of reptiles, birds, monotremes, marsupials, and eutherians, but was not detected in choroid plexus incubation media from amphibians (Harms et al., 1991; Achen et al., 1993; Richardson et al., 1994; Duan et al., 1995a,b; Richardson et al., 1997; Prapunopj et al., 2002) or fish (G. Schreiber, unpublished observations). Examples are shown in Figure 11.6. It appears that the TTR gene in the choroid plexus was turned on once at the stage of the stem-reptiles, the closest common ancestor to reptiles, birds, and mammals. The early reptiles were the first to develop traces of a cerebral neocortex (Kent, 1987), thereby increasing the brain volume. As thyroid hormones are lipophilic and readily partition into cell membranes, the increase in brain size may have been the selection pressure for turning on the TTR gene in the choroid plexus, resulting in TTR assisting in moving thyroid hormones from the blood across the blood–CSF barrier into the brain, and also acting as a thyroid hormone distributor protein in the CSF (Schreiber and Richardson, 1997).

In modern reptiles, amphibians, and fish, TTR is synthesized by the liver during development only (Richardson et al., 2004), whereas homeothermic animals synthesize TTR in their livers throughout life (Richardson et al., 1994). It is most probable that the stem-reptiles had the TTR gene in their genomes, possibly expressed in the liver during development; then a change in specificity of transcription factors could have been all that was required to achieve TTR synthesis in the choroid plexus.

In adult rats, TTR constitutes about 20% of the protein synthesized by the choroid plexus and 50% of the protein secreted by the choroid plexus (Dickson et al., 1986).

The major protein synthesized and secreted by the choroid plexus of amphibians was a lipocalin, most probably prostaglandin D synthetase (PGDS) (Achen et al., 1992), also known as beta-trace (Beuckmann et al., 1999) and Cpl1 (Lepperdinger, 2000) (see Figure 11.6). PGDS is a monomeric 20 kDa protein, which belongs to the lipocalin superfamily of proteins because it forms a calyx. Lipocalins are specialized in binding small molecules (Godovac-Zimmerman, 1988). This raises the question as to whether this lipocalin is the functional precursor to transthyretin in the choroid plexus.

11.6 SCHREIBER'S MODEL OF TTR SYNTHESIZED BY THE CHOROID PLEXUS TRANSPORTING THYROID HORMONES INTO THE BRAIN AND ACTING AS A THYROID HORMONE DISTRIBUTOR PROTEIN IN THE CSF

The first paper reporting TTR mRNA in the choroid plexus but nowhere else in the brain immediately raised the suggestion that TTR synthesized by the choroid plexus could be involved in thyroid hormone transport from blood to brain via the choroid plexus (Dickson et al., 1985). Goodman's group published a paper on TTR mRNA in the brain the same year. They had cut a brain into five regions and found TTR mRNA in all but one region. The choroid plexus was not identified as the site of synthesis, nor was a function for TTR synthesized by the brain suggested (Soprano et al., 1985).

Schreiber's group then went on to demonstrate that TTR synthesis by the choroid plexus was not under negative acute phase regulation (Dickson et al., 1986) and that only the epithelial cells of the choroid plexus synthesize TTR (Stauder et al., 1986) (see Figure 11.7).

The first data demonstrating TTR's possible role in transporting thyroid hormones from blood into the brain were presented by Dickson et al. (1987). Freshly dissected choroid plexus accumulated ^{125}I-T4 and ^{125}I-T3 from surrounding medium

FIGURE 11.7 In situ hybridization followed by autoradiography of choroid plexus with ^{35}S TTR cDNA probe showing silver grains lying only over epithelial cells, demonstrating that only the epithelial cells of the choroid plexus synthesize TTR (From Stauder et al., 1986.)

in a nonsaturable process, and fluorescence-quenching experiments demonstrated that maximal accumulation of thyroid hormone was in the middle of a membrane bilayer. ^{125}I-T4 injected into rats accumulated in the liver and choroid plexus only, whereas ^{125}I-T3 did not accumulate in any specific tissue. Kinetics of intravenously injected ^{125}I-T4 uptake into the brain showed an initial accumulation in the choroid plexus, followed by the striatum, then in the cortex, and finally into the cerebellum (see Figure 11.8). This did not occur following injection of ^{125}I-T3.

The model offered by the authors was the following: T4 in blood partitions into the choroid plexus. De novo synthesized TTR binds the T4 at a site not yet defined (intracellular, near membranes, or extracellular), and the TTR-T4 complex is secreted into the CSF and swept away by the CSF to other parts of the brain. The authors were careful to state that this would not be the sole site of thyroid hormone entry into the brain. That ^{125}I-T4 and not ^{125}I-T3 showed these patterns was explained as being either due to TTR's higher affinity for T4 than T3, or because the molar concentration of free T4 is 7-fold higher than that for free T3 in rats.

The developmental profiles of TTR synthesis by the choroid plexus (discussed in Sec. 11.4) also give strength to the model of TTR synthesis by the choroid plexus being involved in transport of T4 into the brain and acting as a thyroid hormone distributor protein in the CSF.

The next set of experiments producing further data to support Schreiber's model was performed when Malcolm Segal visited Schreiber's laboratory. They perfused a sheep choroid plexus with medium containing ^{14}C-leucine. Highly radioactive TTR was collected from the newly secreted CSF on the surface of the choroid plexus. Radioactive TTR that had been secreted back into the perfusion medium could not

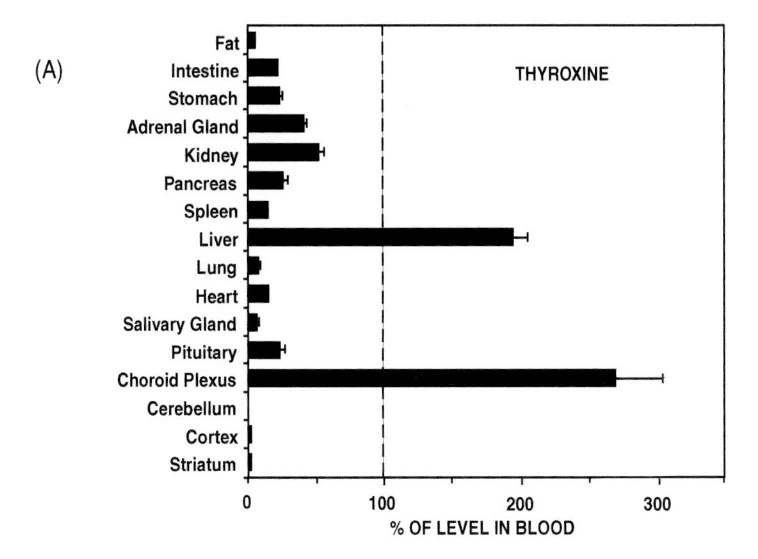

FIGURE 11.8 (A) Tissue distribution of ^{125}I-T4 10 min after injection into the vena cava of rats. Data is presented as radioactivity in 1 mg tissue as a percentage of radioactivity in 1 μL blood. The radioactivity in each tissue has been corrected for blood content. (B) Kinetics of accumulation of ^{125}I-T4 after injection into the vena cava of rats, in choroid plexus (CP), liver (L), blood (B), kidney (K), and pituitary (P). (C) Kinetics of accumulation of ^{125}I-T4 after injection into the vena cava of rats, in striatum, then cortex, then cerebellum. (From Dickson et al., 1987.)

be detected, even after the perfusion medium had been concentrated, indicating that 100% of TTR synthesized by the choroid plexus is secreted into the CSF (and none is secreted into the bloodstream). In addition, ^{125}I-T4 was intravenously injected into rats. First there was an accumulation of ^{125}I-T4 in the choroid plexus, then in the CSF, and thereafter in the striatum and cortex (Dickson et al., 1987).

Two sets of elegant experiments were carried out by Chanoine et al. (1992). In the first set they used N-bromoacetyl-L-^{125}I-T4 covalently bound to TTR and albumin. The second set of experiments involved the use of the synthetic plant flavonoid EMD 21388 (now known as F21388), which inhibits binding of T4 specifically to TTR. They convincingly demonstrated that (1) the TTR-T4 complex is not transported into the CSF (therefore, TTR synthesized by the liver does not contribute to TTR in the CSF), (2) T4 transport from blood to CSF is dependent on the concentration of free T4 in serum and on T4 binding to TTR in the choroid plexus, and (3) T4 in the CSF, which came from serum via the choroid plexus, may contribute to the T4 pool in the brain.

Southwell developed a rat choroid plexus cell culture model in a two-chambered system and demonstrated that TTR accumulated in the fluid in the apical chamber to twice the level in the basal chamber. ^{125}I-T4 added to the basal chamber permeated to the apical chamber to twice the concentration in the basal chamber. Addition of cycloheximide to cease protein synthesis resulted in the ^{125}I-T4 equilibrating between the two chambers to equal concentrations. Competitive inhibition of ^{125}I-T4 binding to TTR by EMD 21388 also prevented accumulation of ^{125}I-T4 in the apical chamber

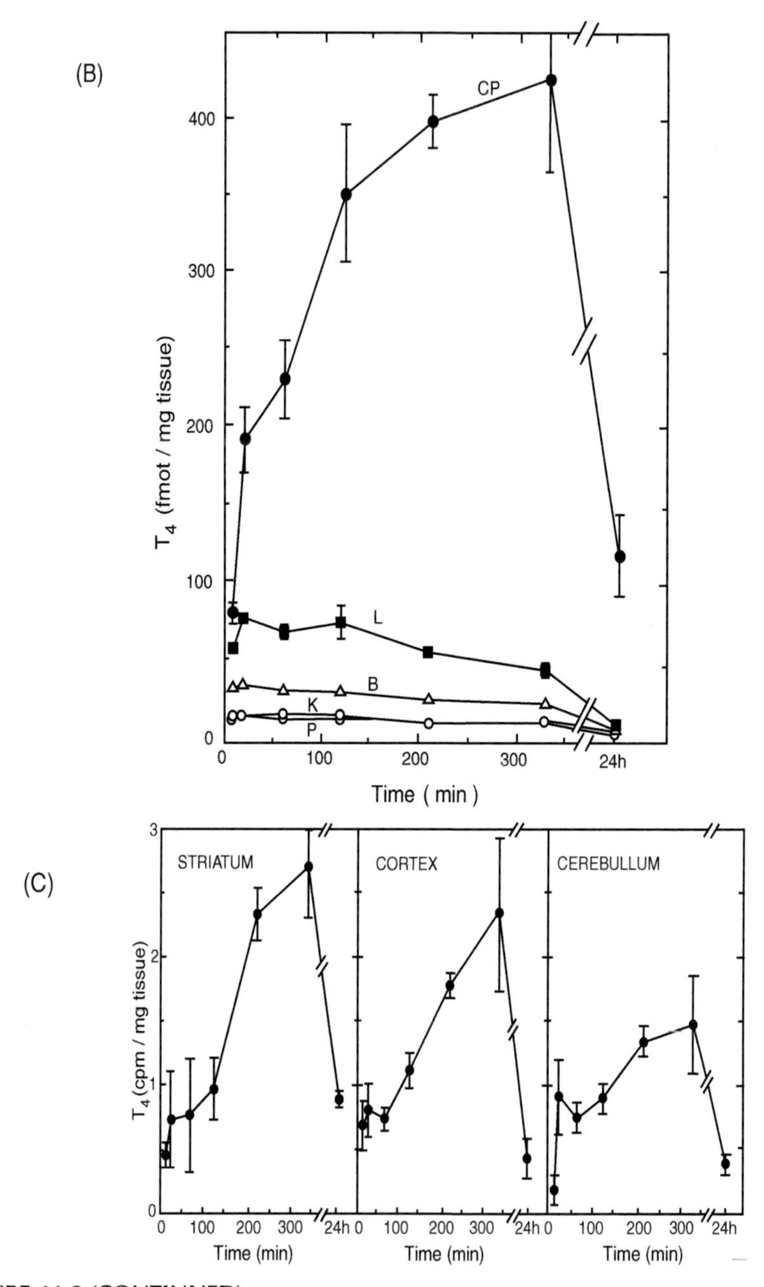

FIGURE 11.8 (CONTINUED)

(Southwell et al., 1993). The original model proposed by Dickson et al. (1987) was slightly modified in Southwell et al. (1993), and this is the currently accepted model (see Figure 11.9). As shown in the lower part of the figure, this model clearly includes access of thyroid hormone to the brain by paths not involving the choroid plexus.

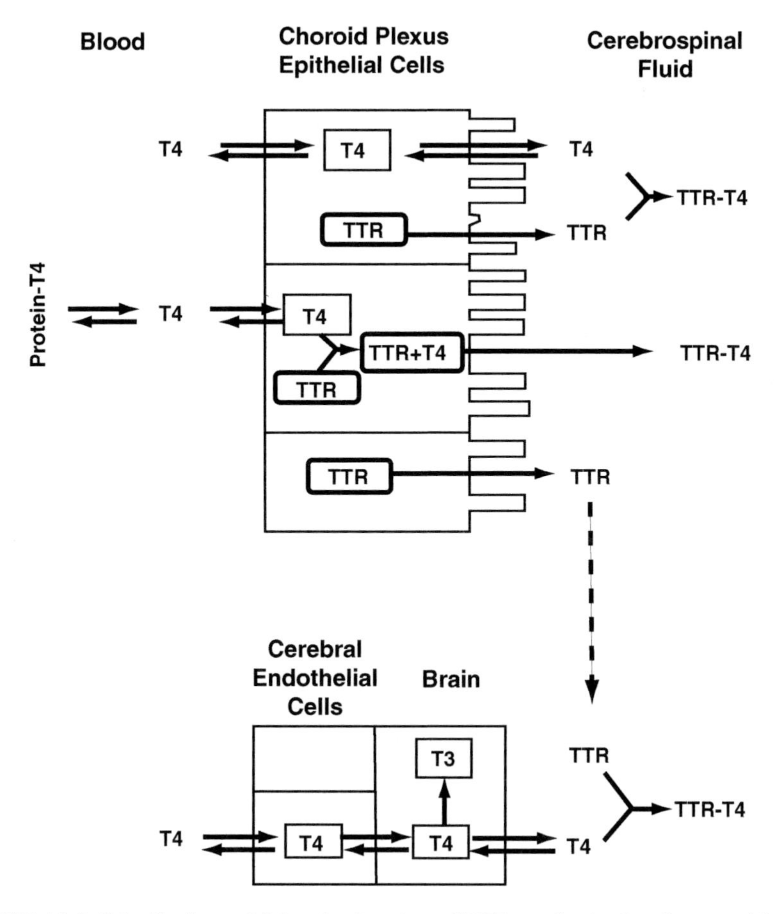

FIGURE 11.9 Schreiber's model for the function of TTR synthesized and secreted by the choroid plexus epithelial cells in the movement of T4 from the blood into the CSF and for TTR to act as a thyroid hormone distributor protein in the CSF. T4 bound to a thyroid hormone distributor protein in blood is in equilibrium with free T4 (depending on the Kd value), which is lipophilic and partitions into the membrane of the choroid plexus. TTR is synthesized by choroid plexus epithelial cells and either is secreted in complex with T4 or binds T4 in the CSF. The complex is washed away by the flow of the CSF. Alternatively, T4 may enter via the cerebral endothelial cells. (From Southwell et al., 1993.)

The next phase in understanding the role of TTR in thyroid hormone transport into the brain came from the results of a comparative study. Semi-quantitative data during the study by Duan et al. (1991) had suggested that chicken bound T4 with lower affinity than did mammalian TTRs. Chang et al. (1999) developed a method for accurately and precisely measuring the affinity of TTRs from a variety of species to T3 and T4. TTRs from 9 species were studied, including three birds (emu, *Dromaius novaehollandiae*; chicken, *Gallus gallus*; pigeon, *Columbia livia*), three marsupials (Southern hairy-nosed wombat, *Lasiorhinus latifrons*; brush-tailed possum, *Trichosurus vulpecula*; tammar wallaby, *Macropus eugenii*), and three

TABLE 11.1
Dissociation Values for T4 and T3 with TTRs from 11 Species

Source TTR	K_dT4 (nM)	K_dT3 (nM)	K_dT3/K_dT4
	Eutherians		
Human	13.6	56.6	4.2
Sheep	11.3	63.5	3.2
Rat	8.0	67.2	8.4
	Marsupials		
Wombat	21.8	97.8	4.5
Possum	15.9	206.1	12.9
Wallaby	13.8	65.3	4.7
	Birds		
Emu	37.4	18.9	0.51
Chicken	28.8	12.3	0.43
Pigeon	25.3	16.1	0.64
	Reptile		
Crocodile[a]	36.7	7.56	0.21
	Amphibian		
Toad[a]	508.0	248.0	0.49

[a] Crocodile and toad TTRs were synthesized in the recombinant yeast expression system *Pichia pastoris*.

Source: Schreiber, 2002.

eutherians (human, *Homo sapiens*; sheep, *Ovis aries*; rat, *Rattus norvegicus*). Whereas eutherian TTRs had higher affinity for T4 than T3, avian TTRs had higher affinity for T3 than T4. TTRs from marsupials had higher affinity for T4 than T3, but these values were intermediate between those of eutherians and birds (see Table 11.1).

Thus, the functional TTR appeared to change from preferentially binding T3 to preferentially binding T4 (Chang et al., 1999). The x-ray crystal structure of chicken TTR indicated that there were no differences in the thyroid hormone-binding sites compared to human TTR (Sunde et al., 1996). Was this change in ligand preference related to the change in structure of the N-termini? If this were true, then reptilian TTRs would bind T3 with higher affinity than T4. As adult reptiles do not synthesize TTR in their liver, it cannot be purified from blood. Purification of TTR from reptilian CSF was not feasible given the amount of TTR required for the binding assays and the number of animals that would need to be killed, so recombinant TTR was synthesized from salt water crocodile (*Crocodylus porosus*) TTR cDNA. Crocodile TTR bound T3 with higher affinity than T4 (Prapunpoj et al., 2002). *Xenopus laevis* TTR was also found to bind T3 with higher affinity than T4 (Prapunpoj et al., 2000b) (see Table 11.1), as was TTR from sea bream (*Sparus aurata*) (Santos and Power, 1999), but this has not yet been quantitated.

It therefore appears that TTR's function changed during vertebrate evolution, from preferentially binding T3 to preferentially binding T4. In this book, however, we are primarily concerned with the role of TTR in the choroid plexus (fish and amphibians do not synthesize TTR in the choroid plexus). When TTR was first synthesized by the choroid plexus, it probably preferentially bound T3. Today, TTRs in the choroid plexus of reptiles and birds preferentially bind T3, and only TTRs in the choroid plexus of mammals preferentially bind T4. Does this mean that in reptiles and birds T3 is transported via the choroid plexus into the CSF, rather than T4? What implications could this have for the evolution of deiodinases in vertebrate brains? The selection pressure leading to the change from TTR preferentially binding T3 to T4 could be that, instead of transporting the "active" form of the hormone, a "precursor" form of the hormone is transported. This would allow greater flexibility and specificity at the local tissue level to either activate the T4 by deiodinating it to T3, or to inactivate the T4 by deiodinating it to rT3. This could be especially true in the brain, as in the rat brain the percentage of T3 due to local deiodination of T4 is very specific to the region of the brain: 65% in the cortex, 51% in the cerebellum, 35% in the pons, 32% in the hypothalamus, 30% in the medulla oblongata, and 22% in the spinal cord (van Doorn et al., 1985).

11.7 OTHER POSSIBLE FUNCTIONS OF TTR SYNTHESIZED BY THE CHOROID PLEXUS

In the mid-1990s, Schwarzman and colleagues discovered that TTR sequesters amyloid β protein, thereby preventing its formation of amyloid in CSF (Schwarzman et al., 1994). Since then several papers have appeared on this topic, reviewed by Stein and Johnson (2003). It does seem ironic that a protein that itself can form amyloid (both FAP and SSA) can sequester and thereby prevent another protein from forming amyloid. Several studies have found that patients suffering from Alzheimer's disease have lowered concentrations of TTR in their CSF and have attributed the onset of the disease to the reduced TTR levels being insufficient to sequester the amyloid β protein (e.g., Riisoen, 1988; Serot et al., 1997).

RBP mRNA has been detected in total RNA prepared from the choroid plexus of some mammals (rats, mice, sheep, cattle) but not others (dog) (Duan and Schreiber, 1992). Of the mammals synthesizing RBP in their choroid plexus, the relative amount of RBP mRNA in total RNA was high for the ruminants, but extremely low for the rodents. This raises the question as to the function of RBP synthesized by the choroid plexus. Vitamin A has a pivotal role in embryogenesis, including the development of the CNS. How vitamin A gains access to the brain is not fully understood. Goodman claimed the function of TTR in the blood was to bind RBP, thereby preventing loss of RBP via filtration in the kidneys (Raz and Goodman, 1969). If the RBP synthesized by the choroid plexus was secreted into the CSF, would TTR binding to RBP have a purpose in the CSF?

11.8 TRANSTHYRETIN NULL MICE

Humans lacking TTR have not been described, although humans lacking the other thyroid hormone distributor proteins (TBG or albumin) are healthy (see Schreiber, 2002). The obvious difference between TTR and TBG or albumin is that TTR is synthesized by the brain as well as the liver. This raised the question as to whether TTR was essential for early life.

One way to answer such a question is to remove TTR and score for a phenotype. Therefore, TTR null mice were investigated for thyroid hormone metabolism. They were found to have reduced levels of retinol in their plasma and reduced concentrations of total T4 and T3, but were otherwise healthy (Episkopou et al., 1993). The next description of TTR null mice included levels of free thyroid hormones, which were reported as being normal, thus, the TTR null mice were considered euthyroid (Palha et al., 1994). A further study exploring the turnover of thyroid hormone pools revealed that the T4 content of the brain was reduced to 36% compared to wild-type mice and had a decreased residence time in the brain compared to wild-type mice. In addition, the kinetic model of rate constants of T4 transfer between compartments the authors constructed did not hold true for the brain (Palha et al., 1997). The next study showed reduced levels of T4 (14%) and T3 (48%) in the choroid plexus (Palha et al., 2000). The overall conclusion at the end of each of these (and other) papers is that TTR is not important for thyroid hormone metabolism. This argument is used by Palha to discredit Schreiber's model.

Schreiber is very careful to state that T4 transport into the brain via TTR synthesis in the choroid plexus is not the only mechanism for thyroid hormone entry into the brain, and he is well aware of the leakiness of the blood–brain barriers to plasma proteins (e.g., Schreiber et al., 1990; Southwell et al., 1993).

The issues of knock-out mice not displaying expected phenotypes, biological redundancy, and biological networks are now old and very familiar. Airplanes have two steering systems so that, if one fails, the other can take over. If one is removed and the airplane keeps flying, does that mean that steering is not important for airplanes? The same logic applies to the number of wheel nuts on car wheels. By the time eutherians evolved, many evolutionary safety nets were in place. It is only worthwhile to "back up" biologically important features with redundancy and networks in order to ensure survival. In the case of TTR null mice, 99.89% of their T4 in blood is bound to protein, compared to 99.94% in wild-type mice. It appears that albumin and TBG are acting as "buffers" in the absence of TTR.

Two areas that should be studied in the TTR null mice are their perinatal development and susceptibility to hypothyroidism. In particular, the rate of brain development around the time when TTR synthesis is maximal in the choroid plexus of wild-type mice should be investigated, as Strait et al. (1992) demonstrated that, by 60 days of age, hypothyroid mice had "caught up" to euthyroid mice in terms of gene regulation in the brain.

11.9 CONCLUSIONS AND FUTURE DIRECTIONS

TTR binds thyroid hormones and RBP in all vertebrate species analyzed. The amino acid sequence and the 3D structure of TTR have remained virtually unchanged during vertebrate evolution, implying a strong selection pressure against change and therefore an important function in vertebrates. The stepwise change in the structure of the N-termini, which may be responsible for the change in ligand preference from T3 to T4, is an example of neo-Darwinian positive selection and may have occurred to result in increased level of regulation of tissues to the effects of T3 on gene transcription. Both classes of hormones transported by TTR (thyroid hormones and retinoids) have extremely important roles in regulation of development. TTR is synthesized by the choroid plexus, the liver, and the retina. The TTR gene is differentially regulated in the choroid plexus compared to the liver. TTR has been synthesized by the choroid plexus since the development of the neocortex, about 320 million years ago, and is the major protein synthesized and secreted by the choroid plexus of reptiles, birds, and mammals. Only the epithelial cells of the choroid plexus synthesize TTR. TTR is synthesized by choroid plexus Anlage cells, and the rate of TTR gene expression in the choroid plexus during development is closely linked with the timing of the development of the brain in both altricial and precocial species. There is a compelling body of evidence that TTR synthesized by the choroid plexus is involved with transport of thyroid hormones from the blood via the blood–CSF barrier into the CSF. This TTR most probably acts as a thyroid hormone distributor protein in the CSF.

The data on the developmental regulation and on the evolution of TTR synthesis by the choroid plexus make it extremely likely that TTR synthesis and secretion by the choroid plexus has a significant function. Most probably this function is described by Schreiber's model, that TTR synthesized by the choroid plexus is involved in transporting thyroid hormones into the brain and acting as a thyroid hormone distributor protein in the CSF.

Mammals, in particular mice and rats, are very convenient models to work with because they are easily stored in animal houses, have short generation times, and can be handled easily. We know quite a lot about the mammalian choroid plexus, but to understand an organ fully we should look at its evolution by taking a comparative approach. One area that should be studied in order to give us a greater understanding of the choroid plexus is thyroid hormone transport across the choroid plexus in nonmammalian species, i.e., in reptiles and birds, whose TTRs preferentially bind T3, and in amphibians and fish, whose choroid plexus do not synthesize TTR.

REFERENCES

Achen MG, Duan W, Pettersson TM, Harms PJ, Richardson SJ, Lawrence MC, Wettenhall REH, Aldred AR, Schreiber G. (1993) Transthyretin gene expression in choroid plexus first evolved in reptiles. *Am J Physiol* 265:R982–9.

Achen MG, Harms PJ, Thomas T, Richardson SJ, Wettenhall REH, Schreiber G. (1992) Protein synthesis at the blood-brain barrier. The major protein secreted by amphibian choroid plexus is a lipocalin. *J Biol Chem* 267:23170–4.

Aldred AR, Prapunpoj P, Schreiber G. (1997) Evolution of shorter and more hydrophilic transthyretin N-termini by stepwise conversion of exon 2 into intron 1 sequences (shifting the 3β splice site of intron 1). *Eur J Biochem* 246:401–9.

Bernal J, Guadano-Ferraz A, Morte B. (2003) Perspectives in the study of thyroid hormone action on brain development and function. *Thyroid* 13:1005–12.

Beuckmann CT, Aoyagi M, Okazaki I, Hiroike T, Toh H, Hayaishi O, Urade Y. (1999) Binding of biliverdin, bilirubin and thyroid hormones to lipocalin-type prostaglandin D synthetase. *Biochemistry* 38:8006–13.

Blake CCF. Geisow MJ, Oatley SJ, Rerat C, Rerat B. (1978) Structure of prealbumin: secondary, tertiary and quaternary interactions determined by Fourier refinement at 1.8 Å. *J Mol Biol* 121:339–56.

Blake CCF. Geisow MJ, Swan IDA, Rerat C, Rerat B. (1974) Structure of human plasma prealbumin at 2.5 Å resolution. A preliminary report on the polypeptide chain conformation, quaternary structure and thyroxine binding. *J Mol Biol* 88:1–12.

Brack CM, Duan W, Hulbert AJ, Schreiber G. (1995) Wallaby transthyretin. *Comp Biochem Physiol* 110B:523–9.

Braverman LE, Utiger RD. (2000) Introduction to hypothyroidism. In:*Werner and Ingbar's The Thyroid. A Fundamental and Clinical Text*, 8th ed., LE Barverman, RD Utiger, eds. Lippincott Williams and Wilkins, Philadelphia, pp 719–20.

Bridges CD, Alvarez RA, Fong SL, Gonzales-Fernandez F, Lam DM, Liou GI. (1984) Visual cycle in the mammalian eye. Retinoid-binding proteins and the distribution of 11-cis retinoids. *Vis Res* 24:1581–94.

Bui BV, Armitage JA, Fletcher EL, Richardson SJ, Schreiber G, Vingrys AJ. (2001) Retinal anatomy and function of the transthyretin null mouse. *Exp Eye Res* 73:651–9.

Calkins E. (1974) Amyloidosis. In: *Harrison's Principles of Internal Medicine*, 7th ed. McGraw-Hill, New York, pp 644–7.

Chang L, Munro SLA, Richardson SJ, Schreiber G. (1999) Evolution of thyroid hormone binding by transthyretins in birds and mammals. *Eur J Biochem* 259:634–42.

Chanoine J-P, Alex S, Fang SL, Stone S, Leonard JL, Köhrle J, Braverman LE. (1992) Role of transthyretin in the transport of thyroxine from the blood to the choroid plexus, the cerebrospinal fluid, and the brain. *Endocrinology* 130:933–8.

Colon W, Kelly JW. (1992) Partial denaturation of transthyretin is sufficient for amyloid fibril formation *in vitro*. *Biochemistry* 31:8654–60.

Connors LH, Richardson AM, Theberg R, Costello CE. (2000) Tabulation of transthyretin (TTR) variants as of 1/1/2000. *Amyloid* 7:54–69.

Costa RH, Grayson DR. (1991) Site-directed mutagenesis of hepatocyte nuclear factor (HNF) binding sites in the mouse transthyretin (TTR) promoter reveal synergistic interactions with its enhancer region. *Nucleic Acids Res* 19:4139–45.

Costa RH, van Dyke TA, Yan C, Kuo F, Darnell JE. (1990) Similarities in transthyretin gene expression and differences in transcription factors: liver and yolk sac compared to choroid plexus. *Proc Natl Acad Sci USA* 87:6589–93.

Craik DJ, Duggan BM, Munro SLA. (1996) Conformations and binding interactions of thyroid hormone analogues In: *Biological Inhibitors*, Vol 2, *Studies in Medicinal Chemistry.* MI Choudhary, ed. Harwood Academic Publishers, Amsterdam.

Dickson PW, Aldred AR, Marley PD, Bannister D, Schreiber G. (1986) Rat choroid plexus specializes in the synthesis and the secretion of transthyretin (prealbumin)—regulation of transthyretin synthesis in choroid plexus is independent from that in liver. *J Biol Chem* 261:3475–8.

Dickson PW, Aldred AR, Menting JGT, Marley PD, Sawyer WH, Schreiber G. (1987) Thyroxine transport in choroid plexus. *J Biol Chem* 262:13907–15.

Dickson PW, Howlett GJ, Schreiber G. (1995) Rat transthyretin (prealbumin). Molecular cloning, nucleotide sequence, and gene expression in liver and brain. *J Biol Chem* 260:8214–9.

Duan W, Achen MG, Richardson SJ, Lawrence MC, Wettenhall REH, Jaworowski A, Schreiber G. (1991) Isolation, characterisation, cDNA cloning and gene expression of an avian transthyretin: implications for the evolution of structure and function of transthyretin in vertebrates. *Eur J Biochem* 200:679–87.

Duan W, Cole TJ, Schreiber G. (1989) Cloning and nucleotide sequence of transthyretin (prealbumin) cDNA from rat choroid plexus and liver. *Nucleic Acids Res* 17:3979.

Duan W, Richardson SJ, Babon JJ, Heyes RJ, Southwell BR, Harms PJ, Wettenhall REH, Dziegielewska KM, Selwood L, Bradley AJ, Brack CM, Schreiber G. (1995a) Evolution of transthyretin in marsupials. *Eur J Biochem* 227:396–406.

Duan W, Richardson SJ, Köhrle J, Chang L, Southwell BR, Harms PJ, Brack CM, Pettersson TM, Schreiber G. (1995b) Binding of thyroxine to pig transthyretin, its cDNA structure, and other properties. *Eur J Biochem* 230:977–86.

Duan W, Schreiber G. (1992) Expression of retinol-binding protein mRNA in mammalian choroid plexus. *Comp Biochem Physiol* 101B:399–406.

Dussault JH, Ruel J. (1987) Thyroid hormones and brain development. *Annu Rev Physiol* 49:321–34.

Dwork AJ, Cavallaro T, Martone RL, Goodman DS, Schon EA, Herbert J. (1990) Distribution of transthyretin in the rat eye. *Invest Opthalmol Vis Sci* 31:489–96.

Episkopou V, Maeda S, Nishigushi S, Shimada K, Gaitanaris GA, Gottesman ME, Robertson EJ. (1993) Disruption of the transthyretin gene results in mice with depressed levels of plasma retinol and thyroid hormone. *Proc Natl Acad Sci USA* 90:2375–9.

Farsetti A, Mitsuhashi T, Desvergne B, Robbins J, Nikodem VM. (1991) Molecular basis of thyroid hormone regulation of myelin basic protein gene expression in rodent brain. *J Biol Chem* 266:23226–32.

Fisher DA, Polk DH. (1989) Maturation of thyroid hormone actions. In: *Research in Congenital Hypothyroidism*, F Delange, DA Fisher, D Glinoer, eds. Plenum Press, New York, pp 61–77.

Folli C, Pasquato N, Ramazzina I, Battistutta, Zanotti G, Berni R. (2003) Distinctive binding and structural properties of piscine transthyretin. *FEBS Lett* 555:279–84.

Fung WP, Thomas T, Dickson PW, Aldred AR, Milland J, Dziadek M, Power B, Hudson P, Schreiber G. (1988) Structure and expression of the rat transthyretin (prealbumin) gene. *J Biol Chem* 263:480–8.

Godovac-Zimmerman J. (1988) The structural motif of β-lactoglobulin and retinol-binding protein: a basic framework for binding and transport of small hydrophobic molecules? *Trends Biochem Sci* 13:64–66.

Harms PJ, Tu GF, Richardson SJ, Aldred AR, Jaworowski A, Schreiber G. (1991) Transthyretin (prealbumin) gene expression in choroid plexus is strongly conserved during evolution of vertebrates. *Comp Biochem Physiol* 99B:239–49.

Hatterer JA, Herbert J, Hidaka C, Roose SP, Gorman JM. (1993) CSF transthyretin in patients with depression. *Am J Psychiatry* 150:813–5.

Herbert J, Cavallaro T, Martone R. (1991) The distribution of retinol-binding protein and its mRNA in the rat eye. *Invest Opthalmol Vis Sci* 32:302–9.

Ingbar S. (1958) Prealbumin: a thyroxine-binding protein of human plasma. *Endocrinology* 63:256–9.

Ingenbleek Y, Young V. (1994) Transthyretin (prealbumin) in health and disease: nutritional implications. *Annu Rev Nutr* 14:495–533.

Irikara D, Kanaoka Y, Urade Y. Bovine TTR sequence submitted to EMBL/GenBank/DDBJ databases, December 1997.

Jacobsen M. (1978) *Developmental Neurobiology.* Plenum Press, New York.

Kabat EA, Landow H, Moore DH. (1942a) Electrophoretic patterns of concentrated cerebrospinal fluid. *Proc Soc Exp Bio Med* 49:260–3.

Kabat EA, Moore DH, Landow H. (1942b) An electrophoretic study of the protein components in cerebrospinal fluid and their relationship to the serum proteins. *J Clin Invest* 21:571–7.

Kent GC. (1987) *Comparative Anatomy of the Vertebrates.* Time Mirror/Mosby College, St Louis, p 542.

Köhrle J. (2002) Iodothyronine deiodinases. *Methods Enzymol.* 347:125–67.

Köhrle J, Irmscher K, Hesch RD. (1988) Flavonoid effects on transport, metabolism and action of thyroid hormones. In: *Plant Flavonoids in Biology and Medicine II: Biochemical, Cellular and Medicinal Properties,* Cody V, Harborne JB, Beretz A, eds. Alan R. Liss, New York, pp 323–40.

Lepperdinger G. (2000) Amphibian choroid plexus lipocalin, Cpl1. *Biochim Biophys Acta* 1482:119–26.

Mendel CM. (1989) The free hormone hypothesis: a physiologically based mathematical model. *Endocr Rev* 10:232–74.

Mendel CM, Weisiger RA. (1990) Thyroxine uptake by perfused rat liver. No evidence for facilitation by five different thyroxine-binding proteins. *J Clin Invest* 86:1840–7.

Mendel CM, Weisiger RA, Jones AL, Cavalieri RR. (1987) Thyroid hormone-binding proteins in plasma facilitate uniform distribution of thyroxine within tissues: a perfused rat liver study. *Endocrinology* 120:1742–9.

Miller SR, Sekijima Y, Kelly JW. (2004) Native state stabilization by NSAIDs inhibits transthyretin amyloidogenesis from the most common familial disease variants. *Lab Invest* 84:545–52.

Mita S, Maeda S, Shimada K, Araki S. (1984) Cloning and sequence analysis of cDNA for human prealbumin. *Biochem Biophys Res Commun* 124:558–64.

Mizuno R, Cavallaro T, Herbert J. (1992) Temporal expression of the transthyretin gene in the developing rat eye. *Invest Opthalmol Vis Sci* 33:341–9.

Monaco H, Rizzi M, Coda A. (1995) Structure of a complex of two plasma proteins: transthyretin and retinol-binding protein. *Science* 268:1039–41.

Morreale de Escobar G, Obregon MJ, Escobar del Rey F. (1987) Fetal and maternal thyroid hormones. *Horm Res* 26:12–27.

Munro SL, Lim C-F, Hall JG, Barlow JW, Craik DJ, Topliss DJ, Stockigt JR. (1989) Drug competition for thyroxine binding to transthyretin (prealbumin): comparison with effects on thyroxine-binding globulin *J Clin Endocrinol Metab* 68:1141–7.

Nomenclature Committee of the IUB (NC-IUB) IUB-IUPAC Joint Commission on Biochemical Nomenclature (JCBN) Newsletter 1981. (1981) *J Biol Chem* 256:12–14.

Ong DE, Davis JT, O'Day WT, Bok D. (1994) Synthesis and secretion of retinol-binding protein and transthyretin by cultured retinal pigment epithelium. *Biochemistry* 33:1835–42.

Palha JA. (2002) Transthyretin as a thyroid hormone carrier: function revisited. *Clin Chem Lab Med* 40:1292–1300.

Palha JA, Episkopou V, Maeda S, Shimada K, Gottesman ME, Saraiva MJM. (1994) Thyroid hormone metabolism in a transthyretin null mouse strain. *J Biol Chem* 269:33135–9.

Palha JA, Fernandes R, Morreale de Escobar GM, Episkopou V, Gottesman M, Saraiva MJ. (2000) Transthyretin regulates thyroid hormone levels in the choroid plexus, but not in the brain parenchyma: study in a transthyretin-null mouse model. *Endocrinology* 141:3267–72.

Palha JA, Hays MT, Morreale de Escobar G, Episkopou V, Gottesman ME, Saraiva MJ. (1997) Transthyretin is not essential for thyroxine to reach the brain and other tissues in transthyretin-null mice. *Am J Physiol* 272:E485–93.

Prapunpoj P, Richardson SJ, Fugamalli L, Schreiber G. (2000a) The evolution of the thyroid hormone distributor protein transthyretin in the order Insectivora, class Mammalia. *Mol Biol Evol* 17:1199–1209.

Prapunpoj P, Richardson SJ, Schreiber G. (2002) Crocodile transthyretin: structure, function and evolution. *Am J Physiol* 283:R885–96.

Prapunpoj P, Yamauchi K, Nishiyama N, Richardson SJ, Schreiber G. (2000b) Evolution of structure, ontogeny of gene expression and function of *Xenopus laevis* transthyretin. *Am J Physiol* 279:R2026–41.

Qian X, Samadani U, Porcella A, Costa RH. (1995) Decreased expression of hepatocyte nuclear factor 3α during the acute-phase response influences transthyretin gene transcription. *Mol Cell Biol* 15:1364–76.

Raz A, Goodman DS. (1969) The interaction of thyroxine with plasma prealbumin and with prealbumin-retinol-binding protein complex. *J Biol Chem* 244:3230–7.

Richardson SJ. (2002) The evolution of transthyretin synthesis in vertebrate liver, in primitive eukaryotes and in bacteria. *Clin Chem Lab Med* 40:1191–9.

Richardson SJ, Bradley AJ, Duan W, Wettenhall REH, Harms PJ, Babon JJ, Southwell BR, Nicol S, Donnellan SC, Schreiber G. (1994) Evolution of marsupial and other vertebrate thyroxine-binding plasma proteins. *Am J Physiol* 266:R1359–70.

Richardson SJ, Hunt JL, Aldred AR, Licht P, Schreiber G. (1997) Abundant synthesis of transthyretin in the brain, but not in the liver, of turtles. *Comp Biochem Physiol* 117B:421–9.

Richardson SJ, Monk JA, Shepherdley CA, Ebbesson LOE, Sin F, Power DM, Frappell PB, Köhrle J, Renfree MB. (2004) Developmentally regulated thyroid hormone distributor proteins in marsupials, a reptile and fishes. (submitted).

Richardson SJ, Wettenhall REH, Schreiber G. (1996) Evolution of transthyretin gene expression in the liver of *Didelphis virginiana* and other American marsupials. *Endocrinology* 137:3507–12.

Riisoen H. (1988) Reduced prealbumin (transthyretin) in CSF of severely demented patients with Alzheimer's disease. *Acta Neurol Scand* 78:455–9.

Robbins J. (2000) Thyroid hormone transport proteins and the physiology of hormone binding. In: *Werner and Ingbar's The Thyroid*, 8th ed., Braverman LE, Utiger RD, eds. Lippincott Williams and Wilkins, Philadelphia, pp 105–20.

Robbins J, Rall JE. (1957) The interaction of thyroid hormones and protein in biological fluids. *Recent Progr Horm Res* 13:161–202.

Samadani U, Qian X, Costa RH. (1996) Identification of a transthyretin enhancer site that selectively binds the hepatic nuclear factor-3 beta isoform. *Gene Expr* 6:23–33.

Santos CRA, Power DM. (1999) Identification of transthyretin in fish (*Sparus aurata*): cDNA cloning and characterisation. *Endocrinology* 140:2430–3.

Saunders NR, Habgood MD, Dziegielewska KM. (1999) Barrier mechanisms in the brain, I. Adult brain. *Clin Exp Pharm Physiol* 26:11–19.

Schreiber G. (1987) Synthesis, processing and secretion of plasma proteins by the liver and other organs and their regulation. In: *The Plasma Proteins*, Vol 5, FW Putnam, ed. Academic Press, New York, pp 292–363.

Schreiber G. (2002) Beyond carrier proteins. The evolutionary and integrative roles of transthyretin in thyroid hormone homeostasis. *J Endocrinol* 175:61–73.

Schreiber G, Aldred AR. (1993) Extrahepatic synthesis of acute phase proteins. In: *Acute Phase Proteins. Molecular Biology, Biochemistry and Clinical Applications*, A Mackiewicz, I Kusher, H Baumann, eds. CRC Press, Boca Raton, FL, pp 39–76.

Schreiber G, Aldred AR, Jaworowski A, Nilsson C, Achen MG, Segal MB. (1990) Thyroxine transport from blood to brain via transthyretin synthesis in choroid plexus. *Am J Physiol* 258:R338–45.

Schreiber G, Howlett GJ. (1983) Synthesis and secretion of acute-phase proteins. In: *Plasma Protein Secretion by the Liver*, H Glaumann, T Peters Jr, C Redman, eds. Academic Press, London, pp 423–49.

Schreiber G, Richardson SJ. (1997) The evolution of gene expression, structure and function of transthyretin. *Comp Biochem Physiol* 116B:137–60.

Schwarzman AL, Gregori L, Vitek MP, Lyubski S, Strittmatter WJ, Enghilde JJ, Bhasin R, Silverman J, Weisgraber KH, Coyle PK, Zagorski MG, Talafous J, Eisenberg M, Saunders AM, Roses AD, Goldgaber D. (1994) Transthyretin sequesters amyloid β protein and prevents amyloid formation. *Proc Natl Acad Sci USA* 91:8368–72.

Serot JM, Christmann D, Dubost T, Couturier M. (1997) Cerebrospinal fluid transthyretin: aging and late onset Alzheimer's disease. *J Neurol Neurosurg Psychiatry* 63:506–8.

Siebert FB, Nielson JW. (1942) Electrophoretic study of the blood protein response in tuberculosis. *J Biol Chem* 143:29–38.

Soprano DR, Herbert J, Soprano KJ, Schon EA, Goodman DS. (1985) Demonstration of transthyretin mRNA in the brain and other extrahepatic tissues in the rat. *J Biol Chem* 260:11793–8.

Southwell BR, Duan W, Alcorn D, Brack C, Richardson SJ, Köhrle J, Schreiber G. (1993) Thyroxine transport to the brain: role of protein synthesis by the choroid plexus. *Endocrinology* 133:2116–26.

Southwell BR, Duan W, Tu GF, Schreiber G. (1991) Ontogenesis of transthyretin gene expression in chicken choroid plexus and liver. *Comp Biochem Physiol* 100:329–38.

Stauder AJ, Dickson PW, Aldred AR, Schreiber G, Mendelsohn FAO, Hudson P. (1986) Synthesis of transthyretin (prealbumin) mRNA in choroid plexus epithelial cells, localized by in situ hybridization in the rat brain. *J Histochem Cytochem* 34:949–52.

Strait KA, Zou L, Oppenheimer JH. (1992) β1 isoform-specific regulation of a triiodothyronine-induced gene during cerebellar development. *Mol Endocrinol* 6:1874–80.

Stein TD, Johnson JA. (2003) Genetic programming by the proteolytic fragments of the amyloid precursor protein: somewhere between confusion and clarity. *Rev Neurosci* 14:317–41.

Sturrock RR. (1979) A morphometric study of the development of the mouse choroid plexus. *J Anat* 129:777–93.

Sullivan GM, Hatterer JA, Herbert J, Chen X, Roose SP, Attia E, Mann J, Marangell LB, Goetz RR, Gorman JM. (1999) Low levels of transthyretin in the CSF of depressed patients. *Am J Psychiatry* 156:710–5.

Sunde M, Richardson SJ, Chang L, Pettersson TM, Schreiber G, Blake CCF. (1996) The crystal structure of transthyretin from chicken. *Eur J Biochem* 236:491–9.

Sundelin J, Melhus H, Das S, Eriksson U, Lind P, Trägårdh L, Peterson PA, Rask L. (1985) The primary structure of rabbit and rat prealbumin and a comparison with the tertiary structure of human prealbumin. *J Biol Chem* 260:6481–7.

Tan Y, Costa RH, Kovesdi I, Reichel RR. (2001) Adenovirus-mediated increase of HNF-3 levels stimulates expression of transthyretin and sonic hedgehog, which is associated with F9 cell differentiation toward the visceral endoderm lineage. *Gene Expr* 9:237–48.

Thomas T, Power B, Hudson P, Schreiber G, Dziadek M. (1988) The expression of transthyretin mRNA in the developing rat brain. *Dev Biol* 128:415–28.

Tsuzuki T, Mita S, Maeda S, Araki S, Shimada K. (1985) Structure of the human prealbumin gene. *J Biol Chem* 260:12224–7.

Tu G-F, Cole T, Duan W, Schreiber G. (1989) The nucleotide sequence of transthyretin cDNA isolated from a sheep choroid plexus cDNA library. *Nucleic Acids Res* 17:6384.

Tu G-F, Cole T, Southwell BR, Schreiber G. (1990) Expression of the genes for transthyretin, cystatin C and βA4 amyloid precursor protein in sheep choroid plexus during development. *Dev Brain Res* 55:203–8.

van Doorn J, Roelfsema F, van der Heide D. (1985) Concentration of thyroxine and 3,5,3′-triiodothyronine at 34 different sites in euthyroid rats as determined by isotopic equilibrium technique. *Endocrinology* 117:1201–8.

van Jaarsveld P, Branch WT, Robbins J, Morgan FJ, Kanda Y, Canfield RE. (1973) Polymorphism of rhesus monkey serum prealbumin. *J Biol Chem* 248:7898–7930.

Wakasugi S, Maeda S, Shimada K. (1986) Structure and expression of the mouse prealbumin gene. *J Biochem (Tokyo)* 100:49–58.

Wakasugi S, Maeda S, Shimada K, Nakashima H, Migita S. (1985) Structural comparisons between mouse and human prealbumin. *J Biochem Tokyo* 98:1707–14.

Wei S, Episkopou V, Piantedosi R, Maeda S, Shimada K, Gottesman ME, Blaner WS. (1995) Studies on the metabolism of retinol and retinol-binding protein in transthyretin-deficient mice produced by homologous recombination. *J Biol Chem* 270:866–70.

Wojczak A. (1997) Crystal structure of rat transthyretin at 2.5 Å resolution: first report on a unique tetrameric structure. *Acta Biochim Pol* 44:505–18.

Wojczak A, Cody V, Luft JR, Pangborn W. (2001) Structure of rat transthyretin (rTTR) complex with thyroxine at 2.5 Å resolution: first non-biased insight into thyroxine binding reveals different hormone orientation in two binding sites. *Acta Cryst* D57:1061–70.

Yamauchi K, Takeuchi H, Overall M, Dziadek M, Munro SLA and Schreiber G. (1998) Structure characteristics of bullfrog (*Rana catesbeiana*) transthyretin and its cDNA: comparison of its pattern of expression during metamorphosis with that of lipocalin. *Eur J Biochem* 256:287–96.

Yan C, Costa RH, Darnell JE Jr, Chen JD and van Dyke TA. (1990) Distinct positive and negative elements control the limited hepatocyte and choroid plexus expression of transthyretin in transgenic mice. *EMBO J* 9:869–78.

Zheng W, Deane R, Redzic Z, Preston JE and Segal MB. (2003) Transport of L-[^{125}I]thyroxine by in situ perfused ovine choroid plexus: inhibition by lead exposure. *J Toxicol Environ Health* 66:435–51.

Section Three

Blood–CSF Barrier in Diseases

12 Choroid Plexus and CSF in Alzheimer's Disease: Altered Expression and Transport of Proteins and Peptides

Conrad E. Johanson, Gerald D. Silverberg, John E. Donahue, John A. Duncan, and Edward G. Stopa
Brown Medical School, Rhode Island Hospital, Providence, Rhode Island, U.S.A.

CONTENTS

12.1 INTRODUCTION: MODIFIED CHOROIDAL FUNCTIONS IN AGING AND IN ALZHEIMER'S DISEASE

In advanced stages of Alzheimer's disease (AD), there is severely disrupted homeostasis of the extracellular fluid environment of neurons. This is due partly to structural and functional deficits in the blood–cerebrospinal fluid (CSF) and blood–brain barriers. The brain, in its daily housekeeping, normally produces a number of peptides and proteins, such as amyloid beta-peptides (Aβs) and tau protein, which can self-assemble at higher than normal concentrations and become toxic. Removal of such potentially harmful macromolecules depends upon the normal function of a number of macromolecular clearance pathways. These include transport of such molecules from the interstitial fluid (ISF) into the bloodstream across the capillary endothelium, passage into the CSF by diffusion and into the venous circulation by bulk flow of CSF, in situ enzymatic degradation and scavenger cell uptake. Aging and disease take a toll on all of these systems.

The pivotal role of the choroid plexus (CP) in maintaining normal CSF composition has been established, but the impact of large-cavity (ventricular and subarachnoid) CSF on brain ISF dynamics in disease states is less well known.The CSF originates mainly from the choroid plexuses (Johanson, 2003); therefore it is clinically and pharmacologically useful to learn how the operation of this secretory epithelium affects the function of neuronal systems distant from the cerebral ventricles and how this process is modified by disease (Strazielle and Ghershi-Egea, 2000).

In several ways the CP-CSF system has a bearing on the "metabolic milieu" and fluid balance in the CNS. Transport, secretory, and permeability properties of the blood–CSF interface determine the distribution of materials between the choroidal blood and ventricular CSF. One can view the CPs as dynamic tissues that regulate the bidirectional fluxes of a wide array of substances to ensure a special, stable biochemical environment for the neurons. Accordingly, choroidal epithelial functions include barrier protection, inward transport, synthesis/secretion, reabsorptive transport, and fluid formation. As a preface to more detailed discussion later in the chapter, a brief overview of the above-mentioned functions is given here in the context of neurodegenerative disease:

1. CSF, and consequently the brain, is protected from potentially toxic elements in the blood by a "barrier" in the CP. Tight junctions between the epithelial cells constitute the anatomical substrate of the blood–CSF barrier, as compared to the capillary bed of the brain, the blood–brain barrier, wherein the tight junctions occur between the vascular endothelial cells. The permeability of tight junctions to water-soluble molecules is regulated by hormones, growth factors, and cytokines, the titers of which in CSF and plasma can be altered in neurodegeneration. Tight junctions are also vulnerable to toxicants (Zheng et al., 2003).
2. In health, the blood-to-CSF transporters in the basolateral and apical membranes move micronutrients, like vitamins B and C, as well as ions, organic solutes, and hormones into ventricular fluid for bulk flow delivery to neurons and glia (Spector and Johanson, 1989). Any serious diminution in choroidal epithelium viability in AD dementia, e.g., due to severe ischemia, choroidal amyloid deposition, and so on, can reduce the capacity of the plexus-CSF system to translocate inorganic ions, e.g., Ca, and peptidergic substances to target cells in the brain parenchyma.
3. Choroidal epithelial cells have transcripts for transthyretin, cystatin C, ceruloplasmin, transferrin, and growth factors; they manufacture and secrete these proteins into the CSF (Dickson et al., 1985; Knuckey et al., 1996; Schreiber, 2002). Mutations in proteins can alter their functional capabilities. CSF protein levels are often increased or decreased in AD, with implications for brain interstitial plaque development.
4. On the apical membrane of the choroid epithelium, there are protein transporters that remove numerous types of organic molecules from the ventricular CSF (Chodobski and Szmydynger-Chodobska, 2001; Ghershi-Egea and Strazielle, 2002). Fragments of Aβ peptides, for example, may be actively removed from the CSF by apically located transporters. Pathological factors that alter the choroidal removal of amyloid fragments and other peptides can cause changes in their CSF concentration; this, in turn, may affect plaque formation in the brain parenchyma.
5. Continual CSF formation by active transport processes in the CP sets up the beneficial CSF "sink" action on small as well as larger molecules in the brain interstitium (Parandoosh and Johanson, 1982). When CSF production in AD is reduced as the result of failing physiological mechanisms (Silverberg et al., 2001), the metabolic waste products and xenobiotics (Strazielle and Ghershi-Egea, 1999) build up in brain as the result of a decrease in the CSF turnover, i.e., the CSF formation rate ÷ the CSF volume. Such CSF stagnation in dementia would be expected to exacerbate the course of neurodegeneration (Silverberg et al., 2003). Clearly, there is a compelling need for a quantitative assessment of CSF dynamics in aging and age-related dementia, especially in regard to the clearance of protein/peptide fragments and other organic solutes from the disabled CNS with its reduced CSF turnover.

12.2 PATHOPHYSIOLOGICAL PERSPECTIVES ON THE BLOOD–CSF BARRIER

This chapter focuses on altered transport and expression of proteins and peptides in the CP-CSF system stressed by AD. Previous publications have treated the ultrastructural pathophysiology of the choroidal epithelium and the subjacent basement membrane and stroma in a variety of dementias and in hydrocephalus (Serot et al., 1997a, 2000, 2003; Weaver et al., 2004). Moreover, in an excellent review of the CP as a target and source of polypeptides, Chodobski and Szmydynger-Chodobska (2001) have discussed the functional significance of amyloid fragment clearance from the CSF of AD patients. It is also useful to analyze peptide transport data from blood–brain barrier investigations (Banks et al., 1988; Shibata et al., 2000) in order to eventually integrate such information on AD into CSF models.

When interpreting CSF malfunction in neurodegenerative states, it is essential to relate it to the appropriate baseline (untreated or disease-free) parameters that accompany normal aging. This is especially true for assessments of AD dementia, which is age-related and is progressively exacerbated by advancing age. Preston (2001) and Serot et al. (2001b) have thoroughly summarized the characteristics of the aging CP-CSF system in mammals. In order to improve comparisons of dementia states, it is also important to distinguish, whenever feasible, between the various disease phases, such as Braak stages I-II, III-IV, and V-VI in humans, and mild vs. moderate vs. severe pathological changes in experimental animals. The standardization of stages should also help to evaluate future studies of the CP and ventricular pathology in transgenic mice with various AD phenotypes.

12.3 CHOROID PLEXUS PROTEINS AND PEPTIDES IN HEALTH AND DISEASE

There are marked differences as well as similarities in regard to protein transport and expression by the parenchymal cells or barrier elements of the blood–CSF and blood–brain interfaces (Pollay and Roberts, 1980). In AD there may be simultaneous injury to both the CPs and the cerebral capillary endothelial cells, especially at an advanced stage of the disease. While it is beyond the scope of the present treatise to include the blood–brain barrier disruption in AD, we do discuss the modified synthesis/secretion by the plexus of protein and peptide compounds in several classes, i.e., heat shock proteins, growth factors, fluid-regulating hormones, carriage proteins, and protease inhibitors; the transport of amyloid peptide fragments into and out of CSF has been discussed elsewhere (Chodobski and Szmydynger-Chodobska, 2001) and so will not be covered herein. The molecular distribution model involving the CP-CSF nexus has been useful, in the context of volume transmission (Johanson et al., 2003), to describe functional relationships between large-cavity CSF and adjacent brain. When possible, the CP information compiled below is related to corresponding findings for CSF. More attention needs to be paid to modified volume transmission of CSF-borne substances when the choroidal epithelium is stressed by degenerative processes.

12.3.1 Choroid Plexus, Oxidative Stress, and Neurodegeneration

The central nervous system (CNS) transport-interfaces and meninges are well equipped biochemically to remove (by clearance transport) and metabolize molecules detrimental to brain and choroidal function. In aging and in neurodegenerative states, there is an increased burden of free radicals and oxidants in the CSF and cerebral ISF. Potentially deleterious molecules, e.g., some organic anions and cations, are actively removed by the choroidal epithelium for enzymatic breakdown or eventual disposal into the venous blood draining the plexuses.

A variety of antioxidant systems in the choroidal epithelium help to reduce or catabolize harmful substrates generated in the CNS by normal as well as pathological processes associated with the progression of AD and other diseases. Ghersi-Egea and Strazielle (2001) have reviewed the plethora of enzymes in the CP that inactivate xenobiotic compounds and harmful metabolites, thereby restricting their access to the brain. Although it is beyond the scope of this chapter to survey all the degradation pathways in the choroidal epithelium of the blood–CSF barrier, several of the major antioxidant systems involved in redox balance are summarized in Table 12.1.

Because, the CP is the site of major trafficking of a wide range of proteins, peptides, and other organic solutes into and out of the CSF, it is essential that this choroidal epithelial system be able to protect itself, as well as the brain, from injurious by-products of metabolism. A high level of glutathione in the plexus promotes detoxification—by way of conjugation and glutathione S-transferase activity (Otieno et al., 1997). Glutathione confers protection against reactive oxygen species. Plasma-borne glutathione is filtered across the permeable choroidal capillaries and, following epithelial uptake, is available for transport into the CSF; upon accessing the ventricles, it is moved by bulk flow to target cells throughout the brain to aid in defending against oxygen-related toxicities (Shikama et al., 1995). Unfortunately, however, under certain circumstances, glutathione may become a toxin,

TABLE 12.1
Antioxidant Gene Expression in Choroid Plexus Epithelial Cells

Protectors	Localization	Analysis	Species	Investigators
Glutathione S-transferase	Nucleus of epithelium	Immunostain	Rat	Otieno et al., 1997
Paraoxonases (PON1)	Apical membrane of epithelium	Immunostain	Rat	Rodrigo et al., 2001
Metallothionein	Epithelium	mRNA	Rat	Rojas et al., 1996
Thioredoxin	Epithelium	mRNA	Rat	Lippoldt et al., 1995
Superoxide dismutase	Epithelium	Immunostain	Human	Uchino et al., 1994

i.e., some glutathione-xenobiotic conjugates may introduce toxicity to the CNS (Monks et al., 1999).

Several other enzymes in the plexus inactivate or reduce oxidative stress reactions (see Table 12.1). Paraoxonases (PONs) have been identified in a variety of tissues due to interest in their ability to prevent peroxidative damage to phospholipids and cholesteryl-esters in cell membranes (Rodrigo et al., 2001). PON1, expressed in the proximal renal tubule, is also found on the ventricular surface of the CP, where it likely helps to attenuate oxidative stress. Moreover, mRNA transcripts for metallothionein I and thioredoxin have been found in choroidal epithelium. Metallothionein regulates intracellular redox potential; when oxidative stressor agents are injected into the cerebral ventricles, there is a compensatory upregulation of metallothionein I in the plexus and circumventricular regions (Rojas et al., 1996). Thioredoxin catalyzes protein disulfide reductions and is expressed in CP as well as in many neurons; a role for thioredoxin in brain regeneration, following injury and oxidative stress, has been proposed by Lippoldt et al. (1995). Thioredoxin expression in AD awaits elucidation.

Superoxide dismutase (SOD) is differentially expressed in neurons and the CP in certain neurodegenerative diseases. In amyotrophic lateral sclerosis (ALS), for example, there is sparse staining of neurons with both anti-manganese and copper/zinc antibodies, but heavy immunoreactivity is seen in the choroidal epithelium (Uchino et al., 1994). Due to the importance of dismutase activity in reducing superoxide accumulation in cells, there is a need for further investigation into possible reactive (compensatory) expression of SOD at barrier interfaces in AD.

The preponderance of antioxidant capabilities in the plexus, as surveyed above, undoubtedly serves this tissue well in the early stages of neurodegenerative disease. At some point, however, aging and degenerative changes, as seen in advanced AD, lead to progressively increasing oxidative stress loads that may overwhelm these barrier mechanisms, leading to a compromised ability of the choroidal epithelial cells to maintain homeostasis. The end result of structural deficits and failing energy metabolism would be a curtailed manufacture and secretion of stress proteins, growth factors, and other protective peptides for the CP-CSF system. Choroidal reabsorptive mechanisms, e.g., for organic anions, may also be less efficient when CSF inflammatory processes accompany neurodegeneration (Strazielle et al., 2003).

12.3.2 Heat Shock Proteins

Not usually expressed in healthy tissues, the various heat shock proteins (HSPs) are synthesized in response to homeostatic disruptions caused by injury, disease, or temperature fluctuation. Known as stress proteins, the HSPs are markers of cell injury. HSP expression is linked, at least in part, to oxidative stress. To minimize the build-up of stress-damaged proteins in the cell, HSPs act as "molecular chaperones" to inhibit aggregation of denatured proteins, promote proper conformational refolding, and assist in degradation (Yoo et al., 2001). HSPs can be divided into families on the basis of molecular size: small, intermediate, and large molecular weights (see Table 12.2). As expected, there are substantial changes in brain expression of HSPs as the result of perturbed neuronal and glial functions in AD. Although

TABLE 12.2
Heat Shock Protein Immunostaining of Human Choroid Plexus in Alzheimer's Disease[a]

HSP Type	HSP Family	Compartments	Staining Intensity	Expression (vs. Control)
HSP 32 (HO-1)	Small	Stroma and epithelium	3.8	↓ ($p < 0.05$)
Ubiquitin	Small	Epithelium	0	No change
Alpha-B-crystalline	Small	Epithelium	0	No change
HSP 60	Intermediate	Stroma/epithelium	3.2	No change
HSP 70	Intermediate	Stroma	0.7	No change
GRP 78	Intermediate	Epithelium	2.4	↓ ($p < 0.05$)
HSP 90	Large	Epithelium	3.1	↑ ($p < 0.05$)
GRP 94	Large	Stroma/epithelium	1.4	↓ ($p < 0.05$)

[a] Plexus tissues were from lateral ventricles. Staining degree was evaluated on a scoring system of 0–5 (see Anthony et al., 2003).

there is limited information on HSP profiles of CP epithelium in AD, it is evident that the choroidal response cannot be explained simply as the consequence of oxidative stress per se.

12.3.2.1 Small HSPs

HSP 32, otherwise known as heme oxygenase-1 (HO-1), is one of the smallest stress proteins, with a molecular weight of 32 kDa. HO pathways defend cells against oxidant challenge. Relative to age-matched controls, there was substantially less expression of HSP 32 in AD choroidal epithelial cells (see Figure 12.1) (Anthony et al., 2003). This is consistent with a 40% reduction in the CSF concentration of HO-1 in AD patients, i.e., to 19 ng/mL (Schipper et al., 2000). Two other small HSPs, ubiquitin and alpha-B-crystalline, do not undergo expression changes in the AD plexus (see Table 12.2). Overall, the absence of a profile of upregulated small HSPs in the oxidatively stressed AD plexus raises the question of what other factors might regulate HSPs in this secretory tissue.

The plexus, like the brain, is stressed by cytokine and amyloid burdens in AD. Interestingly, neurons and glia exposed to cytokine and Aβ toxicities in AD respond to oxidative stress by enhancing HO-1 expression (Schipper et al., 1995). CP responds oppositely. Thus, the lack of upregulated small HSPs (see Table 12.2) may be due to factors other than oxidative stress, e.g., plasma components, that differentially modulate the plexus and brain. One such possibility is the presence of HO-1 suppressor activity in AD patients' plasma (Kravitz et al., 2002). Water-soluble factors in plasma readily penetrate the permeable choroidal capillaries but not their cerebral counterparts, which have tight junctions. Therefore, choroidal cell expression patterns are complexly influenced by both systemic (basolateral membrane exposure) and central (CSF or apical membrane) modulation.

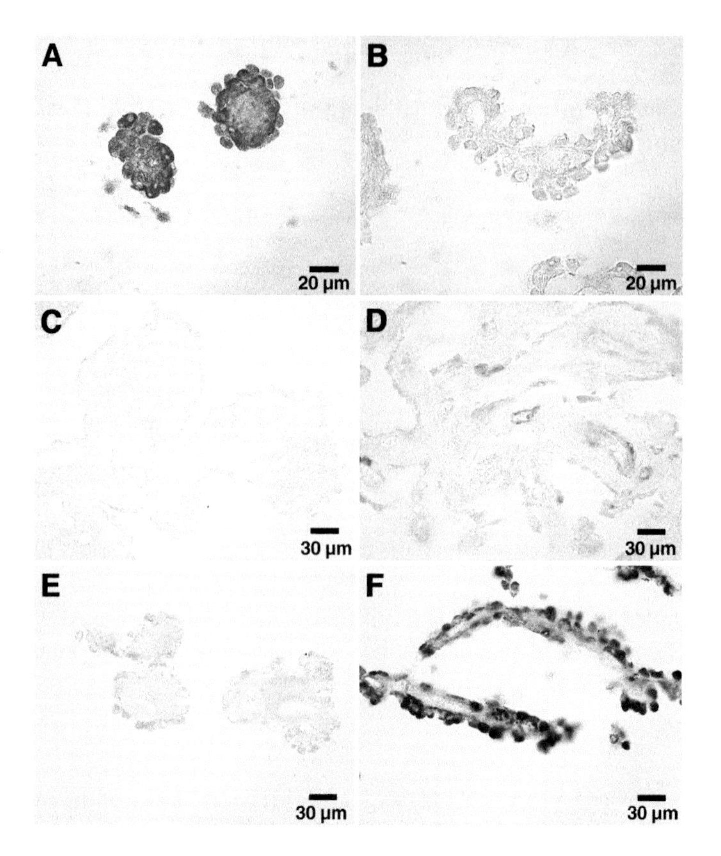

FIGURE 12.1 Expression of heat shock proteins in human choroid plexus. Controls are on left (A, C, and E) and corresponding AD specimens (mainly Braak stage VI) are on the right side of the figure (B, D, and F). See Anthony et al. (2003) for description of the immunostaining procedures. (A) Stress protein GRP 94 immunostaining in cytoplasm and stroma of control lateral ventricle plexus. (B) Lower level of GRP 94 staining in AD plexus. (C) Negligible staining of HSP 27 in control choroidal tissue. (D) Enhanced staining of HSP 27 mainly in the endothelial cells of the AD plexus. (E) Low level of HSP 90 expression in control plexus tissue from aged controls. (F) Robust staining of HSP mainly in the cytoplasm of the epithelium from an AD specimen (Panels A, B, E, and F are reproduced with permission from the *Journal of Alzheimer's Disease*.)

Moreover, the lack of small HSP induction in the AD plexus can be analyzed from another point of view involving two diametrically opposed, but equally valid, interpretations: (1) the CP in AD experiences relatively low levels of oxidative stress and therefore needs not upregulate the small HSPs, or (2) the CP may be incapable of mounting an appropriate small HSP response to oxidative stress, thereby rendering the CP more vulnerable to oxidative damage (relative to brain parenchyma). To differentiate between these possibilities, one would need to measure direct indices of oxidative damage.

12.3.2.2 Intermediate HSPs

In AD plexuses, compared to appropriate controls, there are no significant alterations in the expression of HSP 60 and HSP 70 (see Table 12.2). However, a stress protein that is glucose regulated, GRP 78, displayed a 40% reduction in immunoreactivity in human choroidal tissues (Anthony et al., 2003). This is in contrast to AD brain, in which augmented levels of GRP 78 were observed in hippocampal neurons that were cytologically normal (Hamos et al., 1991). While oxidative nucleic acid damage or general impairment of protein biosynthesis may curtail the expression of GRP 78 in human CP, it seems more likely that a specific derangement in glucose or calcium metabolism explains this abnormality. More information is needed to clarify why, for instance in AD, the plexus is evidently less capable than the brain parenchyma of upregulating specific HSPs in response to stress. One possibility is that the choroidal epithelial cells sustain earlier and greater damage in AD/aging than many of their brain parenchymal cell counterparts.

12.3.2.3 Large HSPs

The largest HSPs have molecular weights of 90 kDa or more. As is the case for the smaller GRP 78, the immunostaining of GRP 94 in the CP of AD is also less than in controls (see Table 12.2). The GRPs usually respond to a more restricted range of stimuli, such as glucose deprivation or calcium ionophores, rather than to widespread intracellular oxidative stress. Diminished GRP expression has been attributed to specific malfunctions in signal transduction pathways, as well as to the more general cellular damage involving nucleic acid and protein syntheses; however, exact mechanisms await elucidation. It is noteworthy that of eight stress proteins analyzed in human AD plexus by Anthony et al. (2003), there was significantly enhanced immunostaining only for HSP 90 (see Table 12.2). The ability of HSP 90 to form intracellular complexes with several protein kinases points to possible extensive downregulation of gene transcription in the AD plexus. Because HSP 90 functions as a steroid receptor chaperone protein, its overexpression in AD could pathologically interfere with glucocorticoid modulation of CSF formation.

12.3.2.4 Overview of HSP Findings

In CP specimens from AD patients, analyzed immunohistochemically, the level of many HSPs is not augmented as it is generally in the brain parenchyma. The interpretation of HSP expression responses by the CP, as well as the degree of HSP immunostaining observed, suggests a number of complicating factors. Unlike the brain, the CP sees injury signals and regulatory molecules at both sides of its epithelial surfaces, i.e., from the plasma and CSF. Due to the secretory nature of the blood–CSF interface, the choroidal cell levels of proteins are determined by their relative rates of synthesis versus release; hence, caution is in order in interpreting immunostaining phenomena. Other variables in comparing brain versus CP in AD include possible differences in cellular damage and redox enzyme profiles. CP experimentation that includes transcript analyses and cultured cells (Zheng and Zhao, 2002) can shed more light on HSP function at the blood–CSF barrier.

12.3.3 Growth Factors

Growth factors have a major role in the development, maintenance, and repair of the CNS. There is abundant growth factor expression in the choroidal epithelium (Johanson et al., 2000). Many growth factors synthesized by the plexus are secreted into ventricular CSF, from which they are taken up by the brain to promote ontogenic processes (Miyan et al., 2003). Growth factor expression by choroidal epithelium is extensive during fetal development, wanes in young healthy adults, but can be substantially upregulated in later life following trauma, stroke, or neurodegenerative disease. Growth factors subserve multiple functions, including angiogenesis (Luciano et al., 2001), mitosis (subventricular zone), brain fluid balance, and neuroprotection (preventing programmed cell death). There are specific spectra associated with the various growth factors pertaining to effects exerted on neurons, endothelium, and epithelium.

12.3.3.1 Basic Fibroblast Growth Factor (FGF-2)

FGF-2 immunostaining of the plexus epithelial cells occurs in the nucleus and cytoplasm; in both control human and rat tissues (Stopa et al., 2001; Hayamizu et al., 2001), staining is most intense at the apical membrane, where the peptide is thought to be released into the CSF (Szmydynger-Chodobska et al., 2002). In the human AD plexus, prominent FGF-2 immunoreactivity has been noted in the epithelium as well as in the choroidal interstitium. Moreover, in AD there is enhanced punctate staining, near membranes, for FGF receptors in human CP epithelium. Curiously, FGF-2 is sequestered in Aβ plaques. Even though it is established that FGF-2 levels are increased in the AD brain (Stopa et al., 1990), more information is needed on the FGF-2 titers in AD CSF at various Braak stages and the associated kinetics of FGF-2 transfer between CSF and cerebral ISF. This may provide insight into FGF-2's presence in Aβ plaques in the brain.

12.3.3.2 Transforming Growth Factor Beta (TGFβ)

In AD there is an elevation of TGFβ isoforms in the CSF (Chao et al., 1994; Tarkowski et al., 2002). As it has a role in repair at brain injury sites, TGFβ in CSF can gain access to many regions in the CNS. The CP manufactures TGFβ isoforms 1, 2, and 3 (Knuckey et al., 1996) and therefore is a source of these peptides for the CSF. Moreover, Vawter et al. (1996) have offered evidence that in neurodegenerative disease there is leakage of TGFβ from plasma to ventricular CSF. Thus, the increased levels of TGFβ in CSF may be due to passive permeation of the peptide at the blood–CSF barrier as well as to upregulated synthesis of TGFβ by the choroidal epithelium in response to brain injury (Logan et al., 1992; Knuckey et al., 1996). Whereas TGFβ confers protection against ischemia and immunologically mediated neuronal injury, it may also promote amyloidogenesis (Tarkowski et al., 2002). Beneficial versus detrimental actions should be analyzed in regard to TGFβ expression in aging and neurodegeneration, particularly concerning the effects of peptide concentrations and exposure times.

12.3.3.3 Vascular Endothelial Growth Factor (VEGF)

Both the endothelial and epithelial cells in the plexus express VEGF (Sandstrom et al., 1999; Chodobski et al., 2003). The expression of VEGF in the epithelial compartment of the human plexus in AD is similar to that seen in patients without dementia (Stopa et al., 2001). This comparable expression of VEGF was determined by both immunohistochemistry and ELISA. Still, in an extensive Swedish study, elevated levels of VEGF were found in the CSF of AD and vascular dementia patients (Tarkowski et al., 2002). Augmented immunoreactivity of VEGF in AD neocortex has been attributed to compensatory angiogenic mechanisms to counter insufficient vascularity or reduced perfusion (Kalaria et al., 1998). Delivery of the *VEGF* gene into the subarachnoid space results in significant improvement in cerebral blood flow in a stroke model (Yoshimura et al., 2002), suggesting a possible application to the cerebral hypoperfusion seen in AD. The relationship in AD between elevated levels of VEGF in CSF and in brain deserves further exploration, as does the source of VEGF in CSF, i.e., choroidal or extrachoroidal. We postulate that the CP, via bulk flow of CSF, furnishes VEGF to the CNS to help repair regions insulted by AD degeneration.

12.3.3.4 Hepatocyte Growth Factor (HGF)

Choroidal epithelial cells express HGF in response to injury (Hayashi et al., 1998). Upon its secretion into CSF, HGF can exert angiogenic and neurotrophic effects on the brain parenchyma. There is a significantly higher titer of HGF in the CSF of AD patients than in controls (Tsuboi et al., 2003). Similarly, there is a greater concentration of HGF in cerebral cortex and white matter of AD specimens than in nondemented counterparts (Fenton et al., 1998); the HGF immunoreactivity is associated with senile plaques but not neurofibrillary tangles. The high CSF levels of HGF in AD correlate with the extent of white matter damage (Tsuboi et al., 2003). Given the ability of intrathecal HGF to promote tissue vascularization, blood flow, and neurotrophic effects generally, it is tempting to hypothesize that the greater amounts of HGF in the CSF and CNS in AD is a homeostatic response to tissue damage incurred in the dementia.

HGF can counter fibrosis, e.g., as induced by TGFβ in the kidney (Gong et al., 2003). It is therefore important to ascertain whether HGF can prevent or minimize the CNS fibrosis, e.g., in the arachnoid membrane, that is induced by CSF-borne FGF-2 and TGFβ (Johanson et al., 1999b; Whitelaw et al., 2004). This is of significance because the elevation of both FGF-2 and TGFβ in the AD brain may disrupt CNS fluid homeostasis due to interference with CSF dynamics, particularly CSF absorption. CSF growth factor excess may also contribute to the ventriculomegaly (Johanson et al., 1999b) that commonly occurs in AD patients.

12.3.4 Fluid-Regulating Neuropeptides and Hormones

A key issue in AD progression is a reduced CSF renewal. The decrease in the CSF turnover, defined as the volume of CSF produced in 24 hours ÷ the volume of the CSF space, has a dual cause: diminished CSF production by the plexus and enlargement of the ventricles and subarachnoid space into which the choroidal fluid is secreted (Silverberg et al., 2001, 2003). Because of the histopathological changes in the CP with advancing age (Preston 2001), which are coincidental with the progressive ex vacuo hydrocephalus in AD neurodegeneration, there is a continual diminution in CSF secretion and turnover and a decrease in the "sink action" important in clearing catabolites from the brain ISF. Indeed, in elderly patients with hydrocephalus, another disease characterized by decreased CSF production and enlarged ventricles, there is a much higher than expected coincidence of AD pathology (Silverberg et al., 2003). Another factor in the curtailed CSF production during aging and dementia is the altered titers of CSF hormones and growth factors capable of modulating fluid balance.

12.3.4.1 Basic Fibroblast Growth Factor

Recently, FGF-2 has been implicated in fluid-regulatory functions mediated by the CP and the hypothalamic-pituitary axis (Gonzalez et al., 1994). FGF-2 expression in the choroidal epithelium and pituicytes, for example, is markedly upregulated in response to dehydration (Johanson et al., 2001), a disorder that occurs both in aging and in AD. In both in vivo and in vitro experiments, exogenously administered FGF-2 results in decreased CSF formation (Johanson et al., 1999b; Hakvoort and Johanson, 2000). Information is lacking, however, for CSF levels of FGF-2 at the various Braak stages of AD. Interestingly, there is a FGF-binding protein (FGF-BP) in the CSF of AD and control patients. The FGF-BP probably has roles in transporting, sequestering and/or delivering FGF peptides to target cells in the CNS (Haaneken et al., 1995). The procurement of additional data for CNS FGF-2 and its binding protein in dementia syndromes is a useful goal for translational research in the context of CSF volume transmission and reabsorption. Another area requiring more analysis is the co-localization and putative functional coupling of FGF-2 with arginine vasopressin (Szmydynger-Chodobska et al., 2002), especially in relation to compensatory adjustments in CSF formation (Faraci et al., 1990; Johanson et al., 1999a).

12.3.4.2 Arginine Vasopressin (AVP)

The CP is a likely source of AVP in the CSF (Chodobski et al., 1997, 1998b), although the hypothalamus is probably also a contributor. The presence of vasopressin-associated neurophysin in the choroidal epithelium is consistent with a hypothalamic-like biosynthesis of AVP from its prohormone. Centrally secreted or administered AVP in animal models leads to reduced CSF formation, an effect that is mediated by V1 receptors in the plexus (Faraci et al., 1990; Chodobski et al., 1998a).

AD patients are vulnerable to dehydration. Therefore, their AVP response to water imbalance is of interest. Upregulation of AVP-binding sites in CP specimens obtained at autopsy from patients with AD has been reported by Korting et al. (1996). Such choroidal upregulation, i.e., an elevated B_{max} for V1a-antagonist binding, is consistent with AD reductions in CSF levels of AVP (Mazurek et al., 1986; Raskind et al., 1986) and the associated neurophysin (North et al., 1992). Reduced levels of serum AVP in response to thirst following dehydration from fluid restriction has been noted in AD patients (Albert et al., 1989); individuals with advanced cognitive impairment seem to be particularly susceptible to dehydration due to altered endocrine responses to thirst (Albert et al., 1994).

Because the cholinergic system is a modulator of vasopressin-producing neurons, it is of interest to ascertain the effect of acetylcholinesterase inhibitors on AVP secretion. However, in patients with AD, physostigmine did not alter the AVP in CSF or plasma or the response of the human subjects to hypertonic saline infusion (Peskind et al., 1995). There may be changing titers of AVP in the CSF during early versus later stages of AD neurodegeneration (Jolkkonen et al., 1989). More information is needed to elucidate the CP-CSF vasopressin interactions in AD. It would be useful, therefore, to quantify CSF AVP concentrations at various Braak stages. Moreover, it would also be instructive to evaluate aging per se in the relationship between plexus V1 receptor expression and CSF levels of AVP.

12.3.4.3 Melatonin

Neuroendocrine regulation is diversely modified by AD pathophysiology. Melatonin, like AVP, functions via neuroendocrine pathways that span the circumventricular organs, hypothalamus, and CSF. Reduced levels of melatonin in the CSF occur early in AD, even before the onset of clinical symptoms (Zhou et al., 2003). Postmortem analyses of ventricular fluid reveal an inverse relation between CSF melatonin concentration and Braak stage. Individuals with even a few neurofibrillary tangles and neuritic plaques in the temporal cortex have greatly decreased titers of melatonin in their ventricular CSF (Zhou et al., 2003). Moreover, patients with apolipoprotein E 4/4 alleles have lower levels of CSF melatonin than those with the 3/4 genotype (Liu et al., 1999). Clearly, the most severe cases of AD are associated with substantial melatonin deficiency.

CSF melatonin comes primarily from the pineal gland, the activity of which is suppressed in aging and AD (Wu et al., 2003). The pineal gland is subject to progressive age-related changes and pronounced early calcification. Even in cognitively intact subjects, i.e., Braak stages I-II, CSF melatonin levels are low compared to age-matched controls. An early decrease in CSF melatonin may lead to oxidative injury to the CP. Alterations in both the concentration and circadian rhythm of CSF melatonin, occurring in both preclinical and advanced AD, are due to disrupted monoaminergic regulatory systems in the pineal gland. Because sleep disturbances in both aged and demented patients are attributable to a shortfall in CSF melatonin, it is important to pursue initial observations that medically supplemented melatonin can ameliorate the nightly sleep disturbances attending AD.

It is clinically relevant to consider the relationship between CSF neurohumoral agents, such as melatonin, and CSF dynamics. Choroidal tissues are stimulated by exogenous melatonin, pointing to a greater secretory activity by the epithelial cells (Decker and Quay, 1982). Human CSF formation is elevated at night (Nilsson et al., 1994) at a time when CSF melatonin levels are also augmented (Kanematsu et al., 1989; Bruce et al., 1991). Thus, CSF-borne melatonin and other substances probably have a key role in promoting the volume transmission that mediates physiological mechanisms like sleep and wakefulness, which require facilitated movement of neuroendocrine signals between CSF and hypothalamic nuclei. Consequently, an interference with melatonin distribution pathways, related to compromised CP-CSF dynamics, may partly explain the sleep disorders that often accompany AD and other dementias.

Another facet of melatonin's wide spectrum of biological actions is its antioxidant capability. Normally, melatonin is secreted into the CSF at a concentration higher than in blood. In healthy adults, melatonin is transported down the CSF neuraxis, where it undoubtedly has beneficial antioxidant effects on the CP and on brain tissue surrounding the ventricles. In the CSF system of AD-afflicted individuals, the combined deficiencies in melatonin concentration and fluid turnover are thought to result in drastically reduced delivery of melatonin to periventricular neurons. One working hypothesis is that an inadequate CSF melatonin level in AD enables hydroxyl radicals to damage neuronal mitochondria in ventricle-bordering regions like the hippocampus (Maurizi, 2001). However, the CNS has numerous antioxidant systems (see Table 12.1); therefore, although deficient melatonin availability may contribute to AD pathophysiology, it is highly likely that it is the collective paucity of various antioxidants that leads to damage of the neuropil and the CP.

12.3.5 Vitamin Transporters

The brain's need for a steady supply of water-soluble vitamins having antioxidant capability is largely met by transport via the CP-CSF system. Because cerebral capillaries are impervious to plasma ascorbate, the brain is highly dependent upon CP secretion for its supply of vitamin C (Rice, 2000). Facilitated and active transporters in the basolateral and apical membranes of the choroidal epithelium translocate vitamins B and C from the blood into ventricular fluid (Spector and Johanson, 1989).

CSF vitamin depletion in AD subjects (Ikeda et al., 1990) begs the question of whether the problem is limited substrate availability (from plasma) or choroidal vitamin transport disability. At least in early AD, the CSF level of ascorbate is stable (Paraskevas et al., 1997). In a study by Quinn et al. (2003), the enhanced CSF/plasma ratio of vitamin C in AD patients, compared to controls, was interpreted as furnishing more antioxidant capability for injured brain tissue. Other investigations have revealed decreased ascorbate in the CSF of AD subjects, which was restored to normal by vitamin C supplementation (Kontush et al., 2001b). Therefore, the evidence points to functional Na-ascorbate transport in the CP of some AD-afflicted patients. Consequently, CNS vitamin C deficiency in AD may be correctable by dietary supplements.

Folate concentration is maintained close to normal in the CSF of mild and moderate senile-onset AD patients. In fact, the CSF/plasma folate ratio is actually increased in AD subjects (4.20:1) compared to control (3.13:1); however, this augmented ratio mainly reflects a 36% reduction in the plasma concentration of folate (Abalan et al., 1996). A more recent study of CSF folate concentrations in late-onset

AD subjects furnished evidence for decreased (by 18%) CSF concentration compared to elderly controls (Serot et al., 2001a). This reduced level of CSF folate was offered as support for compromised transport of folate by the CP in AD patients; such a conclusion would be consistent with the choroidal epithelial atrophy noted in AD. However, a plasma folate reduction also has to be considered as a potential contributor to the modest decrement in CSF folate seen in AD. Overall, it appears that CSF vitamins are fairly well maintained even in moderate AD. Information is needed on the homeostasis, or lack thereof, of the CSF vitamin content in the most severe stages of AD.

12.3.6 Iron-Transporting Proteins

In order to control iron passage from blood to CP and CSF, the blood–CSF barrier makes use of at least three transporters: the transferrin receptor (TfR) protein, the divalent metal transport 1 protein (DMT1), and the hemochromatosis protein (Hfe) (see also Chapter 17). TfR was the first and most extensively studied iron-transporting protein (Giometto et al., 1990), but the more recently identified DMT1 and Hfe proteins also have roles in regulating iron fluxes. Iron transfer analyses of choroidal tissues, including the adhering Kolmer cells (Lu et al., 1995), are important because too little or too much iron in the brain leads to neuronal malfunctioning and injury. CNS iron buildup probably causes significant oxidative damage to neurons, and iron excess in the CSF–brain system may contribute to the progression of AD and other neurological diseases. It is of interest to note that Aβ peptides may ameliorate the free radical damage to neurons caused by iron (Atwood et al., 2002).

12.3.6.1 Transferrin Receptor (TfR)

Most cells utilize the TfR to take up iron that is bound to transferrin (Tf). TfR distribution in the CNS matches the regional uptake of ^{59}Fe (Morris et al., 1992; Deane et al., 2004). The CP is a prominent site of receptors for Tf. An iron delivery or carriage protein, Tf is both transported and synthesized by the CP epithelium. The plexuses in the lateral and 3rd ventricles, but not the 4th, have a rich expression of Tf mRNA (Bloch et al., 1987). The choroidal epithelium also expresses TfR (Moos, 1996). Plasmaborne Tf diffuses out of the fenestrated capillaries in the plexus and is actively taken up by epithelial TfRs. Deane et al. (2004) perfused rat brains with Tf-bound ^{59}Fe and, by rate constant analysis, determined that the plexus initially sequesters Tf from the basolateral side of the epithelium and then slowly releases it across the apical membrane into CSF. In addition, Tf mRNA (Dickson et al., 1985) and protein (Kissel et al., 1998) are present in the epithelial cells; this choroidal manufacturing is another source of Tf for the CSF. Thus, it is generally presumed that most of the Tf in the CSF originates in the plexus tissues (Connor et al., 2001a). Tf transfer across the blood–CSF barrier is greater than that of albumin (Moos and Morgan, 2000). As an iron mobilization protein, then, Tf facilitates the distribution of iron from the plexus to CSF to periventricular brain (Moos, 2003).

Neurons, with their preponderance of TfRs, are the target cells for Tf-bound and free iron that is transported by TfRs in the CP as well as the blood–brain barrier

(Moos, 1996). An increased frequency of the C2 variant form of Tf has been linked to AD onset (van Landeghem et al., 1998). The finding of Tf in senile plaques (Connor et al., 1992) by immunostaining has prompted genetic analyses of Tf alleles in AD cohorts. Interestingly, in late-onset AD, not only is the C2 variant frequency greater than in age-matched controls, but also the C2 allele occurrence is twice as high in apo-E 4,4 individuals compared to those with no 4 alleles (Namekata et al., 1997). The age of onset of AD is earlier, by six to seven years, in patients having the C2 and E4 alleles than in AD individuals without both C2 and E4 (van Rensburg et al., 2000). However, E4 interaction with C2 in AD may be more complex than early studies indicated (Zambenedetti et al., 2003). At any rate, possible alterations in brain iron transport need to be studied in order to explain the significant associations between C2, Tf, and neurodegenerative disorders.

12.3.6.2 Divalent Metal Transporter-1 (DMT1)

DMT1 is a proton-gradient–driven iron pump (Gunshin et al., 1997) that is expressed at the choroidal epithelium (Connor et al., 2001a; Moos and Morgan, 2004). This divalent metal ion transporter is expressed at the major transport interfaces (see Figure 12.2), pointing to its key transport role in CNS iron homeostasis. This idea is supported by the observation of brain iron deficiency in the Belgrade rat, a strain with a defective DMT1 gene (Burdo et al., 1999). Moos and Morgan (2004), using antibodies targeted to different sites on DMT1, consistently found DMT1 expression in CP and in neurons, but not in the endothelium of cerebral capillaries. Studies of Sprague-Dawley rats have shown that DMT1 expression in CP at five days postnatal precedes TfR immunostaining by at least a week (Siddappa et al., 2002). The DMT1 reactive expression to altered iron levels in early development is more sensitive in neurons than in the parenchymal cells of the barriers (Siddappa et al., 2003).

Expression of DMT1 and TfR should also be assessed in aging and in AD in the context of iron mismanagement and oxidative stress. A step in this direction is the development of the heavy-chain ferritin null mutant, which mimics the iron milieu of the AD brain (Thompson et al., 2003). These mice, having enhanced expression of DMT1 and TfR and greater oxidative stress, should be useful in devising strategies for antioxidant therapy in AD neurodegeneration.

12.3.6.3 Hemochromatosis Protein (Hfe)

A third iron-transporting protein is Hfe, the gene involved in hemochromatosis. Normally Hfe acts as a modulating brake on TfR-mediated uptake of Tf by cells. However, in hemochromatosis, the mutation of Hfe renders the protein inactive, releasing the braking effect by enhancing the affinity between Tf and TfR. A considerable number of AD patients carry the Hfe mutation. The occurrence in males of an Hfe mutation together with the apo-E 4,4 alleles is associated with an even higher risk for AD (Moalem et al., 2000).

Hfe is expressed in the CP, ependyma, and cerebral vessels (Connor et al., 2001b). When mutated, Hfe is thought to cause disordered iron movement at these CNS transport interfaces, resulting in an augmented iron burden for the CNS.

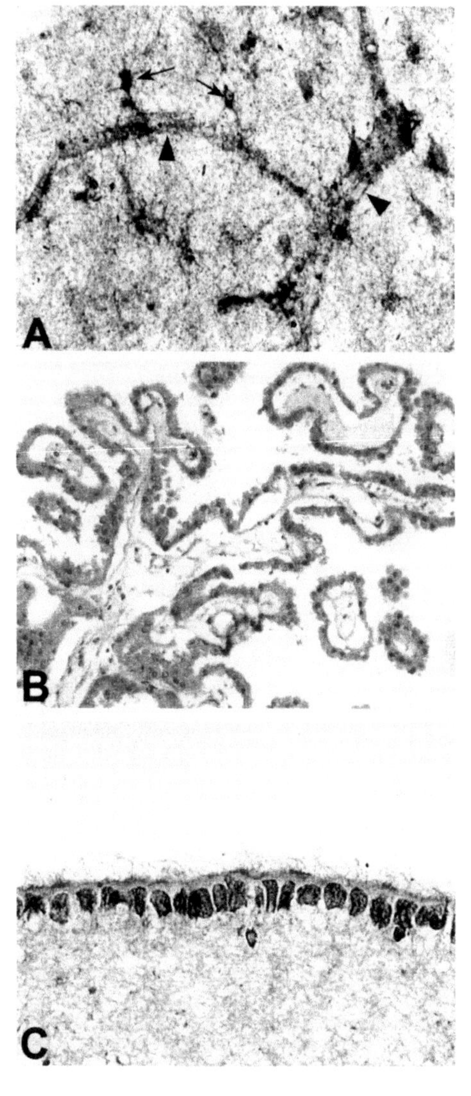

FIGURE 12.2 The divalent metal transport protein (DMT1) has recently been identified in human tissue by Connor et al. (2001a). (A) DMT1 in the cerebral cortical blood vessels (arrowheads point to endothelial cells). DMT1 is also present in astrocyte processes surrounding the blood vessels (arrows). (B) Strong DMT1 staining by the choroid plexus epithelial cells. Nuclei are counterstained. (C) Immunostaining for DMT1 in the ventricular ependymal lining. DMT1 is markedly concentrated at the apical membrane, suggesting metal uptake from CSF. (All panels are reproduced with the permission of Dr. James Connor, *Pediatric Neurology* and Elsevier.)

Consequently, in the presence of mutant Hfe, more iron can be transported across the barrier cells and subsequently taken up by neuronal TfRs to a greater degree. Such enhanced Tf-mediated iron flux increases the risk of iron-induced oxidative damage in the brain of Hfe mutants. It would be interesting to analyze human CSF samples to ascertain whether the iron level is elevated in the Hfe-mutated condition. Another factor in iron accumulation by the human brain may be the relatively long human life span, compared to animal models.

12.3.6.4 Melanotransferrin (MTf)

Yet another iron-transporting protein at the CNS barrier interfaces is MTf. This protein also has significance in AD. MTf levels in CSF are elevated in AD, which makes it a useful disease marker. It has been hypothesized that MTf, which is associated with senile plaques, originates at least in part from reactive microglia (Yamada et al., 1999). MTf mRNA has also been localized to human brain capillary endothelium. The involvement of the CP-CSF nexus in the distribution dynamics of MTf needs to be assessed.

12.3.6.5 Iron Influx Versus Efflux

Whereas considerable progress has been made in understanding the roles of various transporters and pumps in the uptake and distribution of iron in the CNS, there is far less appreciation of the efflux mechanisms for removing iron. In the healthy adult at steady state, the iron influxes at the blood–CSF and blood–brain barriers must be balanced by CSF and interstitial outflow mechanisms to regulate brain iron at the appropriate level. With the exception of a study by Banks et al. (1988), there is a paucity of information on reabsorptive transport of Tf and iron across the CNS barrier systems. Nor has there been sufficient quantification of brain iron turnover in either the normal or neurodegenerative states. Intraventricular injections of iron and Tf are cleared within hours by volume transmission of CSF into venous blood (Moos and Morgan, 1998). That CSF bulk flow is the primary elimination route for Tf is indicated by the demonstration that intraventricularly administered [^{125}I]Tf and [^{131}I]albumin are "washed out" of the CSF system at a similar rate.

Is the CSF a "source" or a "sink" for brain iron? The answer is both, depending upon factors such as CNS developmental stage or disease progression. During postnatal maturation, the rat brain needs a substantial amount of iron. Moos and Morgan (2000) demonstrated that an intraventricular injection of radiolabeled Fe and transferrin led to a higher degree of penetration of infant rat brain than the adult counterpart. The authors attributed the greater uptake by postnatal brain to a lower CSF turnover rate in infant rats (Johanson and Woodbury, 1974), thereby enabling a sustained concentration of tracer in CSF; thus, the larger concentration gradient, from CSF to brain, would promote a greater net uptake of test marker. In healthy adult rats, however, as the CSF formation rate increases and the barriers tighten, the CSF would exert an opposite effect, i.e., a "sink action" on the brain iron (Moos, 2002; Deane et al., 2004).

In severe AD, the CNS barriers become less restrictive, and the CSF turnover decreases (Silverberg et al., 2001, 2003). Such alterations can attenuate the CSF sink action and probably promote iron retention in the brain, especially if homeostatic clearance mechanisms, such as CSF bulk flow, are compromised. Brain iron overloading is a consistent finding in AD (Connor et al., 2001b). Investigators of brain iron burdens, including models for aging and AD phenotypes, should consider assessing CSF dynamics in relation to iron retention. Because excessive iron can generate harmful levels of reactive oxygen, it seems pertinent to continue to investigate iron regulation by the CP-CSF system in neurodegenerative diseases.

12.3.7 Protease Inhibitors

The choroidal epithelium actively secretes cystatin C into CSF at a concentration exceeding that in plasma. Cystatin C, a cysteine protease inhibitor, is a highly conserved protein with abundant mRNA in the plexus, implying important functions in both brain development and repair (Tu et al., 1992). Cystatin C has been associated with neuronal repair following insults to the hippocampus (Palm et al., 1995). As a protease inhibitor that is carried by CSF bulk flow, cystatin C likely has a role in neuronal repair or regenerative attempts following AD-related injury.

Cystatin C colocalizes with Aβ in senile plaques (Levy et al., 2001). Therefore, there is considerable interest in the nature and source of the cystatin C that envelops Aβ. Cystatin C is expressed in reactive astrocytes and in some neurons, thus the CSF is not the only source of this protease inhibitor in the CNS. Still, the CSF is of interest in regard to being a possible supplier of cystatin C to the neuropil. There is a dearth of information about cystatin C levels in the CSF and serum of AD patients. Apparently only Kalman et al. (2000) have reported cystatin C concentrations in CSF samples from demented patients; they found no differences between controls and individuals with late-onset AD or vascular dementia.

Cystatin C mutations, however, have been genetically linked to AD (Goddard et al., 2004). Moreover, in transgenic mice expressing a mutated amyloid precursor protein, there are high levels of cystatin C in astrocytes that encompass Aβ plaques and a discrete layer of cystatin C adhering to the plaque cores (Steinhoff et al., 2001). Accordingly, the interaction between Aβ and cystatin C is of primary interest, especially in regions susceptible to plaque formation, e.g., cortical layers III and V (Deng et al., 2001). In one homozygous variant of cystatin C, i.e., the B/B, which predisposes to AD development, there is an impairment of cystatin C secretion (by fibroblasts) due to inefficient cleavage of the signal peptide (Benussi et al., 2003). This begs the question about possible reductions in choroidal secretion of cystatin C. Overall, more attention should be focused on cystatin C variants in the CSF of AD patients and how the concentration and distribution of cystatin C throughout the CNS could be affected by altered CSF dynamics in aging and age-related dementias.

Another protease inhibitor synthesized and secreted by the CP that has relevance to AD pathology is α_1-antichymotrypsin (ACT) (Aldred et al., 1995). ACT is a serine protease inhibitor with high affinity for cathepsin G. Immunostaining for ACT is particularly prominent at the apical surface of the choroidal epithelium (Justice et al., 1987), suggesting potential secretion of this peptide into ventricular CSF. CSF levels of ACT do not significantly change in aging (Harigaya et al., 1995), but are elevated in AD (DeKosky et al., 2003), especially in late-onset patients. CSF titers of ACT reflect disease severity; therefore, this peptide may be a useful biomarker to evaluate interventions and monitor inflammation in AD.

ACT is present in CSF in monomer as well as polymer form. Polymerization of ACT, e.g., by prostaglandin E_2, decreases its protease inhibitory activity (Licastro et al., 1994). There is an increase in ACT polymers in AD patients carrying the apo-E 4 allele. It is possible that inflammatory molecules inactivate protease inhibitors like ACT, promoting amyloid deposition. Because ACT has been found in the core of Aβ plaques, there has been longstanding interest in the role of this peptide in AD plaque formation (Abraham et al., 1990). The intimate association between ACT and Aβ in AD pathology has prompted geneticists to analyze risk factors associated with ACT mutations. Recent findings by Wang et al. (2002) indicate that the ACT gene has variable sites that are associated with either greater or less risk of AD. There apparently has been no CSF analysis of ACT variants in AD patient cohorts.

12.3.8 Estrogen and Its Receptors

Several studies have demonstrated that estrogen replacement therapy reduces the risk of AD or delays its onset. Cognitive function in some AD females is enhanced by estrogen supplements (Cholerton et al., 2002). Conversely, it is remarkable that in AD, both the increased Aβ burden and the reduced metabolism in the hippocampus are associated with decreased concentrations of estradiol in the CSF (Schonknecht et al., 2001, 2003). This prompts the question of whether estrogen transport at the blood–CSF interface involves estrogen receptors, the function of which may affect the concentration of this sex steroid in CSF. Estrogen receptors alpha and beta (ERα and ERβ) are present at lower densities in the CP of AD patients than in nonAD patients (Hong-Goka and Chang, 2004). Does this imply reduced transport of estrogen from blood into the CSF of some AD patients? Oxidative stress regulates the expression of ERα and ERβ (Tamir et al., 2002); therefore, it is timely to evaluate how the altered ER density in the AD-stressed CP might affect estrogen modulation and transport at the blood–CSF barrier.

CSF-borne estrogen is carried by bulk flow throughout the ventriculo-subarachnoid system and can thereby exert widespread neurotrophic effects on the AD brain. Estrogen (17β-estradiol) confers protection against Aβ toxicity by activating protein kinase C in cultured cortical neurons (Cordey et al., 2003). In addition to exerting beneficial effects directly on the neuronal metabolism of Aβ, estrogen may also attenuate neuronal Aβ toxicity indirectly by enhancing the removal of Aβ from ISF by way of increased CSF turnover. It is interesting that CSF estrogen concentration is augmented in pseudotumor cerebri (Toscano et al., 1991), a condition in which CSF dynamics are altered. Direct analyses of estrogen effects on choroid plexus transport activity (Lindvall-Axelsson and Owman, 1990) and CSF turnover can place in perspective the functional relationships among estrogen modulation, CSF dynamics, and Aβ clearance from the CNS.

12.4 TRANSTHYRETIN: ITS ROLE IN AMYLOIDOGENESIS AND ANTIAMYLOIDOGENESIS

Transthyretin (TTR), also known as prealbumin, is the hallmark protein secreted by the choroidal epithelium into the CSF. The plexus is a major factory for manufacturing TTR; indeed, about one eighth of total choroidal protein synthesis is TTR (Schreiber et al., 2002). TTR likely has several functions, e.g., interacting with retinol-binding protein and serving as a transport protein for thyroxine in the CNS. But TTR-null mice appear to have unimpaired retinal and cerebral functions. Therefore, other critical functions of TTR have been considered, among them its ability to prevent the formation of Aβ plaques in the brain.

TTR was initially discovered in the CSF (Robbins, 2002). Its prominence in CSF is due primarily to the substantial secretion of TTR by CP. The CSF/serum ratio of 0.04 for TTR (Vatassery et al., 1991) is considerably higher than the corresponding ratio for albumin. The plexus expresses the highest concentration of TTR mRNA in the body (Schreiber, 2002). Because about 25% of the total protein in CSF is TTR, it is of interest to investigate the normal roles for TTR in the CNS and possible pathophysiological aspects of TTR in neurodegenerative diseases. In addition to TTR acting as a carriage or ligand-transporting protein, it may be that TTR plays a modulating role in stabilizing the conformation and solubility of other extracellular proteins, such as Aβ (Palha, 2002).

TTR exists as a soluble, symmetrical tetramer. Mutations spanning the entire TTR protein, however, can alter its structure. The altered molecule may then enhance amyloidogenesis. Over 80 genetic variants have been documented. Patients harboring TTR mutations present with syndromes characterized by amyloid depositions in the peripheral nervous system (Saraiva, 1995). The pathology associated with TTR variants and systemic amyloidogenesis has sparked interest in a possible role of CSF-borne TTR in amyloid modification in AD.

TTR can itself form amyloid fibrils. Amyloidosis in the CP and ependyma, in the form of curly fibers and tangles, occurs as a conjugate of TTR with amyloid peptides; such deposits may be different from true β-sheet amyloidosis seen in the neuritic plaques of AD (Miklossy et al., 1999). Still, this well-described choroidal amyloidosis may preclude the optimal transfer of molecules across the blood–CSF barrier. Because the plexus supplies many trophic molecules to the brain and removes harmful ones, any serious amyloidosis disrupting the integrity of the choroidal epithelium and interstitium could conceivably exacerbate the progression of AD and other dementias (Silverberg et al., 2001, 2003). Therefore, the nature of this choroidal amyloidosis needs to be more fully explored.

A popular hypothesis is that choroidally derived TTR has another role, i.e., in attenuating the formation of insoluble Aβ in the brain. This working model has been put forth on the basis that TTR favorably modulates Aβ conformation both in vivo and in vitro (see Table 12.3). TTR availability promotes Aβ stabilization in solution in CSF incubation experiments (Wisniewski et al., 1993; Schwarzman et al., 1994).

TABLE 12.3
Evidence for Transthyretin Stabilization of Aβ

Methodology	Species	Tissue/ Fluid	Observations	Investigators
In vitro incubation	Human	CSF	Aβ 1–40 homolog fibril formation inhibited by CSF	Wisniewski et al., 1993
In vitro incubation	Human	CSF	TTR sequestered Aβ into stable complexes	Schwarzman et al., 1994
In vivo (engineering)	*C. elegans*	Muscle	Coexpression of TTR with Aβ reduced amyloid deposits	Link, 1995
Culture	Human; Dog	Muscle	TTR blocked E3/E4 induction of cytoplasmic Aβ	Mazur-Kolecka et al., 1995
Culture	Dog (aged)	Muscle	CSF, but not serum, reduced Aβ deposits	Mazur-Kolecka et al., 1997
Biochemical analysis	Human	CSF	Lower CSF TTR concentration in Alzheimer's disease	Serot et al., 1997b
Autopsy	Human	Kidney	Amyloid deposit not observed when TTR colocalized with Aβ	Tsuzuki et al., 1997
Biochemical	Human (AD)	CSF	↑ CSF [TTR] correlated with ↓ in amyloid plaques	Merched et al., 1998
Recombinant TTR variants	N/A	PBS	Some TTR variants unable to prevent Aβ polymerization	Schwarzman et al., 2004

The greater effectiveness of CSF compared to serum (with a higher TTR concentration) in reducing Aβ deposits in cell cultures (Mazur-Kolecka et al., 1997) is an interesting phenomenon. In kidney and muscle, the co-expression of TTR with amyloid tends to minimize deposition of the latter (Link, 1995; Tsuzuki et al., 1997). The CSF of AD patients contains less TTR when compared to controls (Davidsson et al., 1997; Serot et al., 1997b; Dorta-Contreras et al., 2000). On the other hand, amyloid-mutant transgenic mice do not develop senile plaques when TTR is expressed in the hippocampus, where TTR is not normally synthesized (Stein and Johnson, 2002). In a wide spectrum of analyses, then, TTR availability is inversely related to Aβ insolubility, plaque formation, and dementia (Riisoen, 1988).

Many mutations of TTR have been described. TTR variant analysis is important because it may shed light on the ability of native TTR to maintain Aβ solubility in the brain. Nearly four dozen variants of TTR have been produced and analyzed for their ability to disaggregate Aβ (Schwarzman et al., 2004). Most of the TTR variants were able to bind to Aβ and inhibit aggregation; however, a few mutations disabled such abilities. Evidently, the β-amlyoidogenic potential of TTR variants is related to their tendency to aggregate in solution (Colon and Kelly, 1992; Quintas et al., 1997). Thus, additional in vitro analyses of TTR and Aβ interactions, as well as epidemiological studies of TTR variant occurrence, seem warranted in order to gain a deeper insight on the effects of CSF TTR on brain Aβ composition and deposition.

It would be instructive to characterize amyloid deposits in the CP to determine if they mimic the systemic amyloidosis seen in familial amyloid polyneuropathy and cardiomyopathy and to ascertain how they might affect AD neurodegeneration by way of altered CSF secretion.

12.5 CHOROID PLEXUS AND ALZHEIMER'S DISEASE: FUTURE RESEARCH DIRECTIONS

With advancing age, the extracellular fluid environment of the neurons is subject to unfavorable changes in composition. In AD, particularly during the late stages, the brain ISF compartment deteriorates even further. Part of the decline in cerebral function in AD dementia can be attributed to a failure of the CP-CSF system to carry out its homeostatic roles. As the choroidal epithelium becomes less able to generate fluid into the ventricles, the CSF turnover rate begins to decline. Although it has been shown that amyloid is deposited in the CP in an age-dependent fashion (Eriksson and Westermark, 1990), more systematic studies are needed in both animal models and humans to quantify progressive alterations in CSF formation rate and ventricular volume and to determine how these parameters correlate with changes in CSF chemical composition and plaque formation (Silverberg et al., 2001, 2003).

Another area that deserves more attention is the clearance of macromolecules from the ISF and CSF and the removal of noxious constituents from the CNS. Little is known about aging and AD phenomena at the extrachoroidal aspects of the blood–CSF barrier, i.e., the arachnoid membrane. It is becoming increasingly clear that neurodegenerative processes are exacerbated by compromised functions of the choroidal epithelium, ependyma, and meninges. An important challenge is to find ways to minimize the deterioration of CP and CSF functions both in aging and in states of neurodegeneration.

AD is a multifactorial disease with a multitude of demonstrable physiological and metabolic abnormalities. Some may be causal and others reactive. The above discussion is not exhaustive but serves to describe a number of choroidal and CSF abnormalities seen in AD —what we know and what we do not know. It is intriguing that many proteins secreted by CP co-localize with Aβ senile plaques. It is also of interest that the CSF in AD has augmented levels of growth factors but a deficiency of certain vitamins and hormones. Curiously, the ApoE 4 genotype occurs with mutations in some iron-transporting proteins and protease inhibitors. Moreover, given the data supporting the idea that a decrease in CSF turnover might increase the contact time between macromolecules in the CNS, and thereby lead to greater glycation of proteins (Shuvaev et al., 2001), it seems pertinent to focus more attention on CSF glycation products in AD. Much research will be necessary to unravel the complex causation and possible therapy of AD.

12.6 ACKNOWLEDGMENTS

The authors are grateful for support from the NIH (NS R01 27601), the Department of Neurosurgery at Rhode Island Hospital, and *Lifespan*. Appreciation is also extended to our CSF research colleagues in the Department of Clinical Neurosciences at the Brown Medical School. We thank Drs. Hyman Schipper, Torben Moos, Michael Vitek, Gilbert Faure, Jean Marie Serot, and James Connor for helpful comments about the manuscript.

REFERENCES

F. Abalan, J. Zittoun, C. Boutami, G. Dugrais, G. Manciet, A. Decamps and B. Antoniol, Plasma, red cell, and cerebrospinal fluid folate in Alzheimer's disease, *Encephale* 22, 1996, 430–4.

C.R. Abraham, T. Shirahama and H. Potter, Alpha 1-antichymotrypsin is associated solely with amyloid deposits containing the beta-protein. Amyloid and cell localization of alpha 1-antichymotrypsin, *Neurobiol. Aging* 11, 1990, 123–9.

S.G. Albert, B.R. Nakra, G.T. Grossberg and E.R. Caminal, Vasopressin response to dehydration in Alzheimer's disease, *J. Am. Geriat. Soc.*, 37, 1989, 843–7.

S.G. Albert, B.R. Nakra, G.T. Grossberg and E.R. Caminal, Drinking behavior and vasopressin responses to hyperosmolality in Alzheimer's disease, *Int. Psychogeriat.* 6, 1994, 79–86.

A.R. Aldred, C.M. Brack and G. Schreiber, The cerebral expression of plasma protein genes in different species, *Comp. Biochem. Physiol. B Biochem. Mol. Biol.* 111, 1995, 1–15.

S.G. Anthony, H.M. Schipper, R. Tavares, V. Hovanesian, S.C. Cortez, E.G. Stopa and C.E. Johanson, Stress protein expression in the Alzheimer-diseased choroid plexus, *J. Alzheimer's Dis.* 5, 2003, 171–7.

C.S. Atwood, S.R. Robinson and MA. Smith, Amyloid-beta: redox-metal chelator and antioxidant, *J. Alzheimer's Dis.* 4, 2002, 203–14.

W.A. Banks, A.J. Kastin, M.B. Fasold, C.M. Barrera and G. Augereau, Studies of the slow bidirectional transport of iron and transferrin across the blood-brain barrier, *Brain Res. Bull.* 21, 1988, 881–5.

L. Benussi, R. Ghidoni, T. Steinhoff, A. Alberici, A. Villa, F. Mazzoli, et al., Alzheimer disease-associated cystatin C variant undergoes impaired secretion, *Neurobiol. Dis.*, 13, 2003, 15–21.

B. Bloch, T. Popovici, S. Chouham, M.J. Levin, D. Tuil and A. Kahn, Transferrin gene expression in choroid plexus of the adult rat brain, *Brain Res. Bull.* 18, 1987, 573–6.

J. Bruce, L. Tamarkin, C. Riedel, S. Markey and E. Oldfield, Sequential cerebrospinal fluid and plasma sampling in humans: 24-hour melatonin measurements in normal subjects and after peripheral sympathectomy, *J. Clin. Endocrinol. Metab.* 72, 1991, 819–23.

J.R. Burdo, J. Martin, S.L. Menzies, K.G. Dolan, M.A. Romano, R.J. Fletcher, M.D. Garrick, L.M. Garrick and J.R. Connor, Cellular distribution of iron in the brain of the Belgrade rat, *Neurosci.* 93, 1999, 1189–96.

C.C. Chao, S. Hu, W.H. Frey, 2nd, T.A. Ala, W.W. Tourtellotte and P.K. Peterson, Transforming growth factor beta in Alzheimer's disease, *Clin. Diagn. Lab. Immunol.* 1, 1994, 109–10.

V.P. Chauhan, I. Ray, A. Chauhan and H.M. Wisniewski, Binding of gelsolin, a secretory protein, to amyloid beta-protein, *Biochem. Biophys. Res. Commun.* 258, 1999, 241–6.

A. Chodobski, I. Chung, E. Kozniewska, T. Ivanenko, W. Chang, J.F. Harrington, J.A. Duncan and J. Szmydynger-Chodobska, Early neutrophilic expression of vascular endothelial growth factor after traumatic brain injury, *Neurosci.* 122, 2003, 853–67.

A. Chodobski, Y.P. Loh, S. Corsetti, J. Szmydynger-Chodobska, C.E. Johanson, Y.P. Lim and P.R. Monfils, The presence of arginine vasopressin and its mRNA in rat choroid plexus epithelium, *Brain Res. Mol. Brain Res.* 48, 1997, 67–72.

A. Chodobski and J. Szmydynger-Chodobska, Choroid plexus: target for polypeptides and site of their synthesis, *Microsc. Res. Techn.* 52, 2001, 65–82.

A. Chodobski, J. Szmydynger-Chodobska and C.E. Johanson, Vasopressin mediates the inhibitory effect of central angiotensin II on cerebrospinal fluid formation, *Eur. J. Pharmacol.* 347, 1998a, 205–9.

A. Chodobski, B.E. Wojcik, Y.P. Loh, K.A. Dodd, J. Szmydynger-Chodobska, C.E. Johanson, D.M. Demers, Z.G. Chun and N.P. Limthong, Vasopressin gene expression in rat choroid plexus, *Advances Exp. Med. Biol.* 449, 1998b, 59–65.

B. Cholerton, C.E. Gleason, L.D. Baker and S. Asthana, Estrogen and Alzheimer's disease: the story so far, *Drugs Aging* 19, 2002, 405–27.

W. Colon and J.W. Kelly, Partial denaturation of transthyretin is sufficient for amyloid fibril formation *in vitro*, *Biochem.* 31,1992, 8654–60.

J.R. Connor, S.L. Menzies, J.R. Burdo and P.J. Boyer, Iron and iron management proteins in neurobiology, *Pediatr. Neurol.* 25, 2001a, 118–29.

J.R. Connor, S.L. Menzies, S.M. St. Martin and E.J. Mufson, A histochemical study of iron, transferrin, and ferritin in Alzheimer's diseased brains, *J. Neurosci. Res.* 31, 1992, 75–83.

J.R. Connor, E.A. Milward, S. Moalem, M. Sampietro, P. Boyer, M.E. Percy, C.N. Vergani, R.J. Scott and M. Chorney, Is hemochromatosis a risk factor for Alzheimer's disease?, *J. Alzheimer's Dis.* 3, 2001b, 471–477.

M. Cordey, U. Gundimeda, R. Gopalakrishna and C.J. Pike, Estrogen activates protein kinase C in neurons: role in neuroprotection, *J. Neurochem.* 84, 2003, 1340–8.

D.A. Cottrell, E.L. Blakely, M.A. Johnson, P.G. Ince and D.M. Turnbull, Mitochondrial enzyme-deficient hippocampal neurons and choroidal cells in AD, *Neurol.* 24, 2001, 260–4.

P. Davidsson, R. Ekman and K. Blennow, A new procedure for detecting brain-specific proteins in cerebrospinal fluid, *J. Neural Transm.* 104, 1997, 711–20.

R. Deane, W. Zheng and B.V. Zlokovic, Brain capillary endothelium and choroid plexus epithelium regulate transport of transferrin-bound and free iron into the rat brain, *J. Neurochem.* 88, 2004, 813–20.

J.F. Decker and W.B. Quay, Stimulatory effects of melatonin on ependymal epithelium of choroid plexuses in golden hamsters, *J. Neural Transm.* 55, 1982, 53–6.

S.T. DeKosky, M.D. Ikonomovic, X.Wang, M. Farlow, S. Wisniewski et al., Plasma and cerebrospinal fluid alpha 1-antichymotrypsin levels in Alzheimer's disease: correlation with cognitive impairment, *Ann. Neurol.* 53, 2003, 81–90.

A. Deng, M.C. Irizarry, R.M. Nitsch, J.H. Growdon and G.W. Rebeck, Elevation of cystatin C in susceptible neurons in Alzheimer's disease, *Am. J. Pathol.* 159, 2001, 1061–8.

P.W. Dickson, A.R. Aldred, P.D. Marley, G.F. Tu, G.J. Howlett and G. Schreiber, High prealbumin and transferrin mRNA levels in the choroid plexus of rat brain, *Biochem. Biophys. Res. Commun.* 127, 1985, 890–5.

A.J. Dorta-Contreras, M. Barshatzky, E. Noris-Garcia and T. Serrano-Sanchez, Transthyretin levels in serum and cerebrospinal fluid sustain the nutrio-viral hypothesis of the Cuban epidemic neuropathy, *Rev. Neurol.*, 31, 2000, 801–4.

L. Eriksson and P. Westermark, Amyloid inclusions in choroid plexus epithelial cells. A simple autopsy method to rapidly obtain information on the age of an unknown dead person, *Forensic Sci. Int.* 48, 1990, 97–102.

F.M. Faraci, W.G. Mayhan and D.D. Heistad, Effect of vasopressin on production of cerebrospinal fluid: possible role of vasopressin (V1)-receptors, *Am. J. Physiol.* 258, 1990, R94–8.

H. Fenton, P.W. Finch, J.S. Rubin, J.M. Rosenberg, W.G. Taylor, V. Kuo-Leblanc, M. Rodriguez-Wolf, A. Baird, H.M. Schipper and e.g., Stopa, Hepatocyte growth factor (HGF/SF) in Alzheimer's disease, *Brain Res.* 779, 1998, 262–70.

H.M. Fillit, The role of hormone replacement therapy in the prevention of Alzheimer's disease, *Arch. Intern. Med.* 162, 2002, 1934–42.

J.F. Ghersi-Egea and N. Strazielle, Brain drug delivery, drug metabolism, and multidrug resistance at the choroid plexus, *Microsc. Res. Techn.* 52, 2001, 83–8.

J.F. Ghersi-Egea and N. Strazielle, Choroid plexus transporters for drugs and other xenobiotics, *J. Drug Targeting* 10, 2002, 353–7.

B. Giometto, F. Bozza, V. Argentiero, P. Gallo, S. Pagni, M.G. Piccinno and B. Tavolato, Transferrin receptors in rat central nervous system. An immunocytochemical study, *J. Neurol. Sci.* 98, 1990, 81–90.

K.A. Goddard, J.M. Olson, H. Payami, M. Van Der Voet, H. Kuivaniemi and G. Tromp, Evidence of linkage and association on chromosome 20 for late-onset Alzheimer's disease, *Neurogenetics*, 5, 2004, 121–8.

R. Gong, A. Rifai, E.M. Tolbert, J.N. Centracchio and L.D. Dworkin, Hepatocyte growth factor modulates matrix metalloproteinases and plasminogen activator/plasmin proteolytic pathways in progressive renal interstitial fibrosis, *J. Am. Soc. Nephrol.* 14, 2003, 3047–60.

A.M. Gonzalez, A. Logan, W. Yin, D.A. Lappi, M. Berry and A. Baird, Fibroblast growth factor in the hypothalamic-pituitary axis: differential expression of fibroblast growth factor-2 and a high affinity receptor, *Endocrinol.* 134, 1994, 2289–97.

H. Gunshin, B. Mackenzie, U.V. Berger, Y. Gunshin, M.F. Romero, W.F. Boron, S. Nussberger, J.L. Gollan and M.A. Hediger, Cloning and characterization of a mammalian proton-coupled metal-ion transporter, *Nature* 388, 1997, 482–8.

A. Hakvoort and C.E. Johanson, Growth factor modulation of CSF formation by isolated choroid plexus: FGF-2 vs. TGF-beta1, *Eur. J. Pediatr. Surg.* 10, 2000, S 44–6.

J.E. Hamos, B. Oblas, D. Pulaski-Salo, W.J. Welch, D.G. Bole and D.A. Drachma, Expression of heat shock proteins in Alzheimer's disease, *Neurol.* 41, 1991, 345–50.

A. Hanneken, S. Frautschy, D. Galasko and A. Baird, A fibroblast growth factor binding protein in human cerebral spinal fluid, *Neuroreport* 19, 1995, 886–8.

T.F. Hayamizu, P.T. Chan and C.E. Johanson, FGF-2 immunoreactivity in adult rat ependyma and choroid plexus: responses to global forebrain ischemia and intraventricular FGF-2, *Neurol. Res.* 23, 2001, 353–8.

T. Hayashi, K. Abe, M. Sakurai and Y. Itoyama, Inductions of hepatocyte growth factor and its activator in rat brain with permanent middle cerebral artery occlusion, *Brain Res.* 799, 1998, 311–6.

Y. Harigaya, M. Shoji, T. Nakamura, E. Matsubara, K. Hosoda and S. Hirai, Alpha 1-antichymotrypsin level in cerebrospinal fluid is closely associated with late onset Alzheimer's disease, *Intern. Med.* 34, 1995, 481–4.

B.C. Hong-Goka and F.L. Chang, Estrogen receptors alpha and beta in choroid plexus epithelial cells in Alzheimer's disease, *Neurosci. Lett.* 360, 2004, 113–6.

T. Ikeda, Y. Furukawa, S. Mashimoto, K. Takahashi and M. Yamada, Vitamin B12 levels in serum and cerebrospinal fluid of people with Alzheimer's disease, *Acta Psychiatr. Scand.* 82, 1990, 327–9.

C.E. Johanson, The choroid plexus-CSF nexus: gateway to the brain. In: *Neurosci. Med.*, 2nd ed., P.M. Conn, ed., Humana Press, Totowa, NJ, 2003, pp. 165–95.

C.E. Johanson, P.N. McMillan, D.E. Palm, E.G. Stopa, C.E. Doberstein and J.A. Duncan,Volume transmission-mediated protective impact of choroid plexus-CSF growth factors on forebrain ischemic injury. In: *Blood-Spinal Cord Brain Barriers Health Dis.*, H.S. Sharma and J. Westman, eds., Academic Press, San Diego, CA, 2003, pp. 361–384.

C.E. Johanson, A.M. Gonzalez and E.G. Stopa, Water-imbalance-induced expression of FGF-2 in fluid-regulatory centers: choroid plexus and neurohypophysis, *Eur. J. Pediatr. Surg.* 11, 2001, S37–8.

C.E. Johanson, D.E. Palm, M.J. Primiano, P.N. McMillan, P. Chan, N.W. Knuckey and E.G. Stopa, Choroid plexus recovery after transient forebrain ischemia: role of growth factors and other repair mechanisms, *Cell. Mol. Neurobiol.* 20, 2000, 197–216.

C.E. Johanson, J.E. Preston, A. Chodobski, E.G. Stopa, J. Szmydynger-Chodobska and P.N. McMillan, AVP V1 receptor-mediated decrease in Cl-efflux and increase in dark cell number in choroid plexus epithelium, *Am. J. Physiol.* 276, 1999a, C82–90.

C.E. Johanson, J. Szmydynger-Chodobska, A. Chodobski, A. Baird, P. McMillan and E.G. Stopa, Altered formation and bulk absorption of cerebrospinal fluid in FGF-2-induced hydrocephalus, *Am. J. Physiol.* 277, 1999b, R263–71.

C.E. Johanson and D.M. Woodbury, Changes in CSF flow and extracellular space in the developing rat. In: *Drugs and the Developing Brain*, A. Vernadakis and N. Weiner, eds., Plenum Press, New York, 1974, pp. 281–87.

J. Jolkkonen, E.L. Helkala, R. Kutvonen, M. Lehtinen and P.J. Riekkinen, Vasopressin levels in CSF of Alzheimer patients: correlations with monoamine metabolites and neuropsychological test performance, *Psychoneuroendocrinol.* 14, 1989, 89–95.

D.L. Justice, R.H. Rhodes and Z.A. Tokes, Immunohistochemical demonstration of proteinase inhibitor alpha-1-antichymotrypsin in normal human central nervous system, *J. Cell Biochem.* 34, 1987, 227–38.

R.N. Kalaria, D.L. Cohen, D.R. Premkumar, S. Nag, J.C. LaManna and W.D. Lust, Vascular endothelial growth factor in Alzheimer's disease and experimental cerebral ischemia, *Brain Res. Mol. Brain Res.* 62, 1998, 101–5.

J. Kalman, J. Marki-Zay, A. Juhasz, A. Santha, L. Dux and Z. Janka, Serum and cerebrospinal fluid cystatin C levels in vascular and Alzheimer's dementia, *Acta Neurol. Scand.* 101, 2000, 279–82.

N. Kanematsu, Y. Mori, S. Hayashi and K. Hoshino, Presence of a distinct 24-hour melatonin rhythm in the ventricular cerebrospinal fluid of the goat, *J. Pineal Res.* 7, 1989, 143–52.

K. Kissel, S. Hamm, M. Schulz, A. Vecchi, C. Garlanda and B. Engelhardt, Immunohistochemical localization of the murine transferrin receptor (TfR) on blood-tissue barriers using a novel anti-TfR monoclonal antibody, *Histochem. Cell. Biol.* 110, 1998, 63–72.

N.W. Knuckey, P. Finch, D.E. Palm, M.J. Primiano, C.E. Johanson, K.C. Flanders and N.L. Thompson, Differential neuronal and astrocytic expression of transforming growth factor beta isoforms in rat hippocampus following transient forebrain ischemia, *Brain Res. Mol. Brain Res.* 40, 1996, 1–14.

A. Kontush, N. Donarski and U. Beisiegel, Resistance of human cerebrospinal fluid to *in vitro* oxidation is directly related to its amyloid-beta content, *Free Radical Res.* 35, 2001a, 507–17.

A. Kontush, U. Mann, S. Arlt, A. Ujeyl, C. Luhrs, T. Muller-Thomsen and U. Beisiegel, Influence of vitamin E and C supplementation on lipoprotein oxidation in patients with Alzheimer's disease, *Free Radical Biol. Med.* 31, 2001b, 345–54.

B. Korting, E.J. van Zwieten, G.J. Boer, R. Ravid and D.F. Swaab, Increase in vasopressin binding sites in the human choroid plexus in Alzheimer's disease, *Brain Res.* 706, 1996, 151–4.

S. Kravitz, Y. Mawal, D.J. Sahlas, A. Liberman, H.M. Chertkow, H. Bergman and H.M. Schipper, Heme oxygenase-1 suppressor activity in Alzheimer's plasma, *Ann. Neurol.* 52, S31, 2002.

E. Levy, M. Sastre, A. Kumar, G. Gallo, P. Piccardo, B. Ghetti and F. Tagliavini, Codeposition of cystatin C with amyloid-beta protein in the brain of Alzheimer disease patients, *J. Neuropathol. Exp. Neurol.* 60, 2001, 94–104.

M.D. Li, J.K. Kane, S.G. Matta, W.S. Blaner and B.M. Sharp, Nicotine enhances the biosynthesis and secretion of transthyretin from the choroid plexus in rats: implications for beta-amyloid formation, *J. Neurosci.* 20, 2000, 1318–23.

F. Licastro, L.J. Davis, M.C. Morini, D. Cucinotta and G. Savorani, Cerebrospinal fluid patients with senile dementia of Alzheimer's type show an increased inhibition of alpha-chymotrypsin, *Alzheimer Dis. Assoc. Disord.* 8, 1994, 241–9.

M. Lindvall-Axelsson and C. Owman, Actions of sex steroids and corticosteroids on rabbit choroid plexus as shown by changes in transport capacity and rate of cerebrospinal fluid formation, *Neurol. Res.* 12, 1990, 181–6.

C.D. Link, Expression of human beta-amyloid peptide in transgenic *Caenorhabditis elegans*, *Proc. Natl. Acad. Sci. USA* 92, 1995, 9368–72.

A. Lippoldt, C.A. Padilla, H. Gerst, B. Andbjer, E. Richter, A. Holmgren and K. Fuxe, Localization of thioredoxin in the rat brain and functional implications, *J. Neurosci.* 15, 1995, 6747–56.

R.Y. Liu, J.N. Zhou, J. van Heerikhuize, M.A. Hofman and D.F. Swaab, Decreased melatonin levels in postmortem cerebrospinal fluid in relation to aging, Alzheimer's disease, and apolipoprotein E-epsilon4/4 genotype, *J. Clin. Endocrinol. Metab.* 84, 1999, 323–7.

A. Logan, S.A. Frautschy, A.M. Gonzalez, M.B. Sporn and A. Baird, Enhanced expression of transforming growth factor beta1 in the rat brain after a localized cerebral injury, *Brain Res.* 587, 1992, 216–25.

B.J. Lu, C. Kaur, and E.A. Ling, Expression and upregulation of transferrin receptors and iron uptake in the epiplexus cells of different aged rats injected with lipopolysaccharide and interferon-gamma, *J. Anat.* 187, 1995, 603–11.

M.G. Luciano, D.J. Skarupa, A.M. Booth, A.S. Wood, C.L. Brant and M.J. Gdowski, Cerebrovascular adaptation in chronic hydrocephalus, *J. Cereb. Blood Flow Metab.* 21, 2001, 285–94.

N. Matsumoto, H. Kitayama, M. Kitada, K. Kimura, M. Noda and C. Ide, Isolation of a set of genes expressed in the choroid plexus of the mouse using suppression subtractive hybridization, *Neurosci.* 117, 2003, 405–15.

C.P. Maurizi, Alzheimer's disease: roles for mitochondrial damage, the hydroxyl radical, and cerebrospinal fluid deficiency of melatonin, *Medical Hypoth.* 57, 2001, 156–60.

B. Mazur-Kolecka, J. Frackowiak, R.T. Carroll and H.M. Wisniewski, Accumulation of Alzheimer amyloid-beta peptide in cultured myocytes is enhanced by serum and reduced by cerebrospinal fluid, *J. Neuropathol. Exp. Neurol.* 56, 1997, 263–72.

B. Mazur-Kolecka, J. Frackowiak and H.M. Wisniewski, Apolipoproteins E3 and E4 induce, and transthyretin prevents accumulation of the Alzheimer's beta-amyloid peptide in cultured vascular smooth muscle cells, *Brain Res.* 698, 1995, 217–22.

M.F. Mazurek, J.H. Growdon, M.F. Beal and J.B. Martin, CSF vasopressin concentration is reduced in Alzheimer's disease, *Neurol.* 36, 1986, 1133–7.

A. Merched, J.M. Serot, S. Visvikis, D. Aguillon, G. Faure and G. Siest, Apolipoprotein E, transthyretin and actin in the CSF of Alzheimer's patients: relation with the senile plaques and cytoskeleton biochemistry, *FEBS Lett.* 425, 1998, 225–8.

J. Miklossy, K. Taddei, R. Martins, G. Escher, R. Kraftsik, O. Pillevuit, D. Lepori and M. Campiche, Alzheimer disease: curly fibers and tangles in organs other than brain, *J. Neuropathol. Exp. Neurol.* 58, 1999, 803–14.

J.A. Miyan, M. Nabiyouni and M. Zendah, Development of the brain: a vital role for cerebrospinal fluid, *Can. J. Physiol. Pharmacol.* 81, 2003, 317–28.

S. Moalem, M.E. Percy, D.F. Andrews, T.P. Kruck, S. Wong, A.J. Dalton, P. Mehta, B. Ferdo and A.C. Warren, Are hereditary hemochromatosis mutations involved in Alzheimer's disease?, *Am. J. Medical Genet.* 93, 2000, 58–68.

T.J. Monks, J.F. Ghersi-Egea, M. Philbert, A.J. Cooper and E.A. Lock, Symposium overview: the role of glutathione in neuroprotection and neurotoxicity, *Toxicol. Sci.* 51, 1999,161–77.

T. Moos, Immunohistochemical localization of intraneuronal transferrin receptor immunoreactivity in the adult mouse central nervous system, *J. Comp. Neurol.* 375, 1996, 675–92.

T. Moos, Brain iron homeostasis, *Dan. Medical Bull.* 49, 2002, 279–301.

T. Moos, Delivery of transferrin and immunoglobulins to the ventricular system of the rat, *Frontiers Biosci.* 8, 2003, 102–9.

T. Moos and E.H. Morgan, Kinetics and distribution of [59Fe-125I] transferrin injected into the ventricular system of the rat, *Brain Res.* 790, 1998, 115–28.

T. Moos and E.H. Morgan, Transferrin and transferrin receptor function in brain barrier systems, *Cell Mol. Neurobiol.* 20, 2000, 77–95.

T. Moos and E.H. Morgan, The significance of the mutated divalent metal transporter (DMT1) on iron transport into the Belgrade rat brain, *J. Neurochem.* 88, 2004, 233–45.

C.M. Morris, A.B. Keith, J.A. Edwardson and R.G.Pullen, Uptake and distribution of iron and transferrin in the adult rat brain, *J. Neurochem.* 59, 1992, 300–6.

K. Namekata, M. Imagawa, A. Terashi, S. Ohta, F. Oyama and Y. Ihara, Association of transferrin C2 allele with late-onset Alzheimer's disease, *Hum. Genet.* 101, 1997, 126–9.

C. Nilsson, F. Stahlberg, P. Gideon, C. Thomsen and O. Henriksen, The nocturnal increase in human cerebrospinal fluid production is inhibited by a beta 1-receptor antagonist, *Am. J. Physiol.* 267,1994, R1445–8.

W.G. North, R. Harbaugh and T. Reeder, An evaluation of human neurophysin production in Alzheimer's disease: preliminary observations, *Neurobiol. Aging* 13, 1992, 261–5.

M.A. Otieno, R.B. Baggs, J.D. Hayes and M.W. Anders, Immunolocalization of microsomal glutathione S-transferase in rat tissues, *Drug Metab. Dispos.* 25, 1997, 12–20.

J.A. Palha, Transthyretin as a thyroid hormone carrier: function revisited, *Clin. Chem. Lab. Med.* 40, 2002, 1292–1300.

D. E. Palm, N.W. Knuckey, M.J. Primiano, A.G. Spangenberger and C.E. Johanson, Cystatin C, a protease inhibitor, in degenerating rat hippocampal neurons following transient forebrain ischemia, *Brain Res.* 691,1995, 1–8.

Z. Parandoosh and C.E. Johanson, Ontogeny of blood-brain barrier permeability to, and cerebrospinal fluid sink action on, [14C]urea, *Am. J. Physiol.* 243, 1982, R400–7.

G.P. Paraskevas, E. Kapaki, G. Libitaki, C. Zournas, I. Segditsa and C. Papageorgiou, Ascorbate in healthy subjects, amyotrophic lateral sclerosis and Alzheimer's disease, *Acta Neurol. Scand.* 96, 1997, 88–90.

E.R. Peskind, D. Wingerson, M. Pascualy, L. Thal, R.C. Veith, D.M. Dorsa, S. Bodenheimer and M.A. Raskind, Oral physostigmine in Alzheimer's disease: effects on norepinephrine and vasopressin in cerebrospinal fluid and plasma, *Biol. Psychiatry* 38, 1995, 532–8.

M. Pollay and P.A. Roberts, Blood-brain barrier: a definition of normal and altered function, *Neurosurgery* 6, 1980, 675–85.

J.E. Preston, Ageing choroid plexus-cerebrospinal fluid system, *Microsc. Res. Techn.* 52, 2001, 31–7.

J. Quinn, J. Suh, M.M. Moore, J. Kaye, and B. Frei, Antioxidants in Alzheimer's disease — vitamin C delivery to a demanding brain, *J. Alzheimer's Dis.* 5, 2003, 309–13.

A. Quintas, M.J. Saraiva and R.M. Brito, The amyloidogenic potential of transthyretin variants correlates with their tendency to aggregate in solution, *FEBS Lett.* 418, 1997, 297–300.

M.A. Raskind, E.R. Peskind, T.H. Lampe, S.C. Risse, G.J. Taborsky Jr. and D. Dorsa, Cerebrospinal fluid vasopressin, oxytocin, somatostatin, and beta-endorphin in Alzheimer's disease, *Arch. Gen. Psychiatry* 43, 1986, 382–8.

M.E. Rice, Ascorbate regulation and its neuroprotective role in the brain, *Trends Neurosci.* 23, 2000, 209–16.

H. Riisoen, Reduced prealbumin (transthyretin) in CSF of severely demented patients with Alzheimer's disease, *Acta Neurol. Scand.*, 78,1988, 455–9.

J. Robbins, Transthyretin from discovery to now, *Clin. Chem. Lab. Med.* 40, 2002, 1–8.

L. Rodrigo, A.F. Hernandez, J.J. Lopez-Caballero, F. Gil and A. Pla, Immunohistochemical evidence for the expression and induction of paraoxonase in rat liver, kidney, lung and brain tissue. Implications for its physiological role, *Chem. Biol. Interact.* 137, 2001, 123–37.

P. Rojas, D.R. Cerutis, H.K. Happe, L.C. Murrin, R. Hao, R.F. Pfeiffer and M. Ebadi, 6-Hydroxydopamine-mediated induction of rat brain metallothionein I mRNA, *Neurotoxicol.* 17, 1996, 323–34.

M. Sandstrom, M. Johansson, J. Sandstrom, A.T. Bergenheim and R. Henriksson, Expression of the proteolytic factors, tPA and uPA, PAI-1 and VEGF during malignant glioma progression, *Int. J. Developmental Neurosci.* 17, 1999, 473–81.

M. Saraiva, Transthyretin mutations in health and disease, *Hum. Mutat.* 5, 1995, 191–192.

H.M. Schipper, H. Chertkow, K. Mehindate, D. Frankel, C. Melmed and H. Bergman, Evaluation of hemeoxygenase-1 as a systemic biological marker of sporadic AD, *Neurol.* 54, 2000, 1297–304.

H.M. Schipper, S. Cisse and E.G. Stopa, Expression of heme oxygenase-1 in the senescent and Alzheimer-diseased brain, *Ann. Neurol.* 37, 1995, 758–68.

P. Schonknecht, M. Henze, A. Hunt, K. Klinga, U. Haberkorn and J. Schroder, Hippocampal glucose metabolism is associated with cerebrospinal fluid estrogen levels in postmenopausal women with Alzheimer's disease, *Psychiatry Res.* 124, 2003, 125–7.

P. Schonknecht, J. Pantel, K. Klinga, M. Jensen, T. Hartmann, B. Salbach and J. Schroder, Reduced cerebrospinal fluid estradiol levels are associated with increased beta-amyloid levels in female patients with Alzheimer's disease, *Neurosci. Lett.* 307, 2001, 122–4.

G. Schreiber, The evolution of transthyretin synthesis in the choroid plexus, *Clin. Chem. Lab. Med.* 40, 2002, 1200–1210.

A.L. Schwarzman, L. Gregori, M.P, Vitek, S. Lyubuski, W.J. Strittmatter, J.J. Enghilde, R. Bhasin, J. Silverman, K.H. Weisgraber, J.K. Coyle, et al., Transthyretin sequesters amyloid beta protein and prevents amyloid formation, *Proc. Natl. Acad. Sci.* 91, 1994, 8368–72.

A.L. Schwarzman, M. Tsiper, H. Wente, A. Wang, M.P. Vitek, V. Vasilev and D. Goldgaber, Amyloidogenic and anti-amyloidogenic properties of recombinant transthyretin variants, *Amyloid* 11, 2004, 1–9.

J.M. Serot, M.C. Bene and G.C. Faure, Choroid plexus, ageing of the brain, and Alzheimer's disease, *Frontiers Biosci.* 8, 2003, 515–21.

J.M. Serot, M.C. Bene, B. Foliguet and G.C. Faure, Altered choroid plexus basement membrane and epithelium in late-onset Alzheimer's disease: an ultrastructural study, *Ann. NY Acad. Sci.* 826, 1997a, 507–9.

J.M. Serot, M.C. Bene, B. Foliguet and G.C. Faure, Morphological alterations of the choroid plexus in late-onset Alzheimer's disease, *Acta Neuropathol. (Berl.)* 99, 2000, 105–8.

J.M. Serot, D. Christmann, T. Dubost, M.C. Bene and G.C. Faure, CSF-folate levels are decreased in late-onset AD patients, *J. Neural. Transm.* 108, 2001a, 93–9.

J. M. Serot, D. Christmann, T. Dubost and M. Couturier, Cerebrospinal fluid transthyretin: aging and late onset Alzheimer's disease, *J. Neurol. Neurosurg. Psychiatry* 63, 1997b, 506–8.

J.M. Serot, B. Foliguet, M.C. Bene and G.C. Faure, Choroid plexus and ageing in rats: a morphometric and ultrastructural study, *Eur. J. Neurosci.* 14, 2001b, 794–8.

M. Shibata, S. Yamada, S.R. Kumar, M. Calero, J. Bading, B. Frangione, D.M. Holtzman, C.A. Miller, D.K. Strickland, J. Ghiso and B.V. Zlokovic, Clearance of Alzheimer's amyloid-ss(1–40) peptide from brain by LDL receptor-related protein-1 at the blood-brain barrier, *J. Clin. Inv.* 106, 2000, 1489–99.

H. Shikama, A. Ohta, A. Iwai, H. Koutoku, M. Umeda, K. Noguchi, M. Takeda and I. Ohhata, Transport and metabolism of glutathione isopropyl ester in cerebrospinal fluid, *Res. Commun. Mol. Pathol. Pharmacol.* 88, 1995, 349–57.

V.V. Shuvaev, I. Laffont, J.M. Serot, J. Fujii, N. Taniguchi and G. Siest, Increased protein glycation in cerebrospinal fluid of Alzheimer's disease, *Neurobiol. Aging* 22, 2001, 397–402.

A.J. Siddappa, R.B. Rao, J.D. Wobken, K. Casperson, E.A. Leibold, J.R. Connor and M.K. Georgieff, Iron deficiency alters iron regulatory protein and iron transport protein expression in the perinatal rat brain, *Pediatr. Res.* 53, 2003, 800–7.

A.J. Siddappa, R.B. Rao, J.D. Wobken, E.A. Leibold, J.R. Connor and M.K. Georgieff, Developmental changes in the expression of iron regulatory proteins and iron transport proteins in the perinatal rat brain, *J. Neurosci. Res.* 68, 2002, 761–75.

G.D. Silverberg, G. Heit, S. Huhn, R. Jaffe, S.D. Chang, H. Bronte-Stewart, E. Rubenstein, K. Possin and T.A. Saul, The cerebrospinal fluid production rate is reduced in dementia of the Alzheimer's type, *Neurol.* 57, 2001, 1763–6.

G.D. Silverberg, M. Mayo, T. Saul, E. Rubenstein and D. McGuire, Alzheimer's disease, normal-pressure hydrocephalus, and senescent changes in CSF circulatory physiology: a hypothesis, *Lancet Neurol.* 2, 2003, 506–11.

R. Spector and C.E. Johanson, The mammalian choroid plexus, *Sci. Am.* 261, 1989, 68–74.

T.D. Stein and J.A. Johnson, Lack of neurodegeneration in transgenic mice overexpressing mutant amyloid precursor protein is associated with increased levels of transthyretin and the activation of cell survival pathways, *J. Neurosci.* 22, 2002, 7380–8.

T. Steinhoff, E. Moritz, M.A. Wollmer, M.H. Mohajeri, S. Kins and R.M. Nitsch, Increased cystatin C in astrocytes of transgenic mice expressing the K670N–M671L mutation of the amyloid precursor protein and deposition in brain amyloid plaques, *Neurobiol. Dis.* 8, 2001, 647–54.

E.G. Stopa, T.M. Berzin, S. Kim, P. Song, V. Kuo-LeBlanc, M. Rodriguez-Wolf, A. Baird and C.E. Johanson, Human choroid plexus growth factors: What are the implications for CSF dynamics in Alzheimer's disease?, *Exp. Neurol.* 167, 2001, 40–7.

E.G. Stopa, A.M. Gonzalez, R. Chorsky, R.J. Corona, J. Alvarez, E.D. Bird and A. Baird, Basic fibroblast growth factor in Alzheimer's disease, *Biochem. Biophys. Res. Commun.* 171, 1990, 690–6.

N. Strazielle and J.F. Ghersi-Egea, Demonstration of a coupled metabolism-efflux process at the choroid plexus as a mechanism of brain protection toward xenobiotics, *J. Neurosci.* 19, 1999, 6275–89.

N. Strazielle and J.F. Ghersi-Egea, Choroid plexus in the central nervous system: biology and physiopathology, *J. Neuropathol. Exp. Neurol.* 59, 2000, 561–74.

N. Strazielle, S.T. Khuth, A. Murat, A. Chalon, P. Giraudon, M.F. Belin and J.F. Ghersi-Egea, Pro-inflammatory cytokines modulate matrix metalloproteinase secretion and organic anion transport at the blood-cerebrospinal fluid barrier, *J. Neuropathol. Exp. Neurol.* 62, 2003, 1254–64.

J. Szmydynger-Chodobska, Z.G. Chun, C.E. Johanson and A. Chodobski, Distribution of fibroblast growth factor receptors and their co-localization with vasopressin in the choroid plexus epithelium, *Neuroreport* 13, 2002, 257–9.

S. Tamir, S. Izrael and J. Vaya, The effect of oxidative stress on ERalpha and ERbeta expression, *J. Steroid Biochem. Mol. Biol.* 81, 2002, 327–32.

E. Tarkowski, R. Issa, M. Sjogren, A. Wallin, K. Blennow, A. Tarkowski and P. Kumar, Increased intrathecal levels of the angiogenic factors VEGF and TGF-beta in Alzheimer's disease and vascular dementia, *Neurobiol. Aging* 23, 2002, 237–43.

K. Thompson, S. Menzies, M. Muckenthaler, F.M. Torti, T. Wood, S.V. Torti, M.W. Hentze, J. Beard and J. Connor, Mouse brains deficient in H-ferritin have normal iron concentration but a protein profile of iron deficiency and increased evidence of oxidative stress, *J. Neurosci. Res.* 71, 2003, 46–63.

V. Toscano, G. Sancesario, P. Bianchi, C. Cicardi, D. Casilli and P. Giacomini, Cerebrospinal fluid estrone in pseudotumor cerebri: a change in cerebral steroid hormone metabolism?, *J. Endocrinol. Inv.* 14, 1991, 81–86.

Y. Tsuboi, K. Kakimoto, M. Nakajima, H. Akatsu, T. Yamamoto, K. Ogawa, T. Ohnishi, Y. Daikuhara and T. Yamada, Increased hepatocyte growth factor level in cerebrospinal fluid in Alzheimer's disease, *Acta Neurol. Scand.* 107, 2003, 81–6.

K. Tsuzuki, R. Fukatsu, Y. Hayashi, T. Yoshida, N. Sasaki, Y. Takamaru, H. Yamaguchi, M. Tateno, N. Fujii and N. Takahata, Amyloid beta protein and transthyretin, sequestrating protein colocalize in normal human kidney, *Neurosci. Lett.* 222, 1997, 163–6.

G.F. Tu, A.R. Aldred, B.R. Southwell and G. Schreiber, Strong conservation of the expression of cystatin C gene in choroid plexus, *Am. J. Physiol.* 263, 1992, R195–200.

M. Uchino, Y. Ando, Y. Tanaka, T. Nakamura, E. Uyama, S. Mita, T. Murakami and M. Ando, Decrease in Cu/Zn- and Mn-superoxide dismutase activities in brain and spinal cord of patients with amyotrophic lateral sclerosis, *J. Neurol. Sci.* 127, 1994, 61–7.

G.F. van Landeghem, C. Sikstrom, L.E. Beckman, R. Adolfsson and L. Beckman, Transferrin C2, metal binding and Alzheimer's disease, *Neuroreport* 9, 1998, 177–9.

S.J. van Rensburg, F.C. Potocnik, J.N. De Villiers, M.J. Kotze and J.J. Taljaard, Earlier age of onset of Alzheimer's disease in patients with both the transferrin C2 and apolipoprotein E-epsilon 4 alleles, *Ann. NY Acad. Sci.* 903, 2000, 200–3.

G.T. Vatassery, H.T. Quach, W.E. Smith, B.A. Benson and J.H. Eckfeldt, A sensitive assay of transthyretin (prealbumin) in human cerebrospinal fluid in nanogram amounts by ELISA, *Clin. Chim. Acta* 197, 1991, 19–25.

M.P. Vawter, O. Dillon-Carter, W.W. Tourtellotte, P. Carvey and W.J. Freed, TGFbeta1 and TGFbeta2 concentrations are elevated in Parkinson's disease in ventricular cerebrospinal fluid, *Exp. Neurol.* 142, 1996, 313–22.

X. Wang, S.T. DeKosky, E. Luedecking-Zimmer, M. Ganguli and I.M. Kamboh, Genetic variation in alpha(1)-antichymotrypsin and its association with Alzheimer's disease, *Hum. Genet.* 110, 2002, 356–65.

C.E. Weaver, J.A. Duncan, E.G. Stopa and C.E. Johanson, Hydrocephalus disorders: their biophysical and neuroendocrine impact on the choroid plexus epithelium. In: *Non-Neuronal Cells of the Nervous System: Function and Dysfunction*, L. Hertz, ed., Elsevier Press, New York, 2004, 269–294.

A. Whitelaw, S. Cherian, M. Thoresen and I. Pople, Posthaemorrhagic ventricular dilatation: new mechanisms and new treatment, *Acta Paediatr. Supplement* 93, 2004, 11–4.

T. Wisniewski, E. Castano, J. Ghiso and B. Frangione, Cerebrospinal fluid inhibits Alzheimer beta-amyloid fibril formation *in vitro*, *Ann. Neurol.* 34, 1993, 631–3.

Y.H. Wu, M.G. Feenstra, J.N. Zhou, R.Y. Liu, J.S. Torano, H.J. Van Kan, D.F. Fischer, R. Ravid and D.F. Swaab, Molecular changes underlying reduced pineal melatonin levels in Alzheimer's disease: alterations in preclinical and clinical stages, *J. Clin. Endocrinol. Metab.* 88, 2003, 5898–906.

L.J. Wu, A.G. Leenders, S. Cooperman, E. Meyron-Holtz, S. Smith, W. Land, R.Y. Tsai,
U. Berger, Z.H. Sheng and T.A. Rouault, Expression of the iron transporter ferroportin in vesicles and the blood-brain barrier, *Brain Res.* 1001, 2004, 108–17.

T. Yamada, Y. Tsujioka, J. Taguchi, M. Takahashi, Y. Tsuboi, I. Moroo, J. Yang, and W.A. Jefferies, Melanotransferrin is produced by senile plaque-associated reactive microglia in Alzheimer's disease, *Brain Res.* 845, 1999, 1–5.

B.C. Yoo, S.H. Kim, N. Cairns, M. Fountoulakis and G. Lubec, Deranged expression of molecular chaperones in brains of patients with Alzheimer's disease, *Biochem. Biophys. Res. Commun.* 280, 2001, 249–58.

S. Yoshimura, R. Morishita, K. Hayashi, J. Kokuzawa, M. Aoki, K. Matsumoto, T. Nakamura, T. Ogihara, N. Sakai and Y. Kaneda, Gene transfer of hepatocyte growth factor to subarachnoid space in cerebral hypoperfusion model, *Hypertension* 39, 2002, 1028–34.

S.W. Yun, U. Gartner, T. Arendt and S. Hoyer, Increase in vulnerability of middle-aged rat brain to lead by cerebral energy depletion, *Brain Res. Bull.* 52, 2000, 371–8.

P. Zambenedetti, G. De Bellis, I. Biunno, M. Musicco and P. Zatta, Transferrin C2 variant does confer a risk for Alzheimer's disease in Caucasians, *J. Alzheimer's Dis.* 5, 2003, 423–7.

W. Zheng, M. Aschner and J.F. Ghershi-Egea, Brain barrier systems: a new frontier in metal neurotoxicological research, *Toxicol. Applied Pharmacol.* 192, 2003, 1–11.

W. Zheng and Q. Zhao, Establishment and characterization of an immortalized Z310 choroidal epithelial cell line from murine choroid plexus, *Brain Res.* 958, 2002, 371–80.

W. Zheng, Q. Zhao, V. Slavkovich, M. Aschner and J.H. Graziano, Alteration of iron homeostasis following chronic exposure to manganese in rats, *Brain Res.* 833, 1999,125–32.

J.N. Zhou, R.Y. Liu, W. Kamphorst, M.A. Hofman and D.F Swaab, Early neuropathological Alzheimer's changes in aged individuals are accompanied by decreased cerebrospinal fluid melatonin levels, *J. Pineal Res.* 35, 2003, 125–30.

13 The Blood–CSF Barrier and Cerebral Ischemia

Richard F. Keep, Steven R. Ennis and Jianming Xiang
University of Michigan,
Ann Arbor, Michigan, U.S.A.

CONTENTS

13.1 INTRODUCTION

About 600,000 people suffer a stroke in the United States each year. It is the third leading cause of death. Even in those who survive, it can result in major mental and physical disabilities. There are currently about 4.4 million people who have suffered a stroke. The predominant cause of stroke is cerebral ischemia, where a reduction or cessation of blood flow to an area of the brain causes tissue damage. There are also hemorrhagic strokes involving rupture of a blood vessel. This latter, often fatal, stroke subtype accounts for about 15% of strokes in the United States and a greater percentage in Japan and China (Qureshi et al., 2001).

Recent decades have seen an explosion in our knowledge of the effects of a stroke on the brain. Thus, although the underlying cause of ischemic brain damage is reduced oxygen and glucose delivery to the affected area, this initial event triggers a series of changes that may further exacerbate brain damage (secondary brain injury). As yet, the only therapeutic modality for ischemic stroke is prompt restoration of blood flow through administration of tissue plasminogen activator (tPA) to reopen the occluded vessel (National Institute of Neurological Disorders and Stroke, 1995). However, there are strict limitations on the use of this therapy, as it can result in cerebral hemorrhage (hemorrhagic transformation) due to reperfusion of damaged blood vessels. In addition, delayed restoration of blood flow can actually exacerbate brain injury (reperfusion injury) because of events triggered by the resupply of oxygen and other factors to the injured brain.

The primary focus of much of the research into cerebral ischemia has been neuronal. Less well studied are the effects of cerebral ischemia on other cell types, e.g., astrocytes, oligodendrocytes, and endothelial cells. The effects of cerebral ischemia on the choroid plexus have, in particular, been the subject of very few studies, and these will be the subject of this review. The choroid plexus epithelial cells and their linking tight junctions are the site of the blood–CSF barrier (BCSFB). The effects of cerebral ischemia on BCSFB function have also received much less attention than effects on the blood–brain barrier (BBB) formed by the endothelial cells of the cerebral capillaries.

Although there has been understandable interest in developing therapeutic strategies to prevent the mechanisms that induce or exacerbate ischemic brain damage, considerable evidence has been derived on endogenous mechanisms that limit such damage. Factors derived from the choroid plexuses may play a part in this (Johanson et al., 2000, 2003). In addition, there is rapidly expanding interest in repair or recovery of function following stroke (from adoption of function by noninjured tissue to neurogenesis) and how to modulate these processes. Again, evidence suggests that the choroid plexuses may play a role in these processes (Li et al., 2002).

This chapter will concentrate on the lateral ventricle choroid plexuses, which have been the subject of most experimental studies. In addition, any prolonged ischemia that affects the hindbrain (the site of the 4th ventricle choroid plexus) is lethal because of the respiratory and cardiovascular centers that reside in that area.

This review will examine the following broad topics:

- Normal blood flow to the choroid plexuses
- The effect of cerebral ischemia on choroid plexus function
- Mechanisms of cell death and therapeutics
- Occurrence of choroid plexus atrophy/repair
- Hemorrhagic stroke
- The role of the choroid plexus in parenchymal injury
- An overview addressing gaps in our knowledge

13.2 CHOROID PLEXUS BLOOD FLOW

The choroid plexuses have a very rich blood supply and appear as "glistening, bright red, distended structures" within the ventricles (Bradbury, 1979). In humans, lateral ventricle choroid plexus blood supply is from the anterior and posterior choroidal arteries. The blood drains into the choroidal vein. The anterior choroidal artery is usually a branch off the internal carotid. The posterior choroidal artery is a branch of the posterior cerebral artery, which in turn branches from the basilar artery. The posterior choroidal artery also supplies the third ventricle choroid plexus. The dual blood supply to the lateral ventricle choroid plexuses probably offers some protection from focal ischemic events (e.g., blockage to one of the choroidal arteries).

The choroid plexuses have a high blood flow compared to other areas of the brain. Thus, in rabbit and dog, choroid plexus blood flow is about 10-fold higher than that in the cerebral cortex (Faraci et al., 1988). In rat, choroid plexus blood flows are about 3- to 5-fold higher than that in the cerebral cortex with flows of about 3 to 5 ml/g/min being reported for choroid plexus (Williams et al., 1991a, 1991b; Szmydynger-Chodobska et al., 1994). There are a number of potential underlying reasons for the high blood flow, a reflection of the different functions of the choroid plexus. An understanding of those reasons is important, since it will determine how reductions in flow (ischemia) affect the choroid plexus.

In the brain parenchyma, blood flow matches metabolic demand (i.e., areas of high activity have high flow). This raises the question as to whether the choroid plexuses have a high metabolic demand compared to other brain structures. A number of studies have used [^{14}C]deoxyglucose to assess choroid plexus energy metabolism. Those studies generally show a level similar to cortical structures (Tyson et al., 1982; Vitte et al., 1989; Gross et al., 1992). There are, however, difficulties in interpreting such data. Methodologically there are concerns with label washout from the choroid plexus, which could cause an underestimate of metabolism. In addition, while glucose utilization is an accurate measurement of metabolic rate in cortex (Clarke and Sokoloff, 1999), there is a question over whether this is the case in the choroid plexus. For example in a number of cell types, including at least some epithelial cells, glutamine is a major substrate used to generate ATP (Windmueller and Spaeth, 1980; Spolarics et al., 1991). An alternate approach to examining the metabolic demand of the choroid plexus is to compare mitochondrial density. Choroid plexus epithelial cells are richly endowed with mitochondria, and they occupy about 8% of cell volume (Keep et al., 1986; Keep and Jones, 1990b). This is greater than the ~6% found in cerebral cortex (Keep and Jones, 1990a), although the disparity does not appear to be as great as might be expected from blood flow. Further evidence that the high blood flow to the choroid plexus is not to meet the metabolic needs of the choroid plexus comes from experiments where intraventricular endothelin administration was found to decrease blood flow (~40%) but increase glucose metabolism in choroid plexus (Gross et al., 1992).

The choroid plexuses are involved in CSF secretion and the movement of compounds from blood to CSF and from CSF to blood (Davson et al., 1987; Johanson, 1989; Gao and Meier, 2001; Johanson, 2003). It is possible that one of these properties necessitates a high blood flow. By analogy, the extremely high flow to the cerebral endothelium (~100 ml/g endothelium/min) reflects the requirement for oxygen and glucose delivery to brain and CO_2 removal from brain rather than the metabolic requirements of the endothelium. In the rat, CSF secretion rate is about 2 to 3 μl/min, and the choroid plexuses are the major source (Davson et al., 1987). Blood flow to the lateral, third, and fourth ventricle choroid plexuses is about 24 μl/min (assuming a combined choroid plexus weight of 6 mg and a flow of 4 ml/g/min). This raises the question of whether blood flow is normally a limiting factor for CSF secretion. Indeed, endothelin can cause marked reductions in choroid plexus blood flow with only a minor reduction in CSF secretion (Schalk et al.,

1992a), while atrial natriuretic peptide causes increase in flow with no change in CSF secretion (Schalk et al., 1992b). It should be noted, however, that marked reductions in choroidal blood flow must induce reductions in CSF secretion rate (e.g., at a 20% blood flow, plasma flow to the choroid plexus would be approximately equal to the normal CSF production rate).

The choroid plexuses are involved in the transport of compounds from blood to CSF. Overall, the compounds that are required in the greatest amounts (oxygen and glucose) are primarily transported across the blood–brain barrier situated at the cerebral capillaries. However, some compounds (e.g., ascorbate and folate) are selectively transported across the blood–CSF barrier (Spector and Eells, 1984), although the amounts of these compounds transported mean that it is unlikely that their delivery to CSF is limited by choroidal blood flow.

The choroid plexuses are also involved in the transport of compounds from CSF to blood. Thus, for example, the choroid plexus epithelium has a wide range or organic anion transporters, organic cation transporters, and multidrug resistance proteins that are involved with the clearance of potentially toxic compounds from CSF (Gao and Meier, 2001). The choroid plexus has been compared to the kidney in terms of its potential role in the clearance of toxic compounds (Spector and Johanson, 1989). The extent to which this clearance function requires a high blood flow is uncertain.

In the brain parenchyma, there are threshold blood flows for induction of permanent tissue damage. Thus, sustained reductions to about 18 ml/100 g/min cause energy failure and tissue death (Jones et al., 1981; Mies et al., 1991). In the rat, this might represent a reduction in flow of about 70 to 80%. The threshold for choroid plexus damage is unknown. If only a fraction of choroidal blood flow is related to meeting direct choroid plexus metabolic requirements, it is possible that greater reductions in flow may be required to cause permanent structural damage. It is, however, possible that smaller reductions in flow could alter certain choroid plexus functions.

13.3 EFFECT OF CEREBRAL ISCHEMIA ON CHOROID PLEXUS FUNCTION

Studies on the effects of cerebral ischemia on choroid plexus structure and function are summarized in Table 13.1. By way of introduction to those results, animal studies of cerebral ischemia are broadly divided into those using global models and those using focal ischemia models. Global models generally are thought to reflect events that occur after a temporary cessation of heart function. Models include those where the heart is stopped and restarted, the gerbil bilateral carotid occlusion model, and the rat 4-vessel occlusion and 2-vessel occlusion + hypotension models. The gerbil has a poor circle of Willis, so bilateral occlusion of the common carotid arteries results in forebrain blood flow dropping to near zero. The rat has a better circle of Willis (although this is strain dependent), so that an extra operation is required to induce ischemic brain damage. This can be either occlusion of the vertebral arteries (4-vessel occlusion) or the addition of hypotension (2-vessel occlusion + hypotension). In both cases, the model results in very low forebrain blood flows

(4-vessel <3% of normal). The models are designed for short durations of ischemia with reperfusion.

Most human strokes are, however, focal in nature (i.e., there is a reduction in blood flow to a limited area of the brain). The most common animal models of focal cerebral ischemia are those based on middle cerebral artery occlusion (MCAO). The occlusion can be generated by permanent ablation by cautery or by insertion of a thread to either permanently or transiently block flow to the MCA territory (Bederson et al., 1986; Longa et al., 1990). There are always concerns as to the relevance of such models, and there has been increased use of embolic models (e.g., Zivin et al., 1985), although these have the disadvantage of some variability in the site of occlusion.

Pulsinelli et al. (1982) first examined the effects of transient forebrain ischemia (4-vessel occlusion with reperfusion of the carotid arteries after 10, 20, or 30 min) on lateral ventricle choroid plexus morphology. They found evidence of necrosis after 6 hours, but the number of animals affected decreased by 24 hours and disappeared by 72 hours. This pattern of injury contrasts to that in nearby brain parenchyma, where there was delayed cell death.

The results of Pulsinelli et al. (1982) demonstrated that very low flows induce choroid plexus damage, that such damage can precede parenchymal injury, and that there is either rapid repair of the damage to the epithelium or the rapid loss of damaged cells. This general pattern has been supported by a number of subsequent morphological studies. Thus, Palm et al. (1995) and Johanson et al. (2000), using 2-vessel occlusion with hypotension in the rat, found early damage to the epithelial brush border, organelles, and nucleus following reperfusion (and restoration of blood pressure) with rapid necrosis. However, by 24 hours necrotic epithelial cells were absent. Similarly, using the same model, Ferrand-Drake and Wieloch (Ferrand-Drake and Wieloch, 1999; Ferrand-Drake, 2001) reported DNA fragmentation in the epithelium after 18 to 24 hours of reperfusion, but very little at 48 hours. Again, DNA fragmentation in the choroid plexus precedes that in adjacent brain (fragmentation in CA1 neurons begins at 36 hours and increases by 48 hours) (Ferrand-Drake, 2001). Following permanent focal cerebral ischemia, Gillardon et al. (1996) found evidence of DNA fragmentation after 6 but not 1.5 hours of MCAO.

This general pattern of morphological changes is reflected in physiological changes. Thus, Nagahiro et al. (1994) used magnetic resonance imaging (MRI) with gadolinium-diethylene triamine pentacetic acid (Gd-DPTA) to examine BCSFB disruption following transient MCAo in the rat. They found that 30 and 60 minutes of occlusion (and, in some animals, 15 min of occlusion) caused BCSFB disruption as assessed after 6 hours of reperfusion. There was, however, recovery of function with longer durations of reperfusion, although the speed of that recovery decreased with the length of ischemia. Thus, with 30 minutes of ischemia, the animals recovered by day 1. With 60 minutes of ischemia, they recovered by day 7. Importantly, the changes in BCSFB seem to precede changes in blood–brain barrier disruption.

Although most studies examining the effects of ischemia on the choroid plexus have been in adult animals, two studies have examined ischemic effects in neonates (Towfighi et al., 1995; Rothstein and Levison, 2002). Both studies used a rat hypoxia-ischemia model with unilateral carotid occlusion and exposure to 8%

TABLE 13.1
Effects of Cerebral Ischemia on Lateral Ventricle Choroid Plexus

Model	Species	Duration	Effect on Epithelium	Ref.
4-vessel occlusion	Rat	10, 20 or 30 min + reperfusion	Necrosis at 6 h, less at 24 h, none at 72 h	Pulsinelli et al., 1982
4-vessel occlusion	Rat	30 min + 0–72 h reperfusion	Enhanced calcium accumulation in choroid plexus; restoration toward normal calcium content after peak at 5 h, loss of choroid plexus weight ~35%	Dienel, 1984
2-vessel occlusion + hypotension	Rat	10 min + reperfusion	Increase in sodium and water and decrease in potassium at 0.5 h; damage to brush border, mitochondria, and necrosis (0.5–12 h), none at 24 h	Palm et al., 1995; Johanson et al., 2000
2-vessel occlusion + hypotension	Rat	15 min + reperfusion	DNA fragmentation at 24 h, diminished by 48 h	Ferrand-Drake and Wieloch, 1999
2-vessel occlusion + hypotension	Rat	15 min + reperfusion	DNA fragmentation at 18 h and then gradually diminishing (24 + 36 h); little at 48 h	Ferrand-Drake, 2001
2-vessel occlusion + hypotension	Rat	15 min + 20 min or 1 or 24 h reperfusion	No enhanced blood to CSF calcium flux	Ohta et al., 1992
2-vessel occlusion	Gerbil	5 min + 7 d reperfusion	DNA fragmentation after 7 d	Kitagawa et al., 1998
2-vessel occlusion	Gerbil	5 min + 4 or 7 d reperfusion	Enhanced blood to CSF calcium flux; morphological changes	Ikeda et al., 1992
Permanent MCAO	Rat	1.5 or 6 h	DNA fragmentation after 6 but not 1.5 h occlusion	Gillardon et al., 1996
Transient MCAO	Rat	5, 15, 30, or 60 min + 6 h, 1 d, and 7 d of reperfusion	BCSFB disruption following 15, 30 and 60 min of ischemia and 6 h of reperfusion; recovery by day 1 except after 60 min ischemia when recovery by day 7	Nagahiro et al., 1994

(continued)

TABLE 13.1 (CONTINUED)
Effects of Cerebral Ischemia on Lateral Ventricle Choroid Plexus

Model	Species	Duration	Effect on Epithelium	Ref.
Hypoxia/ Ischemia Unilateral carotid + 90 min hypoxia	Neonatal rat	0, 2, or 4 h reoxygenation	Damage to brush border, nucleus, swollen organelles (0 h reoxygenation); loss of plasma membrane integrity, necrosis (2–4 h); endothelium normal	Rothstein and Levinson, 2002
Hypoxia/ Ischemia Unilateral carotid + 120 min hypoxia	Neonatal rat	0–3 weeks reoxygenation	Pyknotic nuclei within few hours, necrosis at 24 h; by 4 d necrotic cells disappear, choroid plexus atrophied	Towfighi et al., 1995
Transient MCAO	Rat	2 h + 2–28d of reperfusion	BrdU positive cells in choroid plexus; peak at 7 d	Li et al., 2002
Medial posterior choroidal artery occlusion	Human	Unknown	Infarction	Liebeskind and Hurst, 2004

oxygen. The studies recapitulate findings in the adult. Morphologically, necrotic cells appear soon after reoxygenation (within a few hours). There is then a sloughing of these cells, and the epithelial cell structure appears normal by 4 days.

Most studies have been performed on the rat. However, Ikeda et al. (1992) found that brief global ischemia (bilateral common carotid occlusion) in the gerbil caused prolonged BCSFB disruption (as assessed by ^{45}Ca entry into CSF) as well as morphological changes to the choroid plexus. Kitagawa et al. (1998) found evidence of choroid plexus DNA fragmentation in the same model. Very little is known about the effect of ischemia on choroid plexus function in humans. A recent report, however, did find infarction of a lateral choroid plexus following a posterior choroidal artery stroke (Liebeskind and Hurst, 2004).

13.4 MECHANISMS OF CELL DEATH AND THERAPEUTICS

One of the very first events that follows cerebral ischemia in the parenchyma is a redistribution of ions between the intra- and extracellular compartments (Kristian and Siesjo, 1997). Thus, energy failure results in a loss of parenchymal cell K^+ (due to Na^+/K^+-ATPase inhibition) and an entry of Na^+ and Ca^{2+} into cells down their electrochemical gradients. Cell Na^+ accumulation usually exceeds K^+ loss, and this results in an influx of water into the cells. These events also happen in the choroid plexus epithelium. Palm et al. (1995) have reported an increase in choroid plexus

Na^+ and a loss of K^+ following forebrain ischemia, and Dienel (1984) reported an increase in choroid plexus Ca^{2+} content.

In neurons, the ionic disequilibrium that follows ischemia is exacerbated by the presence of voltage-gated ion channels, which open in response to the depolarization induced by energy failure. In addition, ischemia induces the release of neurotransmitters (secondary to increased intracellular Ca^{2+}). The release of glutamate activates other ion channels [such as *N*-methyl-*D*-aspartate (NMDA) channel], which further increases intracellular Ca^{2+} and Na^+ (so-called excitoxicity). Alterations in Ca^{2+} are thought to play a pivotal role in cell injury through activating proteases, disrupting the cell cytoskeleton, DNA damage, and free radical production. Free radicals can also result from other pathways, such as arachidonic acid metabolism. Free radicals are particularly important during the reoxygenation that occurs during reperfusion. In addition, neuronal injury can result from ischemic effects on other cell types. Thus, for example, alterations in glutamate clearance by astrocytes enhance excitoxicity, and an influx of white bloods cells into the brain can also enhance injury (inflammation).

This incomplete description of some of the events that occur in neuronal injury serves to raise several points. First, there is an initial injury linked to energy failure. Second, this is then exacerbated by other events, so-called secondary brain injury. This is exemplified by the fact that delayed reperfusion can actually enhance ischemic damage. Third, injury can be mediated through effects on other cell types. For the choroid plexus, the relative importance of these events is uncertain. Thus, is injury solely the result of energy failure or is there a significant secondary injury component? At the blood–brain barrier there is significant influx of neutrophils and monocytes after cerebral ischemia that may contribute to barrier breakdown. Does this occur at the blood–CSF barrier, and does it contribute to choroid plexus injury? These basic questions are basically unanswered.

In the parenchyma, the core of the infarct undergoes necrosis. However, there is considerable evidence for a role of apoptosis (programmed cell death) in the penumbra (MacManus et al., 1993; Du et al., 1996). The relative importance of apoptosis in cell death after ischemia has been the subject of much debate in the last decade, and it depends, to some extent, on the definition of apoptosis (morphological or biochemical). Indeed evidence indicates that there is a continuum between the two poles of necrosis and apoptosis with some cell death-sharing characteristics of both (Hou and MacManus, 2002). Similarly, there is some disagreement as to the mechanism by which the choroid plexus epithelial cells die following cerebral ischemia. Morphological studies have strongly indicated that cell death is via necrosis (Pulsinelli et al., 1982; Palm et al., 1995; Towfighi et al., 1995; Johanson et al., 2000; Rothstein and Levison, 2002). However, a number of studies have shown that cells display evidence of DNA fragmentation consistent with apoptosis (Ferrand-Drake, 2001; Ferrand-Drake and Wieloch, 1999; Gillardon et al., 1996). It appears in the choroid plexus epithelium that, although cells may undergo some of the biochemical changes associated with apoptosis, the programmed cell death pathway may be superceded by necrosis before the morphological features of apoptosis become present.

Johanson et al. (2000) point out an interesting aspect of ischemia-induced cell death in the choroid plexus epithelium, that is, that ischemia can have profoundly different effects on nearby cells. One might expect regional differences in injury dependent upon the origin of blood supply, but adjacent cells can display normal and necrotic morphology. Although morphologically all choroid plexus epithelial cells appear fairly similar, there is evidence that they may fulfill different functions. For example, during development, very selective epithelial cells have a pathway for the movement of low molecular weight dextran (Saunders et al., 1999). In the adult, light and dark cells can be identified by light microscopy, and this may reflect a difference in water handling (Dohrmann, 1970). Similarly, differences in vasopressin mRNA levels within cells of the choroid plexus have been reported (Chodobski et al., 1997). Whether differences in physiological function underlie altered susceptibility to ischemic damage remains to be investigated.

There have been few studies examining protective strategies that might limit ischemia-induced choroid plexus damage. Ferrand-Drake and Wieloch (1999) found that hypothermia reduced global ischemia-induced DNA fragmentation.

13.5 OCCURRENCE OF CHOROID PLEXUS ATROPHY/REPAIR

Despite the cell death that occurs in the choroid plexus soon after cerebral ischemia, all the morphological studies indicate that choroid plexus epithelial cell structure then returns to normal (Pulsinelli et al., 1982; Palm et al., 1995; Towfighi et al., 1995; Johanson et al., 2000; Ferrand-Drake, 2001). Similarly, BCSFB function and choroid plexus sodium and calcium concentrations return toward normal (Dienel, 1984; Nagahiro et al., 1994; Palm et al., 1995). The question arises as to how this occurs. Two broad (not mutually exclusive) possibilities are that new epithelial cells are produced or that the remaining epithelial cells may be reorganized to form a smaller, but functionally intact, choroid plexus. The evidence for the former is somewhat conflicting (at least in respect to importance). Li et al. (2002) found evidence that populations of choroid plexus cells proliferate after transient MCAO using bromodeoxyuridine (BrdU) labeling. The number of cells peaked at day seven. Such proliferative cells may be involved in either choroid plexus or brain parenchyma repair (some cells expressed either neuronal or glial markers). In contrast, Johanson et al. (2000) found little evidence of cell division after forebrain ischemia (in terms of mitotic cells or ^{3}H-thymidine incorporation), although they postulated that there may be some migration from the stalk zone of the choroid plexus. There is evidence that any cell proliferation is not enough to replace all ischemic cell death. Towfighi et al. (1995) noted ipsilateral choroid plexus atrophy following their neonatal rat hypoxia/ischemia model, whereas Dienel (1984) found a ~35% reduction in lateral ventricle choroid plexus weight following 4-vessel occlusion in adult rats. This suggests that the restoration of a normal choroid plexus appearance and function is primarily associated with a rearrangement of the surviving cells to create a smaller plexus. The long-term effects of such choroid plexus atrophy on surviving brain have not been investigated. Longer-term studies examining whether there is a gradual replacement of choroid plexus epithelium are also required.

13.6 HEMORRHAGIC STROKE

While there have been few studies on the effects of cerebral ischemia on the choroid plexus, there have been almost no studies on the effects of cerebral hemorrhage. This is surprising, considering the importance of intraventricular hemorrhage in preterm infants in relation to the occurrence of cerebral palsy (Wildrick, 1997). Hydrocephalus is a common occurrence after intraventricular hemorrhage (Naff and Tuhrim, 1997), and it would be important to know whether choroid plexus function has been altered following such a hemorrhage. One animal study suggests that thrombolytic therapy to remove the intraventricular hematoma may result in choroid plexus damage (Wang et al., 2002).

Intraventicular hemorrhage is also a risk factor for poor prognosis following intracerebral hemorrhage in adults (Daverat et al., 1991). Extension of an intracerebral hemorrhage into the ventricular system occurs in about 40% of cases (Daverat et al., 1991). Studies on animal models of intracerebral hemorrhage indicate that a number of clot-derived factors cause damage to the surrounding tissue (Xi et al., 2002). Thus, elements of the coagulation cascade (particularly thrombin), complement, and hemoglobin breakdown products have all been implicated in inducing parenchymal damage (Xi et al., 2002). Whether these blood components have similar effects on the choroid plexus epithelium is unknown.

An example of a potential role of the choroid plexus in hemorrhagic brain injury relates to bilirubin and leukotriene C_4. Bilirubin is a degradation product of hemoglobin, and evidence suggests that it may contribute to brain injury after cerebral hemorrhage (Fujii et al., 1989; Huang et al., 2002). Leukotriene C_4 is an eicosanoid produced in the brain after cerebral hemorrhage (White et al., 1990), and evidence suggests that it, too, can produce brain injury (Unterberg et al., 1990). Both of these compounds are substrates for organic anion transport, and a variety of organic anion transporters have been identified at the choroid plexus (OAT1, OAT3, oatp1, oatp2, and MRP1) (see review by Sun et al., 2003). This raises an interesting possibility that these compounds might compete for clearance from CSF to blood at the choroid plexus. We have found that bilirubin (1 μM) and leukotriene C_4 (4 μM) inhibit p-aminohippuric acid uptake (an organic anion transporter substrate) into isolated rat choroid plexus by 75% and 61%, respectively, and that bilirubin (1 μM) inhibits leukotriene C_4 uptake by 45% (J. Xiang and R.F. Keep, unpublished results). Thus, the production of bilirubin after an intraventricular hemorrhage as the clot resolves might inhibit the clearance of leukotriene C_4 from CSF by the choroid plexus, potentially exacerbating injury.

13.7 BCSFB FUNCTION AND ISCHEMIC BRAIN DAMAGE

The normal choroid plexus is the site of production of multiple polypeptides including growth factors (reviewed in (Johanson et al., 2000; Chodobski and Szmydynger-Chodobska, 2001; Johanson et al., 2003). Multiple growth factors (e.g., fibroblast growth factor, insulin-like growth factor, and nerve growth factor) have been shown to be neuroprotective when administered to the brain (Ay et al., 2001; Johanson et al., 2003), raising the possibility that endogenous production of such factors may also protect the brain from ischemic damage. As an example, TGFβ1 is produced by the choroid plexus ischemia (Knuckey et al., 1996). Administration of TGFβ to the CSF limits ischemic brain damage (reviewed in Ay et al., 2001; Johanson et al., 2003), strongly suggesting that endogenous TGFβ1 will have the same effect. Further support for this concept is the finding that choroid plexus TGFβ1 is upregulated following cerebral ischemia (Knuckey et al., 1996).

Other functions of the choroid plexus may also serve to limit brain damage following stroke. The production of CSF and the resultant sink effect (Davson et al., 1987) will serve to clear potentially harmful compounds generated at the site of injury. Similarly, the wide array of transporters present at the choroid plexuses (Gao and Meier, 2001) may serve a similar function as discussed above.

If the stroke extensively damages the choroid plexus, it is possible that the above beneficial effects will be impaired. In addition, disruption of the BCSFB may have consequences for the evolution of brain injury. Thus, Ferrand-Drake (2001) postulated that disruption may allow the entry of a number of plasma components that could have deleterious effects on the brain. For example, evidence indicates that the plasma protein prothrombin via generation of thrombin participates in brain injury after stroke (Xi et al., 2002; Karabiyikoglu et al., 2004). Similarly, BCSFB may allow entry of plasma proteins that are part of the complement system. The complement cascade participates in ischemic and hemorrhagic brain injury (Vasthare et al., 1998; Hua et al., 2000), although there is still a question as to the role of plasma-derived versus brain-derived complement. In contrast, Johanson et al. (2000) have postulated that BCSFB disruption may allow entry of potentially beneficial compounds from blood to CSF such as growth factors. The relative deleterious/beneficial effect of BCSFB has still to be determined.

13.8 AN OVERVIEW—GAPS IN OUR KNOWLEDGE

The choroid plexuses, which are the major component of the BCSFB, are vulnerable to ischemia. They undergo profound morphological changes and cell loss. These morphological changes are reflected in physiological changes (e.g., BCSFB disruption). The degree of damage depends on the duration of ischemia. Surprisingly little is known about the degree of ischemia necessary to induce choroid plexus damage. This makes it difficult to translate animal data to human stroke and to elucidate the underlying mechanisms for choroid plexus damage. We (S.R. Ennis and R.F. Keep, unpublished results) have performed preliminary experiments with permanent MCAO in the rat using the thread occlusion model of Longa et al. (1990). This model caused only a 38 ± 18% reduction in lateral ventricle choroid plexus blood flow in the ipsilateral hemisphere. It did, however, cause significant choroidal edema at 24 hours (water content in the ipsilateral and contralateral lateral ventricle choroid plexuses was 4.72 ± 0.30 and 3.84 ± 0.23 g/g dry weight, respectively). Combining MCAO with ipsilateral common carotid artery occlusion caused a 53 ± 13% reduction in choroidal blood flow. These results suggest that choroid plexus function can be compromised by small reductions in flow compared to parenchymal tissue or that parenchymal damage may indirectly be affecting the choroid plexus.

In models of ischemia/reperfusion, there is rapid restoration of the appearance of the choroid plexus epithelium and a return of barrier function. The rapidity of this restoration is again dependent on the duration of the ischemic insult. Whether this repair involves rearrangement of the surviving epithelial cells or the production of new epithelial cells is, as yet, not totally known. The rapidity of functional restoration and some limited evidence on atrophy suggests that repair is primarily due to rearrangement of surviving cells. However, long-term studies are required to examine whether cell proliferation may have a significant role in delayed recovery. There is also the need to examine the degree to which the balance of restructuring and cell proliferation is model (i.e., blood flow) dependent.

Changes in choroid plexus (morphological and physiological) often precede parenchymal changes (e.g., it precedes delayed cell death in the hippocampus in models of forebrain ischemia). This raises the interesting possibility that altered choroid plexus function might contribute to ischemic damage to the parenchyma. As discussed previously, BCSFB disruption may have detrimental or beneficial effects on parenchymal cell function dependent upon which factors enter the CSF from blood (Johanson et al., 2000; Ferrand-Drake, 2001). The net effect of choroid plexus damage during ischemia will also depend upon how the other functions of the choroid plexus are affected by ischemia (e.g., the carrier-mediated transport of compounds between blood and CSF and the production and secretion of compounds, such as growth factors, by the plexus itself).

A true understanding of the importance of ischemia-induced changes in choroid plexus function in stroke really depends on developing ways of selectively preventing choroid plexus damage. In that regard, we need a much better understanding of the mechanisms underlying such damage. Is it purely due to energy failure, or are there secondary mechanisms? Such mechanisms may be easier to study using in vitro choroid plexus epithelial models. If therapeutic targets are identified, it may be possible to selectively protect the choroid plexus epithelium with gene therapy, as suggested by Johanson et al. (2000). Intraventricular administration of some gene transfer vectors, e.g., adenoviruses, results in preferential gene expression in the choroid plexus (Kitagawa et al., 1998). In addition, promoters of genes selectively expressed in the choroid plexus can be used.

There is a paucity of data on the effects of human stroke on choroid plexus function. Multiple studies have examined changes in CSF composition following stroke, but interpreting whether changes reflect BCSFB hyperpermeability, altered choroid plexus transport, secretion or absorption, changes in the movement of compounds across the blood–brain barrier, or altered parenchymal cell function is extremely difficult. Detailed postmortem morphological studies that take into account and measure potential atrophy are needed. In vivo imaging techniques with improved resolution would, of course, be of great value.

Stroke is primarily a disease of the elderly. The animal studies on ischemia-induced injury to the choroid plexus have not been in aged rats. Whether age affects such injury is unknown, although it does enhance parenchymal cell damage in animals and humans (Howard et al., 1987; Davis et al., 1995). A number of morphological changes occur in the choroid plexus with age in human and rat, including flattening of the epithelium, thickening of the basement membrane, and lipofuscin deposits (Preston, 2001) (see also Chapter 14). Whether these changes with age will affect the response to ischemia or might even reflect past small transient ischemic events is unknown. People do suffer multiple strokes, and it is possible that damage to the choroid plexus from a prior stroke might influence the progression of a subsequent ischemic event.

The effect of cerebral ischemia on the choroid plexus is a neglected topic. Many fundamental questions remain to be addressed, let alone answered.

13.9 ACKNOWLEDGMENTS

This study was supported in part by Grants R01 NS034709 and P01 HL018575 from the National Institutes of Health.

REFERENCES

Ay I, Ay H, Koroshetz WJ, Finklestein SP (2001) Growth factors and cerebral ischemia. In: *Current Review of Cerebrovascular Disease* (Fisher M, Bogousslavsky J, eds.). Philadelphia, Current Medicine Inc., pp. 25–33.

Bederson JB, Pitts LH, Tsuji M, Nishimura MC, Davis RL, Bartkowski H (1986) Rat middle cerebral artery occlusion: evaluation of the model and development of a neurologic examination. *Stroke* 17:472–6.

Bradbury MWB (1979) *The Concept of a Blood-Brain Barrier.* John Wiley & Sons, Chichester.

Chodobski A, Szmydynger-Chodobska J (2001) Choroid plexus: target for polypeptides and site of their synthesis. *Microsc. Res. Tech.* 52:65–82.

Chodobski A, Loh YP, Corsetti S, Szmydynger-Chodoboska J, Johanson CE, Lim Y-P, Monfils PR (1997) *Mol. Brain Res.* 48:67–72.

Clarke DD, Sokoloff L (1999) Circulation and energy metabolism of the brain. In: *Basic Neurochemistry* (Siegel GJ, Agranoff BW, Albers RW, Fisher SK, Uhler MD, eds.). Lippincott-Raven, Philadelphia, pp. 637–669.

Daverat P, Castel JP, Dartigues JF, Orogozo JM (1991) Death and functional outcome after spontaneous intracerebral hemorrhage. *Stroke* 22:1–6.

Davis M, Mendelow AD, Perry RH, Chambers IR, James OF (1995) Experimental stroke and neuroprotection in the aging rat brain. *Stroke* 26:1072–1078.

Davson H, Welch K, Segal MB (1987) *Physiology and Pathophysiology of the Cerebrospinal Fluid.* Churchill Livingstone, Edinburgh.

Dienel G (1984) Regional accumulation of calcium in postischemic rat brain. *J. Neurochem.* 43:913–925.

Dohrmann GJ (1970) Dark and light epithelial cells in the choroid plexus of mammals. *J. Ultrastruct. Res.* 32:268–273.

Du C, Hu R, Csernansky C, Hsu C, Choi D (1996) Very delayed infarction after mild focal cerebral ischemia: a role of apoptosis. *J. Cereb. Blood Flow Metab.* 16:195–201.

Faraci FM, Mayhan WG, Williams JK, Heistad DD (1988) Effects of vasoactive stimuli on blood flow to choroid plexus. *Am. J. Physiol.* 254:H286–H291.

Ferrand-Drake M (2001) Cell death in the choroid plexus following transient forebrain global ischemia in the rat. *Microsc. Res. Tech.* 52:130–136.

Ferrand-Drake M, Wieloch T (1999) The time-course of DNA fragmentation in the choroid plexus and the CA1 region following transient global ischemia in the rat brain: the effect of intra-ischemic hypothermia. *Neuroscience* 93:537–549.

Fujii H, Yamashima T, Higashi S, Hashimoto M, Yamamoto S (1989) Experimental intracranial hyperetension following cisternal injection of various blood components. *Neurochirurgia* 32:165–167.

Gao B, Meier PJ (2001) Organic anion transport across the choroid plexus. *Microsc. Res. Tech.* 52:60–64.

Gillardon F, Lenz C, Kuschinsky W, Zimmermann M (1996) Evidence for apoptotic cell death in the choroid plexus following focal cerebral ischemia. *Neurosci. Lett.* 207:113–116.

Gross PM, Zochodne DW, Wainman DS, Ho LT, Espinosa FJ, Weaver DF (1992) Intraventricular endothelin-1 uncouples the blood-flow: metabolism relationship in periventricular structures of the rat brain: involvement of L-type calcium channels. *Neuropeptides* 22:155–165.

Hou ST, MacManus JP (2002) Molecular mechanisms of cerebral ischemia-induced neuronal death. *Int. Rev. Cytol.* 221:93–148.

Howard G, Toole JF, Frye-Pierson J, Hinshelwood LC (1987) Factors influencing the survival of 451 transient ischemic attack patients. *Stroke* 18:552–557.

Hua Y, Xi G, Keep RF, Hoff JT (2000) Complement activation in the brain after experimental intracerebral hemorrhage. *J. Neurosurg.* 92:1016–1022.

Huang F-P, Xi G, Hua Y, Keep RF, Nemoianu A, Hoff JT (2002) Brain edema after experimental intracerebral hemorrhage: Role of hemoglobin breakdown products. *J. Neurosurg.* 96:287–293.

Ikeda J, Mies G, Nowak TS, Joo F, Klatzo I (1992) Evidence for increased calcium influx across the choroid plexus following brief ischemia of gerbil brain. *Neurosci. Lett.* 142:257–259.

Johanson CE (1989) Potential for pharmacologic manipulation of the blood-cerebrospinal fluid barrier. In: *Implications of the Blood-Brain Barrier and Its Manipulation.* (Neuwelt EA, ed.). New York, Plenum, pp. 223–260.

Johanson CE (2003) The choroid plexus-cerebrospinal fluid nexus: gateway to the brain. In: *Neuroscience in Medicine* (Conn PM, ed.). Totowa, Humana Press, pp. 1–31.

Johanson CE, Palm DE, McMillan PN, Stopa EG, Doberstein CE, Duncan JA (2003) Volume transmission-mediated protective impact of choroid plexus-CSF growth factors on forebrain ischemic injury. In: *Blood-Spinal Cord and Brain Barriers in Health and Disease.* (Sharma HS, ed.). San Diego, Academic Press, pp. 361–384.

Johanson CE, Palm DE, Primiano MJ, McMillan PN, Chan P, Knuckey NW, Stopa EG, (2000) Choroid plexus recovery after transient forebrain ischemia: role of growth factors and other repair mechanisms. *Cell. Mol. Neurobiol.* 20:197–216.

Jones TH, Morawetz RB, Crowell RM, Marcoux FW, FitzGibbon SJ, DeGirolami U (1981) Threshold of focal cerebral ischemia in awake monkeys. *J. Neurosurg.* 54:773–782.

Karabiyikoglu M, Hua Y, Keep RF, Ennis SR, Xi G (2004) Intracerebral hirudin injection attenuates ischemic tissue damage and neurological deficits without altering local cerebral blood flow. *J. Cereb. Blood Flow Metab.* 24:159–166.

Keep RF, Jones HC (1990a) Cortical microvessels during brain development. A morphometric study in the rat. *Microvasc. Res.* 40:412–426.

Keep RF, Jones HC (1990b) A morphometric study on the development of the lateral choroid plexus, choroid plexus capillaries and ventricular ependyma in the rat. *Dev. Brain Res.* 56:47–53.

Keep RF, Jones HC, Cawkwell RD (1986) A morphometric analysis of the development of the fourth ventricle choroid plexus in the rat. *Dev. Brain Res.* 27:77–85.

Kitagawa H, Setoguchi Y, Fukuchi Y, Mitsumoto Y, Koga N, Mori T, Abe K (1998) DNA fragmentation and HSP72 gene expression by adenovirus-mediated gene transfer in postischemic gerbil hippocampus and ventricle. *Metab. Brain Dis.* 13:211–223.

Knuckey NW, Finch P, Palm DE, Primiano MJ, Johanson CE, Flanders KC, Thompson NL (1996) Differential neuronal and astrocytic expression of transforming growth factor beta isoforms in rat hippocampus following transient forebrain ischemia. *Mol. Brain Res.* 40:1–14.

Kristian T, Siesjo BK (1997) Changes in ionic fluxes during cerebral ischaemia. *Int. Rev. Neurobiol.* 40:27–45.

Li Y, Chen J, Chopp M (2002) Cell proliferation and differentiation from ependymal, subependymal and choroid plexus cells in response to stroke in rats. *J. Neurol. Sci.* 193:137–146.

Liebeskind DS, Hurst RW (2004) Infarction of the choroid plexus. *Am. J. Neuroradiol.* 25:289–290.

Longa EZ, Weinstein PR, Carlson S, Cummins R (1990) Reversible middle cerebral artery occlusion without craniectomy in rats. *Stroke* 20:84–91.

MacManus JP, Buchan AM, Hill I.E., Rasquinha I, Preston E (1993) Global ischemia can cause DNA fragmentation indicative of apoptosis in rat brain. *Neurosci. Lett.* 164:89–92.

Mies G, Ishimaru S, Xie Y, Seo K, Hossmann K-A (1991) Ischemic threholds of cerebral protein synthesis and energy state following middle cerebral artery occulusion in rat. *J. Cereb. Blood Flow Metab.* 11:753–761.

Naff NJ, Tuhrim S (1997) Intraventricular hemorrhage in adults: complications and treatment. *New Horizons* 5:359–363.

Nagahiro S, Goto S, Korematsu K, Sumi M, Takahashi M, Ushio Y (1994) Disruption of the blood-cerebrospinal fluid barrier by transient cerebral ischemia. *Brain Res.* 663:305–311.

National Institute of Neurological Disorders and Stroke rt-PA Stroke Study Group (1995) Tissue plasminogen activator for acute ischemic stroke. *N. Engl. J. Med.* 333:1581–1587.

Palm D, Knuckey N, Guglielmo M, Watson P, Primiano M, Johanson CE (1995) Choroid plexus electrolytes and ultrastructure following transient forebrain ischemia. *Am. J. Physiol.* 269:R73–R79.

Preston JE (2001) Ageing choroid plexus-cerebrospinal fluid system. *Microsc. Res. Tech.* 52:31–37.

Pulsinelli WA, Brierley JB, Plum F (1982) Temporal profile of neuronal damage in a model of transient forebrain ischemia. *Ann. Neurol.* 11:491–498.

Qureshi AI, Turhim S, Broderick JP, Batjer HH, Hondo H, Hanley DF (2001) Spontaneous intracerebral hemorrhage. *N. Engl. J. Med.* 344:1450–1460.

Rothstein RP, Levison SW (2002) Damage to the choroid plexus, ependyma and subependyma as a consequence of perinatal hypoxia/ischemia. *Dev. Neurosci.* 24:426–436.

Saunders NR, Dzigielewska KM, Ek J, Mollgard K (1999) Morphological and physiological aspects of barriers in the developing brain. In: *Brain Barrier Systems; Alfred Benzon Symposium 45* (Paulson O, Knudsen GM, Moos T, eds.). Copenhagen, Munksgaard.

Schalk KA, Faraci FM, Heistad DD (1992a) Effect of endothelin on production of cerebrospinal fluid in rabbits. *Stroke* 23:560–563.

Schalk KA, Faraci FM, Williams JL, VanOrden D, Heistad DD (1992b) Effect of atriopeptin on production of cerebrospinal fluid. *J. Cereb. Blood Flow Metab.* 12:691–696.

Spector R, Eells J (1984) Deoxynucleoside and vitamin transport into the central nervous system. *Fed. Proc.* 43:196–200.

Spector R, Johanson CE (1989) The mammalian choroid plexus. *Sci. Amer.* 261:68–74.

Spolarics Z, Lang CH, Bagby GJ, Spitzer JJ (1991) Glutamine and fatty acid oxidation are the main sources of energy for Kupffer and endothelial cells. *Am. J. Physiol.* 261:G185–90.

Sun H, Dai H, Shaik N, Elmquist WF (2003) Drug efflux transporters in the CNS. *Adv. Drug Deliv. Rev.* 55:83–105.

Szmydynger-Chodobska J, Chodobski A, Johanson CE (1994) Postnatal developmental changes in blood flow to choroid plexuses and cerebral cortex of the rat. *Am. J. Physiol.* 266:R1488–R1492.

Towfighi J, Zec N, Yager J, Housman C, Vannucci RC (1995) Temporal evolution of neuropathologic changes in an immature rat model of cerebral hypoxia: a light microscopic study. *Acta Neuropathol.* 90:375–386.

Tyson G, Kelly P, McCulloch J, Teasdale G (1982) Autoradiographic assessment of choroid plexus blood flow and glucose utilization in the unanesthetized rat. *J. Neurosurg.* 57:543–547.

Unterberg A, Schmidt W, Wahl M, Baethmann A (1990) Role of leukotrienes as mediator compounds in brain edema. *Adv. Neurol.* 52:211–214.

Vasthare US, Barone FC, Sarau HM, Rosenwasser RH, DiMartino M, Young WF, Tuma RF (1998) Complement depletion improves neurological function in cerebral ischemia. *Brain Res. Bull.* 45:413–419.

Vitte P-A, Brun J, Lestage P, Claustrat B, Bobillier P (1989) The effects of melatonin and pinealectomy upon local cerebral glucose utilization in awake unrestrained rats are restricted to a few specific regions. *Brain Res.* 489:273–282.

Wang Y-C, Lin C-W, Shen C-C, Lai S-C, Kuo J-S (2002) Tissue plasminogen activator for the treatment of intraventricular hematoma: the dose-effect relationship. *J. Neurol. Sci.* 2002:35–41.

White RP, Leffler CW, Bada HS (1990) Eicosanoid levels in CSF of premature infants with posthemorrhagic hydrocephalus. *Am. J. Med. Sci.* 299:230–235.

Wildrick D (1997) Intraventricular hemorrhage and long-term outcome in the premature infant. *J. Neurosci. Nurs.* 29:281–289.

Williams JL, Jones SC, Bryan RM (1991a) Vascular responses of choroid plexus during hypercapnia in rats. *Am. J. Physiol.* 260:R1066–R1070.

Williams JL, Shea M, Furlan AJ, Little JR, Jones SC (1991b) Importance of freezing time when iodoantipyrine is used for measurement of cerebral blood flow. *Am. J. Physiol.* 261:H252–H256.

Windmueller HG, Spaeth AE (1980) Respiratory fuels and nitrogen metabolism *in vivo* in small intestine of fed rats. *J. Biol. Chem.* 255:107–112.

Xi G, Keep RF, Hoff JT (2002) Pathophysiology of brain edema formation. *Neurosurg. Clin. North Am.* 13:371–383.

Zivin JA, Fisher M, DeGirolami U, Hemenway CC, Stashak JA (1985) Tissue plasminogen activator reduces neurological damage after cerebral embolism. *Science* 230:1289–1292.

14 Aging of the Choroid Plexus and CSF System: Implications for Neurodegeneration

Jane E. Preston
King's College London
London, United Kingdom

Michael R. Wilson
Imperial College School of Medicine
Chelsea and Westminster Hospital
London, United Kingdom

Ruo Li Chen
King's College London
London, United Kingdom

CONTENTS

The function of physiological systems in later life has tended to be dominated by the investigation of specific diseases, and the choroid plexus (CP) is no exception. Biogerontologists interested in nondiseased aging states ask the question: What makes old cells more vulnerable to disease than young cells? It is argued that this approach has the potential to illuminate multiple disease states as well as the process of "healthy" aging (Hayflick, 2000). Studies of aging CP can be broadly approached at three levels: molecular damage, cellular changes, and the CP–cerebrospinal fluid (CSF) functional level.

14.1 MOLECULAR DAMAGE

Oxidative damage to molecular and cellular components by the action of reactive oxygen species (ROS) is a ubiquitous finding in aged tissue. Evidence of oxidative nuclear DNA damage was shown by Nakae and colleagues (2000), with formation of the modified base 8-hydroxydeoxyguanosine (8-OHdG) apparent in two-week-old Fischer rats and at all ages tested up to two years. Of the 15 organs and tissues sampled, 8-OHdG was detected earlier only in the cerebellum, brain stem, and kidney. While the presence of 8-OHdG does not necessarily correlate with DNA mutation rate, it does result in GC-TA transversions (Shibutani et al., 1991), which, if unrepaired, will form mutations affecting gene expression. Mitochondrial DNA is more susceptible to oxidative damage due to its close proximity to ROS generation via the electron transport chain and its paucity of repair mechanisms and protective histones. In human CP, Cottrell and colleagues (2001a) studied the presence of the electron transport chain complex IV, cytochrome *c* oxidase (CcO), subunits of which are encoded by mitochondrial DNA (Anderson et al., 1981). They found the percentage of CcO-deficient cells increased from 0.18% at age 15 to 4.16% at age 87 (Cottrell et al., 2001a), around four times the level seen in hippocampal neurons, despite the parallel age-related increase in hippocampus. These cells were not deficient in complex II, encoded by the nuclear DNA, indicating that the deficit was at the level of the mitochondria, with resultant potential impairment in ATP production, elevated production of ROS, and breach of the epithelial cell layer. Similar CcO deficiency of CP is seen in the mitochondrial diseases Leigh syndrome (Tanji et al., 2001) and mitochondrial encephalomyopathy, lactic acidosis, and stroke-like episodes (MELAS) (Cottrell et al., 2001b), and considered important to the neuropathological outcomes of these syndromes. It is also interesting that old CP CcO-negative cells were larger than their neighbors (Cottrell et al., 2001a), also seen in MELAS (Cottrell et al., 2001b), likely due to excess proliferation of mitochondria in an attempt to rescue the ATP-deficient cell. Aged rat CP is also deficient in dehydrogenase enzymes of the Krebs cycle and glycolysis (Ferrante and Amenta, 1987), further compromising energy production. Induction of ATP depletion in young guinea pig CP by addition of DNP and 2-deoxyglucose or incubation at 27°C results in excess accumulation of β-amyloid (Aβ) peptide 1–40 (Wilson et al., 2001), mirroring the accumulations seen in Alzheimer's disease (AD) and, to a lesser extent, normal aged CP (Cortez et al., 1995; Premkumar and Kalaria, 1996; Preston et al., 1999). In old rat CP incubation with excess 1 μM Aβ 1–40 for 1 hour led to reduced dehydrogenase activity (MTT assay) and reduced Aβ degradation compared to young

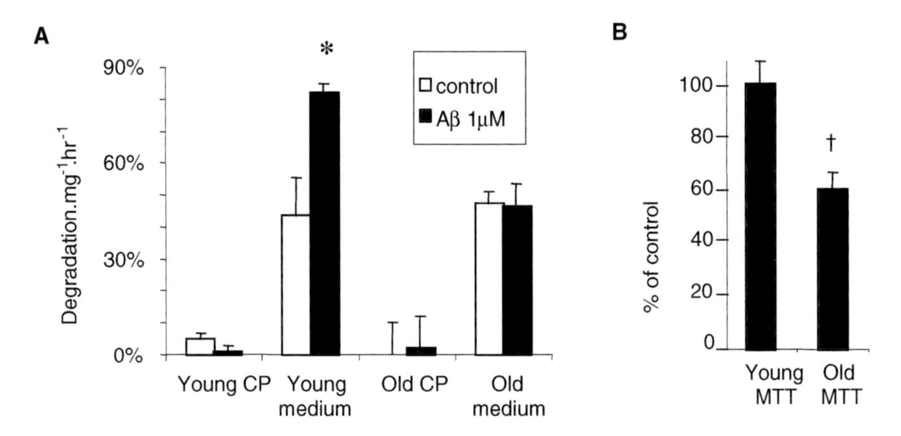

FIGURE 14.1 (A) Degradation of ^{125}I-amyloidβ 1–40 by old and young rat CP after 1-hour incubation in artificial CSF medium, with or without excess peptide (1 μM Aβ 1–40). (B) MTT assay of tissue dehydrogenase activity after addition of 1 μM Aβ 1–40. *Note:* Lateral ventricle choroid plexuses from young (6 months) and old (20–22 months) Wistar rats were incubated for 1 hour in 1 mL artificial CSF (37°C, gassed 95% O_2 5% CO_2) containing 0.5μCi ^{125}I-amyloidβ 1–40 after the method of Smith and Johanson (1991). Extracellular marker was ^{14}C-mannitol. The two plexuses from each rat were paired, one placed in control medium, the other in medium containing 1 μM Aβ 1–40. Tissues were solubilized (triton X), homogenized, 0.5 mL trichloroacetic acid added, and centrifuged, all at 4 °C. Degradation was estimated from the TCA-soluble dpm in the supernatant from the CP and incubating medium samples, corrected for initial TCA solubility of 10–15%. MTT assay was used to determine the dehydrogenase activity. Values are mean ± SEM of 5 old and 4 young plexuses; $*p < 0.05$, difference from control, unpaired t-test; $\dagger p < 0.01$, difference from control, paired *t*-test.

rats (see Figure 14.1) (unpublished data). Taken together, this data indicates a circular trend whereby poor energy transduction facilitates peptide accumulation, which further compromises cellular function in old tissue.

The effects of excess ROS on both nuclear and mitochondrial genomes are known to include mutations and epigenetic effects, inhibiting DNA methylation, for example, and altering gene expression, predisposing cells to either transformation or replicative senescence (Poirier, 1994; von Zglinicki, 2003). A candidate gene for premature senescence, *klotho* (Nabeshima, 2002), has been suggested as such a target (Nakae et al., 2000) since the Klotho protein is highly expressed in mice CP. Transgenic Klotho mice lacking the gene have a greatly shortened life span and are unable to regulate calcium adequately, posited to result in increased calcium-dependent protease, degradation of cytoskelatal proteins, and cell death (Tsujikawa et al., 2003). Whether this impacts on CP aging remains to be determined.

An example of lipid oxidative damage in many long-lived tissues is accumulation of intracellular lipofuscin, also seen in human CP (Wen et al., 1999; Serot et al., 2000). Whether this is a significant factor in CP aging is not clear, although it is suggested that retinal pigment epithelium lipofuscin contributes to age-related macular degeneration (Katz, 2002). Lipofuscin is rarely seen in aged rat CP (Serot et al., 2001), and despite accumulation in human CP, CSF samples from healthy human donors do not indicate age-related increase in the lipid peroxidation end product F2-isoprostane (Montine et al., 2002).

In the context of oxidative stress, the CP has abundant antioxidant defense compared with neuronal tissues. In particular, glutathione peroxidase has 10 times greater activity in CP than in cerebrum, and this remains constant in aged rats (Tayarani et al., 1989). CP also avidly transports ascorbic acid (Angelow et al., 2003), elevating CSF levels above those in plasma (Reiber et al., 1993; Alho et al., 1998). CSF superoxide dismutase is increased with age in humans and rats (Bracco et al., 1991; Hiramatsu et al., 1992). However, the concentration of the potent antioxidant melatonin declines in CSF with increasing age as well as AD (Lui et al., 1999), and this is postulated to add to increased oxidative stress caused by elevation of reactive oxygen species $OH^{\bullet}$ (Maurizi, 2001) and predispose to pathogenesis.

14.2 CELLULAR AND MORPHOLOGICAL CHANGES

Several studies (see Serot et al., 2003, for review) have provided important insights into the morphological changes of CP epithelium with both age and disease. The cells reduce in height by some 16% between 6 and 30 months (Serot et al., 2001) and 10% between infancy and 88 years in rats and humans (Serot et al., 2000). The microvilli in both mice and rats are shortened (Sturrock, 1988; Serot et al., 2001) suggesting a diminution of the apical surface area. Increasing calcification of the CP is evident in humans, with an incidence of 0.5% in the first decade rising to 86% in the eighth decade (Modic et al., 1980). The nucleus becomes irregular and flattened (Serot et al., 2000) and elongated in rats (Serot et al., 2001).

Like other secretory epithelia, such as in the kidney and seminiferous tubules, the epithelial basement membrane becomes significantly thicker in old age, more than doubling in 30-month-old rats (Serot et al., 2001), and increasing by around 22% in elderly humans (Serot et al., 2000). Fibrosis of the stroma is a common feature (Shuangshoti and Netsky, 1970; Serot et al., 2000, 2001) and is clearly seen in low-power light photomicrographs (see Figure 14.2) of CP from 22-month-old rats (Figure 14.2B) compared to 3-month-old CP (see Figure 14.2A). A likely cause is non-enzymic protein glycation, which resists proteolytic turnover (Koschinsky et al., 1997). The direct consequences of these changes on CP function remain unknown, but in the kidney, at least, thickening of the basement membrane is thought to contribute to reduced fluid filtration in the renal tubules (Vlassara, 1994), and it is reasonable to suggest that there are similar consequences for plasma filtration in the aging CP with a subsequent decrease in CSF formation.

Common cellular inclusions include the Biondi bodies, tightly packed linear or ring amyloid-associated structures seen in aged human CP and, to a greater extent, in Alzheimer's disease (Eriksson and Westermark, 1990; Miklossy et al., 1998; Wen et al., 1999). The cause of such fibrillary aggregation is not clear, but ATP deficiency certainly plays a role in neuronal accumulation of the structurally related neurofibrillary tangles (Roder et al., 1993), and lack of ATP predisposes the CP to amyloid accumulation (Wilson et al., 2001).

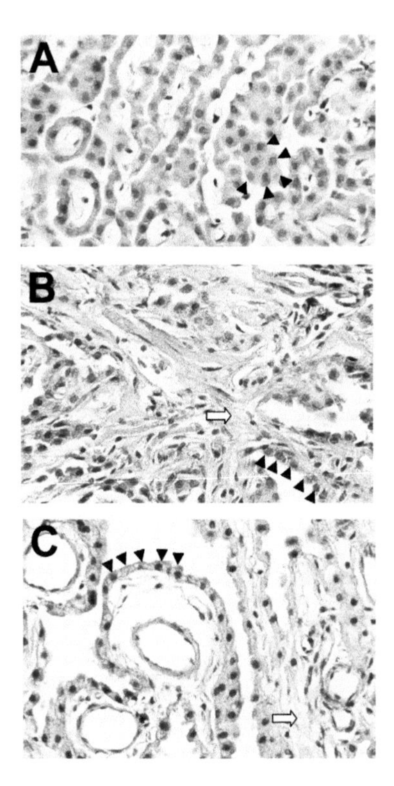

FIGURE 14.2 Light photomicrographs of lateral ventricular CPs from Sprague-Dawley rats: (A) 3 months old; (B, C) 22 months old. (Magnification 100x). Increased fibrosis of the stroma can be seen in B and C (open arrows), with flattening of the epithelial cell layer and darkening of the cytoplasm (filled arrow heads). (Images kindly provided by Dr. Adam Chodobski and Dr. Joanna Szmydynger-Chodobska.)

14.3 CSF SECRETION AND CP FUNCTION

14.3.1 ION TRANSPORT

There is evidence that ion transport is compromised with age. An 18% decline in blood-to-CSF sodium transfer was seen in 34 month-old Fischer rats (Smith et al., 1982), and ouabain sensitive Na-K-ATPase activity declined by 50% in rats between ages 8 and 26 months (Kvitnitskaia-Ryzhova and Shkapenko, 1992). Chloride efflux rate also declined by nearly one third between 3 and 26 months (Preston, 1999). In aged sheep, attempts to upregulate K/Cl cotransport with *N*-ethylmaleimide (NEM) failed to elicit any change in either Cl transport or CSF secretion, suggesting deficits in ion transport capacity, while young sheep responded with transient doubling in both measures (Chen et al., 2004). An interesting area for further research is the role of nitric oxide (NO) in the aged phenotype. Bovine CP Na-K-ATPase is inhibited by NO (Ellis et al., 2000) released following cholinergic activity. While there are no direct lines of evidence for elevation in CP NO with age, after infection iNOS mRNA is notably induced in CP within 4 hours (McCann et al., 1998), and the authors suggest that "recurrent infections over the lifespan play a significant role in producing aging changes." In the CP, that would manifest as reduced Na-K-ATPase activity and subsequent compromise of CSF secretion.

14.3.2 CSF Secretion

Decline in CSF secretion with age has been demonstrated using a variety of techniques in different species including human, rat, and sheep (Cutler et al., 1969; May et al., 1990; Preston, 2001; Chen et al., 2003). Such age-related decline is also seen in patients with Parkinson's disease and acute hydrocephalus (Silverberg et al., 2002), neither of which appears to differ from measurements in healthy subjects (May et al., 1990). Measurements in humans are not without controversy (Preston, 2001), not least because they compare cohorts, rather than longitudinal follow-up design and employ invasive techniques. Indeed, some MRI studies imply no change with age in humans (Gideon et al., 1994). Despite such controversy, the bulk of evidence suggests that CSF secretion rate is reduced in later life, which is consistent with the molecular and cellular evidence.

14.3.3 Vasopressin

There is increasing evidence that vasopressin (VP) regulates CP function (Szmydynger-Chodobska et al., 2002), and some interesting clues for involvement in CP aging are emerging. VP receptors (V1) have been described in many brain regions, including the CP, and activation of V1 receptors results in decreased blood flow, reduced CSF secretion (Faraci et al., 1990), the appearance of "dark" cells, and slower chloride efflux from isolated tissues (Johanson et al., 1999). In aging mice, the number of dark epithelial cells increases, in parallel with other age-related morphological changes (Sturrock, 1988), and mirror the VP-induced dark cells. With aging, there are suggestions that VP secretion increases, and this combined with greater numbers of VP containing fibers penetrating the 3rd ventricle and elevation of VP in rat CSF and human plasma (Frolkis et al., 1999) point to an additional mechanism by which CSF secretion is attenuated.

14.3.4 Insulin-Like Growth Factor

A general decline in systemic and central levels of growth hormone and its anabolic mediator insulin-like growth factor-I (IGF-I) is common in healthy aging (see Khan et al., 2002, for review), and these factors hold an interest, too, for CP function. The choroid plexus synthesizes and secretes into CSF the related insulin-like growth factor-II (IGF-II) and its binding protein IGFBP-2 (Holm et al., 1995). Under normal circumstances, and in health, these peptides act in concert with IGF-I to stimulate DNA synthesis, cell growth, and myelin synthesis in the brain (Lenoir, 1983; Mill, 1985; McMorris, 1991). In addition, a major role for IGF-II is postulated in brain injury (Beilharz et al., 1998), when IGF-II and IGFBP-2, presumed to be of CP origin, increase in CSF after injury, then later at the site of brain injury in the absence of any local IGF-II mRNA expression (Walter et al., 1999). IGF-II is also thought to play a role in promoting choroidal epithelial cell growth (Nilsson et al., 1996). Production of both IGFs is linked to the presence of circulating growth hormone (Cohen et al., 1992), and in later life, in addition to a decline in plasma growth hormone (Arnold et al., 1999), there is also reduced density of growth hormone binding to CP (Lai et al., 1993). IGF-II mRNA expression seems also to be regulated by circulating IGF-I since, in diabetic rats, intravenous injection of IGF-I increases brain IGF-II mRNA (Armstrong et al., 2000), which the authors suggest results from IGF-I action on the CP. Thus, an age-related decline in plasma IGF-I could further affect CP IGF-II synthesis. To date there are no studies quantifying IGF-II synthesis with advanced age, but the circumstantial evidence suggests that maintenance of synthesis would be in the face of several obstacles. Any reduction in IGF-II activity or content in the CP could have significant consequences for CP repair and cell turnover and also in neuronal survival after injury or ischemia, consistent with an evolving theory for the role of CP-derived growth factors in the protection of the brain from ischemic injury (Johanson et al., 2000; Stopa et al., 2001).

14.4 CSF TURNOVER AND DRAINAGE

As CSF secretion rate declines, there is a parallel increase in the proportion of the intracranial volume occupied by CSF (Narr et al., 2003). MRI studies estimate that the CSF intracranial space is 20 to 33% in 71- to 80-year-olds, while in young adults it is around 12% (Courchesne et al., 2000). This equates to an increase in CSF volume from around 150 to 350 mL by age 80. Longitudinal studies of healthy elderly patients in their eighth decade also showed increased CSF volume over a 5-year period (Wahlund et al., 1996). The cause is widely held to be due to neuronal atrophy, but the result is a significant reduction in the rate of turnover of CSF, which would be further exacerbated by decline in CSF production. In humans, CSF is replaced approximately four times each day (Silverberg et al., 2003) while with aging or disease this may occur only one to two times each day (Rubenstein, 1998; Silverberg et al., 2003). In Fischer rats, CSF volume increased from 0.16 to 0.31 mL between 3 and 30 months of age, with corresponding turnover rates falling from 12 to 3 times per day (Preston, 2001). Additional contributors to turnover are mechanisms for CSF drainage and include drainage of fluid into the sagittal venous sinus at the arachnoid villi. Increased resistance to drainage is associated with decreased CSF secretion rate in older people (Czosnyka et al., 2001) and may be due in part to effects at the level of the CP since in experimental hydrocephalus in the rat with elevated intracranial pressure, CP chloride efflux is reduced (Knuckey et al., 1993). Resistance to drainage increases with age (Albeck et al., 1998), the arachnoid membrane thickens in humans (Bellur et al., 1980), and in many older people, central venous pressure increases along with generalized vascular disease, all contributing to the slow turnover (Rubenstein, 1998). The reduced ability to replace CSF will contribute to build-up of molecules or metabolites that rely on CSF drainage for removal; these may enter the fluid, either by penetration across the CP, de novo CP synthesis, or by bulk flow from brain interstitial fluid. Thus, many proteins of plasma origin, especially those of large molecular size, have increased

CSF/plasma ratios with age, for example, albumin in humans and sheep and IgG in humans (Garton et al., 1991; Blennow et al., 1993; Reiber, 2001; Chen et al., 2003). Radioiodinated serum albumin (RIHSA) is cleared from the CNS very slowly in older humans after lumbar injection, with most remaining in the brain one to two days after administration (Henriksson and Voight, 1976). This contrasts with younger age groups, where most RIHSA had been cleared by this time. Brain-derived proteins have a more varied pattern (Reiber, 2001) presumably because their removal depends on interstitial fluid flow and the effect of the local blood–brain barrier as well as CSF drainage. The effect of turnover on clearance of compounds is illustrated by studies of β-amyloid clearance from healthy rat CSF. In young animals radiolabeled ^{125}I-Aβ40 is rapidly cleared from CSF after intraventricular injection (Ghersi-Egea et al., 1996). However, in aged Fischer rats the rate of clearance fell from 10.44 $\mu L.min^{-1}$ to 0.7 $\mu L.min^{-1}$ between 3 and 30 months of age (Preston, 2001), with most of the decline occurring by 19 months. This was accompanied by an age-related increase in brain ^{125}I accumulation from 7% to 49% over the 90 minutes of the experiment (Preston, 2001). Similarly, in the senescence accelerated mouse (SAMP8), efflux of radiolabeled Aβ42 from brain after intraventricular injection was significantly slower in old SAMP8 than young SAMP8 or controls (Banks et al., 2003). In a study of patients with Alzheimer's disease, establishment of a shunt to increase CSF drainage stabilized the usual cognitive deficits, so that, at 12 months after shunt, no change in Mattis Dementia rating scale was seen. In the nonshunted group cognitive decline followed its expected course (Silverberg et al., 2003). Such changes in CSF protein could, and have, been interpreted as an increase in CP permeability. However, studies on both sheep and rats show no significant change, at least for large compounds (Preston, 2001), whereas the documented elevation of plasma proteins in CSF is wholly consistent with reduced CSF turnover (Reiber, 2001).

14.4.1 Transthyretin

In addition to mediating protein transfer between blood and CSF, the CP epithelium has an exclusive or contributory role in synthesizing protein for CSF distribution. The most abundant protein synthesized is transthyretin (TTR) (Southwell et al., 1993), which transports thyroid hormones—principally thyroxine (T4)—in blood and CSF of mammals, reptiles, birds, and marsupials. It is suggested that systemically produced T4 binds to CP TTR, and this complex is transported unidirectionally into CSF providing the CSF supply of T4 and contributing to brain levels (Chanoine et al., 1992; Southwell et al., 1993; Kuchler-Bopp et al., 2000; Schreiber, 2002). Importantly TTR has other functions in relation to amyloidosis and binding of Aβ and can prevent amyloid aggregation (Wisniewski et al., 1993; Schwarzman et al., 1994), a hallmark of Alzheimer's disease pathogenesis. Levels of TTR in aging human CSF have been variously described as elevated or stable (Kleine et al., 1993; Serot et al., 1997; Kunicki et al., 1998). However, given the effect of CSF turnover on protein concentrations, absolute concentrations may give a misleading impression of the capacity of the CP to synthesize TTR. For example Serot and colleagues (1997) show CSF albumin levels rise from 113 to 217 $mg.L^{-1}$ between 10 and 76 years, an increase of some 90%, whereas the TTR increase was, only around 30%, from 15.5 to 20 mg.L^{-1}. Reiber (2001) suggests that protein changes be corrected for the albumin ratio to normalize the CSF turnover effects, and we might, therefore, expect that TTR production has in real terms declined. In the aged sheep (Chen et al., 2003) CSF TTR was two-thirds the level found in young sheep, despite normal plasma TTR levels (derived from liver). Since thyroxine is essential to maintain neuronal metabolism and chelate amyloid peptides, it is suggested that lack of sufficient TTR contributes to age-related cognitive decline (Rubenstein, 1998). Clinical studies have shown that nicotine increases synthesis and secretion of TTR in brain, presumed to be of CP origin (Li et al., 2000), and may provide a mechanistic link to the epidemiological observations of an association between cigarette smoking and protection from dementia (Graves et al., 1991). It is of interest that the CP has recently been shown to express mRNA for gelsolin (Matsumoto et al., 2003), which inhibits Aβ fibrillogenesis and defibrillizes preformed Aβ 1–40 and 1–42 fibrils (Ray et al., 2000), and gelsolin expression increases in aged rat brain (Ahn et al., 2003). Thus TTR may be one of a concert of proteins found in the CP-CSF axis, along with gelsolin, apolipoprotein E, and clusterin (DeMattos et al., 2004), which are involved in CNS regulation of amyloid peptide.

14.5 IMPLICATIONS FOR NEURODEGENERATIVE DISEASE

The question initially asked is what makes the old CP-CSF system more vulnerable to disease than the young system, and the multiple factors to consider are summarized in Table 14.1. The CP is a highly metabolically active tissue, and loss of any capacity for energy transduction or gene expression will manifest in reduced capacity to resist stressors as well as maintain normal homeostatic mechanisms. It would be interesting to evaluate the extent of oxidative protein damage since this occurs in at least 30% of brain proteins (Stadtmann and Levine, 2003), rendering them inactive and reducing enzyme activity. This may be one explanation for reduced amyloid degradation and increased accumulation with age, which itself confers toxicity (see Figure 14.1). A generalized reduction in the ability to upregulate protein and growth factor production as part of normal cell function or in response to brain injury would again compromise the CP and the wider CNS. The combination of morphological changes, particularly basement membrane thickening, and reduced ion transport ability contribute to impaired capacity for the CP to secrete CSF and to the gradual slowing of CSF turnover, which has been linked not only to Alzheimer's disease but, through TTR, to cognitive decline with age. The CP must also cope with age-related changes in the rest of the body, and these cannot be ignored as potential contributors to CP-CSF vulnerability with age, for example, the increased prevalence of type II diabetes with associated increase in protein glycation and thickening of basement membranes; increased cardiovascular disease and heart failure associated with reduced CP perfusion (Rubenstein, 1998); endocrine senescence resulting in elevated VP reducing CSF secretion; and declining circulating growth hormone reducing IGF-II synthesis. All these factors contribute to senescence of the CP in particular and the consequent decline in protection afforded to the CNS by this gatekeeper of the brain.

TABLE 14.1
Age-Related Changes in CP and CSF

	Species	Ref.
In CP		
Nuclear DNA damage, elevated 8-OHdG	Rat	Nakae et al., 2000
Mitochondrial DNA damage	Human	Cottrell et al., 2001a
Lipofuscin accumulation	Human	Serot et al., 2000; Wen et al., 1999
Krebs cycle, glycolysis enzyme deficit	Rat	Ferrante and Amenta, 1987
Biondi bodies accumulate	Human	Eriksson and Westermark, 1990; Miklossy et al., 1998
Na-K-ATPase decline	Rat	Kvitnitskaia-Ryzhova and Shkapenko, 1992
Chloride efflux decline	Rat	Preston, 1999
Flattened epithelium	Human, rat	Serot et al., 2000, 2001
Shortened microvilli	Rats, mouse	Serot et al., 2001; Sturrock, 1988
Calcification	Human	Modic et al., 1980
Thickened basement membrane	Human, rat	Serot et al., 2001; Serot et al., 2000
Presence of "dark cells"	Mouse	Sturrock, 1988
In CSF		
Decline in secretion rate	Human, rat, sheep	Cutler et al., 1969; May et al., 1990; Silverberg, 2002; Preston, 2001; Chen et al., 2004
Intracranial CSF volume increase	Human, rat	Courchesne et al., 2000; Preston, 2001
Slow CSF turnover	Human, rat	Rubenstein, 1998, Silverberg et al., 2003; Preston, 2001
Increased resistance to drainage	Human	Albeck et al., 1998
Elevated CSF/plasma protein ratio	Human, sheep	Garton et al., 1991; Reiber, 2001; Chen et al., 2003
Transthyretin elevated	Human	Serot et al., 1997
Transthyretin stable	Human	Kunicki et al., 1998
Transthyretin reduced	Sheep	Chen et al., 2003
Reduced CSF melatonin	Human	Lui et al., 1999
Elevated CSF vasopressin	Human	Frolkis et al., 1999
Reduced clearance of amyloid peptides	Rat, SAMP (senescence accelerated prone) mouse	Preston, 2001; Banks et al., 2003

14.6 ACKNOWLEDGMENTS

We would like to thank K. Doctor, N. Kassem, A. Chodobski, J. Szmydynger-Chodobska, and M. Segal. The work was funded in part by the Biotechnology and Biological Research Council, the Trustees of Guy's and St. Thomas' Hospitals, Sir Jules Thorn Charitable Trust, and the Wellcome Trust.

REFERENCES

J.S. Ahn, I.S. Jang, D.I. Kim, K.A. Cho, Y.H. Park, K. Kim, C.S. Kwak, S.C. Park, Ageing-associated increase of gelsolin for apoptosis resistance, *Biochem Biophys Res Commun* 312, 2003, 1335–41.

M.J. Albeck, C. Skak, P.R. Nielsen, K.S. Olsen, S.E. Børgesen, F. Gjerris, Age dependency of resistance to cerebrospinal fluid outflow, *J Neurosurg* 89, 1998, 275–8.

H. Alho, J.S. Leionen, M. Erhola, K. Lonnrot, R. Aejmelaeus, Assay of antioxidant capacity of human plasma and CSF in aging and disease, *Restor Neurol Neurosci* 12, 1998, 159–65.

S. Anderson, A.T. Bankier, B.G Barrell, M.H, de Bruijn, A.R. Coulson, J. Drouin, I.C. Eperon, D.P. Nierlich, B.A. Roe, F. Sanger, P.H. Schreier, A.J. Smith, R. Staden, I.G. Young, Sequence and organization of the human mitochondrial genome, *Nature* 290, 1981, 457–65.

S. Angelow, M. Haselbach, H-J. Galla, Functional characterisation of the active ascorbic acid transport into cerebrospinal fluid using primary cultured choroid plexus cells, *Brain Res* 988, 2003, 105–13.

C.S. Armstrong, L. Wuarin, D.N. Ishii, Uptake of circulating insulin-like growth factor-I into the cerebrospinal fluid of normal and diabetic rats and normalization of IGF-II mRNA content in diabetic rat brain, *J Neurosci Res* 59, 2000, 649–60.

P.M. Arnold, J.Y. Ma, B.A. Citron, B.W. Festoff, Insulin-like growth factor binding proteins in cerebrospinal fluid during human development and aging, *Biochem Biophys Res Commun* 264, 1999, 652–6.

W.A. Banks, S.M. Robinson, S. Verma, J.E. Morley, Efflux of human and mouse amyloid beta proteins 1–40 and 1–42 from brain: impairment in a mouse model of Alzheimer's disease, *Neurosci* 121, 2003, 487–92.

E.J. Beilharz, V.C. Russo, G. Butler, N.L. Baker, B. Connor, E.S. Sirimanne, M. Dragunow, G.A. Werther, P.D. Gluckman, C.E. Williams, A. Scheepens, Co-ordinated and cellular specific indication of the components of the IGF/IGFBP axis in the rat brain following hypoxic-ischemic injury, *Mol Brain Res* 59, 1998, 119–34.

S.N. Bellur, V. Chandra, L.W. McDonald, Arachnoidal cell hyperplasia: its relationship to aging and chronic renal failure, *Arch Pathol Lab Med* 104, 1980, 414–6.

K. Blennow, P. Fredman, A. Wallin, C.G. Gottfries, I. Karlsson, G. Langstrom, I. Skoog, L. Svennerholm, C. Wikkelso, Protein analysis in cerebrospinal fluid. II. Reference values derived from healthy individuals 18–88 years of age, *Eur Neurol* 33, 1993, 129–33.

L.N. Boggs, K.S. Fuson, M. Baez, L. Churgay, D. McClure, G. Becker, P.C. May, Clusterin (Apo J) protects against *in vitro* amyloid-beta (1–40) neurotoxicity, *J Neurochemistry* 67, 1996, 1324–7.

F. Bracco, M. Scarpa, A. Rigo, L. Battistin, Determination of superoxide dismutase activity by the polarographic method of catalytic currents in the cerebrospinal fluid of aging brain and neurological degenerative disease, *Proc Soc Exp Biol Med* 196, 1991, 36–41.

J-P. Chanoine, S. Alex, S.L. Fang, S. Stone, J.L. Leonard, J. Korhle, L.E. Braverman, Role of transthyretin in the transport of thyroxine from blood to the choroids plexus, the cerebrospinal fluid and the brain, *Endocrinol* 130, 1992, 933–8.

R.L. Chen, N. Kassem, M.B. Segal, J.E. Preston, N-Ethylmaleimide increases cerebrospinal fluid secretion in young but not old sheep, *J Physiol* 555P, 2004, PC43.

R.L. Chen, M.B. Segal, J.E. Preston, Age-related decline in cerebrospinal fluid transthyretin (thyroid binding protein), *Biochem Soc Trans* 31, 2003, 63.

P. Cohen, H.C. Graves, D.M. Peehl, M. Kamarei, L.C. Giudice, R.G. Rosenfeld, Prostate-specific antigen (PSA) is an insulin-like growth factor binding protein-3 protease found in seminal plasma, *J Clin Endocrinol Metab* 75, 1992, 1046–53.

S. Cortez, C.E. Johanson, V. Kuo-LeBlanc, M. Rodriguez-Wolf, A. Baird, A.M. Gonzalez, A. Sneddon, P. Bohlen, E.G. Stopa, Heparan-binding growth factors in control and Alzheimer's choroid plexus, *Soc Neurosci* (abstr.) 25, 1995, 741.

D.A. Cottrell, E.L. Blakely, M.A. Johnson, P.G. Ince, G.M. Borthwick, D.M. Turnbull, Cytochrome c oxidase deficient cells accumulate in the hippocampus and choroid plexus with age, *Neurobiol Aging*, 22, 2001a, 265–72.

D.A. Cottrell, P.G. Ince, T.M. Wardell, D.M. Turnbull, M.A. Johnson, Accelerated aging changes in the choroid plexus of a case with multiple mitochondrial DNA deletions, *Neuropathol Appl Neurobiol* 27, 2001b, 206–24.

E. Courchesne, H.J. Chisum, J. Townsend, A. Cowles, J. Covington, B. Egaas, M. Harwood, S. Hinds, G.A. Press, Normal brain development and aging: quantitative analysis at *in vivo* MR Imaging in healthy volunteers, *Neuroradiol* 216, 2000, 672–82.

R.W. Cutler, L. Page, J. Galicich, G.V. Watters, Formation and absorption of cerebrospinal fluid in man, *Brain* 91, 1969, 707–20.

M. Czosnyka, Z.H. Czosnyka, P.C. Whitfield, T. Donovan, J.D. Pickard, Age dependence of cerebrospinal pressure-volume compensation in patients with hydrocephalus, *J Neurosurg* 94, 2001, 482–6.

R.B. DeMattos, J.R. Cirrito, M. Parsadanian, P.C. May, M.A. O'Dell, J.W. Taylor, J.A.K. Harmony, B.J. Aronow, K.R. Bales, S.M. Paul, D.M. Holtzman, ApoE and clusterin cooperatively suppress Aβ levels and deposition: evidence that ApoE regulates Aβ metabolism *in vivo*, *Neuron* 41, 2004, 193–202.

D.Z. Ellis, J.A. Nathanson, K.J. Sweadner, Carbacol inhibits Na^+-K^+-ATPase activity in choroid plexus via stimulation of the NO/cGMP pathway, *Am J Physiol* 279, 2000, C1685–93.

L. Eriksson, P. Westermark, Characterisation of intracellular amyloid fibrils in the human choroid plexus epithelial cells, *Acta Neuropathol* 80, 1990, 597–603.

F.M. Faraci, W.G. Mayhan, D.D. Heisted, Effect of vasopressin on cerebrospinal fluid production: possible role of vasopressin V1 receptors, *Am J Physiol* 258, 1990, R94–9.

F. Ferrante, F. Amenta, Enzyme histochemistry of the choroid plexus in old rats, *Mech Ageing Dev* 41, 1987, 65–72.

V.V. Frolkis, T. Kvitnitskaia-Ryzhova, T.A. Dubiley, Vasopressin, hypothalmo-neurohypophyseal system and aging, *Arch Gerontol Geriatr* 29, 1999, 109–214.

M.J. Garton, G. Keir, M.V. Lakshmi, E.J. Thompson, Age-related changes in cerebrospinal fluid protein concentrations, *J Neurolol Sci* 104, 1991, 74–80.

J-F Ghersi-Egea, PD Gorevic, J Ghiso, B Frangione, CS Patlak, JD Fenstermacher, Fate of cerebrospinal fluid-borne amyloid beta-peptide: rapid clearance into blood and appreciable accumulation by cerebral arteries, *J Neurochemistry* 67, 1996, 880–3.

P. Gideon, C. Thomsen, F. Stahlberg, O. Henriksen, Cerebrospinal fluid production and dynamics in normal aging: a MRI phase-mapping study, *Acta Neurol Scand* 89, 1994, 362–6.

A.B. Graves, C.M. van Duijn, V. Chandra, L. Fratiglioni, A. Heyman, A.F. Jorm, E. Komen, K. Kondo, J.A. Mortimer, W.A. Rocca, Alcohol and tobacco consumption as risk factors for Alzheimer's disease: a collaborative re-analysis of case-control studies, *Int J Epidemiol* 20, 1991, S48–S57.

L. Hayflick, The future of ageing, *Nature* 408, 2000, 37–9.

L. Henriksson, K. Voigt, Age-dependent differences of distribution and clearance patterns in normal RIHSA cisternograms, *Neuroradiol* 12, 1976, 103–7.

M. Hiramatsu, M. Kohno, R. Edmatsu, K. Mitsuta, A. Mori, Increased superoxide dismutase activity in aged human cerebrospinal fluid and rat brain determined by electron spin resonance spectrometry using the spin trap method, *J Neurochemistry* 58, 1992, 1160–8.

N.R. Holm, L.B. Hansen, C. Nilsson, S. Gammeltoft, Gene expression and secretion of insulin-like growth factor-II and insulin-like growth factor binding protein-2 from cultured sheep choroid plexus epithelial cells, *Mol Brain Res*, 21, 1994, 67–74.

C.E. Johanson, D.E. Palm, M.J. Primiano, P.N. McMillan, P. Chan, N.W. Knuckey, E.G. Stopa, Choroid plexus recovery after transient forebrain ischemia: role of growth factors and other repair mechanisms, *Cell Mol Neurobiol* 20, 2000, 197–216.

C.E. Johanson, J.E. Preston, A. Chodobski, E. Stopa, J. Szmydynger-Chodobska, P.N. McMillan, AVP V1 receptor-mediated decrease in Cl efflux and increase in dark cell number in choroid plexus epithelium, *Am J Physiol* 276, 1999, C82–90.

M.L. Katz, Potential role of retinal pigment epithelial lipofuscin accumulation in age-related macular degeneration, *Arch Gerontol Geriatr* 34, 2002, 359–70.

A.S. Khan, D.C. Sane, T. Wannenburg, W.E. Sonntag, Growth hormone, insulin-like growth factor-1 and the aging cardiovascular system, *Cardiovasc Res* 54, 2002, 25–35.

T.O. Kleine, R. Hackler, P. Zofel, Age-related alterations of the blood-brain barrier (BBB) permeability to protein molecules of different size, *Zeitschr fur Gerontol* 26, 1993, 256–9.

N.W. Knuckey, J.E Preston, D. Palm, M.H. Epstein, C.E. Johanson, Hydrocephalus decreases chloride efflux from the choroid-plexus epithelium, *Brain Res* 618, 1993, 313–7.

T. Koschinsky, C. He, T. Mitsuhashi, R. Bucala, C. Liu, C. Buenting, K. Heitmann, H. Vlassara, Orally absorbed reactive glycation products: an environmental risk factor in diabetic nephropathy, *Proc Natl Acad Sci USA* 94, 1997, 6474–9.

S. Kuchler-Bopp, J.B. Dietrich, M. Zaepfel, J.P. Delaunoy, Receptor-mediated endocytosis of transthyretin by ependymoma cells, *Brain Res* 870, 2000, 185–194.

S. Kunicki, J. Richardson, P.D. Mehta, K.S. Kim, E. Zorychta, The effects of age, apolipoprotein E phenotype and gender on the concentration of amyloid-b (Ab) 40, Ab42, apolipoprotein E and transthyretin in human cerebrospinal fluid, *Clin Biochemistry* 31, 1998, 409–15.

T. Kvitnitskaia-Ryzhova, A.L. Shkapenko, A comparative ultracytochemical and biochemical study of the ATPases of the choroid plexuses in aging, *Tsitologiia* 34, 1992, 81–7.

Z. Lai, P. Roos, O. Zhai, Y. Olsson, K. Fhölenhag, C. Larsson, F. Nyberg, Age-related reduction of human growth hormone-binding sites in the human brain, *Brain Res* 621, 1993, 260–6.

D. Lenoir, P. Honnegger, Insulin-like growth factor I (IGF-I) stimulates DNA synthesis in fetal rat brain cell cultures, *Dev Brain Res* 7, 1983, 205–13.

M.D. Li, J.K. Kane, S.G. Matta, W.S. Blaner, B.M. Sharp, Nicotine enhances the biosynthesis and secretion of transthyretin from the choroid plexus in rats: implications for β-amyloid formation, *J Neurosci* 20, 2000, 1318–23.

A. Logan, A-M. Gonzalez, D.J. Hill, M. Berry, N.A. Gregson, A. Baird, Coordinated pattern of expression and localisation of insulin-like growth factor-II (IGF-II) and IGF-binding protein-2 in the adult rat brain, *Endocrinol* 135, 1994, 2255–64.

R. Lui, J. Zhou, J. van Heerikhuize, M.A. Hofman, D.F. Swaab, Decreased melatonin levels in post-mortem cerebrospinal fluid in relation to aging, Alzheimer's disease and apolipoprotein E-e4/4, *J Clin Endocrinol Metab* 84, 1999, 323–7.

N. Matsumoto, H. Kitayama, M. Kitada, K. Kimura, M. Noda, C. Ide, Isolation of a set of genes expressed in the choroid plexus of the mouse using suppression subtractive hybridisation, *Neurosci* 117, 2003, 405–15.

C.P. Maurizi, Alzheimer's disease: roles for mitochondrial damage, the hydroxyl radical, and cerebrospinal fluid deficiency of melatonin, *Medical Med* 57, 2001, 156–60.

C. May, J.A. Kaye, J.R. Atack, M.B. Schapiro, R.P. Friedland, S.I. Rapoport, Cerebrospinal fluid production is reduced in healthy aging, *Neurol* 40, 1990, 500–3.

S.M. McCann, J. Licino, M.L. Wong, W.H. Yu, S. Karanth, V. Rettorri, The nitric oxide hypothesis of aging, *Exp Gerontol* 33, 1998, 813–26.

M.S. McLachlan, The aging kidney, *Lancet*, 2, 1978, 143–5.

F.A. McMorris, R.W. Furlanetto, R.L. Mozell, M.J. Carson, D.W. Raible, Regulation of oligodendrocyte development by insulin-like growth factors and cyclic nucleotides, *Ann NY Acad Sci* 605, 1991, 101–9.

J. Miklossy, R. Kraftsik, O. Pillevuit, D. Leori, C. Genton, F. Bosman, Curly fibre and tangle-like inclusions in the ependyma and choroid plexus—a pathogenic relationship with the cortical Alzheimer-type changes?, *J Neuropathol Exp Neurol* 57, 1998, 1202–12.

J.F. Mill, M.V. Chao, D.N. Ishii, Insulin, insulin-like growth factor II and nerve growth factor effects on tubulin mRNA levels and neurite formation, *Proc Natl Acad Sci USA* 82, 1985, 7126–30.

M.T. Modic, M.A. Weinstein, A.D. Rothner, G. Erenberg, P.M. Duchesneau, B. Kaufman, Calcification of the choroid plexus visualized by computed tomography, *Radiol* 135, 1980, 369–72.

T.J. Montine, M.D. Neely, J.F. Quinn, M.F. Beal, W.R. Marksberry, L.J. Roberts, J.D. Morrow Lipid peroxidation in aging brain and Alzheimer's disease, *Free Radical Biol Med* 33, 2002, 620–6.

Y. Nabeshima, Klotho: a fundamental regulator of aging, *Aging Res Rev* 1, 2002, 627–38.

D. Nakae, H. Akai, H. Kishida, O. Kusuoka, M. Tsutsumi, Y. Konishi, Age and organ dependent spontaneous generation of nuclear 8-hydroxydeoxyguanosine in male Fischer 344 rats, *Lab Invest* 80, 2000, 249–61.

K.L. Narr, T. Sharma, R.P. Woods, P.M. Thompson, E.R. Sowell, D. Rex, S. Kim, D. Asuncion, S. Jang, J. Mazziotta, A.W. Toga, Increases in regional subarachnoid CSF without apparent cortical grey matter deficits in schizophrenia: modulating effects of sex and age, *Am J Psychiatry* 160, 2003, 2169–80.

C. Nilsson, F. Stahlberg, C. Thomsen, O. Henriksen, M. Herning, C. Owman, Circadian variation in human cerebrospinal fluid production measured by magnetic resonance imaging, *Am J Physiol* 262, 1996, R20–4.

L.A. Poirier, Methyl group deficiency in hepatocarcinoma genesis, *Drug Metab Rev* 26, 1994, 185–9.

D.R.D. Premkumar, R.N. Kalaria, Altered expression of amyloid β precursor mRNAs in cerebral vessels, meninges and choroid plexus in Alzheimer's disease, *Ann NY Acad Sci* 777, 1996, 288–92.

J.E. Preston, Aging of the choroid plexus-cerebrospinal fluid system, *Microsc Res Technique* 52, 2001, 31–7.

J.E. Preston, Age-related reduction in rat choroid plexus chloride efflux and CSF secretion rate, *Soc Neurosci* (abstr.) 25, 1999, P698.7.

J.E. Preston, M.R. Wilson, N.J. Abbott, M.B. Segal, Uptake of ^{125}I-labelled β-amyloid(1–40) from blood into rat CSF, choroid plexus and brain increases with age, using *in situ* brain perfusion, *J Physiol (London)* 515P, 1999, 5P.

I. Ray, A. Chauhan, J. Wegiel, V.P.S. Chauhan, Gelsolin inhibits the fibrillization of amyloid beta-protein, and also defribrillizes its preformed fibrils, *Brain Res* 853, 2000, 344–51.

H. Reiber, Dynamics of brain-derived proteins in cerebrospinal fluid, *Clin chim Acta* 310, 2001, 173–86.

H. Reiber, M. Ruff, M. Uhr, Ascorbate concentration in human cerebrospinal fluid (CSF) and serum. Intrathecal accumulation and CSF flow rate, *Clin chim Acta* 217, 1993, 163–73.

H. Roder, P. Eden, V. Ingram, Brain protein kinase PK40 converts TAU into a PHF-like form as found in Alzheimer's disease, *Biochem Biophys Res Commun* 193, 1993, 639–47.

E. Rubenstein, Relationship of senescence of cerebrospinal fluid circulatory system to dementias of the aged, *Lancet* 351, 1998, 283–5.

G. Schreiber, The evolutionary and integrative roles of transthyretin in thyroid hormone homeostasis, *J Endocrinol* 175, 2002, 61–73.

A.L Schwarzman, L. Gregori, M.P. Vitek, S. Lyubski, W.J. Strittmatter, J.J. Enghilde, R. Bhasin, J. Silverman, K.H. Weisgraber, P.K. Coyle, Transthyretin sequesters amyloid beta protein and prevents amyloid formation, *Proc Natl Acad Sci USA* 91, 1994, 8368–72.

J-M. Serot, M.C. Béné, G.C. Faure, Choroid plexus, aging of the brain, and Alzheimer's disease, *Frontiers Biosci* 8, 2003, s515–21.

J-M. Serot, M.C. Béné, B. Foliguet, G.C. Faure, Morphological alterations of the choroid plexus in late-onset Alzheimer's disease, *Acta Neuropathol* 99, 2000, 105–8.

J-M. Serot, D. Christmann, T. Dubost, M. Couturier, Cerebrospinal fluid transthyretin aging and late onset Alzheimer's disease, *J Neurol, Neurosurg Psychiatry* 63, 1997, 506–8.

J-M. Serot, B. Foliguet, M.C. Béné, G.C. Faure, Choroid plexus and aging in rats: a morphometric and ultrastructural study, *Eur J Neurosci* 14, 2001, 794–8.

S. Shuangshoti, M.G. Netsky, Human choroid plexus—morphologic and histochemical alterations with age, *Am J Anat* 128, 1970, 73–95.

S. Shubutani, M. Takeshita, A.P. Grollman, Insertion of bases during DNA synthesis past the oxidation-damaged base 8-oxodG, *Nature* 349, 1991, 431–4.

G.D. Silverberg, M. Mayo, T. Saul, E. Rubenstein, D. McGuire, Alzheimer's disease, normal-pressure hydrocephalus and senescent changes in CSF circulatory physiology: a hypothesis, *Lancet Neurol* 2, 2003, 506–11.

G.D. Silverberg, S. Huhn, R.A. Jaffe, S.D. Chang, T. Saul, G. Heit, A. von Essen, E. Rubenstein, Down regulation of cerebrospinal fluid production in patients with chronic hydrocephalus, *J Neurosurg* 97, 2002, 1271–5.

Q.R. Smith, C.E. Johanson, Chloride efflux from isolated choroid plexus, *Brain Res* 25, 1991, 306–10.

Q.R. Smith, Y. Takasato, S.I. Rapoport, Age-associated decrease in the rate of cerebrospinal fluid uptake of Na in the Fischer-344 rat, *Soc Neurosci* (abstr.) 8, 1982, 443.

B.R Southwell, W. Duan, D. Alcorn, C. Brack, S.J. Richardson, J. Kohrle, G. Schreiber, Thyroxine transport to the brain: role of protein synthesis by the choroid plexus, *Endocrinol* 133, 1993, 2116–26.

E.R. Stadtman, R.L. Levine, Free radical-mediated oxidation of free amino acids and amino acid residues in proteins, *Amino Acids* 25, 2003, 207–18.

E.G. Stopa, M.T. Berzin, S. Kim, P. Song, V. Kuo-LeBlanc, M. Rodriguez-Wolf, A. Baird, C.E. Johanson, Human choroid plexus growth factors: What are the implications for CSF dynamics in Alzheimer's disease?, *Exp Neurol* 167, 2001, 40–7.

R.R. Sturrock, An ultrastructural study of the choroid plexus of aged mice, *Anatom Anz* 65, 1988, 379–85.

J. Szmydynger-Chodobska, Z.G. Chun, C.E. Johanson, A. Chodobski, Distribution of fibroblast growth factor receptors and their co-localization with vasopressin in the choroid plexus epithelium, *Neuroreport* 11, 2002, 257–9.

K. Tanji, T. Kunimatsu, T.H. Vu, E. Bonilla, Neuropathological features of mitochondrial disorders, *Cell Dev Biol* 12, 2001, 429–39.

I. Tayarani, I. Cloez, M. Clement, J-M. Bourre, Antioxidant enzyme and related trace elements in aging brain capillaries and choroid plexus, *J Neurochem* 53, 1989, 817–24.

H. Tsujikawa, Y. Kurotaki, T. Fujimori, K. Fukuda, Y. Nabeshima, Klotho, a gene related to a syndrome resembling human premature aging, functions in a negative regulatory circuit of vitamin D endocrine system, *Mol Endocrinol* 17, 2003, 2393–403.

H. Vlassara, Recent progress in the biologic and clinical significance of advanced glycosylation end-products, *J Lab Clin Med* 124, 1994, 19–30.

T. von Zglinicki, Replicative senescence and the art of counting, *Exp Gerontol* 38, 2003, 259–64.

L-O. Wahlund, O. Almkvist, H. Basun, P. Julin, MRI in successful aging, a 5-year follow-up study from the eighth to ninth decade of life, *Magn Reson Imag* 14, 1996, 601–8.

H.J. Walter, M. Berry, D.J. Hill, S. Cwyfan-Hughes, J.M.P. Holly, A. Logan, Distinct sites of insulin-like growth factor (IGF)-II expression and localization in lesioned rat brain: possible roles of IGF binding proteins (IGFBPs) in the mediation of IGF-II activity, *Endocrinol* 140, 1999, 520–32.

G.Y. Wen, H.M. Wisniewski, R.J. Kascsak, Biondi ring tangles in the choroid plexus of Alzheimer's disease and normal aging brains: a quantitative study, *Brain Res* 832, 1999, 40–6.

M.R. Wilson, J.E. Preston, M.B. Segal, Accumulation of Alzheimer's associated beta-amyloid peptide by the guinea-pig choroid plexus, *Mech Ageing Dev* 122, 2001, 34.

T. Wisniewski, E. Castano, J. Ghiso, B. Frangione, Cerebrospinal fluid inhibits Alzheimer beta-amyloid fibril formation *in vitro*, *Ann Neurol* 34, 1993, 631–3.

15 The Role of the Choroid Plexus in HIV-1 Infection

Carol K. Petito
University of Miami School of Medicine
Miami, Florida, U.S.A.

CONTENTS

15.1 INTRODUCTION

The choroid plexus (CPx) plays an important role in the pathogenesis of human immunodeficiency virus type I (HIV-1) infection. It contains HIV-1–infected cells in a little over half of all AIDS patients at the time of death and immune complexes in approximately 75% of acquired immunodeficiency syndrome (AIDS) patients (Falangola et al., 1994, 1995). The immune complexes may be viral-protective in the CPx as they are postulated to be in systemic sites (Tacchetti et al., 1997). This chapter will review briefly the characteristics of the blood–cerebrospinal fluid (CSF) barrier of the CPx, which may contribute to its role in HIV-1 neuropathogenesis, detail the specific pathology and neurovirology of the choroid plexus in HIV-1 infection and its experimental models, and summarize the evidence that suggests that the CPx is important in viral dissemination to the central nervous system (CNS) and participates as a site for viral sanctuary or reservoir.

15.2 VIROLOGY AND NEUROVIROLOGY

Specific characteristics of HIV-1 contribute to its neuropathogenesis and infectivity. This retrovirus has an inner core of viral RNA, reverse transcriptase, and structural core proteins that is surrounded by an outer bilipid membrane composed of viral transmembrane (gp41) and external membrane (gp120) glycoproteins plus components of the host plasma membrane (Hasseltine and Wong-Staal, 1989). Two different host cell receptors participate in viral entry. First, host cell CD4 receptor couples with viral gp120, producing a conformational change in the gp120, which then permits the protein to complex with host cell HIV-1 chemokine coreceptor. Sequences in the V3 loop of gp120 determine coreceptor usage (Border and Collman, 1997). T lymphocyte–tropic viral strains preferentially utilize the CXCR4 receptor, whereas macrophage (M)–tropic strains preferentially utilize the CCR5 receptor. Dual-tropic viruses and multiple chemokine receptors on a particular host cell are common.

Viral strains obtained from the CNS differ from those obtained from peripheral tissues. Brain isolates are M-tropic, and their gene sequences in the V3 loop of gp120 differ from those in systemic compartments (Korber et al., 1994; Ball et al., 1994). Paradoxically, neurons and astrocytes are most vulnerable to the T-tropic viral strains, as discussed later in this chapter. This selective vulnerability is related to the high expression levels of neuronal and astrocyte CXCR4 when compared to CCR5.

HIV-1 is cultured from the CSF during the first few months after the initial systemic infection (Chiodi et al., 1992; McArthur et al., 1998), but productive brain infection and encephalitis is delayed until the onset of immune suppression and AIDS (Bell et al., 1993; Sinclair et al., 1994; Kibayashi et al., 1996). Before the onset of effective antiretroviral therapy, up to 30% of AIDS patients had HIV-1 encephalitis (HIVE) at the time of death. This infection is characterized by multifocal collections of inflammation and perivascular lymphocytes that predominate in the basal ganglia and cerebral white matter. Cortical involvement generally is confined to the more severe cases of HIVE, and its inflammatory lesions often appear in subpial areas (Petito et al., 1986). Actual viral replication takes place in monocyte-lineage cells (Koenig et al., 1986; Wiley et al., 1986), although restricted or latent viral infection may be present in astrocytes and neurons. Parallel studies with simian immunodeficiency virus (SIV) reveal a similar relationship between viral isolation in the CSF and brain (Sharma et al., 1992; Smith et al., 1995). Between one and two weeks after systemic inoculation, a time at which plasma viral load is high and anti-SIV antibodies have not yet developed, CSF inflammation and viral DNA are common but decrease with the subsequent systemic production of anti-SIV antibodies. Like HIVE, SIV encephalitis is not seen until the late stages of infection and requires systemic or cerebral inoculation of M-tropic viral strains.

Brain infection occurs via entry of infected cells—the so-called Trojan horse theory of brain infection. Infected monocytes in the circulation are the likely culprits to carry virus into the brain since productive viral infection in brain is confined to monocyte-lineage cells and since brain-derived virus is M-tropic. Alternatively, infected T lymphocytes could enter the CNS and secondarily infect its monocytic cells. T lymphocytes are the principal infectious cell type in the circulation of end-stage AIDS patients (Connor et al., 1997; van der Ende et al., 1999), and both $CD4^+$ and $CD8^+$ T cells are increased in brains with HIVE (Petito et al., 2002). While there is little evidence for endothelial infection in vivo, it is readily found in experimental models of HIVE and in experimental blood–brain barrier (BBB) models in tissue culture (see, for example, Mankowsie et al., 1994; Steffan et al., 1994; Moses et al., 1994).

The importance of CNS HIV-1 is threefold. First, CNS infection—namely, HIVE—is the underlying cause of HIV-1–associated dementia (Cherner et al., 2002), a slowly progressive disorder that is an important cause of morbidity and mortality in patients with AIDS (Mayeux et al., 1992; Ellis et al., 1997). Second, the CNS may be a sanctuary for viral production during the period of immune competency or in the setting of highly active antiretroviral therapy (HAART). Evidence for this is found in the occasional patient whose plasma HIV RNA levels fall to undetectable levels with HAART exposure, but whose CSF HIV RNA levels remain elevated (Pialoux et al., 1997). Third, the CNS may be a reservoir for ongoing viral production, permitting reseeding of virus from the CNS to the periphery because of the unique properties of the BBB and the blood–cerebrospinal fluid barrier that modify drug and immune cell entry.

15.3 BARRIERS OF THE BRAIN

In normal brain, the BBB limits entry of inflammatory cells, restricts or prevents macromolecular transport (Betz et al., 1989), and reduces the efficiency of drug penetration, including antiretroviral agents. Structurally, BBB capillaries have a paucity of pinocytotic vesicles and tight junctions between adjacent cells (Reese and Karnovsky, 1967). The CPx lies outside the normal BBB since its endothelial cells are connected by gap junctions (Tennyson and Pappas, 1969) that facilitate passage of macromolecules from blood to CPx stroma. The blood–CSF (B-CSF) barrier exists at the level of the CPx epithelial cells (see, for review, Strazielle and Ghersi-Egea, 2000; Chodobski and Szmydynger-Chodobska, 2001). It is formed by impermeable tight junctions between adjacent epithelial cells, thus forming the second of the two barriers between the brain and systemic circulation.

Both the BBB and the B-CSF barriers contribute to the immune system of the CNS in resting conditions and with infection and inflammation. Normal brain contains a small number of activated T cells (Hickey et al., 1991) that are rapidly eliminated by apoptosis once they leave the perivascular spaces and enter the brain parenchyma (Bauer et al., 1998). Normal CPx stroma also contains a small number of T lymphocytes, monocytes, and dendritic cells in humans (Nathanson and Chun, 1989; Hickey et al., 1991; Biermacki et al., 2001) and experimental animals (Strazielle and Ghersi-Egea, 2000). Differential expression of cell adhesion molecules on endothelium and epithelium in the CPx versus brain causes selective recruitment of immune cells to the CPx stroma, whence they can migrate into the CSF. Choroid plexus and meningeal endothelium express P- and E-selectin but not intracellular adhesion molecule-1 (ICAM-1) or vascular intracellular adhesion molecule (VCAM), whereas the reverse is true for brain endothelium (Deckert-Schluter et al., 1994; Steffen et al., 1996; Kivisakk et al., 2003). Choroid plexus expression of

P- and E-selectin also is important in human brain since CSF lymphocytes display high expression levels of the P-selectin ligand PSGP-1 (Kivisakk et al., 2003). Choroid plexus epithelial cells, on the other hand, express ICAM-1 and VCAM-1 but not P- and E-selectin, and thus have a cell adhesion profile more closely resembling that of brain endothelium.

As in the brain, CPx cell adhesion molecules, major histocompatability (MHC) antigens, and inflammatory cells are upregulated in infection and inflammation (Werkerle et al., 1986; Hickey et al., 1991; Masopust et al., 2001; Biermacki et al., 2001; Carrithers et al., 2000). In AIDS patients, enhanced vascular permeability in brains (Petito and Cash, 1992) as well as an increase in brain of lymphocytes and monocytes in AIDS and SIV (Lane et al., 1996a; Birdsell et al., 1997; Petito et al., 2002) are indicative of the prominent alterations of the normal neuroimmune system in these infections.

15.4 T-CELL TRAFFICKING IN THE CPX AND BRAIN

The choroid plexus accumulates large numbers of T lymphocytes in inflammation and disease (Engelhardt et al., 2001). In order to compare T cell accumulation in CPx versus brain of the same animal, we acutely activated systemic $CD4^+$ and $CD8^+$ TCR Vβ8-bearing T cells by injecting mice with an intraperitoneal injection of the superantigen *Staphylococcus* enterotoxin B (SEB) (Murphy et al., 1990) and counting the numbers of T cells per unit area in CPx and brain. In controls, number of T cells per unit area was higher in the CPx than in brain. Choroid plexus T cells averaged 0.41 ± 0.79, 0.24 ± 0.48, and 0.02 ± 0.04 per 0.01 mm^2 in lateral, third, and fourth ventricular CPx but only 0.0008 ± 0.01 per 0.01 mm^2 in brain. T cells accumulated in CPx as well as brain following systemic immune activation, reaching their maximum at three days after SEB injections (see Figure 15.1). The CPx showed the greatest increase at three days post–SEB injection, especially in the fourth ventricular CPx that had a 150-fold increase over controls. Brain T cells also increased although the relative increase over controls was no more than fourfold higher. The accumulated cells tended to cluster near the vascular stalk of the CPx and the epithelial-ependymal junction (see Figure 15.2). A small number of $CD3^+$ T cells were also seen in adjacent subependymal brain (see Figure 15.2) and on the supra-ependymal surfaces.

Flow cytometry of cell suspensions from pooled spleen and CPx (Adkins et al., 1993) in controls and at three days after SEB injection confirmed the increases in $CD4^+$ and $CD8^+$ Vβ8 cells in both spleen and CPx (see Figure 15.3), The ratio of Vβ8 $CD8^+$-to-$CD4^+$ T cells was higher in the SEB CPx than in SEB spleen, suggesting preferential accumulation of activated $CD8^+$ Tcells to the CPx. Carson et al. (1999) suggest that the presence of antigen-presenting cells (APCs) leads to preferential accumulation of $CD8^+$ T cells, an observation made following local brain injections of APCs. The CPx contains abundant numbers of stromal MHC type II–expressing dendritic cells (see Serot et al., 1997; Hanly and Petito, 1998) that might explain the selective $CD8^+$ T-cell recruitment to this structure.

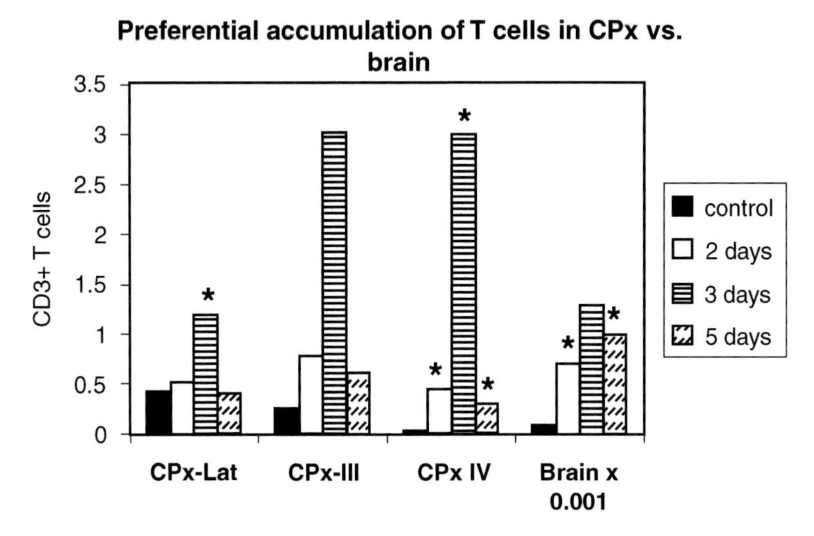

FIGURE 15.1 Normal CPx contains threefold more T cells per unit area than brain. Following peripheral immune activation, the number of T cells significantly increases in both brain and CPx, reaching the maximum at three days after injection with the superantigen. The largest absolute and relative increase is in the CPx from the fourth ventricle. $CD3^+$ T cells: number per 0.01 mm^2. Note: brain values are ×0.001. CPx-Lat: lateral ventricular choroid plexus (CPx). * $p < 0.05$.

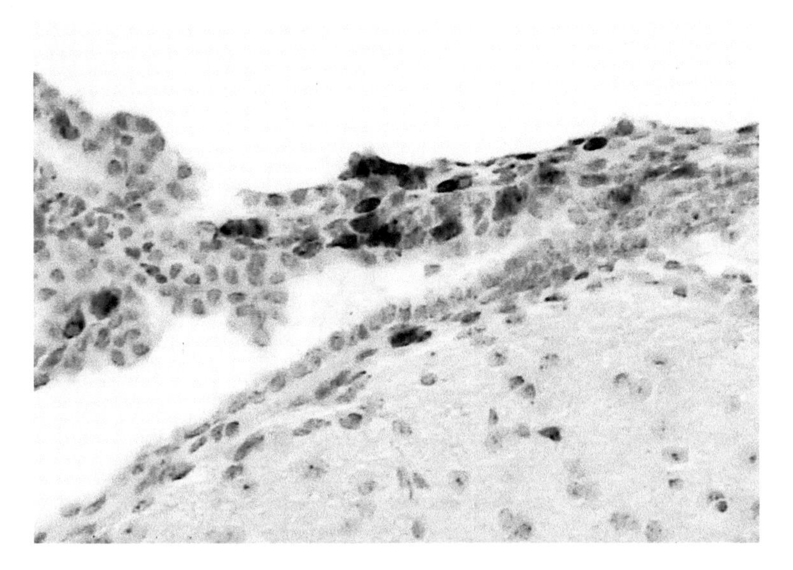

FIGURE 15.2 $CD3^+$ T cells accumulate in the CPx three days after intraperitoneal injection of superantigen. Note associated T-cell increase in the subependymal brain. Hematoxylin counterstain; original magnification × 400.

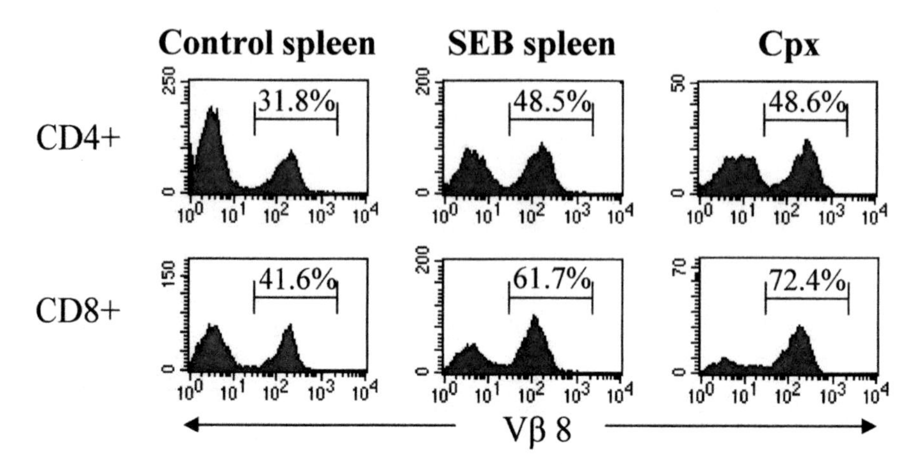

FIGURE 15.3 Flow cytometry of spleen and choroid plexus cell suspensions at 3 days following intraperitoneal injection of streptocccal antigen B. Activated $CD4^+$ and $CD8^+$ T cells increase in both tissues, with a relatively greater accumulation of $CD8^+$ T cells in the CPx.

15.5 CHOROID PLEXUS AND HIV-1 INFECTION

Several years ago, we tested the hypothesis that the CPx is an important site for HIV dissemination to the CNS. We reasoned that the permeable capillaries of the CPx would facilitate entry of immune cells into the CPx and thence into the CSF and that the distribution of the inflammatory lesions of HIVE was highly consistent with a CSF dissemination of virus. The concept that hematogenous dissemination of infectious organisms occurs first in the CPx is not novel (Levine, 1986; see, for example, Falangola and Petito, 1993; Pron et al., 1997). Experimental animal models showing initial infection of the CPx include parasites (*Trypanosoma cruzi*), bacteria (*Niesseria meningitidis*), and viruses (lymphocyte choriomeningitis virus). Although the importance of CPx is less readily examined in clinical situations, the CPx is the sole or principal site of infection in patients dying in early stages of Chagas' disease, AIDS patients with cerebral toxoplasmosis, and infants with bacterial meningitis.

We found that 29% of asymptomatic, HIV-1–infected (ASY) patients had HIV-immunoreactive cells in the CPx, but their brains were negative for both immunoreactive cells and HIVE (Petito et al., 1999). We also found that CPx viral sequences were common (80%), whereas brain viral sequences were infrequent (20%). To do these studies, we used formalin-fixed, paraffin-embedded tissue sections of autopsy CPx, brain, and spleen, removed the paraffin, extracted total DNA, and amplified HIV *env* sequences by the polymerase chain reaction (PCR).

We also examined the relative frequency of HIV-1 in CPx versus brains of AIDS patients and its relationship to HIV-1 encephalitis (see Figure 15.4A). First, we found AIDS-related increases in stromal dendritic cells, T cells, and monocytes (see Figure 15.4 B,C). Second, the incidence of CPx HIV-1 immunoreactive cells increased (see Figure 15.4D) to 57% (see Table 15.1), whereas HIVE was encountered in only 30% of the same brains (Falangola et al., 1994). The majority of infected cells in the CPx stroma had the morphology of tissue monocytes or dendritic cells and

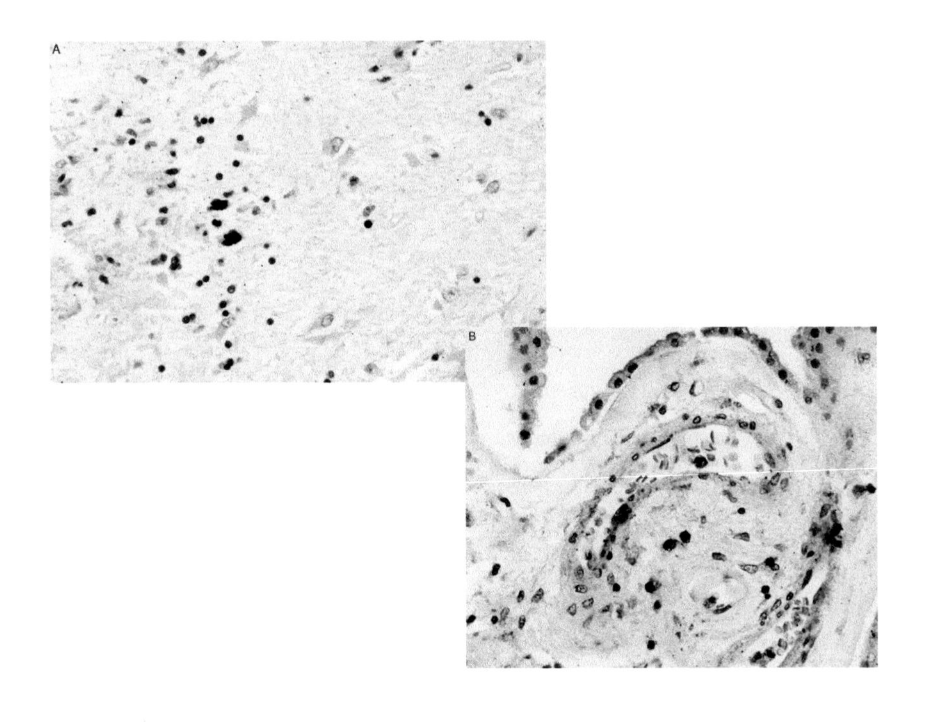

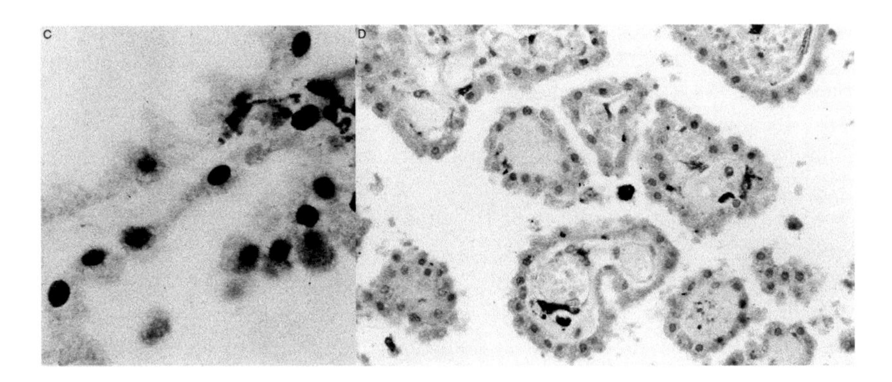

FIGURE 15.4 (A) HIV-1 encephalitis in AIDS patient showing a microglial inflammatory nodule in the basal ganglia that contains monocytes, lymphocytes, and the multinucleated giant cells that are characteristic of this infection. A few reactive astrocytes in the region are consistent with brain injury. Hematoxylin-eosin; original magnification × 400. (B–D) AIDS choroid plexus containing $CD3^+$ T lymphocytes (B), $LN3^+$ dendritic cells (C), and HIV $gp41^+$ cells (D) that have a morphology suggestive of monocytes and dendritic cells. Hematoxylin counterstain; original magnification × 400.

TABLE 15.1
Distribution of HIV-1 in Choroid Plexus, Brain, and Spleen

		CPx	Brain	Spleen
ASY	IHC (n = 7)	29%	0	ND
	Viral isolation (n = 31)	0	0	0
AIDS	IHC (n = 14)	57%	30%	ND
	Viral isolation (n = 31)	12%	29%	18%
	Tropism	M	M	M,T
Control	IHC	0	0	ND
	Viral isolation (n = 6)	0	0	0

ASY: Asymptomatic HIV-1–infected patients without immune suppression; AIDS: acquired immunodeficiency syndrome; CPx: choroid plexus; IHC: immunohistochemistry; ND: not done.

colabeled for immunoreactive monocyte and dendritic cell markers with serial section immunohistochemistry (Falangola et al., 1994; Hanly and Petito, 1998). Rarely, we detected co-labeling for T-cell and HIV-1 gp41 protein. We found no evidence of HIV-1 infection in the CPx epithelial cells by immunohistochemistry. When we removed epithelial cells from tissue sections using laser capture microdissection, we were unable to amplify HIV-1 gene sequences even though housekeeping genes were readily amplified.

We theorized that CPx viral strains would include blood-related sequences if its infection were derived from hematogenous sources. Accordingly, we extracted DNA from paraffin-embedded or fresh-frozen brain, CPx, and spleen sections of AIDS patients and amplified HIV-1 *env* sequences by PCR. We cloned the amplified sequences and examined the relationship among the sequences from different tissues within the same patients. Four AIDS patients were so studied in our first report (Chen et al., 2000), and three additional cases in a recent study (Burkala et al., 2004). In all seven cases, brain and splenic viral isolates clustered according to tissues, with tissue-specific distributions for brain and for spleen. In six of the seven cases, CPx sequences were an admixture of brain and spleen sequences. Only in one of the seven patients did the CPx viral sequences cluster only with the brain sequences.

We evaluated the phenotypic differences in virus from CPx versus brain and spleen by determining the viral cell tropism and chemokine coreceptor usage in viruses isolated from all three tissue compartments (see Table 15.1) (Burkala et al., 2004). Positive viral cultures were confined to AIDS patients and correlated with tissue of origin. We found positive cultures in 29% of basal ganglia samples, 18% of splenic samples, and 12% of CPx samples; it was absent in all samples of frontal cortex. Neuropathological changes in the six AIDS cases with viable virus included two with HIVE, two with meningitis, one with lymphoma, and one with gliosis. Cultures were negative in three ASY patients and six HIV-1–negative controls (see Table 15.1).

All brain and CPx viruses were macrophage-tropic and preferentially utilized the CCR5 coreceptor. All splenic isolates had dual tropism for macrophages and T lymphocytes and utilized both CCR5 and CXCR4 coreceptors as major coreceptors. All tissue isolates had minor usage for CCR3. However, the CPx viral isolates, but none of the brain isolates, also had minor usage for CXCR4.

15.6 PERSPECTIVES

Pro-inflammatory states, induced by systemic injections of nonseptic, low-dose lipopolysaccaride (Quan et al., 1999a), trypanosomal parasites (Quan et al., 1999b), or pro-inflammatory cytokines (Strazielle et al., 2003) upregulate pro-inflammatory cytokines (tumor necrosis factor-α and interleukin-1β) and metalloproteinases in the CPx and circumventricular organs prior to changes in brain parenchyma. This preferential expression of chemoattractant cytokines in CPx over brain may be responsible for the massive accumulation of T lymphocytes in the CPx when compared to brain that we detected shortly after acute systemic immune activation (Petito and Adkins, unpublished observations). Similar mechanisms may operate in HIV-1–infected patients whose systemic pro-inflammatory cytokine levels are elevated. Preferential recruitment of infected immune cells to the CPx, combined with a protective environment induced by local immune complex deposition, may enable this structure to serve as a site for hematogenous dissemination of virus from blood to CNS and a site for viral protection during the immune-competent period and during effective antiretroviral therapy.

The CPx is a likely site for viral entry and a sanctuary for viral production. It supports HIV replication in vitro (Harouse et al., 1989). As described above, ASY and AIDS patients have a higher frequency of virus in the CPx than in the brain, which represent admixtures of blood as well as brain viral strains. Choroid plexus infection is not confined to HIV-1 but also occurs with its SIV and feline immunodeficiency virus (FIV) animal models. In macaques, 50% of fetal animals contained SIV in the CPx versus 29% with SIV infection in brain parenchyma (Lane et al., 1996). CPx FIV develops in cats infected with FIV (Bragg et al., 2002). Explants of feline CPx exhibit low levels of infection upon exposure to FIV in culture (Bragg et al., 2002), and the infected cell type was consistent with CPx stromal macrophages.

What are the clinical significances of CPx HIV-1 and its M-tropism and mixed chemokine receptor usage? First, infected CPx cells can enter the CSF and either travel to the brain and initiate subsequent brain infection or could spread to the regional cervical lymph nodes and reinfect the periphery. As summarized by Johanson (1993), the CNS has four bulk flow and diffusional pathways. These include (1) bulk flow of CSF from ventricles to subarachnoid space and thence into the dural venous sinuses, (2) bi-directional bulk flow and diffusion from brain to intraventricular spaces, (3) drainage of CSF fluid into cervical lymph nodes, and (4) bulk flow of interstitial fluids within the brain itself. These fluid flows bring with them particulate matter, macromolecules, and cells, including small nondividing lymphocytes and HIV gp120 protein (Oehmichen et al., 1982; Seabrook et al., 1998; Cashion et al., 1999).

Second, the specific finding of the more neurotoxic CXCR4-using virus in the choroid plexus in our studies, as well as the recent finding of a CXCR4-using virus in the CSF of an AIDS patient (Yi et al., 2003), provides a means whereby the brain could be exposed to this viral more neurotoxic strain. Both neurons and astrocytes are more vulnerable to cell injury and death when exposed to CXCR4-using than CCR5-using virus in vitro (Nath et al., 1995; Hesselgesser et al., 1998; McCarthy et al., 1998). This selective vulnerability may be related to the higher expression levels of CXCR4 than of CCR5 (Lavi et al., 1997; McManus et al., 2000; Petito et al., 2001). This relationship is enhanced in patients with AIDS and HIVE. In these conditions, neuronal expression of CXCR4 increases and neuronal expression of CCR5 decreases (van der Meer, 2000; Petito et al., 2001), thereby rendering them more vulnerable to the effects of CXCR4-using viruses.

15.7 ACKNOWLEDGMENTS

The author gratefully acknowledges the continued support of her colleagues, Drs. B. Adkins, E. Burkala, H. Chen, F. Falangola, and C. Wood, as well as the technical and secretarial assistance of Ms. Cynthia Carrasco and Ms. Gloria Diaz. This work was supported by the National Institutes of Health, NS RO1-39177 (to CKP).

REFERENCES

Adkins, B., Ghanei, A., et al. (1993) Developmental regulation of IL-4, IL-2 and IFN-gamma production by murine peripheral T lymphocytes. J Immunol 151:6617–6626.

Ball, J.K., Holmes, E.C., Whitwell, H. and Desselberger, U. (1994) Genomic characterization of human immunodeficiency virus type 1 (HIVE): molecular analyses of HIVE in sequential blood samples and various organs obtained at autopsy. J Gen Virol 75:67–79.

Bauer, J., Bradl, M., Hickey, W.F., Forrs-Perrer, S., Breitschopf, H., Linington, C., Wekerle, H., Lassman, H. (1998) T-cell apoptosis in inflammatory brain lesions: Destruction of T-cell does not depend on antigen recognition. Am J Pathol 153:715–724.

Bell, J.E., Busuttil, A., Ironside, J.W., Rebus, S., Donaldson, Y.K., Simmonds, S., Peuther, J.R. (1993) Human immunodeficiency virus and the brain: investigation of virus load and neuropathological changes in pre-AIDS subjects. J Infect Dis 168:818–824.

Betz, L.A., Goldstein, G.W. and Latzman, R. (1989) Blood-brain-cerebrospinal fluid barriers. *Basic Neurochemistry: Molecular and Cellular and Medical Aspects.* Ed. G.J. Siegal, NY. Raven Press, pp. 591–606.

Biernacki, K., Prat, A., Blain, M. and Antel JP. (2001) Regulation of Th1 and Th2 lymphocyte migration by human adult brain endothelial cells. J Neuropathol Exp Neurol 60:1127–1136.

Birdsall, H.H., Trial, J., Lin, H.J., Green, D.M., Sorrentino, G.W., Siwak, E.B., de Jong, A.L. and Rossen, R.D. (1997) Transendothelial migration of lymphocytes from HIV-1 infected donors: a mechanism for extravascular dissemination of HIV-1. J Immunol 158:5968–5977.

Bragg, D.C., Childers, T.A., Tompkins, W.A. and Meeker, R.B. (2002) Infection of the choroid plexus by feline immunodeficiency virus. J NeuroVirol 8:211–244.

Border, C.C. and Collman, R.G. (1997) Chemokine receptors and HIV. J Leukocyte Biol 62:20–29.

Burkala, E., West, J.T., He, J., Petito, C.K. and Wood, C. The choroid plexus and viral entry into the brain. In: *The Neurology of AIDS*. Eds. H.E. Gendelman, I.P. Everall, I. Grant, S.A. Lipton, S. Swindells. Oxford University Press, 2004.

Carrithers, M.D., Visintin, I., Kang, S.J. and Janeway, C.A. (2000) Differential adhesion molecule requirement for immune surveillance and inflammatory recruitment. Brain 123:1092–1101.

Carson, M.J., Reilly, C.R., Sutcliffe, J.G. and Lo, D. (1999) Disproportionate recruitment of CD8+ T cells into the central nervous system by professional antigen-presenting cells. Am J Pathol 154(2):481–494.

Cashion, M.F., Banks, W.A., Bost, K.L. and Kastin, A.J. (1999) Transmission routes of HIV1 gp120 from brain to lymphoid tissues. Brain Res 822:26–33.

Chen, H., Wood, C. and Petito, C.K. (2000) Comparisons of HIV-1 viral sequences in brain choroid plexus and spleen: potential role of choroid plexus in the pathogenesis of HIV-1 encephalitis. J Neuropathol Exp Neurol 6:498–506.

Cherner, M., Masliah, E., Ellis, R.J., Marcotte, T.D., Moore, D.J., Grant, I. and Heaton, R.K. (2002) Neurocognitive dysfunction predicts postmortem findings of HIVE. Neurology 59:1563–1567.

Chiodi, F., Keys, B., Albert, J., Hagberg, L., Lundeberg, J., Uhlen, M., Fenyo, E.M. and Norkrans, G. (1992) Human immunodeficiency virus type 1 is present in the cerebrospinal fluid of a majority of infected individuals. J Clin Microbiol,7:1768–1771.

Chodobski, A., Szmydynger-Chodoska, J. (2000) Choroid plexus: target for polypeptides and site of their synthesis. Microsc Res Tech 52:65–82.

Connor, R.I., Sheridan, K.E., Ceradini, D., Choe, S. and Landau, N.R. (1997) Change in co-receptor use correlates with disease progression in HIV-1 infected individuals. J Exp Med 185:621–628.

Deckert-Schluter, M., Schluter, D., Hof, H., Wiestler, O.K. and Lassmann H. (1994) Differential expression of ICAM-1, VCAM-1 and their ligands FLA-1, Mac-1, CD43, VLA-4 and MHC class II antigens in murine toxoplasma encephalitis. A light and microscopic and ultrastructural immunohistochemical study. J Neuropath Exp Neurol 53:457–468.

Ellis, R.J., Deutch, R., Heaton, R.K., et al. (1997) Neurocognitive impairment is an independent risk factor for death in HIV infection. Arch Neurol 54:416–424.

Falangola, M.F., Hanly, A., Galvao-Castro, B. and Petito, C.K. (1995) HIV infection of human choroid plexus: a possible mechanism of viral entry into the CNS. J Neuropathol Exp Neurol 54:497–503.

Falangola, M.F., Castro, B., Filha, B.G. and Petito, C.K. (1994) Immune complex deposition in the choroid plexus of AIDS patients. Ann Neurol 36:437–440.

Falangola, M.F. and Petito, C.K. (1993) Choroid plexus infection in AIDS patients with cerebral toxoplasmosis. Neurology 43:2085–2040.

Hanly, A. and Petito, C.K. (1998) HLA-DR-positive dendritic cells of the normal human choroid plexus: a potential reservoir of HIV in the central nervous system. Hum Pathol 29(1):88–93.

Harouse, J.M., Wroblewska, Z., Laughlin, M.A., Hickey, W.F., Schonwetter, B.S., and Gonzalez-Scarano, F. (1989) Human choroid plexus cells can be latently infected with human immunodeficiency virus. Ann Neurol 25:406–411.

Haseltine, W.A. and Wong-Staal, F. (1988). The molecular biology of the AIDS virus. Sci Am 259:52–62.

Hesselgesser, J., Taub, D., Baskar, P., Greenberg, M., Hoxie, J., Kolson, D.L. and Horuk, R. (1998) Neuronal apoptosis induced by HIVE gp120 and the chemokine SDF-1 is mediated by the chemokine receptor CXCR4. Curr Biol 7:8(10):595–598.

Hickey, W.F., Hsu, B.L. and Kimura, H. (1991) T lymphocyte entry into the central nervous system. J Neurosci Res 28:254–260.

Johanson, C.E. (1993) Tissue barriers: diffusion, bulk flow and volume transmission of proteins and peptides within the brain. In: *Biological Barriers to Protein Delivery*, Eds. K.L. Audus and T.J. Raub. Plenum Press, NY. pp. 467–486.

Kibayashi, K., Masti, A.R. and Hirsch, C.S. (1996) Neuropathology of human immunodeficiency virus at different disease stages. Hum Pathol 27:637–642.

Kivisakk, P., Mahad, D.J., Callahan, M.K., Trebvst, C., Tucky, B., Wei, T., Wu, L., Baekkevold, E.S., Lassmann, H., Staugaitis, S.M., Campbell, J.J. and Ranmsohoff, R.M. (2003) Human cerebrospinal fluid central memory $CD4^+$ T cells: evidence for trafficking through choroid plexus and meninges via P-selectin. Proc Natl Acad Sci USA 100:8389–8394.

Koenig, S., Gendelman, H.E. and Orenstein, J.F. (1986) Detection of AIDS virus in macrophages in brain tissue from AIDS patient with encephalopathy. Science 233:1089–1093.

Korber, B.T., Kunstman, K.J., Patterson, B.K., Furtado, M., McEvilly, M.M., Levy, R. and Wolinsky, S.M. (1994) Genetic differences between blood- and brain-derived viral sequences from human immunodeficiency virus type 1-infected patients: evidence of conserved elements in the V3 region of the envelope protein of brain-derived sequences. J Virol 68:7467–7481.

Lane, J.H., Sasseville, V.G., Smith M.O., Vogel, P., Pauley, D.R., Heyes, M.P. and Lackner, A.A. (1996a) Neuroinvasion by simian immunodeficiency virus coincides with increased numbers of perivascular macrophages/microglia and intrathecal immune activation. J NeuroVirol 2(6):423–432.

Lane, J.H., Tarantal, A.F., Pauley, D., Marthas, M., Miller, C.J. and Lacknew, A.A. (1996b) Localization of simian immunodeficiency virus nucleic acid and antigen in brains of fetal macaques inoculated *in utero*. Am J Pathol 149:1097–1104.

Lavi, E., Strizki, J.M., Ulrich, A.M., Zhang, W., Fu, L., Wang, Q., O'Connor, M., Hoxie, J.A. and Gonzalez-Scarano, F. (1997) CXCR-4 (Fusin), a co-receptor for the type 1 human immunodeficiency virus (HIV-1), is expressed in the human brain in a variety of cell types, including microglia and neurons. Am J Pathol 151:1035–1042.

Levine, S. (1987) Choroid plexus: target for systemic disease and pathway to the brain (editorial). Lab Invest 56:231–233.

Mankowski, J.L., Spelman, J.P., Ressetar, H.G., et al. (1994) Neurovirulent simian immunodeficiency virus replicates productively in endothelial cells of the central nervous system *in vivo* and *in vitro*. J Virol 68:8202–8208.

Masopust, D., Vezys, V., Marzo, A.L. and Lefvreancoid, L. (2001) Preferential localization of effector memory cells in nonlymphoid tissue. Science 291:2413–2417.

Mayeux, R., Stern, Y., Tana, M.X., et al. (1993) Mortality risks in gay men with human immunodeficiency virus infection and cognitive impairment. Neurology 43:173.

McArthur, J.C., Cohen, B.A., Farzedegan, H., et al. (1998) Cerebrospinal fluid abnormalities in homosexual men with and without neuropsychiatric findings. Ann Neurol 23(suppl):S34–S37.

McManus, C.M., Weidenheim, K., Woodman, S.E., Nunez, J., Hesselgesser, J., Nath, A., Berman, J.W. (2000) Chemokine and chemokine-receptor expression in human glial elements. Am J Pathol 156:1441–1453.

Moses, A.V. and Nelson, J.A. (1994) HIV infection of human brain capillary endothelial cells—implications for AIDS dementia. Adv. Neuroimmunol 4:239–247.

Murphy, K.M., Heimberger, A.B. and Loh, D.Y. (1990) Induction by antigen of intrathymic apoptosis of $CD4^{+}CD8^{+}TCR^{1o}$.Thymocytes *in vivo*. Science 250:1720–1723.

Nath, A., Harloper, V., Furer, M. and Fowke, K. (1995) Infection of human fetal astrocytes with HIV-1: Viral tropism and the role of cell to cell contact in viral transmission. J Neuropathol Exp Neurol 54:320–330.

Nathanson, J.A. and Chun, L.L.Y. (1989) Immunological function of blood–cerebrospinal fluid barrier. Proc Natl Acad Sci 86:1684–1688.

Petito, C.K., Adkins, B., McCarthy, M., Roberts, B. and Khamis, I. (2003) $CD4^{+}$ and $CD8^{+}$ cells accumulate in the brains of acquired immunodeficiency syndrome patients with human immunodeficiency virus encephalitis. J NeuroVirol 9:36–44.

Petito, C.K. and Cash, K.S. (1992) Blood-brain barrier abnormalities in the acquired immunodeficiency syndrome dementia: immunohistochemical localization of serum proteins in post-mortem brain. Ann Neurol,32:658–660.

Petito, C.K., Chen, H., Mastri, A., Torres-Muñoz, J., Roberts, B. and Wood, C. (1999) HIV Infection of choroid plexus in AIDS and asymptomatic HIV-infected patients suggests that the choroid plexus may be an important reservoir of productive infection. J Neurovirol 5:670–677.

Petito, C.K., Cho, E.S., Lemann, W., Navia, B.A. and Price, R.W. (1986) Neuropathology of acquired immunodeficiency syndrome (AIDS): an autopsy review. J Neuropathol Exp Neurol 45:635–646.

Petito, C.K., Roberts, B., Cantando, J.D., Rabinstein, A., Duncan, R. (2001) Hippocampal injury and alterations in neuronal chemokine co-receptor expression in patients with AIDS. J Neuropathol Exp Neurol 60:377–38510.

Pialoux, G., Fournier, S., Moulignier, A., Poveda, J.D., Clavel, F. and DuPont, B. (1997) Central nervous system as a sanctuary for HIV-1 infection despite treatment with zidovudine and indinavir. AIDS 11:1302–1303.

Pron, B., Than, M.K., Rombaud, C., Fournet, J.C., Pattey, N., Monnet, J.P., Musilek, M., Beretti, J.L., and Nassif, X. (1997) Interactions of *Neisseria meningitidis* with the components of the blood-brain barrier correlate within increase expression of PilC. J Infect Dis 176:1285–1292.

Quan, N., Mhlanga, J.D.M., Whiteside, M.B., McCoy, A.N., Kristensson, K. and Herkenham, M. (1999a) Chronic overexpression of proinflammatory cytokines and histopathology in the brains of rats infected with *Trypanosoma brucei*. J Comp Neurol 414:114–130.

Quan, N., Stern, E.L., Whiteside, M.B., Herkenham, M. (1999b) Induction of pro-inflammatory cytokine mRNAs in the brain after peripheral injection of subseptic doses of lipopolysaccharide in the rat. J Neuroimmunol 93:72–80.

Reese, T.S. and Karnovsky, M.J. (1967) Fine structural localization of a blood-brain barrier to exogenous peroxidase. J Cell Biol 34:207–217.

Sactor, N. (2002) The epidemiology of human immunodeficiency virus associated neurological disease in the era of HAART therapy. J NeuroVirol 8 (suppl) 2:115–121.

Seabrook, T.J., Johnston, M. and Hay, J.B. (1998) Cerebral spinal fluid lymphocytes are part of the normal recirculating lymphocyte pool. J Neuroimmunol 91:100–107.

Serot, J.M., Foliguet, B., Bene, M.C. and Faure, G.C. (1997) Ultrastructural and immunohistological evidence for dendritic-like cells within human choroid plexus epithelium. Neuroreport 8:1995–1998.

Sharma, D.P., Zinc, M.C., Anderson, M., et al. (1992) Derivation of neurotropic simian immunodeficiency virus from exclusively lymphocytotropic parental virus: pathogenesis of infection in macaques. J Virol 66:3550–3556.

Sinclair, W., Gray, F., Leitner, T. and Chiodi, F. (1994) Immunohistochemical changes and PCR detection of HIV provirus DNA in brains of asymptomatic HIV-positive patients. J Neuropathol Exp Neurol 53:43–50.

Smith, M.O., Heyes, M.P., Lackner, A. and Gonzalez, R.G. (1995) Early intrathecal events in rhesus macaques infected with pathogenic and nonpathogenic molecular clones of simian immunodeficiency virus. Lab Invest 72:547–558.

Steffan, A.M., Lafon, M.E., Gendrault, J.L., et al. (1994) Feline immunodeficiency virus can productively infect cultured endothelial cells from cat brain microvessels. J Gen Virol 75:3647–3653.

Steffen, B.J., Breier, G., Bucher, E.C., Schulz, M., and Engelhardt, B. (1996) ICAM-1, VCAM-1 and Mod CAM-1 are expressed on choroid plexus epithelium but not endothelium and moderate binding of lymphocytes *in vitro*. Am J Pathol 148:1819–1938

Strazielle, N. and Ghersi-Egea, J.F. (2000) Choroid plexus in the central nervous system biology and physiopathology. J Neuropath Exp Neurol 59:561–574.10.

Strazielle, N., Khuth, S.T., Murat, A., Chalon, A., Giraudon, P., Belin, M.F. and Ghersi Edea, J.F. (2003) Pro-inflammatory cytokines modulate matrix metalloproteinase secretion and organic anion transport at the blood–cerebrospinal fluid barrier. J Neuropath Exp Neurol 62(12):1254–1264.

Tacchetti, C., Favre, A., Moresco, L. et al. (1997) HIV is trapped and masked in the cytoplasm of lymph node follicular dendritic cells. Am J Path 150(2):533–542.

Tennyson, V.M. and Pappas, G.D. (1969) The fine structure of the choroid plexus: adult and developmental stages. In: *Brian Barrier Systems*. Eds. A., Lajtha, D.H. Ford. Amsterdam: Elsevier 63–86.

Van der Ende, M.E., Schutten, M., Raschdorff, B., Grossschupff, G., Racz, P., Osterhaus, A.D.M. and Tenner-Racz, K. (1999) CD4 T cells remain the major source of HIV-1 during end stage disease. AIDS 13:1015–1019.

Werkerle, H., Linnington, C., Lassmann, H. and Meyermann, R. (1986) Cellular immune reactivity within the CNS. Trends Neurosci 9:271–277.

Wiley, C.A., Schrier, R.B., Nelson, J.A., Lampert, P.W. and Oldstone, M.B.A. (1986) Cellular localization of human immunodeficiency virus infection within the brains of acquired immune deficiency syndrome patients. Proc Natl Acad Sci USA 93:7089–7093.

Yi, Y., Chen, W., Frank, I., Cutilli, J., Singh, A., Starr-Spires, L., Sulcove, J., Kolson, D.L. and Collman, R.G. (2003) An unusual syncytia-inducing human immunodeficiency virus type 1 primary isolate from the central nervous system that is restricted to CXCR4 replicates efficiently in macrophages, and induces neuronal apoptosis. J NeuroVirol 9:432–441.

16 Choroid Plexus and Drug Therapy for AIDS Encephalopathy

Julie E. Gibbs and Sarah A. Thomas
King's College London, Guy's Campus,
London, United Kingdom

CONTENTS

16.1 INTRODUCTION

Human immunodeficiency virus (HIV) is a retrovirus that causes acquired immunodeficiency syndrome (AIDS). In 2003, the global HIV/AIDS epidemic killed more than three million people, and an estimated five million acquired HIV. This brings the number of people living with the virus around the world to 40 million (UNAIDS: AIDS Epidemic Update 2003). HIV attacks the immune system by targeting CD4+ T lymphocytes, and as the number of these cells falls, immune deficiency occurs. Without treatment, the AIDS patient becomes susceptible to opportunistic infections, which, in due course, leads to death. With the introduction of highly active antiretroviral therapy (HAART), whereby three or more anti-HIV drugs are used in parallel, there have been dramatic reductions in mortality and morbidity associated with HIV in resource-rich countries (Sabin 2002). Since AIDS involves encephalopathy, the question as to how to design the structure of therapeutic drugs so that they can more efficiently gain access to the brain has become a major subject in the treatment of HIV-induced brain damage. This chapter will start with a brief introduction of HIV in the central nervous system (CNS) and currently used anti-AIDS drugs. The focus will then be directed to treatment of HIV infection within the CNS and the significance of achieving therapeutic levels of drugs within the brain and cerebrospinal fluid (CSF), with particular emphasis on the role of the choroid plexus.

16.1.1 HIV in the CNS

HIV-1 is a neurotropic retrovirus and enters the CNS early on in the course of infection. HIV antibodies can be detected in the CSF in the early stages of HIV infection (Diederich et al., 1988), and HIV-1 nucleic acid has been detected in the brain just 15 days after infection (Davis et al., 1992). Macrophages are the major cell types in which HIV replicates within the brain. There are a number of different macrophage populations in the CNS, including microglia, macrophages of the choroid plexus, macrophages of the meninges, and perivascular macrophages. While there is evidence to indicate that neurons and astrocytes may become infected with HIV (Wiley et al., 1986; Takahashi et al., 1996; An et al., 1999; Trillo-Pazos et al., 2003), the microglia and the perivascular macrophages are thought to be the major populations of infected cells within the CNS (Wiley et al., 1986; Takahashi et al., 1996; Fischer-Smith et al., 2001; Williams et al., 2001). Consequently it is important that anti-HIV drugs can target these infected cells. However, entry of anti-HIV drugs into the CNS is partly limited by the blood–brain barrier (BBB) located at the level of the cerebral capillary endothelium, and the blood–CSF barrier found at the level of the choroid plexuses and the arachnoid membrane.

16.1.2 Anti-HIV Drugs

Currently four classes of anti-HIV drugs are available in the United Kingdom: nucleoside analogues or nucleoside reverse transcriptase inhibitors (NRTIs), nucleotide reverse transcriptase inhibitors (NtRTIs), nonnucleoside reverse transcriptase inhibitors (NNRTIs), and protease inhibitors (PIs) (Table 16.1). There are several more classes in development, for example, the fusion inhibitors and the integrase inhibitors (Gulick, 2003). HAART is currently the recommended treatment strategy for HIV infection. There have been no definitive controlled trials to demonstrate the clinical superiority of any one HAART regime; however, it is thought that for each patient a regime should be established to achieve the best potency, adherence, and tolerability possible, to minimize potential toxicity, and to avoid any likely drug interactions. An audit of HIV treatment published in 2003 revealed that of the HIV-positive patients receiving antiretroviral therapy, 97.6% were receiving combinations of three or more drugs (HAART), with a majority of patients taking nucleoside analogue-based regimes (Curtis et al., 2003). However, it has become evident that even with the range of drugs available (see Figure 16.1), total eradication of HIV-1 is still not possible (Finzi et al., 1999). This is due to the presence of virus in host cellular and anatomical reservoirs that are inaccessible to HAART and that are

TABLE 16.1
U.K.- and U.S.-Approved Anti-HIV Drugs[a]

NRTIs	NNRTIs	NtRTIs	PIs
Zidovudine (3'-azido-3'-deoxythymidine; AZT)	Efavirenz (EFV)	Tenofovir (PMPA)	Saquinavir
Zalcitabine (2'3'-dideoxycytidine; ddC)	Nevirapine (NVP)		Ritonavir
Didanosine (2'3'-dideoxyinosine; ddI)	Delaviridine (DLV)		Amprenavir
Stavudine (2'3'-didehydro-3'deoxythymidine; d4T)			Indinavir
Lamivudine ((-) -β-*L*-2'3'-dideoxy-3'-thiacytidine; 3TC)			Nelfinavir
Abacavir ([()-(1*S*,4*R*)-4-[2-amino-6-(cyclopropylamino)-*H*-purin-9-yl]-2-cyclopentene-1-methanol; 1592U89)			Lopinavir

[a] Several anti-HIV agents are in development, some of which have alternative mechanisms of action, including entry inhibitors and inhibitors of HIV integrase (Gulick, 2003).

consequently a source of viral rebound to the plasma and CSF if therapy is discontinued or inadequate (Wong et al., 1997; Schrager and D'Souza 1998; Blankson et al., 2002).

A number of different cellular reservoirs of HIV have been identified, including CD4$^+$ T cells (Chun and Fauci 1999; Finzi et al., 1999; Siliciano and Siliciano 2000), macrophages (Crowe et al., 1990; Aquaro et al., 2002), and follicular dendritic cells (Smith et al., 2001). The key anatomical viral reservoirs are the CNS, lymphoid organs (Gunthard et al., 2001), and the genitourinary tract (Schrager and D'Souza 1998; Blankson et al., 2002).

16.2 ROLE OF CHOROID PLEXUS IN HIV INFECTION/TREATMENT

The presence of a drug in the CSF relates to blood–CSF barrier permeability and does not necessarily reflect BBB permeability or brain drug concentrations (Groothuis and Levy, 1997; Thomas and Segal, 1998). Table 16.2 confirms the complexity of drug movement into and within the brain, CSF, and choroid plexuses, with varying rates of drug uptake being seen for each tissue. This is related to the differing paracellular permeability and transporter characteristics of the brain barriers (Bouldin and Krigman, 1975) and the fact that exchange of molecules between the brain and CSF compartments is relatively slow and limited (Groothuis and Levy, 1997). The ability of drugs to cross the choroid plexuses and directly reach the CSF from the blood is of special interest in the treatment of HIV infection. This is because drugs that have crossed the choroid plexuses and are present in the ventricular CSF

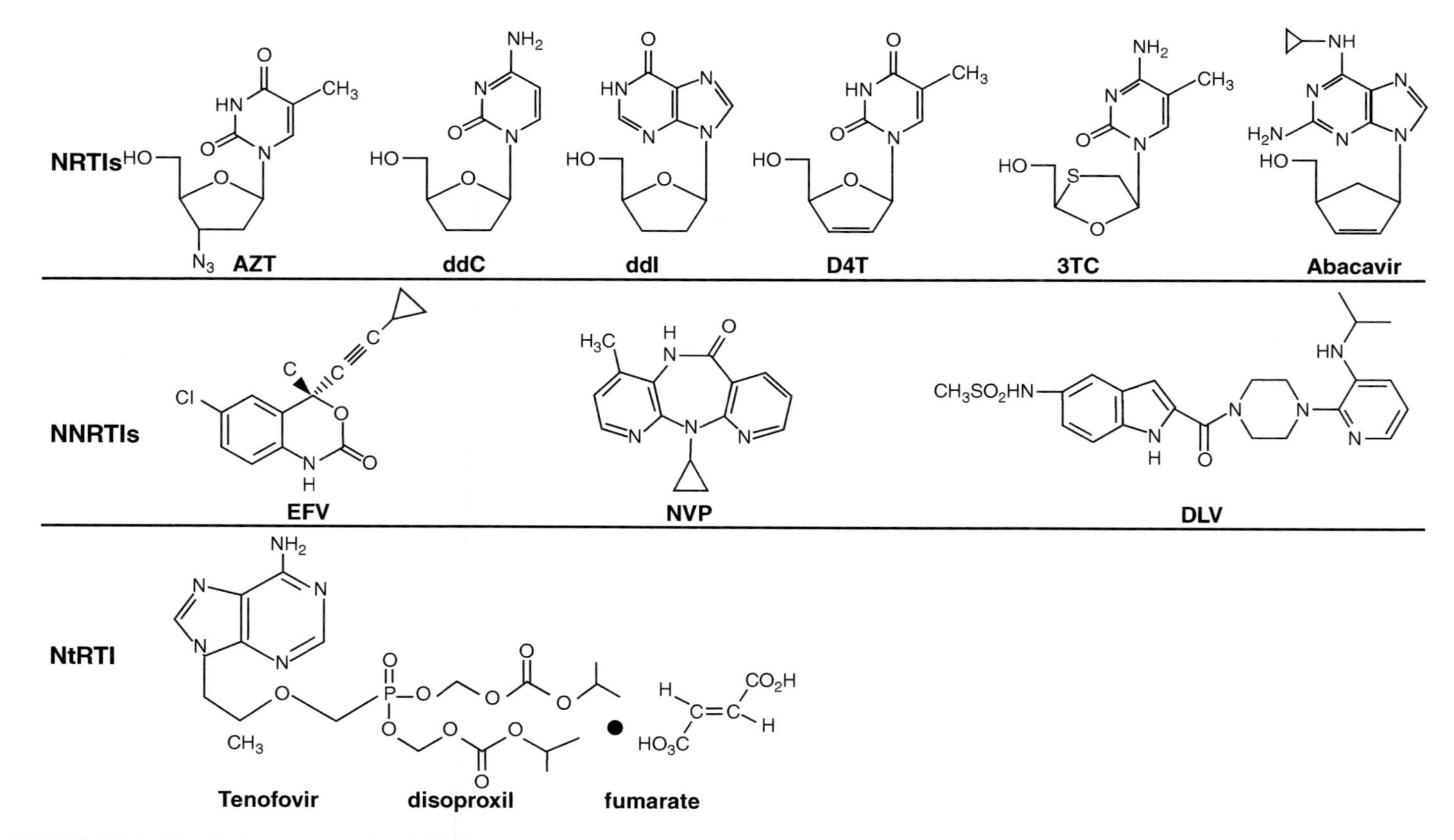

FIGURE 16.1 Chemical structures of anti-HIV drugs.

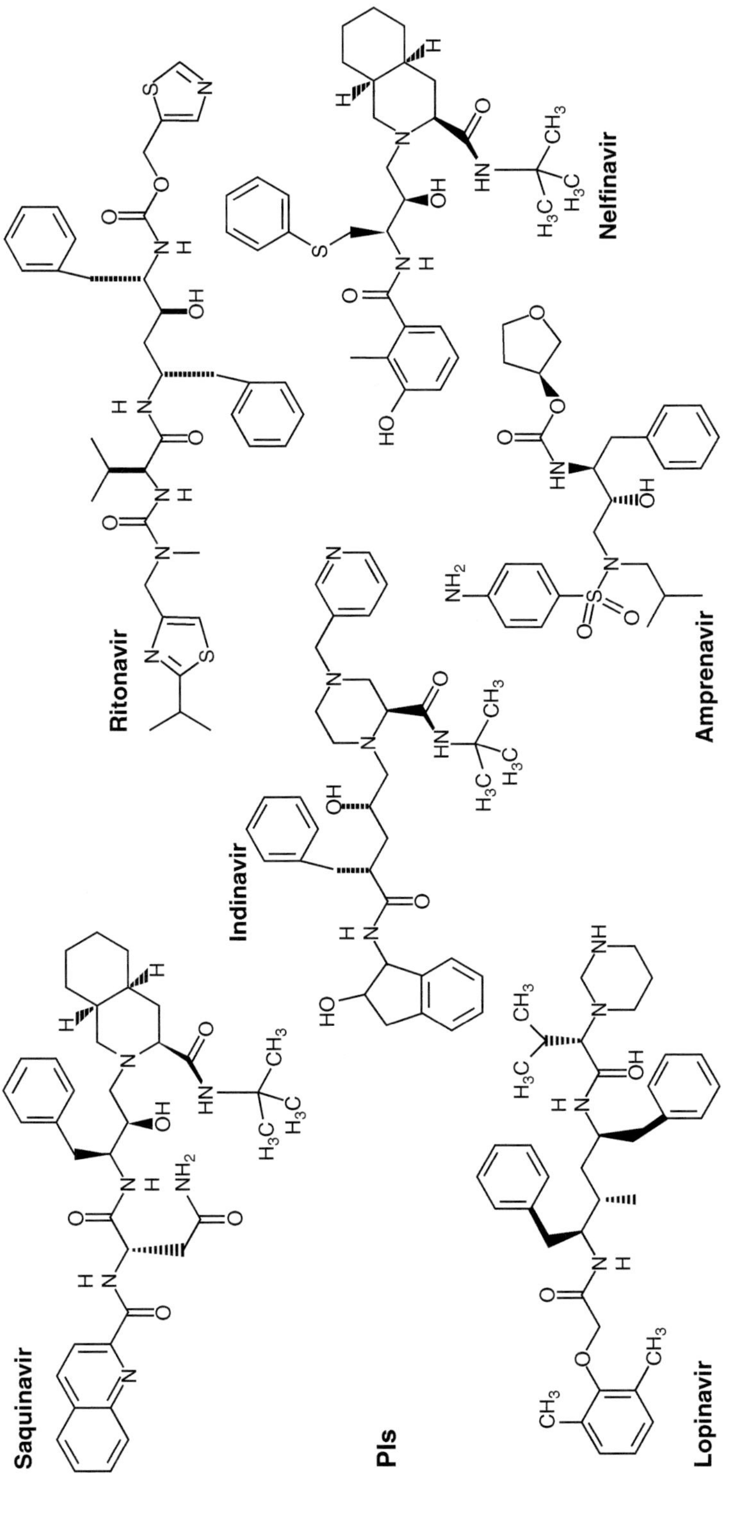

Saquinavir

Ritonavir

Indinavir

PIs

Nelfinavir

Lopinavir

Amprenavir

FIGURE 16.1B Chemical structures of anti-HIV drugs.

TABLE 16.2
Unidirectional Transfer Constants (K_{in} [μmicroL/min/g]) Calculated for Anti-HIV Drugs Tested in Bilateral In Situ Brain Perfusion Technique in the Guinea Pig

NRTIs	Cerebrum	Choroid Plexus	CSF	Ref.
[^{3}H]AZT	0.8 ± 0.1[a]	1.8 ± 1.0[b(10min)]	0.5 ± 0.1[b(10min)]	Thomas and Segal, 1997
[^{3}H]ddC	0.5 ± 0.1[a]	7.1 ± 0.7[b(30min)]	0.5 ± 0.2[b(30 min)]	Gibbs and Thomas, 2002
[^{3}H]ddI	0.9 ± 0.1[a]	18.4 ± 3.4[b]	0.8 ± 0.1[b]	Gibbs et al., 2003a
[^{3}H]d4T	0.3 ± 0.1[a]	5.8 ± 1.1[b(10min)]	0.8 ± 0.2[b(10min)]	Thomas and Segal, 1998
[^{3}H]3TC	0.7 ± 0.1[a]	20.3 ± 2.8[b]	0.3 ± 0.1[b]	Gibbs et al., 2003b
[^{14}C]abacavir	9.3 ± 0.6[b(10min)]	not determined	4.5 ± 1.4[b(10min)]	Thomas et al., 2001
NtRTI				
[^{3}H]PMPA	0.2 ± 0.1[b]	9.0 ± 2.8[b]	2.4 ± 0.9[b]	Anthonypillai et al.[c]
PIs				
[^{3}H]ritonavir	3.2 ± 0.4[a]	78.6 ± 13.4[b]	0.2 ± 0.1[b]	Anthonypillai et al., 2004
[^{3}H]amprenavir	9.8 ± 2.2[b(15min)]	29.8 ± 4.1[b(15min)]	1.0 ± 0.1[b(15min)]	Parsons et al.[c]

[a]Multiple-time uptake.
[b]Single-time uptake at 20 min; if different, time stated in parentheses. All values are vascular space corrected.
[c]Preliminary observations.

will have rapid access to the infected perivascular and meningeal macrophages (Rennels et al., 1985; Ghersi-Egea et al., 1996).

On the other hand, the choroid plexus is thought to play a role in HIV infection of the CNS. Certain characteristics of the choroid plexus make it a potential site for HIV to gain entry into the CNS. First, the permeable nature of the choroid plexus capillaries and the exclusion of this tissue from the protection of the BBB make it a potential route of entry for virus into the CNS from the periphery. Second, the choroid plexus stroma contains T lymphocytes and monocytes derived from circulation, hence it is a prospective site for infected lymphocytes and monocytes to enter the CSF from the blood and perhaps gain access to the brain parenchyma. Examination of the viral sequence of HIV in the choroid plexus reveals that it is a mixture of systemic and brain sequences (Chen et al., 2000), which supports the view that the choroid plexus may be a site of viral entry into the CNS from the blood.

The first evidence for the choroid plexus involvement in HIV infection came from studies by Harouse and co-workers. They demonstrate that HIV can latently infect cells isolated from human choroid plexus tissue (Harouse et al., 1989). Further to this finding, HIV-infected cells have been found in postmortem choroid plexus tissue from patients who died with AIDS, within the stroma, supra-epithelial area (and, rarely, the capillary endothelium) (Falangola et al., 1995; Petito et al., 1999).

On the basis of immunoreactivity for viral proteins, it is suggested that the infected cells are monocytes and dendritic cells (Petito et al., 1999). However, previous studies have also suggested that the epithelial cells of the choroid plexus become infected with HIV (Bagasra et al., 1996). Taken together, these studies demonstrate that the choroid plexus is a reservoir for HIV within the CNS and thus must also be an important target for anti-HIV drugs.

16.3 THE IMPORTANCE OF TREATING HIV INFECTION WITHIN THE CNS

The presence of HIV within the brain and CSF is directly associated with the development of a syndrome called HIV-associated dementia (HAD), which is characterized by a collection of cognitive, motor, and behavioral symptoms (McArthur et al., 1997). In order to reduce the occurrence of HAD, therapeutic concentrations of antiviral drugs must reach the brain, choroid plexus, and CSF. In support, it has been demonstrated that those drugs that are able to reach the CSF can improve CNS function (as measured by psychomotor testing, a sensitive predictor of HAD) in HIV-infected individuals (Arendt et al., 2001; von Giesen et al., 2002). Furthermore, subtherapeutic levels of anti-HIV agents within the CNS may permit the evolution of drug-resistant viral strains in the CSF, which have the potential to reinfect the periphery (Schrager and D'Souza, 1998).

Several studies have provided evidence for the development of drug-resistant HIV strains in the CSF independently of the plasma (Cunningham et al., 2000; Venturi et al., 2000; Stingele et al., 2001). The presence of drug-resistant variants of HIV is a major clinical concern as it is associated with therapy failure, disease progression, and death. Consequently, a clear understanding of the ability of anti-HIV drugs to reach effective concentrations within the CNS, including the choroid plexus, is needed if we are to prevent the development of the neurological symptoms of HIV infection and reduce the risk of peripheral reinfection.

16.4 EVALUATION OF CNS EFFICACY OF ANTI-HIV DRUGS

16.4.1 CSF STUDIES

As stated earlier, free drug movement into the CNS is limited by the innate characteristics of the blood–brain and blood–CSF barriers and the physical nature of the drugs themselves. Many clinical trials have examined whether anti-HIV drugs can be detected within the CSF (see Table 16.3); however, differences between patient groups, disease progression, dosing, number, and timing of samples make direct comparisons of the drug efficacy difficult, if not impossible. In addition, human CSF studies, which are obtained only from lumbar puncture, may not reflect the ventricular and upper brain subarachnoid CSF. This has been shown in non-human primates for 3TC for which the concentration in the lumbar CSF was fourfold higher ($41 \pm 23\%$) than in the ventricular CSF ($7.9 \pm 4.7\%$) (Blaney et al., 1995). Furthermore, evaluation of the CNS distribution of one anti-HIV drug is now complicated by the clinical need

TABLE 16.3
Human CSF/Plasma Ratios for Anti-HIV Drugs

	CSF/Plasma Ratios	Ref.
NRTIs		
AZT	60%	Rolinski et al., 1997
ddC	9–37%	Yarchoan et al., 1988
ddI	21–27%	Burger et al., 1995; Hartman et al., 1990
d4T	40%	Haworth et al., 1998
3TC	10%	Foudraine et al., 1998
abacavir	30%	McDowell et al., 1999
NNRTIs		
NVP	40%	Enzensberger and von Giesen 1999
EFV	0.6–1.2%	Tashima et al., 1999
PIs		
Ritonavir	0–0.5%	Gisolf et al., 2000; Kravcik et al., 1999
Indinavir	7%	Solas et al., 2003; Zhou et al., 2000
Saquinavir	0–2%	Gisolf et al., 2000; Shafer and Vuitton 1999
Amprenavir	Not detectable	Sadler et al., 2001
Lopinavir	Not detectable	Solas et al., 2003
Nelfinavir	Not detectable	Aweeka et al., 1999

to give multiple drugs at the same time. Hence experimental models in animals are necessary if we are to further our understanding of both the potential CNS efficacy of the different drugs and the role of the choroid plexus in drug distribution.

Brain microdialysis, brain uptake index (BUI) methods, whole body autoradiography, and cultured cerebral microvascular endothelial cells have all been used to examine the distribution of certain antiretroviral agents in the brain (see Sawchuk and Yang, 1999). Direct CNS efficacy has also been examined by means of a severe combined immune deficiency (SCID) mouse model of HIV-1 encephalitis, which assays the effect of certain drugs on viral replication in brain macrophages (Limoges et al., 2000, 2001).

16.4.2 Choroid Plexus Studies

Several research groups have assessed the movement of individual anti-HIV drugs into and from the mammalian CSF using intravenous infusion/injection (Anderson et al., 1990; Galinsky et al., 1990, 1991; Sawchuk and Hedaya, 1990; Hoesterey et al., 1991; Tuntland et al., 1994; Takasawa et al., 1997b), intracerebroventricular injection (Takasawa et al., 1997b, 1997c), ventriculo-cisternal perfusion (Thomas and Segal, 1997), isolated choroid plexus (whole and sliced) incubation studies (Takasawa et al., 1997a), and primary cultures of choroidal epithelial cells (Strazielle et al., 2003). However, not many groups directly examine the role of the choroid plexus in this distribution for more than one class of anti-HIV drugs (Thomas, 2004).

The bilateral in situ brain perfusion technique is an ideal model to examine the distribution of anti-HIV drugs into the choroid plexus, brain, and CSF simultaneously (Thomas nee Williams and Segal, 1996). In short, adult animals are anesthetized and both carotid arteries are cannulated with silicon tubing connected to fluid pumps. The perfusion fluid consists of an artificial plasma containing the radiolabeled anti-HIV drug. The perfusion can be maintained for periods up to 30 minutes while still maintaining normal physiological parameters, which enables the movement of slowly moving molecules to be measured (Thomas nee Williams and Segal, 1996; Gibbs and Thomas, 2002). Both jugular veins are sectioned at the start of perfusion to enable fluid outflow. The perfusion of both cerebral hemispheres has an added advantage over unilateral in situ brain perfusion methods (Zlokovic et al., 1986), as it enables drug distribution into the brain, choroid plexuses and cisternal CSF to be examined simultaneously (Thomas nee Williams and Segal, 1996; Gibbs and Thomas, 2002). The uptake of radiolabeled drug into the brain, CSF and choroid plexuses can be expressed as a percentage of that in the artificial plasma. Although it is appreciated that in in vivo studies, passage across the blood–brain and blood–CSF barriers cannot be separated, comparison of the drug uptake into the brain, CSF, and choroid plexus enables assumptions to be made concerning the predominant route for CNS entry (Thomas and Segal, 1998).

The permeability of the brain barriers for the anti-HIV drug can be expressed by the unidirectional transfer constant, K_{in}, which can be calculated from multiple-time (2.5 to 30 min) or single-time uptake data. Both calculations have been previously described in detail (Thomas nee Williams and Segal, 1996; Williams et al., 1996). Table 16.2 summarizes the anti-HIV drug data obtained from a bilateral in situ brain perfusion technique in the anesthetized guinea pig. These studies clearly demonstrate that there is consistently a lower rate of drug uptake into the cisternal CSF when compared to that measured into the choroid plexuses. This suggests that the choroid plexus regulates the distribution of each of the tested anti-HIV drugs into the CSF from the plasma. In some cases this discrepancy has been demonstrated to be indicative of an efflux transport process at the choroid plexus, which limits drug accumulation in the CSF (Gibbs and Thomas, 2002; Gibbs et al., 2003a).

In an attempt to clarify the role of the choroid plexuses in this distribution, several studies have examined the removal of anti-HIV drugs from the CSF into the choroid plexuses by means of the isolated incubated choroid plexus model (see Table 16.4). This method predominantly examines the accumulation of molecules from an artificial CSF across the apical (CSF facing) membrane of the incubated choroid plexus (Pritchard et al., 1999) and provides further evidence for the involvement of transport systems in the removal of anti-HIV drugs from the CSF (Takasawa et al., 1997a; Gibbs and Thomas, 2002).

16.4.3 Drug Transporters

Several drug transporters are expressed by the choroid plexus (see Figure 16.2); for example, members of the multidrug resistance (MDR), multidrug resistance-associated protein (Mrp), organic anion-transporting polypeptide (Oatp), organic anion transport (Oat), organic cation transporter (OCT), and concentrative (CNT)

TABLE 16.4
Accumulation of Test Drugs by the Isolated Incubated Choroid Plexus of the Guinea Pig[a]

	Total NRTI or PI (mL/g)	Extracellular Space Marker (mL/g)	Corrected NRTI or PI (mL/g)	Ref.
NRTIs				
[^{3}H]AZT	1.2 ± 0.1	0.6 ± 0.1	0.6 ± 0.1	Anthonypillai et al.[b]
[^{3}H]ddC	1.3 ± 0.1	0.5 ± 0.1	0.9 ± 0.1	Gibbs and Thomas, 2002
[^{3}H]ddI	1.3 ± 0.1	0.6 ± 0.1	0.7 ± 0.1	Gibbs et al., 2003a
[^{3}H]d4T	1.1 ± 0.1	0.5 ± 0.1	0.5 ± 0.1	Ham et al.[b]
[^{3}H]3TC	1.1 ± 0.1	0.5 ± 0.1	0.6 ± 0.1	Gibbs et al., 2003b
PIs				
[^{3}H]ritonavir	5.5 ± 1.1	0.6 ± 0.1	4.9 ± 1.0	Anthonypillai et al., 2004
[^{3}H]amprenavir	1.7 ± 0.1	0.7 ± 0.1	1.0 ± 0.1	Parsons et al.[b]

[a]Length of isotope incubation was 10 min for NRTIs and 5 min for PIs. The extracellular space marker was *D*-[^{14}C]mannitol for the NRTIs and [^{14}C]sucrose for the PIs (n = 3–6).

[b]Preliminary observations.

and equilibrative (ENT) nucleoside transporter families have all been found at the choroid plexus (Choudhuri et al., 2003). The NRTI AZT was the first drug to be used against HIV infection. Although initially there was some debate as to whether AZT used a transporter mechanism for nucleosides to cross the brain barriers and enter the CNS (Collins et al., 1988), many studies now point toward a nonfacilitated transfer of AZT into the CSF and brain (Masereeuw et al., 1994; Tuntland et al., 1994; Thomas and Segal, 1997). However, there is considerable evidence supporting the active removal of AZT across both the blood–brain and blood–CSF barriers by a probenecid sensitive transport system (Sawchuk and Hedaya 1990; Dykstra et al., 1993; Takasawa et al., 1997b; Sawchuk and Yang, 1999). Probenecid is a nonspecific inhibitor of organic anion transport and can inhibit Mrp, Oatp, and Oat transporters. The CNS removal process of AZT has also been shown to be sensitive to *p*-aminohippurate (PAH) (Takasawa et al., 1997c), which is a model substrate for Oat (Sekine et al., 1997; Sweet et al., 1997). OAT1 and OAT3, which are both expressed in the rat choroid plexus, are strong candidates for mediating the removal of AZT from the CNS (Pritchard et al., 1999; Wada et al., 2000; Nagata et al., 2002; Strazielle et al., 2003). A recent study has further confirmed that AZT is removed by Oat1 or Oat3 across the in vitro model of the blood–CSF interface cultured choroid plexus (Strazielle et al., 2003). The possible involvement of Mrp in mediating AZT efflux transport has also been indicated at the brain endothelial cells (Sawchuk and Yang, 1999), but not at the basolateral membrane of the choroidal epithelial cells (Strazielle et al., 2003).

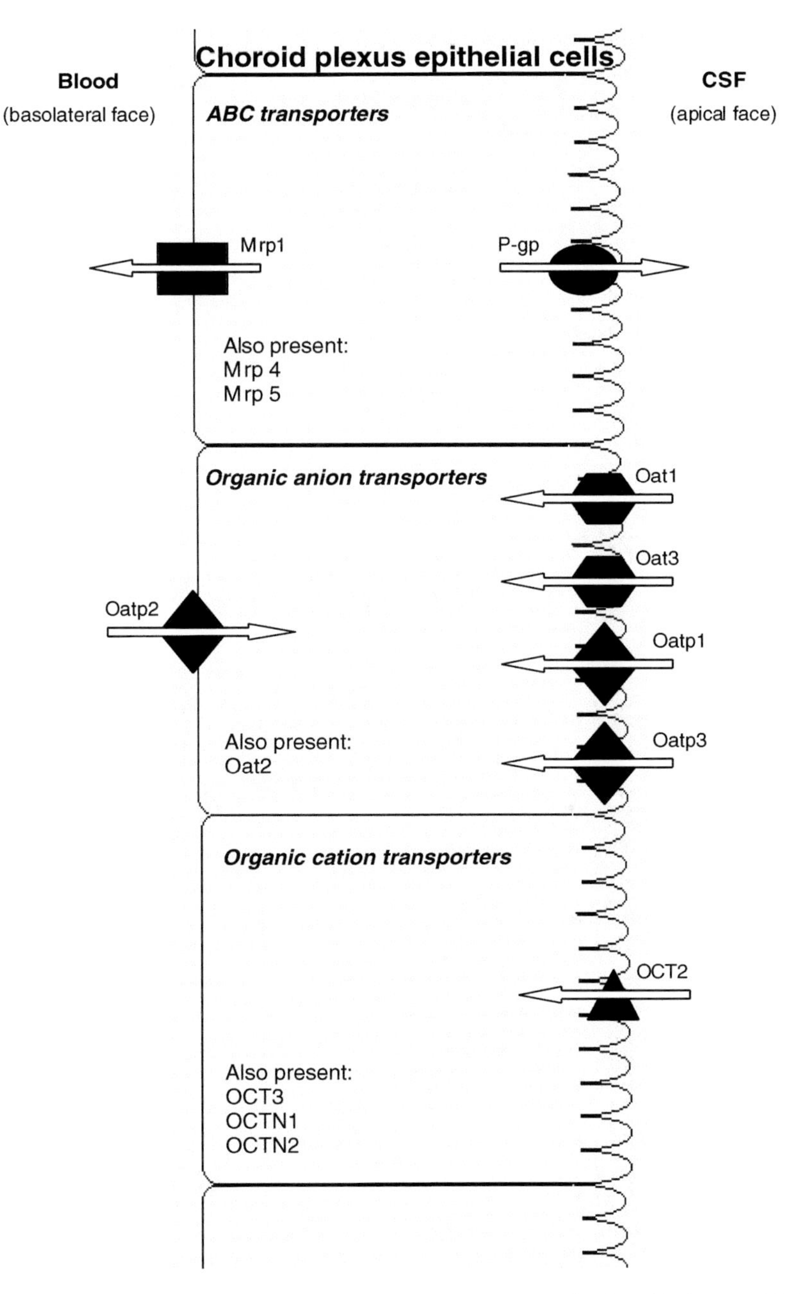

FIGURE 16.2 The expression and function of drug transporters in choroid plexus epithelial cells. P-gp (P-glycoprotein), Mrp (multidrug-resistance proteins), Oat (organic anion transporter), Oatp (organic anion–transporting polypeptide), and OCT and OCTN (organic cation transporters). Where the transporters are not assigned a membrane face, their localization within the epithelial cells is unknown. (Adapted from Gibbs, 2003.)

TABLE 16.5
Summary of Octanol-Saline (pH 7.4) Partition Coefficients for Tested Anti-HIV Drugs

	Octanol/Saline Partition Coefficients	Ref.
NRTIs		
[^{3}H]AZT	1.02 ± 0.02	Thomas and Segal, 1997
[^{3}H]ddC	0.047 ± 0.001	Gibbs and Thomas, 2002
[^{3}H]ddI	0.05 ± 0.01	Gibbs et al., 2003a
[^{3}H]d4T	0.167 ± 0.003	Thomas and Segal, 1998
[^{3}H]3TC	0.108 ± 0.006	Gibbs et al., 2003b
[^{14}C]abacavir	6.6 ± 0.2	Thomas et al., 2001
NtRTIs		
[^{3}H]PMPA	0.013 ± 0.00003	Anthonypillai et al.[a]
PIs		
[^{3}H]ritonavir	22.2 ± 0.7	Anthonypillai et al., 2004
[^{3}H]amprenavir	36.7 ± 1.6	Parsons et al.[a]

[a]Preliminary observations.

Using the bilateral in situ brain perfusion technique and the isolated incubated choroid plexus method in parallel, the transport of another NRTI, ddC, from the blood into the CSF was found to be limited, although a significant accumulation was noted in the choroid plexus (Gibbs and Thomas, 2002). The low lipophilicity of this NRTI may be partly responsible for the slow movement across the blood–CSF barrier (see Table 16.5); however, further studies revealed that [^{3}H]ddC was effluxed out from the choroid plexuses toward the blood in a saturable fashion. The inhibition characteristics of this transport suggests that a member of the Oat family (possibly Oat1-like or Oat3-like) and OATP1-like or OATPA-like isoforms are likely involved in the limited distribution of ddC to the CNS (Gibbs and Thomas, 2002).

The protease inhibitors, [^{3}H]ritonavir and [^{3}H]amprenavir have also been studied using the bilateral in situ brain perfusion and isolated choroid plexus techniques in the guinea pig (Anthonypillai et al., 2004). Both drugs are lipophilic (see Table 16.5) and have substantial rates of uptake into the brain and choroid plexuses (see Table 16.2). However, the uptake of ritonavir and amprenavir into the CSF is rather limited and corresponds to findings in the clinical situation (see Table 16.3). These studies confirm the importance of the blood–CSF barrier in controlling drug entry into the CNS, although further studies have not been able to elucidate the identity of any transporter involved in the distribution of these PIs (Anthonypillai et al., 2004).

Overall, the limited CNS distribution of PIs is thought to be linked to the use of efflux pumps, specifically MDR1 transporter P-glycoprotein (P-gp) (Kim et al., 1998; Polli et al., 1999; Choo et al., 2000; Huisman et al., 2000; Huisman et al., 2001) and possibly MRP (Srinivas et al., 1998; Miller et al., 2000; Huisman et al., 2002; van der Sandt et al., 2001). Using P-gp knockout mice, it has been demonstrated that P-gp limits the brain accumulation of several anti-HIV protease inhibitors (Kim et al., 1998; Polli et al., 1999; Huisman et al., 2003). P-gp is expressed at both the BBB (Cordon-Cardo et al., 1989; Jette et al., 1993; Jette et al., 1995; Schinkel 1999; Demeule et al., 2002) and choroid plexus epithelium (Rao et al., 1999; Warren et al., 2000). Within the choroidal epithelial cells, P-gp has a predominantly subapical distribution (although areas of intracellular and basolateral foci have been observed) (Rao et al., 1999). Interestingly, this localization suggests that P-gp in the choroid plexus epithelial cells functions to actively influx substances into the CSF. This discrepancy between the differential functions of transporters, depending on their location, further highlights the complexity of understanding drug movement within the CNS.

An interesting aspect of the use of transporters by anti-HIV drugs is that certain combinations of anti-HIV drugs could interact for transport at the brain barriers. For example, the concentration of ddI in the guinea pig choroid plexus was altered in the presence of the other NRTIs—abacavir, 3TC, and AZT (Gibbs et al., 2003a). In addition, ritonavir accumulation by the perfused choroid plexus was significantly reduced by NVP and abacavir (Anthonypillai et al., 2004). This could ultimately result in subtherapeutic concentrations of drugs within the brain/choroid plexus/CSF, which would enable viral replication to continue and drug resistant strains of HIV to be selected. Thus, a clear understanding of the use of transport mechanisms by the anti-HIV drugs would aid in the design of more effective drug regimens.

In addition, specific inhibitors of P-gp and MRP 1 are being developed in an attempt to enhance the distribution of CNS-active drugs, and it is thought they may prove beneficial as an adjuvant therapy in the treatment of HIV-infected patients who are receiving PIs (Taylor 2002). The presence of an efflux process for AZT has also indicated that systemic co-administration with an efflux inhibitor (e.g., probenecid) may produce more favorable CNS effects. Although probenecid administered in combination with AZT is not clinically ideal due to the development of cutaneous adverse events, benzbromarone has been shown to inhibit AZT efflux across the choroid plexus and has been suggested by Strazielle and colleagues (2003) as an alternative adjuvant therapy to potentiate the CNS activity of AZT.

16.5 BLOOD–BRAIN BARRIER IN HIV INFECTION

Another aspect that further complicates our understanding of antiretroviral drug delivery to the HIV-infected CNS is the fact that the disease may alter the overall integrity of the brain barriers. Findings from a study that examined tissue from AIDS patients who were suffering from HIV encephalitis (the biological correlate of HAD) demonstrated that the vascular permeability of the BBB was increased in approximately 50% of these patients (Petito and Cash, 1992). Interestingly, in another study, 15% of the HIV-1–infected patients who were neuro-asymptomatic had increased levels of albumin in their CSF (Andersson et al., 2001). Although this was attributed to elevated BBB permeability, it must be pointed out that CSF albumin levels also reflect changes in the permeability of the blood–CSF barrier (i.e., choroid plexus). Interestingly, following treatment with AZT, the levels of neopterin and β_2-microglobulin in the CSF were significantly decreased, and this change in protein levels was correlated with a decrease in CSF HIV RNA levels (Gisslen et al., 1997).

However, no correlation was found between HIV RNA plasma and CSF levels (Brouwers et al., 1997; Gisslen et al., 1997). These results suggest either that the integrity of the blood–brain and blood–CSF interfaces is maintained in terms of virus exchange or that the damage in the barriers is reversible upon treatment. How these barrier changes affect the CNS distribution of anti-HIV drugs is unknown.

A downregulation in the cerebral expression of P-gp has also been noted in HIV patients both with and without neurological complications. This reflects work undertaken with peripheral blood mononuclear cells (PBMCs), where expression of P-gp in the PBMCs of patients with HIV was significantly lower than in uninfected individuals (Meaden et al., 2001). Altered P-gp expression in the BBB of HIV-infected patients could be associated with a reduced ability of the brain to protect itself from putative toxins and may also affect the cerebral distribution of anti-HIV drugs. These concepts should be considered when evaluating the ability of anti-HIV drugs to penetrate the HIV-infected CNS.

16.6 SUMMARY

A clear understanding of drug distribution across the blood–CSF interface could significantly contribute to developing anti-HIV drug treatment regimens that improve the neurological deficits associated with HIV infection and prolong patient survival. As discussed, there has been considerable research in this field, but further studies need to be undertaken if we are to maximize viral suppression and improve the clinical benefit that could be offered by HAART.

16.7 ACKNOWLEDGMENTS

The authors' work described in this chapter was funded by a Wellcome Trust grant awarded to S. A. Thomas (grant code RCDF 057254).

REFERENCES

An, S.F., Groves, M., Gray, F. and Scaravilli, F. (1999) Early entry and widespread cellular involvement of HIV-1 DNA in brains of HIV-1 positive asymptomatic individuals, *J Neuropathol Exp Neurol* 58: 1156–1162.

Anderson, B.D., Hoesterey, B.L., Baker, D.C. and Galinsky, R.E. (1990) Uptake kinetics of 2',3'-dideoxyinosine into brain and cerebrospinal fluid of rats: intravenous infusion studies, *J Pharmacol Exp Ther* 253: 113–118.

Andersson, L.M., Hagberg, L., Fuchs, D., Svennerholm, B. and Gisslen, M. (2001) Increased blood-brain barrier permeability in neuro-asymptomatic HIV-1-infected individuals—correlation with cerebrospinal fluid HIV-1 RNA and neopterin levels, *J Neurovirol* 7: 542–547.

Anthonypillai, C., Sanderson, R.N., Gibbs, J.E. and Thomas, S.A. (2004) The distribution of the HIV protease inhibitor, ritonavir, to the brain, cerebrospinal fluid, and choroid plexuses of the guinea pig, *J Pharmacol Exp Ther* 308: 912–920.

Aquaro, S., Bagnarelli, P., Guenci, T., De Luca, A., Clementi, M., Balestra, E., Calio, R. and Perno, C.F. (2002) Long-term survival and virus production in human primary macrophages infected by human immunodeficiency virus, *J Med Virol* 68: 479–488.

Arendt, G., von Giesen, H.J., Hefter, H. and Theisen, A. (2001) Therapeutic effects of nucleoside analogues on psychomotor slowing in HIV infection, *AIDS* 15: 493–500.

Aweeka, F., Jayewardene, A., Staprans, S., Bellibas, S.E., Kearney, B., Lizak, P., Novakovic-Agopian, T. and Price, R.W. (1999) Failure to detect nelfinavir in the cerebrospinal fluid of HIV-1—infected patients with and without AIDS dementia complex, *J AIDS Hum Retrovirol* 20: 39–43.

Bagasra, O., Lavi, E., Bobroski, L., Khalili, K., Pestaner, J.P., Tawadros, R. and Pomerantz, R.J. (1996) Cellular reservoirs of HIV-1 in the central nervous system of infected individuals: identification by the combination of in situ polymerase chain reaction and immunohistochemistry, *AIDS* 10: 573–585.

Blaney, S.M., Daniel, M.J., Harker, A.J., Godwin, K. and Balis, F.M. (1995) Pharmacokinetics of lamivudine and BCH-189 in plasma and cerebrospinal fluid of nonhuman primates, *Antimicrob Agents Chemother* 39: 2779–2782.

Blankson, J.N., Persaud, D. and Siliciano, R.F. (2002) The challenge of viral reservoirs in HIV-1 infection, *Annu Rev Med* 53: 557–593.

Bouldin, T.W. and Krigman, M.R. (1975) Differential permeability of cerebral capillary and choroid plexus to lanthanum ion, *Brain Res* 99: 444–448.

Brouwers, P., Hendricks, M., Lietzau, J.A., Pluda, J.M., Mitsuya, H., Broder, S. and Yarchoan, R. (1997) Effect of combination therapy with zidovudine and didanosine on neuropsychological functioning in patients with symptomatic HIV disease: a comparison of simultaneous and alternating regimens, *AIDS* 11: 59–66.

Burger, D.M., Kraayeveld, C.L., Meenhorst, P.L., Mulder, J.W., Hoetelmans, R.M., Koks, C.H. and Beijnen, J.H. (1995) Study on didanosine concentrations in cerebrospinal fluid. Implications for the treatment and prevention of AIDS dementia complex, *Pharm World Sci* 17: 218–221.

Chen, H., Wood, C. and Petito, C.K. (2000) Comparisons of HIV-1 viral sequences in brain, choroid plexus and spleen: potential role of choroid plexus in the pathogenesis of HIV encephalitis, *J Neurovirol* 6: 498–506.

Choo, E.F., Leake, B., Wandel, C., Imamura, H., Wood, A.J., Wilkinson, G.R. and Kim, R.B. (2000) Pharmacological inhibition of P-glycoprotein transport enhances the distribution of HIV-1 protease inhibitors into brain and testes, *Drug Metab Dispos* 28: 655–660.

Choudhuri, S., Cherrington, N.J., Li, N. and Klaassen, C.D. (2003) Constitutive expression of various xenobiotic and endobiotic transporter mRNAs in the choroid plexus of rats, *Drug Metab Dispos* 31: 1337–1345.

Chun, T.W. and Fauci, A.S. (1999) Latent reservoirs of HIV: obstacles to the eradication of virus, *Proc Natl Acad Sci USA* 96: 10958–10961.

Collins, J.M., Klecker, R.W., Jr., Kelley, J.A., Roth, J.S., McCully, C.L., Balis, F.M. and Poplack, D.G. (1988) Pyrimidine dideoxyribonucleosides: selectivity of penetration into cerebrospinal fluid, *J Pharmacol Exp Ther* 245: 466–470.

Cordon-Cardo, C., O'Brien, J.P., Casals, D., Rittman-Grauer, L., Biedler, J.L., Melamed, M.R. and Bertino, J.R. (1989) Multidrug-resistance gene (P-glycoprotein) is expressed by endothelial cells at blood-brain barrier sites, *Proc Natl Acad Sci USA* 86: 695–698.

Crowe, S.M., Mills, J., Kirihara, J., Boothman, J., Marshall, J.A. and McGrath, M.S. (1990) Full-length recombinant CD4 and recombinant gp120 inhibit fusion between HIV infected macrophages and uninfected CD4-expressing T-lymphoblastoid cells, *AIDS Res Hum Retroviruses* 6: 1031–1037.

Cunningham, P.H., Smith, D.G., Satchell, C., Cooper, D.A. and Brew, B. (2000) Evidence for independent development of resistance to HIV-1 reverse transcriptase inhibitors in the cerebrospinal fluid, *AIDS* 14: 1949–1954.

Curtis, H., Sabin, C.A. and Johnson, M.A. (2003) Findings from the first national clinical audit of treatment for people with HIV, *HIV Med* 4: 11–17.

Davis, L.E., Hjelle, B.L., Miller, V.E., Palmer, D.L., Llewellyn, A.L., Merlin, T.L., Young, S.A., Mills, R.G., Wachsman, W. and Wiley, C.A. (1992) Early viral brain invasion in iatrogenic human immunodeficiency virus infection, *Neurology* 42: 1736–1739.

Demeule, M., Regina, A., Jodoin, J., Laplante, A., Dagenais, C., Berthelet, F., Moghrabi, A. and Beliveau, R. (2002) Drug transport to the brain: key roles for the efflux pump P-glycoprotein in the blood-brain barrier, *Vasc Pharmacol* 38: 339–348.

Diederich, N., Ackermann, R., Jurgens, R., Ortseifen, M., Thun, F., Schneider, M. and Vukadinovic, I. (1988) Early involvement of the nervous system by human immune deficiency virus (HIV). A study of 79 patients, *Eur Neurol* 28: 93–103.

Dykstra, K.H., Arya, A., Arriola, D.M., Bungay, P.M., Morrison, P.F. and Dedrick, R.L. (1993) Microdialysis study of zidovudine (AZT) transport in rat brain, *J Pharmacol Exp Ther* 267: 1227–1236.

Enzensberger, W. and von Giesen, H.J. (1999) Antiretroviral therapy (ART) from a neurological point of view. German Neuro-AIDS study group (DNAA), *Eur J Med Res* 4: 456–462.

Falangola, M.F., Hanly, A., Galvao-Castro, B. and Petito, C.K. (1995) HIV infection of human choroid plexus: a possible mechanism of viral entry into the CNS, *J Neuropathol Exp Neurol* 54: 497–503.

Finzi, D., Blankson, J., Siliciano, J.D., Margolick, J.B., Chadwick, K., Pierson, T., Smith, K., Lisziewicz, J., Lori, F., Flexner, C., Quinn, T.C., Chaisson, R.E., Rosenberg, E., Walker, B., Gange, S., Gallant, J. and Siliciano, R.F. (1999) Latent infection of $CD4^+$ T cells provides a mechanism for lifelong persistence of HIV-1, even in patients on effective combination therapy, *Nat Med* 5: 512–517.

Fischer-Smith, T., Croul, S., Sverstiuk, A.E., Capini, C., L'Heureux, D., Regulier, E.G., Richardson, M.W., Amini, S., Morgello, S., Khalili, K. and Rappaport, J. (2001) CNS invasion by $CD14^+/CD16^+$ peripheral blood-derived monocytes in HIV dementia: perivascular accumulation and reservoir of HIV infection, *J Neurovirol* 7: 528–541.

Foudraine, N.A., Hoetelmans, R.M., Lange, J.M., de Wolf, F., van Benthem, B.H., Maas, J.J., Keet, I.P. and Portegies, P. (1998) Cerebrospinal-fluid HIV-1 RNA and drug concentrations after treatment with lamivudine plus zidovudine or stavudine, *Lancet* 351: 1547–1551.

Galinsky, R.E., Flaharty, K.K., Hoesterey, B.L. and Anderson, B.D. (1991) Probenecid enhances central nervous system uptake of 2',3'-dideoxyinosine by inhibiting cerebrospinal fluid efflux, *J Pharmacol Exp Ther* 257: 972–978.

Galinsky, R.E., Hoesterey, B.L. and Anderson, B.D. (1990) Brain and cerebrospinal fluid uptake of zidovudine (AZT) in rats after intravenous injection, *Life Sci* 47: 781–788.

Ghersi-Egea, J.F., Finnegan, W., Chen, J.L. and Fenstermacher, J.D. (1996) Rapid distribution of intraventricularly administered sucrose into cerebrospinal fluid cisterns via subarachnoid velae in rat, *Neuroscience* 75: 1271–1288.

Gibbs, J.E. (2003) Transport of anti-HIV nucleoside analogues across the blood-brain and blood-CSF barriers., Ph.D. thesis. University of London.

Gibbs, J.E., Jayabalan, P. and Thomas, S.A. (2003a) Mechanisms by which 2',3'-dideoxyinosine (ddI) crosses the guinea-pig CNS barriers; relevance to HIV therapy, *J Neurochem* 84: 725–734.

Gibbs, J.E., Rasheed, T. and Thomas, S.A. (2003b) Effect of transport inhibitors and additional anti-HIV drugs on the movement of lamivudine (3TC), across the guinea-pig brain-barriers., *J Pharm Exp Ther* 306: 1035–1041.

Gibbs, J.E. and Thomas, S.A. (2002) The distribution of the anti-HIV drug, 2'3'-dideoxycytidine (ddC), across the blood-brain and blood-cerebrospinal fluid barriers and the influence of organic anion transport inhibitors, *J Neurochem* 80: 392–404.

Gisolf, E.H., Enting, R.H., Jurriaans, S., de Wolf, F., van der Ende, M.E., Hoetelmans, R.M., Portegies, P. and Danner, S.A. (2000) Cerebrospinal fluid HIV-1 RNA during treatment with ritonavir/saquinavir or ritonavir/saquinavir/stavudine, *AIDS* 14: 1583–1589.

Gisslen, M., Norkrans, G., Svennerholm, B. and Hagberg, L. (1997) The effect on human immunodeficiency virus type 1 RNA levels in cerebrospinal fluid after initiation of zidovudine or didanosine, *J Infect Dis* 175: 434–437.

Groothuis, D.R. and Levy, R.M. (1997) The entry of antiviral and antiretroviral drugs into the central nervous system, *J Neurovirol* 3: 387–400.

Gulick, R.M. (2003) New antiretroviral drugs, *Clin Microbiol Infect* 9: 186–193.

Gunthard, H.F., Havlir, D.V., Fiscus, S., Zhang, Z.Q., Eron, J., Mellors, J., Gulick, R., Frost, S.D., Brown, A.J., Schleif, W., Valentine, F., Jonas, L., Meibohm, A., Ignacio, C.C., Isaacs, R., Gamagami, R., Emini, E., Haase, A., Richman, D.D. and Wong, J.K. (2001) Residual human immunodeficiency virus (HIV) type 1 RNA and DNA in lymph nodes and HIV RNA in genital secretions and in cerebrospinal fluid after suppression of viremia for 2 years, *J Infect Dis* 183: 1318–1327.

Harouse, J.M., Wroblewska, Z., Laughlin, M.A., Hickey, W.F., Schonwetter, B.S. and Gonzalez-Scarano, F. (1989) Human choroid plexus cells can be latently infected with human immunodeficiency virus, *Ann Neurol* 25: 406–411.

Hartman, N.R., Yarchoan, R., Pluda, J.M., Thomas, R.V., Marczyk, K.S., Broder, S. and Johns, D.G. (1990) Pharmacokinetics of 2',3'-dideoxyadenosine and 2',3'-dideoxyinosine in patients with severe human immunodeficiency virus infection, *Clin Pharmacol Ther* 47: 647–654.

Haworth, S.J., Christofalo, B., Anderson, R.D. and Dunkle, L.M. (1998) A single-dose study to assess the penetration of stavudine into human cerebrospinal fluid in adults, *J AIDS Hum Retrovirol* 17: 235–238.

Hoesterey, B.L., Galinsky, R.E. and Anderson, B.D. (1991) Dose dependence in the plasma pharmacokinetics and uptake kinetics of 2',3'-dideoxyinosine into brain and cerebrospinal fluid of rats, *Drug Metab Dispos* 19: 907–912.

Huisman, M.T., Smit, J.W., Crommentuyn, K.M., Zelcer, N., Wiltshire, H.R., Beijnen, J.H. and Schinkel, A.H. (2002) Multidrug resistance protein 2 (MRP2) transports HIV protease inhibitors, and transport can be enhanced by other drugs, *AIDS* 16: 2295–2301.

Huisman, M.T., Smit, J.W. and Schinkel, A.H. (2000) Significance of P-glycoprotein for the pharmacology and clinical use of HIV protease inhibitors, *AIDS* 14: 237–242.

Huisman, M.T., Smit, J.W., Wiltshire, H.R., Beijnen, J.H. and Schinkel, A.H. (2003) Assessing safety and efficacy of directed P-glycoprotein inhibition to improve the pharmacokinetic properties of saquinavir coadministered with ritonavir, *J Pharmacol Exp Ther* 304: 596–602.

Huisman, M.T., Smit, J.W., Wiltshire, H.R., Hoetelmans, R.M., Beijnen, J.H. and Schinkel, A.H. (2001) P-glycoprotein limits oral availability, brain, and fetal penetration of saquinavir even with high doses of ritonavir, *Mol Pharmacol* 59: 806–813.

Jette, L., Pouliot, J.F., Murphy, G.F. and Beliveau, R. (1995) Isoform I (mdr3) is the major form of P-glycoprotein expressed in mouse brain capillaries. Evidence for cross-reactivity of antibody C219 with an unrelated protein, *Biochem J* 305 (Pt 3): 761–766.

Jette, L., Tetu, B. and Beliveau, R. (1993) High levels of P-glycoprotein detected in isolated brain capillaries, *Biochim Biophys Acta* 1150: 147–154.

Kim, R.B., Fromm, M.F., Wandel, C., Leake, B., Wood, A.J., Roden, D.M. and Wilkinson, G.R. (1998) The drug transporter P-glycoprotein limits oral absorption and brain entry of HIV-1 protease inhibitors, *J Clin Invest* 101: 289–294.

Kravcik, S., Gallicano, K., Roth, V., Cassol, S., Hawley-Foss, N., Badley, A. and Cameron, D.W. (1999) Cerebrospinal fluid HIV RNA and drug levels with combination ritonavir and saquinavir, *J AIDS* 21: 371–375.

Limoges, J., Persidsky, Y., Poluektova, L., Rasmussen, J., Ratanasuwan, W., Zelivyanskaya, M., McClernon, D.R., Lanier, E.R. and Gendelman, H.E. (2000) Evaluation of anti-retroviral drug efficacy for HIV-1 encephalitis in SCID mice, *Neurology* 54: 379–389.

Limoges, J., Poluektova, L., Ratanasuwan, W., Rasmussen, J., Zelivyanskaya, M., McClernon, D.R., Lanier, E.R., Gendelman, H.E. and Persidsky, Y. (2001) The efficacy of potent anti-retroviral drug combinations tested in a murine model of HIV-1 encephalitis, *Virology* 281: 21–34.

Masereeuw, R., Jaehde, U., Langemeijer, M.W., de Boer, A.G. and Breimer, D.D. (1994) In vitro and *in vivo* transport of zidovudine (AZT) across the blood–brain barrier and the effect of transport inhibitors, *Pharm Res* 11: 324–330.

McArthur, J.C., McClernon, D.R., Cronin, M.F., Nance-Sproson, T.E., Saah, A.J., St Clair, M. and Lanier, E.R. (1997) Relationship between human immunodeficiency virus-associated dementia and viral load in cerebrospinal fluid and brain, *Ann Neurol* 42: 689–698.

McDowell, J.A., Chittick, G.E., Ravitch, J.R., Polk, R.E., Kerkering, T.M. and Stein, D.S. (1999) Pharmacokinetics of [(14)C]abacavir, a human immunodeficiency virus type 1 (HIV-1) reverse transcriptase inhibitor, administered in a single oral dose to HIV-1-infected adults: a mass balance study, *Antimicrob Agents Chemother* 43: 2855–2861.

Meaden, E.R., Hoggard, P.G., Maher, B., Khoo, S.H. and Back, D.J. (2001) Expression of P-glycoprotein and multidrug resistance-associated protein in healthy volunteers and HIV-infected patients, *AIDS Res Hum Retroviruses* 17: 1329–1332.

Miller, D.S., Nobmann, S.N., Gutmann, H., Toeroek, M., Drewe, J. and Fricker, G. (2000) Xenobiotic transport across isolated brain microvessels studied by confocal microscopy, *Mol Pharmacol* 58: 1357–1367.

Nagata, Y., Kusuhara, H., Endou, H. and Sugiyama, Y. (2002) Expression and functional characterization of rat organic anion transporter 3 (rOat3) in the choroid plexus, *Mol Pharmacol* 61: 982–988.

Petito, C.K. and Cash, K.S. (1992) Blood-brain barrier abnormalities in the acquired immunodeficiency syndrome: immunohistochemical localization of serum proteins in post-mortem brain, *Ann Neurol* 32: 658–666.

Petito, C.K., Chen, H., Mastri, A.R., Torres-Munoz, J., Roberts, B. and Wood, C. (1999) HIV infection of choroid plexus in AIDS and asymptomatic HIV-infected patients suggests that the choroid plexus may be a reservoir of productive infection, *J Neurovirol* 5: 670–677.

Polli, J.W., Jarrett, J.L., Studenberg, S.D., Humphreys, J.E., Dennis, S.W., Brouwer, K.R. and Woolley, J.L. (1999) Role of P-glycoprotein on the CNS disposition of amprenavir (141W94), an HIV protease inhibitor, *Pharm Res* 16: 1206–1212.

Pritchard, J.B., Sweet, D.H., Miller, D.S. and Walden, R. (1999) Mechanism of organic anion transport across the apical membrane of choroid plexus, *J Biol Chem* 274: 33382–33387.

Rao, V.V., Dahlheimer, J.L., Bardgett, M.E., Snyder, A.Z., Finch, R.A., Sartorelli, A.C. and Piwnica-Worms, D. (1999) Choroid plexus epithelial expression of MDR1 P glycoprotein and multidrug resistance-associated protein contribute to the blood–cerebrospinal fluid drug-permeability barrier, *Proc Natl Acad Sci USA* 96: 3900–3905.

Rennels, M.L., Gregory, T.F., Blaumanis, O.R., Fujimoto, K. and Grady, P.A. (1985) Evidence for a 'paravascular' fluid circulation in the mammalian central nervous system, provided by the rapid distribution of tracer protein throughout the brain from the subarachnoid space, *Brain Res* 326: 47–63.

Rolinski, B., Bogner, J.R., Sadri, I., Wintergerst, U. and Goebel, F.D. (1997) Absorption and elimination kinetics of zidovudine in the cerebrospinal fluid in HIV-1-infected patients, *J AIDS Hum Retrovirol* 15: 192–197.

Sabin, C.A. (2002) The changing clinical epidemiology of AIDS in the highly active antiretroviral therapy era, *AIDS* 16 (Suppl) 4: S61–S68.

Sadler, B.M., Chittick, G.E., Polk, R.E., Slain, D., Kerkering, T.M., Studenberg, S.D., Lou, Y., Moore, K.H., Woolley, J.L. and Stein, D.S. (2001) Metabolic disposition and pharmacokinetics of [14C]-amprenavir, a human immunodeficiency virus type 1 (HIV-1) protease inhibitor, administered as a single oral dose to healthy male subjects, *J Clin Pharmacol* 41: 386–396.

Sawchuk, R.J. and Hedaya, M.A. (1990) Modeling the enhanced uptake of zidovudine (AZT) into cerebrospinal fluid. 1. Effect of probenecid, *Pharm Res* 7: 332–338.

Sawchuk, R.J. and Yang, Z. (1999) Investigation of distribution, transport and uptake of anti-HIV drugs to the central nervous system, *Adv Drug Deliv Rev* 39: 5–31.

Schinkel, A.H. (1999) P-Glycoprotein, a gatekeeper in the blood-brain barrier, *Adv Drug Deliv Rev* 36: 179–194.

Schrager, L.K. and D'Souza, M.P. (1998) Cellular and anatomical reservoirs of HIV-1 in patients receiving potent antiretroviral combination therapy, *JAMA* 280: 67–71.

Sekine, T., Watanabe, N., Hosoyamada, M., Kanai, Y. and Endou, H. (1997) Expression cloning and characterization of a novel multispecific organic anion transporter, *J Biol Chem* 272: 18526–18529.

Shafer, R.W. and Vuitton, D.A. (1999) Highly active antiretroviral therapy (HAART) for the treatment of infection with human immunodeficiency virus type 1, *Biomed Pharmacother* 53: 73–86.

Siliciano, J.D. and Siliciano, R.F. (2000) Latency and viral persistence in HIV-1 infection, *J Clin Invest* 106: 823–825.

Smith, B.A., Gartner, S., Liu, Y., Perelson, A.S., Stilianakis, N.I., Keele, B.F., Kerkering, T.M., Ferreira-Gonzalez, A., Szakal, A.K., Tew, J.G. and Burton, G.F. (2001) Persistence of infectious HIV on follicular dendritic cells, *J Immunol* 166: 690–696.

Solas, C., Lafeuillade, A., Halfon, P., Chadapaud, S., Hittinger, G. and Lacarelle, B. (2003) Discrepancies between protease inhibitor concentrations and viral load in reservoirs and sanctuary sites in human immunodeficiency virus-infected patients, *Antimicrob Agents Chemother* 47: 238–243.

Srinivas, R.V., Middlemas, D., Flynn, P. and Fridland, A. (1998) Human immunodeficiency virus protease inhibitors serve as substrates for multidrug transporter proteins MDR1 and MRP1 but retain antiviral efficacy in cell lines expressing these transporters, *Antimicrob Agents Chemother* 42: 3157–3162.

Stingele, K., Haas, J., Zimmermann, T., Stingele, R., Hubsch-Muller, C., Freitag, M., Storch-Hagenlocher, B., Hartmann, M. and Wildemann, B. (2001) Independent HIV replication in paired CSF and blood viral isolates during antiretroviral therapy, *Neurology* 56: 355–361.

Strazielle, N., Belin, M.-F. and Gersi-Egea, J.-F. (2003) Choroid plexus controls brain availability of anti-HIV nucleoside analogs via pharmacologically inhibitable organic anion transporters, *AIDS* 17: 1473–1485.

Sweet, D.H., Wolff, N.A. and Pritchard, J.B. (1997) Expression cloning and characterization of ROAT1. The basolateral organic anion transporter in rat kidney, *J Biol Chem* 272: 30088–30095.

Takahashi, K., Wesselingh, S.L., Griffin, D.E., McArthur, J.C., Johnson, R.T. and Glass, J.D. (1996) Localization of HIV-1 in human brain using polymerase chain reaction/in situ hybridization and immunocytochemistry, *Ann Neurol* 39: 705–711.

Takasawa, K., Suzuki, H. and Sugiyama, Y. (1997a) Transport properties of 3'-azido-3'-deoxythymidine and 2',3'-dideoxyinosine in the rat choroid plexus, *Biopharm Drug Dispos* 18: 611–622.

Takasawa, K., Terasaki, T., Suzuki, H., Ooie, T. and Sugiyama, Y. (1997b) Distributed model analysis of 3'-azido-3'-deoxythymidine and 2',3'-dideoxyinosine distribution in brain tissue and cerebrospinal fluid, *J Pharmacol Exp Ther* 282: 1509–1517.

Takasawa, K., Terasaki, T., Suzuki, H. and Sugiyama, Y. (1997c) In vivo evidence for carrier-mediated efflux transport of 3'-azido-3'-deoxythymidine and 2',3'-dideoxyinosine across the blood-brain barrier via a probenecid-sensitive transport system, *J Pharmacol Exp Ther* 281: 369–375.

Tashima, K.T., Caliendo, A.M., Ahmad, M., Gormley, J.M., Fiske, W.D., Brennan, J.M. and Flanigan, T.P. (1999) Cerebrospinal fluid human immunodeficiency virus type 1 (HIV-1) suppression and efavirenz drug concentrations in HIV-1-infected patients receiving combination therapy, *J Infect Dis* 180: 862–864.

Taylor, E.M. (2002) The impact of efflux transporters in the brain on the development of drugs for CNS disorders, *Clin Pharmacokinet* 41: 81–92.

Thomas, S.A. and Segal, M.B. (1996) Identification of a saturable uptake system for deoxyribonucleosides at the blood-brain and blood-cerebrospinal fluid barriers, *Brain Res* 741: 230–239.

Thomas, S.A. (2004) Anti-HIV drug distribution to the central nervous system., *Curr Pharm Des* 10:1313–1324.

Thomas, S.A., Bye, A. and Segal, M.B. (2001) Transport characteristics of the anti-human immunodeficiency virus nucleoside analog, abacavir, into brain and cerebrospinal fluid, *J Pharmacol Exp Ther* 298: 947–953.

Thomas, S.A. and Segal, M.B. (1997) The passage of azidodeoxythymidine into and within the central nervous system: does it follow the parent compound, thymidine?, *J Pharmacol Exp Ther* 281: 1211–1218.

Thomas, S.A. and Segal, M.B. (1998) The transport of the anti-HIV drug, 2',3'-didehydro-3'-deoxythymidine (D4T), across the blood-brain and blood-cerebrospinal fluid barriers, *Br J Pharmacol* 125: 49–54.

Trillo-Pazos, G., Diamanturos, A., Rislove, L., Menza, T., Chao, W., Belem, P., Sadiq, S., Morgello, S., Sharer, L. and Volsky, D.J. (2003) Detection of HIV-1 DNA in microglia/macrophages, astrocytes and neurons isolated from brain tissue with HIV-1 encephalitis by laser capture microdissection, *Brain Pathol* 13: 144–154.

Tuntland, T., Ravasco, R.J., al Habet, S. and Unadkat, J.D. (1994) Efflux of zidovudine and 2',3'-dideoxyinosine out of the cerebrospinal fluid when administered alone and in combination to *Macaca nemestrina*, *Pharm Res* 11: 312–317.

Van der Sandt, I., Vos, C.M., Nabulsi, L., Blom-Roosemalen, M.C., Voorwinden, H.H., de Boer, A.G. and Breimer, D.D. (2001) Assessment of active transport of HIV protease inhibitors in various cell lines and the *in vitro* blood–brain barrier, *AIDS* 15: 483–491.

Venturi, G., Catucci, M., Romano, L., Corsi, P., Leoncini, F., Valensin, P.E. and Zazzi, M. (2000) Antiretroviral resistance mutations in human immunodeficiency virus type 1 reverse transcriptase and protease from paired cerebrospinal fluid and plasma samples, *J Infect Dis* 181: 740–745.

von Giesen, H.J., Koller, H., Theisen, A. and Arendt, G. (2002) Therapeutic effects of nonnucleoside reverse transcriptase inhibitors on the central nervous system in HIV-1-infected patients, *J AIDS* 29: 363–367.

Wada, S., Tsuda, M., Sekine, T., Cha, S.H., Kimura, M., Kanai, Y. and Endou, H. (2000) Rat multispecific organic anion transporter 1 (rOAT1) transports zidovudine, acyclovir, and other antiviral nucleoside analogs, *J Pharmacol Exp Ther* 294: 844–849.

Warren, K.E., Patel, M.C., McCully, C.M., Montuenga, L.M. and Balis, F.M. (2000) Effect of P-glycoprotein modulation with cyclosporin A on cerebrospinal fluid penetration of doxorubicin in non-human primates, *Cancer Chemother Pharmacol* 45: 207–212.

Wiley, C.A., Schrier, R.D., Nelson, J.A., Lampert, P.W. and Oldstone, M.B. (1986) Cellular localization of human immunodeficiency virus infection within the brains of acquired immune deficiency syndrome patients, *Proc Natl Acad Sci USA* 83: 7089–7093.

Williams, K., Alvarez, X. and Lackner, A.A. (2001) Central nervous system perivascular cells are immunoregulatory cells that connect the CNS with the peripheral immune system, *Glia* 36: 156–164.

Williams, S.A., Abbruscato, T.J., Hruby, V.J. and Davis, T.P. (1996) Passage of a delta-opioid receptor selective enkephalin, [D-penicillamine2,5] enkephalin, across the blood-brain and the blood-cerebrospinal fluid barriers, *J Neurochem* 66: 1289–1299.

Wong, J.K., Hezareh, M., Gunthard, H.F., Havlir, D.V., Ignacio, C.C., Spina, C.A. and Richman, D.D. (1997) Recovery of replication-competent HIV despite prolonged suppression of plasma viremia, *Science* 278: 1291–1295.

Yarchoan, R., Perno, C.F., Thomas, R.V., Klecker, R.W., Allain, J.P., Wills, R.J., McAtee, N., Fischl, M.A., Dubinsky, R., McNeely, M.C. et al. (1988) Phase I studies of 2',3'-dideoxycytidine in severe human immunodeficiency virus infection as a single agent and alternating with zidovudine (AZT), *Lancet* 1: 76–81.

Zhou, X.J., Havlir, D.V., Richman, D.D., Acosta, E.P., Hirsch, M., Collier, A.C., Tebas, P. and Sommadossi, J.P. (2000) Plasma population pharmacokinetics and penetration into cerebrospinal fluid of indinavir in combination with zidovudine and lamivudine in HIV-1-infected patients, *AIDS* 14: 2869–2876.

Zlokovic, B.V., Begley, D.J., Djuricic, B.M. and Mitrovic, D.M. (1986) Measurement of solute transport across the blood-brain barrier in the perfused guinea pig brain: method and application to N-methyl-alpha-aminoisobutyric acid, *J Neurochem* 46: 1444–1451.

17 Blood–CSF Barrier in Iron Regulation and Manganese-Induced Parkinsonism

Wei Zheng
Purdue University
West Lafayette, Indiana, U.S.A.

CONTENTS

17.1 INTRODUCTION

The role of iron (Fe) in the etiology of several neurodegenerative diseases has been recognized (Connor, 1997). Particularly regarding idiopathic Parkinson's disease (IPD), cumulative clinical and laboratory evidence has suggested that malfunction of brain iron may initiate free radical reactions, diminish mitochondrial energy production, and provoke oxidative cytotoxicity (Beard and Connor, 2003). The participation of iron in neuronal cell death is especially intriguing, in that iron acquisition and regulation in the brain are highly conservative and therefore vulnerable to interference by other metals bearing similar chemical reactivity. In addition to cellular iron regulatory mechanisms, the blood–brain barrier and blood–cerebrospinal fluid (CSF) barrier in the choroid plexus partake in the regulation of brain iron homeostasis by a restricted transport of iron via a number of iron-binding/transport proteins such as transferrin, transferrin receptor (TfR), divalent metal transporter-1 (DMT1), melanotransferrin (p97), and lactoferrin (Jefferies et al., 1996; Berg et al., 2001; Ke and Qian 2003). Efflux en route to the choroid plexus has also recently been proposed (Deane et al., 2004).

Exposure to manganese causes syndromes similar to Parkinsonism (Crossgrove and Zheng, 2004). Recent interest in manganese neurotoxicity stems from increased use of unleaded gasoline using an antiknock agent, methylcyclopentadienyl Mn tricarbonyl (MMT). MMT has been associated with increased health problems in heavily air polluted areas (Cooper, 1984; Abbott, 1987; Loranger and Zayed, 1995; Sierra et al., 1995). The neurodegenerative damage caused by manganese is usually irreversible and permanent. Similar to iron, brain homeostasis of manganese is regulated by both barriers. The influx of manganese at the blood–brain barrier appears to be associated with one or more transporter proteins. A common theory is that manganese binds to plasma transferrin. This complex is then transported into brain via a TfR-mediated mechanism at the blood–brain barrier. Noticeably, this process competes directly with iron transport, or vice versa (Goddard et al., 1997; Aschner et al., 1999; Conrad et al., 2000; Crossgrove and Yokel, 2004). Some investigators have also suggested that the divalent metal transporter-1 (DMT1) may be involved in manganese influx into the brain (Goddard et al., 1997; Gunshin et al., 1997; Conrad et al., 2000); however, a more recent study failed to reveal a significant impact of the lack of functional DMT1 in knockout rats on brain influx of manganese ion or manganese-transferrin complex (Crossgrove and Yokel, 2004). Moreover, some investigators (Moos and Morgan, 2004) have shown that DMT1 may not even exist in brain capillary endothelial cells, which argues against the involvement of DMT1 in manganese transport into the brain. In addition, evidence has suggested that manganese may cross the cell membranes via voltage-gated calcium (Ca) channels, the Na/Ca exchanger, the Na/Mg antiporter, or mitochondrial active Ca uniporter (Takeda, 2003).

This chapter deals with the current understanding of the role of brain barriers, particularly the choroid plexus, in management of iron and manganese in the central nervous system (CNS). For each metal, the primary utilities of the metal in the human body will be briefly reviewed, followed by a discussion of the biochemical mechanisms underlying metal transport by the blood–brain and blood–CSF barriers.

Inasmuch as the chemical/biochemical properties of manganese are very similar to iron, the interaction of these two metals at the molecular, cellular, and systemic levels and the relevance of these interactions to consequential neurodegenerative disorders will be discussed. Finally, future research needed to reveal the molecular mechanisms of metal transport at the brain barriers is discussed. For the role of the choroid plexus in neurotoxicities induced by other metals, readers are referred to other reviews (Zheng, 1996, 2001; Zheng et al., 2003).

17.2 IRON AND MANGANESE IN HEALTH AND DISEASE

17.2.1 Essential Functions of Iron and Its Homeostasis

As a transition metal, iron exists in a variety of oxidation states from Fe(−II) to Fe(VI). The multiple oxidation states enable iron to readily participate in oxidation-reduction reactions and therefore to carry out a wide range of biological functions. Iron functions mainly to transport oxygen in hemoglobin and myoglobin. Iron is also required for the normal function of cytochromes, cytochrome oxidase, peroxidase, and catalase. In the brain, not only is iron required for DNA synthesis and mitochondrial respiration, it also participates in the biosynthesis of neurotransmitters, axonal growth, and receptor-mediated postsynaptic signal transduction. Iron-sulfur proteins in mitochondrial respiration chain deliver the electrons generated from the tricarboxylic acid cycle to ADP molecules for energy production, thereby complying with the high demand of energy consumption of neuronal cells.

The total quantity of iron in the body averages 4 to 5 g, 65% of which is in the form of hemoglobin and 4% in the form of myoglobin. Another 15 to 30% is stored in the reticuloendothelial system and liver parenchymal cells, primarily in the form of ferritin. The metabolism and systemic balance of iron in the body are precisely regulated by several macromolecules. Once iron enters the blood circulation from the small intestine, iron binds to a β-globulin, apotransferrin, to form transferrin. Transferrin carries iron in the blood and serves as the major vehicle for iron transport in the body. Upon arriving at the target cells, transferrin binds strongly with transferrin receptors (TfR) on the outer surface of the cell membrane. By endocytosis, transferrin carries iron to the cytoplasm and delivers iron directly to the mitochondria for heme synthesis. In the cytoplasm, the released free iron ions are either utilized in metabolic processes or conjugated with the large molecular weight protein apoferritin (460,000 daltons) to form ferritin. The latter acts as a storage vehicle for iron.

17.2.2 Role of Iron in Parkinsonism and Mechanism of Iron Cytotoxicity

Abnormal iron homeostasis, both systemically and subcellularly, is believed to be associated with the etiology of idiopathic Parkinson's disease (IPD) and chemical-induced parkinsonism (Connor and Benkovic, 1992; Jenner et al., 1992; Youdim et al., 1993; Logroscino et al., 1997). In the substantia nigra of IPD patients, high levels of total iron, decreased ferritin, oxidative stress, and deficiency in mitochondrial functions have been observed repeatedly (Dexter et al., 1991; Sofic et al., 1991;

Jenner et al., 1992; Griffiths and Crossman, 1993; Youdim et al., 1993; Mann, 1994; Loeffler et al., 1995; Ye et al., 1996). In a case-control study, serum iron concentrations are significantly altered in IPD patients as compared to controls (Logroscino et al., 1997). Serum ferritin, transferrin, total Fe-binding capacity (TIBC), and % Fe saturation are all reduced in IPD patients. These findings suggest a generalized "defect" in iron systemic metabolism in IPD.

Studies using animal models also reveal a general increase in the concentrations of iron in the substantia nigra following treatment with 1-methyl-4-phenyl-1,2,3,6-tetrahydropyridine (MPTP), 6-hydroxydopamine (6-OHDA), ibotenic acid infusion (to destroy GABAergic neurons), and direct intranigral iron infusion (Sengstock et al., 1992, 1994; Mochizuki et al., 1994; Sastry and Arendash, 1995; He et al., 1996). It is still debatable whether iron overload in the substantia nigra is the cause or consequence of neuronal degeneration. For instance, the increased iron could result directly from the intervention of toxicants on iron regulatory processes, which is followed by neuronal cell death. It is also possible that the increased iron may be a secondary event in the course of cell injury. In the latter case, the initiation of cell impairment may not necessarily be related to iron status. Despite these arguments, cellular iron overload in the substantia nigra may indeed represent the hallmark of the IPD disease process. Accordingly, iron-mediated oxidative stress would lead to further degeneration of nigrostriatal dopamine neurons in IPD patients or experimental animals.

It is still uncertain why and how iron accumulates in the specific area of IPD brain. Conceivably, an uneven distribution of iron transporters at brain barriers may contribute to the specific brain regional accumulation of iron.

About 70 to 80% of iron in the brain is found in the myelin fraction, and histochemically it is localized in oligodendrocytes. The rich iron content renders the oligodendrocytes particularly susceptible to oxidative injury. Loss of myelin not only interferes with neuronal firing, it also leads to increased oxidative stress in neurons. A high cellular iron level is directly toxic to the striatal GABAergic neurons in IPD (Gabrielsson et al., 1986; Palmeira et al., 1993; Sohn and Yoon, 1998; Beard and Connor, 2003).

Biochemically, iron is known to initiate the cascade of free radical reactions. A free radical is a molecule or molecular fragment that contains one or more unpaired electrons in its outer orbital. They are unstable and highly reactive. Free Fe(II) ions can catalyze the Fenton reaction, which generates highly toxic hydroxyl radical (OH) from superoxide radical (O_2^-) and hydrogen peroxide (H_2O_2):

$$\cdot O_2^- + Fe^{3+} \ddagger O_2 + Fe^{2+} \quad (17.1)$$

$$Fe^{2+} + H_2O_2 \ddagger Fe^{3+} + OH^- + OH \quad (17.2)$$

or directly catalyze the decomposition of lipid hydroperoxides:

$$Fe^{2+} + LOOH \ddagger Fe^{3+} + OH^- + LO\cdot \quad (17.3)$$

Free Fe(II) ions and cytotoxic OH radicals promote the oxidative reaction in polypeptides of proteins, nucleotides in DNA structure, or phospholipids in membrane. The ensuing oxidation damage (i.e., oxidative stress) to cell membrane, enzyme activities, and nutrient production leads to the ultimate cell death of dopamine neurons in the substantia nigra of IPD.

Excess iron can interfere with protein phosphorylation by altering protein kinase C (PKC) activity in the progress of cell death. PKC is activated as a consequence of receptor-dependent increases in intracellular [Ca^{2+}] and diacylglycerol (DAG). Activation leads to translocation of the enzyme from the soluble to membrane-associated particulate component of the cell. The enzyme plays a critical role in transduction of cellular signals, regulation of membrane functions, and control of cell proliferation. A study from this laboratory has demonstrated that toxic metal such as lead (Pb) activates PKC and promotes its translocation from cytosol to membrane in cultured choroidal epithelial cells (Zhao et al., 1998). Iron can upregulate the expression of PKC (Alcantara et al., 1994) and facilitate the oxidation of PKC subunits to activate directly the PKC pathway (Taher et al., 1993), which leads to cellular oxidative damage. Noticeably, iron metabolism itself is partly regulated by PKC. PKC can catalyze the phosphorylation of the [Fe-S] cluster in iron regulatory protein 1 (IRP1), thereby regulating its availability, being reversible as an IRE-binding protein or cytosolic aconitase (Eisenstein et al., 1993; Schalinske et al., 1997; Brown et al., 1998). Conceivably, modulation of PKC activity by other metals in the choroid plexus could subsequently affect cellular iron balance.

17.2.3 Essential Functions of Manganese in Human Body

The outer electron shell of manganese can donate up to 7 electrons. Thus, manganese can assume 11 different oxidation states. In living organisms, manganese has been found as Mn(II), Mn(III), and Mn(IV). The ability of manganese to assume various valence states allows it to act as either a prooxidant or an antioxidant in biological matrices.

In humans, food is the main source of manganese, and the usual daily intake ranges from 1 to 5 mg per day. As an essential nutrient, manganese actively participates in bone mineralization, protein and energy metabolism, metabolic regulation, cellular protection from damaging free radical species, and formation of glucosaminoglycans (Wedler, 1994). Manganese can activate numerous enzymes involved with either a catalytic or regulatory function such as transferase, decarboxylases, hydrolases, dehydrogenase, synthetases, and lyases. Mitochondrial superoxide-dismutase, pyruvate carboxylase, and liver arginase are known manganese metalloenzymes. Manganese is also required for many neuronal activities. The regulation of brain manganese depends largely on the blood–brain barrier and blood–CSF barrier. A primary transport form of manganese in blood (in the divalent oxidation state) crosses the blood–brain barrier via specific carriers at a rate far slower than in other tissues (Aschner et al., 1999).

17.2.4 Manganese-Induced Parkinsonism

Occupational exposure to manganese can lead to the symptoms resembling Parkinson's disease (Tepper, 1961; Mena et al., 1967, 1970; Chandra et al., 1981; Barbeau, 1985; Gorell et al., 1997). Among welders exposed to airborne manganese during welding, serum levels of manganese and iron in welders were 4.3-fold and 1.9-fold, respectively, higher than those of controls (Li et al., 2004). Evidence suggests that occupational exposure to welding fumes disturbs the homeostasis of trace elements in systemic circulation and induces oxidative stress. The other common source of manganese is found in the street drug called "Bazooka," a cocaine-based drug contaminated with manganese-carbonate from free-base preparation methods (Ensing, 1985).

Manganese-induced neurologic lesions are located in the globus pallidus and striatum of the basal ganglia. The pallidus and striatum display a marked decrease in myelinated nerve fibers, accompanied by depletion of striatal dopamine (Mena et al., 1967, 1970; Neff et al., 1969; Bonilla, 1980; Yamada et al., 1986; Eriksson et al., 1987; Ingersoll et al., 1995). It is generally accepted that manganese specifically affects the striatum and pallidus, whereas IPD preferentially affects dopaminergic neurons of the substantia nigra (Inoue and Makita, 1996). The question as to whether this rather selective metal targeting in the brain is due to the sensitivity of a particular neuronal pathway to the metal or due to brain region-specific metal transport remains unsolved. It is noted that manganese uptake by the choroid plexus is several orders of magnitude higher than that by other brain regions (Murphy et al., 1991). Readers are referred to a recent review for details on manganese toxicity (Crossgrove and Zheng, 2004).

17.3 REGULATION OF IRON TRANSPORT AT BRAIN BARRIERS

17.3.1 Transferrin Receptor-Mediated Transport

Under normal physiological conditions, the brain stringently regulates iron balance by three well-orchestrated systems: (1) the influx of iron into brain, which is regulated by transferrin receptor-mediated transport at brain barriers; (2) the storage of iron in which the cellular sequestration is largely dependent upon availability of ferritin; and (3) the efflux of iron, whose rate is controlled by bulk CSF flow to the blood circulation (Connor and Benkovic, 1992; Jefferies et al., 1996; Bradbury, 1997). The critical proteins involved in iron metabolism and their functions are listed in Table 17.1. At brain barriers, the iron-transferrin complex is taken up by endocytosis into cerebral capillary endothelia (and possibly into choroidal epithelia), where the molecules subsequently dissociate. A dissociated apotransferrin is recycled to the blood compartment; the released iron crosses the abluminal membrane of the barriers into the cerebral compartment by binding to brain transferrin derived discretely from oligodendrocytes and choroid plexus epithelia. The cerebral transferrin-bound iron thus becomes available for neurons expressing transferrin receptors (Beard and Connor, 2003). Some investigators suggest that iron-transferrin can be

TABLE 17.1
Critical Proteins Involved in Iron Metabolism

Protein	Chromosomal Site of Gene	MW	Structure	Function
Transferrin (Tf)	3q31-qter	79,570	Single-chain glycoprotein with 2 Fe-binding sites	Fe transport in plasma and extracellular fluid
Transferrin receptor (TfR)	3q26.2-qter	185,000	Transmembrane glycoprotein dimer with 2 Tf-binding sites	Receptor-mediated endocytosis of ferric-Tf; recycled
Ferritin	H subunit: 11	440,000	Spherical protein of 24 subunits, binds up to 4500 Fe atoms	Fe storage
	L subunit: 19			
Cytosolic aconitase (IRE-BP or IRP1)	9	90,000	Member of family of [4Fe-4s] cluster proteins; homology with mitochondrial aconitase	Coordinate regulation of translation of ferritin, TfR, and Δ-ALA synthase
Erythroid-specific Δ-aminolevulinic acid synthase	Xp21-q21	59,500	Single-chain peptide of 579 amino acids with presequence of 56 highly basic residues	Catalyzes condensation of glycine and succinyl CoA to form Δ-ALA; regulate heme synthesis

transported into the brain by means of transcytosis through the blood–brain barrier (Moos and Morgan, 2000).

In a study conducted in this laboratory, a brain perfusion technique was used to investigate the role of transferrin in iron (^{59}Fe) transport into cerebral capillaries, various brain regions, choroid plexus, and CSF (Deane et al., 2004). The brain capillaries were enriched by a capillary depletion technique, which separates the brain parenchyma from capillary fraction by a Dextran-formed gradient. When either transferrin-bound or free ^{59}Fe was infused into rat brain, the amounts of ^{59}Fe uptake by the brain capillary far exceeded those into brain parenchyma, with the brain tissue being only 1% of that in the capillaries. Evidently, only a very small fraction of blood iron, no matter if it is bound or free, is capable of entering brain tissues by crossing the endothelial barrier.

There is a regional difference in brain iron uptake for both free and transferrin-bound iron species (see Figure 17.1). In the presence of ^{59}Fe-transferrin, the order of uptake rate is striatum > hippocampus > cerebellum > frontal cortex > brain stem. This order appears to be in agreement with the regional difference in capillary distribution of TfR, as the maximum binding (B_{max}) and the affinity constant (K_D) of

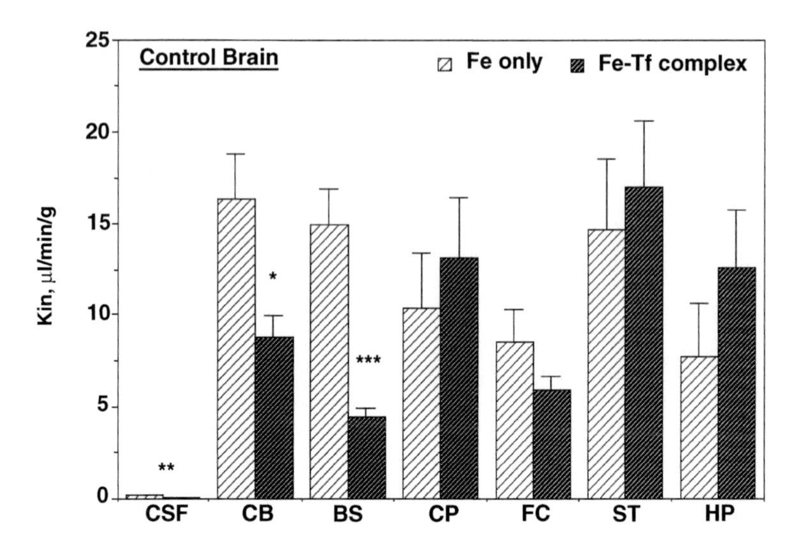

FIGURE 17.1 Influx of ^{59}Fe to various brain regions in the absence or presence of transferrin. Brains of control rats were infused with a Ringer solution containing ^{59}Fe only or ^{59}Fe-transferrin complex for 15 minutes. Cerebral capillaries were removed by capillary depletion method. Data represent mean + SEM; n = 3–4. CSF: cerebrospinal fluid, CB: cerebellum, BS: brain stem, CP: choroid plexus, FC: frontal cortex, ST: striatum, HP: hippocampus. TfR-mediated transport was higher in ST, HP, and CP than in other regions, whereas non–TfR-mediated transport was higher in CB, ST, and BS.

TfR are higher in the capillaries of hippocampus and striatum than in the capillaries of frontal cortex. Thus, a higher transferrin-dependent iron uptake into striatum and hippocampus may reflect the greater density of TfR in these regions.

Similar observations were obtained in the choroid plexus. The kinetics of iron uptake into the choroid plexus did not change whether the iron in the perfusate was bound or free; the rate of uptake by the choroid plexus was much faster than the rate into the CSF. There was a significant difference, however, between bound and free iron species in iron influx to the CSF. The binding of iron to transferrin reduced the influx of iron into the CSF nearly 2-fold in comparison to the results from a free iron experiment. It seems likely that the transport of iron via the blood–CSF barrier may be facilitated by a nontransferrin-dependent mechanism.

Overall, from both iron-only and iron-transferrin studies, the rate of iron uptake shows the following order, brain capillaries > choroid plexus > brain parenchyma > CSF. Therefore, it is reasonable to postulate that the mechanisms at the brain barriers for transport of transferrin-bound iron and for delivery of free iron appear to play an equally important role in transport of blood iron to the CNS. This data also suggest that the blood–brain barrier and blood–CSF barrier serve as the major iron regulatory sites to limit blood iron to entry into brain, no matter how iron concentrations in the blood may vary. There appeared to be a gradient for iron to be transported, by bulk flow, from the interstitial fluid of brain tissues to the CSF.

The choroid plexus and oligodendrocytes are the only two cell types in brain capable of producing transferrin. It has long been postulated that the CSF transferrin may function to carry iron in the CSF to extracellular fluids between neurons and glial cells, serving as an alternative iron transport mechanism in brain. However, Moos and Morgan (1998b) reported that after injecting radiolabeled transferrin directly into rat lateral cerebral ventricle, transferrin in the CSF was rapidly washed and incapable of penetrating further into brain parenchyma. While the authors questioned the importance of brain-derived transferrin in the transport of CSF iron in adult animals, they suggested that CSF transferrin may serve such a function in young animals.

17.3.2 Nontransferrin-Mediated Transports

Nontransferrin-mediated iron transport to neurons and glial cells has been proposed by other investigators (Ueda et al., 1993; Bradbury, 1997; Moos and Morgan, 1998a; Burdo et al., 2001). DMT1, also called Nramp2 (natural resistance-associated macrophage protein), was originally cloned in 1997 by two different strategies and found to play a role in the transport of Fe(II) through biological membranes (Fleming et al., 1997; Gunshin et al., 1997). Mammalian DMT1 has two isoforms resulting from alternative splicing of the 3'-terminal exons. The isoform I mRNA contains an iron-responsive element (IRE) in its 3'-noncoding region, whereas the isoform II mRNA does not have the IRE structure (Tchernitchko et al., 2002). DMT1 is a proton-driven transporter, transporting one proton and one atom of ferrous iron in the same direction. The divalent ions such as iron, lead, manganese, zinc, cobalt, cadmium, copper, and nickel are transported at pH 5.5, the optimum pH for DMT1 operation (Gunshin et al., 1997). In brain, DMT1 exists in the choroid plexus (Gunshin et al., 1997).

The other nontransferrin iron transport pertains to a newly discovered iron transport protein, metal transport protein 1 (MTP1). MTP1, also known as ferroportin1 or IREG1 (Donovan et al., 2000; McKie et al., 2000), functions to facilitate iron export and to remove iron from the cells. The function of this protein is consistent with the cell's role in iron delivery to the rest of the organ or the body.

While there is no question about the presence of DMT1 in the blood–CSF barrier in the choroid plexus (Gunshin et al., 1997; Siddappa et al., 2002; Choudhuri et al., 2003; Moos and Morgan, 2004), the dispute remains as to whether DMT1 exists in the blood–brain barrier. Burdo et al. (2001) studied the expression of DMT1 in brain. In the normal rat, DMT1 is indeed expressed in ependymal cells lining the third ventricle and vascular cells throughout the brain, with the highest expression in neurons of striatum, cerebellum, and thalamus. The authors explicitly showed a strong DMT1 immunofluorescence in an isolated blood vessel, which colocates with the immunofluorescence signal of TfR. The evidence by immunostaining was further strengthened by a Western blot study in which an immunoblot of DMT1 was identified in brain endothelial cells. Clearly, the staining in the ependymal cells and endothelial cells indicates an important role of DMT1 in iron transport into the brain. In the same rat brains, MTP1 staining was identified in most brain regions. While

MTP1 expression in the brain is robust in pyramidal neurons of the cerebral cortex, it is not detected in the vascular endothelial cells and ependymal cells. It remains unknown if MTP1 presents in the choroid plexus. It has been suggested that DMT1 and MTP1 are involved in brain iron transport, but both transporters may not colocate as in the intestine. Therefore, the involvement of DMT1 and MTP1 in brain iron transport is both brain region and cell type specific.

The work by Siddappa et al. (2002) supports the presence of DMT1 at both the blood–brain barrier and blood–CSF barrier. They reported that iron regulatory proteins (IRP1 and IRP2) and DMT1 were partially expressed in the choroid plexus epithelial cells at postnatal days 5 and 10 and fully expressed at day 15. In contrast, the cerebral blood vessels and ependymal cells strongly expressed IRP1, IRP2, and DMT1 as early as postnatal day 5. The authors suggested that the different time period of brain development may demand iron in a different way, which may render brain structure in a special period more vulnerable to iron deficiency or iron overload.

The existence of DMT1 at the blood–brain barrier was challenged by a recent study by Moos and Morgan (2004). Using antibodies targeting different sites on DMT1 molecules, the authors reported that DMT1 can be consistently detected in neurons and choroid plexus, but not in brain capillary endothelial cells, macroglial cells, or microglial cells. It is noted that this study was performed in mice, whereas the reports by others were mainly conducted in rats.

The existence of DMT1 in the choroid plexus is consistent with the results from our own brain perfusion study (Deane et al., 2004). A high influx of iron into the CSF following perfusion with nontransferrin-bound iron suggests that the choroid plexus may serve as the primary site for nontransferrin-bound iron to enter the CSF. It is possible that the multiple status of iron in the blood circulation, either as bound or free, may be modulated differently at brain barriers. Hypothetically, the blood–brain barrier transports those iron species that bind to transferrin via a TfR-mediated mechanism, whereas the blood–CSF barrier may transport free, unbound iron species by DMT1-associated transport. This hypothesis is, of course, subject to extensive experimental verification.

17.4 REGULATION OF MANGANESE TRANSPORT AT BRAIN BARRIERS

17.4.1 Transferrin Receptor-Mediated Transport

The homeostasis of brain manganese appears to be controlled by both the blood–brain barrier and blood–CSF barrier. However, each barrier may play a different role in response to different blood concentrations of manganese. At normal plasma concentrations, manganese appears to be transported primarily across the endothelial cells of brain capillaries. Several mechanisms have been proposed, including facilitated diffusion (Rabin et al., 1993), active transport (Murphy et al., 1991; Rabin et al., 1993; Aschner and Gannon, 1994), as well as transferrin (Tf)-dependent transport (Aschner and Gannon, 1994).

TfR-mediated transport may be the primary mechanism for Mn(III) transport at brain barriers. The normal blood and serum total manganese concentrations are ~200 and ~20 nM, respectively. Most manganese ions in the serum exist in the form of Mn(II), as albumin-bound species (84%), hydrated ions (6.4%), and in 1:1 complexes with bicarbonate (5.8%) and citrate (2.0%) (Harris and Chen, 1994). Mn(III) presents in a small quantity (1.8% of blood manganese) mainly in the small molecular weight (MW) species, most of which is believed to be bound to transferrin (Aisen et al., 1969; Harris and Chen, 1994). Based on animal data, manganese distributes under normal physiological conditions in brain regions in the following order: substantia nigra > striatum > hippocampus > frontal cortex in a concentration range of 0.3 to 0.7 μg/g of wet tissue weight (Zheng et al., 1998). Similar to iron distribution in brain, the order of manganese distribution is consistent with the distribution pattern of TfR in cerebral endothelial capillaries collected from various brain regions (see discussions above) (Deane et al., 2004).

The distribution of TfR in brain parenchyma may also contribute to selective manganese accumulation in brain regions. The thalamic nuclei, the pallidum, and the substantia nigra contain the highest manganese concentrations (Barbeau, 1985). Noticeably, these structures also contain high levels of iron (Hill and Switzer, 1984). Iron and manganese may compete for the same TfR carrier transport system. When the plasma iron level is elevated, manganese uptake into the brain across the blood–brain barrier is significantly decreased. For example, intravenous administration of ferric-hydroxide dextran complex significantly inhibits the net uptake of manganese into brain (Diez-Ewald et al., 1968). In contrast, a deficiency in systemic iron is associated with an increased CNS burden of manganese (Mena et al., 1974; Aschner and Aschner, 1990). All this corroborates a potential relationship between iron and manganese transport by the brain barriers.

Transport of manganese via TfR-mediated mechanism is also supported by the evidence from in vitro studies. Transferrin, whose primary function is to carry iron, can bind to a variety of other metals and may function in vivo to transport diverse metals. The binding of manganese to transferrin is time-dependent (Keefer et al., 1970; Scheuhammer and Cherian, 1985; Aschner and Aschner, 1990). Manganese in transferrin complex is exclusively present in the trivalent oxidation state (Aisen et al., 1969). At normal plasma iron concentrations (0.9 to 2.8 μmg/mL), the normal plasma concentration of transferrin is about 3 mg/mL. The apotransferrin (MW 77,000) has two metal ion–binding sites per molecule. Considering the normal iron-binding capacity (2.5 to 4 μg/mL), only 30% of the transferrin is occupied by Fe(III), which leaves about 50 μmol/L of unoccupied binding sites available for Mn(III).

At the blood–brain barrier, Mn(III) enters the endothelial cells as a transferrin complex. The metal is subsequently released from the complex in the endothelial cell interior by endosomal acidification. The apotransferrin is then returned to the luminal surface (Morris et al., 1992a,b). Manganese released within the endothelial cells is further transported to the abluminal cell surface for release into the extracellular fluid. The metal is carried by brain-derived transferrin for extracellular transport and uptake at neurons, oligodendrocytes, and astrocytes for usage and storage. Suarez and Eriksson (1993) have provided direct evidence to support receptor-mediated endocytosis of the manganese-transferrin complex in cultured neuroblastoma cells.

17.4.2 Involvement of DMT1

Several studies have suggested that DMT1 may be involved in manganese influx into the brain (Goddard et al., 1997; Gunshin et al., 1997; Conrad et al., 2000); however, recent results have shown that the lack of functional DMT1 in knockout rats had no apparent effect on brain influx of manganese either as free ions or conjugated with transferrin (Crossgrove and Yokel, 2004).

17.4.3 Transport at Choroid Plexus

When the blood concentration of manganese exceeds normal levels, the transport of manganese may take place at the blood–CSF barrier of the choroid plexus. For example, acute bolus injections of manganese result in a rapid appearance and persistent elevation of manganese in the choroid plexus (London et al., 1989; Takeda et al., 1994a, b, 1995; Ingersoll et al., 1995), which is consistent with its transport across the choroid plexus in conditions of high manganese plasma concentrations (Murphy et al., 1991; Rabin et al., 1993; Takeda et al., 1994a, b, 1995). The data from this laboratory also demonstrated that following exposure of rats to 6 mg Mn/kg as $MnCl_2$, ip, once daily for 30 consecutive days, manganese concentrations in the choroid plexus were increased by 5-fold (see Figure 17.2), which was in line with the 3-fold increase in the CSF (Zheng et al., 1998).

17.4.4 Efflux of Manganese

In contrast to the evidence for brain manganese influx, much less is known about manganese movement out of the brain into blood. Once manganese enters the brain, it persists there for a relatively long time. The half-life in monkey brain would be expected to exceed 100 days (Dastur et al., 1971). Such a long half-life could be due to a tight tissue binding of metal ions or to slow clearance of manganese from the cerebral compartment. Some evidence suggests that the brain efflux of manganese across the BBB does not occur through a transporter, but rather through slow diffusion (Yokel et al., 2003).

Manganese in the CSF appears to equilibrate with manganese in brain tissues. In an in vivo study by Zheng et al. (1998), when rats were injected once daily for 30 days with Mn(II) as $MnCl_2$ via the ip route, manganese concentrations in the striatum, substantia nigra, hippocampus, and frontal cortex were all increased. However, the elevation of Mn in all selected brain regions (range between 3.1- and 3.9-fold) in this chronic exposure model was similar in magnitude to that in CSF (3.1-fold) rather than in serum (6.1-fold). These observations suggest that manganese within the CSF may serve as a "sink" for manganese deposition in brain tissues.

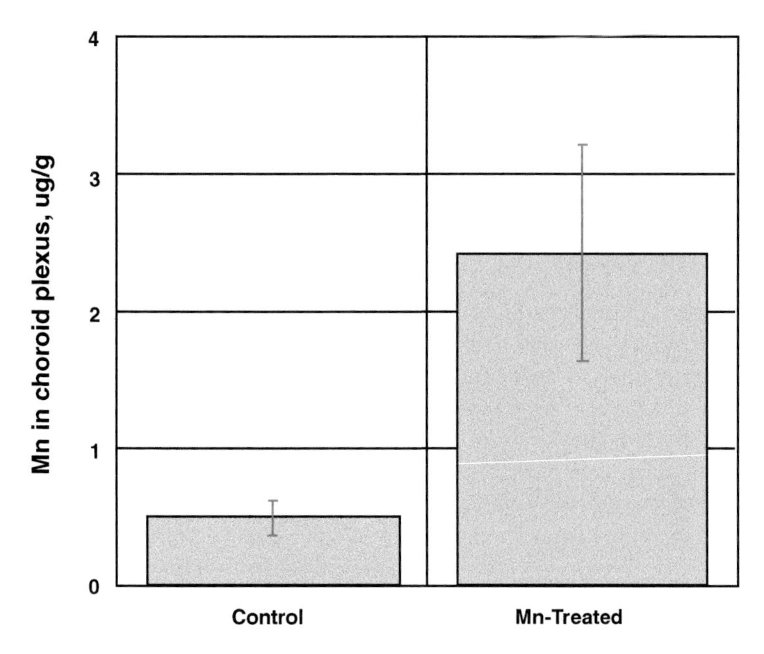

FIGURE 17.2 Accumulation of manganese in the choroid plexus following chronic manganese exposure. Rats received ip injection of 6 mg/kg Mn (as $MnCl_2$) once daily for 30 consecutive days. Mn concentrations were determined by flameless AAS. Data represent means ± SD; $n = 3–4$ for controls and 6 for Mn-treated group.* $p < 0.05$ as compared to control.

17.5 INTERACTION BETWEEN MANGANESE AND IRON AT BRAIN BARRIERS

17.5.1 Chemical and Biochemical Basis of Interactions between Manganese and Iron

Chemically and biochemically, manganese shares numerous similarities with iron:

1. Both metals are transition elements adjacent to each other in the periodic table. Manganese has the atomic number 25, iron 26.
2. Both metals have similar ionic radius. The electron configurations for manganese and iron are -8-13-2 and -8-14-2, respectively.
3. Both metals carry similar valencies (2+ and 3+) in physiological conditions.
4. In the body, both metals strongly bind transferrin (Aschner and Aschner, 1990; Suarez and Eriksson, 1993; Ueda et al., 1993).
5. Intracellularly, both preferentially accumulate in mitochondria (Gavin et al., 1990, 1992).

Because of these similarities, it is not surprising that manganese can interact directly with iron at the cellular and subcellular levels, particularly on the enzymes that require iron as a cofactor in their active catalytic center, such as aconitase, NADH-ubiquinone reductase (Complex I, Cpx-I), and succinate dehydrogenase (SDH).

17.5.2 Manganese Alters Systemic and Cellular Iron Homeostasis

Chronic Mn exposure in animals alters the systemic homeostasis of iron. Studies in this laboratory indicate that manganese appears to favor the influx of iron from the systemic circulation to the CSF (Zheng et al., 1999). Groups of rats received intraperitoneal injections of $MnCl_2$ at a dose of 6 mg Mn/kg/day or an equal volume of saline for 30 days. Exposure to manganese resulted in a 32% decrease in plasma iron and no changes in plasma total iron-binding capacity (TIBC). Surprisingly, the iron concentration in the CSF of the same rats did not decline as it did in plasma. Instead, it increased by more than three-fold the control values. The ratio of CSF to plasma iron ($Fe_{CSF/plasma}$) was increased from 0.16 in the control rats to 0.72 in the treated rats, reflecting an influx of iron from the systemic circulation to the CSF.

In the same animal model, the choroid plexus was sampled to determine the iron concentration. Manganese exposure caused a 26% reduction of iron in the choroid plexus (see Figure 17.3), suggesting that accumulation of manganese in the choroid plexus may alter the iron regulatory mechanism, possibly to reduce iron storage in the choroid plexus.

This pattern of plasma iron, while unexpected, is consistent with the observations in IPD patients whose circulating iron, ferritin, transferrin, as well as TIBC are all significantly lower than those of control subjects (Logroscino et al., 1997). The explanation for this overall deficiency in iron metabolism in IPD patients remains uncertain; however, manganese-induced diminished plasma iron may be attributed to manganese-enhanced intracellular distribution of iron. For example, Chua and Morgan (1996) observed that manganese supplementation in food led to an increased ^{59}Fe uptake by the brain, liver, and kidneys of rats. The authors concluded that manganese and iron interact during transfer from the plasma to the brain as well as other organs in a synergistic rather than competitive manner. Seligman et al. (1988) also reported that incubation of cultured human HL60 cells with a manganese-transferrin complex dramatically increased cellular uptake of ^{59}Fe. In vivo study in monkeys shows that manganese intoxication causes an elevated iron deposition in the globus pallidus and substantia nigra pars reticulata of experimental animals (Olanow et al., 1996).

To obtain further evidence, this laboratory performed an iron uptake study in three cell types, namely, neuronal type PC12 cells, cultured primary astrocytes, and primary choroidal epithelial cells. Exposure of PC12 cells to manganese caused a significant increase in cellular net uptake of ^{59}Fe. The cellular uptake is characterized by (1) a rapid uptake in the early stage followed by a slow decline of cellular iron in both manganese-treated and control groups and (2) a much higher ^{59}Fe level in manganese-treated cells than in controls in the later phase. The promoting effect of

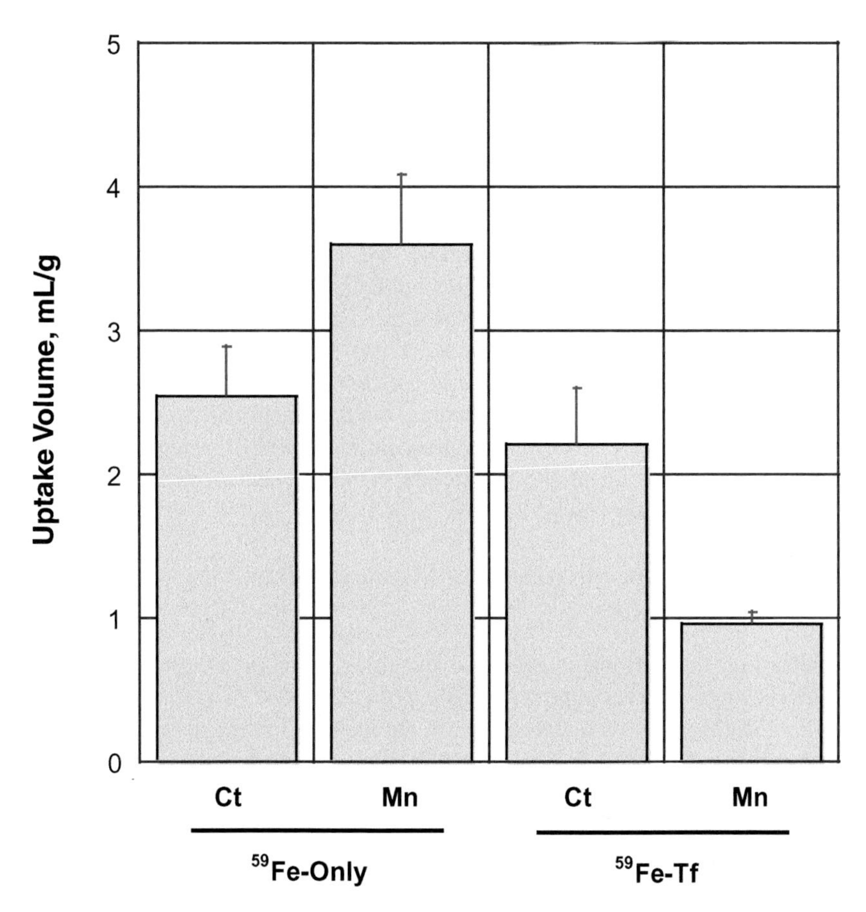

FIGURE 17.3 Effect of in vivo manganese exposure on iron concentrations in the choroid plexus. Rats received ip injection of 6 mg/kg Mn (as $MnCl_2$) once daily for 30 consecutive days. Fe concentrations were determined by flameless AAS. Data represent means ± SD, $n = 3–4$ for controls and 6 for Mn-treated group;* $p = 0.0768$.

manganese on cellular iron uptake is manganese concentration–dependent within the range of 50 to 200 μM Mn in culture medium (Zheng and Zhao, 2001).

When cultured astrocytes were treated with manganese under the same conditions, no significant difference in cellular net uptake of ^{59}Fe was observed between manganese-treated and control groups. The results demonstrate a different sensitivity in cellular iron net uptake between neuronal type PC12 cells and neuroglial astrocytes. Studies by Northern blot of TfR mRNA showed that astrocytes express TfR mRNA, but much more weakly than PC12 or choroidal epithelial cells. The rather low base level of TfR may partially explain the insensitivity of astrocytes to iron cytotoxicity, although this cell type is known to accumulate as much as 80% of manganese in the brain. Thus, it seems likely that astrocytes in the brain may act as a metal depot for manganese ions (Zheng and Zhao, 2001).

17.5.3 Manganese Alters Iron Transport in a Transwell Model

A Transwell model with the primary culture of choroidal epithelial cells was also used to evaluate the trans-epithelial transport of iron. Upon formation of an impermeable monolayer at day 7–8, trace amounts of ^{59}Fe were added to the outer (donor) chamber. The rate of appearance of ^{59}Fe radioactivity in the inner (acceptor) chamber thus represents transepithelial transport of ^{59}Fe. Exposure to manganese promoted the transport of ^{59}Fe across the choroidal epithelial barrier by 35% (Li et al., 2004). This observation may explain the elevated iron concentration in the CSF following in vivo manganese exposure (Zheng et al., 1999). More interestingly, the same study shows a vast reduction in ^{59}Fe storage in the choroidal epithelial cells as compared to controls. The in vitro results corroborate the data from in vivo experiments (see Figure 17.3). Presumably, manganese may interfere with iron regulatory proteins, resulting in suppression of the production of ferritin and decreased cellular iron storage (see discussion below).

17.5.4 Molecular Mechanism of Manganese–Iron Interaction

Cellular iron homeostasis is regulated by iron regulatory proteins (IRPs), one of which is cytoplasmic aconitase (see Table 17.1). Aconitase, whose structure contains a [4Fe-4S] cluster in its active center, is an enzyme present in both mitochondria and cytoplasm (Henson and Cleland, 1967; Kennedy et al., 1983). Cytosolic aconitase (ACO1) regulates intracellular iron metabolism, while mitochondrial aconitase (ACO2) is responsible for the conversion of citrate to isocitrate in the TCA cycle.

Through its [4Fe-4S] cluster, ACO1 binds to IREs, which fold stereoscopically as a stem loop and locate in the 5'- or 3'-untranslated region (UTR) of the mRNA sequence (see Figure 17.4). In iron-replete cells, ACO1 secures iron in the form of a [4Fe-4S] cluster. While this form of ACO1 binds poorly with mRNA, it can enzymatically catalyze the conversion of bound citrate to isocitrate. When cellular iron levels are insufficient, however, ACO1 assumes a [3Fe-4S] configuration and is then transformed into an mRNA-binding protein. At this state, the enzyme binds with high affinity to IRE-containing mRNAs, inhibits translation of those whose IREs are 5' (e.g., ferritin, SDH, mitochondrial aconitase), and stimulates the expression of those whose IREs are 3' (e.g., TfR). The net result of this RNA-protein interaction is an increase in cellular iron uptake (due to increased TfR) and a decrease in iron storage (due to reduced ferritin) (Kennedy et al., 1983; Henderson, 1996).

Since manganese shares many chemical similarities with iron, manganese may replace iron in ACO1s [4Fe-4S] and therefore alter cellular iron regulatory mechanism. In an in vitro study using purified aconitase, incubation of the enzyme with Mn(II) reduced the activity of aconitase by 81% (Zheng et al., 1998). The Mn(III) species is more efficient than Mn(II) in the inhibition of aconitase (Chen et al., 2001). The enzyme activity was nearly entirely inhibited by Mn(III) at 10 μM in vitro. In vivo chronic exposure of rats to manganese (6 mg Mn/kg, ip, for 30 days) led to a region-specific reduction in total aconitase (ACO1 + ACO2) activity in

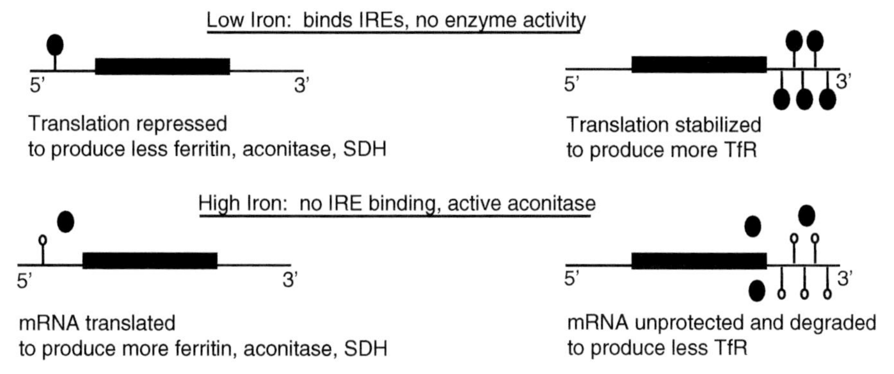

FIGURE 17.4 Control of intracellular Fe homeostasis by IRP-I. Ovals represent IRP-I. Graph A on the left represents the mRNA encoding ferritin, and so on with one clustered IRE at the 5'-cap site. Graph B on the right represents the mRNA encoding Tf receptor with five clustered IREs at the 3' untranslated region.

brain tissue. The inhibitory effect of manganese on aconitase activity reached the maximum in striatum (60% of controls), followed by a 48.5% reduction in frontal cortex ($p < 0.01$), 20.6% in substantia nigra ($p < 0.139$), and 19.3% in hippocampus ($p < 0.323$) (Zheng et al., 1998).

Since the coordination chemistry of manganese closely resembles that of iron, it appears likely that manganese may insert itself into the fourth, highly labile iron-binding site of both mitochondrial and cytosolic aconitase. The action, while suppressing the enzyme's catalytic function, may increase its binding affinity to the mRNAs encoding major proteins in iron metabolism, such as ferritin and transferrin receptor (see Figure 17.5).

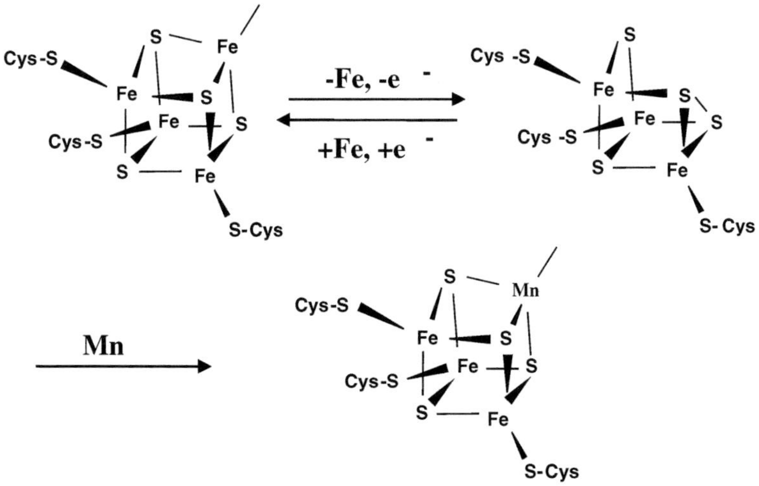

FIGURE 17.5 Possible competition between Mn and Fe for the binding site in aconitase.

Studies with Northern blot were performed to quantify mRNAs encoding TfR in primary culture of choroid plexus cells. When the choroidal epithelia were exposed to manganese (100 μM) for 14 hours, the expression of TfR mRNA was more abundant in manganese-treated groups than in controls (Zheng et al., 1999). Thus, it seems likely that the choroid plexus participates in the regulation of cerebral iron homeostasis, and this function can be modulated by excess manganese presented in the blood or accumulated in the choroid plexus.

Taken together, these in vitro and in vivo data are likely to illustrate the hypothesized sequential events following manganese exposure. Chronic manganese exposure results in the accumulation of manganese in the choroid plexus and brain tissues. Excess cellular manganese may stimulate the expression of TfR in the choroid plexus and at brain capillary endothelia. The overexpression of TfR at brain barriers and the ensuing facilitated iron transport from blood to the cerebral compartment would explain a compartmental shift of iron from the blood to the CSF.

This hypothesis, however, was recently challenged by our own brain perfusion experiment on a subacute manganese exposure model (Deane et al., 2002). In that experiment, rats received ip injections of 6 mg Mn/kg once daily for 14 days. Brains were then dual-perfused with a Ringer solution containing traces of ^{59}Fe with or without transferrin via common carotid arteries. The cerebral capillaries were separated by a capillary-depletion method. In the absence of perfusate transferrin, manganese exposure significantly increased the influx of ^{59}Fe to the choroid plexus by 2.3-fold but did not affect influx to the other brain regions examined. In the presence of transferrin, however, the influxes of ^{59}Fe to most brain regions were unexpectedly reduced in manganese-exposed animals except for a nearly 10-fold increase into the CSF compared to controls (Deane et al., 2002). Thus, manganese exposure appears to affect ^{59}Fe uptake at the cerebral capillaries by upregulating nonTfR-mediated iron transport and downregulate TfR-mediated processes. An increased influx of iron to the CSF in manganese-treated animals is likely due to enhanced transport by DMT1 or other nontransferrin-related transport mechanisms at both brain barriers.

17.6 SUMMARY

The brain barrier system clearly plays a fundamental role in restricting blood iron flushing into the brain and therefore regulating iron homeostasis in the brain. Two barriers, however, appear to have specific functions. Transport of iron at the endothelial interface of the blood–brain barrier is mainly TfR-mediated, while at the epithelial interface of the blood–CSF barrier transport is likely via a nontransferrin-mediated mechanism. A gradient iron distribution from the barrier cells to brain parenchyma and further to the CSF may suggest a pathway for iron dynamic movement in the cerebral compartment. Various transporters, specific or nonspecific, transferrin-mediated or nontransferrin-related, during the path of iron flow could affect the homeostasis of iron regulation. The choroid plexus, with regard to its vast surface area in the brain ventricles, should play a significant role in iron homeostasis in the CSF. Since many iron regulatory proteins and transporters have not been studied in the choroid plexus, research should be directed to understanding the direction of iron fluxes at this barrier. The choroid plexus is a known target for metal sequestration in the brain. How other metals interfere with iron regulation and the ensuing consequences in CSF iron deserve in-depth investigation.

Manganese transport into the brain relies on a transferrin-mediated mechanism at both the blood–brain barrier and blood–CSF barrier. While it remains disputable whether DMT1 is involved in manganese transport via the blood–brain barrier, the abundant presence of DMT1 in the choroid plexus suggests that manganese may be transported via this nonselective metal transporter. It remains uncertain if MTP1 exits in the choroid plexus, where the MTP1 moiety is subcellularly located, and how MTP1, in coordinating with DMT1, contributes to transport of manganese and iron by the blood–CSF barrier.

Manganese exposure alters iron homeostasis systemically and subcellularly. The interaction of manganese with iron appears to happen at the cellular iron regulatory mechanism, of which a [4Fe-4S]-containing iron regulatory protein (IRP1) may serve as the critical action site. Excess manganese may replace iron from the fourth binding site in IRP1, stabilize the expression of TfR, and enhance the cell surface TfR in neurons and choroidal epithelial cells. Elevated TfR, in turn, facilitates cellular uptake of iron and at the blood–CSF barrier may assist iron transport into the CSF. Manganese exposure, however, increases nontransferrin-mediated uptake by both the blood–brain barrier and blood–CSF barrier. Whether this is due to DMT1 or another transport mechanism and how manganese alters these unknown mechanisms is an exciting research subject for further investigation.

17.7 ACKNOWLEDGMENT

The author's research has been supported by NIH Grants RO1 ES-07042, RO1 ES-08146, PO1 ES-09089, Calderone Foundation, Johnson & Johnson Focused Giving Funds, Lily Research Funds, and Burroughs Wellcome Foundation.

REFERENCES

Abbott PJ (1987) Methylcyclopentadienyl manganese tricarbonyl (MMT) in petrol: the toxicological issues. *Sci Total Environ* 67: 247–255.

Aisen P, Aasa R, Redfield AG (1969) The chromium, manganese, and cobalt complexes of transferrin. *J Biol Chem* 244: 4628–4633.

Alcantara O, Obeid L, Hannun Y, Ponka P, Boldt DH (1994) Regulation of protein kinase C (PKC) expression by iron: effect on different iron compounds on PKC-b and PKC-alpha gene expression and role of the 5'-flanking region of the PKC-beta gene in the response to ferric transferrin. *Blood* 84: 3510–3517.

Aschner M, Aschner JL (1990) Manganese transport across the blood-brain barrier: relationship to iron homeostasis. *Brain Res Bull* 24: 857–860.

Aschner M, Gannon M (1994) Manganese (Mn) transport across the blood-brain barrier: saturable and transferrin-dependent transport mechanisms. *Brain Res Bull* 33: 345–349.

Aschner M, Vrana KE, Zheng W (1999) Manganese uptake and distribution in the central nervous system (CNS). *Neurotoxicology* 20: 173–180.

Barbeau A (1985) Manganese and extrapyramidal disorders. *Neurotoxicology* 5: 13–16.

Beard JL, Connor JR (2003) Iron status and neural functioning. *Ann Rev Nutr* 23: 41–58.

Berg D, Gerlach M, Youdim, MBH, Douible KL, Zecca L, Riederer P, Becker G (2001) Brain iron pathways and their relevance to Parkinson's disease. *J Neurochem* 79: 225–236.

Bonilla E (1980) *L*-Tyrosine hydroxylase activity in the rat brain after chronic oral administration of manganese chloride. *Neurobehav Toxicol* 2: 37–41.

Bradbury MW (1997) Transport of iron in the blood-brain-cerebrospinal fluid system. *J Neurochem* 69: 443–454.

Brown NM, Anderson SA, Steffen DW, Carpenter TB, Kennedy MC, Walden WE, Eisenstein RS (1998) Novel role of phosphorylation in Fe-S cluster stability revealed by phosphomimetic mutations at Ser-138 of iron regulatory protein 1. *Proc Natl Acad Sci USA* 95: 15235–15240.

Burdo JR, Menzies SL, Simpson IA, Garrick LM, Garrick MD, Dolan KG, Haile DJ, Beard JL, Connor JR (2001) Distribution of divalent metal transporter 1 and metal transport protein 1 in the normal and Belgrade rat. *J Neurosci Res* 66: 1198–1207.

Chandra SV, Shukla GS, Srivastawa RS, Singh H, Gupta VP (1981) An exploratory study of manganese exposure to welders. *Clin Toxicol* 18: 407–418.

Chen JY, Tsao G, Zhao Q, Zheng W (2001) Differential cytotoxicity of Mn(II) and Mn(III): special reference to mitochondrial [Fe-S] containing enzymes. *Toxicol Appl Pharmacol* 175: 160–168.

Choudhuri S, Cherrington NJ, Li N, Klaassen CD (2003) Constitutive expression of various xenobiotic and endobiotic transporter mRNAs in the choroid plexus of rats. *Drug Metab Disp* 31: 1337–1345.

Chua AC, Morgan EH (1996) Effects of iron deficiency and iron overload on manganese uptake and deposition in the brain and other organs of the rat. *Biol Trace Elem Res* 55: 39–54.

Connor JR (1997) Evidence for iron mismanagement in the brain in neurological disorders. In: *Metals and Oxidative Damage in Neurological Disorders* (Connor JR, ed.). Plenum Press, New York, pp. 23–39.

Connor JR, Benkovic SA (1992) Iron regulation in the brain: histochemical, biochemical, and molecular considerations. *Ann Neurol* 32(suppl): S51–61.

Conrad ME, Umbreit JN, Moore EG, Hainsworth LN, Porubcin M, Simovich MJ, Nakada MT, Dolan K, Garrick MD (2000) Separate pathways for cellular uptake of ferric and ferrous iron. *Am J Physiol Gastrointest Liver Physiol* 279: G767–774.

Cooper WC (1984) The health implications of increased manganese in the environment resulting from the combustion of fuel additives: a review of the literature. *J Toxicol Environ Health* 14: 23–46.

Crossgrove JS, Yokel RA (2004) Manganese distribution across the blood–brain barrier III. The divalent metal transporter-1 is not essential for brain manganese uptake. *NeuroToxicology* 25: 451–460.

Crossgrove JS, Zheng W (2004) Review of manganese toxicity upon overexposure. *NMR Biomed* 17:1–10.

Dastur DK, Manghani DK, Raghavendran KV (1971) Distribution and fate of 54Mn in the monkey: studies of different parts of the central nervous system and other organs. *J Clin Invest* 50: 9–20.

Deane D, Kong B, Pilsner R, Zheng W (2002) Brain regional influx of 59.Fe: Effect of *in vivo* manganese exposure. *Toxicol Sci* 66(1-S): 1004.

Deane R, Zheng W, Zlokovic BV (2004) Brain capillary endothelium and choroid plexus epithelium regulate transport of transferrin-bound and free iron into the rat brain. *J Neurochem* 88: 813–820.

Dexter DT, Carayon A, Javoy-Agid F, Agid Y, Wells FR, Daniel SE, Lees AJ, Jenner P, Marsden CD (1991) Alterations in the levels of iron, ferritin and other trace metals in Parkinson's disease and other neurodegenerative diseases affecting the basal ganglia. *Brain* 114: 1953–1975.

Diez-Ewald M, Weintraub LR, Crosby WH (1968) Interrelationship of iron and manganese metabolism. *Proc Soc Exp Biol Med* 129: 448–451.

Donovan A, Brownlie A, Zhou Y, Shepard J, Pratt SJ, Moynihan J, Paw BH, Drejer A, Barut B, Zapata A, Law TC, Brugnara C, Lux SE, Pinkus GS, Pinkus JL, Kingsley PD, Palis J, Fleming MD, Andrews NC, Zon LI (2000) Positional cloning of zebrafish ferroportin1 identifies a conserved vertebrate iron exporter. *Nature* 403: 711–713.

Eisenstein RS, Tuazon PT, Schalinske KL, Anderson SA, Traugh JA (1993) Iron-responsive element-binding protein. Phosphorylation by protein kinase C. *J Biol Chem* 268: 27363–27370.

Ensing JG (1985) Bazooka: cocaine-base and manganese carbonate. *J Anal Toxicol* 9: 45–46.

Eriksson H, Magiste K, Plantin LO, Fonnum F, Hedstrom KG, Norheim-Theodorsson E, Kristensson K, Stalberg E, Heibronn E (1987) Effects of manganese oxide on monkeys as revealed by a combined neurochemical, histological and neurophysiological evaluation. *Arch Toxicol* 61: 46–52.

Fleming MD, Trenor CC Su MA, Foernzler D, Beier DR, Dietrich WF, Andrews NC (1997) Microcytic anemia mice have a mutation in Nramp2, a candidate iron transporter gene. *Nat Genet* 16: 383–386.

Gabrielsson B, Robson T, Norris D, Chung SH (1986) Effects of divalent metal ions on the uptake of glutamate and GABA from synaptosomal fractions. *Brain Res* 384: 218–223.

Gavin CE, Gunter KK, Gunter TE (1990) Manganese and calcium efflux kinetics in brain mitochondria. Relevance to manganese toxicity. *Biochem J* 266: 329–334.

Goddard WP, Coupland K, Smith JA, Long RG (1997) Iron uptake by isolated human enterocyte suspensions *in vitro* is dependent on body iron stores and inhibited by other metal cations. *J Nutr* 127: 177–183.

Gorell JM, Johnson CC, Rybicki BA, Peterson EL, Kortsha GX, Brown GG, Richardson RJ (1997) Occupational exposures to metals as risk factors for Parkinson's disease. *Neurology* 48: 650–658.

Griffiths PD, Crossman AR (1993) Distribution of iron in the basal ganglia and neocortex in postmortem tissue in Parkinson's disease and Alzheimer's disease. *Dementia* 4: 61–65.

Gunshin H, Mackenzie B, Berger UV, Gunshi Y, Romero MF, Boron WF, Nussberger S, Gollan JL, Hediger MA (1997) Cloning and characterization of a mammalian protein-coupled metal-ion transporter. *Nature (London)* 388: 482–488.

Harris WR, Chen Y (1994) Electron paramagnetic resonance and difference ultraviolet studies of Mn2+. binding to serum transferrin. *J Inorg Biochem* 54: 1–19.

He Y, Thong PS, Lee T, Leong SK, Shi CY, Wong PT, Yuan SY, Watt F (1996) Increased iron in the substantia nigra of 6OHDA induced parkinsonian rats: a nuclear microscopy study. *Brain Res* 735: 149–153.

Henderson BR (1996) Iron regulatory proteins 1 and 2. *BioEssays* 18: 739–746.

Henson CP, Cleland WW (1967) Purification and kinetic studies of beef liver cytoplasmic aconitase. *J Biol Chem* 17: 3833–3838.

Hill JM, Switzer RC (1984) The regional distribution and cellular localization of iron in the rat brain. *Neuroscience* 11: 595–603.

Ingersoll RT, Montgomery EB, Aposhian HV (1995) Central nervous system toxicity of manganese. I. Inhibition of spontaneous motor activity in rats after intrathecal administration of manganese chloride. *Fundam Appl Toxicol* 27: 106–113.

Inoue N, Makita Y (1996) Neurological aspects in human exposures to manganese. In: Toxicology of Metals (Chang LW, ed.). CRC Press, Boca Raton, FL, pp. 415–421.

Jefferies WA, Gabathuler R, Rotherberger S, Food M, Kennard ML (1996) Pumping iron in the '90s. *Trend Cell Biol* 6, 223–228.

Jenner P, Schapira AH, Marsden CD (1992) New insights into the cause of Parkinson's disease. *Neurology* 42: 2241–2250.

Ke Y, Qian ZM (2003) Iron misregulation in the brain: a primary cause of neurodegenerative disorders. *Lancet Neurology* 2: 246–253.

Keefer RC, Barak AJ, Boyett, JD. (1970) Binding of manganese and transferrin in rat serum. *Biochim Biophys Acta* 221: 390–393.

Kennedy MC, Emptage MH, Dreyer JL, Beinert H (1983) The role of iron in the activation-inactivation of aconitase. *J Biol Chem* 258: 11098–11105.

Li GJ, Zhang L, Lu L, Wu P, Zheng W (2004) Occupational exposure to welding fume among welders: alterations of manganese, iron, zinc, copper, lead in body fluids and the oxidative stress status. *J Occup Environ Med* 46: 241–248.

Li GJ, Zhao Q, Zheng W (2004) Alteration at translational but not transcriptional level of transferrin receptor expression following manganese exposure at the blood–CSF barrier *in vitro*. *Toxicol Appl Pharmacol* 201(3).

Loeffler DA, Connor JR, Juneau PL, Snyder BS, Kanaley L, DeMaggio AJ, Nguyen H, Brickman CM, LeWitt PA (1995) Transferrin and iron in normal, Alzheimer's disease, Parkinson's disease brain regions. *J Neurochem* 65: 710–724.

Logroscino G, Marder K, Graziano JH, Freyer G, Slavkovich V, LoIacono N, Cote L, Mayeux R (1997) Altered systemic iron metabolism in Parkinson's disease. *Neurology*, 49: 714–717.

London RE, Toney G, Gabel SA, Funk A (1989) Magnetic resonance imaging studies of the brains of anesthetized rats treated with manganese chloride. *Brain Res Bull* 23: 229–235.

Loranger S, Zayed J (1995) Environmental and occupational exposure to manganese: a multimedia assessment. *Int Arch Occ Env Health* 67: 101–110.

Mann VM, Cooper JM, Daniel SE, Srai K, Jenner P, Marsden CD, Schapira AH (1994) Complex I, iron, and ferritin in Parkinson's disease substantia nigra. *Ann Neurol* 36: 876–881.

McKie AT, Marciani P, Rolfs A, Brennan K, Wehr K, Barrow D, Miret S, Bomford A, Peters TJ, Farzaneh F, Hediger MA, Hentze MW, Simpson RJ (2000) A novel duodenal iron-regulated transporter, IREG1, implicated in the basolateral transfer of iron to the circulation. *Mol Cell* 5: 299–309.

Mena I, Marin O, Fuenzalida S, Cotzias GC (1967) Chronic manganese poisoning: clinical picture and manganese turnover. *Neurology* 17: 128–136.

Mena I, Court J, Fuenzalida S, Papavasiliou PS, Cotzias GC (1970) Modification of chronic manganese poisoning. Treatment with L-dopa or 5-OH tryptophan. *N Engl J Med* 282: 5–10.

Mena I, Horiuchi K, Lopez G (1974) Factors enhancing entrance of manganese into brain: iron deficiency and age. *J Nuclear Med* 15: 516.

Mochizuki H, Imai H, Endo K, Yokomizo K, Murata Y, Hattori N, Mizuno Y (1994) Iron accumulation in the substantia nigra of 1-methyl-4-phenyl-1,2,3,6-tetrahydropyridine (MPTP)-induced hemiparkinsonian monkeys. *Neurosci Lett* 168: 251–253.

Moos T, Morgan EH (1998a). Evidence for low molecular weight, non-transferrin-bound iron in rat brain and CSF. *Nerosci Res.* 54: 486–494.

Moos T, Morgan EH (1998b). Kinetics and distribution of [59Fe-125I]transferrin injected into the ventricular system of the rat. *Brain Res* 790: 115–128.

Moos T, Morgan EH (2000) Transferrin and transferrin receptor function in brain barrier systems. *Cell Mol Neurobiol* 20: 77–95.

Moos T, Morgan EH (2004) The significance of the mutated divalent metal transporter (DMT1) on iron transport into the Belgrade rat brain. *J Neurochem* 88: 233–245.

Morris CM, Keith AB, Edwardson JA, Pullen, RGL (1992a). Uptake and distribution of iron and transferrin in the adult brain. *J Neurochem* 59: 300–306.

Morris CM, Candy JM, Keith AB, Oakley A, Taylor G, Pullen, RGL, Bloxham CA, Gocht A, Edwardson JA (1992b). Brain iron homeostasis. *J Inorganic Biochem* 47: 257–265.

Murphy VA, Wadhwani KC, Smith QR, Rapoport SI (1991) Saturable transport of manganese (II) across the rat blood-brain barrier. *J Neurochem* 57: 948–954.

Neff NH, Barrett RE, Costa E (1969) Selective depletion of caudate nucleus dopamine and serotonin during chronic manganese dioxide administration to squirrel monkeys. *Experientia* 25: 1140–1141.

Olanow CW, Good PF, Shinotoh H, Hewitt KA, Vingerhoets F, Snow BJ, Beal MF, Calne DB, Perl DP (1996) Manganese intoxication in the rhesus monkey: a clinical, imaging, pathologic, and biochemical study. *Neurology* 46: 492–498.

Palmeira CM, Santos MS, Carvalho AP, Oliveira CR (1993) Membrane lipid peroxidation induces changes in gamma-[3H]aminobutyric acid transport and calcium uptake by synaptosomes. *Brain Res* 609: 117–123.

Rabin O, Hegedus L, Bourre JM, Smith QR (1993) Rapid brain uptake of manganese(II) across the blood-brain barrier. *J Neurochem* 61: 509–517.

Sastry S, Arendash GW (1995) Time-dependent changes in iron levels and associated neuronal loss within the substantia nigra following lesions within the neostriatum/globus pallidus complex. *Neuroscience* 67: 649–66.

Schalinske KL, Anderson SA, Tuazon PT, Chen OS, Kennedy MC, Eisenstein RS (1997) The iron-sulfur cluster of iron regulatory protein 1 modulates the accessibility of RNA binding and phosphorylation sites. *Biochemistry* 36: 3950–3958.

Scheuhammer AM, Cherian MG (1985) Binding of manganese in human and rat plasma. *Biochim Biophys Acta* 840: 163–169.

Seligman PA, Chitambar C, Vostrejs M, Moran PL (1988) The effects of various transferrins on iron utilization by proliferating cells. *Ann NY Acad Sci* 526: 136–140.

Sengstock GJ, Olanow CW, Dunn AJ, Arendash GW (1992) Iron induces degeneration of nigrostriatal neurons. *Brain Res Bull* 28: 645–649.

Sengstock GJ, Olanow CW, Dunn AJ, Barone S, Arendash GW (1994) Progressive changes in striatal dopaminergic markers, nigral volume, and rotational behavior following iron infusion into the rat substantia nigra. *Exp Neurol* 130: 82–94.

Siddappa AJ, Rao RB, Wobken JD, Leibold EA, Connor JR, Georgieff MK (2002) Developmental changes in the expression of iron regulatory proteins and iron transport proteins in the perinatal rat brain. *J Neurosci Res* 68: 761–775.

Sierra P, Loranger S, Kennedy G, Zayed J (1995) Occupational and environmental exposure of automobile mechanics and nonautomotive workers to airborne manganese arising from the combustion of methylcyclopentadienyl manganese tricarbonyl (MMT). *Am Ind Hyg Assoc J* 56: 713–716.

Sofic E, Paulus W, Jellinger K, Riederer P, Youdim MB (1991) Selective increase of iron in substantia nigra zona compacta of parkinsonian brains. *J Neurochem* 56: 978–982.

Sohn J, Yoon YH (1998) Iron-induced cytotoxicity in cultured rat retinal neurons. *Korean J Ophthalmol* 12: 77–84.

Suarez N, Eriksson H (1993) Receptor-mediated endocytosis of a manganese complex of transferrin into neuroblastoma (SHSY5Y) cells in culture. *J Neurochem* 61: 127–131.

Taher MM, Garcia JG, Natarajan V (1993) Hydroperoxide-induced diacylglycerol formation and protein kinase C activation in vascular endothelial cells. *Arch Biochim Biophys* 303: 260–266.

Takeda A (2003) Manganese action in brain function. *Brain Res Brain Res Rev* 41: 79–87.

Takeda A, Akiyama T, Sawashita J, Okada S (1994a). Brain uptake of trace metals, zinc and manganese, in rats. *Brain Res* 640: 341–344.

Takeda A, Sawashita J, Okada S (1994b). Localization in rat brain of the trace metals, zinc and manganese, after intracerebroventricular injection. *Brain Res* 658: 252–254.

Takeda A, Sawashita J, Okada S (1995) Biological half-lives of zinc and manganese in rat brain. *Brain Res* 695: 53–58.

Tchernitchko D, Bourgeois M, Martin ME, Beaumont C (2002) Expression of the two mRNA isoforms of the iron transporter Nramp2/DMT1 in mice and function of the iron responsive element. *Biochem J* 363: 449–455.

Tepper LB (1961) Hazards to health: manganese. *N Engl J Med* 264: 347–348.

Ueda F, Raja KB, Simpson RJ, Trowbridge IS, Bradbury, MWB (1993) Rate of ^{59}Fe uptake into brain a cerebrospinal fluid and the influence thereon of antibodies against the transferrin receptor. *J Neurochem* 60: 106–113.

Wedler FC (1994) Biochemical and nutritional role of manganese: an overview. In: *Manganese in Health and Disease* (Klimis-Tavantzis DJ, ed). CRC Press, Boca Raton, FL, pp. –36.

Yamada M, Ohno S, Okayasu I, Hatakeyama S, Watanabe H, Ushio K, Tsukagoshi H (1986) Chronic manganese poisoning: a neuropathological study with determination of manganese distribution in the brain. *Acta Neuropathol (Berl)* 70: 273–278.

Ye FQ, Allen PS, Martin WR (1996) Basal ganglia iron content in Parkinson's disease measured with magnetic resonance. *Mov Disorders* 11: 243–249.

Yokel RA, Crossgrove JS, Bukaveckas BL (2003) Manganese distribution across the blood-brain barrier II. Manganese efflux from the brain does not appear to be carrier mediated. *NeuroToxicology* 24: 15–22.

Youdim, MBH, Ben-Shachar D, Riederer P (1993) The possible role of iron in the etiopathology of Parkinson's disease. *Mov Disord* 8: 1–12.

Zhao Q, Slavkovich V, Zheng W (1998) Lead exposure promotes translocation of protein kinase C activity in rat choroid plexus *in vitro*, but not *in vivo*. *Toxicol Appl Pharmacol* 149: 99–106.

Zheng W (1996) Choroid plexus and metal toxicity. In: *Toxicology of Metals* (Chang LW, ed). CRC Press, Boca Raton, FL, pp. 605–622.

Zheng W (2001) Neurotoxicology of the brain barrier system: new implications. *J Toxicol Clin Toxicol* 39(7):711–719.

Zheng W, Zhao Q (2001) Iron overload following manganese exposure in cultured neuronal, but not neuroglial cells. *Brain Res* 897: 175–179.

Zheng W, Aschner M, Ghersi-Egea JF (2003) Brain barrier systems: a new frontier in metal neurotoxicological research. Invited review. *Toxicol Appl Pharmacol* 192: 1–11.

Zheng W, Ren S, Graziano JH (1998) Manganese inhibits mitochondrial aconitase: a mechanism of manganese neurotoxicity. *Brain Res* 799: 334–342.

Zheng W, Zhao Q, Slavkovich V, Aschner M, Graziano H (1999) Alteration of iron homeostasis following chronic exposure to manganese in rats. *Brain Res* 833: 125–132.

18 Involvement of the Choroid Plexus and the Cerebrospinal Fluid in Immune Molecule Signaling in the Central Nervous System

Miles Herkenham
National Institute of Mental Health
Bethesda, Maryland, U.S.A.

CONTENTS

18.1 THE CEREBROSPINAL FLUID FLOW SYSTEM

The choroid plexus (ChPlx) resides in the cerebral ventricles and is the source of cerebrospinal fluid (CSF). The CSF flows rapidly through the ventricles, subarachnoid cisterns, and perivascular spaces (Rennels et al., 1985) and diffuses more slowly in the interstitial spaces (Brightman, 1965) into the parenchyma, reaching every location in the brain (Ghersi-Egea et al., 1996; Proescholdt et al., 2000). The CSF flow system serves an important function as a drainage route for clearance of degraded proteins and metabolites. In addition, bioactive molecules or live white blood cells can enter the CSF stream at any point along its flow route, and they can travel to nearby or more distant sites in the brain to affect or interact with other responsive cells. The ChPlx, as the source of the circulating fluid, lies "upstream." At the "downstream" end, both fluid and its contents exit the brain into blood and lymph along several described routes (Cserr et al., 1992; Kida et al., 1993). Sampling contents of CSF at points along the flow path can give important insights into its information-carrying potential, but identifying the original sources of the informational molecules and cells can be very difficult. Because the ChPlx is upstream, one can surmise that it is an important source of the informational contents, but this is not easy to demonstrate. It has been shown that molecules produced by the ChPlx, or cells or molecules passing through the ChPlx into the ventricular CSF, have the potential to reach target sites throughout the brain by the aforementioned flow routes. We have calculated that the tracer molecule [^{14}C]inulin injected into the lateral venticle of rats is evenly distributed throughout the parenchyma at about 1% of the starting ventricular concentration by 4 hours (Proescholdt et al., 2000).

18.2 THE BLOOD–BRAIN BARRIER AND THE BLOOD–CSF BARRIER

The blood–brain barrier (BBB) for the most part comprises the vascular endothelia and the tight junctions between them that effectively prevent movement of molecules and cells out of the blood and into the brain compartment. That essential arrangement is modified in the circumventricular organs (CVOs) and ChPlx, where the blood vessels are fenestrated. Important CVOs related to immune signaling pathways are the organum vasculosum of the lamina terminalis (OVLT), lying at the rostral tip of the third ventricle in the anterior preoptic area; the subfornical organ (SFO), situated below the ventral hippocampal commissure at the foramen of Monro; the median eminence (ME), comprising the floor of the third ventricle at the level of the hypothalamic arcuate nucleus and infundibular stalk; and the area postrema (AP), lying at the dorsum of the medulla, above the nucleus of the solitary tract, at the origin of the central canal of the spinal cord. CVOs, by virtue of their circumventricular locations, and the ChPlx, as the thin boundary separating blood from the ventricular CSF, both seem to reside in ideal positions to enable movement of molecules or cells directly from blood, across the fenestrated vessels, into the large fluid compartments of the brain. But this simple passage is prevented. In the CVOs, passage from blood vessels into the parenchymal spaces occurs, but easy movement out of the organ and into the CSF compartment is blocked by tight junctions along the ependymal walls of the CVOs (Krisch et al., 1978; Petrov et al., 1994). In the ChPlx, which is a sandwich composed of vascular endothelia, stroma, and choroidal epithelia, tight junctions reside between the epithelial cells (see Chapters 2 and 3), and access of molecules and cells from the blood to the ventricular CSF is effectively blocked. Nature thus appears to work in mysterious ways by placing CVOs and ChPlx in an ideal position to provide molecules access to the upstream portions of the CSF flow routes, but then restricts access in both locations.

18.3 THE BRAIN'S IMMUNE RESPONSE

The brain and the immune system communicate with each other via pathways and mechanisms that are much studied but not completely understood. Brain-to-immune signaling is achieved by neural efferent projections arising from brainstem, spinal cord, and autonomic ganglia that innervate immune organs like the thymus and spleen. These connections mediate central nervous system (CNS) control of immune organ functioning (Katafuchi et al., 1993; Hori et al., 1995). In addition, circulating hormones and neurotransmitters under CNS control powerfully affect immune system function.

In immune-to-brain signaling, immune system molecules such as cytokines and pathogen-associated molecular patterns (PAMPs) (Dalpke and Heeg, 2002) travel in the blood to bind to specific receptors on target cells with important consequences, depending in large part on the particular cell type bearing the respective receptors. Receptors on nerve endings in afferent peripheral nerves, such as in the vagus nerve, trigger signals that ascend to the brainstem (Goehler et al., 2000). Alternatively, receptor-bearing cells at the blood–brain interface or in brain areas where the molecules have direct access to the brain can trigger responsive signals.

Immune signaling to the brain can invoke a CNS response whereby CNS-derived immune system molecules are produced and released to act as paracrine signals. The various mechanisms and issues relevant to immune-to-brain signaling by cytokines and PAMPs have been reviewed (Quan and Herkenham, 2002; Rivest, 2003). Finally, cells of the immune system travel in the blood, and they pass into the brain both normally and in response to infection, injury, degeneration, or disease. The purpose of this chapter is to review these processes in the context of the potential roles played by the ChPlx and CSF.

18.4 IMMUNE-TO-BRAIN SIGNALING INVOLVING THE ChPlx AND CSF

Pathways for immune-to-brain communication have been thoroughly explored and discussed. Cytokines and PAMPs are considered the primary information-bearing molecules signaling the presence of peripheral infection or injury (Watkins et al., 1995). Possible cytokine/PAMP communication pathways include two humoral routes—direct entry into the brain across the BBB mediated by a saturable transport mechanism (Banks et al., 1995) and entry at CVOs where the BBB is leaky (Broadwell and Sofroniew, 1993; Lee et al., 1998)—and a neural route, namely activation in the gut organs of peripheral nerve afferents, prominently the vagus nerve, leading to transsynaptic neuronal signaling ascending from the medulla to the hypothalamus (Ericsson et al., 1997). A fourth mechanism of immune-to-brain signaling is signal transduction at the BBB itself; by this model, an inflammatory molecule (cytokine or PAMP) in the blood interacts with receptors on barrier cells, and these cells in turn synthesize inflammatory molecules, notably prostaglandins (Matsumura et al., 1998), which are released on the brain side of the BBB (Ek et al., 2001), from where they diffuse to target cells responsible for producing fever, sleep, anorexia, activation of the hypothalamic-pituitary-adrenal (HPA) axis stimulating corticosterone secretion, and other behavioral and hormonal changes associated with the acute sickness response. The endothelial response to the bloodborne signals is found all over the brain's vasculature, so it is not known whether or not there is a critical site for this transduction event. In addition, the diffusional pathways on the brain side of the BBB have not been explored. However, the critical role of the CVOs, notably AP, ME, SFO, and OVLT, in production of fever and HPA axis response has been indicated by results of lesion and tract-tracing studies (Takahashi et al., 1997; Lee et al., 1998; Romanovsky et al., 2003). These findings diminish the possibility of a role for ChPlx and CSF in these aspects of the acute phase response.

Nevertheless, histochemical studies have shown that cells of the ChPlx are highly responsive to peripheral immune challenges. The ChPlx has receptors for cytokines, e.g., the type 1 interleukin-1 receptor (IL-1R1), demonstrated by ligand-binding studies and by in situ hybridization (Ban, 1994; Ericsson et al., 1995; Haour et al., 1995; Nadeau and Rivest, 1999) (see Figure 18.1a, b). Recently, receptors for PAMPs have been identified and mapped, and Toll-like receptor-4 (TLR4), the receptor for the bacterial cell wall component lipopolysaccharide (LPS), has been found to be constitutively expressed in low numbers on ChPlx cells (see Figure 18.1c, d) (Laflamme and Rivest, 2001). The distributions of the other TLRs (1–9), have not yet been mapped, though one, TLR2, is a highly inducible receptor that shows strong mRNA expression in the ChPlx following LPS challenge (Laflamme et al., 2003).

Following peripheral administration of the PAMP LPS or the proinflammatory cytokine interleukin-1 (IL-1), the brain has been shown to mount its own inflammatory response. These inflammatory challenges induce the expression of cytokines (IL-1α, IL-1β, IL-6) (Van Dam et al., 1992; Vallières and Rivest, 1997; Quan et al., 1998a, 1999a; Eriksson et al., 2000; Garabedian et al., 2000), immune-related receptors and transcription factors (CD14, IκBα) (Quan et al., 1997; Lacroix et al., 1998), chemokines (monocyte chemoattractant protein, MCP-1) (Thibeault et al., 2001), and general inflammatory molecules (cyclooxygenase-2, Cox-2, and inducible nitric oxide synthase, iNOS, are markers of production of prostaglandins and nitric oxide, respectively) (Wong et al., 1996; Lacroix and Rivest, 1998; Quan et al., 1998b) in cells at or near the BBB, especially the ChPlx and CVOs. Most of these transcripts have no constitutive expression in the absence of a specific immune challenge, but upregulation following peripheral administration of cytokines or LPS is rapid and transient, lasting 12 to 24 hours. As an example, Figure 18.2 shows the pattern of induced IL-1β mRNA expression at three time points following LPS administration (Quan et al., 1998a). At 0.5 hour after intravenous LPS administration in rats, both vascular and epithelial cells express message (see Figure 18.2a). The expression reaches maximal levels at two to four hours (see Figure 18.2b), then declines somewhat at eight hours. At 12 hours, a second wave of mRNA expression can be seen throughout the brain. In the ChPlx, most cells expressing mRNA appear to be displaced cells, probably therefore epiplexus cells (see Figure 18.2c). In double-label studies, they stain positively for the macrophage marker ED1 at 24 hours after LPS injection (Eriksson et al., 2000).

Most of the IL-1β mRNA measured in brain following peripheral LPS injection derives from meninges and ChPlx (Garabedian et al., 2000). The induction of IL-1β

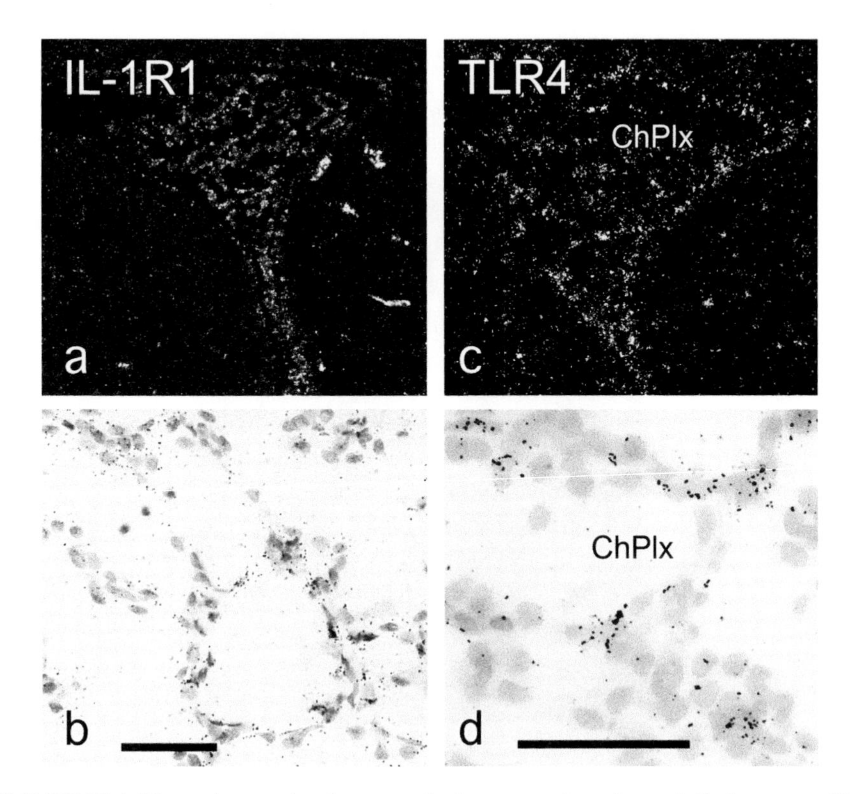

FIGURE 18.1 Photomicrographs show constitutive expression of type 1 IL-1 receptor (IL-1R1) (a, b) and TLR4 (LPS receptor). (c, d) mRNA expression in ChPlx and ventricular ependyma of rat (IL-1R1) and mouse (TLR4) at low (a, c) (darkfield illumination) and high (b, d) (brightfield illumination) magnifications. The labeling appears as silver grains overlying both ependymal and epithelial cells for both cRNA ribonucleotide probes. There is strong vascular labeling of IL-1R1 in the brain parenchyma as well (a). Magnification bars = 100 μm for b and d. (IL-1R1 from N. Quan; TLR4 from S. Chakravarty.)

mRNA has a temporal course that parallels measures of bioactive IL-1 in the CSF and brain parenchyma (Quan et al., 1994). Similarly, selective induction of other immune system molecules in the ChPlx and meninges suggests that the ChPlx is a major source of elevated cytokines in the CSF measured six hours after peripheral LPS administration (Quan et al., 1994), though this origin has not been definitively demonstrated. The observations have nonetheless led many investigators to study the consequences of cytokine administration into the CSF (see below).

In one study we used c-fos mRNA as a general cellular activation marker and found that cells of the ChPlx have induced c-fos mRNA expression shortly after intravenous IL-1β administration in rats (Herkenham et al., 1998). We also used a marker for NF-κB transcription factor activity to show more specifically immune-related responses in the CNS to an immune challenge (Quan et al., 1997). NF-κB is activated by cytokines, chemokines, PAMPs, viruses, heat shock, and other stressors impinging on cell-surface receptors. Following its translocation from the cytoplasm to the nucleus and binding to its response element on the DNA, NF-κB can

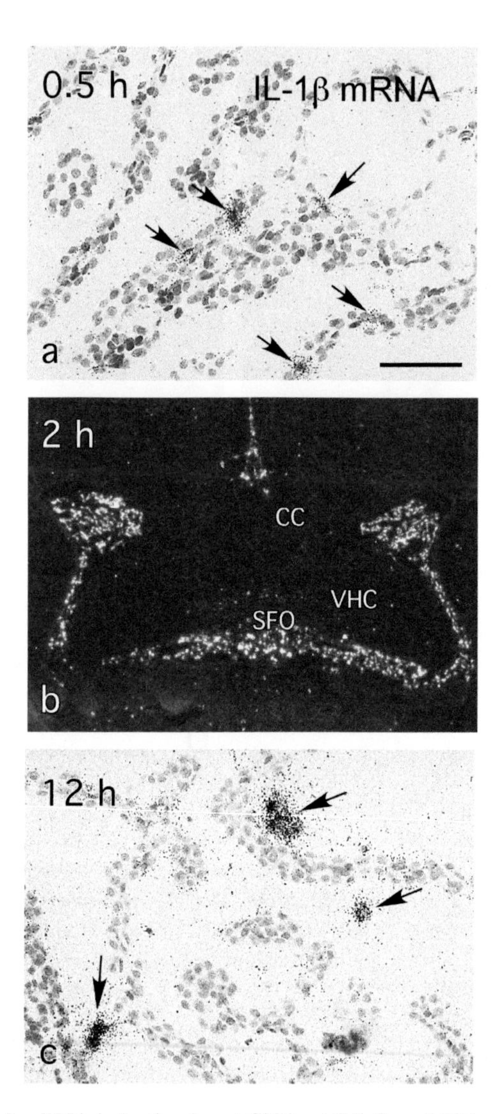

FIGURE 18.2 IL-1β mRNA induction in rat ChPlx at 0.5, 2, and 12 hours after intravenous LPS injection in rats. In unchallenged animals, there is no detectable IL-1β mRNA expression (data not shown). High-magnification photomicrographs show progression of labeled cell type from endothelial and epithelial (a) to epiplexus (c). Peak of mRNA induction at two to four hours is represented at low magnification (darkfield) in b, and the labeling is dense in the ChPlx, vasculature of the interhemispheric fissure, and the subfornical organ (SFO), a CVO. Abbreviations: cc, corpus callosum; VHC, ventral hippocampal commissure. Bar = 100 μm for a and c. (From Quan et al., 1998a.)

induce the transcription of hundreds of genes, most of which are in the immune family or are associated with cell death or survival pathways (Barkett and Gilmore, 1999; Li and Verma, 2002). One mRNA transcribed is the NF-κB inhibitor IκB, which serves to reset the transcription factor activity. The induction of IκBα mRNA parallels NF-κB activation in a time-dependent manner, and IκBα is an immediate-early gene that generally reflects immune-related activity in many cell types (Quan et al., 1997).

A temporal evolution of IκBα mRNA expression can be documented in the ChPlx following peripheral LPS administration in rats, whereby induced expression of IκBα occurs first in vascular endothelial cells at 0.5 hour (see Figure 18.3a, b) and at later times (2 to 12 hours) in epithelial and perhaps also epiplexus cells (see Figure 18.3c,d). Expression of IκBα mRNA in the ventricular ependyma at 12 hours (see Figure 18.3e, f) supports the speculation that a secondary immunologically active molecule has been generated by the LPS stimulation and subsequently secreted into the CSF.

Likely candidates for secondary inflammatory molecules are prostaglandins because they appear to be synthesized all along the BBB axis, including but not limited to the ChPlx, as indicated by mRNA induction of the Cox-2 enzyme gene (Lacroix and Rivest, 1998). However, due to their instability and lipophilia, they are better suited for exerting a local paracrine effect on target cells bearing prostaglandin receptors. Another candidate inflammatory signaling molecule is nitric oxide (NO), which can be produced following induction of iNOS in barrier locations following sepsis-like stimuli or infectious conditions. Markers of NO can be measured in the CSF following peripheral LPS (Wong et al., 1996), but its origins are not known.

18.5 IMMUNE CELLS IN THE ChPlx

Macrophage-like cells with significant phagocytic abilities are found throughout the brain. These cells include the microglia in the brain parenchyma proper, supraependymal cells lining the ventricles, epiplexus cells of the ChPlx, meningeal macrophages of the meninges, and perivascular cells throughout the brain (Jordan and Thomas, 1988). The perivascular cell, also called the perivascular macrophage or perivascular microglial cell (Hickey et al., 1992), is hematogenic and has been shown to have a slow, steady turnover rate (Bechmann et al., 2001; Vallières and Sawchenko, 2003). Presumably all the other cells have similar turnover properties, trafficking in and out of brain. These cells are the major antigen-presenting cells of the brain, and they serve as scavengers for foreign debris and other molecules in the CSF (Angelov et al., 1998). It has been suggested that they help clear the CSF of foreign molecules that have breached the BBB (Lu et al., 1993; Ling et al., 1998).

Cells immunologically identified as bone marrow-derived because they immunostain for CD45, also called leukocyte common antigen (LCA), are found normally in the ChPlx, both in the stromal portion of the structure and in the locations occupied by epiplexus cells, at the margins of the epithelial cell wall, and protruding into the ventricular compartment (see Figure 18.4a). The epiplexus cell (Kolmer cell) has many characteristics of a macrophage, staining for the markers OX42, OX1 (CD45), OX18, and ED1 and showing slow turnover, derived from source hematogenic monocytes (Kaur et al., 1996; Ling et al., 1998).

Cells staining positively for ED2 typically lie along the epithelial margins (Jordan and Thomas, 1988). The ED2 antibody is a classic marker for mature tissue macrophages and perivascular cells elsewhere in the brain (Graeber et al., 1989).

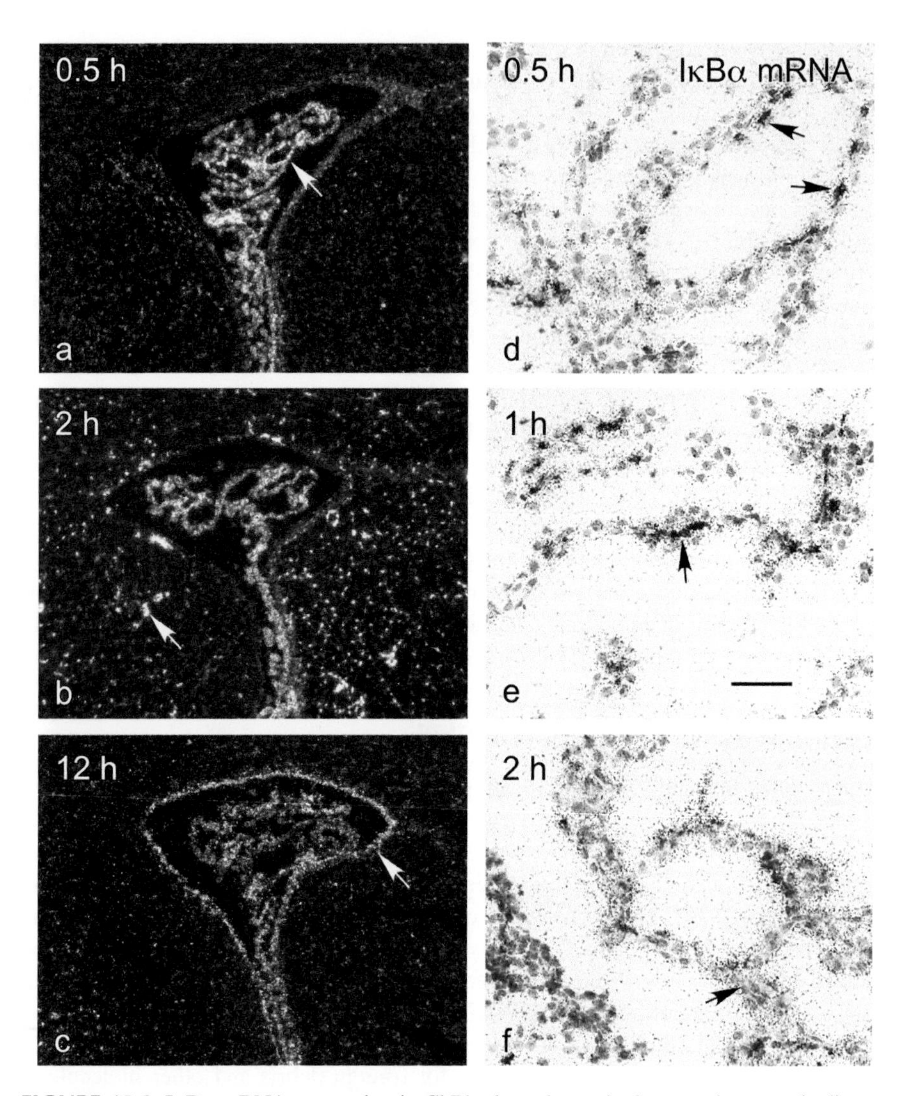

FIGURE 18.3 IκBα mRNA expression in ChPlx, lateral ventricular ependyma, and adjacent parenchyma (septum, striatum, corpus callosum) is shown in darkfield low magnification (a–c) and brightfield high magnification (d–f) at 0.5 (a, d), 1 (e), 2 (b, f), and 12 hours (c) post–intravenous LPS administration. Arrows in a and d point to endothelial cell labeling at 0.5 hour. Arrows in e and f point to epithelial labeling at 1 to 2 hours. Arrow in b points to endothelial cell labeling in the parenchyma. Arrow in c points to ventricular ependymal labeling at 12 hours. Bar = 500 μm and 100 μm for a–c and d–f, respectively. (From Quan et al., 1997.)

Constitutive expression of the cell adhesion molecule ICAM-1 on the outer epithelial surface of the choroidal epithelia may serve to anchor these cells (Engelhardt et al., 2001) (see Figure 18.4b). Curiously, ICAM-1 staining has been reported to be restricted to endothelial cells in human ChPlx (Kivisakk et al., 2003). It is noteworthy

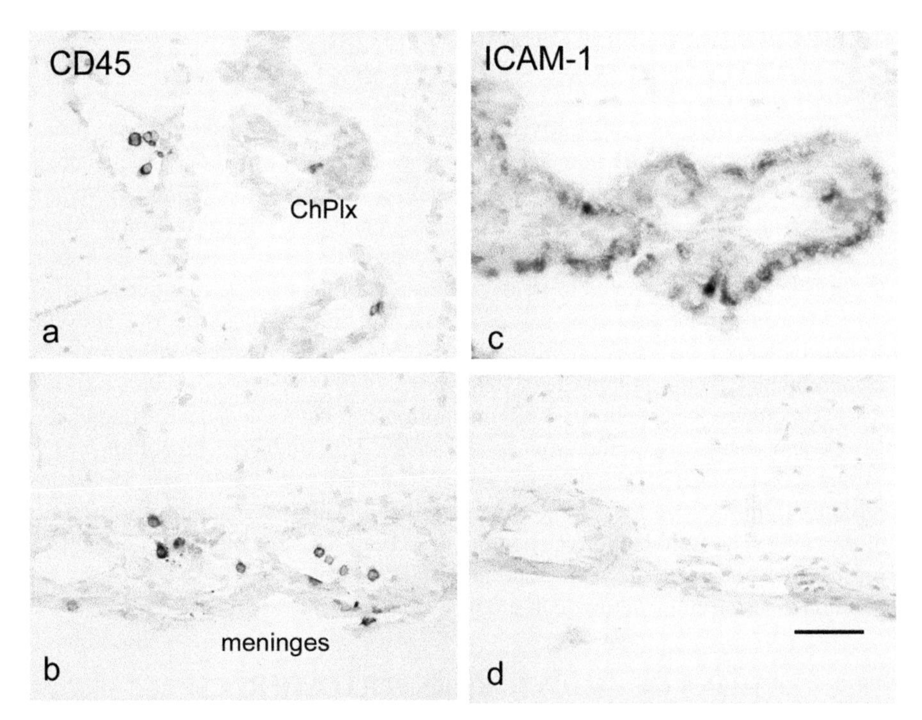

FIGURE 18.4 CD45-immunostained leukocytes normally appear in ChPlx (a) and meninges (b) of untreated rats. In the same animals, ICAM-1 immunostaining is constitutively present on epithelial side of the ChPlx (c) and is almost absent in the meninges (d). Bar = 50 μm.

that CD45-positive leukocytes in other periepithelial locations exist in the absence of ICAM-1 immunostaining (see Figure 18.4c,d), suggesting another role for constitutive ICAM-1 staining in the ChPlx.

Dendritic cells, which are type II major histocompatibility class (MHC)-positive, as defined by OX6 and OX62 immunostaining, and show "dendriform" morphology, have been shown to be abundantly distributed in the ChPlx stroma in whole-mount preparations (McMenamin, 1999). Dendritic cells are also present in the dura mater and leptomeninges but are not found in the brain parenchyma in normal conditions. They are usually found in low numbers in the CSF, but in inflammatory conditions they appear in large numbers in the CSF (Pashenkov et al., 2003). The origin of the additional cells is not known, but chemokine-based recruitment is suggested. Dendritic cells in the CSF serve another important function for immune-brain communication; they can exit the brain into the cervical lymph nodes, where they can present CNS-derived antigens to T lymphocytes for elaboration of an acquired immune response directed against the brain (Carson et al., 1999).

In addition, lymphocytes enter the brain in normal conditions, where they serve as surveillance cells (Hickey et al., 1991). $CD3^{+}$ T lymphocytes have been found in normal human ChPlx stroma (Kivisakk et al., 2003). Similarly, P-selectin–stimulated "pioneer" T lymphocytes have been marked in mouse ChPlx stroma (Carrithers et al., 2000, 2002).

18.6 ChPlx IMMUNE CELLS AND CYTOKINES FOLLOWING IMMUNE CHALLENGE

In addition to bone marrow-derived immune cells residing in brain under normal conditions, leukocyte entry into the brain can be induced by inflammatory challenges. Depending on the nature of the stimulus, cells with signature CD immunostaining characteristics are found in the CSF, ChPlx, meninges, and perivascular spaces. Discussion of such responses is beyond the scope of this chapter because the cellular trafficking is usually targeted to brain regions other than the ChPlx and may involve the ChPlx only minimally or as part of its role as a brain barrier structure. However, some challenges do cause significant leukocyte entry into the ChPlx and CSF compartments, so the examples are noteworthy.

For example, peripheral LPS administration can induce leukocyte adherence to ChPlx and meninges (Andersson et al., 1992). LPS induces chemokine mRNA expression in the ChPlx and the vasculature (Thibeault et al., 2001). ICAM-1 immunostaining intensity in the ChPlx is increased following LPS administration (Endo et al., 1998). The combination of chemokine and cell adhesion molecule upregulation presumably enhances the adherence and entry of leukocytes into the ChPlx and meninges. These cells, in turn, can produce cytokines that are released into the CSF.

18.7 ACTIONS OF CSF CYTOKINES FOLLOWING CSF FLOW ROUTES

Leukocytes that gain access to the subarachnoid spaces (as in meningitis) themselves secrete inflammatory molecules that move throughout the brain along established CSF flow routes. In an attempt to show this humoral flow pathway, we mapped the spread using [^{14}C]inulin, an inert, large molecular weight tracer molecule (Proescholdt et al., 2000). Following its injection into the lateral ventricle of awake rats, the radioactivity spread rapidly along the ventricular path caudalward from the third to the fourth ventricle, after which it entered the cisterna magna and moved forward on the outside of the brain, within the cisterns and subarachnoid spaces. All along this route, tracer diffused at a slower rate into the brain parenchyma, filling the extracellular spaces in a time-dependent manner. Figure 18.5a–c shows the pattern of tracer distribution at 0.5 hour after injection. Spread into perivascular spaces within the parenchyma was seen along major blood vessels both with [^{14}C]inulin and with a more sensitive (but not quantitative) marker, Alexa-488–labeled dextran, similarly injected into the lateral ventricle and microscopically visualized in brain sections with fluorescence illumination (see Figure 18.5g, h).

We next administered the cytokine IL-1β into the lateral ventricle, using the same experimental conditions, and measured induced expression of IκBα mRNA as a marker for immune activation (Proescholdt et al., 2002). Induced signal was strong all along the CSF flow routes and was particularly evident in the ChPlx, meninges, and ventricular walls. It was also evident, in a gradient of expression density, in vascular and other parenchymal cells (not neurons) throughout the brain. This mRNA induction pattern matched the inulin spread pattern but was dramatically amplified along the vasculature (see Figure 18.5d–f). This can be explained by the

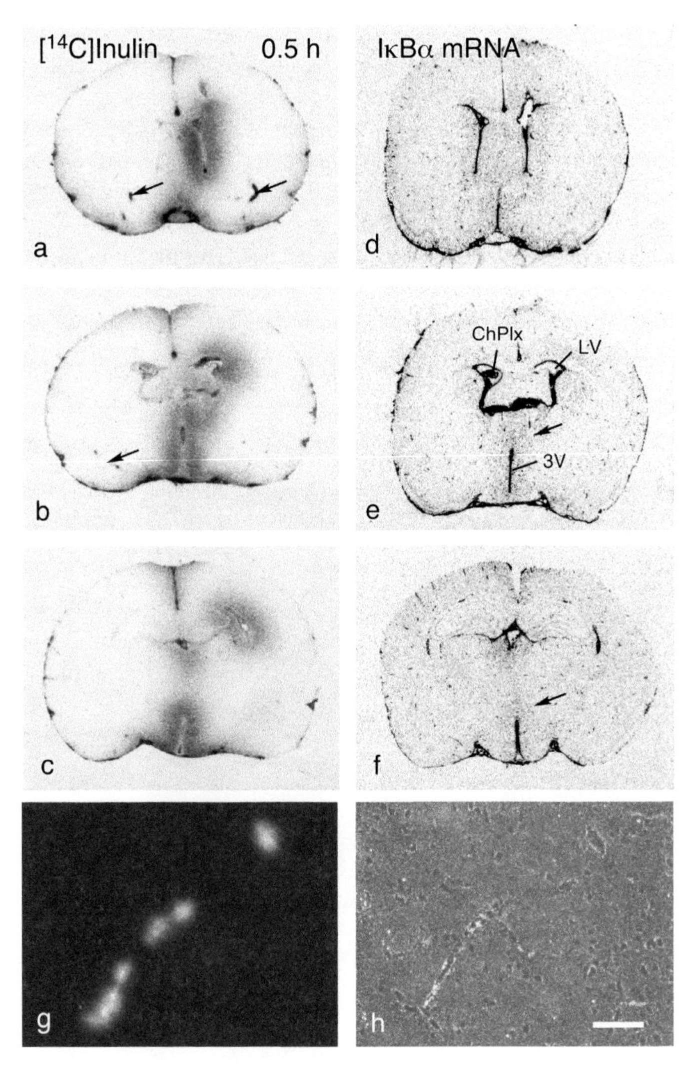

FIGURE 18.5 Film images show CSF flow patterns tracked by [^{14}C]inulin in the brain at 0.5 hour after lateral ventricular injection in awake rats (a–c). Following rapid spread through the ventricles and subarachnoid spaces (seen at 5-minute survival), the tracer spreads more slowly into the adjacent brain parenchyma, as shown. Some perivascular spread is also noticeable (arrows in a and b). High magnification of fluorescent image of Alexa-488 conjugated to dextran (Molecular Probes) shows perivascular spread of the CSF flow marker in the Virchow-Robin spaces in the diencephalon (g). The same field viewed in darkfield illumination (unstained section) shows the blood vessel in the corresponding location (h). Bar = 100 μm for g, h. The tracer distribution pattern is compared with the early pattern of induction of IκBα mRNA in blood vessels following lateral ventricular IL-1β administration (d–f). Arrows in e and f point to areas of relatively greater IκBα mRNA expression, matching the [^{14}C]inulin spread. However, elsewhere there is widespread vascular labeling that is attributed to IκBα mRNA induction in endothelial cells that express IL-1R1. Note also dense labeling of the ChPlx. (From Proescholdt et al., 2000, 2002.)

preexisting strong presence of type 1 IL-1 receptors (IL-1R1) on vascular endothelia (Ericsson et al., 1995). Thus, whereas only a small amount of injected IL-1β gained access to these locations via flow in the perivascular space, the response by the receptor-bearing cells was dramatic. As a result of the downstream activation of the NF-κB transcription factor, the endothelial cells began to express cell adhesion molecules (ICAM-1) and chemokines (MCP-1) (Proescholdt et al., 2002), signaling bloodborne leukocytes to attach to the vascular walls and cross into the brain parenchyma several hours later (see Figure 18.6a). The locations of the infiltrating leukocytes matched both the pattern of [^{14}C]inulin spread and the location of the earlier IκBα mRNA induction in the endothelial cells. Interestingly, when IL-1β was administered intravenously, the vascular response still occurred (IκBα mRNA induction), but there was no subsequent ICAM-1 or MCP-1 induction and no leukocyte entry (Proescholdt et al., 2002).

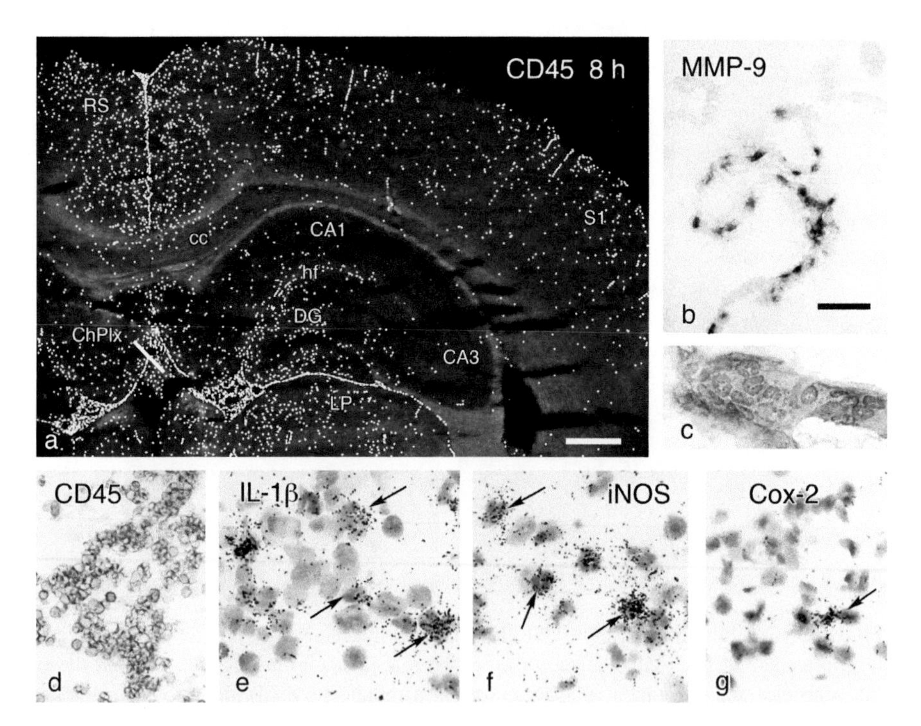

FIGURE 18.6 (a) CD45-immunostained leukocytes infiltrating the brain eight hours after intracerebroventricular injection of IL-1β. The cells are numerous in the ChPlx, meninges, and perivascular locations. Their overall pattern of localization mirrors the earlier distribution on the administered IL-1β. High-magnification photomicrographs show that the leukocytes stain for MMP-9 (b, c), have neutrophil morphology (c), and express CD45 (d) and mRNA for IL-1β (e), iNOS (f), and Cox-2 (g). Bar in a = 1 mm. Bar = 100 μm for b and e–g, 20 μm for c, and 200 μm for d. Abbreviations: CA1, CA3, hippocampal CA fields 1 and 3; cc, corpus callosum; DG, dentate gyrus; hf, hippocampal fissure; LP, lateral posterior thalamic nucleus; RS, retrosplenial cortex; SI, primary somatosensory cortex. (From Proescholdt et al., 2002.)

The dramatic induction and infiltration pattern underscores the significance of the CSF distribution system for bioactive molecules like IL-1β. Another spatiotemporal pattern of IκBα mRNA induction was seen in the brain following administration of tumor necrosis factor-α (TNF-α) into the lateral ventricle of rats (Nadeau and Rivest, 2000), though in that case the spread of induced message was restricted to areas closely adjacent to the ventricles and, therefore, more closely matched the pattern of [^{14}C]inulin movement. A proposed difference between the distributions induced by the two cytokines is that receptors for TNFα are in low abundance constitutively and are locally induced by the circulating TNFα, enabling the response to transpire in the more proximal locations (Nadeau and Rivest, 2000).

These widespread responses generated as a consequence of the CSF-mediated distribution of the inflammatory molecules can be contrasted with the very regionally restricted response seen when immune stimuli are injected intraparenchymally. Holmin et al. (2000) showed leukocyte entry regionally restricted to the site of administration following IL-1β injection into the cerebral cortex. Apparently in this case, the IL-1β diffused and acted locally and did not enter the CSF flow pathways in amounts significant enough to trigger the widespread response.

The intracerebroventricular (icv) administration of IL-1β elicited massive leukocyte entry into subarachnoid spaces and ChPlx as well as the brain parenchyma (see Figure 18.6a–d). The leukocytes expressed mRNA for IL-1β, iNOS, Cox-2, and IκBα (see Figure 18.6e–g). They also stained intensely for matrix metalloproteinase-9 (MMP-9) (see Figure 18.6b), which may have contributed to the BBB opening. The infiltrating leukocytes, mostly neutrophils based on their polymorphonuclear appearance (see Figure 18.6c), could be an alternate source of matrix metalloproteinases measured in the CSF following inflammatory events, in addition to ChPlx cells themselves (Strazielle et al., 2003). In the ventricular IL-1β administration model, a final outcome of the induced cascade of proinflammatory events was a brainwide glial activation involving both astrocytes and microglia. Both mRNA transcription and protein immunostaining of GFAP (astrocyte marker) and iba1 (microglial marker) were elevated, as was nonneuronal c-fos mRNA expression (Proescholdt et al., 2002). These effects dramatically demonstrate how an immune signal originating in the ChPlx could have widespread effects on brain activity.

18.8 ChPlx IN IMMUNE DISEASE—TRYPANOSOMIASIS

There are several diseases in which the ChPlx appears to be a preferred target for infectious agents (Levine, 1987; Petito, 2004). Harboring of parasites, viral particles, or other microbial agents in the ChPlx stroma may afford them a degree of protection from immune system attack. Some of the specific disease examples are discussed elsewhere in this book (see Chapter 15), so only one is presented here, which is caused by a trypanosome parasite.

African sleeping sickness is caused by the parasite *Trypanosoma brucei gambiense* (Haller et al., 1986). The disease involves significant CNS degeneration (Pentreath, 1995). Its counterpart in small animals is caused by the *Trypanosoma brucei*

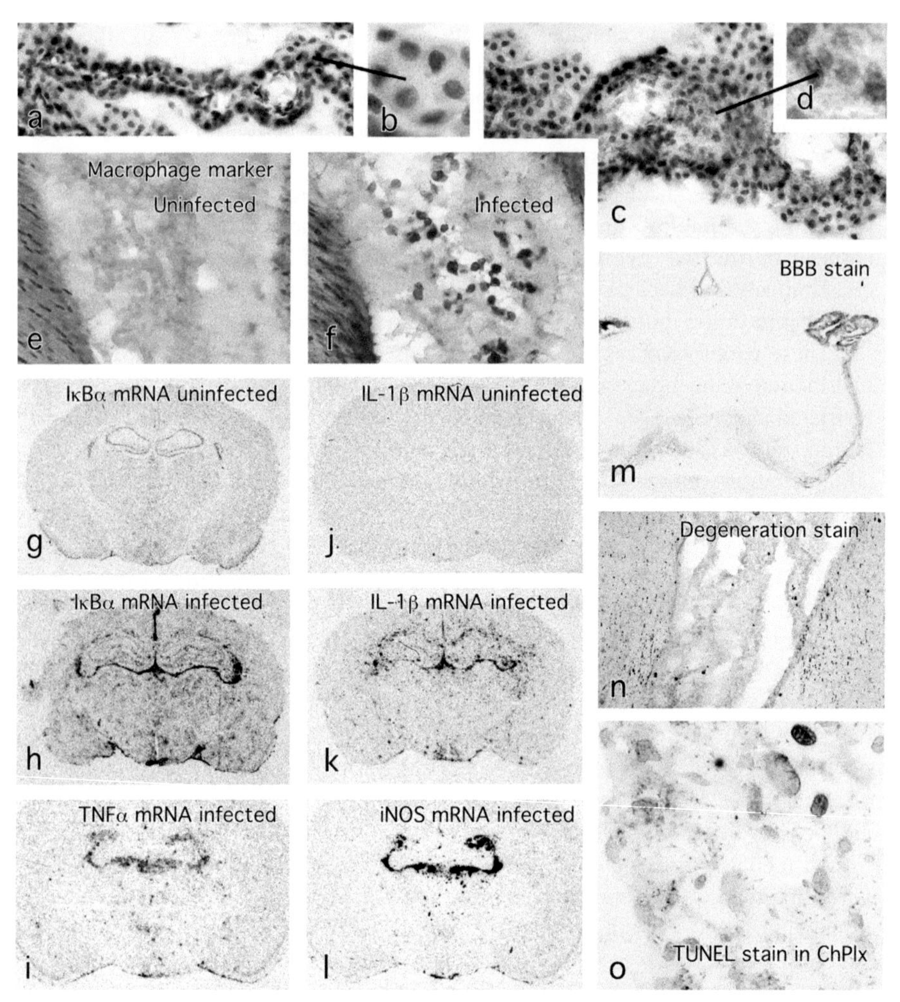

FIGURE 18.7 The CNS consequences of infection in rats by the trypanosome parasite are shown. Several weeks after infection, the parasites become heavily lodged in the ChPlx (a–d), and macrophages invade the ChPlx (e, f) (silver stains of macrophages by NeuroScience Associates, NSA). In parallel, there is a time course of cytokine mRNA induction in ChPlx, CVOs, and adjacent brain areas (g–l), a breakdown of the blood–CSF barrier (shown by immunoglobulin staining in m), and neurodegeneration in nearby fiber tracts (n, silver stained by NSA). The TUNEL stain for fragmented DNA (o) shows staining of parasites (small cells) and macrophages (large cells filled with debris). (From Quan et al., 1999b.)

brucei parasite (Keita et al., 1997). The trypanosome is an extracellular organism that travels in the blood and lodges in various organs. It cannot cross the BBB, but it resides in the interstices of the ChPlx (see Figure 18.7 a–c) and CVOs (Poltera et al., 1980; Schultzberg et al., 1988). Because it does not enter the brain in significant numbers until the late stages of the disease, a possible mechanism by which neurodegeneration occurs is by the release of toxic substances into the CSF pathways. We used this chronic inflammation model to demonstrate that a spatiotemporal sequence of events occurs, similar to that following peripheral LPS administration, but protracted over weeks instead of hours (Quan et al., 1999b). We showed the early appearance of cytokine mRNA induction in the ChPlx and CVOs (see Figure 18.7g–i). The ChPlx itself was engorged with parasites, and this region became a battleground characterized by the presence of macrophages (see Figure 18.7e,f), breakdown of the blood–CSF barrier (see Figure 18.7m), apoptosis and phagocytosis of nuclear debris as seen with a TUNEL stain (see Figure 18.7o), and strong induction of iNOS mRNA (see Figure 18.7l). These observations support the hypothesis that toxic substances can gain access to the CSF flow pathways. The appearance of cytokine mRNA-positive cells in the vicinities of the ventricles (see Figure 18.7g–i), and of neurodegeneration in areas close to the ventricular and meningeal surfaces (see Figure 18.7n), supports this hypothesis. In a recent study, inactivation of TNF-α in the CSF by icv introduction of TNF antibodies was shown to reduce the severity of neurodegeneration and of general disease indicators (Quan et al., 2003).

The model generated by this example is of inflammatory and possibly toxic molecules moving from sources in the ChPlx and meninges throughout the CSF flow routes. Several diseases appear to be related in some way to the CSF flow routes, occasionally indicating an involvement of the ChPlx. Injury-, cytokine- and LPS-induced prostaglandins [PGE_2, PGD_2, and PGD synthase (β-trace)] (Van Dam et al., 1993; Beuckmann et al., 2000; Kunz et al., 2002; Takano et al., 2003) produced in and released from the ChPlx and meninges may regulate sleep cycles (Urade and Hayaishi, 1999; Hayaishi, 2000). Leukotriene C4 synthase is exclusively produced in the ChPlx under normal conditions (Soderstrom et al., 2003). It was suggested that leukotriene C4 in the CSF might alter the permeability of blood vessels throughout the brain in response to injury. Other functions need to be elucidated in various inflammation models. Inflammation-induced choroidal release of MMP-2 and MMP-9 could reduce blood–CSF barrier integrity (Strazielle et al., 2003).

Upregulation of cell adhesion molecules in ChPlx occurs in experimental allergic encephalomyelitis (EAE), a model of the immune disease multiple sclerosis, and there are structural changes in the choroidal epithelial cells (Engelhardt et al., 2001). In human multiple sclerosis, the brain lesions localized by magnetic resonance imaging predominate in periventricular white matter (Lee et al., 1999), suggesting that substances flowing in the CSF contribute to lesion generation and are more relevant to final lesion location than initial sites of BBB breakdown.

18.9 SUMMARY AND CONCLUSIONS

The ChPlx and the CVOs are uniquely sensitive to circulating cytokines and pathogen-associated inflammatory molecules. The cells of the ChPlx respond to peripheral LPS or cytokine administration by producing a variety of inflammatory molecules that are released into the CSF and distributed to distant receptor-bearing target cells. Similar kinds of responses occur in the CVOs and perhaps to a lesser extent in the vascular BBB cells throughout the brain, and the actions of the transduced inflammatory molecules in these locations may be more local.

Although the precise sources and destinations of the inflammatory molecules produced in the ChPlx following peripheral immune challenge are not precisely known, a pattern of responses can be demonstrated by experimentally introducing the cytokines directly into the CSF. In these studies, widespread vascular and glial responses can be observed. In chronic disease models, patterns of induced cytokine expression in the brain occur along with related neurodegeneration.

In addition, cellular trafficking across the ChPlx occurs normally, and bone marrow-derived phagocytic macrophages and lymphocytes serve scavenging and surveillance roles. These cells are activated and recruited following specific immune challenges or during disease, and their responses are not unique but rather resemble those of similar cells found throughout the brain.

Further work is needed in this field to demonstrate that the blood-to-CSF transfer of immune molecules and cells occurs in a functionally relevant way as proposed by the experimental demonstrations that are typically phenomenological. Work with tagged or modified molecules or genetically engineered cells may aid in their tracking across the blood–CSF barrier and assessment of their effect on brain processes. Work with transgenic animals will help to disclose the key molecular signals involved. Experiments can be designed to assess the role of this trafficking system in the elaboration of injury, disease, or other processes involving inflammatory signals.

REFERENCES

Andersson PB, Perry VH, Gordon S (1992) The acute inflammatory response to lipopolysaccharide in CNS parenchyma differs from that in other body tissues. Neuroscience 48:169–186.

Angelov DN, Walther M, Streppel M, Guntinas-Lichius O, Neiss WF (1998) The cerebral perivascular cells. Adv Anat Embryol Cell Biol 147:1–87.

Ban EM (1994) Interleukin-1 receptors in the brain: characterization by quantitative in situ autoradiography. Immunomethods 5:31–40.

Banks WA, Kastin AJ, Broadwell RD (1995) Passage of cytokines across the blood-brain barrier. Neuroimmunomodulation 2:241–248.

Barkett M, Gilmore TD (1999) Control of apoptosis by Rel/NF-κB transcription factors. Oncogene 18:6910–6924.

Bechmann I, Kwidzinski E, Kovac AD, Simburger E, Horvath T, Gimsa U, Dirnagl U, Priller J, Nitsch R (2001) Turnover of rat brain perivascular cells. Exp Neurol 168:242–249.

Beuckmann CT, Lazarus M, Gerashchenko D, Mizoguchi A, Nomura S, Mohri I, Uesugi A, Kaneko T, Mizuno N, Hayaishi O, Urade Y (2000) Cellular localization of lipocalin-type prostaglandin D synthase (β-trace) in the central nervous system of the adult rat. J Comp Neurol 428:62–78.

Brightman MW (1965) The distribution within the brain of ferritin injected into cerebrospinal fluid compartments. II. Parenchymal distribution. Am J Anat 117:193–219.

Broadwell RD, Sofroniew MV (1993) Serum proteins bypass the blood-brain fluid barriers for extracellular entry to the central nervous system. Exp Neurol 120:245–263.

Carrithers MD, Visintin I, Kang SJ, Janeway CA, Jr. (2000) Differential adhesion molecule requirements for immune surveillance and inflammatory recruitment. Brain 123:1092–1101.

Carrithers MD, Visintin I, Viret C, Janeway CS, Jr. (2002) Role of genetic background in P selectin-dependent immune surveillance of the central nervous system. J Neuroimmunol 129:51–57.

Carson MJ, Reilly CR, Sutcliffe JG, Lo D (1999) Disproportionate recruitment of $CD8^+$ T cells into the central nervous system by professional antigen-presenting cells. Am J Pathol 154:481–494.

Cserr HF, Harling-Berg CJ, Knopf PM (1992) Drainage of brain extracellular fluid into blood and deep cervical lymph and its immunological significance. Brain Pathol 2:269–276.

Dalpke A, Heeg K (2002) Signal integration following Toll-like receptor triggering. Crit Rev Immunol 22:217–250.

Ek M, Engblom D, Saha S, Blomqvist A, Jakobsson PJ, Ericsson-Dahlstrand A (2001) Inflammatory response: pathway across the blood-brain barrier. Nature 410:430–431.

Endo H, Sasaki K, Tonosaki A, Kayama T (1998) Three-dimensional and ultrastructural ICAM-1 distribution in the choroid plexus, arachnoid membrane and dural sinus of inflammatory rats induced by LPS injection in the lateral ventricles. Brain Res 793:297–301.

Engelhardt B, Wolburg-Buchholz K, Wolburg H (2001) Involvement of the choroid plexus in central nervous system inflammation. Microsc Res Tech 52:112–129.

Ericsson A, Arias C, Sawchenko PE (1997) Evidence for an intramedullary prostaglandin-dependent mechanism in the activation of stress-related neuroendocrine circuitry by intravenous interleukin-1. J Neurosci 17:7166–7179.

Ericsson A, Liu C, Hart RP, Sawchenko PE (1995) Type 1 interleukin-1 receptor in the rat brain: distribution, regulation, and relationship to sites of IL-1-induced cellular activation. J Comp Neurol 361:681–698.

Eriksson C, Nobel S, Winblad B, Schultzberg M (2000) Expression of interleukin 1α and β, and interleukin 1 receptor antagonist mRNA in the rat central nervous system after peripheral administration of lipopolysaccharides. Cytokine 12:423–431.

Garabedian BV, Lemaigre-Dubreuil Y, Mariani J (2000) Central origin of IL-1β produced during peripheral inflammation: role of meninges. Mol Brain Res 75:259–263.

Ghersi-Egea JF, Finnegan W, Chen JL, Fenstermacher JD (1996) Rapid distribution of intraventricularly administered sucrose into cerebrospinal fluid cisterns via subarachnoid velae in rat. Neuroscience 75:1271–1288.

Goehler LE, Gaykema RP, Hansen MK, Anderson K, Maier SF, Watkins LR (2000) Vagal immune-to-brain communication: a visceral chemosensory pathway. Auton Neurosci 85:49–59.

Graeber MB, Streit WJ, Kreutzberg GW (1989) Identity of ED2-positive perivascular cells in rat brain. J Neurosci Res 22:103–106.

Haller L, Adams H, Merouze F, Dago A (1986) Clinical and pathological aspects of human African trypanosomiasis (*T. b. gambiense*) with particular reference to reactive arsenical encephalopathy. Am J Trop Med Hyg 35:94–99.

Haour F, Marquette C, Ban E, Crumeyrolle-Arias M, Rostene W, Tsiang H, Fillion G (1995) Receptors for interleukin-1 in the central nervous and neuroendocrine systems. Role in infection and stress. Ann Endocrinol (Paris) 56:173–179.

Hayaishi O (2000) Molecular mechanisms of sleep-wake regulation: a role of prostaglandin D2. Philos Trans R Soc Lond B Biol Sci 355:275–280.

Herkenham M, Lee HY, Baker RA (1998) Temporal and spatial patterns of *c-fos* mRNA induced by intravenous interleukin-1: a cascade of non-neuronal cellular activation at the blood-brain barrier. J Comp Neurol 400:175–196.

Hickey WF, Hsu BL, Kimura H (1991) T-lymphocyte entry into the central nervous system. J Neurosci Res 28:254–260.

Hickey WF, Vass K, Lassmann H (1992) Bone marrow-derived elements in the central nervous system: an immunohistochemical and ultrastructural survey of rat chimeras. J Neuropathol Exp Neurol 51:246–256.

Holmin S, Mathiesen T (2000) Intracerebral administration of interleukin-1β and induction of inflammation, apoptosis, and vasogenic edema. J Neurosurg 92:108–120.

Hori T, Katafuchi T, Take S, Shimizu N, Niijima A (1995) The autonomic nervous system as a communication channel between the brain and the immune system. Neuroimmunomodulation 2:203–215.

Jordan FL, Thomas WE (1988) Brain macrophages: questions of origin and interrelationship. Brain Res 472:165–178.

Katafuchi T, Ichijo T, Take S, Hori T (1993) Hypothalamic modulation of splenic natural killer cell activity in rats. J Physiol 471:209–221.

Kaur C, Singh J, Lim MK, Ng BL, Yap EP, Ling EA (1996) Studies of the choroid plexus and its associated epiplexus cells in the lateral ventricles of rats following an exposure to a single non-penetrative blast. Arch Histol Cytol 59:239–248.

Keita M, Bouteille B, Enanga B, Vallat J-M, Dumas M (1997) *Trypanosoma brucei brucei:* a long-term model of human African trypanosomiasis in mice, meningo-encephalitis, astrocytosis, and neurological disorders. Exp Parasitol 85:183–192.

Kida S, Pantazis A, Weller RO (1993) CSF drains directly from the subarachnoid space into nasal lymphatics in the rat. Anatomy, histology and immunological significance. Neuropathol Appl Neurobiol 19:480–488.

Kivisakk P, Mahad DJ, Callahan MK, Trebst C, Tucky B, Wei T, Wu L, Baekkevold ES, Lassmann H, Staugaitis SM, Campbell JJ, Ransohoff RM (2003) Human cerebrospinal fluid central memory $CD4^+$ T cells: evidence for trafficking through choroid plexus and meninges via P-selectin. Proc Natl Acad Sci USA 100:8389–8394.

Krisch B, Leonhardt H, Buchheim W (1978) The functional and structural border between the CSF- and blood-milieu in the circumventricular organs (organum vasculosum laminae terminalis, subfornical organ, area postrema) of the rat. Cell Tissue Res 195:485–497.

Kunz T, Marklund N, Hillered L, Oliw EH (2002) Cyclooxygenase-2, prostaglandin synthases, and prostaglandin H2 metabolism in traumatic brain injury in the rat. J Neurotrauma 19:1051–1064.

Lacroix S, Rivest S (1998) Effect of acute systemic inflammatory response and cytokines on the transcription of the genes encoding cyclooxygenase enzymes (COX-1 and COX-2) in the rat brain. J Neurochem 70:452–466.

Lacroix S, Feinstein D, Rivest S (1998) The bacterial endotoxin lipopolysaccharide has the ability to target the brain in upregulating its membrane CD14 receptor within specific cellular populations. Brain Pathol 8:625–640.

Laflamme N, Rivest S (2001) Toll-like receptor 4: the missing link of the cerebral innate immune response triggered by circulating gram-negative bacterial cell wall components. FASEB J 15:155–163.

Laflamme N, Echchannaoui H, Landmann R, Rivest S (2003) Cooperation between toll-like receptor 2 and 4 in the brain of mice challenged with cell wall components derived from gram-negative and gram-positive bacteria. Eur J Immunol 33:1127–1138.

Lee HY, Whiteside MB, Herkenham M (1998) Area postrema removal abolishes stimulatory effects of intravenous interleukin-1β on hypothalamic-pituitary-adrenal axis activity and *c-fos* mRNA in the hypothalamic paraventricular nucleus. Brain Res Bull 46:495–503.

Lee MA, Smith S, Palace J, Narayanan S, Silver N, Minicucci L, Filippi M, Miller DH, Arnold DL, Matthews PM (1999) Spatial mapping of T2 and gadolinium-enhancing T1 lesion volumes in multiple sclerosis: evidence for distinct mechanisms of lesion genesis? Brain 122:1261–1270.

Levine S (1987) Choroid plexus: target for systemic disease and pathway to the brain. Lab Invest 56:231–233.

Li Q, Verma IM (2002) NF-κB regulation in the immune system. Nat Rev Immunol 2:725–734.

Ling EA, Kaur C, Lu J (1998) Origin, nature, and some functional considerations of intraventricular macrophages, with special reference to the epiplexus cells. Microsc Res Tech 41:43–56.

Lu J, Kaur C, Ling EA (1993) Uptake of tracer by the epiplexus cells via the choroid plexus epithelium following an intravenous or intraperitoneal injection of horseradish peroxidase in rats. J Anat 183:609–617.

Matsumura K, Cao C, Ozaki M, Morii H, Nakadate K, Watanabe Y (1998) Brain endothelial cells express cyclooxygenase-2 during lipopolysaccharide-induced fever: light and electron microscopic immunocytochemical studies. J Neurosci 18:6279–6289.

McMenamin PG (1999) Distribution and phenotype of dendritic cells and resident tissue macrophages in the dura mater, leptomeninges, and choroid plexus of the rat brain as demonstrated in wholemount preparations. J Comp Neurol 405:553–562.

Nadeau S, Rivest S (1999) Effects of circulating tumor necrosis factor on the neuronal activity and expression of the genes encoding the tumor necrosis factor receptors (p55 and p75) in the rat brain: a view from the blood-brain barrier. Neuroscience 93:1449–1464.

Nadeau S, Rivest S (2000) Role of microglial-derived tumor necrosis factor in mediating CD14 transcription and nuclear factor κB activity in the brain during endotoxemia. J Neurosci 20:3456–3468.

Pashenkov M, Teleshova N, Link H (2003) Inflammation in the central nervous system: the role for dendritic cells. Brain Pathol 13:23–33.

Pentreath VW (1995) Trypanosomiasis and the nervous system. Pathology and immunology. Trans Roy Soc Trop Med Hyg 89:9–15.

Petito CK (2004) Human immunodeficiency virus type 1 compartmentalization in the central nervous system. J Neurovirol 10 (suppl 1):21–24.

Petrov T, Howarth AG, Krukoff TL, Stevenson BR (1994) Distribution of the tight junction-associated protein ZO-1 in circumventricular organs of the CNS. Brain Res Mol Brain Res 21:235–246.

Poltera AA, Hochmann A, Rudin W, Lambert PH (1980) Trypanosoma brucei brucei: a model for cerebral trypanosomiasis in mice—an immunological, histological and electron-microscopic study. Clin Exp Immunol 40:496–507.

Proescholdt MG, Hutto B, Brady LS, Herkenham M (2000) Studies of cerebrospinal fluid flow and penetration into brain following lateral ventricle and cisterna magna injections of the tracer [^{14}C]inulin in rat. Neuroscience 95:577–592.

Proescholdt MG, Chakravarty S, Foster JA, Foti SB, Briley EM, Herkenham M (2002) Intracerebroventricular but not intravenous interleukin-1β induces widespread vascular-mediated leukocyte infiltration and immune signal mRNA expression followed by brain-wide glial activation. Neuroscience 112:731–749.

Quan N, Herkenham M (2002) Connecting cytokines and brain: a review of current issues. Histol Histopathol 17:273–288.

Quan N, Sundar SK, Weiss JM (1994) Induction of interleukin-1 in various brain regions after peripheral and central injections of lipopolysaccharide. J Neuroimmunol 49:125–134.

Quan N, Whiteside M, Herkenham M (1998a) Time course and localization patterns of interleukin-1β messenger RNA expression in brain and pituitary after peripheral administration of lipopolysaccharide. Neuroscience 83:281–293.

Quan N, Whiteside M, Herkenham M (1998b) Cyclooxygenase 2 mRNA expression in rat brain after peripheral injection of lipopolysaccharide. Brain Res 802:189–197.

Quan N, He L, Lai W (2003) Intraventricular infusion of antagonists of IL-1 and TNFα attenuates neurodegeneration induced by the infection of *Trypanosoma brucei*. J Neuroimmunol 138:92–98.

Quan N, Whiteside M, Kim L, Herkenham M (1997) Induction of inhibitory factor κBα mRNA in the central nervous system after peripheral lipopolysaccharide administration: an in situ hybridization histochemistry study in the rat. Proc Natl Acad Sci USA 94:10985–10990.

Quan N, Stern EL, Whiteside MB, Herkenham M (1999a) Induction of pro-inflammatory cytokine mRNAs in the brain after peripheral injection of subseptic doses of lipopolysaccharide in the rat. J Neuroimmunol 93:72–80.

Quan N, Mhlanga JDM, Whiteside MB, McCoy AN, Kristensson K, Herkenham M (1999b) Chronic over-expression of pro-inflammatory cytokines and histopathology in the brains of rats infected with *Trypanosoma brucei*. J Comp Neurol 414:114–130.

Rennels ML, Gregory TF, Blaumanis OR, Fujimoto K, Grady PA (1985) Evidence for a 'paravascular' fluid circulation in the mammalian central nervous system, provided by the rapid distribution of tracer protein throughout the brain from the subarachnoid space. Brain Res 326:47–63.

Rivest S (2003) Molecular insights on the cerebral innate immune system. Brain Behav Immun 17:13–19.

Romanovsky AA, Sugimoto N, Simons CT, Hunter WS (2003) The organum vasculosum laminae terminalis in immune-to-brain febrigenic signaling: a reappraisal of lesion experiments. Am J Physiol Regul Integr Comp Physiol 285:R420–428.

Schultzberg M, Ambatsis M, Samuelsson EB, Kristensson K, van Meirvenne N (1988) Spread of *Trypanosoma brucei* to the nervous system: early attack on circumventricular organs and sensory ganglia. J Neurosci Res 21:56–61.

Soderstrom M, Engblom D, Blomqvist A, Hammarstrom S (2003) Expression of leukotriene C4 synthase mRNA by the choroid plexus in mouse brain suggests novel neurohormone functions of cysteinyl leukotrienes. Biochem Biophys Res Commun 307:987–990.

Strazielle N, Khuth ST, Murat A, Chalon A, Giraudon P, Belin MF, Ghersi-Egea JF (2003) Pro-inflammatory cytokines modulate matrix metalloproteinase secretion and organic anion transport at the blood-cerebrospinal fluid barrier. J Neuropathol Exp Neurol 62:1254–1264.

Takahashi Y, Smith P, Ferguson A, Pittman QJ (1997) Circumventricular organs and fever. Am J Physiol 273:R1690–1695.

Takano M, Horie M, Yayama K, Okamoto H (2003) Lipopolysaccharide injection into the cerebral ventricle evokes kininogen induction in the rat brain. Brain Res 978:72–82.

Thibeault I, Laflamme N, Rivest S (2001) Regulation of the gene encoding the monocyte chemoattractant protein 1 (MCP-1) in the mouse and rat brain in response to circulating LPS and proinflammatory cytokines. J Comp Neurol 434:461–477.

Urade Y, Hayaishi O (1999) Prostaglandin D2 and sleep regulation. Biochim Biophys Acta 1436:606–615.

Vallières L, Rivest S (1997) Regulation of the genes encoding interleukin-6, its receptor, and gp130 in the rat brain in response to the immune activator lipopolysaccharide and the proinflammatory cytokine interleukin-1β. J Neurochem 69:1668–1683.

Vallières L, Sawchenko PE (2003) Bone marrow-derived cells that populate the adult mouse brain preserve their hematopoietic identity. J Neurosci 23:5197–5207.

Van Dam A-M, Brouns M, Louisse S, Berkenbosch F (1992) Appearance of interleukin-1 in macrophages and in ramified microglia in the brain of endotoxin-treated rats: a pathway for the induction of non-specific symptoms of sickness? Brain Res 588:291–296.

Van Dam A-M, Brouns M, Man AHW, Berkenbosch F (1993) Immunocytochemical detection of prostaglandin E2 in microvasculature and in neurons of rat brain after administration of bacterial endotoxin. Brain Res 613:331–336.

Watkins LR, Maier SF, Goehler LE (1995) Cytokine-to-brain communication: a review and analysis of alternative mechanisms. Life Sci 57:1011–1126.

Wong ML, Rettori V, al-Shekhlee A, Bongiorno PB, Canteros G, McCann SM, Gold PW, Licinio J (1996) Inducible nitric oxide synthase gene expression in the brain during systemic inflammation. Nat Med 2:581–584.

19 Experimental Models of Hydrocephalus: Relation to Human Pathology

Charles E. Weaver and John A. Duncan
Brown University School of Medicine
Providence, Rhode Island, U.S.A.

CONTENTS

19.1 INTRODUCTION

Hydrocephalus is the abnormal accumulation of cerebrospinal fluid (CSF) within the cranial cavity accompanied by expansion of the cerebral ventricles. Treatment often involves surgical insertion of a shunt into the ventricles to divert excess CSF to another bodily cavity, usually the peritoneum. Hydrocephalus, if untreated, can result in permanent injury to the brain and death. In the United States, hydrocephalus accounts for nearly 70,000 hospital admissions per year, and 25,000 shunt-related surgeries are performed in this country each year with 18,000 related to initial placements for hydrocephalus (Bondurant and Jimenez, 1995). The treatment and management of patients with hydrocephalus has been one of the most significant contributions and successes of modern neurosurgery. Congenital hydrocephalus is present at birth (0.1 to 3.5/1000 live births) or develops in the first months of life and accounts for 50% of patients with hydrocephalus (Dignan and Warkany, 1974; Stein et al., 1981; Lemire, 1988). Many congenital diseases and syndromes have hydrocephalus as a significant component of their presentation (see Table 19.1). This suggests that the development of hydrocephalus involves multiple genes and their interactions. Many animal models have been examined in an attempt to gain insight into the pathophysiology underlying the development of hydrocephalus and to develop better treatment strategies for humans suffering from hydrocephalus. In some cases the underlying molecular and genetic mechanisms of hydrocephalus development have been determined. In this review we will examine the spontaneously occurring and genetically induced animal models of congenital hydrocephalus currently available and the relevance of these models to congenital hydrocephalus in humans.

The number of terms that have been used to further describe the clinical finding of dilated ventricles in hydrocephalus indicates the complexity of this clinical entity. Hydrocephalus can be acute or chronic, compensated or uncompensated, normal pressure or high pressure, communicating or noncommunicating, and obstructive or nonobstructive. Acute hydrocephalus develops over hours to days and may require neurosurgical intervention, whereas chronic hydrocephalus progresses slowly over months to years and may or may not be accompanied by clinical symptoms. Normal and high pressure hydrocephalus refer to the measured pressure within the system. Communicating or noncommunicating refer to the movement of CSF between the ventricular system and the subarachnoid space. Obstructive and nonobstructive refer to the presence or absence of a physical blockage to the flow of CSF. Inactive, arrested, and compensated refer to hydrocephalus without symptoms and imply the absence of need for CSF diversion. Ex vacuo hydrocephalus is a term used to describe

TABLE 19.1
Association of Hydrocephalus with Congenital Disease

Strong Association	Occasional Association
Aqueductal stenosis	Achondroplasia
X-linked recessive hydrocephalus	Acrodysostosis
Familial type Dandy-Walker malformation	Apert's disease
Spinal bifida cystica	Basal cell nevus
Osteopetrosis (Albers-Schönberg disease)	Hurler's disease
Intraventricular hemorrhage	Immotile-cilia syndrome (Kartagener's syndrome)
Infectious meningitis	Incontinentia pigmenti
Trauma/subarachnoid hemorrhage	Linear sebaceous nevus sequence
Tumor	Meckel-Gruber syndrome
	Oral-facial-digital syndrome
	Osteogenesis imperfecta
	Riley-Day syndrome
	Thanotophoric dwarfism
	Triploidy
	Trisomy 13, 18

Source: Adapted from McLaurin et al., 1989.

ventricular dilation as a result of brain volume loss as in the case of atrophy, stroke, or surgical resection. Congenital hydrocephalus is present during the perinatal period or develops during the first months of life.

For simplicity, hydrocephalus can be divided into two subtypes: noncommunicating and communicating. Noncommunicating hydrocephalus arises from obstruction of the ventricular system proximal to the exit sites from the fourth ventricle and leads to dilation proximal to the obstruction. Obstruction can be caused by congenital malformations such as aqueductal stenosis, tumors within or extending into the ventricular system, and scar due to infection or trauma. Communicating hydrocephalus implies that no obstruction to the flow of CSF into the subarachnoid space is present. Classically, communicating hydrocephalus was felt to be secondary to obstruction of CSF circulation at the level of arachnoid granulations. There is a growing realization that other factors such as alterations in CSF pulsations may play a role in ventricular dilation (Egnor et al., 2002).

Regardless of the etiology of ventricular expansion, the clinical findings in hydrocephalus arise from the mechanical stresses placed on important neural structures as a result of ventricular deformations and the often-associated increase in intracranial pressure (ICP). The pathogenesis of the brain injury caused by hydrocephalus is multifactorial and includes ischemia in white matter due to changes in intracranial pressure and decreased vascular perfusion, stretch-related damage to periventricular axons leading to demyelination and disconnection of neurons, and changes in the composition of CSF and extracellular fluid. The global neuropathophysiologic changes due to hydrocephalus have been reviewed elsewhere (Akai et al., 1987; Jensen, 1979; Del Bigio, 1993; McAllister and Chovan, 1998).

19.2 PRIMARY MODELS OF COMMUNICATING HYDROCEPHALUS

19.2.1 Neural Cell Adhesion Molecule L1 Knockout Mouse

Neural cell adhesion molecule L1 (L1CAM) is a transmembrane glycoprotein of the immunoglobulin superfamily (IgSF). The structure consists of an extracellular component with six immunoglobulin domains (Ig) and five fibronectin type III-like domains (FN), a transmembrane domain that passes once through the lipid bilayer, and a short cytoplasmic C-terminal tail (Moos et al., 1988; Hlavin and Lemmon, 1991). A subgroup of IgSF with this characteristic domain structure is now known as the L1 subfamily (Grumet et al., 1991). L1CAM and other members of the subgroup are expressed predominantly in the developing central and peripheral nervous system (Martini and Schachner 1986, Bartsch et al., 1989). The human *L1CAM* gene has 28 exons. An alternative splice variant of this gene, in which exons 2 and 27 are removed, is expressed in nonneuronal cells (Kowitz et al., 1992; Reid and Hemperly, 1992; Jouet et al., 1995; Takeda et al., 1996; Debiec et al., 1998). During development of the nervous system L1CAM plays a role in the interactions between adjacent neurons and between neurons and Schwann cells (Faissner et al., 1984; Rathjen and Schachner, 1984; Persohn and Schachner, 1987). Through these interactions L1CAM is involved in axon pathfinding and neuronal migration (Lindner et al., 1983; Fischer et al., 1986; Lagenaur and Lemmon, 1987; Asou et al., 1992), neurite growth and fasciculation (Stallcup and Beasley, 1985; Fischer et al., 1986; Chang et al., 1987; Lagenaur and Lemmon, 1987; Kunz et al., 1996; Yip et al., 1998), myelination (Wood et al., 1990), and the synaptic plasticity involved in long-term potentiation (Luthi et al., 1994; Rose, 1995). These functions are all dependent on homophilic interactions between L1CAM molecules on adjacent cells (Lemmon et al., 1989) and heterophilic interactions between L1CAM and other molecules of the IgSF (Kuhn et al., 1991; Brummendorf et al., 1993; DeBernardo and Chang, 1996; Kunz et al., 1996) as well as components of the extracellular matrix (Friedlander et al., 1994; Milev et al., 1994). In addition to its role as an adhesion molecule, L1CAM may also be involved in intracellular signaling via second messengers such as Ca^{2+} and inositol phosphate (Schuch et al., 1989; Von Bohlen Und Halbach et al., 1992), interactions with intracellular kinases (Ignelzi et al., 1994), direct interactions with the cytoskeleton (Davis and Bennett, 1993, 1994; Dahlin-Huppe et al., 1997; Sammar et al., 1997), and interactions with other molecules within the cell membrane such as axonin-1 and fibroblast growth factor receptor (Buchstaller et al., 1996; Doherty and Walsh, 1996; Saffell et al., 1997).

In 1949 Bickers and Adams described a family in which male siblings died at birth from congenital hydrocephalus (Bickers and Adams, 1949). The syndrome became know as hydrocephalus due to stenosis of the aqueduct of Sylvius (HSAS) because of the narrowed cerebral aqueduct found at postmortem examination. The recessive X-linked inheritance pattern identified in families with this syndrome also led to the term X-linked hydrocephalus (Shannon and Nadler, 1968). In the early 1990s linkage analysis placed the gene for human X-liked hydrocephalus in the Xq28 region (Willems et al., 1990, 1992; Lyonnet et al., 1992; Jouet et al., 1993). Specific mutations were eventually identified in one of the genes of the Xq28 region in HSAS family cohorts. The gene identified was *L1CAM*, and to date 142 mutations in the *L1CAM* gene in 153 families have been identified. (An updated list of *L1CAM* mutations is maintained at the *L1CAM* mutation Web page: http://www.uia.ac.be/dnalab/l1/.)

Mutations in the *L1CAM* gene cause a spectrum of diseases, including HSAS, MASA (mental retardation, aphasia, shuffling gait, adducted thumbs), SP-1 (complicated spastic paraplegia type *1*), ACC (agenesis of corpus callosum), and MR-CT (mental retardation with clasped thumbs). CRASH syndrome (corpus callossal hypoplasia, mental retardation, adducted thumbs, spastic paraplegia, and hydrocephalus) is now used to refer to the entire spectrum of diseases caused by mutations in the *L1CAM* gene (Fransen et al., 1995). The phenotype of the diseases in humans and the degree of associated hydrocephalus are dependent upon the part of the gene that is altered by mutation (Yamasaki et al., 1997; Fransen et al., 1998b). Patients with mutations that affect only the L1CAM cytoplasmic domain, termed class-1 mutations by Yamasaki et al. (1997), rarely develop ventricular dilation and seldom require CSF diversion. Patients with class-1 mutations also tend to have a lower mortality and less severe cognitive deficits. Patients with mutations causing nonsense or frame shift mutations, termed class-3 mutations by Yamasaki, almost always have severe hydrocephalus requiring CSF diversion. These patients also have the highest mortality and most severe cognitive deficits. Patients with missense or point mutations in the extracellular domain, termed class-2 by Yamaski, have disease phenotypes dependent on the particular residue involved. Mutations in residues responsible for maintaining the extracellular Ig or FN domain structure lead to severe phenotypes, whereas mutations causing little change in domain structure cause less severe phenotypes (Bateman et al., 1996; Michaelis et al., 1998).

Two independently created *L1cam* knockout mouse lines have been shown to display a spectrum of abnormalities similar to those seen in human CRASH patients (Cohen et al., 1997; Dahme et al., 1997; Fransen et al., 1998a). The mutations in both mouse lines correspond to class-3 mutations that are associated with the most severe disease phenotype in humans, including massive and progressive hydrocephalus. *L1cam* knockout mice have dilated ventricular systems. Interestingly, the degree of ventricular dilation is dependent on the background strain (Dahme et al., 1997). This suggests that other modifier genes are involved in the final disease phenotype and may explain both the intra- and interfamily variability of hydrocephalus severity noted in humans. The number of *L1cam* knockout offspring born is only 50% of that predicted by Mendelian genetics, suggesting intrauterine death (Cohen et al., 1997; Dahme et al., 1997; Fransen et al., 1998a), and the life spans of knockout mice are shorter than that of their wild-type littermates (Dahme et al., 1997), reproducing the increased mortality seen in human patients with *L1CAM* mutations. Performance in a Morris water maze task, testing spatial memory, is impaired in *L1cam* knockout mice (Fransen et al., 1998a) reminiscent of the cognitive impairments noted in human patients with *L1CAM* mutations. Most human patients with *L1CAM* mutations are hindered by spastic lower extremity paraplegia, and knockout mice were found to be impaired in tests of motor function particularly regarding the hind limbs (Dahme et al., 1997; Fransen et al., 1998a). These motor impairments are most likely explained by the abnormalities noted in the corticospinal tract of *L1cam* knockout mice. In *L1cam* knockout mice many corticospinal neurons fail to cross at the level of the pyramidal decussation and instead pass into the ipsilateral dorsal column (Cohen et al., 1997).

The role of *L1CAM* mutations in the development of hydrocephalus is unclear. Stenosis of the aqueduct of Sylvius was initially thought to be the primary cause of hydrocephalus due to *L1CAM* mutation. Further observations beyond the initial human autopsy specimens have not supported this conclusion (Landrieu et al., 1979; Varadi et al., 1987). The aqueduct in *L1cam* knockout mice has been described as normal in size but altered in shape (Fransen et al., 1998a), dilated (Demyanenko et al., 1999), and to have no consistent morphologic difference compared to wild type (Rolf et al., 2001). Rolf et al. propose that secondary occlusion of the cerebral aqueduct can occur in *L1cam* knockout mice. These authors found complete occlusion of the aqueduct in *L1cam* knockout mice that develop severe hydrocephalus, whereas animals with moderate ventricular dilation have normal-appearing aqueducts (Rolf et al., 2001). These authors further propose that severe dilation of the ventricles in some *L1cam* knockout mice can lead to deformation of the brain and secondary compression of the aqueduct and a conversion from communicating to noncommunicating hydrocephalus. Secondary aqueductal stenosis has also been proposed as a mechanism for the observed occlusion in human patients with X-linked hydrocephalus (Landrieu et al., 1979; Renier et al., 1982) and has been observed in other congenital hydrocephalus models (Borit and Sidman 1972; Raimondi et al., 1973; Jimenez et al., 2001). Ventricular dilation my lead to secondary aqueductal occlusion, but the role of L1CAM in the initial communicating hydrocephalus leading to ventricular dilation in human patients and knockouts is less clear.

Kamiguchi et al. (1998), building upon the theory of Van Essen, proposed a novel explanation for the ventricular dilation seen in *L1cam* knockout animals and human CRASH patients. Van Essen (1997) postulated that the mechanical tension generated by axons, dendrites, and glial processes plays a role in CNS morphogenesis. Kamiguchi et al. noted that mutations in the extracellular domain of L1CAM are more frequently associated with ventricular dilation than mutations in the intracellular domain, suggesting that a loss of cell adhesion, through L1CAM interactions with adjacent cells, is critical for the development of ventricular dilation (Yamasaki et al., 1997; Fransen et al., 1998b). Since ventricular size is reliant upon the balance between intraventricular CSF pressure and brain compliance, Kamiguchi et al. sugested that a loss of cell-to-cell adhesion would increase brain compliance and lead to ventricular dilation. The loss of axon cables in the corpus callosum and corticospinal tract seen in *L1cam* mutant mice would similarly contribute to increased brain compliance.

19.2.2 Congenital Hydrocephalus Mutant Mouse (ch) and Forkhead/Winged Helix Gene (Mf1) Knockout Mouse

Gruneberg first described the *ch* mutant mouse in 1943 (Gruneberg, 1943). It is a recessive form of hydrocephalus in which homozygous animals die at birth with massively dilated ventricles. Gruneberg also noted that these animals have defects in the skull bones as well as the axial skeleton and initially attributed these phenotypic changes to a primary problem in the differentiation of prochondrogenic mesenchyme (Gruneberg, 1953). Green (1970) and subsequently Gruneberg and Wickramaratne (1974) noted defects in other mesenchymal tissues including the urogenital system as well as the ectodermally derived parotid and nasal glands. Green proposed that the hydrocephalus in *ch* mice was due to delayed differentiation of the meninges causing impaired flow of cerebrospinal fluid into the arachnoid granulations. The *ch* locus was subsequently mapped to mouse chromosome 13.

Winged Helix (WH) proteins belong to a family of DNA-binding proteins (Clark et al., 1993; Kaufmann and Knochel, 1996). These proteins are components of transcriptional complexes involved in the regulation of tissue-specific gene expression (Qian and Costa, 1995; Cirillo et al., 1998). Spontaneous and targeted mutations in WH proteins are associated with proliferation and differentiation defects in specific cell populations in embryonic mice (Ang and Rossant, 1994; Nehls et al., 1994; Xuan et al., 1995; Hatini et al., 1996; Dou et al., 1997; Kaestner et al., 1997; Labosky et al., 1997; Winnier et al., 1997).

Homzygous *Mf1* null mutant mice were created to examine the role of this winged helix protein in development and the animals demonstrate a phenotype identical to the spontaneously occurring *ch* mutant mouse (Kume et al., 1998). Homozygous null *Mf1* mice die at birth with hydrocephalus, eye defects, and multiple skeletal abnormalities. The importance of the *Mf1* gene in *ch* mutant mice was revealed when Kume et al. identified a point mutation in the *Mf1* gene of *ch* mutants. The predicted structure of the ch mutant protein is truncated and lacks the DNA-binding domain of the wild-type protein. Kume et al. also demonstrated that the Mf1 protein is highly expressed in the developing meninges. The meninges of the arachnoid are disorganized in the *Mf1* knockout mice, leading to a thinner arachnoid layer with more closely packed cells compared to wild-type mice. The observations of Kume et al. in *Mf1* knockout mice are similar to those made by Green with regard to the meninges of the spontaneous *ch* mutant mouse and suggest that the cause for hydrocephalus in both *ch* and *Mf1* mutants is impaired flow of CSF through the arachnoid space. It is important to note that no obstruction at the level of the cerebral aqueduct is seen in *ch* mutant mice (Green, 1970) or *Mf1* mutants (Kume et al., 1998).

Kume et al. also provide convincing evidence that the human gene *FREAC3* (chromosome 6p25), having greater than 91% predicted amino acid identity, is the human homolog to the mouse *Mf1* gene. Interestingly, when patients with cytologically visible deletions involving the 6p25 region were examined for deletions of the *FREAC3* gene, a correlation was found between loss of a single copy of the gene and hydrocephalus (Kume et al., 1998). Hydrocephalus has also been reported in human patients with deletions of 6p25 as well as ring chromosome 6 (Levin et al., 1986; Chitayat et al., 1987; Kelly et al., 1989; Zurcher et al., 1990; Alashari et al., 1995; Walker et al., 1996; Davies et al., 1999; Urban et al., 2002). These observations taken together suggest that the *FREAC3* gene may be involved in some cases of congenital hydrocephalus in humans. Screening for alterations in the *FREAC3* gene in human patients with hydrocephalus would add further support for a role of this gene in human disease.

19.2.3 Hydrocephalus with Hop Gait Mutant Mouse

Hydrocephalus with hop gait (*hyh*) is a spontaneous autosomal recessive mouse mutation causing congenital hydrocephalus that arose in the C57BL/10J strain (Lane, 1985). Homozygous newborn mice demonstrate a characteristic hopping gate and moderate hydrocephalus with dilated lateral and third ventricles and a patent but narrowed rostral aqueduct (Bronso'n and Lane, 1990). Detailed anatomic studies have demonstrated that the earliest detectable changes in *hyh* mutant mice is damage and sloughing of the ventral ependymal lining of the ventricular system occurring between embryonic days 12 and 16 (E12-E16) (Jimenez et al., 2001; Wagner et al., 2003). This ependymal denudation precedes the first significant increase in ventricular size in *hyh* mutants. Ventricular dilation is not noted until E15 in the third ventricle, followed by expansion of the cerebral aqueduct beginning at E16 (Jimenez et al., 2001; Wagner et al., 2003). At birth *hyh* mutant mice have moderate dilation of the lateral and third ventricles and stenosis of the rostral end of the cerebral aqueduct (Perez-Figares et al., 1998; Jimenez et al., 2001; Wagner et al., 2003). The fourth ventricle remains unchanged in size compared to nonhydrocephalic littermates throughout the development of hydrocephalus. Progressive dilation of the lateral ventricles develops from the time of birth and by postnatal day five the caudal aqueduct becomes occluded and leads to severe hydrocephalus and eventually death by two months of age (Bronson and Lane, 1990; Perez-Figares et al., 1998; Wagner et al., 2003).

Abnormalities of the subcommissural organ (SCO), a circomventricular organ located in the posterior roof of the third ventricle at the opening of the cerebral aqueduct, have been suggested to contribute to the development of hydrocephalus in *hyh* mutant mice (Perez-Figares et al., 1998). The SCO normally secretes glycoproteins into the CSF that assemble into a fibrous structure known as Reissner's fiber. The Reissner's fiber normally grows and extends into the aqueduct, 4th ventricle, and central canal of the spinal cord (Rodriguez et al., 1992, 1998). It has been suggested that the secretions of the SCO prevent closure of the aqueduct during fetal development and that dysfunction of the SCO may cause aqueductal stenosis and congenital hydrocephalus (Overholser, 1954). In support of this theory, immunologic blockade of Reissner's fiber formation during fetal and postnatal life in rats has been shown to cause aqueductal stenosis and hydrocephalus (Vio et al., 2000). The SCO of *hyh* mutant mice unlike nonhydrocephalic littermates secretes a material that does not form Reissner's fiber (Perez-Figares et al., 1998). Therefore, SCO dysfunction may play a role in development of hydrocephylus in the *hyh* mutant mouse. The MT/HokIdr mouse strain provides additional support for the role of the SCO in the development of hydrocephalus. The SCO is completely absent from the brains of these mice, in which congenital hydrocephalus is likely to be due at least in part to stenosis of the midportion of the cerebral aqueduct (Takeuchi et al., 1987).

The gene mutation causing the *hyh* phenotype has recently been reported by two groups (Chae et al., 2004; Hong et al., 2004). Both groups report that *hyh* mutant mice contain a G-to-A missense mutation in the Napa gene coding for soluble *N*-ethylmaleimide-sensitive factor attachment protein-α (α-SNAP). The mutation leads to an isoleucine for methionine substitution at residue 105. *Hyh* mutant mice express the α-SNAP mRNA and protein but at significantly lower levels than heterozygous or wild-type mice, possibly due to mRNA (Chae et al., 2004) or protein (Hong et al., 2004) instability. α-SNAP is an important component of the molecular machinery responsible for membrane fusion events such as Golgi transport (Rothman and Wieland, 1996), neuronal exocytosis (Sollner et al., 1993; Whiteheart et al., 1993; Rothman, 1994) and is therefore involved in protein trafficking. The ependymal lining of the ventricles is a strongly polarized epithelium and, as discussed, is implicated in the development of hydrocephalus in the *hyh* mutant. Chae et al. (2004) have found abnormalities in the cellular distribution of proteins (F-actin, α-catenin, and E-cadherin), important in adherens junctions within the ependymal layer of *hyh* mutant mice. These abnormalities in apical protein distribution may contribute to dysfunction of the specialized epithelium of the SCO (Perez-Figares et al., 1998) and the ependymal denudation noted to precede the development of hydrocephalus in the *hyh* mutant mice (Jimenez et al., 2001; Wagner et al., 2003).

19.2.4 Obstructive Hydrocephalus Mutant Mouse

First described by Borit and Sidman (1972), obstructive hydrocephalus (*oh*) is an autosomal recessive mutation in the mouse. Animals develop generalized ventricular dilation within a few days of birth. Similar to *hy-3* mutants, the cerebral aqueduct appears to be secondarily obstructed in *oh* mice due to compression by the expanding lateral ventricles and cerebral hemispheres. Aqueduct obstruction occurs by two weeks of age, and animals die before four weeks of age.

19.2.5 Transforming Growth Factor-β1 Gene

Growth factors have been shown to play an important role in the regulation of CSF dynamics. CSF retention and associated ventricular dilation occur when levels of transforming growth factor-β1 (TGF-β1) or basic fibroblast growth factor (FGF2) are raised in the CNS extracellular fluid, by direct infusion of growth factors into the CSF spaces (Johanson et al., 1999; Takizawa et al., 2001), as a consequence of subarachnoid hemorrhage in adults (Flood et al., 2001; Takizawa et al., 2001) or intraventricular hemorrhage in premature babies (Whitelaw et al., 1999). The resultant growth factor-induced fibrosis of CSF drainage pathways interferes with CSF flow and reabsorption resulting in communicating hydrocephalus. Not surprisingly, transgenic mice engineered to overexpress TGF-β1 in the CNS also develop neurological abnormalities including hydrocephalus (Galbreath et al., 1995; Wyss-Coray et al., 1995).

Transgenic animals generated by Wyss-Coray et al. expressing high levels of TGF-β1 developed severe communicating hydrocephalus, seizures, motor difficulties, and early runting. These authors have also demonstrated an upregulation of the extracellular matrix proteins laminin and fibronectin in proximity to the perivascular astrocytes shown to express TGF-β1. The transgenic animals produced by Galbreath et al. also developed severe hydrocephalus accompanied by spasticity and limb tremors. MRI studies suggested CSF obstruction within the subarachnoid space at the exit of the 4th ventricle. This was confirmed by examination at the gross and light microscopic level (Cohen et al., 1999). Immunohistochemical studies have localized TGF-β1 and its receptor primarily to the meninges and subarachnoid space in both the transgenic mice produced by Galbreath et al. as well as normal controls. Presumably elevated levels of cytokines such as TGFβ1 play a role in other models of hydrocephalus. However, a decreased expression of TGF-β1 was observed in the brains of H-Tx rats (Cai et al., 1999). If elevated levels of TGF-β1 are associated with ventricular dilation, why do H-Tx rats have decreased expression of this cytokine? The authors speculate that the decreased levels of TGF-β1, as well as epidermal growth factor (EGF), seen in hydrocephalic H-Tx rats is the result of negative feedback. This would require elevation of an as yet unidentified downstream gene that is normally regulated by TGF-β1 and EGF. While this may be the case, H-tx mutant rats do not develop hydrocephalus because of alterations in the subarachnoid space and increased resistance to CSF absorption (Jones and Bucknall, 1987). Hydrocephalus in H-Tx rats is most likely due to primary stenoisis or occlusion of the cerebral aqueduct. Therefore, it is not surprising that TGF-β1 expression is decreased in the brains of hydrocephalic H-Tx rats. It would be interesting to examine the levels of these cytokines in animal models where hydrocephalus is known to be nonobstructive or communicating.

19.3 MODELS WITH OBSTRUCTIVE HYDROCEPHALUS

19.3.1 SUMS/NP Mutant Mouse

Congenital hydrocephalus develops in the SUMS/NP mutant mouse as an autosomal recessive trait (Jones, 1984). The specific gene or genes involved have not been determined. Hydrocephalus is noted by the third postnatal day, and affected animals develop domed heads and die soon after weaning (Jones et al., 1987; Bruni et al., 1988b). Ventricular and cisternal infusion experiments evaluate the resistance to flow and absorption from various levels of the ventricular system. The resistance to CSF absorption from the lateral ventricles is greater in hydrocephalic than normal littermates, whereas there is no difference in resistance to absorption from the cisterna magna (Jones, 1984, 1985). These findings suggest that hydrocephalic SUMS/NP mice have a resistance to the flow of CSF across the cerebral aqueduct and normal absorption of CSF from the subarachnoid space. Detailed anatomic studies of developing SUMS/NP mice demonstrate that the development of hydrocephalus in this mutant strain begins in the prenatal period. The first difference detected in the CSF pathway is dilation of the lateral and third ventricles at E18. Ventriculomegaly is due to a decreased cross-sectional area or obstruction of the cerebral aqueduct (Jones et al., 1987; Bruni et al., 1988b). Abnormal development of the diencephalons or the rostral midbrain or both has been proposed as a possible cause of distortions in the anatomy of the cerebral aqueduct (Bruni et al., 1988b). These studies suggest that aqueductal stenosis/occlusion is the primary event in the development of hydrocephalus in the SUMS/NP mutant.

19.3.2 Hydrocephalus Texas (H-Tx) Rat

The H-Tx rat is a strain in which hydrocephalus develops at a high frequency in the offspring of normal adults. The strain arose spontaneously and was first described by Kohn et al. as a congenital communicating hydrocephalus (Kohn et al., 1981). Further anatomic studies have demonstrated that aqueductal stenosis plays a significant role in the development of hydrocephalus in this strain. Some controversy exists as to whether aqueductal stenosis is the primary event leading to ventricular dilation in the H-Tx rat or if communicating hydrocephalus leads to secondary obstruction of the cerebral aqueduct (Jones, 1997; Oi, 1998). Jones and Bucknall examined H-Tx animals from E16 through one day after birth and first detected hydrocephalus at E18–E20 in the lateral and sometimes third ventricles (Jones and Bucknall, 1988). The cerebral aqueduct is either occluded or significantly smaller in animals with ventricular dilation. Some animals without hydrocephalus are also noted to have occluded aqueducts, suggesting that events at the aqueduct precede development of hydrocephalus. Jones and Bucknall consider their anatomic observations evidence for aqueductal stenosis or occlusion as the primary event leading to ventricular dilation and hydrocephalus in the H-Tx rat. The observations of Pourghasem et al. (2001) regarding the anatomy of the developing ventricular system of H-Tx rats also support occlusion of the aqueduct prior to the onset of ventricular dilation. Consistent with these anatomic studies, ventricular infusion experiments demonstrated that resistance to CSF flow across the cerebral aqueduct is normal in all E19 fetuses but increases from this point on in animals that develop hydrocephalus (Jones and Bucknall, 1987). In addition, the resistance to absorption of CSF from the cisterna magna is not different between hydrocephalic and normal H-Tx littermates, demonstrating that no impairment to CSF absorption is present in H-Tx rats. Therefore, Jones and Bucknall (1988) and Pourghasem et al. (2001) believe that aqueductal stenosis precedes or coincides with the development of hydrocephalus in the H-Tx rat. The observations of Oi et al. (1996), on the other hand, appear to contradict these observations. These authors also examined H-Tx rats in the perinatal period and reported the development of mild ventriculomegaly on E17 prior to the development of any aqueductal occlusion. They observed that the aqueduct tends to be patent at E17 and that aqueductal narrowing or occlusion does not occur until the perinatal period. When the aqueduct did become occluded, it was felt to be due to hypercellular changes in the midbrain, especially in the periaqueductal region. It should be noted that in the report of Oi et al. the methodology for quantification of ventricular size and aqueductal stenosis is not clear, and differences in interpretation may be due to differences in the preparation of tissue and techniques of measurement. It is also possible that the differences noted relate to true differences between the strains examined. The strains used by Jones et al. and Oi et al. were derived from animals obtained from D. F. Kohn. The strains have subsequently been bred and maintained independently, and divergences in the phenotype of hydrocephalus observed may have developed. The origin of the H-Tx animals used in the studies of Pourghasem et al. is unclear. Environmental factors have been shown to influence the frequency of hydrocephalus in H-Tx rats (Jones et al., 2002), and environmental differences regarding the care and handling of the animals in the different colonies may have influenced the hydrocephalus phenotype.

The mode of inheritance for hydrocephalus in the H-Tx rat is complex. Linkage studies suggest that inheritance is likely polygenetic involving multiple gene loci (Jones et al., 2000, 2001b). The overall penetration of hydrocephalus within the H-Tx strain is reported to be 40% (Jones et al., 2000). Linkage analysis on H-Tx rats crossed to a nonhydrocephalic strain suggests that the number of hydrocephalus susceptibility loci correlates with the severity of hydrocephalus seen (Jones et al., 2001a). The mapping of specific candidate genes to the loci involved in hydrocephalus susceptibility in the H-Tx rat awaits further investigation.

19.3.3 LEW/jms Rat

The LEW/jms inbred rat strain was originally developed and first described in Japan (Sasaki et al., 1983). Anatomic studies have suggested that aqueductal stenosis followed by occlusion beginning at E17 is the primary event leading to ventricular dilation (Yamada et al., 1991; Oi et al., 1996). Affected pups are identifiable soon after birth by domed heads, and they die 10 to 20 days after birth. Transmission of the hydrocephalus trait was originally reported to be through simple autosomal recessive inheritance (Sasaki et al., 1983). Recent linkage studies have suggested that the inheritance may be far more complex. Experiments involving crosses of LEW/jms rats to a nonhydrocephalic strain suggest that the inheritance may be semidominant or involve more than one gene and, because of a male predominance of affected offspring, involve linkage to the X chromosome (Jones et al., 2003). Fifty percent of the nonhydrocephalic littermates develop mild ventricular dilation as adults, but the presence or absence of ventricular dilation in adults does not appear to affect the frequency of severely affected offspring (Jones et al., 2003). The association of this milder form of hydrocephalus to the severe form in neonates is unknown. Delayed presentation of aqueductal stenosis is occasionally seen in human patients. Further examination of the LEW/jms rat may help explain why some patients present early in the perinatal period with symptoms due to aqueductal stenosis whereas others survive into adulthood before presenting with symptoms of hydrocephalus.

19.4 MODELS OF EX VACUO HYDROCEPHALUS

19.4.1 Hypoxia Inducible Factor-1α Knockout Mouse

To investigate the role of hypoxia inducible factor-1α (HIF-1α) in CNS development, knockout mice deficient in CNS HIF-1α expression were created (Tomita et al., 2003). Hypoxia brings about a wide variety of responses in mammalian tissues, including alterations in the expression of many genes. HIF-1α appears to be a common regulatory element involved in the response to oxygen deprivation including angiogenesis, erythropoiesis, glycolysis, and cell survival (Semenza, 1999; Schipani et al., 2001; Hofer et al., 2002). Adult *Hif1α*-deficient mice show no differences in external appearance from control animals. When brains were examined, however, there was pronounced atrophy of the cerebral cortex with decreased numbers of neurons and bilateral dilation of the lateral ventricles (Tomita et al., 2003). These authors further demonstrated that the apoptotic loss of neurons was due to impaired angiogenesis. Behavioral testing in a radial maze test shows that mice lacking HIF-1α have impaired memory function. The available data would suggest that the ventricular dilation seen in *Hif1α*-deficient mice is hydrocephalus ex vacuo. It is unclear, however, if the behavioral impairment observed in these mice is due to the neuronal loss incurred during development or the mechanical forces exerted on axonal pathways by ventricular dilation. Systematic measurement of ventricular pressure during development would help to determine if hydrocephalus itself plays a role in neuronal death or functional impairment.

19.5 MODELS OF UNCLEAR ETIOLOGY COMMUNICATING VERSUS NONCOMMUNICATING HYDROCEPHALUS

19.5.1 E2F Transcription Factor-5 Knockout Mouse

E2F transcription factor 5 (E2F-5) is a member of a larger family of transcription factors implicated in the regulation of growth controlling genes. Other E2F transcription factors have been shown to be important in cell cycle regulation (Johnson et al., 1993; Qin et al., 1994; Sardet et al., 1995), neoplastic transformation (Adams and Kaelin, 1996), and apoptosis (Qin et al., 1994; Fan and Steer, 1999). In situ hybridization studies have shown increased E2F-5 expression in embryonic epithelial tissues (Dagnino et al., 1997), including the choroid plexus, ventricular ependyma, and subependymal areas (Lindeman et al., 1998). *E2f5* knockout mice have normally appearing brains and ventricular systems at birth but develop progressive hydrocephalus shortly after birth (Lindeman et al., 1998). Knockout animals developed hydrocephalus independent of the background strain used for breeding, and no abnormalities were noted in heterozygous littermates. The ependymal lining of the lateral ventricles in these animals is disrupted at multiple sites, and the ventricles are frequently filled with blood, suggesting a significant elevation of intracranial pressure. The third ventricle is moderately dilated and the 4th ventricle is normal (Lindeman et al., 1998). Free flow of dye from the lateral ventricle to the subarachnoid space suggests a communicating hydrocephalus, but a normal appearance of the 4th ventricle with proximal ventricular dilation suggests some level of obstruction at the cerebral aqueduct. Unfortunately, detailed measurements of the aqueduct in knockout mice are lacking. Interestingly, electron microscopy of the choroid plexus in *E2f5* knockout animals demonstrates an abundance of electron lucent cells (Lindeman et al., 1998). This light cell morphology in the choroid plexus is thought to represent a cell in a more hydrated state and has been interpreted to represent a cell actively involved in CSF secretion (Dohrmann, 1970). An increased number of light choroid plexus cells in knockout mice compared to wild-type littermates may suggest excessive production of CSF as a contributing factor in the development of hydrocephalus. Excessive production of CSF has been shown to be a cause of hydrocephalus in choroid plexus papilloma (Milhorat et al., 1976). Detailed measurements of the rate of CSF production in knockout compared to wild-type animals should be performed to determine if hydrocephalus in *E2f5* knockout mice is due to an overproduction of CSF and communicating hydrocephalus. Detailed examination of the ventricular system and the cerebral aqueduct during the perinatal period may determine if primary or secondary aqueductal stenosis contributes to the hydrocephalus seen in this model.

19.5.2 Apolipoprotein B-Modified Mice

Apolipoprotein B (apoB) is an important component of lipoproteins involved in the transport of lipids, cholesterol, and lipid-soluble vitamins, such as vitamin E, in the circulation. Patients with hypobetalipoproteinemia produce a truncated apoB protein and have low circulating concentrations of these plasma components. In an attempt to create an animal model of hypobetalipoproteinemia, transgenic mice were created with targeted modifications of the *apoB* gene such that homozygous mice produce a similarly truncated apoB protein (Homanics et al., 1993; Farese et al., 1996; Kim et al., 1998). Surprisingly, there is a strong association with exencephalus and hydrocephalus in animals with the truncated apoB protein. Hydrocephalus is apparent at three to six weeks of age, with animals developing dome-shaped heads. The brains analyzed at this point demonstrate dilation of the lateral and third ventricles (Homanics et al., 1993). This pattern of dilation suggests obstruction at the level of the cerebral aqueduct. Details regarding the time course of hydrocephalus development and anatomy of the ventricular system and cerebral aqueduct are lacking. Therefore, it is difficult to determine whether ventricular dilation develops due to primary obstruction at the level of the aqueduct or secondary occlusion caused by progressive ventricular dilation. Prenatal vitamin E deficiency is associated with developmental defects in the CNS similar to those seen in *apoB* modified mice (Verma and Wei King, 1967). It is possible that the exencephalus and hydrocephalus seen in this transgenic mouse is due to deficiencies in lipid soluble vitamins indirectly caused by alterations in apoB. A similar association of hydrocephalus or neural tube defects in humans with hypobetalipoproteinemia has not been reported; therefore, the role of apoB in human CNS development remains unclear.

19.5.3 Orthodenticle Homolog-2 Knockout Mice

The Orthodentide homolog-2 (*Otx2)* gene is a homeobox gene that has been shown to be important in head and brain organization (Simeone et al., 1992; Frantz et al., 1994; Matsuo et al., 1995). *Otx2* is expressed in the cerebral cortex, the cerebellum, and the choroid plexus during development (Boncinelli et al., 1993; Frantz et al., 1994). Homozygous knockout mice lack a rostral head, forebrain, and midbrain and die during the embryonic period (Matsuo et al., 1995). Thirteen percent of heterozygous animals develop hydrocephalus in the postnatal period with edema in the periventricular white matter suggesting elevated CSF pressure (Makiyama et al., 1997). Detailed histopathological studies have not been reported to enable determination of the nature of hydrocephalus development in this animal model. These observations will be important to determine whether the hydrocephalus is of a communicating or noncommunicating type. If the hydrocephalus is of a communicating type, what is the role of the *Otx2* gene in the development and anatomy of the subarachnoid space and arachnoid granulations? If, however, the hydrocephalus is noncommunicating, what is the level of the obstruction within the ventricular system and the role of the *Otx2* gene in bringing about this obstruction? It is conceivable that this gene plays a regulatory role in the function of the choroid plexus and ependyma leading to hydrocephalus by an increase in CSF production. All of these questions await further investigation.

19.6 MODELS INVOLVING ALTERATIONS IN CILIARY FUNCTION

The axoneme is the bundle of microtubular filiments that constitutes the central core of a cilium or flagellum. Electron microscopy demonstrates the incredibly conserved organization of nine microtubular doublets around two central microtubules. Other associated proteins include radial spokes that connect outer doublets to the inner pair, nexins that connect outer doublets to each other, and light, intermediate, and heavy axonemal dyneins that assemble into the inner and outer connecting arms between the outer nine doublets and provide the mechanical force for movement. Primary ciliary dyskinesia (PCD), also known as immotile cilia syndrome or Kartagener's syndrome, is a heterogeneous group of genetic disorders in humans leading to disruption of this complicated structure and defective axonemal function. Mutations in three specific genes involved in axonemal structure and function have been identified in human patients with PCD: (1) dynein axonemal intermediate chain-1 (Pennarun et al., 1999; Guichard et al., 2001), (2) dynein axonemal heavy chain-5 (Olbrich et al., 2002), and (3) dynein axonemal heavy chain-11 (Bartoloni et al., 2002). The resulting disease in humans is characterized by randomization of left-right asymmetry, recurrent respiratory infections, and infertility. There is also a less common association with congenital hydrocephalus (al-Shroof et al., 2001; Jabourian et al., 1986; De Santi et al., 1990; Picco et al., 1993; Zammarchi et al., 1993; Greenstone et al., 2001; Wessels et al., 2003). Hydrocephalus is a prominent feature associated with induced mutations in mice (Sapiro et al., 2002; Davy and Robinson, 2003) and spontaneous mutations in mice (Davy and Robinson, 2003), rats (Torikata et al., 1991), dogs (Edwards et al., 1983; Daniel et al., 1995), and pigs (Roperto et al., 1991; Roperto et al., 1993) that affect axonemal function. This heterogeneous group of animal models of PCD demonstrates many of the findings of the human disease. It is unclear why such a strong association between ciliary dysfunction and hydrocephalus exists in animal models, whereas relatively few humans with PCD develop hydrocephalus. Rats treated for seven days with continuous intraventricular infusion of metavanadate, an inhibitor of ciliary movement develop ventricular dilation (Nakamura and Sato, 1993). This finding demonstrates that ciliary immobility alone can lead to ventricular dilation. Ciliated ependymal cells can be demonstrated to contribute to fluid motion as evidenced by the ability of explanted segments of cerebral aqueduct to move latex microbeads in fluid suspension (Nakamura and Sato, 1993). The ciliary immobility may be of relatively greater importance in the rodent brain and may explain the frequent association of hydrocephalus in rats and mice with ciliary dysfunction. Other indirect effects may also contribute to ventricular dilation. The specialized ciliated ependyma of the SCO has been implicated in the maintenance of a patent cerebral aqueduct and development of hydrocephalus in rodents (Takeuchi et al., 1987, 1988; Yamada et al., 1992; Perez-Figares et al., 1998; Rodriguez et al., 1998; Vio et al., 2000; Jimenez et al., 2001; Perez-Figares et al., 2001; Wagner et al., 2003), and ciliary dyskinesia may impair the function of the SCO. Cilia directly interact with the cytoskeleton, and alterations in these interactions may potentially lead to a decrease in brain compliance and therefore ventricular dilation. The association between ciliary dysfunction and hydrocephalus is by no means absolute given that transgenic mice lacking the dynein axonemal heavy chain 11 (*Dnahc11*) gene have loss of left-right asymmetry and ciliary dysfunction without associated hydrocephalus (Okada et al., 1999). This suggests a more complicated link between ciliary dysfunction and hydrocephalus. It is interesting to note that in a human study of ciliary function subjects older than 40 years of age had significantly slower ciliary beat frequencies and longer nasal mucociliary clearance times than younger subjects (Ho et al., 2001). The slower ciliary beat frequencies were suggested to be associated with increased incidence of respiratory infections in the older population. A tantalizing hypothesis is that a similar age-related decrease in the function of ventricular cilia may be involved in the development of normal pressure hydrocephalus (NPH) in some humans. This hypothesis might be tested in the following way: ventricular and respritory ciliary function could be measured directly in aged and young animals. If an age-related decline in ciliary function is found in both tissues, measurement of respiratory cilia function might be used as an indirect measure of ventricular cilia function. This indirect measure could then be used to determine if a correlation exists between the development of NPH and ciliary dysfunction in humans.

19.6.1 Hydrocephalus-3 (Hy-3) Mutant Mouse

Hydrocephalus-1 (Clark, 1932) and hydrocephalus-2 (Zimmerman, 1933) were spontaneously occurring recessive mutations in house mice that are now extinct (Bruni et al., 1988a). Hydrocephalus-3 (*Hy-3*), also a spontaneously occurring autosomal recessive mutation in the mouse, causes lethal communicating hydrocephalus with an onset in the perinatal period. The *Hy-3* mutation was first described by Gruneberg (1943). Newborn mice homozygous for the *Hy-3* mutation are indistinguishable from wild-type littermates at birth. Homozygous pups develop progressive hydrocephalus during the first week of life and die within three to seven weeks of life. Similar to the *ch* mutant mouse, the first histological changes in *Hy-3* mice are in the meninges of the subarachnoid space (Raimondi et al., 1973). These findings suggest obstruction at the level of the subarachnoid space leading to a defect in CSF absorption and communicating hydrocephalus. Raimondi et al. (1973) described hydrocephalus in *Hy-3* mutants as developing in three stages. First there is dilation of the lateral ventricles with an open cerebral aqueduct. This is followed by a second stage involving development of edema in the periventricular white matter with later extension of edema into the gray matter. In the third and final stage the expanding ventricles and edematous cortex cause compression of the quadrageminal plate leading to secondary aqueductal stenosis.

The gene responsible for the *Hy-3* mutant phenotype has recently been identified and characterized by Robinson et al. (2002). The transgenic mouse line OVE459, created by insertional mutation into chromosome 8, demonstrates autosomal recessive hydrocephalus. Mating of heterozygous *Hy-3* and OVE459 animals produces offspring with hydrocephalus in a proportion suggesting that the two mutations are allelic (Robinson et al., 2002). The insertion site in OVE459 is within a novel gene that has been named *Hydin* (hydrocephalus inducing) (Davy and Robinson, 2003). In *Hy-3* mutant mice the *Hydin* gene contains a single CG base pair deletion creating a frame shift mutation that results in a premature stop signal two codons downstream of the insertion site. The mutation results in a predicted loss of the majority of the *Hydin* gene transcript (Davy and Robinson, 2003). The *Hydin* gene contains 87 exons encoding a putative 5099-amino-acid protein. In situ hybridization studies in wild-type mice show that *Hydin* is normally expressed in the choroid plexus, the ependymal lining of the cerebral ventricles, the bronchi of the lung, spermatocytes, and the lining of the oviduct (Davy and Robinson, 2003). Davy and Robinson note that the expression pattern of *Hydin* in wild-type mice suggests a role in the formation, function, or maintenance of ciliated epithelia and structures with components common to cilia (flagella). These authors report that a portion of the putative *Hydin* protein has amino acid similarity to caldesmon, an actin-binding protein important for cytoskeleton structure and smooth muscle contraction (Huber, 1997), suggesting a role in cell motility.

The first description of *Hy-3* mutant mice notes nasal discharge as a peculiar feature of the hydrocephalic pups (Gruneberg, 1943). This observation may be related to bronchociliary dysfunction similar to that seen in humans with PCD and animal models with ciliary dysfunction. It is likely that detailed studies of ciliated epithelia in *Hy-3* and OVE459 mice would demonstrate ciliary dyskinesia. Unlike PCD, left-right asymmetry remains intact in *Hy-3* and OVE459 mice, suggesting that the different functional roles of cilia and related structures during development can become dissociated.

19.6.2 Dynein Axonemal Heavy Chain 5 Knockout Mouse

Transgenic mice with a functional null mutation of the dynein axonemal heavy chain 5 (*Mdnah5)* gene have recently been created and characterized by Ibanez-Tallon et al. (2002). Homozygous mice express many of the features seen in humans with PCD. The mutation was created when a transgene insertion interrupted the *Mdnah5* gene creating a stop codon at the start of the transgene insertion site. The predicted protein of the fusion gene lacks all of the functional domains of the native protein. Electron microscopy of ciliated epithelia confirms the disruption of axonemal structure with absence of outer dynein arms, and examination of tissue slice cultures reveals that the cilia are completely immotile. Homozygous mice demonstrate randomization of left-right asymmetry, recurrent respiratory infections, and hydrocephalus. Ibanez-Tallon et al. report that hydrocephalus is detected as early as three to five days after birth. From this point severe dilation of the lateral ventricles develops, causing thinning of the cortex, multiple hemorrhages, and compression of the cerebellum. Detailed anatomic data regarding the development of hydrocephalus is not available but the data presented by Ibanez-Tallon et al. suggests isolated lateral ventricle dilation and obstruction distally with significant elevation of interventricular presure. Detailed anatomic evaluation of the brain and ventricular system through development may shed light on the role of ciliary motion in the development of hydrocephalus in this model.

19.6.3 Oak Ridge Polycystic Kidney Mutant Mouse (Tg737orpk)

Mutations in the *Tg737* gene encoding the Polaris protein cause a wide spectrum of problems, including randomization of left-right asymmetry (Murcia et al., 2000), polycystic kidney, liver and pancreatic defects, and skeletal abnormalities (Moyer et al., 1994), as well as hydrocephalus (Taulman et al., 2001). Western blot analysis and immunohistochemistry with antibody to Polaris protein demonstrate that *Tg737* expression is concentrated in ciliated epithelium and that the Polaris protein is localized to the area occupied by the basal bodies within the axoneme of cilia and flagella including the cilia of the ventricular ependyma (Taulman et al., 2001). The Polaris protein has also been shown to be required for assembly of cilia in renal cells (Yoder et al., 2002). The spontaneous mouse mutant *Tg737orpk* exhibits polycystic kidney, liver, and pancreas defects, skeletal abnormalities, and hydrocephalus (Moyer et al., 1994; Taulman et al., 2001). Hydrocephalus is present in all adult *Tg737orpk* animals; however, detailed anatomic data regarding the ventricular anatomy and changes during the course of development is lacking. It is not known whether the dramatic loss of ependymal cilia seen in *Tg737orpk* mice is caused by hydrocephalus or if the loss of cilia leads to ventricular dilation. Evaluations of ciliary function in *Tg737orpk* mice correlated with the temporal development of ventricular dilation and changes in key anatomic structures such as the cerebral aqueduct would lead to a better understanding of the role of the Polaris protein in the function of cilia and development of hydrocephalus in this mutant mouse.

19.6.4 DNA Polymerase λ Knockout Mouse

DNA polymerase λ is a new member of the larger polymerase X family (Garcia-Diaz et al., 2000; Nagasawa et al., 2000; Kobayashi et al., 2002). DNA polymerases are involved in DNA replication, repair, recombination, and editing. To better understand the function of this newly identified polymerase knockout mice lacking this gene were created (Kobayashi et al., 2002). The phenotype of homozygous knockout mice includes situs inversus in some animals, evidence of chronic sinusitis, infertility in males due to sperm immotility, and hydrocephalus. Kobayashi et al. recognized the similarity of this phenotype to PCD, and electron microscopy of cilia from both the respiratory epithelium and the ventricular ependyma revealed absence of inner dynein arms, suggesting ciliary dysfunction as a key factor in the mutant phenotype. Kobayashi et al. (2002) report that hydrocephalus is present by two to four weeks of age, leading to doming of the head, and that death occurs in the majority of animals by nine weeks of age. Detailed analysis of the ventricular system at different ages of development demonstrates that ventricular dilation does not begin until postnatal day one when slight enlargement of the lateral ventricles is observed. At four weeks of age severe dilation of the lateral ventricles and moderate dilation of the third ventricle is seen, accompanied by ependymal hyperplasia of the third ventricle and cerebral aqueduct (Kobayashi et al., 2002). Alterations in the aqueduct in combination with the pattern of ventricular dilation suggest obstruction at the level of the aqueduct and noncommunicating-type hydrocephalus. No difference in the number of apoptotic cells was seen between homozygous mutants and wild-type littermates, suggesting that ventricular dilation is not due to excessive neuronal death (Kobayashi et al., 2002). Kobayashi et al. also note that DNA polymerase λ contains a BRCT motif thought to be involved in protein–protein interactions (Zhang et al., 1998). They propose that DNA polymerase lambda may be involved in cilia formation either through direct interactions with axonemal proteins or indirectly through interactions with regulatory proteins leading to modulation of transcription of important axonemal proteins.

19.6.5 Sperm-Associated Antigen-6 (Spag 6) Knockout Mouse

Spag6 is the murine orthologue of PF16, a gene in the green algae *Chlamydomonas*, which when inactivated causes paralysis of flagella (Zhang et al., 1998). Fifty percent of homozygous *Spag6* knockout mice develop dilation of the lateral and third ventricles and die within eight weeks of age (Sapiro et al., 2002). Sapiro et al. also noted male sterility due to severe defects in sperm motility in knockout males that survive to adulthood. Impaired ciliary function is suspected in this model, but direct observations have not been made. The pattern of ventricular dilation reported would suggest obstruction at the level of the cerebral aqueduct. Detailed examination of the ventricular and aqueductal anatomy during perinatal development would help determine if primary stenosis at the aqueduct is indeed responsible for hydrocephalus or if a component of communicating hydrocephalus is also present.

19.6.6 Hepatocyte Nuclear Factor 4 (Forkhead Box J1) Knockout Mouse

Hepatocyte nuclear factor 4 (HFH-4), also referred to as forkhead box J1 (foxj1), is a member of the winged helix transcription factor family. This family of transcription factors has been shown to play an important role in cell-specific gene expression (Kaufmann and Knochel, 1996). HFH-4 is expressed in the lung, spermatids, oviduct, kidney, and choroid plexus of both mice and humans (Hackett et al., 1995; Lim et al., 1997; Pelletier et al., 1998). Homozygous *Hfh4* knockout mice demonstrate randomization of left-right asymmetry, sterility in both sexes, and hydrocephalus (Chen et al., 1998). Chen et al. report that ventricular dilation is present in 50% of the animals examined at greater than one week of age. Detailed analysis of the ventricular anatomy in hydrocephalus is lacking, but the epithelia of the respritory tract, oviduct, and choroid plexus are completely devoid of cilia in the homozygous $Hfh^{4-/-}$ mice.

19.6.7 WIC-Hyd Rat

The WIC-Hyd rat strain arose spontaneously from the Wistar-Imamichi rat. Spontaneous hydrocephalus develops in approximately one third of the offspring produced by mating of normal male and hydrocephalic female WIC-Hyd rats and was first described by Koto et al. (1987). Male rats are more severely affected than females, developing outward signs of hydrocephalus within seven days of birth and death by three to four weeks of age. Hydrocephalic males exhibit randomization of left-right asymmetry. Females develop only moderate hydrocephalus and survive to mating age. No evidence for developmental abnormality has been identified in the ventricular system and the hydrocephalus is felt to be of a communicating type (Koto et al., 1987). The trait appears to be transmitted with an X-linked dominant pattern of inheritance. Ciliary motion is impaired on the cells of the ventricular ependyma, and the degree of ciliary dyskinesia is correlated to the extent of ventricular dilation (Nakamura and Sato, 1993). Electron microscopy demonstrates absence of both inner and outer dynin arms of cilia and a random organization of cilia and basal bodies, suggesting that the gene involved relates either directly or indirectly to axonemal structure and function (Torikata et al., 1991).

19.6.8 Hop-Sterile (hop) and Hop-hpy (hpy) Mutant Mice

Hop and *hpy* are allelic variants (Handel et al., 1985) of a gene located on mouse chromosome 6 (Hollander, 1976). *Hpy* was first described in descendents of X-irradiated mice (Hollander, 1966), and *hop* was first identified as a spontaneous mutation (Johnson and Hunt, 1971). The pattern of inheritance is autosomal recessive, and homozygous animals show preaxial polydactyly, scoliosis, male sterility, a characteristic hopping gait, and nonobstructive hydrocephalus. The hydrocephalus is associated with destruction of the ependymal cells over wide areas of the ventricular system, with no evidence of obstruction of the cerebral aqueduct, and affected animals die within 14 days of birth (Bryan et al., 1977). In addition to the defects in sperm tail development, presumed to cause the observed male sterility, global defects are also seen in the axoneme ultrastructure of ciliated epithelia, including loss of inner dynein arms (Bryan, 1983). The defects in the cilia of the ventricular apendymal cells are presumed to be involved in the development of hydrocephalus.

19.7 MODELS OF MYELOMENINGOCELE

In the United States spina bifida aperta or spina bifida cystica (spina bifida with meningocele or myelomeningocele) is the most common major birth defect. Hydrocephalus develops in 65 to 85% of these children, but only 5 to 10% have clinically overt hydrocephalus at birth (Stein and Schut, 1979). The majority of newborns with myelomeningocele (MMC), therefore, have normal or small ventricular systems. Pregressive ventriculomegaly and elevated intercranial pressure classically develop after myelomeningocele repair. The majority of MMC patients have caudal dislocation of the cervicomedullary junction, pons, fourth ventricle, and medulla (Type II Chiari, or Arnold-Chiari malformation). The prevailing theory behind the Chiari II malformation is that CSF egress from the open neural tube defect leads to decreased CSF backpressure and that this pressure is required for normal formation of the posterior fossa contents and CSF pathways (McLone and Knepper 1989). Early closure of myelomeningocele defects by fetal surgery in theory should decrease the formation of the Chiari II defect and perhaps decrease the number of children that require CSF diversion. Preliminary studies of fetal surgery in humans have suggested that this is the case (Farmer et al., 2003; Sutton et al., 2003).

No useful model of spontaneous myelomenigocele has yet been identified. A promising model involving the surgical creation of an open lumbar defect in fetal sheep has been utilized to better understand the potential benefits of fetal surgical repair of MMC and its impact on Chiari II malformation and neurological outcome (Bouchard et al., 2003). The defect is created on gestational day 75 and includes a laminectomy, opening of the dura, and a midline myelotomy to open the central canal of the spinal cord. Bouchard et al. report that animals develop myelomeningocele defects and posterior fossa herniation very similar to those seen in human MMC patients. The ventricular system is normal or decreased in size, and no animals with the surgically created defect are born with overt hydrocephalus. In utero surgical repair in this model effectively reverses hindbrain herniation and restores the gross anatomy of the cerebellar vermis (Bouchard et al., 2003). It is not clear what the natural history would be for animals if the surgically created defect were repaired postnatally. A reasonable hypothesis would be that animals so treated would develop hydrocephalus after the repair. Therefore, an important aspect of this model to explore further would be to repair the surgically created defect postpartum, as has classically been done in humans, to determine if these animals would develop hydrocephalus with the same high rate seen in humans. The importance of this model would be increased significantly if it could be shown to mimic the human disease in this regard. It is also possible that animals repaired postnatally would not develop hydrocephalus. If this were the case, it would suggest that the events leading to hydrocephalus in human MMC patients occur earlier in fetal development than the surgical defect is created. Further exploration of this issue could define critical time periods during gestation when repair of the MMC defect should be done to prevent later development of hydrocephalus.

19.8 LOOKING TO THE FUTURE: UNANSWERED QUESTIONS AND REMAINING CHALLENGES

The causes of hydrocephalus are multiple and varied. It is clear from our review of the many available animal models of congenital hydrocephalus that ventricular dilation can develop due to alterations of multiple genes alone or in combination. Screening for mutations known to carry a significant risk for hydrocephalus could help select individuals that may benefit from specific early surgical or possibly medical interventions. At some point it may be possible through the developing field of gene therapy to repair or replace a defective gene and prevent hydrocephalus and the secondary damage it causes.

At present the key treatment modality for children with congenital hydrocephalus involves surgical insertion of a shunt into the ventricles to divert excess CSF to another bodily cavity, usually the peritoneum. This often produces good short-term results, but more than 40% of shunts fail within two years (Kestle et al., 2000) and 3 to 15% become infected (Kulkarni et al., 2001). In an effort to reduce shunt-related complications and the cumulative risk of repeated shunt revisions, several modifications to shunt systems have been made, including (1) anti-siphon and flow-limiting devices to prevent overdrainage and subdural hematoma (Sainte-Rose et al., 1987; Horton and Pollay, 1990); (2) programmable valve mechanisms that obviate the need to repeat surgery if a change in pressure setting is required (Black et al., 1994; Aschoff et al., 1995; Belliard et al., 1996); (3) antibiotic impregnation of catheters to prevent infection (Bayston and Lambert, 1997; Stanton et al., 1999; Hampl et al., 2003; Kohnen et al., 2003); (4) neuroendoscopy and neuronavigation to aid in the precise placement of ventricular catheters (McCallum, 1997; Villavicencio et al., 2003), and (5) endoscopic third ventriculostomy (ETV) performed to create an internal shunt and avoid the need for shunting hardware (Griffith, 1975; Fukushima, 1978; Vries, 1978). The results of a randomized trial of different shunt valve designs failed to demonstrate a significant advantage of newer designs, including a valve with an antisiphon device or a flow-limiting device, over a standard differential pressure valve (Drake et al., 1998). The primary endpoint, namely the time to shunt failure, did not significantly differ among the valves, and long-term follow-up also failed to demonstrate a clear advantage of any valve (Kestle et al., 2000). Equivalent failure rates have also been found in a randomized trial comparing conventional valves and a programmable device (Pollack et al., 1999). However, the increased opening pressure control provided by a programmable device and the ability to change this pressure without the need for additional surgery may lead to an advantage over conventional valves. The efficacy of antibacterial impregnated catheters to reduce the rate of shunt infection remains to be proven in a randomized controlled trial.

Careful characterization of animal models of congenital hydrocephalus with regard to the underlying etiology of ventricular dilation and the relation to human hydrocephalus (see Table 19.2) may aid in the selection of particular modes of treatment. In situations where communicating hydrocephalus is the primary cause of ventricular dilation, conventional shunting systems may be the treatment of choice. In noncommunicating hydrocephalus, where the subarachnoid space and absorptive apparatus are intact, treatments such as endoscopic third ventriculostomy may be the better approach. Failure rates for ETV range from 20 to 40%, and have been reported to be significantly higher in the pediatric population presumably due to poor absorption of CSF in neonates (Kelly, 1991; Jones et al., 1996; Teo and Jones, 1996). Mohanty et al. (2002) recently reported a series of 72 ETV procedures in which 13 patients failed. Cine-MRI or endoscopic exploration demonstrated an open stoma in 8 of the 13 ETV failures. This finding would suggest that in this study the majority of ETV failures were due to an unrecognized component of communicating

TABLE 19.2
Primary Etiology in Experimental Models of Congenital Hydrocephalus

Model	Human Disease Correlate	Noncommunicating		Communicating	Ciliary Dysfunction	Ex vacuo Changes	Increased CSF Production
		Primary Aqueductal Stenosis	Secondary Aqueductal Stenosis				
oh	CH, AS	—	+	+	?	?	?
ch	CH	—	—	+	?	?	?
MF1	CH	—	—	+	?	?	?
E2F-5	AS	+	—	?	?	?	+
L1CAM	CRASH syndrome, CH, AS	—	+	+	?	?	?
Hif-1alpha	CH	—	—	—	?	?	?
H-Tx	AS	+	?	?	—	?	?
LEW/jms	AS	+	—	—	?	?	?
ApoB	Hypolipoproteinemia	+	?	?	?	?	?
hyh	AS, CH	+	—	+	?	?	?
Otx2	AS	+	—	—	?	?	?
SUMS/NP	AS	+	—	—	?	?	?
Hy-3	AS, CH, PCD	—	+	+	+	?	?
OVE459	AS, CH, PCD	—	+	+	+	?	?
Mdnah5	AS, PCD	+	—	—	+	?	?
DNA lambda	AS, PCD	+	—	—	+	—	?
Tg737orpk	PCD	?	?	?	+	?	?
Spag-6	AS, PCD	+	?	?	+	?	?
HFH-4	PCD	?	?	?	+	?	?
WIC-Hyd	CH, PCD	?	?	+	+	?	?
Hop-sterile/hop-hpy	CH, PCD	—	—	+	+	?	?
TGFb1	CH	—	—	+	?	?	?

+, positive evidence available; −, negative evidence available; ?, no positive or negative evidence available; AS, aqueductal stenosis; CH, communicating hydrocephalus; PCD, primary ciliary dyskinesia.

hydrocephalus. At present the most reliable predictor of ETV failure is a history of intracerebral infection, and this is most likely due to obliteration of the subarachnoid space and absorptive apparatus (Fukuhara et al., 2000). Patients that fail ETV require placement of a conventional shunt. The ability to define the population that will not respond to ETV will enable these patients to avoid an unnecessary surgery. It is clear from a number of the animal models examined (oh, L1CAM, hyh, hy-3, OVE459, hop-sterile, hop-hpy) that primary or secondary aqueductal stenosis can exist on a background of communicating hydrocephalus. It is therefore conceivable that genetic screening may eventually be able to determine the mutation(s) responsible for hydrocephalus in an individual patient and its association with communicating or noncommunicating hydrocephalus. This information taken together with the patient's age and the risk of conventional shunt complications would determine the most appropriate surgical intervention.

REFERENCES

Adams, P. D., and Kaelin, W. G., Jr. (1996) The cellular effects of E2F overexpression. *Curr Top Microbiol Immunol* **208**, 79–93.

Akai, K., Uchigasaki, S., Tanaka, U., and Komatsu, A. (1987) Normal pressure hydrocephalus. Neuropathological study. *Acta Pathol Jpn* **37**, 97–110.

Alashari, M., Chen, E., and Poskanzer, L. (1995) Partial deletion of chromosome 6p: autopsy findings in a premature infant and review of the literature. *Pediatr Pathol Lab Med* **15**, 941–7.

al-Shroof, M., Karnik, A. M., Karnik, A. A., Longshore, J., Sliman, N. A., and Khan, F. A. (2001) Ciliary dyskinesia associated with hydrocephalus and mental retardation in a Jordanian family. *Mayo Clin Proc* **76**, 1219–24.

Ang, S. L., and Rossant, J. (1994) HNF-3 beta is essential for node and notochord formation in mouse development. *Cell* **78**, 561–74.

Aschoff, A., Kremer, P., Benesch, C., Fruh, K., Klank, A., and Kunze, S. (1995) Overdrainage and shunt technology. A critical comparison of programmable, hydrostatic and variable-resistance valves and flow-reducing devices. *Child Nerv Syst* **11**, 193–202.

Asou, H., Miura, M., Kobayashi, M., and Uyemura, K. (1992) The cell adhesion molecule L1 has a specific role in neural cell migration. *Neuroreport* **3**, 481–4.

Bartoloni, L., Blouin, J. L., Pan, Y., Gehrig, C., Maiti, A. K., Scamuffa, N., Rossier, C., Jorissen, M., Armengot, M., Meeks, M., Mitchison, H. M., Chung, E. M., Delozier-Blanchet, C. D., Craigen, W. J., and Antonarakis, S. E. (2002) Mutations in the DNAH11 (axonemal heavy chain dynein type 11) gene cause one form of situs inversus totalis and most likely primary ciliary dyskinesia. *Proc Natl Acad Sci USA* **99**, 10282–6.

Bartsch, U., Kirchhoff, F., and Schachner, M. (1989) Immunohistological localization of the adhesion molecules L1, N-CAM, and MAG in the developing and adult optic nerve of mice. *J Comp Neurol* **284**, 451–62.

Bateman, A., Jouet, M., MacFarlane, J., Du, J. S., Kenwrick, S., and Chothia, C. (1996) Outline structure of the human L1 cell adhesion molecule and the sites where mutations cause neurological disorders. *EMBO J* **15**, 6050–9.

Bayston, R., and Lambert, E. (1997) Duration of protective activity of cerebrospinal fluid shunt catheters impregnated with antimicrobial agents to prevent bacterial catheter-related infection. *J Neurosurg* **87**, 247–51.

Belliard, H., Roux, F. X., Turak, B., Nataf, F., Devaux, B., and Cioloca, C. (1996) The Codman Medos programmable shunt valve. Evaluation of 53 implantations in 50 patients. *Neurochirurgie* **42**, 139–45; discussion 145–6.

Bickers, D., and Adams, R. (1949) Hereditary stenosis of the aqueduct of Sylvius as a cause of congenital hydrocephalus. *Brain* **72**, 246–262.

Black, P. M., Hakim, R., and Bailey, N. O. (1994) The use of the Codman-Medos Programmable Hakim valve in the management of patients with hydrocephalus: illustrative cases. *Neurosurgery* **34**, 1110–3.

Boncinelli, E., Gulisano, M., and Broccoli, V. (1993) Emx and Otx homeobox genes in the developing mouse brain. *J Neurobiol* **24**, 1356–66.

Bondurant, C. P., and Jimenez, D. F. (1995) Epidemiology of cerebrospinal fluid shunting. *Pediatr Neurosurg* **23**, 254–8; discussion 259.

Borit, A., and Sidman, R. L. (1972) New mutant mouse with communicating hydrocephalus and secondary aqueductal stenosis. *Acta Neuropathol (Berl)* **21**, 316–31.

Bouchard, S., Davey, M. G., Rintoul, N. E., Walsh, D. S., Rorke, L. B., and Adzick, N. S. (2003) Correction of hindbrain herniation and anatomy of the vermis after *in utero* repair of myelomeningocele in sheep. *J Pediatr Surg* **38**, 451–8; discussion 451–8.

Bronson, R. T., and Lane, P. W. (1990) Hydrocephalus with hop gait (hyh): a new mutation on chromosome 7 in the mouse. *Brain Res Dev Brain Res* **54**, 131–6.

Brummendorf, T., Hubert, M., Treubert, U., Leuschner, R., Tarnok, A., and Rathjen, F. G. (1993) The axonal recognition molecule F11 is a multifunctional protein: specific domains mediate interactions with Ng-CAM and restrictin. *Neuron* **10**, 711–27.

Bruni, J. E., Del Bigio, M. R., Cardoso, E. R., and Persaud, T. V. (1988a) Hereditary hydrocephalus in laboratory animals and humans. *Exp Pathol* **35**, 239–46.

Bruni, J. E., Del Bigio, M. R., Cardoso, E. R., and Persaud, T. V. (1988b) Neuropathology of congenital hydrocephalus in the SUMS/NP mouse. *Acta Neurochir (Wien)* **92**, 118–22.

Bryan, J. H. (1983) The immotile cilia syndrome. Mice versus man. *Virchows Arch A Pathol Anat Histopathol* **399**, 265–75.

Bryan, J. H., Hughes, R. L., and Bates, T. J. (1977) Brain development in hydrocephalic-polydactyl, a recessive pleiotropic mutant in the mouse. *Virchows Arch A Pathol Anat Histol* **374**, 205–14.

Buchstaller, A., Kunz, S., Berger, P., Kunz, B., Ziegler, U., Rader, C., and Sonderegger, P. (1996) Cell adhesion molecules NgCAM and axonin-1 form heterodimers in the neuronal membrane and cooperate in neurite outgrowth promotion. *J Cell Biol* **135**, 1593–607.

Cai, X., Pattisapu, J. V., Tarnuzzer, R. W., Fernandez-Valle, C., and Gibson, J. S. (1999) TGF-beta1 expression is reduced in hydrocephalic H-Tx rat brain. *Eur J Pediatr Surg* **9 (suppl 1)**, 35–8.

Chae, T. H., Kim, S., Marz, K. E., Hanson, P. I., and Walsh, C. A. (2004) The hyh mutation uncovers roles for alphaSnap in apical protein localization and control of neural cell fate. *Nat Genet* **36**, 264–270.

Chang, S., Rathjen, F. G., and Raper, J. A. (1987) Extension of neurites on axons is impaired by antibodies against specific neural cell surface glycoproteins. *J Cell Biol* **104**, 355–62.

Chen, J., Knowles, H. J., Hebert, J. L., and Hackett, B. P. (1998) Mutation of the mouse hepatocyte nuclear factor/forkhead homologue 4 gene results in an absence of cilia and random left-right asymmetry. *J Clin Invest* **102**, 1077–82.

Chitayat, D., Hahm, S. Y., Iqbal, M. A., and Nitowsky, H. M. (1987) Ring chromosome 6: report of a patient and literature review. *Am J Med Genet* **26**, 145–51.

Cirillo, L. A., McPherson, C. E., Bossard, P., Stevens, K., Cherian, S., Shim, E. Y., Clark, K. L., Burley, S. K., and Zaret, K. S. (1998) Binding of the winged-helix transcription factor HNF3 to a linker histone site on the nucleosome. *EMBO J* **17**, 244–54.

Clark, F. (1932) Hydrocephalus, a hereditary character in the house mouse. *Proc Natl Acad Sci USA* **18**, 654–656.

Clark, K. L., Halay, E. D., Lai, E., and Burley, S. K. (1993) Co-crystal structure of the HNF-3/fork head DNA-recognition motif resembles histone H5. *Nature* **364**, 412–20.

Cohen, A. R., Leifer, D. W., Zechel, M., Flaningan, D. P., Lewin, J. S., and Lust, W. D. (1999) Characterization of a model of hydrocephalus in transgenic mice. *J Neurosurg* **91**, 978–88.

Cohen, N. R., Taylor, J. S., Scott, L. B., Guillery, R. W., Soriano, P., and Furley, A. J. (1997) Errors in corticospinal axon guidance in mice lacking the neural cell adhesion molecule L1. *Curr Biol* **8**, 26–33.

Dagnino, L., Fry, C. J., Bartley, S. M., Farnham, P., Gallie, B. L., and Phillips, R. A. (1997) Expression patterns of the E2F family of transcription factors during murine epithelial development. *Cell Growth Differ* **8**, 553–63.

Dahlin-Huppe, K., Berglund, E. O., Ranscht, B., and Stallcup, W. B. (1997) Mutational analysis of the L1 neuronal cell adhesion molecule identifies membrane-proximal amino acids of the cytoplasmic domain that are required for cytoskeletal anchorage. *Mol Cell Neurosci* **9**, 144–56.

Dahme, M., Bartsch, U., Martini, R., Anliker, B., Schachner, M., and Mantei, N. (1997) Disruption of the mouse L1 gene leads to malformations of the nervous system. *Nat Genet* **17**, 346–9.

Daniel, G. B., Edwards, D. F., Harvey, R. C., and Kabalka, G. W. (1995) Communicating hydrocephalus in dogs with congenital ciliary dysfunction. *Dev Neurosci* **17**, 230–5.

Davies, A. F., Mirza, G., Sekhon, G., Turnpenny, P., Leroy, F., Speleman, F., Law, C., van Regemorter, N., Vamos, E., Flinter, F., and Ragoussis, J. (1999) Delineation of two distinct 6p deletion syndromes. *Hum Genet* **104**, 64–72.

Davis, J. Q., and Bennett, V. (1993) Ankyrin-binding activity of nervous system cell adhesion molecules expressed in adult brain. *J Cell Sci Suppl* **17**, 109–17.

Davis, J. Q., and Bennett, V. (1994) Ankyrin binding activity shared by the neurofascin/L1/NrCAM family of nervous system cell adhesion molecules. *J Biol Chem* **269**, 27163–6.

Davy, B. E., and Robinson, M. L. (2003) Congenital hydrocephalus in hy3 mice is caused by a frameshift mutation in Hydin, a large novel gene. *Hum Mol Genet* **12**, 1163–70.

De Santi, M. M., Magni, A., Valletta, E. A., Gardi, C., and Lungarella, G. (1990) Hydrocephalus, bronchiectasis, and ciliary aplasia. *Arch Dis Child* **65**, 543–4.

DeBernardo, A. P., and Chang, S. (1996) Heterophilic interactions of DM-GRASP: GRASP-NgCAM interactions involved in neurite extension. *J Cell Biol* **133**, 657–66.

Debiec, H., Christensen, E. I., and Ronco, P. M. (1998) The cell adhesion molecule L1 is developmentally regulated in the renal epithelium and is involved in kidney branching morphogenesis. *J Cell Biol* **143**, 2067–79.

Del Bigio, M. R. (1993) Neuropathological changes caused by hydrocephalus. *Acta Neuropathol (Berl)* **85**, 573–85.

Demyanenko, G. P., Tsai, A. Y., and Maness, P. F. (1999) Abnormalities in neuronal process extension, hippocampal development, and the ventricular system of L1 knockout mice. *J Neurosci* **19**, 4907–20.

Dignan, P., and Warkany, J. (1974) Congenital malformations: hydrocephaly. *Mental Retardation* **6**, 44–83.

Doherty, P., and Walsh, F. S. (1996) CAM-FGF receptor interactions: a model for axonal growth. *Mol Cell Neurosci* **8**, 99–111.

Dohrmann, G. J. (1970) Dark and light epithelial cells in the choroid plexus of mammals. *J Ultrastruct Res* **32**, 268–73.

Dou, C., Ye, X., Stewart, C., Lai, E., and Li, S. C. (1997) TWH regulates the development of subsets of spinal cord neurons. *Neuron* **18**, 539–51.

Drake, J. M., Kestle, J. R., Milner, R., Cinalli, G., Boop, F., Piatt, J., Jr., Haines, S., Schiff, S. J., Cochrane, D. D., Steinbok, P., and MacNeil, N. (1998) Randomized trial of cerebrospinal fluid shunt valve design in pediatric hydrocephalus. *Neurosurgery* **43**, 294–305.

Edwards, D. F., Patton, C. S., Bemis, D. A., Kennedy, J. R., and Selcer, B. A. (1983) Immotile cilia syndrome in three dogs from a litter. *J Am Vet Med Assoc* **183**, 667–72.

Egnor, M., Zheng, L., Rosiello, A., Gutman, F., and Davis, R. (2002) A model of pulsations in communicating hydrocephalus. *Pediatr Neurosurg* **36**, 281–303.

Faissner, A., Kruse, J., Nieke, J., and Schachner, M. (1984) Expression of neural cell adhesion molecule L1 during development, in neurological mutants and in the peripheral nervous system. *Brain Res* **317**, 69–82.

Fan, G., and Steer, C. J. (1999) The role of retinoblastoma protein in apoptosis. *Apoptosis* **4**, 21–9.

Farese, R. V., Jr., Veniant, M. M., Cham, C. M., Flynn, L. M., Pierotti, V., Loring, J. F., Traber, M., Ruland, S., Stokowski, R. S., Huszar, D., and Young, S. G. (1996) Phenotypic analysis of mice expressing exclusively apolipoprotein B48 or apolipoprotein B100. *Proc Natl Acad Sci USA* **93**, 6393–8.

Farmer, D. L., von Koch, C. S., Peacock, W. J., Danielpour, M., Gupta, N., Lee, H., and Harrison, M. R. (2003) In utero repair of myelomeningocele: experimental pathophysiology, initial clinical experience, and outcomes. *Arch Surg* **138**, 872–8.

Fischer, G., Kunemund, V., and Schachner, M. (1986) Neurite outgrowth patterns in cerebellar microexplant cultures are affected by antibodies to the cell surface glycoprotein L1. *J Neurosci* **6**, 605–12.

Flood, C., Akinwunmi, J., Lagord, C., Daniel, M., Berry, M., Jackowski, A., and Logan, A. (2001) Transforming growth factor-beta1 in the cerebrospinal fluid of patients with subarachnoid hemorrhage: titers derived from exogenous and endogenous sources. *J Cereb Blood Flow Metab* **21**, 157–62.

Fransen, E., D'Hooge, R., Van Camp, G., Verhoye, M., Sijbers, J., Reyniers, E., Soriano, P., Kamiguchi, H., Willemsen, R., Koekkoek, S. K., De Zeeuw, C. I., De Deyn, P. P., Van der Linden, A., Lemmon, V., Kooy, R. F., and Willems, P. J. (1998a) L1 knockout mice show dilated ventricles, vermis hypoplasia and impaired exploration patterns. *Hum Mol Genet* **7**, 999–1009.

Fransen, E., Van Camp, G., D'Hooge, R., Vits, L., and Willems, P. J. (1998b) Genotype-phenotype correlation in L1 associated diseases. *J Med Genet* **35**, 399–404.

Fransen, E., Lemmon, V., Van Camp, G., Vits, L., Coucke, P., and Willems, P. J. (1995) CRASH syndrome: clinical spectrum of corpus callosum hypoplasia, retardation, adducted thumbs, spastic paraparesis and hydrocephalus due to mutations in one single gene, L1. *Eur J Hum Genet* **3**, 273–84.

Frantz, G. D., Weimann, J. M., Levin, M. E., and McConnell, S. K. (1994) Otx1 and Otx2 define layers and regions in developing cerebral cortex and cerebellum. *J Neurosci* **14**, 5725–40.

Friedlander, D. R., Milev, P., Karthikeyan, L., Margolis, R. K., Margolis, R. U., and Grumet, M. (1994) The neuronal chondroitin sulfate proteoglycan neurocan binds to the neural cell adhesion molecules Ng-CAM/L1/NILE and N-CAM, and inhibits neuronal adhesion and neurite outgrowth. *J Cell Biol* **125**, 669–80.

Fukuhara, T., Vorster, S. J., and Luciano, M. G. (2000) Risk factors for failure of endoscopic third ventriculostomy for obstructive hydrocephalus. *Neurosurgery* **46**, 1100–1111.

Fukushima, T. (1978) Endoscopic biopsy of intraventricular tumors with the use of a ventriculofiberscope. *Neurosurgery* **2**, 110–3.

Galbreath, E., Kim, S. J., Park, K., Brenner, M., and Messing, A. (1995) Overexpression of TGF-beta 1 in the central nervous system of transgenic mice results in hydrocephalus. *J Neuropathol Exp Neurol* **54**, 339–49.

Garcia-Diaz, M., Dominguez, O., Lopez-Fernandez, L. A., de Lera, L. T., Saniger, M. L., Ruiz, J. F., Parraga, M., Garcia-Ortiz, M. J., Kirchhoff, T., del Mazo, J., Bernad, A., and Blanco, L. (2000) DNA polymerase lambda (Pol lambda), a novel eukaryotic DNA polymerase with a potential role in meiosis. *J Mol Biol* **301**, 851–67.

Green, M. C. (1970) The developmental effects of congenital hydrocephalus (ch) in the mouse. *Dev Biol* **23**, 585–608.

Greenstone, M. A., Jones, R. W., Dewar, A., Neville, B. G., and Cole, P. J. (1984) Hydrocephalus and primary ciliary dyskinesia. *Arch Dis Child* **59**, 481–2.

Griffith, H. B. (1975) Technique of fontanelle and persutural ventriculoscopy and endoscopic ventricular surgery in infants. *Child Brain* **1**, 359–63.

Grumet, M., Mauro, V., Burgoon, M. P., Edelman, G. M., and Cunningham, B. A. (1991) Structure of a new nervous system glycoprotein, Nr-CAM, and its relationship to subgroups of neural cell adhesion molecules. *J Cell Biol* **113**, 1399–412.

Gruneberg, H. (1943a) Congenital hydrocephalus in the mouse, a case of spurious pleotropism. *J Genet* **45**, 1–21.

Gruneberg, H. (1943b) Two new mutant genes in the house mouse. *J Genet* **45**, 22–28.

Gruneberg, H. (1953) Genetical studies on the skeleton of the mouse VII. Congenital hydrocephalus. *J Genet* **51**, 327–358.

Gruneberg, H., and Wickramaratne, G. A. (1974) A re-examination of two skeletal mutants of the mouse, vestigial-tail (vt) and congenital hydrocephalus (ch). *J Embryol Exp Morphol* **31**, 207–22.

Guichard, C., Harricane, M. C., Lafitte, J. J., Godard, P., Zaegel, M., Tack, V., Lalau, G., and Bouvagnet, P. (2001) Axonemal dynein intermediate-chain gene (DNAI1) mutations result in situs inversus and primary ciliary dyskinesia (Kartagener syndrome). *Am J Hum Genet* **68**, 1030–5.

Hackett, B. P., Brody, S. L., Liang, M., Zeitz, I. D., Bruns, L. A., and Gitlin, J. D. (1995) Primary structure of hepatocyte nuclear factor/forkhead homologue 4 and characterization of gene expression in the developing respiratory and reproductive epithelium. *Proc Natl Acad Sci USA* **92**, 4249–53.

Hampl, J. A., Weitzel, A., Bonk, C., Kohnen, W., Roesner, D., and Jansen, B. (2003) Rifampin-impregnated silicone catheters: a potential tool for prevention and treatment of CSF shunt infections. *Infection* **31**, 109–11.

Handel, M., Park, C., and Sotomayor, R. (1985) Allelism of hop and hpy. *Mouse News Lett* **72**, 124.

Hatini, V., Huh, S. O., Herzlinger, D., Soares, V. C., and Lai, E. (1996) Essential role of stromal mesenchyme in kidney morphogenesis revealed by targeted disruption of Winged Helix transcription factor BF-2. *Genes Dev* **10**, 1467–78.

Hlavin, M. L., and Lemmon, V. (1991) Molecular structure and functional testing of human L1CAM: an interspecies comparison. *Genomics* **11**, 416–23.

Ho, J. C., Chan, K. N., Hu, W. H., Lam, W. K., Zheng, L., Tipoe, G. L., Sun, J., Leung, R., and Tsang, K. W. (2001) The effect of aging on nasal mucociliary clearance, beat frequency, and ultrastructure of respiratory cilia. *Am J Respir Crit Care Med* **163**, 983–8.

Hofer, T., Wenger, H., and Gassmann, M. (2002) Oxygen sensing, HIF-1alpha stabilization and potential therapeutic strategies. *Pflugers Arch* **443**, 503–7.

Hollander, W. (1966) Hydrocephalic-polydactyl, a recessive pleotropic mutant in the mouse. *Am Zool* **6**, 588–589.

Hollander, W. (1976) Hydrocephalic-polydactyl, a recessive pleotropic mutant in the mouse, and its location on chromosome 6. *Iowa State J Res* **51**, 13–23.

Homanics, G. E., Smith, T. J., Zhang, S. H., Lee, D., Young, S. G., and Maeda, N. (1993) Targeted modification of the apolipoprotein B gene results in hypobetalipoproteinemia and developmental abnormalities in mice. *Proc Natl Acad Sci USA* **90**, 2389–93.

Hong, H. K., Chakravarti, A., and Takahashi, J. S. (2004) The gene for soluble N-ethylmaleimide sensitive factor attachment protein alpha is mutated in hydrocephaly with hop gait (hyh) mice. *Proc Natl Acad Sci USA* **101**, 1748–53.

Horton, D., and Pollay, M. (1990) Fluid flow performance of a new siphon-control device for ventricular shunts. *J Neurosurg* **72**, 926–32.

Huber, P. A. (1997) Caldesmon. *Int J Biochem Cell Biol* **29**, 1047–51.

Ibanez-Tallon, I., Gorokhova, S., and Heintz, N. (2002) Loss of function of axonemal dynein Mdnah5 causes primary ciliary dyskinesia and hydrocephalus. *Hum Mol Genet* **11**, 715–21.

Ignelzi, M. A., Jr., Miller, D. R., Soriano, P., and Maness, P. F. (1994) Impaired neurite outgrowth of src-minus cerebellar neurons on the cell adhesion molecule L1. *Neuron* **12**, 873–84.

Jabourian, Z., Lublin, F. D., Adler, A., Gonzales, C., Northrup, B., and Zwillenberg, D. (1986) Hydrocephalus in Kartagener's syndrome. *Ear Nose Throat J* **65**, 468–72.

Jensen, F. (1979) Acquired hydrocephalus. III. A pathophysiological study correlated with neuropathological findings and clinical manifestations. *Acta Neurochir (Wien)* **47**, 91–104.

Jimenez, A. J., Tome, M., Paez, P., Wagner, C., Rodriguez, S., Fernandez-Llebrez, P., Rodriguez, E. M., and Perez-Figares, J. M. (2001) A programmed ependymal denudation precedes congenital hydrocephalus in the hyh mutant mouse. *J Neuropathol Exp Neurol* **60**, 1105–19.

Johanson, C. E., Szmydynger-Chodobska, J., Chodobski, A., Baird, A., McMillan, P., and Stopa, E. G., (1999) Altered formation and bulk absorption of cerebrospinal fluid in FGF-2-induced hydrocephalus. *Am J Physiol* **277**, R263–71.

Johnson, D. G., Schwarz, J. K., Cress, W. D., and Nevins, J. R. (1993) Expression of transcription factor E2F1 induces quiescent cells to enter S phase. *Nature* **365**, 349–52.

Johnson, D. R., and Hunt, D. M. (1971) Hop-sterile, a mutant gene affecting sperm tail development in the mouse. *J Embryol Exp Morphol* **25**, 223–36.

Jones, H. C. (1984) The development of congenital hydrocephalus in the mouse. *Z Kinderchir* **39 (suppl 2)**, 87–8.

Jones, H. C. (1985) Cerebrospinal fluid pressure and resistance to absorption during development in normal and hydrocephalic mutant mice. *Exp Neurol* **90**, 162–72.

Jones, H. C. (1997) Aqueduct stenosis in animal models of hydrocephalus. *Child Nerv Syst* **13**, 503–4.

Jones, H. C., and Bucknall, R. M. (1987) Changes in cerebrospinal fluid pressure and outflow from the lateral ventricles during development of congenital hydrocephalus in the H-Tx rat. *Exp Neurol* **98**, 573–83.

Jones, H. C., and Bucknall, R. M. (1988) Inherited prenatal hydrocephalus in the H-Tx rat: a morphological study. *Neuropathol Appl Neurobiol* **14**, 263–74.

Jones, H. C., Dack, S., and Ellis, C. (1987) Morphological aspects of the development of hydrocephalus in a mouse mutant (SUMS/NP). *Acta Neuropathol (Berl)* **72**, 268–76.

Jones, H. C., Lopman, B. A., Jones, T. W., Carter, B. J., Depelteau, J. S., and Morel, L. (2000) The expression of inherited hydrocephalus in H-Tx rats. *Child Nerv Syst* **16**, 578–84.

Jones, H. C., Carter, B. J., Depelteau, J. S., Roman, M., and Morel, L. (2001a) Chromosomal linkage associated with disease severity in the hydrocephalic H-Tx rat. *Behav Genet* **31**, 101–11.

Jones, H. C., Depelteau, J. S., Carter, B. J., Lopman, B. A., and Morel, L. (2001b) Genome-wide linkage analysis of inherited hydrocephalus in the H-Tx rat. *Mamm Genome* **12**, 22–6.

Jones, H. C., Depelteau, J. S., Carter, B. J., and Somera, K. C. (2002) The frequency of inherited hydrocephalus is influenced by intrauterine factors in H-Tx rats. *Exp Neurol* **176**, 213–20.

Jones, H. C., Carter, B. J., and Morel, L. (2003) Characteristics of hydrocephalus expression in the LEW/Jms rat strain with inherited disease. *Child Nerv Syst* **19**, 11–8.

Jones, R. F., Kwok, B. C., Stening, W. A., and Vonau, M. (1996) Third ventriculostomy for hydrocephalus associated with spinal dysraphism: indications and contraindications. *Eur J Pediatr Surg* **6 (suppl 1)**, 5–6.

Jouet, M., Feldman, E., Yates, J., Donnai, D., Paterson, J., Siggers, D., and Kenwrick, S. (1993) Refining the genetic location of the gene for X linked hydrocephalus within Xq28. *J Med Genet* **30**, 214–7.

Jouet, M., Rosenthal, A., and Kenwrick, S. (1995) Exon 2 of the gene for neural cell adhesion molecule L1 is alternatively spliced in B cells. *Brain Res Mol Brain Res* **30**, 378–80.

Kaestner, K. H., Silberg, D. G., Traber, P. G., and Schutz, G. (1997) The mesenchymal winged helix transcription factor Fkh6 is required for the control of gastrointestinal proliferation and differentiation. *Genes Dev* **11**, 1583–95.

Kamiguchi, H., Hlavin, M. L., and Lemmon, V. (1998) Role of L1 in neural development: what the knockouts tell us. *Mol Cell Neurosci* **12**, 48–55.

Kaufmann, E., and Knochel, W. (1996) Five years on the wings of fork head. *Mech Dev* **57**, 3–20.

Kelly, P. C., Blake, W. W., and Davis, J. R. (1989) Tandem Y/6 translocation with partial deletion 6 (p23–pter). *Clin Genet* **36**, 204–7.

Kelly, P. J. (1991) Stereotactic third ventriculostomy in patients with nontumoral adolescent/adult onset aqueductal stenosis and symptomatic hydrocephalus. *J Neurosurg* **75**, 865–73.

Kestle, J., Drake, J., Milner, R., Sainte-Rose, C., Cinalli, G., Boop, F., Piatt, J., Haines, S., Schiff, S., Cochrane, D., Steinbok, P., and MacNeil, N. (2000) Long-term follow-up data from the Shunt Design Trial. *Pediatr Neurosurg* **33**, 230–236.

Kim, E., Cham, C. M., Veniant, M. M., Ambroziak, P., and Young, S. G. (1998) Dual mechanisms for the low plasma levels of truncated apolipoprotein B proteins in familial hypobetalipoproteinemia. Analysis of a new mouse model with a nonsense mutation in the Apob gene. *J Clin Invest* **101**, 1468–77.

Kobayashi, Y., Watanabe, M., Okada, Y., Sawa, H., Takai, H., Nakanishi, M., Kawase, Y., Suzuki, H., Nagashima, K., Ikeda, K., and Motoyama, N. (2002) Hydrocephalus, situs inversus, chronic sinusitis, and male infertility in DNA polymerase lambda-deficient mice: possible implication for the pathogenesis of immotile cilia syndrome. *Mol Cell Biol* **22**, 2769–76.

Kohn, D. F., Chinookoswong, N., and Chou, S. M. (1981) A new model of congenital hydrocephalus in the rat. *Acta Neuropathol (Berl)* **54**, 211–8.

Kohnen, W., Kolbenschlag, C., Teske-Keiser, S., and Jansen, B. (2003) Development of a long-lasting ventricular catheter impregnated with a combination of antibiotics. *Biomaterials* **24**, 4865–9.

Koto, M., Miwa, M., Shimizu, A., Tsuji, K., Okamoto, M., and Adachi, J. (1987) Inherited hydrocephalus in Csk: Wistar-Imamichi rats; Hyd strain: a new disease model for hydrocephalus. *Jikken Dobutsu* **36**, 157–62.

Kowitz, A., Kadmon, G., Eckert, M., Schirrmacher, V., Schachner, M., and Altevogt, P. (1992) Expression and function of the neural cell adhesion molecule L1 in mouse leukocytes. *Eur J Immunol* **22**, 1199–205.

Kuhn, T. B., Stoeckli, E. T., Condrau, M. A., Rathjen, F. G., and Sonderegger, P. (1991) Neurite outgrowth on immobilized axonin-1 is mediated by a heterophilic interaction with L1(G4). *J Cell Biol* **115**, 1113–26.

Kulkarni, A. V., Drake, J. M., and Lamberti-Pasculli, M. (2001) Cerebrospinal fluid shunt infection: a prospective study of risk factors. *J Neurosurg* **94**, 195–201.

Kume, T., Deng, K. Y., Winfrey, V., Gould, D. B., Walter, M. A., and Hogan, B. L. (1998) The forkhead/winged helix gene Mf1 is disrupted in the pleiotropic mouse mutation congenital hydrocephalus. *Cell* **93**, 985–96.

Kunz, S., Ziegler, U., Kunz, B., and Sonderegger, P. (1996) Intracellular signaling is changed after clustering of the neural cell adhesion molecules axonin-1 and NgCAM during neurite fasciculation. *J Cell Biol* **135**, 253–67.

Labosky, P. A., Winnier, G. E., Jetton, T. L., Hargett, L., Ryan, A. K., Rosenfeld, M. G., Parlow, A. F., and Hogan, B. L. (1997) The winged helix gene, Mf3, is required for normal development of the diencephalon and midbrain, postnatal growth and the milk-ejection reflex. *Development* **124**, 1263–74.

Lagenaur, C., and Lemmon, V. (1987) An L1-like molecule, the 8D9 antigen, is a potent substrate for neurite extension. *Proc Natl Acad Sci USA* **84**, 7753–7.

Landrieu, P., Ninane, J., Ferriere, G., and Lyon, G. (1979) Aqueductal stenosis in X-linked hydrocephalus: a secondary phenomenon? *Dev Med Child Neurol* **21**, 637–42.

Lane, P. W. (1985) Hydrocephaly with hop gait (hyh). *Mouse News Letter* **73**, 18.

Lemire, R. J. (1988) Neural tube defects. *Jama* **259**, 558–62.

Lemmon, V., Farr, K. L., and Lagenaur, C. (1989) L1-mediated axon outgrowth occurs via a homophilic binding mechanism. *Neuron* **2**, 1597–603.

Levin, H., Ritch, R., Barathur, R., Dunn, M. W., Teekhasaenee, C., and Margolis, S. (1986) Aniridia, congenital glaucoma, and hydrocephalus in a male infant with ring chromosome 6. *Am J Med Genet* **25**, 281–7.

Lim, L., Zhou, H., and Costa, R. H. (1997) The winged helix transcription factor HFH-4 is expressed during choroid plexus epithelial development in the mouse embryo. *Proc Natl Acad Sci USA* **94**, 3094–9.

Lindeman, G. J., Dagnino, L., Gaubatz, S., Xu, Y., Bronson, R. T., Warren, H. B., and Livingston, D. M. (1998) A specific, nonproliferative role for E2F-5 in choroid plexus function revealed by gene targeting. *Genes Dev* **12**, 1092–8.

Lindner, J., Rathjen, F. G., and Schachner, M. (1983) L1 mono- and polyclonal antibodies modify cell migration in early postnatal mouse cerebellum. *Nature* **305**, 427–30.

Luthi, A., Gahwiler, B. H., and Gerber, U. (1994) Potentiation of a metabotropic glutamatergic response following NMDA receptor activation in rat hippocampus. *Pflugers Arch* **427**, 197–202.

Lyonnet, S., Pelet, A., Royer, G., Delrieu, O., Serville, F., le Marec, B., Gruensteudel, A., Pfeiffer, R. A., Briard, M. L., Dubay, C., et al. (1992) The gene for X-linked hydrocephalus maps to Xq28, distal to DXS52. *Genomics* **14**, 508–10.

Makiyama, Y., Shoji, S., and Mizusawa, H. (1997) Hydrocephalus in the Otx2± mutant mouse. *Exp Neurol* **148**, 215–21.

Martini, R., and Schachner, M. (1986) Immunoelectron microscopic localization of neural cell adhesion molecules (L1, N-CAM, and MAG) and their shared carbohydrate epitope and myelin basic protein in developing sciatic nerve. *J Cell Biol* **103**, 2439–48.

Matsuo, I., Kuratani, S., Kimura, C., Takeda, N., and Aizawa, S. (1995) Mouse Otx2 functions in the formation and patterning of rostral head. *Genes Dev* **9**, 2646–58.

McAllister, J. P., 2nd, and Chovan, P. (1998) Neonatal hydrocephalus. Mechanisms and consequences. *Neurosurg Clin North Am* **9**, 73–93.

McCallum, J. (1997) Combined frameless stereotaxy and neuroendoscopy in placement of intracranial shunt catheters. *Pediatr Neurosurg* **26**, 127–9.

McLaurin, R., Venes, J., Schut, L., and Epstein, F. (1989) *Pediatric Neurosurgery.* W.B. Saunders Company, Philadelphia.

McLone, D. G., and Knepper, P. A. (1989) The cause of Chiari II malformation: a unified theory. *Pediatr Neurosci* **15**, 1-12.

Michaelis, R. C., Du, Y. Z., and Schwartz, C. E. (1998) The site of a missense mutation in the extracellular Ig or FN domains of L1CAM influences infant mortality and the severity of X linked hydrocephalus. *J Med Genet* **35**, 901-4.

Milev, P., Friedlander, D. R., Sakurai, T., Karthikeyan, L., Flad, M., Margolis, R. K., Grumet, M., and Margolis, R. U. (1994) Interactions of the chondroitin sulfate proteoglycan phosphacan, the extracellular domain of a receptor-type protein tyrosine phosphatase, with neurons, glia, and neural cell adhesion molecules. *J Cell Biol* **127**, 1703–15.

Milhorat, T. H., Hammock, M. K., Davis, D. A., and Fenstermacher, J. D. (1976) Choroid plexus papilloma. I. Proof of cerebrospinal fluid overproduction. *Child Brain* **2**, 273–89.

Mohanty, A., Vasudev, M. K., Sampath, S., Radhesh, S., and Sastry Kolluri, V. R. (2002) Failed endoscopic third ventriculostomy in children: management options. *Pediatr Neurosurg* **37**, 304–9.

Moos, M., Tacke, R., Scherer, H., Teplow, D., Fruh, K., and Schachner, M. (1988) Neural adhesion molecule L1 as a member of the immunoglobulin superfamily with binding domains similar to fibronectin. *Nature* **334**, 701–3.

Moyer, J. H., Lee-Tischler, M. J., Kwon, H. Y., Schrick, J. J., Avner, E. D., Sweeney, W. E., Godfrey, V. L., Cacheiro, N. L., Wilkinson, J. E., and Woychik, R. P. (1994) Candidate gene associated with a mutation causing recessive polycystic kidney disease in mice. *Science* **264**, 1329–33.

Murcia, N. S., Richards, W. G., Yoder, B. K., Mucenski, M. L., Dunlap, J. R., and Woychik, R. P. (2000) The Oak Ridge Polycystic Kidney (orpk) disease gene is required for left-right axis determination. *Development* **127**, 2347–55.

Nagasawa, K., Kitamura, K., Yasui, A., Nimura, Y., Ikeda, K., Hirai, M., Matsukage, A., and Nakanishi, M. (2000) Identification and characterization of human DNA polymerase beta 2, a DNA polymerase beta-related enzyme. *J Biol Chem* **275**, 31233–8.

Nakamura, Y., and Sato, K. (1993) Role of disturbance of ependymal ciliary movement in development of hydrocephalus in rats. *Child Nerv Syst* **9**, 65–71.

Nehls, M., Pfeifer, D., Schorpp, M., Hedrich, H., and Boehm, T. (1994) New member of the winged-helix protein family disrupted in mouse and rat nude mutations. *Nature* **372**, 103–7.

Oi, S. (1998) Hydrocephalus chronology in the fetus: pathophysiology of aqueductal stenosis and morphological findings in animal models. Reply to the letter from Dr. H. C. Jones. *Child Nerv Syst* **14**, 614–6.

Oi, S., Yamada, H., Sato, O., and Matsumoto, S. (1996) Experimental models of congenital hydrocephalus and comparable clinical problems in the fetal and neonatal periods. *Child Nerv Syst* **12**, 292–302.

Okada, Y., Nonaka, S., Tanaka, Y., Saijoh, Y., Hamada, H., and Hirokawa, N. (1999) Abnormal nodal flow precedes situs inversus in iv and inv mice. *Mol Cell* **4**, 459–68.

Olbrich, H., Haffner, K., Kispert, A., Volkel, A., Volz, A., Sasmaz, G., Reinhardt, R., Hennig, S., Lehrach, H., Konietzko, N., Zariwala, M., Noone, P. G., Knowles, M., Mitchison, H. M., Meeks, M., Chung, E. M., Hildebrandt, F., Sudbrak, R., and Omran, H. (2002) Mutations in DNAH5 cause primary ciliary dyskinesia and randomization of left-right asymmetry. *Nat Genet* **30**, 143–4.

Overholser, M. D. (1954) The ventricular system in hydrocephlaic rat brains produced by a deficiency of vitamin B12 or folic acid in the maternal diet. *Anat Rec* **120**, 917–933.

Pelletier, G. J., Brody, S. L., Liapis, H., White, R. A., and Hackett, B. P. (1998) A human forkhead/winged-helix transcription factor expressed in developing pulmonary and renal epithelium. *Am J Physiol* **274**, L351–9.

Pennarun, G., Escudier, E., Chapelin, C., Bridoux, A. M., Cacheux, V., Roger, G., Clement, A., Goossens, M., Amselem, S., and Duriez, B. (1999) Loss-of-function mutations in a human gene related to Chlamydomonas reinhardtii dynein IC78 result in primary ciliary dyskinesia. *Am J Hum Genet* **65**, 1508–19.

Perez-Figares, J. M., Jimenez, A. J., Perez-Martin, M., Fernandez-Llebrez, P., Cifuentes, M., Riera, P., Rodriguez, S., and Rodriguez, E. M. (1998) Spontaneous congenital hydrocephalus in the mutant mouse hyh. Changes in the ventricular system and the subcommissural organ. *J Neuropathol Exp Neurol* **57**, 188–202.

Perez-Figares, J. M., Jimenez, A. J., and Rodriguez, E. M. (2001) Subcommissural organ, cerebrospinal fluid circulation, and hydrocephalus. *Microsc Res Tech* **52**, 591–607.

Persohn, E., and Schachner, M. (1987) Immunoelectron microscopic localization of the neural cell adhesion molecules L1 and N-CAM during postnatal development of the mouse cerebellum. *J Cell Biol* **105**, 569–76.

Picco, P., Leveratto, L., Cama, A., Vigliarolo, M. A., Levato, G. L., Gattorno, M., Zammarchi, E., and Donati, M. A. (1993) Immotile cilia syndrome associated with hydrocephalus and precocious puberty: a case report. *Eur J Pediatr Surg* **3 (suppl 1)**, 20–1.

Pollack, I. F., Albright, A. L., and Adelson, P. D. (1999) A randomized, controlled study of a programmable shunt valve versus a conventional valve for patients with hydrocephalus. Hakim-Medos Investigator Group. *Neurosurgery* **45**, 1399–1411.

Pourghasem, M., Mashayekhi, F., Bannister, C. M., and Miyan, J. (2001) Changes in the CSF fluid pathways in the developing rat fetus with early onset hydrocephalus. *Eur J Pediatr Surg* **11 (suppl 1)**, S10–3.

Qian, X., and Costa, R. H. (1995) Analysis of hepatocyte nuclear factor-3 beta protein domains required for transcriptional activation and nuclear targeting. *Nucleic Acids Res* **23**, 1184–91.

Qin, X. Q., Livingston, D. M., Kaelin, W. G., Jr., and Adams, P. D. (1994) Deregulated transcription factor E2F-1 expression leads to S-phase entry and p53-mediated apoptosis. *Proc Natl Acad Sci USA* **91**, 10918–22.

Raimondi, A. J., Bailey, O. T., McLone, D. G., Lawson, R. F., and Echeverry, A. (1973) The pathophysiology and morphology of murine hydrocephalus in Hy-3 and Ch mutants. *Surg Neurol* **1**, 50–5.

Rathjen, F. G., and Schachner, M. (1984) Immunocytological and biochemical characterization of a new neuronal cell surface component (L1 antigen) which is involved in cell adhesion. *EMBO J* **3**, 1–10.

Reid, R. A., and Hemperly, J. J. (1992) Variants of human L1 cell adhesion molecule arise through alternate splicing of RNA. *J Mol Neurosci* **3**, 127–35.

Renier, W. O., Ter Haar, B. G., Slooff, J. L., Hustinx, T. W., and Gabreels, F. J. (1982) X-linked congenital hydrocephalus. *Clin Neurol Neurosurg* **84**, 113–23.

Robinson, M. L., Allen, C. E., Davy, B. E., Durfee, W. J., Elder, F. F., Elliott, C. S., and Harrison, W. R. (2002) Genetic mapping of an insertional hydrocephalus-inducing mutation allelic to hy3. *Mamm Genome* **13**, 625–32.

Rodriguez, E. M., Oksche, A., Hein, S., and Yulis, C. R. (1992) Cell biology of the subcommissural organ. *Int Rev Cytol* **135**, 39–121.

Rodriguez, E. M., Rodriguez, S., and Hein, S. (1998) The subcommissural organ. *Microsc Res Tech* **41**, 98–123.

Rolf, B., Kutsche, M., and Bartsch, U. (2001) Severe hydrocephalus in L1-deficient mice. *Brain Res* **891**, 247–52.

Roperto, F., Galati, P., and Rossacco, P. (1993) Immotile cilia syndrome in pigs. A model for human disease. *Am J Pathol* **143**, 643–7.

Roperto, F., Galati, P., Troncone, A., Rossacco, P., and Campofreda, M. (1991) Primary ciliary dyskinesia in pigs. *J Submicrosc Cytol Pathol* **23**, 233–6.

Rose, S. P. (1995) Cell-adhesion molecules, glucocorticoids and long-term-memory formation. *Trends Neurosci* **18**, 502–6.

Rothman, J. E. (1994) Mechanisms of intracellular protein transport. *Nature* **372**, 55–63.

Rothman, J. E., and Wieland, F. T. (1996) Protein sorting by transport vesicles. *Science* **272**, 227–34.

Saffell, J. L., Williams, E. J., Mason, I. J., Walsh, F. S., and Doherty, P. (1997) Expression of a dominant negative FGF receptor inhibits axonal growth and FGF receptor phosphorylation stimulated by CAMs. *Neuron* **18**, 231–42.

Sainte-Rose, C., Hooven, M. D., and Hirsch, J. F. (1987) A new approach in the treatment of hydrocephalus. *J Neurosurg* **66**, 213–26.

Sammar, M., Aigner, S., and Altevogt, P. (1997) Heat-stable antigen (mouse CD24) in the brain: dual but distinct interaction with P-selectin and L1. *Biochim Biophys Acta* **1337**, 287–94.

Sapiro, R., Kostetskii, I., Olds-Clarke, P., Gerton, G. L., Radice, G. L., and Strauss, I. J. (2002) Male infertility, impaired sperm motility, and hydrocephalus in mice deficient in sperm-associated antigen 6. *Mol Cell Biol* **22**, 6298–305.

Sardet, C., Vidal, M., Cobrinik, D., Geng, Y., Onufryk, C., Chen, A., and Weinberg, R. A. (1995) E2F-4 and E2F-5, two members of the E2F family, are expressed in the early phases of the cell cycle. *Proc Natl Acad Sci USA* **92**, 2403–7.

Sasaki, S., Goto, H., Nagano, H., Furuya, K., Omata, Y., Kanazawa, K., Suzuki, K., Sudo, K., and Collmann, H. (1983) Congenital hydrocephalus revealed in the inbred rat, LEW/Jms. *Neurosurgery* **13**, 548–54.

Schipani, E., Ryan, H. E., Didrickson, S., Kobayashi, T., Knight, M., and Johnson, R. S. (2001) Hypoxia in cartilage: HIF-1alpha is essential for chondrocyte growth arrest and survival. *Genes Dev* **15**, 2865–76.

Schuch, U., Lohse, M. J., and Schachner, M. (1989) Neural cell adhesion molecules influence second messenger systems. *Neuron* **3**, 13–20.

Semenza, G. L. (1999) Regulation of mammalian O_2 homeostasis by hypoxia-inducible factor 1. *Annu Rev Cell Dev Biol* **15**, 551–78.

Shannon, M. W., and Nadler, H. L. (1968) X-linked hydrocephalus. *J Med Genet* **5**, 326–8.

Simeone, A., Acampora, D., Gulisano, M., Stornaiuolo, A., and Boncinelli, E. (1992) Nested expression domains of four homeobox genes in developing rostral brain. *Nature* **358**, 687–90.

Sollner, T., Whiteheart, S. W., Brunner, M., Erdjument-Bromage, H., Geromanos, S., Tempst, P., and Rothman, J. E. (1993) SNAP receptors implicated in vesicle targeting and fusion. *Nature* **362**, 318–24.

Stallcup, W. B., and Beasley, L. (1985) Involvement of the nerve growth factor-inducible large external glycoprotein (NILE) in neurite fasciculation in primary cultures of rat brain. *Proc Natl Acad Sci USA* **82**, 1276–80.

Stanton, C., Bayston, R., and Jellie, S. (1999) Prevention of CSF shunt infection: effect of the use of hybrid shunt systems which include antibacterial catheters. *Eur J Pediatr Surg* **9 (suppl 1)**, 47.

Stein, S. C., Feldman, J. G., Apfel, S., Kohl, S. G., and Casey, G. (1981) The epidemiology of congenital hydrocephalus. A study in Brooklyn, N.Y. 1968—1976. *Child Brain* **8**, 253–62.

Stein, S. C., and Schut, L. (1979) Hydrocephalus in myelomeningocele. *Child Brain* **5**, 413–9.

Sutton, L. N., Adzick, N. S., and Johnson, M. P. (2003) Fetal surgery for myelomeningocele. *Child Nerv Syst* **19**, 587–91.

Takeda, Y., Asou, H., Murakami, Y., Miura, M., Kobayashi, M., and Uyemura, K. (1996) A nonneuronal isoform of cell adhesion molecule L1: tissue-specific expression and functional analysis. *J Neurochem* **66**, 2338–49.

Takeuchi, I. K., Kimura, R., Matsuda, M., and Shoji, R. (1987) Absence of subcommissural organ in the cerebral aqueduct of congenital hydrocephalus spontaneously occurring in MT/HokIdr mice. *Acta Neuropathol (Berl)* **73**, 320–2.

Takeuchi, I. K., Kimura, R., and Shoji, R. (1988) Dysplasia of subcommissural organ in congenital hydrocephalus spontaneously occurring in CWS/Idr rats. *Experientia* **44**, 338–40.

Takizawa, T., Tada, T., Kitazawa, K., Tanaka, Y., Hongo, K., Kameko, M., and Uemura, K. I. (2001) Inflammatory cytokine cascade released by leukocytes in cerebrospinal fluid after subarachnoid hemorrhage. *Neurol Res* **23**, 724–30.

Taulman, P. D., Haycraft, C. J., Balkovetz, D. F., and Yoder, B. K. (2001) Polaris, a protein involved in left-right axis patterning, localizes to basal bodies and cilia. *Mol Biol Cell* **12**, 589–99.

Teo, C., and Jones, R. (1996) Management of hydrocephalus by endoscopic third ventriculostomy in patients with myelomeningocele. *Pediatr Neurosurg* **25**, 57–63.

Tomita, S., Ueno, M., Sakamoto, M., Kitahama, Y., Ueki, M., Maekawa, N., Sakamoto, H., Gassmann, M., Kageyama, R., Ueda, N., Gonzalez, F. J., and Takahama, Y. (2003) Defective brain development in mice lacking the Hif-1alpha gene in neural cells. *Mol Cell Biol* **23**, 6739–49.

Torikata, C., Kijimoto, C., and Koto, M. (1991) Ultrastructure of respiratory cilia of WIC-Hyd male rats. An animal model for human immotile cilia syndrome. *Am J Pathol* **138**, 341–7.

Urban, M., Bommer, C., Tennstedt, C., Lehmann, K., Thiel, G., Wegner, R. D., Bollmann, R., Becker, R., Schulzke, I., and Korner, H. (2002) Ring chromosome 6 in three fetuses: case reports, literature review, and implications for prenatal diagnosis. *Am J Med Genet* **108**, 97–104.

Van Essen, D. C. (1997) A tension-based theory of morphogenesis and compact wiring in the central nervous system. *Nature* **385**, 313–8.

Varadi, V., Csecsei, K., Szeifert, G. T., Toth, Z., and Papp, Z. (1987) Prenatal diagnosis of X linked hydrocephalus without aqueductal stenosis. *J Med Genet* **24**, 207–9.

Verma, K., and Wei King, D. (1967) Disorders of the developing nervous system of vitamin E-deficient rats. *Acta Anat (Basel)* **67**, 623–35.

Villavicencio, A. T., Leveque, J. C., McGirt, M. J., Hopkins, J. S., Fuchs, H. E., and George, T. M. (2003) Comparison of revision rates following endoscopically versus nonendoscopically placed ventricular shunt catheters. *Surg Neurol* **59**, 375–380.

Vio, K., Rodriguez, S., Navarrete, E. H., Perez-Figares, J. M., Jimenez, A. J., and Rodriguez, E. M. (2000) Hydrocephalus induced by immunological blockage of the subcommissural organ-Reissner's fiber (RF) complex by maternal transfer of anti-RF antibodies. *Exp Brain Res* **135**, 41–52.

Von Bohlen Und Halbach, F., Taylor, J., and Schachner, M. (1992) Cell type-specific effects of the neural adhesion molecules L1 and N-CAM on diverse second messenger systems. *Eur J Neurosci* **4**, 896–909.

Vries, J. K. (1978) An endoscopic technique for third ventriculostomy. *Surg Neurol* **9**, 165–8.

Wagner, C., Batiz, L. F., Rodriguez, S., Jimenez, A. J., Paez, P., Tome, M., Perez-Figares, J. M., and Rodriguez, E. M. (2003) Cellular mechanisms involved in the stenosis and obliteration of the cerebral aqueduct of hyh mutant mice developing congenital hydrocephalus. *J Neuropathol Exp Neurol* **62**, 1019–40.

Walker, M. E., Lynch-Salamon, D. A., Milatovich, A., and Saal, H. M. (1996) Prenatal diagnosis of ring chromosome 6 in a fetus with hydrocephalus. *Prenat Diagn* **16**, 857–61.

Wessels, M. W., den Hollander, N. S., and Willems, P. J. (2003) Mild fetal cerebral ventriculomegaly as a prenatal sonographic marker for Kartagener syndrome. *Prenat Diagn* **23**, 239–42.

Whiteheart, S. W., Griff, I. C., Brunner, M., Clary, D. O., Mayer, T., Buhrow, S. A., and Rothman, J. E. (1993) SNAP family of NSF attachment proteins includes a brain-specific isoform. *Nature* **362**, 353–5.

Whitelaw, A., Christie, S., and Pople, I. (1999) Transforming growth factor-beta1: a possible signal molecule for posthemorrhagic hydrocephalus? *Pediatr Res* **46**, 576–80.

Willems, P. J., Dijkstra, I., Van der Auwera, B. J., Vits, L., Coucke, P., Raeymaekers, P., Van Broeckhoven, C., Consalez, G. G., Freeman, S. B., Warren, S. T., et al. (1990) Assignment of X-linked hydrocephalus to Xq28 by linkage analysis. *Genomics* **8**, 367–70.

Willems, P. J., Vits, L., Raeymaekers, P., Beuten, J., Coucke, P., Holden, J. J., Van Broeckhoven, C., Warren, S. T., Sagi, M., Robinson, D., et al. (1992) Further localization of X-linked hydrocephalus in the chromosomal region Xq28. *Am J Hum Genet* **51**, 307–15.

Winnier, G. E., Hargett, L., and Hogan, B. L. (1997) The winged helix transcription factor MFH1 is required for proliferation and patterning of paraxial mesoderm in the mouse embryo. *Genes Dev* **11**, 926–40.

Wood, P. M., Schachner, M., and Bunge, R. P. (1990) Inhibition of Schwann cell myelination *in vitro* by antibody to the L1 adhesion molecule. *J Neurosci* **10**, 3635–45.

Wyss-Coray, T., Feng, L., Masliah, E., Ruppe, M. D., Lee, H. S., Toggas, S. M., Rockenstein, E. M., and Mucke, L. (1995) Increased central nervous system production of extracellular matrix components and development of hydrocephalus in transgenic mice overexpressing transforming growth factor-beta 1. *Am J Pathol* **147**, 53–67.

Xuan, S., Baptista, C. A., Balas, G., Tao, W., Soares, V. C., and Lai, E. (1995) Winged helix transcription factor BF-1 is essential for the development of the cerebral hemispheres. *Neuron* **14**, 1141–52.

Yamada, H., Oi, S., Tamaki, N., Matsumoto, S., and Sudo, K. (1992) Histological changes in the midbrain around the aqueduct in congenital hydrocephalic rat LEW/Jms. *Child Nerv Syst* **8**, 394–8.

Yamada, H., Oi, S. Z., Tamaki, N., Matsumoto, S., and Sudo, K. (1991) Prenatal aqueductal stenosis as a cause of congenital hydrocephalus in the inbred rat LEW/Jms. *Child Nerv Syst* **7**, 218–22.

Yamasaki, M., Thompson, P., and Lemmon, V. (1997) CRASH syndrome: mutations in L1CAM correlate with severity of the disease. *Neuropediatrics* **28**, 175–8.

Yip, P. M., Zhao, X., Montgomery, A. M., and Siu, C. H. (1998) The Arg-Gly-Asp motif in the cell adhesion molecule L1 promotes neurite outgrowth via interaction with the alphavbeta3 integrin. *Mol Biol Cell* **9**, 277–90.

Yoder, B. K., Tousson, A., Millican, L., Wu, J. H., Bugg, C. E., Jr., Schafer, J. A., and Balkovetz, D. F. (2002) Polaris, a protein disrupted in orpk mutant mice, is required for assembly of renal cilium. *Am J Physiol Renal Physiol* **282**, F541–52.

Zammarchi, E., Calzolari, C., Pignotti, M. S., Pezzati, P., Lignana, E., and Cama, A. (1993) Unusual presentation of the immotile cilia syndrome in two children. *Acta Paediatr* **82**, 312–3.

Zhang, X., Morera, S., Bates, P. A., Whitehead, P. C., Coffer, A. I., Hainbucher, K., Nash, R. A., Sternberg, M. J., Lindahl, T., and Freemont, P. S. (1998) Structure of an XRCC1 BRCT domain: a new protein-protein interaction module. *EMBO J* **17**, 6404–11.

Zimmerman, K. (1933) Eine neue Mutation der Hausmaus: "Hydrocephalus." *Z Indukt Abstamm Vererb* **64**, 176–180.

Zurcher, V. L., Golden, W. L., and Zinn, A. B. (1990) Distal deletion of the short arm of chromosome 6. *Am J Med Genet* **35**, 261–5.

20 Clinical Aspects of Disorders of the Choroid Plexus and the CSF Circulation

John D. Pickard, Marek Czosnyka, N. Higgins, B. Owler, S. Momjian, and Alonso Pena
Addenbrooke's Hospital, University of Cambridge, Cambridge, United Kingdom

CONTENTS

20.1 INTRODUCTION

There has been an explosion in experimental data about the choroid plexus (CP), but we understand much less about the human organ, either in health or disease. This chapter describes the limitations of the techniques available for human study and the various conditions that affect the CSF system (production, circulation, and absorption).

20.2 DIRECT VISUALIZATION

Neurosurgical approaches to lesions of the lateral and 3rd ventricles (Apuzzo, 1998) critically depend upon the orientation provided by the CP in the otherwise white, featureless landscape of the lateral ventricles. Modern image guidance techniques have refined such approaches and reduced the size, for example, of incisions through the brain (Paleologus et al., 2001). Once the CP has been identified, it is simply followed anteriorally to the foramen of Monro and hence to the 3rd ventricular pathology.

One inconvenient property of the human CP is its ability to attach to and enter the drainage holes of a ventricular shunt catheter. There is good evidence that the risk of proximal shunt blockage is reduced if the fenestrated end of the ventricular catheter is placed away from the CP (Becker and Nulsen, 1968). Paradoxically, direct visualization of the CP by endoscopic placement of the ventricular catheter does not appear to reduce the risk of proximal blockage (Kestle et al., 2003). Endoscopic coagulation of the CP in the vicinity of the tip of a ventricular catheter may prove helpful (Pople and Ettles, 1995).

20.3 IMAGING

Until the advent of the magnetic resonance (MR) scanner there was no satisfactory way of radiologically imaging of the CP. Plain skull x-rays (Dyke, 1930) and computerized tomography scanning (CT) (Modic et al., 1980) detect the calcification of the CP particularly in the region of the glomus, which progresses with age (75% in the 6th decade) due to psamomma body formation within the CP stroma (Alcolado et al., 1986). Postmortem histology reveals flattening and loss of the secretary epithelium, thickening of the basement membrane, cyst formation, lipid accumulation, fibrosis, and calcification with hyaline-amyloid depositions in the choroidal blood vessels (Silverberg et al., 2003). Mass lesions such as tumors, hematomas, and abscesses may reveal themselves by displacement of a calcified CP.

Both CT and MR imaging (MRI) should reveal major pathology of the CP—cysts, hemorrhages, villous hypertrophy, choroid papillomas, both benign and malignant, with or without any associated hydrocephalus (Atlas, 2002). Villous hypertrophy may cause ventricular distension through hypersecretion of cerebrospinal fluid (CSF) (see below).

The circulation of the CSF may be imaged by tracking the dispersion (using CT scanning or a gamma camera) of a small volume of contrast medium or radioisotope injected either into the ventricle or into the lumbar sac. In communicating hydrocephalus, isotope from the lumbar sac will not enter the cortical subarachnoid space (convexity block) but will enter the ventricles (ventricular reflux). Such patterns, however (Pickard, 1982), are not possible to quantify partly because of partial volume problems (Griffiths, 1973).

Pulsatile brain motion and hence CSF flow may be assessed using dynamic gated MR imaging, which enables analysis of the frequency components at different times during the cardiac cycle within the ventricles, arteries, and venous sinuses (Grietz et al., 1993, 1994; Strik et al., 2002). The flow through the aqueduct may provide a measure of CSF production (see below). Such MR studies have been interpreted by Grietz and his colleagues as showing that CSF flow is not through the cortical subarachnoid space (SAS) with absorption into the superior sagittal sinus (SSS). Such studies, however, depend upon the detection and interpretation of small changes in the MR signal within the few voxels between the curved, convoluted surface of the brain and the skull. Such reports have been based on small series of patients. Although challenging, such interpretations are difficult to reconcile with the enlarged subarachnoid space seen in idiopathic intracranial hypertension, the convexity block and ventricular reflux of communicating hydrocephalus, and the anatomical changes seen in the arachnoid villi with increasing CSF pressure.

20.4 TECHNIQUES FOR CSF PRODUCTION MEASUREMENT IN THE HUMAN

CSF in the primate is derived not only by active secretion from cerebral arterial blood (McComb, 1983; Davson, 1984; Davson et al., 1987) but also from extrachoroidal sources (see below). Neither carbonic anhydrase inhibition nor choroid plexectomy will completely abolish CSF production. With time, CSF production rate appears to return to normal in patients after choroid plexectomy (Milhorat 1974).

In the steady state in all species examined, Davson's equation adequately describes the relationships between CSF pressure, production, and absorption:

$$P_{csf} = (If \times R_{out}) + P_{sss} \tag{20.1}$$

where R_{out} is the resistance to CSF outflow, If = CSF formation rate, and P_{sss} = sagittal sinus pressure.

If P_{csf} is reduced to below P_{sss} by CSF drainage, then all the fluid collected should represent If. If artificial CSF is injected at a certain rate, then the change in (P_{csf} − P_{sss}) should relate to R_{out}, under the assumption that P_{sss} stays constant.

20.4.1 External Drainage

CSF absorption by normal channels must be avoided by maintaining the CSF pressure below P_{sss} for sufficiently long to establish a steady state. Values in humans of 0.24 to 0.37 mL/min have been reported (Cutler et al., 1968; Ekstedt, 1978; Rubin et al., 1966). CSF hypersecretion has certainly been confirmed by this technique in children with choroidal villous hypertrophy (see below). The rate of refilling of a ventricle filled with air during ventriculography indicated a CSF production rate of

0.14 mL/min, which, although almost certainly too low, is remarkably similar to the conventional estimates given the obvious problems with the technique (Deck and Potts, 1969).

20.4.2 Pappenheimer's Ventriculo-Cisternal Perfusion Technique

This method involves the constant infusion of artificial CSF containing a high molecular weight tracer molecule (e.g., Dextran, ^{131}I-labeled serum albumen). Provided that none of the marker diffuses into the brain or bypasses the drainage cannula in the cisterna magna or lumbar sac, its dilution at steady state directly reflects the volume of CSF secreted per unit of time. The drainage pressure should be less than P_{sss}. The method takes a long time to achieve a steady state and is not practical for everyday clinical use. Values of 0.35 to 0.37 mL/min have been recorded together with the effect of diamox (Rubin et al., 1966; Cutler et al., 1968; Milhorat et al., 1976).

20.4.3 Masserman Technique

When a known volume of CSF is withdrawn, P_{csf} falls. The time P_{csf} takes to return to normal is a measure of the CSF production rate. However, the reduction of CSF volume is compensated by an increase in cerebral blood volume. The rate of increase in P_{csf} is a reflection not only of CSF secretion but also of changes in CBV and brain compliance. Furthermore, the time taken is theoretically infinite so that an extrapolation procedure must be used. Finally, the method is accurate only if P_{csf} remains below P_{sss} throughout the testing cycle so that there is no drainage of CSF or suffusion of CSF into the brain parenchyma. Despite this, the results obtained confirm the findings with other techniques (May et al., 1990).

20.4.4 Computerized CSF Infusion Method

The infusion study may be performed via the lumbar CSF space or via a preimplanted ventricular access device. In both cases two needles are inserted (22 g spinal needles for lumbar tests; 25 g butterfly needles for ventricular studies). One needle is connected to a pressure transducer via a stiff saline-filled tube and the other to an infusion pump mounted on a purpose-built trolley containing a pressure amplifier and an IBM-compatible personal computer running software for data analysis (Czosnyka et al., 1990). After 10 minutes of baseline measurement, the infusion of normal saline at a rate of 1.5 or 1 mL/min (if the baseline pressure was higher than 15 mmHg) is started and continued until a steady-state ICP plateau is achieved. If the ICP increases to 40 mmHg, the infusion is stopped. Following cessation of the saline infusion, ICP is recorded until it decreases to steady baseline levels. All compensatory parameters are calculated using computer-supported methods based on physiological models of the CSF circulation. Baseline ICP and R_{csf} characterize static properties of the CSF circulation, while the cerebrospinal elasticity coefficient (E1) and pulse amplitude of ICP waveform (AMP) express the dynamic components of CSF pressure volume compensation.

E1 describes the compliance of the CSF compartment according to the formula:

$$\text{Compliance of CSF space} = C_i = 1/\{E1*(ICP - p_0)\} \qquad (20.2)$$

where p_0 is the unknown reference pressure level representing the hydrostatic difference between the site of ICP measurement and pressure indifferent point of the cerebrospinal axis (Ekstedt, 1978, Czosnyka et al., 1999). Cerebrospinal compliance is inversely proportional to ICP (Marmarou et al., 1978), therefore comparison between different subjects can be made only at the same level of the difference: ICP− p_0. The elastance coefficient E1 is independent of ICP, thus this coefficient is a much more convenient parameter when comparing individual patients. A low value of E1 is specific for a compliant system, while a high value indicates decreased pressure-volume compensatory reserve.

The pulse amplitude of ICP (AMP) increases proportionally when the mean ICP rises. The proportionality ratio (the AMP/P index) characterizes both elastance of the cerebrospinal space and the transmission of arterial pulsations to the CSF compartment.

Using Davson's equation, the sagittal sinus pressure is unknown and cannot be easily measured without increasing the invasiveness of the whole procedure. Consequently, the P_{sss} and $CSF_{formation}$ are estimated jointly using a nonlinear model utilizing the least-squares distance method during the computerized infusion test.

If P_{sss} is assumed to be constant, both CSF formation and R_{out} may be estimated (Vastola, 1980). However, the accuracy of estimation of P_{sss} using the infusion test is not great (Czosnyka et al., 1990), which consequently decreases the accuracy of this method. Although experimentally it is well established that P_{sss} is independent of the pressure in the subarachnoid space over the range of pressures conventionally checked, this may not always be true in the human (see Sec. 7.4). It is becoming clear that there is individual variation in whether an increase in sagittal sinus pressure may be a source of elevated ICP or vice versa (Higgins et al., 2003). Nevertheless, this technique does suggest that CSF formation rate decreases with age (see below).

20.4.5 Dynamic MRI

Cardiac gated and, more recently, echo planar MR imaging of a transaxial slice through the level of the aqueduct of Sylvius has been used to visualize and quantify CSF flow (Gideon et al., 1994a,b; Greitz et al., 1994; Bradley et al., 1996). Mean CSF flow gives very high values for CSF formation rate (e.g., 0.6 ± 0.3 mL/min in volunteers; 1.12 mL/min in IIH and 10.5 mL/min in 5 cases of communicating HC). The method assumes that there is no intraparenchymal suffusion of CSF from the 3rd and lateral ventricles. Different frequencies of oscillations of CSF flow corresponding to cardiac, respiratory, and vasogenic B waves have readily been measured so that the technique continues to evolve (Stric et al., 2002). There remains controversy over the clinical utility of the signal void that may be detected within the aqueduct for distinguishing between cerebral atrophy and active hydrocephalus (Bradley et al., 1996; Krauss et al., 1997). One problem is that MR does not directly measure fluid velocity, and the transaxial technique assumes that the diameter of the aqueduct does not change with systolic expansion of the brain.

20.5 AGE DEPENDENCE OF CSF PRODUCTION

Recently studies have been published referring to the age-related changes in CSF secretion (May et al., 1990; Rubenstein, 1998; Silverberg et al., 2003) using the Masserman technique. Our own finding that the resistance to CSF outflow increases above the 55th year of age (see also Albeck et al., 1998), while baseline ICP remains unaffected, leads to the conclusion that CSF production rate must decrease (Czosnyka et al., 2001) provided that sagittal sinus pressure stays constant (Ekstedt 1978) (see Figure 20.1).

The above agrees with the study of May et al. (1990), who reported a decrease in CSF formation rate by 50% in the elderly (77 years mean age) in comparison to young healthy people (28 years mean age). In contrast, neither Ekstedt (1978) nor Gideon et al. (1994) reported any evidence of age-related decrease in CSF production.

Rubinstein (1998) and Silverberg et al. (2003) have hypothesized that Alzheimer's dementia may be exacerbated by failure of drainage of toxic Abeta-like peptides from the cerebral mantle due to impaired CSF and interstitial fluid (ISF) production. A pilot study using a low-flow Orbis-Sigma V-P shunt has shown that

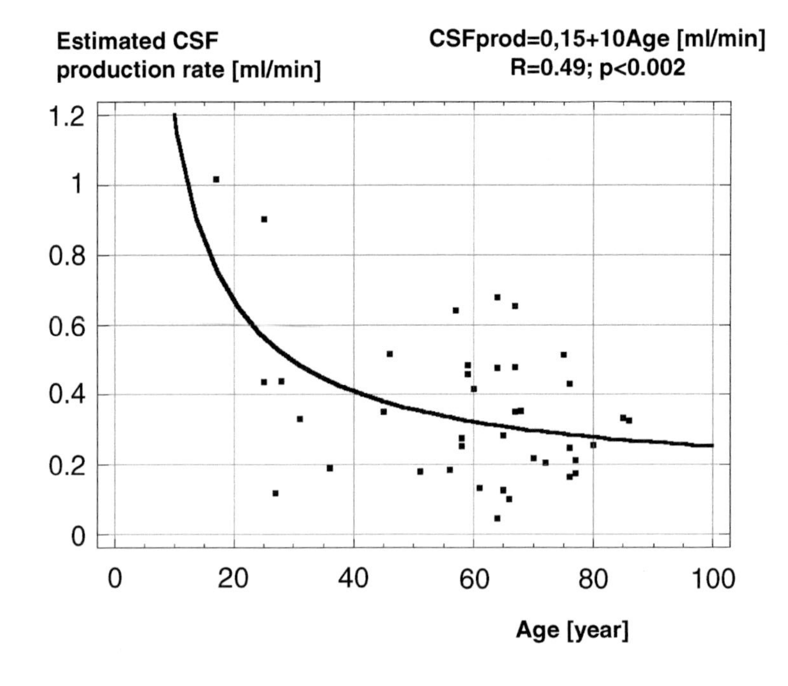

FIGURE 20.1 Estimated CSF production rate (y-axis) depends inversely proportionally on age (x-axis) in patients presenting with normal pressure hydrocephalus (NPH). The method of CSAF estimation was based on infusion test (Vastola, 1980). (From Czosnyka et al., 2001.)

this is a safe procedure and that the CSF levels of Tau and Abeta fall in the shunted group. A larger-scale RCT is currently in progress.

However, one should keep in mind that the estimator of CSF production approximates CSF absorption rather than production rate. If part of the CSF produced leaks into the brain parenchyma, there will be a serious underestimation of the CSF production value. The rate of this leakage may increase with age, as suggested by the age-related increase in the area and intensity of the periventricular lucencies seen on CT scans in patients with NPH. Hence, the age-related decrease in CSF production may signify an increase in CSF parenchymal drainage rather than a decrease in the CSF formation rate. There is no good evidence for lymphatic drainage from the brain in humans despite the findings with lower species.

20.6 THE VASCULAR COMPONENT OF DAVSON'S EQUATION

20.6.1 Evidence for a Vascular Component

The techniques available at the time of the experimental verification of Davson's equation did not permit accurate monitoring of the arterial and CSF pressure waveforms.

Intracranial pressure is derived from the circulation of both the cerebral blood and cerebrospinal fluid and can be expressed as the sum of two components (however, it is not certain whether the operator in the formula below should be represented by a simple addition):

$$ICP = ICP_{vasogenic} + ICP_{csf} \tag{20.3}$$

Hence a vascular term should be added to Davson's equation:

$$P_{csf} = (If \times R_{out}) + P_{sss} + ICP_{vasogenic} \tag{20.4}$$

The vascular component is difficult to express quantitatively. It is probably derived from the pulsation of the cerebral blood volume detected and averaged by nonlinear mechanisms of regulation of cerebral blood volume. More generally, multiple factors including arterial pressure, autoregulation, and cerebral venous outflow all contribute to the vascular component.

Certainly a vascular component exists. Marmarou et al. (1987) measured resistance to CSF outflow, sagittal sinus pressure, and CSF formation rate and revealed that the measured ICP is greater that the value resulting from Davson's equation. Experimental recordings of ICP during cardiac arrest (Czosynka et al., 1999) show a sudden decrease in ICP when the heart stops beating, while the resistance to CSF absorption as assessed by constant rate infusion remains unchanged, at least in the short term (see Figure 20.2). It is very likely that this sudden drop in measured ICP is caused by the disappearance of the vascular component.

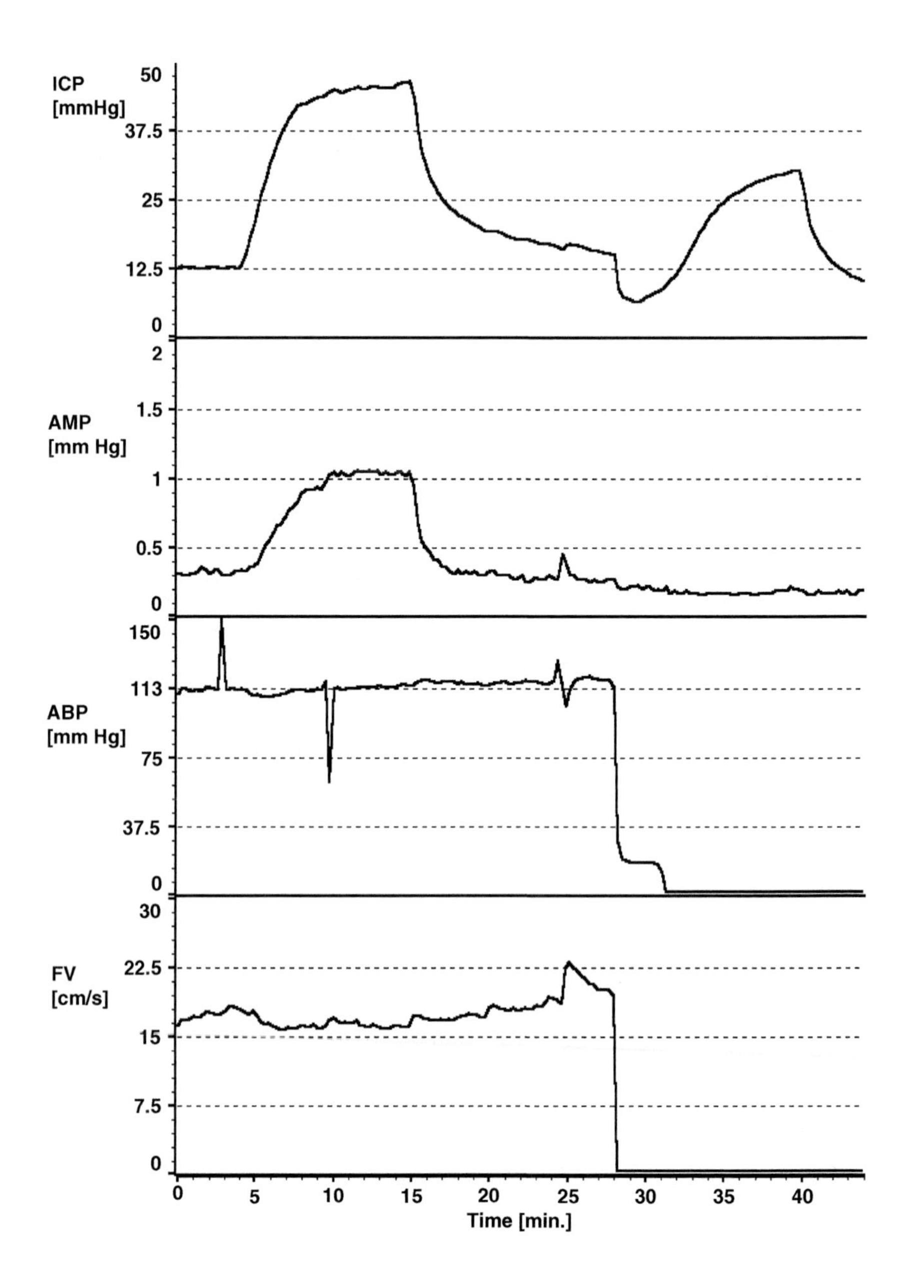

FIGURE 20.2 Time trends of intracranial pressure (ICP), pulse amplitude of ICP (AMP), arterial pressure (ABP), and basilar artery blood flow velocity (FV) during repeated infusion of saline (rate 0.2 mL/min) into subarachnoid space before and after animal was sacrificed (Euthatol 300 mg/kg). Animal sacrificed after ICP elevated during first infusion decreased to the baseline level. Sudden drop of ICP may represent disappearance of vasogenic component. Central venous pressure increased from 4 to 9 mmHg at the same moment. (From Czosnyka et al., 1999.)

20.6.2 Arterial Contribution

The tone of cerebral vessels influences CSF compensatory reserve as described by brain compliance, pressure volume index, brain stiffness, and CSF pulse amplitude. Generally, with cerebral vasodilation, the overall compliance of the cerebrospinal space decreases as with increasing brain edema and expanding mass lesions (Lofgren et al., 1973; Schettini and Walsh, 1988). With modest increases in intracranial pressure produced, for example, by infusion of artificial CSF, hypercapnia dilates parenchymal arterioles and reduces the elastance coefficient and modestly increases R_{csf} (Czosnyka et al., 1999). During arterial hypotension, the tone of all cerebral vessels decreases in autoregulating animals, and the compliance in the arterial walls increases, thereby increasing pulse pressure transmission from arterial blood pressure to intracranial pressure waveforms. This is in contrast to hypercapnia, where a change in arterial pulse transmission is not observed.

With more profound increases in intracranial pressure, provoked experimentally by brain compression with an extradural balloon, hypercapnia may not produce any change in intracranial pressure-volume relationships while arterial hypotension significantly reduces the elastance coefficient (Harper and Glass, 1965; Langfitt et al., 1965).

A modest increase in R_{csf} with hypercapnia and decrease in hypotension may reflect changes in the depth of the cortical subarachnoid space with brain expansion but also undefined effects of arterial venous pressures.

It is difficult to understand why, under steady-state conditions, a vascular bed that is anatomically separated from the CSF compartment may modify mean intracranial pressure. Even if CSF formation is dependent on choroid plexus perfusion pressure, this dependence has never been reported as very significant (Davson et al., 1987). Theoretically, only pulsatile changes in cerebral blood volume may influence mean ICP. Mathematical modeling reveals that simulated ICP in an integrated CBF and CSF circulatory model is always greater than the value resulting from Davson's equation (see Figure 20.3) (Czosnyka et al., 1999). The difference is proportional to the amplitude of arterial blood pressure and nonlinear mechanisms contributing to autoregulation. In all simulations, the vascular component is proportional to the pulsatile arterial blood pressure. The mean ICP decreases abruptly when blood pressure pulsations in the model are simulated to cease without changing mean arterial blood pressure.

Therefore, the hypothesis may be proposed that the continuously fluctuating volume of the arterial bed produces, after detection and filtration by various nonlinear elements of both cerebrovascular autoregulatory and intracerebral compensatory mechanisms, a constant vasogenic ICP component that cannot be explained by Davson's formula. Thus, the resistance to CSF outflow calculated using a constant rate infusion immediately contains a vascular component. This component varies when mean ICP or CPP varies and may explain the hemodynamic-induced changes in R_{csf}.

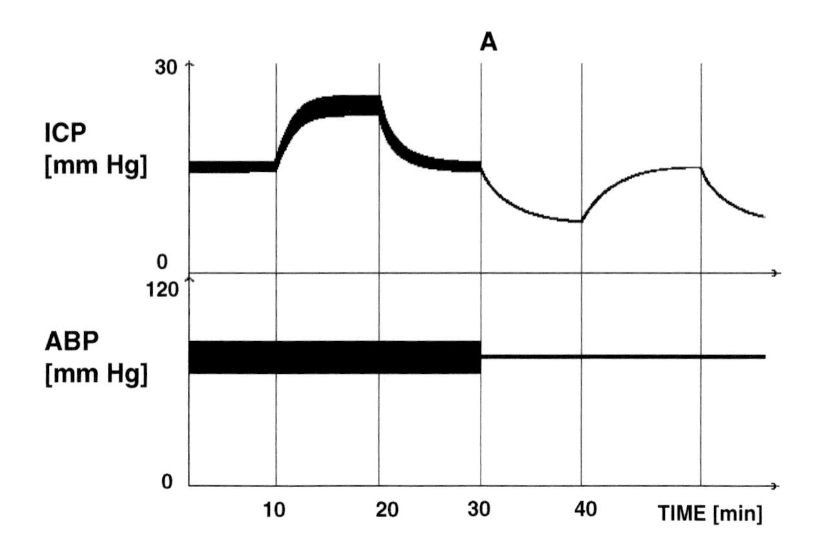

FIGURE 20.3 Results of simulation of ICP during infusion study made in two conditions of pulsatile component of ABP. When pulsations were normal (20 mmHg), baseline ICP was higher than resulting from Davson's equation, and resistance to CSF outflow calculated from the simulated infusion was greater than real parameter (8.7 vs. 6 mmHg/mL/min). When the pulse of ABP was decreased to 0 (point A), mean ICP decreased to a value resulting from Davsons's formula, and measured resistance to CSF outflow calculated from subsequent infusion was exactly equal to the value of real R_{csf}.

20.6.3 Derivation of Indices of Autoregulation from Waveform Analysis of CSF Pressure

Cerebrovascular reactivity may be assessed by observing the response of intracranial pressure to slow spontaneous changes in arterial blood pressure as "interrogated" by the autoregulatory behavior of cerebrovascular smooth muscle, which has a $t_{1/2}$ of 10 to 15 seconds. When the cerebrovascular bed is normally reactive, any change in arterial blood pressure produces inverse changes in cerebral blood volume and hence intracranial pressure. When reactivity is disturbed, changes in arterial blood pressure are passively transmitted to intracranial pressure. The pressure-reactivity index (PRx) is determined by calculating the correlation coefficient between time average data points of intracranial pressure and arterial blood pressure (Czosnyka et al., 1997). A positive PRx signifies a positive gradient of the regression line between the slow components of arterial blood pressure and ICP, which may be associated with the passive behavior of a nonreactive vascular bed. A negative value of PRx reflects a normally reactive vascular bed as ABP waves provoke inversely correlated waves in ICP. This index correlates well with indices of autoregulation based on other techniques including transcranial Doppler ultrasonography and position emission tomography (PET) (Steiner et al., 2003).

Such a technology permits the continuous monitoring of cerebrovascular autoregulation both within neurosurgical neurocritical care and also during various physiological challenges, such as a CSF infusion study. For example, during computerized infusion studies in patients with possible normal pressure hydrocephalus, it was found that global cerebrovascular autoregulation assessed by comparing transcranial Doppler waveforms with arterial blood pressure was inversely related to R_{csf} (Czosnyka et al., 2002). In other words, patients with the purest form of normal pressure hydrocephalus without any cerebrovascular disease have a high R_{csf} and preserved autoregulation. Patients with low R_{csf} and particularly those with white matter ischemia on MR scanning have disturbed autoregulation in the middle cerebral artery territory. More sophisticated techniques based on PET or MR scanning provide only snapshots and of course are not capable of continuous monitoring. The results of the two approaches are consistent (Momjian et al., 2004).

20.6.4 Venous Component

There is increasing evidence that P_{sss} in the human may not always be independent of P_{csf}. For example, in three of the subjects in Martins's study (1974), P_{sss} increased linearly with ICP, probably because the sagittal or transverse/lateral sinuses or both were collapsing. P_{sss} in some individuals may indeed be coupled to ICP (Paztor, 1976; Chazal et al., 1979; Sainte-Rose et al., 1984; Marmarou et al.1987). The venous sinuses may become narrow by various craniofacial anomalies involving the skull base (Taylor et al., 2001). In some patients with idiopathic intracranial hypertension (see Sec. 20.8), intracranial venous sinus pressure may be raised, sometimes because of raised central venous pressure, but more often because of either focal stenotic lesions in the lateral sinuses or extrinsic compression at this level by raised P_{csf} (Higgins et al., 2002, 2003, 2004; King et al., 2002; McGonigal et al., 2004). Some patients with such pathology may complain of pulsatile tinnitus. Under some circumstances, at least theoretically, the venous outflow may be a possible source of spontaneous generation of intracranial slow waves (Piechnik et al., 2001).

20.7 NORMAL PRESSURE HYDROCEPHALUS AND THE CEREBRAL MANTLE

Normal pressure hydrocephalus was first formally described by Hakim and Adams (1965) and consists of a clinical triad of gait disturbance (magnet gait), "dementia," and urinary incontinence with ventricular dilatation and apparently normal CSF pressure under lumbar puncture. However, overnight ICP monitoring may reveal increased CSF pressure and an increased frequency and amplitude of vasogenic waves of ICP. Typically, R_{out} is increased. NPH may be idiopathic or may follow meningitis, subarachnoid haemorrhage, or trauma, among other conditions, often many years later. Patients with aqueduct stenosis may remain asymptomatic for many years but in older life may present with NPH. Characteristically it is a condition of late middle age and the elderly, particularly in its idiopathic form. NPH is important because it is potentially reversible by CSF shunting or, in the subgroup with late presentation of aqueduct stenosis, by third ventriculostomy. The gait disturbance is often improved by shunting, as are many of the problems of higher mental function except for the frontal dysexecutive component (Iddon et al., 1999).

Although considered to be a disorder of the CSF circulation, there is evidence particularly in idiopathic NPH to suggest that the cerebral vasculature may play a role in its pathogenesis. Indeed, it is appropriate to refer to the patients presenting with the Hakim triad as complex dementia as there is often considerable overlap in the same patient between NPH, cerebrovascular disease (Earnest et al., 1974), and Alzheimer's disease (Silverberg et al., 2003). The challenge for the clinician is to identify those patients with a remediable hydrocephalic component that will respond to shunting. Some patients may respond for a few years to a shunt, but then other processes, including Alzheimer's and cerebrovascular disease, in which systemic hypertension is not adequately controlled, may supervene. The situation has been complicated by the hypothesis from the Stanford Group that patients with Alzheimer's disease may be helped by a shunt because it facilitates drainage of the toxic Abeta peptides from the aging cerebral mantle (Silverberg et al., 2003).

The pathology of NPH is poorly defined but may include obliteration by scarring of the subarachnoid space, periventricular gliosis and/or cerebrovascular changes in the periventricular white matter.

Considerable controversy surrounds which investigations provide the most reliable prediction of outcome following shunting, but measurement of R_{csf}, when performed accurately, provides a standard against which other techniques may be judged. CT or MR scanning or both may reveal not only ventriculomegaly but also changes in the white matter, both contiguous and noncontiguous, with the ventricle suggestive of ischemia/infarction, demyelination, and/or gliosis. Until very recently it was not really possible to define global cerebral blood flow and its response to physiological challenges in the periventricular white matter because of limitations of the technology. With modern PET and co-registration with MR, it has now been revealed that global cerebral blood flow is modestly reduced with enlargement of the area of the region of subcortical periventricular low CBF and within the thalamus and basal ganglia. With the increases in P_{csf} provoked by a CSF infusion study, global cerebral blood flow in patients with NPH is modestly reduced, a phenomenon that includes the cerebellum. The CBF is reduced in the thalamus and basal ganglia with a profile of change in regional white matter and cerebral blood flow that suggests a U-shaped relationship with distance from the ventricle. This decrease in white matter cerebral blood flow appears to be maximal in the periventricular watershed region (Owler and Pickard, 2001; Owler et al., 2004a,b; Momjian et al., 2004).

Various hypotheses have been suggested to explain the reduction in periventricular CBF in normal pressure hydrocephalus which are not necessarily mutually exclusive: tissue distortion, watershed ischemia, vasoactive metabolites, vascular disease, and increases in interstitial fluid pressure within the cerebral mantle. The distribution of interstitial parenchymal pressure in NPH has recently been modeled using finite element analysis based on Biot's theory of deformation of a poroelastic medium. Pena and colleagues (2002a,b) have proposed the hypothesis that the chronic ventricular dilatation of communicating hydrocephalus may be explained by a combination of reversal of interstitial fluid flow into the parenchyma and reduced tissue elasticity. This model complements the findings of Momjian et al. (see above). Although it might be argued that any gradient of interstitial tissue pressure will be too low to compress capillaries, the failure of autoregulation and drainage of vaso-active/toxic substances might provide the additional links in the chain of causation of border zone ischemia and tissue damage. There is a spectrum of patients from those with simply an increase of R_{out} and intact cerebrovascular autoregulation to those where cerebrovascular disease predominates and R_{out} is within the normal range or only modestly increased and is accompanied by cerebrovascular dysauto-regulation.

Hence, NPH should not be considered simply as a problem of the CSF circulation, and it is essential to include the properties of the cerebral mantle particularly in the aging brain. The application of the concepts of continuum mechanics as applied to a poroelastic medium are helping to explain various phenomena associated with brain deformation including the effects of brain ventricular shape (Pena et al., 1999), periventricular biomechanics, and the phenomenon of diaschisis in which cerebral blood flow and neurological function may be disturbed at a distance from the primary lesion (Pena et al., 2002).

20.8 IDIOPATHIC INTRACRANIAL HYPERTENSION, PSEUDOTUMOUR CEREBRI, AND BENIGN INTRACRANIAL HYPERTENSION

Much soul searching surrounds the nomenclature of this condition first described by Quincke in 1897. Idiopathic intracranial hypertension (IIH) is characterized by symptoms of increased ICP (headaches, nausea, vomiting, transient visual obscurations, and papilloedema), no localizing neurological signs except for false localizing signs such as a VIth nerve palsy, no deterioration in conscious level, normal CT and MR findings without evidence of dual sinus thrombosis, a raised P_{csf} of over 250 mmH_2O with normal cerebrospinal fluid, cytological and chemical findings, and no other cause for the increased intracranial pressure (see Sussman et al., 1998; Binder et al., 2004 for reviews). There are certainly patients who have IIH without papilloedema. The scan findings are not really normal—the CSF spaces are often enlarged, particularly around the optic nerves and the cortical sulci with an empty sella. CSF protein may actually be reduced from its already low level. Many patients with IIH are obese, and care has to be taken with the circumstances in which P_{csf} is measured because of the effects of obesity on central venous pressure. If left untreated, some patients will become blind. A large list of pathologies and drugs may predispose to IIH.

Various hypotheses have been advanced to explain the pathophysiology. Diffuse cerebral edema is unlikely. It is safe to undertake a lumbar puncture in these patients; their conscious level is not depressed, and the in vivo levels of brain water as measured by diffusion weighted MR are normal as opposed to the water content of biopsies, which are prone to artefact. Hypersecretion of CSF is very unlikely. In infants, choroid villous hypertrophy produces hydrocephalus where there is good evidence for CSF hypersecretion, but IIH does not resemble this condition in any way. CSF secretion rate would have to increase considerably for intracranial pressure to rise to the level seen in IIH if it were the only abnormality. There is no suggestion of hypertrophy of the CP on CT or MR.

The ventricles are normal or even reduced in size in IIH. Changes in cerebral blood flow may play a part, but CBV is difficult to measure in vivo. CSF absorption may certainly be impaired in IIH, but if so, why do the ventricles not increase in size? Where the obstruction to CSF absorption is near or at the arachnoid granulations, then the raised CSF pressure would be equally distributed between the enlarged cortical subarachnoid space and the ventricles so that there would be no gradient of pressure encouraging ventricular dilatation (Foley, 1955). In communicating hydrocephalus, the subarachnoid space is completely or partly obliterated as suggested by the findings on radioisotope cisternography of convexity block and ventricular reflux. If the pathology on IIH is in the cerebral venous drainage, then there will be no tendency for the cerebral ventricles to enlarge (Foley, 1955). As indicated in Sec. 20.6.4, it is now becoming clear that many patients with IIH actually have a problem with cerebral venous outflow, and in some cases this may be the primary pathology. Sagittal sinus thrombosis is a recognized mimic of benign intracranial hyptertension (BIH), but older studies also suggested narrowing in the lateral sinuses in nonthrombotic cases (Chazal et al., 1979; Janny et al., 1981). This was forgotten with the advent of MR venography with focal narrowings and signal gaps in the lateral sinuses regarded as normal variant anatomy or as artefacts of the imaging technique. In fact, if the MR venograms of patients with BIH are compared with those of truly asymptomatic volunteers, two quite different patterns emerge. Normal volunteers usually exhibit uniform signal throughout the lateral sinuses. Patients with BIH usually have bilateral lateral sinus signal gaps. Some stenoses may be due to enlarged arachnoid granulations protruding into the cerebral venous sinus lumen, a phenomenon that increases with age.

Formal venography consists of passing a catheter into the intracranial venous sinuses through the jugular vein, and such studies have recorded raised pressures in the venous sinuses in many patients with IIH (King et al., 1995, 2002; Karahalios et al., 1996; Higgins et al., 2002, 2003). In the minority of cases this is secondary to raised central venous pressure, but it more often appears to be the result of either focal stenotic lesions in the lateral sinuses or of extrinsic compression by raised $P_{csf.}$ Certainly in some patients withdrawal of CSF will reverse these narrowings (Higgins and Pickard, 2004). It has been shown that some patients with IIH may be helped by dilatation of venous stenoses with a stent with reduction in intracranial venous pressure (Higgins et al., 2002, 2003; Owler et al., 2003). Such novel treatments are required because conventional methods of management of IIH have many problems. For the obese, weight reduction may be all that is required. Bariatic surgery for the morbidly obese may be helpful (Sugerman et al., 1995). Of course, any drugs known to cause IIH should be withdrawn. There is no good evidence that drugs such as diamox or diuretics help except by providing a window of opportunity for a patient's condition to resolve spontaneously, which it often does. There was a vogue for corticosteroids, but these should be avoided as they simply induce iatrogenic Cushing's. Frequent lumbar punctures are psychologically traumatizing. Ventriculoperitoneal shunts are difficult to establish because of the small size of the ventricles, with the ventricles often collapsing onto the ventricular catheter and causing intermittent obstruction. A lumboperitoneal shunt is not easy in the obese and may lead to secondary Chiari malformation. Bilateral subtemporal decompressions may be required in a small minority of patients in whom vision is deteriorating and other methods have failed.

20.9 MISCELLANEOUS PATHOLOGY OF THE HUMAN CHOROID PLEXUS

The CP is prone to cyst formation, starting in the fetus (Kraus and Jirash, 2002), where they may be detected by ultrasound (Turner et al., 2003). Periventricular cysts may be associated with serious clinical complications such as congenital viral infections and anomalies (Herini et al., 2003). Intraventricular hemorrhage in neonates often arises from bleeding in to the choroid plexus, particularly the glomus (Graham and Lantos, 2002). Rarely, IVH may result from an arteriovenous malformation of the CP. A rare syndrome of infarction of the CP has been described due to ischemia in the distribution of the medial posterior choroidal artery (Liebeskind et al., 2004).

CP tumors (papilloma, carcinoma) are uncommon and account for only 0.5% of intracranial tumors in all ages, but 3.9% in childhood. They may occur in the fetus. Treatment starts with radical surgery; adjuvant therapy is generally reserved for carcinomas (Wolff et al., 2002; Cavalheiro et al., 2003).

20.10 CONTROL OF CSF PRODUCTION

Despite advances in shunt technology (programmable valves, antisiphon devices, infection-resistant catheters) and third ventriculostomy, significant problems remain with CSF diversion procedures. Reduction in CSF production appears attractive, but given the many functions of the choroid plexus, as outlined elsewhere in this volume, is it wise?

Posthemorrhagic ventricular dilatation is a complication of intraventricular haemorrhage in preterm infants and is associated with a high risk of long-term disability. Furosemide and acetazolamide are used widely in the treatment of this condition as a temporizing measure to avoid the need for placement of a ventriculoperitoneal shunt. However, in a well-conducted randomized controlled trial in 170 infants between 1992 and 1996, such drug treatment increased the risk of death and disability. Neurodevelopmental outcome at one year was also made worse by drug treatment. Hence the use of acetazolamide and furosemide in preterm infants with posthemorrhagic ventricular dilatation was ineffective in reducing the rate of shunt placement and increased neurological morbidity and mortality (Kennedy et al., 2001).

There is no similar Class I evidence for the use of Diamox or diuretics in other CSF disorders. It is increasingly being recognized that drug therapy and multiple lumbar punctures are an imperfect way of managing refractory IIH (Sussman et al., 1998).

Posthemorrhagic ventricular dilatation results initially from multiple small blood clots throughout the CSF channels impeding circulation and reabsorption (Whitelaw et al., 1999). TGF-β released into the CSF may stimulate the laying down of extracellular matrix proteins, which produce permanent obstruction of the CSF pathways. Prolonged raised pressure, pro-inflammatory cytokines, and free radical damage from iron may contribute to periventricular white matter damage and subsequent disability. A phase 1 trial of prevention of hydrocephalus after intraventricular haemorrhage in newborn infants by drainage and irrigation of thrombolytic therapy in 24 infants suggested that shunt surgery was reduced compared with historical controls (Whitelaw et al., 2003). CSF removal in infantile posthemorrhagic hydrocephalus results in significant improvement in cerebral hemodynamics as detected by near infrared spectroscopy (Soul et al., 2004).

20.11 CHOROID PLEXECTOMY

In the rare cases of children with choroid villous hypertrophy and papillomas, hydrocephalus is associated with CSF overproduction (Fujimura et al., 2004). It has been described that ventriculoperitoneal shunting may result in intractable ascites. Choroid plexectomy has been recommended as effective therapy. The role of this operation for the management of conventional hydrocephalus is more controversial. The limited evidence available suggests that CSF production rate returns to normal after choroid plexectomy (Milhorat et al., 1974). The Bristol series of over 100 children suggests that up to one third may achieve long-term control following choroid plexectomy without the need for CSF shunts (Pople and Ettle, 1995). Ventricular size was not significantly reduced, although cranial sulcal markings indicating reduced intracranial pressure became more prominent in all successfully treated patients. Serious morbidity was low in the procedure. It has been recommended for patients with hydranencephaly.

A fascinating recent innovation, still at the in vitro stage, is the development of immunotoxin-mediated destruction of choroid plexus cells by choroid-specific antibodies bound to ricin (Timothy et al., 2004). The long-term goal is to create a viable method for treating hydrocephalus and choroid plexus-derived tumors.

REFERENCES

Albeck MJ, Borgesen SE, Gjerris F, Schmidt JF, Sorensen PS (1991) Intracranial pressure and cerebrospinal fluid outflow conductance in healthy subjects. J Neurosurg 74: 597–600.

Albeck MJ, Skak C, Nielsen PR, Olsen KS, Borgesen SE, Gjerris F (1998) Age dependency of resistance to cerebrospinal fluid outflow. J Neurosurg 89(2): 275–278.

Alcolado JC, Moore IE, Weller RO (1986) Calcification in the human choroids plexus, meningiomas and pineal gland. Neuropathol Appl N Biol 12: 235–250.

Andreasen N, Minthon L, Clarberg A, Davidsson P, Gottfries J, Vanmechelen E, Vaderstichele H, Winbald B, Blennow K. Sensitivity, specificity, and stability of CSF-tau in AD in community-based patient sample. Neurology 1999; 53: 1488–1494.

Apuzzo MLJ (1998) Surgery of the Third Ventricle. Williams and Wilkins, Baltimore.

Artru AA, Hornbein TF. Closed ventriculocisternal perfusion to determine CSF production rate and pressure. Am J Physiol 1986 251(5 pt 2): R996–9.

Artru AA, Nugent M, Michenfelder JD. Closed recirculatory spinal subarachnoid perfusion for determining CSF dynamics. J Neurosurg 1982 56(3): 368–72.

Atlas S (2000) Magnestic Resonance Imaging of the Brain and Spine. 3rd ed. Lippincott Williams and Wilkins, Philadelphia.

Becker DP, Nulsen FE (1968) Control of hydrocephalus by valve-regulated venous shunt: avoidance of complications in prolonged shunt maintenance. J Neurosurg 28: 215–226.

Binder DK, Horton JC, Lawton MT, McDermott MW (2004) Idiopathic intracranial hypertension. Neurosurgery 54: 538–552.

Bradley WG Jr, Scalzo D, Queralt J, Nitz WN, Atkinson DJ, Wong P (1996) Normal-pressure hydrocephalus: evaluation with cerebrospinal fluid flow measurements at MR imaging. Radiology198(2): 523–9.

Cavalheiro S, Moron AF, Hisaba W, et al. (2003) Fetal brain tumours. Childs Nerv Syst 19: 529–36.

Chazal J, Janny P, Georget AM, Colnet G (1979) Benign intracranial hypertension. A clinical evaluation of the CSF absorption mechanism. Acta Neurochir Suppl (Wien) 28: 505–8.

Cutler RWP, Page L, Galicich J, Watters GV (1968) Formation and absorption of cerebrospinal fluid in man. Brain 91: 707–720.

Czosnyka M, Pickard JD (2004) Monitoring and interpretation of intracranial pressure. J Neurol Neurosurg Psychiatry 75: 813–821.

Czosnyka M, Batorski L., Laniewski P, Maksymowicz W, Koszewski W, Zaworski W (1990) A computer system for the identification of the cerebrospinal compensatory model. Acta Neurochir (Wien) 105: 112–116.

Czosnyka M, Czosnyka ZH, Whitfield PC, Donovan T, Pickard JD (2001) Age dependence of cerebrospinal pressure-volume compensation in patients with hydrocephalus. J Neurosurg 94(3): 482–486.

Czosnyka ZH, Czosnyka M, Whitfield PC, Donovan T, Pickard JD (2002) Cerebral autoregulation among patients with symptoms of hydrocephalus. Neurosurgery 50(3): 526–32.

Czosnyka M, Richards HK, Czosnyka Z, Piechnik S, Pickard JD (1999) Vascular components of cerebrospinal fluid compensation. J Neurosurg 90: 752–759.

Czosnyka M, Smielewski P, Kirkpatrick P, Menon DK, Pickard JD (1996) Monitoring of cerebral autoregulation in head-injured patients. Stroke 27: 829–834.

Czosnyka M, Smielewski P, Kirkpatrick P, Laing RJ, Menon D, Pickard JD (1997) Continuous assessment of the cerebral vasomotor reactivity in head injury. Neurosurgery 41: 11–19.

Davson H (1984) Formation and drainage of the CSF in hydrocephalus. In: Hydrocephalus. Shapiro K, Marmarou A, Portnoy H(eds). Raven Press, New York, 112–160.

Davson H, Welch K, Segal MB (1987) The Physiology and Patophysiology of the Cerebrospinal Fluid. Curchill-Livingstone, Edinburgh-New York, 189–246.

Deck MDF, Potts DG (1969) Movements of ventricular fluid levels due to cerebrospinal fluid formation. Am J Roentgenol Radium Ther Nucl Med 106: 354–368.

Dyke CG (1930) Indirect signs of brain tumour as noted in routine Roentgen examination. A survey of 3000 consecutive skull examinations. Amer J Roentgen 23: 598–606.

Earnest MP, Fahu S, Karp JH, Rowland LP (1974) Normal pressure hydrocephalus and hypertensive cerebrovascular disease. Arch Neurol 31: 262–266.

Ekstedt J (1978) CSF hydrodynamic studies in man. 2. Normal hydrodynamic variables related to CSF pressure and flow. J Neurol Neurosurg Psychiatry 41(4): 345–353.

Foley J (1955) Benign forms of intracranial hypertension—'atoxic' and 'otitic' hydrocephalus. Brain 78: 1–41.

Fujimura M, Onvma T, Kameyama M, et al. (2004) Hydrocephalus to cerebrospinal fluid over production by bilateral choroids plexus papillomas. Childs Nerv Syst 20:485–488.

Gideon P, Sorensen PS, Thomsen C, Stahlberg F, Gjerris F, Henriksen O (1994a) Assessment of CSF dynamics and venous flow in the superior sagittal sinus by MRI in idiopathic intracranial hypertension: a preliminary study. Neuroradiology 36(5): 350–354.

Gideon P, Stahlberg F, Thomsen C, Gjerris F, Sorensen PS, Henriksen O (1994b) Cerebrospinal fluid flow and production in patients with normal pressure hydrocephalus studied by MRI. Neuroradiology 36(3): 210–215.

Gideon P, Thomsen C, Stahlberg F, Henriksen O (1994c) Cerebrospinal fluid production and dynamics in normal aging: a MRI phase-mapping study. Acta Neurol Scand. 89(5): 362–366.

Graham DI, Lantos PL, eds. (2002) Greenfield's Neuropathology. 7th ed. Arnold, London.

Greitz D, Franck A, Nordell B (1993) On the pulsatile nature of intracranial and spinal CSF-circulation demonstrated by MR imaging. Acta Radiol 34: 321–328.

Greitz D, Hannerz J, Rahn T, Bolander H, Ericsson A (1994) MR imaging of cerebrospinal fluid dynamics in health and disease. On the vascular pathogenesis of communicating hydrocephalus and benign intracranial hypertension. Acta Radiol 35: 204–211.

Griffiths HB (1973) Transventricular absorption and isotope ventriculography. Arch Neurol 28: 272–273.

Hakim S, Adams RD (1965) The special clinical problem of symptomatic hydrocephalus with normal cerebrospinal fluid pressure. Observation in cerebrospinal fluid hydrodynamics. J Neurol Sci 2: 307–327.

Harper AM, Glass HI (1965) Effect of alterations in the arterial carbon dioxide tension on the blood flow through the cerebral cortex at normal and low arterial blood pressures. J Neurol Neurosurg Psychiatry 28: 449–452.

Herini E, Tsuneishi S, Takada S, et al. (2003) Clinical features of infants with subependymal germinolysis and choroid plexus cysts. Paediatr Int 45: 692–696.

Higgins JN, Pickard JD (2004) Lateral sinus stenoses in idiopathic intracranial hypertension resolving after CSF diversion. Neurology 62: 1907–1908.

Higgins JN, Cousins C, Owler BK, Sarkies N, Pickard JD (2003) Idiopathic intracranial hypertension: 12 cases treated by venous sinus stenting. J Neurol Neurosurg Psychiatry 74(12): 1662–1666.

Higgins JNP, Gillard JH, Owler BK, Harkness K, Pickard JD (2004) MR Venography in idiopathic intracranial hypertension: unappreciated and misunderstood. J Neurol Neurosurg Psychiatry 75: 621–625.

Higgins JNP, Owler BK, Cousins C, Pickard JD (2002) Venous sinus stenting for refractory benign intracranial hypertension. Lancet 359: 228–230.

Iddon JL, Pickard JD, Cross JJ, Griffiths PD, Czosnyka M, Sahakian BJ (1999) Specific patterns of cognitive impairment in patients with idiopathic normal pressure hydrocephalus and Alzheimer's disease: a pilot study. J Neurol Neurosurg Psychiat 67: 723–732.

Janny P, Chazal J, Colnet G, et al. (1981) Benign intracranial hypertension and disorder of CSF absorption. Surg Neurol 15: 168–174.

Kadel KA, Heistad DD, Faraci FM (1990) Effects of endothelin on blood vessels of the brain and choroid plexus. Brain Res 518(1–2): 78–82.

Karaholios DG, Pekate HL, Khayata MH, et al. (1996) Elevated intracranial venous pressure as a universal mechanism in pseudotumor cerebri of varying etiologies. Neurology 46: 198–202.

Kennedy CR, Ayers S, Campbell MJ, et al. (2001) Randomised, controlled trial of acetazolomide and fuxosemide in post haemorrhagic ventricular dilation in infancy: follow-up at 1 year. Paediatrics 108: 597–607.

Kestle JR, Drake JM, Cochrame DD, et al. (2003) Lack of benefit of endoscopic ventriculoperitoneal shunt insertion: a multicenter randomised trial. J Neurosurg 98: 284–290.

King JO, Mitchell PJ, Thomson KR, et al. (1995) Cerebral venography and manometry in idiopathic intracranial hypertension. Neurology 45: 224–228.

King JO, Mitchell PJ, Thomson KR, et al. (2002) Manometry combined with cervical puncture in idiopathic intracranial hypertension. Neurology 58: 26–30.

Kraus I, Jirasek JE. (2002) Some observations of the structure of the choroid plexus and its cysts. Prenat Diagn 22: 1223–1228.

Krauss JK, Regel JP, Vach W, Jungling FD, Droste DW, Wakhloo AK. (1997) Flow void of cerebrospinal fluid in idiopathic normal pressure hydrocephalus of the elderly: can it predict outcome after shunting? Neurosurgery 40(1): 67–74.

Langfitt TW, Weinstein JD, Kassell NF (1965) Cerebral vasomotor paralysis produced by intracranial pressure. Neurology (Minneap) 15: 622–641.

Liebeskind DS, Hurst RW (2004) Infarction of the choroid plexus. Am J Neuroradiol 25: 289–290.

Lofgren J, von Essen C, Zwetnow NN (1973) The pressure-volume curve of the cerebrospinal space in dogs. Acta Neurol Scand 49: 557–574.

Marmarou A, Maset AL, Ward JD, Choi S, Brooks D, Lutz HA, Moulton RJ, Muizelaar JP, DeSalles A, Young HF (1987) Contribution of CSF and vascular factors to elevation of ICP in severely head injured patients. J Neurosurg 66: 883–890.

Martins AN, Kobrine AI, Larsen DL (1974) Pressure in the sagittal sinus during intracranial hypertension in man. J Neurosurg 40: 603–608.

May C, Kaye JA, Atack JR, Schapiro MB, Friedland RP, Rapoport SI (1990) Cerebrospinal fluid production is reduced in healthy aging. Neurology 40(3 pt 1): 500–503.

McComb JG (1983) Recent research into the nature of cerebrospinal fluid formation and absorption. J Neurosurg 59: 369–383.

McGonigal A, Bone I, Teasdale E (2004) Resolution of transverse sinus stenosis in idiopathic intracranial hypertension after L-P shunt. Neurology 62: 514–515.

Milhorat TH (1974) Failure of choroid plexectomy as treatment for hydrocephalus. Surg Gynecol Obstet 139(4): 505–508.

Milhorat TH, Hamnwek MK, Chien T, Davis DA (1976) Normal rate of cerebrospinal fluid formation five years after bilateral choroid plexectomy. J Neurosurg 44: 735–739.

Modic MT, Weinstein MA, Rothner D (1980) Calcification of the choroid plexus visualised by computed tomography. Neuroradiology 135: 369–372.

Momjian S, Owler BK, Czosnyka Z, Czosnyka M, Pena A, Pickard JD (2004) Pattern of white matter regional cerebral blood flow and autoregulation in normal pressure hydrocephalus. Brain 127(pt 5): 965–972.

Owler BK, Pickard JD (2001) Normal pressure hydrocephalus and cerebral blood flow: a review. Acta Neurol Scand 104: 325–342.

Owler BK, Parker G, Helmagyi GM, et al. (2003) Pseudotumor cerebri syndrome: venous sinus obstruction and its treatment with stent placement. J Neurosurg 98: 1045–1055.

Owler BK, Momjian S, Czosnyka Z, Czosnyka M, Pena A, Harris NG, Smielewski P, Fryer T, Donovan T, Coles J, Carpenter A, Pickard JD (2004a) Normal pressure hydrocephalus and cerebral blood flow: a PET study of baseline values. J Cereb Blood Flow Metab. 24(1): 17–23.

Owler BK, Pena A, Momjian S, Czosnyka Z, Czosnyka M, Harris NG, Smielewski P, Fryer T, Donvan T, Carpenter A, Pickard JD (2004b) Changes in cerebral blood flow during cerebrospinal fluid pressure manipulation in patients with normal pressure hydrocephalus: a methodological study. J Cereb Blood Flow Metab 24(5): 579–587.

Paleologus TS, Wadley JP, Kitchen ND, Thomas DG (2001) Interactive image-guided transcallosal microsurgery for anterior third ventricular cysts. Minim. Invasive Neurosurg. 44: 157–162.

Pappenheimer JR, Heisey SR, Jordan EF, Downer JC (1962) Perfusion of the cerebral ventricular system in unanaesthetized goats. Am J Physiol 203: 763–774.

Pasztor E (1976) The effect of increased intracranial pressure on pressure in the superior sagittal sinus. Acta Neurochir (Wien) 34(1–4): 279–283.

Pena A, Bolton MD, Whitehouse H, Pickard JD (1999) Effects of brain ventricular shape on periventricular biomechanics: a finite-element analysis. Neurosurgery 45: 107–116.

Pena A, Harris NG, Bolton MD, Czosnyka M, Pickard JD (2002a) Communicating hydrocephalus: the biomechanics of progressive ventricular enlargement revisited. Acta Neurochir Suppl (Wien) 81: 59–63.

Pena A, Owler BK, Fryer TD, Minhas P, Czosnyka M, Crawford PJ, Pickard JD (2002b) A case study of hemispatial neglect using finite element analysis and positron emission tomography. J Neuroimaging 12: 360–367.

Pickard JD (1982) Adult communicating hydrocephalus. Brit J Hosp Med 27: 35–44.

Pickard JD, Czosnyka M, Steiner LA (2003) Raised intracranial pressure. In: Neurological Emergencies. 4th ed. Hughes RAC (ed.). BMJ Books, 188–246.

Piechnik SK, Czosnyka M, Richards HK, Whitfield PC, Pickard JD (2001) Cerebral venous blood outflow: a theoretical model based on laboratory simulation. Neurosurgery 49: 1214–1223.

Pople IK, Ettles D (1995) The role of endoscopic choroid plexus coagulation in the management of hydrocephalus. Neurosurgery 36: 698–701.

Quincke H (1897) Ueber Meningitis Senose und verwandte Zustände. Dtsch Z Nervenheildke 9: 140–168.

Rubenstein E (1998) Relationship of senescence of cerebrospinal fluid circulatory system to dementias of the aged. Lancet 351: 283–285.

Rubin RC, Henderson ES, Ommaya AK, Walker MD, Roel DP (1966) The production of cerebrospinal fluid in man and its modification by acetazoleamide. J Neurosurg 25: 430–436.

Sainte-Rose C, LaCombe J, Pierre-Kahn A, Renier D, Hirsch JF (1984) Intracranial venous sinus hypertension: cause or consequence of hydrocephalus in infants? J Neurosurg 60(4): 727–736.

Schettini A, Walsh EK (1988) Brain tissue elastic behavior and experimental brain compression. Am J Physiol 255:R799–R805.

Silverberg GD, Heit G, Huhn S, Jaffe RA, Chang SD, Bronte-Stewart H, Rubenstein E, Possin K, Saul TA (2001) The cerebrospinal fluid production rate is reduced in dementia of the Alzheimer's type. Neurol 57(10): 1763–1766.

Silverberg GD, Huhn S, Jaffe RA, Chang SD, Saul T, Heit G, Von Essen A, Rubenstein E (2002) Downregulation of cerebrospinal fluid production in patients with chronic hydrocephalus. J Neurosurg 97(6): 1271–1275.

Silverberg GD, Mayo M, Saul T, Rubenstein E, McGuire D (2003) Alzheimer's disease, normal pressure hydrocephalus and senescent changes in CSF circulatory physiology: a hypothesis. Lancet Neurol 2: 506–511.

Soul JS, Eichenwald E, Walter G, et al. (2004) CSF removal in infantile posthaemorrhagic hydrocephalus results in significant improvement in cerebral haemodynamics. Pediatr Res 55: 872–876.

Steiner LA, Coles JP, Czosnyka M, Minhas PS, Fryer TD, Aigbirhio FI, Clark JC, Smielewski P, Chatfield DA, Donovan T, Pickard JD, Menon DK (2003) Cerebrovascular pressure reactivity is related to global cerebral oxygen metabolism after head injury. J Neurol Neurosurg Psychiatry 74(6): 765–770.

Stric C, Klose U, Kieter C, Grodd W (2002) Slow rhythmic oscillations in intracranial CSF and blood flow: registered by MRI. Acta Neurochir 81(suppl): 139–142.

Sugerman HJ, Felton WL III, Salvant JB (1995) Effects of surgically induced weight loss on idiopathic intracranial hypertension in morbid obesity. Neurology 45: 1655–1659.

Sussman JD, Sarkies N, Pickard JD (1998) Benign intracranial hypertension; pseudotumor cerebri, idiopathic intracranial hypertension. Adv Tech Stand Neurosurg 24: 261–305.

Taylor WJ, Hayward RD, Lasjaunias P, et al. (2001) Enigma of raised intracranial pressure in patients with complex craniosynostosis: the role of abnormal intracranial venous drainage. J Neurosurg 94: 377–385.

Timothy J, Chumas P, Chakraborty A, et al. (2004) Destruction of choroid plexus cells in vitro: a new concept for the treatment of hydrocephalus? Neurosurgery 54: 727–732.

Turner SR, Samei E, Hertzberg BS, et al. (2003) Sonography of fetal choroid plexus cysts: detection depends on cyst size and gestational age. J Ultrasound Med 22: 1219–1227.

Vastola EF (1980) CSF formation and absorption estimates by constant flow infusion method. Arch Neurol 37: 150–154.

Weiss MH, Wertman N (1978) Modulation of CSF production by alterations in cerebral perfusion pressure. Arch Neurol. 35(8): 527–529.

Whitelaw A, Christie S, Pople IK (1999) Transforming growth factor—beta1: a possible signal molecule for post haemorrhagic hydrocephalus. Pediatr Res 46: 576–580.

Whitelaw A, Pople I, Cherian S, et al. (2003) Phase I trial of prevention of hydrocephalus after intraventricular haemorrhage in newborn infants by drainage, irrigation and fibrinolytic therapy. Pediatrics 111: 759–765.

Wolff JE, Sajedi M, Bront R, et al. (2002) Choroid plexus tumours. Br J Cancer 87: 1086–1091.

21 Grafting of the Choroid Plexus: Clinical Implications

Naoya Matsumoto, Yutaka Itokazu, and Masaaki Kitada
Kyoto University Graduate School of Medicine, Kyoto, Japan

Shushovan Chakrabortty
University of Michigan Center for Interventional Pain Medicine, Ann Arbor, Michigan, U.S.A.

Kazushi Kimura
Mie University Faculty of Medicine, Tsu City, Japan

Chizuka Ide
Kyoto University Graduate School of Medicine, Kyoto, Japan

CONTENTS

21.1 INTRODUCTION

The choroid plexus (CP) is composed of epithelial cells and highly vascularized connective tissues (see Figure 21.1a). The epithelial cells are a continuation of the ventricular ependymal cells. Choroid plexus epithelial cells and ventricular ependymal cells are of the same origin developmentally: both are derived from the germinal layer of the neural tube. Choroid plexus epithelial cells can be regarded as a type

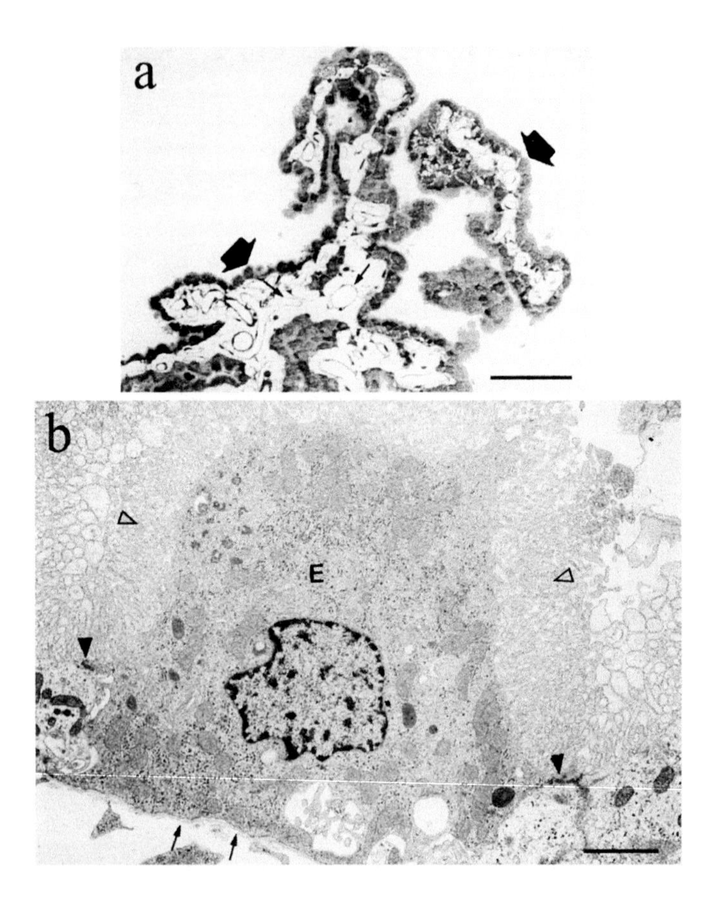

FIGURE 21.1 (a) Choroid plexus from the fourth ventricle. The choroid plexus consists of ependymal (epithelial) cells (thick arrows) and underlying connective tissue containing numerous blood vessels (thin arrows). Bar, 100 μm. (b) By electron microscopy, CPECs have microvilli (empty arrowheads) on the apical surface, and basal lamina (arrows) on the basal side. Filled arrowheads indicate cell junctions. Bar, 2 μm. (From Ide et al., 2001.)

of ependymal cells. Therefore, we call choroid plexus epithelial cells choroid plexus ependymal cells (CPECs). The connective tissue lying beneath the CPECs contains numerous vessels of various diameters, but only a few fibroblasts and nerve fibers. Most blood capillaries are sinusoidal in nature, with fenestrated endothelial cells.

CPECs are characterized by well-developed microvilli and only occasional cilia (see Figure 21.1b). They have junctional complexes at the apical side and are covered by a thin basal lamina at the base. These characteristics are different from those of ventricular ependymal cells, which usually have well-developed cilia and microvilli without basal lamina at the base. From the base of ventricular ependymal cells, cell processes of varying lengths extend into the brain parenchyma.

The choroid plexus is the site of cerebrospinal fluid (CSF) production. The rich vascular networks serve to supply a large amount of blood to the choroid plexus. The choroid plexus functions as the blood–CSF barrier. The CSF circulates from the ventricles to the subarachnoidal space around the brain and spinal cord. The brain and spinal cord can be regarded as being suspended within the CSF. It is known that CSF contains various trophic factors and molecules with particular functions. Although the production of CSF by the choroid plexus has been extensively studied, the roles of the choroid plexus in maintaining normal brain function and protecting the brain from traumatic/pathological damages have not been addressed in detail. We have focused on the role of the choroid plexus in neuronal regeneration in the brain and spinal cord. Here, we describe the effects of CPECs on neurite outgrowth in vitro, and on spinal cord regeneration by transplantation, CPEC differentiation after grafting into injured spinal cord, and the involvement of CPECs in maintaining brain functions by secreting functional molecules into the CSF.

21.2 EFFECTS OF CHOROID PLEXUS EPENDYMAL CELLS ON NEURITE EXTENSION IN VITRO

The influence of CPECs on neurite outgrowth from dorsal root ganglion (DRG) neurons and hippocampal neurons was examined in in vitro experiments in which such neurons were cocultured on monolayers of CPECs. Cocultures of neurons on astrocyte monolayers and cultures of neurons on laminin-coated plates were used for comparison. Since the findings about the effects of CPECs on neurite extension were almost the same for neurons from the DRG and from the hippocampus, only the results with DRG neurons are described.

CPECs were cultured basically according to the method of Weibel et al. (1986). Choroid plexuses were dissected from postnatal day1–10 mice (ddY), mechanically dissociated, and plated in eight-well plastic chamber slides coated with bovine fibronectin, with a final concentration of choroid plexuses from two or three mice per mL of culture medium. The culture medium was composed of alpha minimum essential medium (MEM) supplemented with 10% fatty acid-free bovine serum albumin (0.5 mg/mL), bovine insulin (5 mg/mL), and human transferrin (10 μg/mL). The culture medium was first changed after 2 days and was subsequently changed twice a week. From day 2 through 14, thrombin (0.5 U/mL) was added to the culture medium, and at day 14 alpha MEM was replaced by Waymouth's MB culture medium.

Within 24 hours of being placed in culture, most of the choroid plexus fragments had settled to the bottom of the well. Cells spread from the settled choroid plexus fragments. Two to 3 weeks after plating, they formed cell layers that ranged from sparse to almost complete monolayers. There were mainly two distinct types of ependymal cells in culture: centrally located honeycomb-like aggregates consisted of cells with abundant microvilli on the apical surface, and the peripherally dispersed cells were large flat and polygonal in appearance. The ependymal cells formed junctional complexes with the adjacent cells.

For identifying CPECs in the culture, DiI (5,5'-dibromo-3,3'dioctadecyl-3,3,3',3'-tetraethylindocarbocyanine perchlorate) was injected into the 4th ventricle to stain the CPECs facing the ventricular space 1 day before disection of the choroid plexus for culturing. DiI-stained ependymal cells proliferated in monolayers in the dishes. All DiI-stained cells exhibited strong immunoreactivity of S-100, specific marker for CPECs in the choroid plexus, and were negative for von Willebrand factor, an endothelial cell marker.

For the astrocyte culture, astrocytes were prepared from cerebral cortices of postnatal day 1 mice. Almost pure astrocyte monolayers were obtained after two to three passages. They showed strong positive reactions immunohistochemically for GFAP.

For coculturing, DRG neurons were obtained from mouse fetuses of 14-day gestation. Dissociated DRG neurons were plated on the ependymal cell monolayers, on the astrocyte monolayers or on laminin-coated plates.

After culturing for 4.5 hours, enhanced neurite outgrowth and branching were found with the DRG neurons cocultured on ependymal cell monolayers as compared with those cultured on laminin or on astrocyte monolayers. DRG neurons extended many long neurites with numerous branches on the ependymal cell monolayers (see Figure 21.2). Neurites appeared to prefer to attach to the ependymal cell surfaces. The mean total lengths of all primary neurites (including all branches) of a single DRG neuron were 285 ± 14, 395 ± 15, and 565 ± 12 μm on the laminin substrate, on the astrocyte monolayers, and on the ependymal cell monolayers, respectively (see Figure 21.3A). The mean lengths of individual primary neurites (including all branches) were 126 ± 11, 158 ± 12, and 205 ± 15 μm, respectively (see Figure 21.3B). The mean number of primary neurites per single DRG neuron was 2.26 ± 0.24, 2.50 ± 0.30, and 2.75 ± 0.28, respectively (see Figure 21.3C). The mean number of branches from a primary neurite of a DRG neuron was 1.51 ± 0.30, 2.45 ± 0.32, and 3.30 ± 0.31, respectively (see Figure 21.3D). The differences among the three

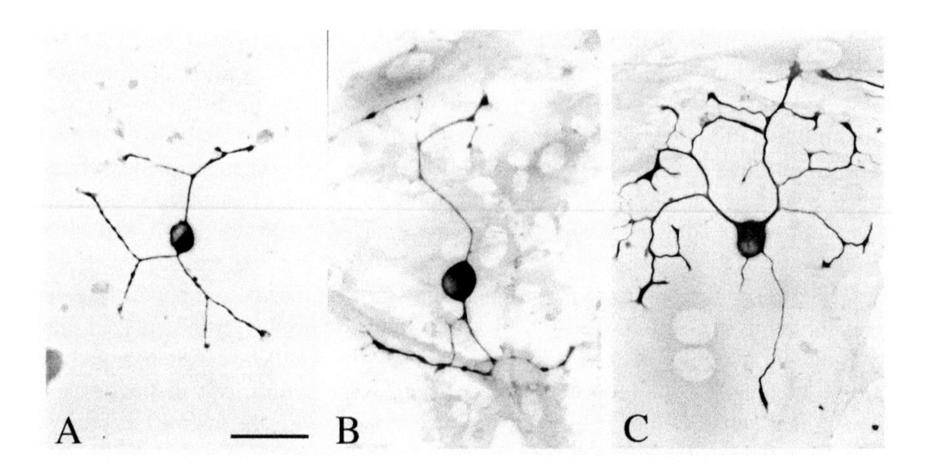

FIGURE 21.2 DRG neurons were cultured on the laminin substrate (A), on the astrocyte monolayers (B), and on the CPEC monolayers (C) for 4.5 hours. Neurons were immunohistochemically stained for neurofilaments. It is clear that the culture on the CPEC monolayers shows vigorous neurite outgrowth with many branches. Bar, 50 μm. (From Chakrabortty et al., 2000.)

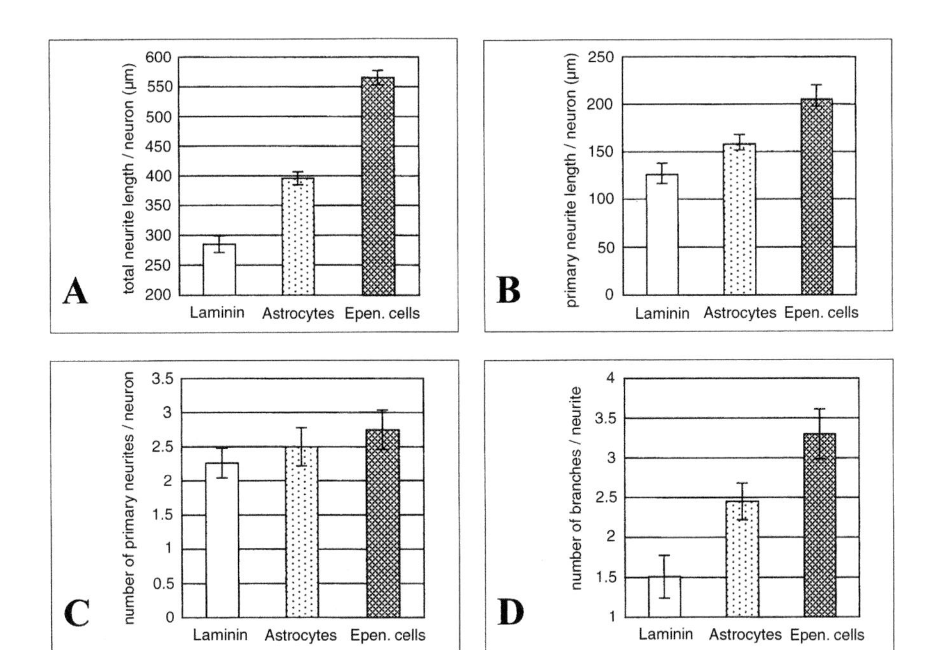

FIGURE 21.3 Graphs showing the mean total length (A), the mean length of individual primary neurites (B), the mean number of primary neurites (C), and the mean number of branches (D) of DRG neurons cultured on the laminin substrate, on the astrocytes monolayers, and on the CPEC monolayers. The values are the mean ± SEM of 50 neurons from each of 10 different cultures. There is a significant difference ($p < 0.05$) in these parameters except the number of primary neurites between the culture on CPEC monolayers and the other cultures on the laminin substrate and astrocyte monolayers. (From Chakrabortty et al., 2000.)

types of substrate were found to be significant by analysis of variance (ANOVA) at a confidence value of $p < 0.05$.

The enhancement of neurite outgrowth by CPECs in culture may be due to some specific molecules expressed on the cytoplasmic membrane of CPECs or secreted from CPECs into the adjacent neighborhood or both.

CPECs produce laminin and fibronectin (Gabrion et al., 1998). In mammals, including humans, ependymal cells express NCAM and E-cadherin (Figarella-Branger et al., 1995), and B-cadherin is expressed at the mRNA and protein levels in the ventricular ependymal cells in the developing chick brain (Murphy-Erdosh et al., 1994). This production of adhesion molecules by ependymal cells may contribute to the enhanced outgrowth of neurites on the cultured ependymal cells. The neurites of DRG neurons were located in most cases on the ependymal cell surfaces. These findings indicate the ability of ependymal cells to act as a substrate providing attachment sites and guidance to neurons and their neurites.

On the other hand, neurotrophic factors from ependymal cells are also considered to participate in the enhancement of neurite extension. Hepatocyte growth factor is normally expressed only in the ventricular ependymal cells and CPECs (Hayashi et al.,

1998). Insulin-like growth factor II (IGF-II) mRNA has been reported to be expressed primarily by the choroid plexus, leptomeninges, and glia (Nilsson et al., 1996). It is reported that IGF-II promotes the survival of neurons and neurite extension from sensory and sympathetic neurons in vitro (Hynes et al., 1988). The choroid plexus expresses mRNAs for neurotrophines and their receptors of the tyrosine kinase family such as trkA, trkB, and trkC (Timmusk et al., 1995). Rodent CPECs produce basic fibroblast growth factor (Cuevas et al., 1994). This production of neurotrophic factors by CPECs may also explain the promotion of neurite extension on the CPECs in the coculture experiments.

21.3 GRAFTING OF CHOROID PLEXUS INTO THE INJURED SPINAL CORD

Neuronal regeneration in the central nervous system (CNS) can be promoted by grafting various tissues and cells, including peripheral nerves (or Schwann cells) (David and Aguayo, 1981; Chen et al., 1996; Xu et al., 1997; Senoo et al., 1998; Menei et al., 1998), embryonic tissue (Iwashita et al., 1994), macrophages (Rapalino et al., 1998), olfactory ensheathing cells (Li et al., 1997), neural stem cells (Takahashi et al.1998; Wu et al., 2001, 2002), and bone marrow stromal cells (Wu et al., 2003; Ohta et al., 2004), or by administration of antibody against inhibitory factors associated with the myelin sheath and oligodendrocytes (Schnell and Schwab, 1990).

As described in the previous chapter, CPECs can promote neurite outgrowth from cocultured DRG and hippocampal neurons in vitro (Chakrabrotty et al., 2000). CPECs are regarded as a type of glial cell, which is associated with well-vascularized connective tissue. The choroid plexus in the 4th ventricle is readily accessible by exposure of the cisterna cerebellomedullaris. These properties prompted us to use the choroid plexus as a graft for the promotion of nerve regeneration in the spinal cord (Ide et al., 2001).

The choroid plexus was excised from the 4th ventricle and kept in Dulbecco's Modified Eagle's Medium (DMEM) for 30 minutes to 1 hour until the recipients were prepared. The spinal cord of recipient rats was exposed at the C2 level, and the dorsal funiculi were cut transversely about 1 to 1.5 mm wide and 1 mm deep with a pair of microscissors. The excised choroid plexus was minced into fragments in PBS and inserted into the spinal cord lesion made as described above.

Within three days after grafting, regenerating axons extended along grafted ependymal cells detached from the basal laminae as mentioned above or along basal laminae freed from ependymal cells. To identify CPECs after grafting, CPECs were labeled by microinjecting DiI into the 4th ventricle of the donor rats 1 day before grafting (see Figure 21.4a). DiI-labeled CPECs were intimately associated with regenerating axons one week after grafting (see Figure 21.4b). Also within the graft there were calcitonin gene-related peptide (CGRP)-containing nerve fibers, which were derived from the dorsal funiculus. CGRP-positive axons were in close contact with DiI-labeled ependymal cells one week after grafting (see Figure 21.4c). Numerous unmyelinated axons were surrounded by presumable CPECs and their cell processes. Ascending fibers from sciatic nerve were labeled by HRP (horseradish

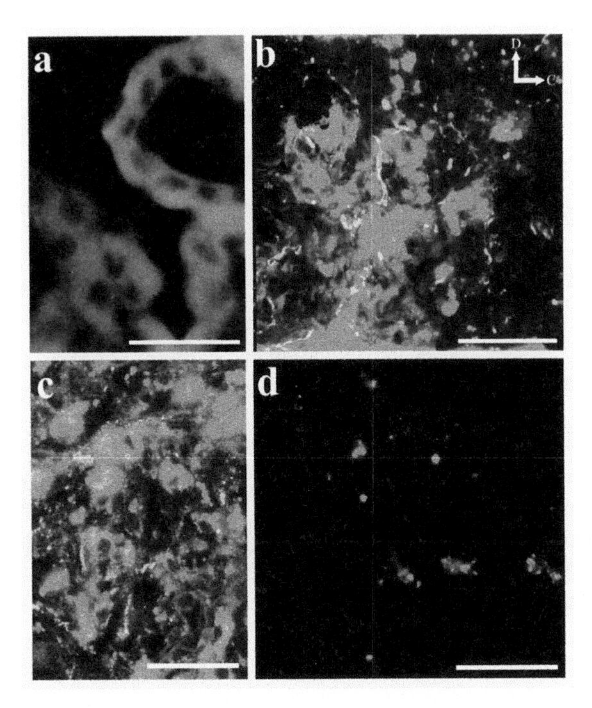

FIGURE 21.4 (See color insert following page 522.) Grafting of DiI-labeled CPECs into the rat spinal cord. (a) Only CPECs were exclusively stained by DiI, which had been injected into the 4th ventricle before grafting. (b) Regenerating axons immunostained for neurofilaments are in close contact with grafted CPECs. D and C: dorsal and caudal direction, respectively. (c) CGRP fibers intimately interact with CPECs. (d) DiI-labeled CPECs were treated repetitively by freezing to kill cellular elements before grafting. No CPECs remained after 7 days in the spinal cord. Some macrophages containing DiI-positive debris in the cytoplasm are seen. Bar, 50 μm. (From Ide et al., 2001.)

peroxidase) in the dorsal funiculus. Numerous HRP-labeled regenerating axons from the fasciculus gracilis extended rostrally into the graft one week after grafting. Unmyelinated axons were surrounded by presumed CPECs and their cell processes in the injury site.

The grafted CPECs closely interacted with regenerating axons for at least one month after grafting. The lesion was well repaired with rich vascularization. There were many myelinated and unmyelinated axons in association with astrocyte processes. Most axons were myelinated by oligodendrocytes, while a few were by Schwann cells.

Eight to ten months after grafting, many HRP-labeled regenerating axons had entered the graft from the fasciculus gracilis (see Figure 21.5a, b). Since the fasciculus gracilis was completely transected by the injury 1 mm deep and 1 to 1.5 mm wide, the HRP-positive axons in the lesion could be considered to regenerate from fibers of the fasciculus gracilis. There was no definite finding indicating elongation of many HRP-labeled fibers further rostrally into the dorsal funiculus. Only a few profiles of HRP-labeled fibers were found in the region immediately rostral to the lesion (see Figure 21.5c, d)

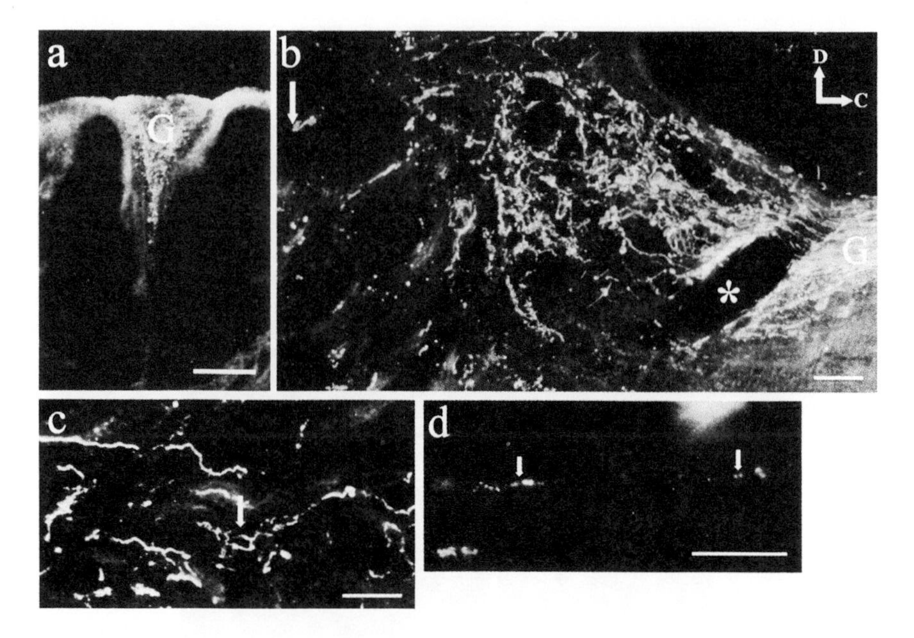

FIGURE 21.5 HRP-labeled regenerating axons eight months after grafting. HRP was injected into the sciatic nerve to label the ascending fibers in the dorsal funiculus of the spinal cord. (a) Transverse section of the spinal cord at the cervical level. HRP-labeled fibers are confined to the fasciculus gracilis. Bar, 200 μm. (b) Sagittal section of the lesion at the cervical level. HRP-labeled fibers extend into the lesion from the fasciculus gracilis (G). D and C: dorsal and caudal direction, respectively. Asterisk indicates a blood vessel. Bar, 100 μm. (c) HRP-labeled axons extend near the rostral border of the lesion. Arrow indicates the finely branched axons. Bar, 100 μm. (d) A few regenerating axons extend into the region of spinal cord rostral to the lesion. Bar, 40 μm. (From Ide et al., 2001.)

Evoked potentials were recorded at a point about 5 mm rostral to the lesion by sciatic nerve stimulation in rats at 8 and 10 months postgrafting. In normal intact rats, sciatic nerve stimulation produced steep evoked potentials of short duration, followed by late slow potentials at a point about 5 mm rostral to the C2 level. The duration (52.3 ± 2.3 ms, $n = 3$) of biphasic evoked potentials in the grafted rats was much longer than that (24.7 ± 2.4 ms, $n = 4$) in the normal intact rats.

It can therefore be concluded that choroid plexus ependymal cells have facilitating effects on nerve regeneration when grafted in the spinal cord.

CPECs have been shown to produce IGF-II (Nilsson et al., 1996) and growth/differentiation factor 15 (GDF-15) (Strelau et al., 2000) and express vascular endothelial growth factor (VEGF) (Chodobski et al., 2003) and adhesion molecules such as N-CAM (Figarella-Branger et al., 1995). It has been shown that CPECs produce several trophic factors, which can be upregulated following injury to the brain (Johanson et al., 2000; Newton et al., 2003). These findings suggest that CPECs may, when transplanted into the CNS, act as a source of trophic factors and as a substrate for the growth of regenerating axons. As described in the previous section, cultured ependymal cells from the choroid plexus can promote the growth of neurites from cocultured dorsal root ganglion neurons more effectively than coculturing with astrocytes (Chakrabortty et al., 2000).

It has been shown that ependymal cells of the spinal cord central canal, following spinal cord injury, proliferate and surround regenerating axons in the rat (Matthews et al., 1979). Our study indicated that the choroid plexus has facilitatory effects on neuronal regeneration when grafted in the spinal cord. It is noteworthy that numerous regenerating axons extended into the lesion and that they were maintained without becoming atrophic in the graft for at least 10 months after grafting.

Although numerous regenerating axons were maintained in the graft for at least 8 to 10 months after grafting, only a few HRP-labeled axons were found extending into the adjacent rostral region. Although CPECs were shown to intimately interact with astrocytes, some functional barriers might be formed at the rostral border between the graft and the host tissue. Such barrier formation is one of the most important problems in neuronal regeneration.

As described below, some CPECs can differentiate into astrocytes when the choroid plexus from green fluorescent protein (GFP)-transgenic mice is grafted into injured spinal cord (Kitada et al., 2001). It is probable that some stem cells might be included in the CPECs (Weiss et al., 1996; Johansson et al., 1999). However, CPECs are considered to function mainly as supporting cells of regenerating axons. In addition, it is probable that immature astrocytes and oligodendrocytes can migrate into the lesion from the host spinal cord tissue. Immature glial cells have been shown to appear at the border of the injured dorsal funiculus in cases of laser irradiation (Sato et al., 1997) and freezing injury (Kitada et al., 1999).

Appropriate angiogenesis is also required for effective repair of the damaged CNS. The choroid plexus is reported to express some angiogenesis-inducing factors, such as angiopoietin-2 (ANG2) (Nourhaghighi et al., 2003), VEGF (Chodobski et al., 2003), and FGF-II (Nilsson et al., 1996). The donor's endothelial cells might contribute, together with the host's counterparts, to the formation of new blood vessels within the lesion. Grafting of the choroid plexus from GFP-transgenic mice into the injured wild-type mice spinal cord showed that donor endothelial cells contributed to the new blood vessel formation. The rich vasculature of the choroid plexus could be beneficial for achieving good tissue repair of the lesion.

21.4 DIFFERENTIATION OF CHOROID PLEXUS EPENDYMAL CELLS INTO ASTROCYTES

To examine whether CPECs grafted into injured CNS can be induced to give rise to other neural cell types, as indicated by morphological and histochemical changes, we transplanted cultured CPECs into pre-lesioned spinal cords of mice (Kitada et al., 2001). For culturing, the choroid plexuses were excised from the 4th ventricle of GFP-transgenic mice, and the dissociated cells were cultured for four to six weeks. CPECs did not exhibit any immunohistochemical characteristics of neurons, astrocytes, and oligodendrocytes in this culture system. These GFP-expressing CPECs were harvested from the monolayer cultures and injected stereotactically into the prelesioned spinal cord of wild-type mice using a syringe. The spinal cord of the recipient mice (8 to 12 weeks old) had been cut transversely using microscissors at a depth of 2 mm from the dorsal surface one week before transplantation.

Three weeks after transplantation, the transplanted cells could be identified as GFP-positive. They formed cell clusters of various sizes in which a small number of GFAP-positive cells were present (see Figure 21.6). The rate of occurrence of double-positive cells for GFAP and GFP was 4.69 ± 0.47% at one week, 5.13 ± 0.68% at two weeks, and 5.00 ± 0.49% at three weeks (mean ± SD, respectively). Some of the GFAP-positive transplanted cells were also positive for GLT-1, a functional marker of astrocytes (Rothstein et al., 2001). All cells that were double-positive for GFAP, and GFP were negative for vimentin, indicating that at least some of the grafted CPECs did not remain as immature astrocytes, but instead differentiated into mature astrocytes. On the other hand, no cells within the clusters displayed immunoreactivity for neurofilament 200k, β-tubulin class III, or MBP.

By immunoelectron microscopy, most of the GFP-positive transplanted cells were found within the pre-lesioned area of the spinal cord (see Figure 21.7A). The cytoplasm and cellular processes of these cells contained bundles of intermediate filaments that were labeled with gold particles for GFAP reactivity (see Figure 21.7B,C). Thus, it was demonstrated that at least a part of grafted CPECs could differentiate into astrocytes in a specialized environment where the spinal cord was traumatically injured.

The fact that CPECs differentiate into astrocytes when transplanted into spinal cord lesions suggests the possibility of using CPECs to supply neural cells to the CNS lesioned site, as in the cases of grafting of neural stem cells (Wu et al., 2001). Thus, choroid plexus grafting might be promising not only as a trophic source or to provide guidance cues for axonal regeneration or neuroprotection, but also as a source to supply cells to reconstruct a newly generated CNS circuit.

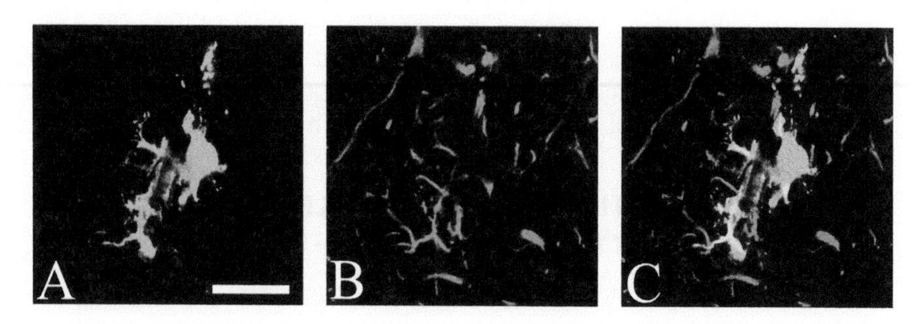

FIGURE 21.6 (See color insert following page 522.) CPECs derived from GFP-positive mice were cultured and grafted 1 week before fixation: (A) grafted GFP-positive CPECs; (B) GFAP-immunostaining; (C) CPECs double-positive for GFP and GFAP are yellow-colored in this merged image. Bar, 20 μm. (From Kitada et al., 2001.)

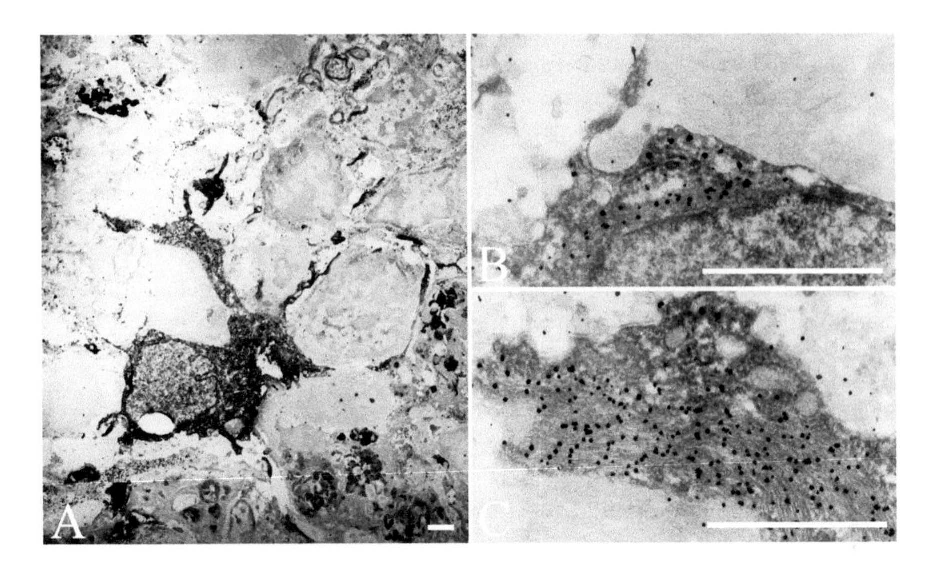

FIGURE 21.7 These electron micrographs show GFP-positive CPECs two weeks after grafting. For electron microscopy, GFP-positive CPECs became electron dense by staining with DAB, and GFAP immunoreactivity was visualized by gold particles. (A) Low magnification. (B, C) Intermediate filaments in the cytoplasm (B) and in the cell process (C) were labeled by gold particles for GFAP immunoreactivity. Bar, 2 μm. (From Kitada et al., 2001.)

21.5 FUNCTIONAL MOLECULES PRODUCED FROM CHOROID PLEXUS EPENDYMAL CELLS

The choroid plexus provides a specialized environment for the CNS through the CSF. Various molecules produced by the choroid plexus should play key roles in maintaining the normal function of the brain, and protecting brain from traumatic and pathological damage. To understand such roles of the choroid plexus, we identified genes predominantly expressed in the mouse choroid plexus using suppression subtractive hybridization (Matsumoto et al., 2003). CPECs secrete various molecules involved in prevention of the fibrilization of amyloid β (Aβ) protein, e.g., transthyretin and gelsolin), lipid metabolism (e.g., phospholipid transfer protein and ABCA8), and detoxification (e.g., androgen-inducible aldehyde reductase).

We applied PCR-based suppression subtractive hybridization to identify genes expressed specifically or abundantly in the choroid plexus. Two types of subtractive cDNA libraries were constructed from mice: (1) choroid plexus (CP) cDNAs from adult mice as the tester and cerebral cortex (CC) cDNAs from adult mice as the driver (CP–CC subtraction), and (2) CP cDNAs from postnatal day 5 (P5) mice as the tester and the CP cDNAs from adult mice as the driver (P5–Ad subtraction). Since the CC is rich in neuronal cells and physically located apart from the CP in the brain, we reasoned that the former subtraction (CP–CC subtraction) would efficiently enrich for transcripts relatively abundant in the choroid plexus. The second subtraction library (P5–Ad) was generated with the aim of identifying genes abundantly expressed in the choroid plexus during brain development, which might play a role during maturation of the brain rather than in the maintenance of the mature brain. Partial sequences of the differentially expressed genes obtained by these two types of subtraction were determined using a DNA sequencer, and a homology search was performed using the GeneBank and dbEST databases.

The identified clones in the CP–CC subtraction and P5–Ad subtraction are summarized in Tables 21.1 and 21.2, respectively. Transthyretin (TTR) (Dickson et al., 1985), nucleotide pyrophosphatase/phosphodiesterase 2 (PDIα) (Narita et al., 1994), and inwardly rectifying potassium channel Kir7.1 (Doring et al., 1998) were identified by the CP-CC subtraction and are molecules known to be expressed abundantly and preferentially in the choroid plexus. This indicates that the subtraction technique employed is reliable.

TABLE 21.1
cDNA Clones Isolated from CP-CC Library

Clone	Identity
1st-a11	Transthyretin
1st-a65	Nucleotide pyrophosphatase/phosphodiesterase2 (NPP2, PDIα)
2nd-1	Kinesin family member 21A (Kif21a)
2nd-10	Inwardly rectifying potassium channel Kir7.1
2nd-27	Novel
2nd-175	Mitochondrial inner membrane translocase component Tim17a
2nd-200	Signal recognition particle 14 kDa (Srp14)
2nd-216	Novel
2nd-265	Cathepsin B
2nd-306	EST (BI731975)

TABLE 21.2
cDNA Clones Isolated from P5-Ad Library

Clone	Identity
FS1	EST (BF784446)
FS7	Syntaxin binding protein 2 (Munc18-2)
FS8	Alpha glucosidase 2, alpha neutral subunit (G2an)
FS12	KIAA0998
FS13	RIKEN cDNA, clone: 4930528H02
FS16	RIKEN cDNA, clone: 1110034007
FS18	Laminin beta1
FS20	Nucleosome assembly protein 1-like 4 (Napl14)
FS25	Vascular cell adhesion molecule-1 truncated form (T-VCAM-1)
FS26	Pyruvate kinase M1 and M2 subunit
FS33	Eukaryotic translation initiator factor 4 gamma, 3 (EIF4G3)

(continued)

TABLE 21.2 (CONTINUED)
cDNA Clones Isolated from P5-Ad Library

Clone	Identity
FS34	Phospholipid transfer protein (PLTP)
FS40	RNA polymerase 1-4 (194 kDa subunit) (Rpo1-4)
Clone	**Identity**
FS42	Apolipoprotein E (apoE)
FS48	Thyroid hormone receptor interactor 6 (Trip6)
FS49	Platelet-derived growth factor receptor, beta polypeptide
FS50	Secreted protein, acidic and rich in cysteine (Sparc)
FS58	Basigin
FS63	Novel
FS64[a]	Novel + ornithine decarboxylase antizyme
FS71	ATP-binding cassette, sub-family A, member 8 (ABCA8)
FS74	Novel
FS80	Splicing factor Srp20
FS81	Ezrin
FS82	K-Cl cotransporter KCC1
FS83	Androgen-inducible aldehyde reductase (AIAR)
FS88	Novel
FS89	RIKEN cDNA, clone: 1110021N07
FS90[a]	ATP synthase, H+ transporting mitchondrial F1 complex, alpha
subunit	(Atp5b) + EST (BI691454)
FS97	Alpha-enolase
FS101	Na+/sulfate cotransporter SUT-1
FS105	KIAA 1576[b]
FS109	Gelsolin
FS114	Dickkopf-3 (dkk-3)
FS116[a]	Microtuble-associated protein, RP/EB family, member 2
(MAPRE2)	+ KIAA1576[b]
FS117	RIKEN cDNA, clone: 2410024K20
FS118	RIKEN cDNA, clone: 2410074K14

[a]FS64, 90, and 116 are cDNA clones fused with two different genes.

[b]KIAA1576 was isolated twice.

The genes expressed in the choroid plexus were analyzed by RNA blot hybridization (see Figure 21.8). The following are the main genes detected in this experiment: (1) apolipoprotein E (ApoE) and secreted protein, acidic and rich in cysteine (SPARC), which have been known to be expressed in the choroid plexus by in situ hybridization; (2) phospholipid transfer protein (PLTP) and androgen-inducible aldehyde reductase (AIAR), which are known to be expressed in the brain; (3) gelsolin, Na^+/sulfate cotransporter SUT-1, and ATP-binding cassette transporter, sub-family A, member 8 (ABCA8) (usually expressed in the peripheral tissues); (4) poorly characterized KIAA1576, RIKEN cDNA clone 2410024K20, and a novel gene,

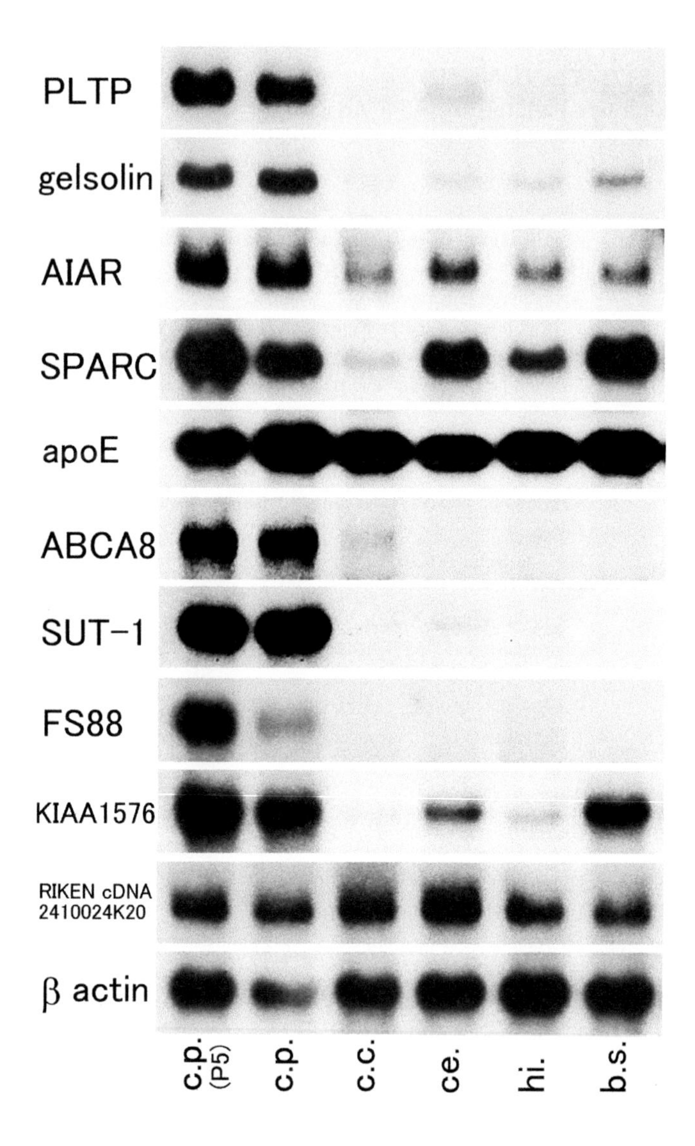

FIGURE 21.8 RNA blot analysis. An aliquot (10 μg) of total RNA extracted from various regions of adult and postnatal day 5 (P5) mouse brains was applied per lane. The probes used are: c.p., choroid plexus; c.c., cerebral cortex; ce., cerebellum; hi., hippocampus; b.s., brain stem. (From Matsumoto et al., 2003.)

FS88. The expression of PLTP, gelsolin, ABCA8, and SUT-1 as well as novel gene FS88 was highly restricted to the choroid plexus. The strong signals for these genes were detected preferentially in the ependymal cells of the choroid plexus by in situ hybridization (Figure 21.9).

PLTP is involved in lipid metabolism regulating the size and composition of high-density lipoprotein (HDL) by the HDL conversion process (Jauhiainen et al.,

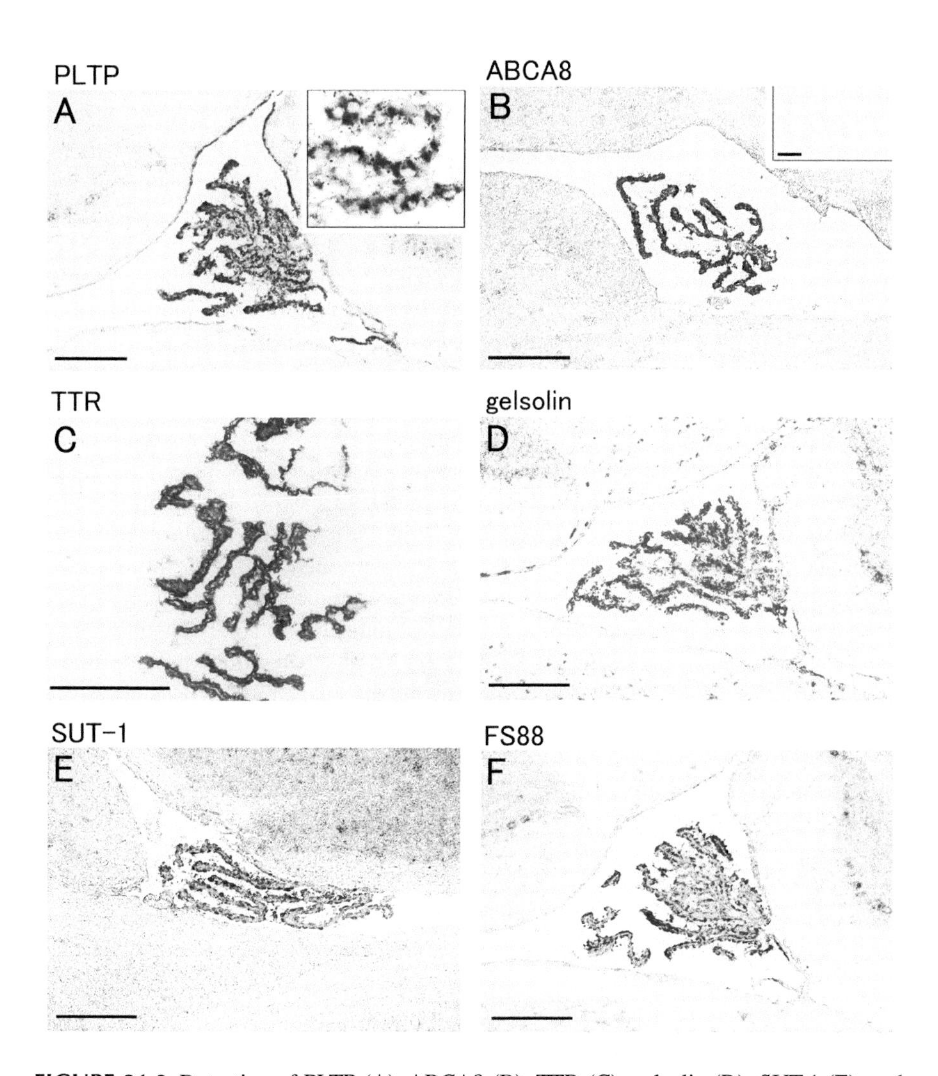

FIGURE 21.9 Detection of PLTP (A), ABCA8 (B), TTR (C), gelsolin (D), SUT-1 (E), and FS88 (F) mRNA in sagittal sections of brain of the adult mouse by in situ hybridization. Strong signals of each mRNA were detected in CPECs of the fourth ventricle. In every case, no significant background signals were detected by hybridization with the corresponding sense probe (see inset in B). TTR (C) was used as positive control because of its restricted expression to CPECs in the brain region. Inset in A: mRNA expression of PLTP is expressed in the cytoplasm of the ependymal cells, but not in the underlying connective tissue. Scale bars, 200 μm. (From Matsumoto et al., 2003.)

1993), controlling HDL levels in the blood plasma (Rao et al., 1997). ABCA1, a prototypic member of the ABC A subfamily of membrane transporters, is a principal regulator of HDL (Drobnik et al., 2001). Since lipoprotein particles cannot be transported across the blood–brain barrier, the synthesis, remodeling, and redistribution of lipids must take place in the brain tissue. CSF was found to contain high levels of PLTP (Demeester et al., 2000). The above data indicate that the choroid plexus serves as a regulatory center for lipid metabolism in the brain. Downregulation of PLTP mRNA in the brains of Down syndrome patients was reported (Krapfenbauer et al., 2001). This is of particular interest since a histopathological feature of Down syndrome is poor myelination, which is composed of a large amount of lipids.

Gelsolin is involved in Aβ metabolism, functioning as an antiamyloidogenic protein in brain (Ray et al., 2000) just as TTR does (Schwarzman et al., 1994). ApoE is also known to interact with Aβ (Naslund et al., 1995) and to play roles in neuronal protection and repair (Mahley and Huang, 1999). It should be noted that CSF inhibits amyloid formation in vitro. The neuronal membrane cholesterol content influences the processing of the amyloid precursor protein (Howland et al., 1998). Thus, the choroid plexus may be involved in Aβ clearance. Aβ is concentrated in senile plaques of Alzheimer's disease and Down syndrome.

PDIα and SPARC are related to cell motility. The membrane glycoprotein PDIα is considered to regulate CSF production (Narita et al., 1994). The identity between PDIα and autotaxin, a tumor cell motility–stimulating factor, has been recognized (Murata et al., 1994). SPARC is a secreted glycoprotein that also influences cell migration (Bradshaw and Sage, 2001). These genes indicate that the choroid plexus might be involved in regulating the growth and motility of neural cells in the brain.

The findings of our studies and of studies from other groups indicate that the choroid plexus produces: (1) trophic or growth factors such as bFGF (Cuevas et al., 1994) and IGF-II (Nilsson et al., 1996), (2) antioxidative factors such as EC-SOD (Fukui et al., 2002), (3) antiamyloidgenic factors (TTR, gelsolin, etc.), (4) lipid metabolism–related factors (PLTP, ABCA8, etc.), (5) cell motility–stimulating factors (SPARC, PDIα, etc.), (6) angiogenesis-inducing factors such as ANG2 (Nourhaghighi et al., 2003), VEGF (Chodobski et al., 2003), and HGF (Hayashi et al., 1998), and (7) detoxification-related factors (AIAR, cathepsin B, etc.). Choroid plexus seems to provide a CNS environment suitable for the development and maintenance of the CNS by producing various factors secreted into the CSF in addition to detoxifying detrimental substrates from the CSF. This inherent ability of the choroid plexus to support the CNS tissue suggests that grafting of choroid plexus would enhance the repair of the CNS tissue following ischemic/traumatic damages. Recently, an immortalized cell line of CPECs was established (Zheng and Zhao, 2000). This will function as the grafting source for the clinical use of CPECs. The enhancement of choroid plexus function by direct manipulation in vivo or transplantation via ex vivo methods is a promising new strategy for the treatment of intractable CNS diseases, including various kinds of injury and degeneration.

21.6 ACKNOWLEDGMENTS

This study was supported in part by Grants-in-Aid for Scientific Research, for Exploratory Research and for Scientific Research on Priority Area from the Japanese Ministry of Education, Culture, Sports, Science and Technology.

REFERENCES

Bradshaw, A.D. and Sage, E.H. (2001) SPARC, a matricellular protein that functions in cellular differentiation and tissue response to injury. *J Clin Invest* 107: 1049–1054.

Chakrabortty, S., Kitada, M., Matsumoto, N., Taketomi, M., Kimura, K. and Ide, C. (2000) Choroid plexus ependymal cells enhance the neurite outgrowth from dorsal root ganglion neurons in vitro. *J Neurocytol* 29: 707–717.

Chen, H., Cao, Y. and Olson, L. (1996) Spinal cord repair in adult paraplegic rats: partial restoration of hind limb function. *Science* 273: 510–513.

Chodobski, A., Chung, I., Kozniewska, E., Ivanenko, T., Chang, W., Harrington, J. F., Duncan, J. A., and Szmydynger-Chodobska, J. (2003) Early neutrophilic expression of vascular endothelial growth factor after traumatic brain injury, *Neuroscience* 122:853–867.

Cuevas, P., Carceller, F., Reimers, D., Fu, X. and Gimenez-Gallego, G. (1994) Immunohistochemical localization of basic fibroblast growth factor in choroid plexus of the rat. *Neurol Res* 16: 310–312

David, S. and Aguayo, A.J. (1981) Axonal elongation in peripheral nervous system "bridges" after central nervous system injury in adult rats. *Science* 214: 931–933.

Demeester, N., Castro, G., Desrumaux, C., De Geitere, C., Fruchart, J.C., Santens, P., Mulleners, E., Engelborghs, S., De Deyn, P.P., Vandekerckhove, J., Rosseneu, M. and Labeur, C. (2000) Characterization and functional studies of lipoproteins, lipid transfer proteins, and lecithin:cholesterol acyltransferase in CSF of normal individuals and patients with Alzheimer's disease. *J Lipid Res* 41: 963–974.

Dickson, P.W., Howlett, G.J. and Schreiber, G. (1985) Rat transthyretin (prealbumin). Molecular cloning, nucleotide sequence, and gene expression in liver and brain. *J Biol Chem* 260: 8214–8219.

Doring, F., Derst, C., Wischmeyer, E., Karschin, C., Schneggenburger, R., Daut, J. and Karschin, A. (1998) The epithelial inward rectifier channel Kir7.1 displays unusual K+. permeation properties. *J Neurosci* 18: 8625–8636.

Drobnik, W., Lindenthal, B., Lieser, B., Ritter, M., Christiansen Weber, T., Liebisch, G., Giesa, U., Igel, M., Borsukova, H., Buchler, C., Fung-Leung, W.P., Von Bergmann, K. and Schmitz, G. (2001) ATP-binding cassette transporter A1 (ABCA1) affects total body sterol metabolism. *Gastroenterology* 120: 1203–1211.

Figarella-Branger, D., Lepidi, H., Poncet, C., Gambarelli, D., Bianco, N., Rougon, G. and Pell, J.F. (1995) Differential expression of cell adhesion molecules(CAM), neural CAM and epithelial cadherin in ependymomas and choroid plexus tumors. *Acta Neuropath.* 89: 248–257.

Fukui, S., Ookawara, T., Nawashiro, H., Suzuki, K. and Shima, K. (2002) Post-ischemic transcriptional and translational responses of EC-SOD in mouse brain and serum. *Free Radic Biol Med* 32: 289–298.

Gabrion, J.B., Herbute, S., Bouille, C., Maurel, D., Kuchiler-Bopp, S., Laabich, A. and Delaunoy, J.P. (1998) Ependymal and choroidal cells in culture: characterization and functional differentiation. *Microsc Res Tech* 41: 124–157.

Hayashi, T., Abe, K., Sakurai, M. and Itoyama, Y. (1998) Induction of hepatocyte growth factor and its activator in rat brain with permanent middle cerebral artery occlusion. *Brain Res* 799: 311–316.

Howland, D.S., Trusko, S.P., Savage, M.J., Reaume, A.G., Lang, D.M., Hirsch, J.D., Maeda, N., Siman, R., Greenberg, B.D., Scott, R.W. and Flood, D.G. (1998) Modulation of secreted beta-amyloid precursor protein and amyloid beta-peptide in brain by cholesterol. *J Biol Chem* 273: 16576–16582.

Hynes, M.A., Brooks, P.J., Van Wyk, J.J. and Lund, P.K. (1988) Insulin-like growth factor II messenger ribonucleic acids are synthesized in the choroid plexus of the rat brain. *Mol Endocrinol* 2: 47–54.

Ide, C., Kitada, M., Chakrabortty, S., Taketomi, M., Matumoto, N., Kikukawa, S., Mizoguchi, A., Kawaguchi, S., Endo, K. and Suzuki, Y. (2001) Grafting of choroid plexus ependymal cells promotes the growth of regenerating axons in the dorsal funiculus of rat spinal cord—a preliminary report. *Exp Neurol* 167: 242–251.

Iwashita, Y., Kawaguchi, A. and Murata, M. (1994) Restoration of function by replacement of spinal cord segments in the rat. *Nature* 367: 167–170.

Jauhiainen, M., Metso, J., Pahlman, R., Blomqvist, S., van Tol, A. and Ehnholm, C. (1993) Human plasma phospholipid transfer protein causes high density lipoprotein conversion. *J Biol Chem* 268: 4032–4036.

Johanson, C.E., Palm, D.E., Primiano, M.J., McMillan, P.N., Chan, P., Knuckey, N.W. and Stopa, E.G. (2000) Choroid plexus recovery after transient forebrain ischemia: role of growth factors and other repair mechanisms. *Cell Mol Neurobiol* 20: 197–216.

Johansson, C.B., Momma, S., Clarke, D.L., Risling, M., Lendahl, U. and Frisen, J. (1999) Identification of a neural stem cell in the adult mammalian central nervous system. *Cell* 96: 25–34.

Kitada, M., Chakrabortty, S., Matsuoto, N., Taketomi, M. and Ide, C. (2001) Differentiation of choroid plexus ependymal cells into astrocytes after grafting into the pre-lesioned spinal cord in mice. *Glia* 36: 364–374.

Kitada, M., Toyama, K. and Ide, C. (1999) Comparison of the axonal and glial reactions between the caudal and rostral border in the cryoinjured dorsal funiculus of the rat spinal cord. *Restor Neurol Neurosci* 14: 251–263.

Krapfenbauer, K., Yoo, B.C., Kim, S.H., Cairns, N. and Lubec, G., (2001) Differential display reveals downregulation of the phospholipid transfer protein (PLTP) at the mRNA level in brains of patients with Down syndrome. *Life Sci* 68: 2169–2179.

Li, Y., Field, P.M. and Raisman, G. (1997) Repair of adult rat corticospinal tract by transplants of olfactory ensheathing cell. *Science* 277: 2000–2002.

Mahley, R.W. and Huang, Y. (1999) Apolipoprotein E: from atherosclerosis to Alzheimer's disease and beyond. *Curr Opin Lipidol* 10: 207–217.

Matsumoto, N., Kitayama, H., Kitada, M., Kimura, K., Noda M. and Ide, C. (2003) Isolation of a set of genes expressed in the choroid plexus of the mouse using suppression subtractive hybridization. *Neuroscience* 117: 405–415.

Matthews, M.A., St. Onge, M.F. and Faciance, C.L. (1979) An electron microscopic analysis of abnormal ependymal cell proliferation and envelopment of sprouting axons following spinal cord transection in the rat. *Acta Neuropathol* 45: 27–36.

Menei, P., Montero-Menei, C., Whittemore, S.R., Bunge, R.P. and Bunge, M.B. (1998) Schwann cell genetically modified to secrete human BDNF promote enhanced axonal regrowth across transected adult rat spinal cord. *Eur J Neurosci* 10: 607–621.

Murata, J., Lee, H.Y., Clair, T., Krutzsch, H.C., Arestad, A.A., Sobel, M.E., Liotta, L.A. and Stracke, M.L. (1994) cDNA cloning of the human tumor motility-stimulating protein, autotaxin, reveals a homology with phosphodiesterases. *J Biol Chem* 269: 30479–30484.

Murphy-Erdosh, C., Napolitano, E.W. and Reichardt, L.F. (1994) The expression of B-cadherin during embryonic chick development. *Dev Biol* 161: 107–125.

Narita, M., Goji, J., Nakamura, H. and Sano, K. (1994) Molecular cloning, expression, and localization of a brain-specific phosphodiesterase I/nucleotide pyrophosphatase (PD-I alpha) from rat brain. *J Biol Chem* 269: 28235–28242.

Naslund, J., Thyberg, J., Tjernberg, L.O., Wernstedt, C., Karlstrom, A.R., Bogdanovic, N., Gandy, S.E., Lannfelt, L., Terenius, L. and Nordstedt, C. (1995) Characterization of stable complexes involving apolipoprotein E and the amyloid beta peptide in Alzheimer's disease brain. *Neuron* 15: 219–228.

Newton, S.S., Collier, E.F., Hunsberger, J., Adams, D., Terwilliger, R., Selvanayagam, E. and Duman, R.S. (2003) Gene profile of electroconvulsive seizures: induction of neurotrophic and angiogenic factors. *J Neurosci* 23: 10841–10851.

Nilsson, C., Hultberg, B.M. and Gammeltoft, S. (1996) Autocrine role of insulin-like growth factor II secretion by the rat choroid plexus. *Eur J Neurosci* 8: 629–635.

Nourhaghighi, N., Teichert-Kuliszewska, K., Davis, J., Stwart, D.J. and Nag, S. (2003) Altered expression of angiopoietins during blood-brain barrier breakdown and angiogenesis. *Lab Invest* 83: 1211–1222.

Ohta, M., Suzuki, Y., Noda, T., Ejiri, Y., Dezawa, M., Kataoka, K., Chou, H., Ishikawa, N., Matsumoto, N., Iwashita, Y., Mizuta, E., Kuno, S. and Ide, C. (2004) Bone marrow stromal cells infused into the cerebrospinal fluid promote functional recovery of the injured rat spinal cord with reduced cavity formation. *Exp Neurol* 187:266–278.

Rao, R., Albers, J.J., Wolfbauer, G. and Pownall, H.J. (1997) Molecular and macromolecular specificity of human plasma phospholipid transfer protein. *Biochemistry* 36: 3645–3653.

Rapalino, O., Lazarov-Spiegler, O., Agranov, E., Velan, G.J., Yoles, E., Fraidakis, M., Solomon, A., Gepstein, R., Katz, A., Belkin, M., Hadani, M. and Schwartz, M. (1998) Implantation of stimulated homologous macrophages results in partial recovery of paraplegic rats. *Nat Med* 4: 814–821.

Ray, I., Chauhan, A., Wegiel, J. and Chauhan, V.P. (2000) Gelsolin inhibits the fibrillization of amyloid beta-protein, and also defibrillizes its preformed fibrils. *Brain Res* 853: 344–351.

Rothstein, J.D., Martin, L., Levey, A.I., Dykes, H.M., Jin, L., Wu, D., Nash, N. and Kuncl, R.W. (2001) Localization of neuronal and glial glutamate transporters. *Neuron* 13: 713–725.

Sato, K., Ohmae, E., Senoo, E., Mase, T., Tohyama, K., Fujimoto, E., Mizoguchi, A. and Ide, C. (1997) Remyelination in the rat dorsal funiculus following demyelination by laser irradiation. *Neurosci Res* 28: 325–335.

Schnell, L. and Schwab, M. (1990) Axonal regeneration in the rat spinal cord produced by an antibody against myelin-associated neurite growth inhibitors. *Nature* 343: 269–272.

Schwarzman, A.L., Gregori, L., Vitek, M.P., Lyubski, S., Strittmatter, W.J., Enghilde, J.J., Bhasin, R., Silverman, J., Weisgraber, K.H., Coyle, P.K., Zagorski, M.G., Talafous, J., Eisenberg, M., Saunders, A.M., Roses, A.D. and Goldgaber, D. (1994) Transthyretin sequesters amyloid beta protein and prevents amyloid formation. *Proc Natl Acad Sci USA* 91: 8368–8372.

Senoo, E., Tamaki, N., Fujimoto, E. and Ide, C. (1998) Effects of prelesioned peripheral nerve graft on nerve regeneration in the rat spinal cord. *Neurosurgery* 42: 1347–1356.

Strelau, J., Sullivan, A., Bottner, M., Lingo, P., Falkenstein, E., Suter-Crazzolara, C., Galter, D., Jaszai, J., Krieglstein, K., and Unsicker, K. (2000) Growth/differentiation factor-15/macrophage inhibitory cytoline-1 is a novel trophic factor for midbrain dopaminergic neurons in vivo. *J Neurosci* 20: 8597–8603.

Takahashi, M., Palmer, T.D., Takahashi, J. and Gage, F.H. (1998) Widespread integration and survival of adult-derived neural progenitor cells in the developing optic retina. *Mol Cell Neurosci* 12: 340–348.

Timmusk, T., Mudo, G., Metsis, M. and Belluardo, N. (1995) Expression of mRNAs for neurotrophins and their receptors in the rat choroid plexus and dura mater. *Neuroreport* 6: 1997–2000.

Weibel, M., Pettmann, B., Artault, J.C., Sensenbrenner, M. and Labourdette, G. (1986) Primary culture of rat ependymal cells in serum-free defined medium. *Brain Res* 390: 199–209.

Weiss, S., Dunne, C., Hewson, J., Wohl, C., Wheatley, M., Peterson, A.C. and Reynolds, B.A. (1996) 'Multipotent CNS stem cells are present in the adult mammalian spinal cord and ventricular neuroaxis' *J Neurosci* 16: 7599–7609.

Wu, S., Suzuki, Y., Noda, T., Bai, H., Kitada, M., Kataoka, K., Chou, H. and Ide, C. (2003) Bone marrow stromal cells enhance differentiation of co-cultured neurosphere cells and promote regeneration of injured spinal cord. *J Neurosci Res* 72: 343–351.

Wu, S., Suzuki, Y., Noda, T., Bai, H., Kitada, M., Kataoka, K., Nishimura, Y. and Ide, C. (2002) Immunohistochemical and electron microscopic study of invasion and differentiation in spinal cord lesion of neural stem cells grafted through cerebrospinal fluid in rat. *J Neurosci Res* 69: 940–945.

Wu, S., Suzuki, Y., Kitada, M., Kitaura, M., Kataoka, K., Takahashi, J., Ide, C. and Nishimura, Y. (2001) Migration integration, and differentiation of hippocampus-derived neurosphere cells after transplantation into injured rat spinal cord. *Neurosci Lett* 312: 173–176.

Xu, X.M., Chen, A., Guenrad, V., Kleitman, N. and Bunge, M.B. (1997) Bridging Schwann cell transplants promote axonal regeneration from both the rostral and caudal stumps of transected adult rat spinal cord. *J Neurocytol* 26: 1–16.

Zheng, W., and Zhao, Q. (2000) Establishment and characterization of an immortalized Z310 choroid epithelial cell line from murine choroid plexus, *Brain Res* 952: 371–380.

Section Four

Current In Vivo *and* In Vitro *Models of the Blood–CSF Barrier*

22 *In Situ* Perfusion Techniques Used in Blood–CSF Barrier Research

Wei Zheng
Purdue University, West Lafayette, Indiana, U.S.A.

Malcolm B. Segal
King's College London, London, United Kingdom

CONTENTS

22.1 INTRODUCTION

The choroid plexus is a relatively independent tissue in the brain: independent blood supply, independent structural support, and independent functional connection to brain parenchyma. It is therefore convenient to distinguish the choroid plexus from the rest of brain structures and to illustrate the tissue, both anatomically and morphologically. Accordingly, techniques that are pertinent to elucidating the distribution of substances at the blood–cerebrospinal fluid (CSF) barrier or to studying the morphological changes of the tissue are most frequently used in choroid plexus research. For example, to see if in vivo metal exposure results in an accumulation of metal in the choroid plexus, the metal can be administered systemically, followed by sampling of the tissue at specified times. By quantifying the amount of the metal in the tissue, either by atomic absorption spectrophotometry or by measuring radioactivity if the radioisotope is used, one can estimate the tissue sequestration of the metal in comparison to other brain regions or other organs. In many cases, a high accumulation of the chemical in the choroid plexus may suggest a route for the chemical to enter or be removed from the CSF compartment.

Alternatively, autoradiography of a radiolabeled chemical in the choroid plexus is another useful tool to locate the transport site of this chemical at the blood–CSF barrier. In this kind of study, radiolabeled compounds can be given according to an acute or chronic dose regimen. At the end of dose administration, the whole brain can be frozen and sectioned for autoradiography. A strong signal from the brain ventricles usually implies a high accumulation of the studied substances in the choroid plexus.

In vivo investigation of the morphological changes of the choroidal epithelia can be accomplished by systemic administration of the study compound, followed by a sequence of procedures to dissect the choroid plexus tissue, to stain or prepare tissue section, and finally to examine by light or electron microscopes.

No great difficulty should be expected for these experiments as long as the investigator understands the location of the choroid plexus. One issue, however, should not be ignored—blood contamination in the choroid plexus tissue. As mentioned in previous chapters, the choroid plexus is rich in blood supply; the flow rate to the choroid plexus is nearly three times faster than that to the rest of the brain. Because of the richness of the blood component, the chemicals present in the blood, if they are not washed out prior to the tissue sampling, could complicate the results. Thus, for the aforementioned studies a rapid brain perfusion procedure is recommended. In an experiment performed in rats, the chest is cut open to expose the heart. An 18 gauge needle attached to a 10 mL syringe is inserted into the left ventricle. Immediately after cutting off the superior vena cava above the heart, 10 mL of ice-cold phosphate-buffered saline (PBS) is injected into the left ventricle. The PBS then circulates through the brain along the blood vessel path; drainage is achieved via the superior vena cava. The method is effective and easily judged for success by eye examination, as the perfused brain is pale in general, without visible blood contamination.

While a location-type study of the choroid plexus is relatively easy to perform, the functional study of the blood–CSF barrier, not at all an easy task. Several factors may contribute to the difficulties of the functional investigation. First, the size or mass of the choroid plexus tissue is rather small. In rats, the wet weight of the choroid plexus is only several milligrams, rendering it nearly impossible for cannulation study. Second, the choroid plexus contacts three fluid compartments: the CSF compartment, the blood compartment, and the choroidal intracellular fluid compartment. Unlike study in the blood–brain barrier, where the transport property can be readily estimated by the appearance of the test compound in brain parenchyma following injection into the bloodstream, the ideal situation to estimate transport at the blood–CSF barrier would be to simultaneously determine the concentration of the test compound in both blood and CSF. Currently no such procedure has been satisfactorily established. The last, but not the least, factor is that sampling of CSF from animals is not straightforward, requiring considerable training and much practice.

Despite all these difficulties, several methods have been developed since the 1960s to study the function of the choroid plexus, particularly regarding the kinetic aspect of material transport by the blood–CSF barrier. This chapter will discuss the in situ techniques used in the perfusion of choroid plexus in sheep and perfusion of the whole brain in rodents.

22.2 IN SITU PERFUSED CHOROID PLEXUS MODEL

22.2.1 Technique Background

In situ perfusion of the choroid plexus in sheep was originally described by Pollay and Kaplan (1972) to study the transport of thiocyanate. The model was further described by Segal's group in King's College London (Blount et al., 1973) and later for investigation of sugar transport via the blood–CSF barrier (Deane and Segal, 1985). Most recently, the sheep model has been used to address the transport kinetics of lipton (Thomas et al., 2001), hypoxanthine (Redzic et al., 2002), and thyroxine (Zheng et al., 2003) in the choroid plexus.

The in situ perfused sheep choroid plexus model offers several advantages as an ideal tool for studying transport of molecules from the blood to the CSF. Unlike the in vivo CSF sampling method, where the CSF concentrations of substances reflect the dynamic balance of transport between the blood–brain barrier and blood–CSF barrier, the in situ perfused choroid plexus excludes interference from the blood–brain barrier and enables the characterization of the kinetic behavior of a given drug at the blood–CSF barrier. Since the tissue under investigation remains intact in the brain and the molecule investigated enters the plexus directly via the blood interface, this preparation provides better similarity to real-life choroid plexus than in vitro incubation of dissected plexus tissues or using cultured choroidal epithelial cells. In addition, application of a dual tracer technique enables one to correct the nonspecific loss of study molecules, making it possible to calculate the unique uptake (or removal) kinetics of the test compound by the choroid plexus.

22.2.2 Animals and Pretreatment

Sheep of either sex can be used for this study. The body weight is usually between 20 and 25 kg. The animals are allowed to have free access to water and food and are quarantined for a period of three to five days prior to experimentation. If in vivo treatment (or exposure) becomes necessary, sheep can be readily i.v. (or otherwise) dosed.

22.2.3 Preparation of In Situ Perfused Sheep Choroid Plexus

Prior to the experiment, each sheep is anesthetized with sodium thiopental (20 mg/kg, i.v.) and heparinized (10,000 units, i.v.). CSF samples (≈ 2 to 5 mL) can be obtained at this time through a butterfly needle (14 gauge) attached to polyethylene tubing. The needle is inserted between the protruberance and the spine of the atlas. CSF samples free of blood can be used for biochemical analysis. Upon CSF sampling, sheep are decapitated, the brains removed rapidly from the skull, and internal carotid arteries cannulated immediately.

The perfusion system can be started at a flow rate of 0.5 to 1.5 mL/min, with the perfusate being directed toward the anterior choroidal arteries by ligation of the other branches on the circle of Willis. The cerebral ventricles are cut open, and the exposed choroid plexus in the lateral ventricle can be readily seen. The ventricles are superperfused with artificial CSF (Preston and Segal, 1990). The perfusion fluid draining from the choroid plexuses is collected by a cannula inserted into the great vein of Galen (Deane and Segal, 1985). A simplified diagram of this in situ model is shown in Figure 22.1.

The perfusate is a modified mammalian Ringer solution (Preston and Segal, 1990) containing 4 g/dL dextran 70 (Gentran) as the colloid osmotic agent and 0.05 g/dL bovine serum albumin to maintain the integrity of the capillary wall. The artificial CSF is protein- and dextran-free. Both solutions are saturated with 95% O_2 and 5% CO_2 at pH 7.4 and warmed to 37°C, with the blood perfusate being filtered and debubbled prior to entering the choroid plexuses. The preparation should be kept warm with a water jacket and by an external heat source; temperature and pressure should be tightly monitored. A preparation like this can stay viable for two to three hours (Preston and Segal, 1990).

22.2.4 Paired Tracer Method for Studying Blood to Plexus Transport

Cellular uptake of radiolabeled chemicals by the choroid plexus from the blood can be determined from the ratio of the recovery of radiolabeled chemicals to the recovery of [^{14}C]mannitol, a nontransported extracellular marker, in choroidal venous outflow. For better understanding of the principle, we use [^{125}I]T_4 (thyroxine) as an example. A calibrated "side loop" of the perfusion system is filled with a 100 μML bolus Ringer containing 0.5 μMCi [^{125}I]T_4 (4.3 fmol), various concentrations of unlabeled T_4, and 2 μMCi of [^{14}C]mannitol (0.1 nmol). By a system of taps, the flow of perfusate can be directed into this side loop and the bolus driven by the flow of perfusate toward either side of the choroid plexus without a rise in perfusion pressure. Approximately 20 seconds later, to enable the bolus to reach the plexus, a run of 20 sequential one-drop samples of venous effluent is collected in about 60 seconds followed by a continuous four-minute collection. Each brain preparation (both sides of the cannulated choroid plexus) can be used for 10 to 15 such runs. The final four-minute collection is usually used to calculate the flow rate. An aliquot (3 mL) of scintillation fluid can be added to each collected drop, 50 μML of final

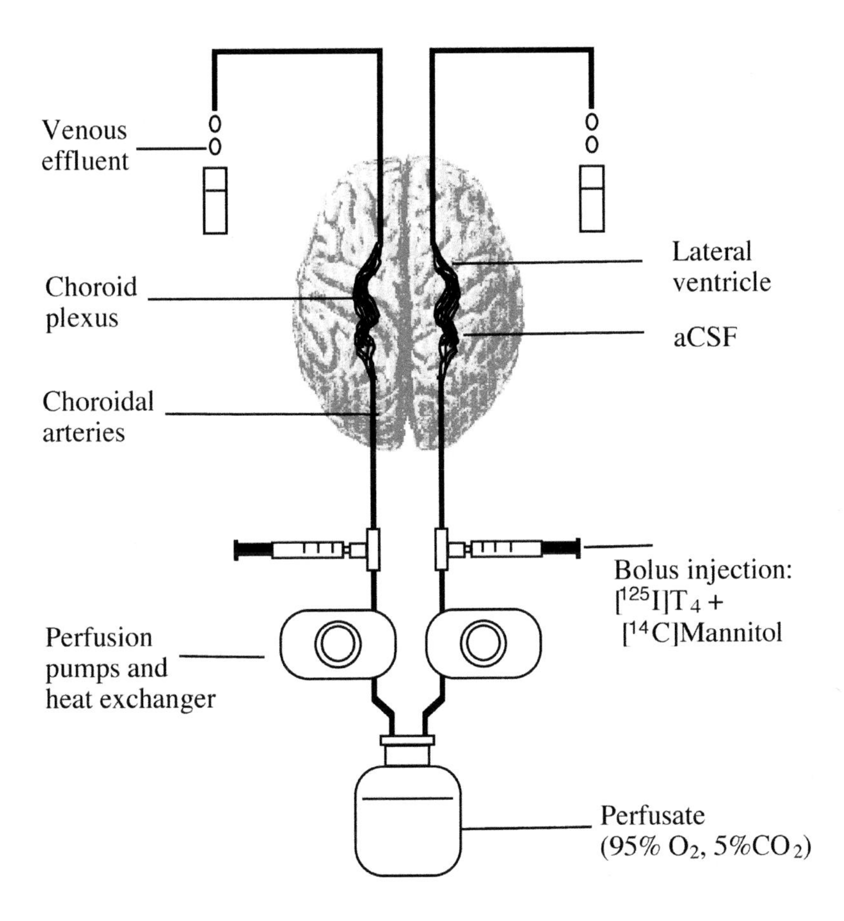

FIGURE 22.1 Illustration of in situ sheep choroid plexus perfusion model. The perfusion system is maintained at a flow rate of 0.5 to 1.5 mL/min at 37°C, with the perfusate being directed toward the anterior choroidal arteries. The cerebral ventricles are exposed, and the choroid plexus in the lateral ventricle is superperfused with an artificial CSF. The perfusion fluid draining from choroid plexuses is collected by a cannula inserted into the great vein of Galen.

four-min effluent, and 10 µML of the remaining injection bolus. Radioactivities will then be counted and expressed as dpm properly established quench curves.

22.2.5 Kinetics Calculation

The recovered [^{14}C]mannitol and [^{125}I]T_4 in each venous drop is expressed as a percentage of the [^{14}C]mannitol and [^{125}I]T_4 injected in the 100 µML bolus (% recovered). For any given drop, the recovery of [^{125}I]T_4 from the choroid plexus is less than the recovery of [^{14}C]mannitol, but peak recovery of both isotopes should be simultaneous. The percentage uptake of [^{125}I]T_4 relative to mannitol, i.e., U%, can be calculated for each drop:

$$U\% ([^{125}I]T_4) = \frac{\%[^{14}C]\text{mannitol recovered} - \%[^{125}I]T4 \text{ recovered}}{\%[^{14}C]\text{mannitol recovered}} \times 100 \qquad (22.1)$$

The U% values for those samples containing the highest levels of recovered radioactivity are averaged to give the maximal cellular uptake (U_{max}) of $[^{125}I]T_4$ for that "run."

In addition, the net uptake (U_{net}) over the whole "run" and final four-minute sample can be calculated:

$$U_{net}\%([^{125}I]T_4) = \frac{\Sigma[^{14}C]\text{mannitol received} - \Sigma[^{125}I]T4 \text{ recovered}}{\Sigma[^{14}C]\text{mannitol recovered}} \times 100 \qquad (22.2)$$

where Σ is the sum of tracer recoveries for the whole "run," plus the final four-minute sample.

The kinetic study of $[^{125}I]T_4$ uptake can be performed under conditions in which the plexuses are perfused with different concentrations of unlabeled T_4. The concentration of unlabeled T_4 in perfusate varies from 0.015 to 20 μM in 100-μL bolus. Under these conditions the 100 μL injected bolus should contain both labeled and unlabeled T_4 and tracer amount of $[^{14}C]$mannitol. The bolus injection mixes with the main perfusion fluid before reaching the plexuses, so that the final concentration of T_4 can be estimated by calculating a dilution factor (usually 5 to 7) based on the passage of $[^{14}C]$mannitol through the plexuses, as described by Segal et al. (1990). For example, a 100 μL injected bolus containing 0.15 μM T_4 would reach the plexuses at an actual concentration of 0.023 μM with an expected dilution factor of 6.5.

The unidirectional flux of $[^{125}I]T_4$ (μmol/min/g) can be calculated, using the U_{max} values, from the following equation:

$$\text{Flux} = -F \times \ln(1 - U_{max}) \times S \qquad (22.3)$$

where F equals the perfusate flow rate (mL/min/g) and S equals unlabeled T_4 concentration (μM). The average weight of sheep choroid plexus is 0.195 ± SD 0.08 g (n = 100), which is taken from many studies previously published (Segal et al., 1990).

To calculate the V_{max} and K_m, the flux values in each run of all experiments are plotted against the actual concentration of the chemical studied, in this case T_4, in each corresponding run. The data sets can be evaluated by any kinetic analysis software package. A concentration-effect math model that simulates the Michaelis-Menten relationship can be used to fit the observed data from which to estimate V_{max} and K_m.

Let us use the paired-tracer method with $[^{125}I]T_4$ as an example to illustrate the steady-state extraction of $[^{125}I]T_4$ relative to $[^{14}C]$mannitol at the basolateral face of the perfused choroid plexus. A typical example of the $[^{125}I]T_4$ fractional extraction measured during one run (a single drop per sample collected for 20 samples) is shown in Figure 22.2. Upon entering the choroidal vessels, the recovery of [^{14}C]mannitol arises quickly to reach the maximum values, suggesting that mannitol diffuses across the fenestrated capillaries into the extracellular space of the choroid plexus. The remainder diffuses back into the circulation and is recovered in the collected drops (see Figure 22.2A). [^{125}I]T_4 injected in the same bolus that accompanies the mannitol is also recovered in the collected drops, but the percentage of recovery of T_4 is smaller than extracellular marker mannitol, although T_4 has access to the same compartments as does the mannitol. This extra loss of T_4 represents the cellular uptake of T_4 by the choroidal epithelial cells. Figure 22.2B shows the uptake of [^{125}I]T_4 into the choroid plexus expressed as the percentage of recovery of [^{125}I]T_4 relative to that of mannitol; this corrects for nonspecific loss of T_4 via the extracellular distribution. The peak values of the uptake, i.e., the region where the greatest recovery occurs, are then averaged (as shown by points joined by a line in Figure 22.2B) to give rise to the value of U_{max}.

One limitation of this technique is that it is difficult to measure how much of a molecule has crossed from the blood into the choroid plexus and then entered the CSF. By collecting newly formed CSF from the surface of the choroid plexus, it is possible to check that the molecule crossing into the CSF is still intact, but it remains difficult to accurately quantify the magnitude of this transport.

22.3 IN SITU BRAIN PERFUSION TECHNIQUE

22.3.1 Technical Background

The technique of brain perfusion via the intracarotid injection of testing compounds has been used since the 1960s (Andjus et al., 1967; Thompson et al., 1968; Takasato et al., 1984; Bradbury et al., 1984). By directly introducing the drug molecules to the brain, one avoids a major problem associated with systemic injection: the hepatic or renal clearance of parent drug molecules. Such clearance (including biotransformation, redistribution, and elimination processes) in many cases makes it difficult to estimate blood–brain transport of substances, since the drug concentration at the site of transport is essentially unknown. Brain perfusion thus enables accurate delivery of a known amount of testing materials to the brain. Based on analysis of drug concentrations in brain parenchyma, which by definition is the brain tissue fractions free of capillaries, the unidirectional influx of drug can then be obtained. A refined method has been described by Takasato et al. (1984) in rat and by Zlokovic et al. (1986) in guinea pig. The same technique can also be applied to a smaller animal such as mouse (Bradbury et al., 1984). Although the method is most frequently used in blood–brain barrier studies, the choroid plexus can also be sampled and thus used for comparison to the brain uptake. The following discussion uses the rat for the purpose of technical description.

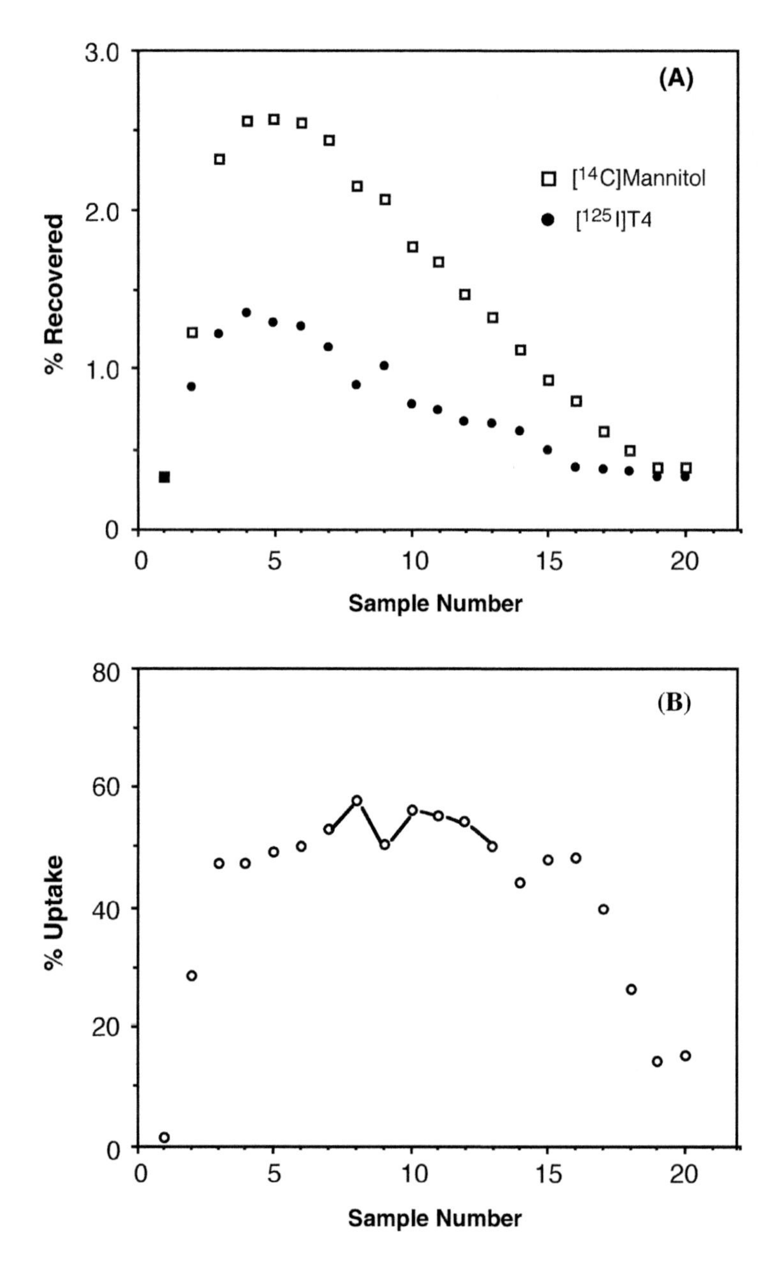

FIGURE 22.2 A typical run of paired-tracer study of T_4 transport by sheep choroid plexus. (A) Recovery of [^{14}C]mannitol and [^{125}I]T_4 in one run of 20 venous samples (as % of radioactivity injected). The low recovery of [^{125}I]T_4 in comparison to [^{14}C]mannitol indicates T_4 uptake at the basolateral face of the choroid plexus. (B) Uptake (%) of [^{125}I]T_4 in each venous sample relative to [^{14}C]mannitol. Samples that contained the higher recovery of isotopes are joined by a line, and these values averaged to estimate U_{max} (%). (Adapted from Zheng et al., 2003.)

22.3.2 Animals

Rats (both sexes) at the time they are used are about 9 to 12 weeks old (280 to 320 g). They are kept in a temperature-controlled, 12-hour light/dark cycle facility and fed ad libitum with a Teklad rat chow. Following a 3- to 5-day quarantine period, rats can be randomly placed into several groups according to the experiment. The rats can be pretreated with testing materials for the duration of the study course.

22.3.3 Perfusion Procedure

At the day of experimentation, rats are anesthetized with 50 mg sodium pentobarbital/kg by ip injection. The skin on the neck is cut open to expose the blood vessels. Both sides of the common carotid arteries are dissected; two surgical threads are imbedded under the carotid arteries. Immediately before cannulation, a node on the carotid arteries toward the heart is made by tightening one string. A cut on the artery is made above the node, followed by insertion of a PE-10 polyethylene tubing (Clay Adams) toward the brain. A second node is then made to secure the tubing inside the blood vessel. Immediately before the start of perfusion, both internal and external jugular veins and the left ventricle of the heart are severed to enable drainage of the perfusate circulating from the brain and to prevent recirculation of the rat blood. By this method, the Ringer solution, both hot and cold, will not mix with rat blood.

Each cannulated artery is initially perfused with a cold (nonradioactive) Ringer solution at 3 to 3.5 mL/min using two independent peristaltic minipumps (Variable-Flow Mini Pump, VWR). The Ringer solution contains (g/L): NaCl, 7.31; KCl, 0.356; $NaHCO_3$, 2.1; KH_2PO_4, 0.166; $MgSO_4.3H_2O$, 0.213; glucose, 1.50, sodium pyruvate, 1 mmol/L, and $CaCl_2$, 2.5 mmol/L, at pH 7.4. The perfusate is circulated through a temperature-controlled water bath system such that the temperature at the tip of the tubing entering the brain is maintained at 37°C. Prior to perfusion, the perfusate must be filtered and presaturated with oxygen. This will enable the extension in perfusion time without anoxia, as reported by Preston et al. (1995). The solution should be continuously oxygenated with 5% CO_2 and 95% O_2 during perfusion. The brain is perfused with nonradioactive Ringer solution for 10 minutes prior to infusion with the hot perfusate (Deane et al., 2004).

The hot perfusate is the Ringer solution containing radioisotopes. The hot perfusate is placed in two syringes and introduced to either side of catheters through two separate three-way tap systems using a slow syringe-driven pump (Harvard Compact Infusion Pump, Model 975). At the time of infusion of the hot perfusate, the peristaltic pumps are turned off, the three-way tap system switched to the syringe pump, and the hot perfusate infused at a flow rate of 0.4 mL/min for up to 30 minutes. Usually [^{3}H]mannitol is used as a reference marker; its distribution volume ($V_d < 5$ μL/g) in the vascular washed brain is subtracted from the total uptake.

About one minute before the end of the timed perfusion period, a cisterna magna CSF sample can be collected by using a 25 gauge butterfly needle (Becton Dickinson, Sandy, UT) inserted between the protuberance and the spine of the atlas. CSF samples (about 100 to 150 μL) free of blood can be used for the biochemical analyses.

At the end of the timed perfusion periods, the syringe pump is turned off, the three-way tap system switched back to peristaltic pumps, and the brain then washed with cold Ringer solution for 30 seconds.

The brain is then removed from the skull, washed in ice-cold saline, and placed on filter paper saturated with saline, which rests on an ice-chilled glass plate. After removal of the meninges, the choroid plexus can be collected. The rest of brain can be further dissected to collect different regions and processed for capillary depletion assay.

22.3.4 Calculations

The uptake is expressed as a distributing volume, V_d, calculated as

$$V_d = \frac{C_{\text{brain or CSF}}}{C_{\text{perfusate}}} = (\text{mL/g}) \tag{22.4}$$

where C_{brain} is dpm/g of tissues, and C_{CSF}, $C_{perfusate}$, dpm/mL of perfusates. The uptake is corrected for residual radioactivity by deducting V_d for [^{3}H]mannitol from the total radioactivity of the studying compound (e.g., ^{59}Fe) distributing volume.

The unidirectional transport rate constant, K_{in} (mL/g/min $\times$ 10^3), corresponding to the slope of the uptake curve, is determined using the linear regression analysis of V_d against the perfusion time, T (min), from Eq. 22.4

$$V_d = K_{in}\,T + V_i \tag{22.5}$$

where V_i is the ordinate intercept of the regression line (Zlokovic et al., 1986; Deane and Bradbury, 1990).

22.4 SUMMARY

In vivo experimentation to study the function of the blood–CSF barrier is generally difficult, as the small size of the choroid plexus limits the capability for surgical operations. In situ choroid plexus perfusion in large animals and brain perfusion in rodents are by far the most frequently used techniques in assessing the choroid plexus functions.

The advantages of these in situ perfusion techniques are obvious: first, the techniques better resemble the in vivo condition than in vitro methods such as choroid plexus incubation study or primary culture of plexus cells. Second, the perfused choroid plexus model minimizes the interference by passage of drug molecules via the blood–brain barrier, making it possible to characterize the unique kinetic behaviors of drug molecules at the blood–CSF barrier. Finally, the in situ brain perfusion model eliminates systemic metabolic interference, enabling comparison of transport kinetics between the blood–brain barrier and blood–CSF barrier. The major disadvantages of these in situ techniques pertain to the technical difficulty in choosing large animals for perfusion of choroid plexus, in surgical procedures, and in using large quantities of radiolabeled materials.

REFERENCES

Andjus RK, Suhara K, Sloviter HA (1967) An isolated, perfused rat brain preparation, its spontaneous and stimulated activity. J Appl Physiol 22: 1033–1039.

Blount R, Foreman P, Harding M, Segal MB (1973) The perfusion of the isolated choroid plexus of the sheep. J Physiol 232: 12P–13P.

Bradbury MWB, Deane R, Rosenberg G (1984) Regional blood flow, EEG, and electrolytes in mouse brain perfused with the perfluorochemical, FC43. J Physiol 355: 31P.

Deane R, Bradbury MWB (1990) Transport of lead-203 at the blood-brain barrier during short cerebrovascular perfusion with saline in the rat. J Neurochem 54: 905–914.

Deane R, Segal MB (1985) The transport of sugars across the perfused choroid plexus of the sheep. J Physiol London 362: 245–260.

Deane R, Zheng W, Zlokovic BV (2004) Brain capillary endothelium and choroid plexus epithelium regulate transport of transferrin-bound and free iron into the rat brain. J Neurochem 88: 813–820.

Pollay M, Kaplan, R (1972) Transependymal transport of thiocyanate. J Neurobiol 3: 339–346.

Preston, JE, Segal, MB (1990) The steady-state amino acid fluxes across the perfused choroid plexus of the sheep. Brain Res 525: 275–279.

Preston JE, Al-Sarraf H, Segal MB (1995) Permeability of the developing blood-brain barrier to mannitol using the rat in situ brain perfusion technique. Dev Brain Res 87: 69–76.

Redzic ZB, Gasic JM, Segal MB, Markovic ID, Isakovic AJ, Rakic MLj, Thomas SA, Rakic LM (2002) The kinetics of hypoxanthine transport across the perfused choroid plexus of the sheep. Brain Res. 925: 169–175.

Segal MB, Preston JE, Collis CS, Zlokovic BV (1990) Kinetics and Na independence of amino acid uptake by the blood side of perfused sheep choroid plexus. Am J Physiol 258: F1288–F1294.

Takasato Y, Rapoport SI, Smith QR (1984) An in situ brain perfusion technique to study cerebrovascular transport in the rat. Am J Physiol 247: H484–H493.

Thomas SA, Preston JE, Wilson MR, Farrell CL, Segal MB. (2001) Leptin transport at the blood–cerebrospinal fluid barrier using the perfused sheep choroid plexus model. Brain Res 895: 283–290.

Thompson AM, Robertson SI, Bauer TA (1968) A rat-head perfusion technique developed for the study of brain uptake of materials. J Appl Physiol 24: 407–411.

Zheng W, Deane R, Redzic Z, Preston JE, Segal MB (2003) Transport of *L*-[125I].thyroxine by in situ perfused ovine choroid plexus: inhibition by lead exposure. J Toxicol Environ Health 66: 435–51.

Zlokovic BV, Begley DJ, Djuricic BM, Mitrovic DM (1986) Measurement of solute transport across the blood-brain barrier in the perfused guinea pig brain: method and application to N-methyl-α-aminoisobutyric acid. J Neurochem 46: 1444–1451.

23 *In Vitro* Investigation of the Blood–Cerebrospinal Fluid Barrier Properties: Primary Cultures and Immortalized Cell Lines of the Choroidal Epithelium

Nathalie Strazielle
Research and Development in Neuropharmacology
INSERM U433, Faculté de Médecine Laennec
Lyon, France

Jean-François Ghersi-Egea
INSERM U433, Faculté de Médecine Laennec
Lyon, France

CONTENTS

23.1 INTRODUCTION

The choroid plexuses (CPs) that form the blood–cerebrospinal fluid (CSF) barrier (BCSFB) are, in conjunction with the blood–brain barrier (BBB), responsible for maintaining the appropriate microenvironment required by the brain. More specifically, the CPs, located in the ventricles of the brain, separate the blood from the ventricular cerebrospinal fluid that they secrete for the most part and, as such, exert a considerable influence on the composition of the medium bathing the central nervous system. The site of the BCSFB is the choroidal epithelium proper (see Figure 23.1).

Defining the role of this epithelium in solute transport between the blood and the CSF or in polarized secretion processes by analyzing CSF in vivo is complicated by the flow and constant renewal of CSF and by the contribution of the BBB and

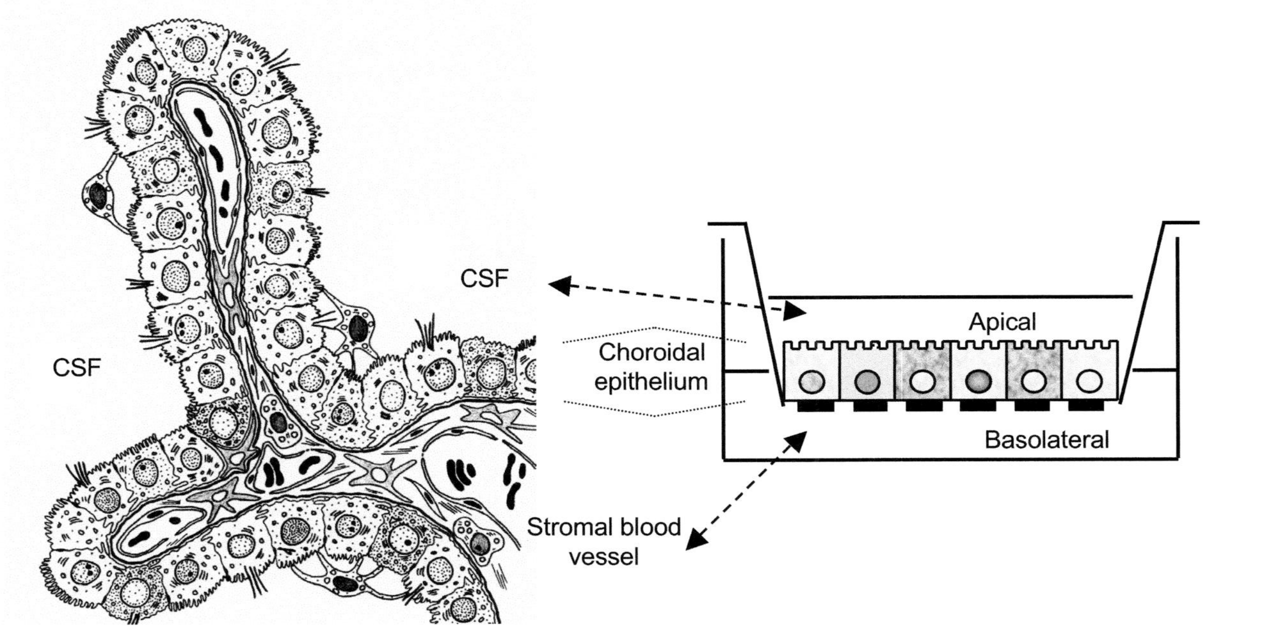

FIGURE 23.1 In vitro model of the BCSFB. Located in the ventricles of the brain and folding into numerous villi, the CPs are formed by a choroidal epithelium delimiting a conjunctive stroma, which contains permeable capillaries and cells of the myeloid lineage. Epiplexuel macrophages, or Kolmer cells, are located at the CSF-facing membrane of the epithelium (left).The site of the BCSFB is the choroidal epithelium proper. The tight regulation of transepithelial permeability required to maintain homeostasis implies a passive blockade of the paracellular pathway and an active control of the transcellular pathway. The former is achieved by the presence of tight junctional complexes sealing the intercellular clefts between epithelial cells and restricting the passage of ions and small lipid-insoluble nonelectrolytes. The control of the transcellular pathway involves both the secretion of fluid, ions, essential nutrients, and hormones and the reabsorption of endogenous toxic metabolites from the CSF. This occurs via a machinery of channels, pumps, and transport systems distributed in a polarized pattern between the apical and basolateral membrane domains of the choroid plexus epithelial (CPE) cells. To reproduce the BCSFB in vitro, choroidal epithelial cells are grown on a permeable support (right). The monolayer separates two compartments, the upper one containing the apical fluid that represents the CSF and the lower chamber containing the basolateral fluid assimilated to the stromal blood.

the nervous tissue to CSF composition. Diffusion from the interstitial fluid into the CSF of material secreted by neural cells or exchanged across the cerebral capillaries occurs across the nonrestrictive permeable ependyma or to a lesser extent across the glia limitans. Alternatively, CSF-borne molecules diffuse into the parenchyma. These different parameters cannot be easily taken into consideration when CP-specific processes are investigated by in vivo techniques such as ventriculocisternal perfusion, in situ brain perfusion, or microdialysis. The development of ex vivo or in vitro techniques was therefore mandatory to study CP functions and physiology without the interference of the BBB and brain parenchyma. The in situ perfused CP enables simultaneous access to both blood and CSF faces of the epithelium and thus can be used to estimate transcellular transport in both directions, from CSF to blood as well as from blood to CSF, providing the delicate recovery of nascent CSF can be carried out (Strazielle and Preston, 2003) (see Chapter 22). However, perfusion of the choroidal vasculature requires a large animal and the technique has been carried out mainly in sheep and goats. The isolated CP and isolated apical and basolateral membrane vesicles prepared from epithelial cells provide (relatively) simpler and more convenient systems. Isolated CPs in particular have been extensively used to characterize choroidal uptake but obviously are not adequate to study transepithelial transfer or polarized transport and secretion mechanisms, as they do not offer easy and separate access to the stromal facing membrane of the choroid plexus epithelial (CPE) cells. Other limitations shared by these different ex vivo/in vitro techniques are (1) the limited survival time of the choroidal tissue, precluding pretreatments and studies related to regulated phenomena or toxicity and (2) the coexistence of other cells in the choroidal tissue that may contribute to protein/polypeptide synthesis and secretion.

For these reasons, establishing in vitro cellular models of the BCSFB retaining the differentiated phenotype of the choroidal epithelium became a challenge a few decades ago. Early attempts at organ culture in the 1950s reported the growth of cystic structures and inverted epithelial vesicles in cultures of intact CP from rat and rabbit embryos near term or from newborn chicks (Cameron, 1953). Having developed a sophisticated chemically defined medium, Agnew was able to generate bulbous vesicular outgrowths from the CP villi explants of juvenile rats (Agnew et al., 1984). After detachment, these vesicles displayed a monolayer structure with the ultrastructural features and polarity of epithelial cells. Stromal cells progressively disappeared to yield a luminal cavity whose diameter could exceed 500 μm, i.e., a size suitable for easy microinjection and microsampling of the luminal fluid. More conventional culture approaches dissociating the epithelial cells prior to culturing aimed at establishing simplified in vitro models, i.e., cell monolayers in primary culture. Methodologies to separate epithelial cells from the CP stroma were developed in the 1970s, but no attempt to maintain the cells in culture was initiated at that time (Nathanson, 1979). With the advent of refined culture methods and permeable culture supports, cultured monolayers of choroidal epithelial cells have now become an appropriate tool to demonstrate transcellular transport and vectorial secretion by the BCSFB (see Figure 23.1). Recently, immortalized cell lines have been generated to overcome the time-consuming aspect inherent in primary culture in general and the difficulty in conducting certain studies such as long-term toxicity studies.

Our aim in this chapter is not to provide the reader with a protocol for the ideal in vitro model of the BCSFB because, in our opinion, a model will always remain a utopia. Rather, our goal is to provide useful information to help investigators set up and characterize their own CPE cell culture system. To that purpose, we will review the latest attempts to develop in vitro models of this interface and discuss the factors critical to isolating cells and selecting culture conditions. We will then focus on primordial specific choroidal features that can be examined within a validation scheme, emphasizing the importance of the methodology and the need for careful interpretation of their significance. Limitations of the different models will be discussed. We will finally survey how these models have served so far to better understand CP physiology and functions.

23.2 CHOROIDAL EPITHELIAL CELLS IN PRIMARY CULTURE

23.2.1 Establishment of the Primary Cultures

Primary cultures of CPE cells have been developed from various animal species by different groups. Since most groups developed their own protocol to establish the culture and investigated different features of the cells, a direct comparison of the various models is difficult. Furthermore, relating differences observed for one given cell characteristic to one particular step or added component is not possible. Nonetheless, some general rules for establishing primary cell cultures of the choroidal epithelium have emerged from all the studies, with several, but not all, steps having met a consensus.

23.2.1.1 Cell Isolation, Seeding, and Culture Conditions

23.2.1.1.1 Cell-Dissociation Techniques

Epithelial cells are closely associated in vivo by junctional complexes, and a general observation in epithelial cell culture is that these cells survive better as clusters, which calls for a rather mild dissociation technique. Crook and colleagues (1981) compared various dissociation conditions (enzymes, concentration, temperature) on adult bovine CPs. They selected pronase over collagenase, dispase, or collagenase/dispase on the basis of cell viability and plating efficiency. Accordingly, collagenase used initially by Tsutsumi and collaborators (1989) on rat choroidal tissue was replaced in further studies of the same group by pronase, which was found to provide better cell yields (Sanders-Bush and Breeding, 1991). Thereafter, pronase (0.5 to 1 mg/mL) has been widely used on CPs from various animal species (rabbit, rat and sheep) (Mayer and Sanders-Bush, 1993; Holm et al., 1994; Villalobos et al., 1997; Plotkin et al., 1997; Zheng et al., 1998; Strazielle and Ghersi-Egea, 1999; Rao et al., 1999; Shu et al., 2002). Further dissociating treatments of the large cell clumps have been added to some protocols, consisting either of trypsin digestion (Tsutsumi et al., 1989; Plotkin et al., 1997; Strazielle and Ghersi-Egea, 1999; Shu et al., 2002) or mechanical trituration (Villalobos et al., 1997; Zheng et al., 1998). Zheng and colleagues found this step crucial for an efficient subsequent plating of rat CPE cells (Zheng and Zhao, 2002a). An alternative to pronase digestion is the use of trypsine (0.25%) alone (Southwell et al., 1993; Ramanathan et al., 1996; Gath et al., 1997). DNAse I is classically added to the dissociating enzymes to digest the genomic DNA released from damaged cells. This prevents an increase in viscosity of the cellular suspension that would interfere with the tissue disaggregation. Owing to their cluster organization, the released epithelial cells can be collected preferentially by enabling them to sediment more rapidly through the more dispersed stromal cells, but the accurate determination of the number of cells in the suspension is thereby impeded. The seeding density can therefore be adjusted on the basis of the initial weight or number of CPs (Gabrion et al., 1988; Gath et al., 1997; Strazielle and Ghersi-Egea, 1999).

23.2.1.1.2 Preparation of Culture Supports

In most studies, CPE cells have been plated directly on plastic or on permeable supports, depending on the purpose of the study. Permeable supports may be selected from a large panel of inserts, available from various manufacturers, made of different materials and offering variable porosities (Strazielle and Preston, 2003). It is worth noting that certain measurements may be influenced by the culture inserts (see Section 23.2.2.3.2). Permeable inserts must be coated beforehand, and the nature of the matrix component has been shown to considerably influence the attachment, growth, and purity of the cultured cells. A number of investigations have designated laminin as the most appropriate matrix component for culturing CPE cells. Rat choroidal cells attached efficiently on collagen-coated plastic support (Zheng et al., 1998) or inserts coated with rat tail collagen (Strazielle and Ghersi-Egea, 1999). However, collagen apparently promoted the adhesion and growth of the fibroblasts present in the cell suspension, thus impairing the purity of the cultures.

Commercial collagen-coated inserts (Transwell-COL™, Costar) also enabled rat epithelial cell attachment, yet the small islands of cells did not spread to form a confluent monolayer (Southwell et al., 1993). By favoring the adhesion of epithelial cells over that of fibroblast-like cells, laminin permitted the formation of rat CPE cell monolayers with good purity on permeable supports (Southwell et al., 1993; Strazielle and Ghersi-Egea, 1999). In a study comparing laminin, collagen, fibronectin, and thrombospondin-1 in the attachment and proliferation rate of porcine epithelial cells, Haselbach and colleagues observed that laminin again offered the highest efficacy for both parameters (Haselbach et al., 2001).

Using primary cultures of fetal CPE cells prepared from mouse early embryos, Dziadek and collaborators (Thomas et al., 1989; Stadler and Dziadek, 1996) investigated in more detail the influence of extracellular matrix composition on cell adhesion, survival, proliferation, and structural organization. Fetal CPE cells attached poorly on collagen I, even in the presence of serum, and formed small aggregates with no evidence of cell spreading and little evidence of cell polarization. Mesenchymal cells were found in some cultures (Thomas et al., 1989). Addition of laminin to collagen I (at a ratio of 9:1) promoted cell adhesion and cell survival, yet did not result in the formation of confluent monolayers (Stadler and Dziadek,

1996). Matrigel™ (BD Biosciences), a commercial matrix composed of a high proportion of laminin with entactin, nidogen, and collagen IV, prevented mesenchymal cell adhesion and induced different epithelial cell organizations depending on its concentration. At low or no dilution, Matrigel promoted the invasion of the CPE cell aggregates into the substratum and the formation of multicellular vesicular structures, consisting of highly polarized cells around a central lumen. In diluted Matrigel, cells lost their invasive capacity and proliferated to form confluent monolayers at the substratum surface, but displayed a similar polarized ultrastructure. These observations highlighted the sensitivity of fetal CPE cells to quantitative changes in matrix composition and also indicated that Matrigel components other than laminin influenced the proliferation of these cells. Surprisingly, tissue specific gene expression was not affected by the degree of cell polarity and the multicellular organization achieved by the fetal CPE cells on different matrix (Thomas et al., 1989).

Deposition of basement membrane components by epithelial cells proper in the course of the culture could conceivably contribute to their proliferation and differentiation. There was no clear evidence of such basement membrane deposition by fetal CPE cells, although the secretion of individual components, not evidenced in the study, cannot be excluded (Thomas et al., 1989). In long-term cultures of mouse choroidal cells enriched in epithelial cells and grown on solid support, fibronectin was detected as a diffused layer under the epithelial cells (Peraldi-Roux et al., 1990). The authors also reported positive staining for laminin within the epithelial cell cytoplasm and under the cells, which persisted in 50-day-old cultures. Immunocytochemical evidence for fibronectin and thrombospondin-1 synthesis and deposition in porcine choroidal cell cultures on permeable support was also obtained (Gath et al., 1997). The decrease in thrombospondin-1 staining in confluent CPE cell monolayers by comparison to subconfluent cultures suggested that it plays a role chiefly in cell proliferation (Haselbach et al., 2001).

23.2.1.1.3 Culture Medium

While CPE cells have been consistently grown in DMEM or a mix of DMEM and Ham's F-12, an important variability in the culture supplements is noticeable among the different in vitro models. Various combinations of the following growth factors, hormones, and culture additives, i.e., hydrocortisone, insulin, selenite, epidermal growth factor, basic fibroblast growth factor, prostaglandin E_1, triiodothyronine, hypoxanthine, forskolin, and progesterone, have been applied to CPE cell cultures. In the absence of a detailed investigation of the influence of each of these compounds taken alone or in combination, it is difficult at present to delineate the exact contribution of individual supplements to the cultured cell characteristics.

By contrast, it is generally agreed that addition of serum to the medium is a prerequisite for efficient adhesion of CPE cells to culture supports. At the time of seeding, and thereafter during proliferation, fetal bovine serum is classically used at a 10% concentration. Calf serum may not always enable attachment, as reported by Tsutsumi and colleagues (1989) for rat cells, in contrast to results from other investigators, who used it successfully (Sanders-Bush and Breeding, 1991; Rao et al., 1999). Nu Serum IV (i.e., diluted fetal bovine serum supplemented with selected factors) has also been used successfully for rat CPE cell cultures (Villalobos et al., 1997). Removal of serum after three to four days has been applied as a technique to limit the proliferation of contaminating fibroblasts (Sanders-Bush and Breeding, 1991; Esterle and Sanders-Bush, 1992). It has also been performed on confluent differentiated rat CPE cell monolayers in specific studies such as protein secretion analysis or to prepare conditioned medium (Hoffmann et al., 1996; Zheng et al., 1998; Strazielle et al., 2003b). This short period of culture in serum-free medium was not reported to affect epithelial cell characteristics. By contrast, serum withdrawal on porcine CPE cell cultures had major morphological and functional consequences and was shown to be necessary for a complete differentiation of the cells (Hakvoort et al., 1998b). These effects will be discussed in more detail later.

23.2.1.1.4 Life Span and Subculture

Another consensus has emerged from the various attempts to culture CPE cells, in that these cells passage poorly. Subculturing rodent choroidal cells resulted in a loss of differentiated functions or rapid cell death (Gabrion et al., 1998; Zheng et al., 1998; N. Strazielle and J.F. Ghersi-Egea, unpublished observations). Furthermore, Crook et al. (1981) reported that fibroblasts could be passaged more easily and suggested that subculture would result in progressive fibroblast enrichement. Repeated subculturing was actually carried out to select fibroblasts and obtain highly enriched cultures of this cell type (Barker and Sanders-Bush, 1993; Ramanathan et al., 1997). As a consequence, CPE cells have been used mostly as primary cultures proper, with the exception of porcine cells, which appear to be subcultured once at the time of seeding on permeable membranes (Hartter et al., 2003). The cells in culture have a life span varying between one and two weeks, during which differentiated functions are retained. Thereafter, dedifferentiation is expected to occur as illustrated by the decrease in Na^+-dependent taurine transport observed in 30-day-old cultures of rabbit CPE cells (Ramanathan et al., 1997). Long-term cultures of CPE cells have been developed from 18-day-old rat and 16-day-old mouse fetuses (Gabrion et al., 1998). The primary cultures could be maintained on plastic supports for up to three months, during which specific features of the epithelial cells such as the apical localization of the Na^+, K^+-ATPase persisted (Peraldi-Roux et al., 1990). The authors reported, however, the presence of fibroblasts, endothelial cells and macrophages intermingled with areas of epithelial cells, within the monolayers. These contaminants could represent up to 25% of the total cell number.

23.2.1.2 Achieving Pure Epithelial Cell Cultures

A recurrent problem encountered in epithelial cell culture in general arises from the contaminating stromal cells, in particular fibroblasts, as the latter usually proliferate more rapidly in serum-supplemented medium and tend to overgrow the epithelial cells. Various physical separation methods and selective culture techniques have been developed over the years to overcome or limit fibroblast contamination. One method used separately may have limited success, but can be of value in combination with other techniques or for relatively short periods of culture. We will focus on the various approaches that have been tested and found efficient for CPE cells.

23.2.1.2.1 Selective Attachment or Differential Attachment

This approach takes advantage of the more or less developed ability that various cell types display to adhere to a given support. It can first be applied to enrich the cell suspension in epithelial cells by promoting the adhesion of stromal elements. Fibroblasts, and to a greater extent macrophages, attach rapidly onto plastic within the first few hours, while epithelial cells, probably because of their greater requirement for extracellular matrix (ECM) regeneration, remain in suspension and are easily recovered for further plating on the appropriate substrate. This technique has proved very useful for CPE cells, in particular for rat cells, and is often used as an enrichment step (Southwell et al., 1993; Ramanathan et al., 1996; Nilsson et al., 1996; Villalobos et al., 1997; Strazielle and Ghersi-Egea, 1999; Shu et al., 2002). This principle of selective attachment is also exploited when permeable filters are coated with laminin, which favors the adhesion of epithelial cells as mentioned above. These rather simple steps proved sufficient to prevent fibroblast contamination for several groups, while others have found it necessary to associate them with selective culture techniques, based on the use of various growth inhibitors. The reason for this difference among laboratories as to the need for growth inhibitor treatment is unclear and is likely to be multifactorial. It may involve differences in animal species and stage of development, as well as small variations in pronase or trypsin concentration and duration of treatment, or else diversity of commercial sources for laminin and coating procedures.

23.2.1.2.2 Selective Culture

D-Valine-selective culture medium, based on substitution of *D*-valine for *L*-valine, was shown to selectively inhibit human and rodent fibroblast proliferation (Gilbert and Migeon, 1975). This effect is based upon the lack in fibroblast cells of *D*-amino acid oxidase activity, which in other cell types, such as epithelial cells, can convert the *D*-amino acid into its essential *L*-isomer. It is important to note that using this technique requires one to dialyze the serum or to omit it from the culture medium because the serum contains endogenous *L*-valine that negates the advantage of *D*-valine–selective medium. While *D*-valine does not result in fibroblast death, it strongly impedes their overgrowth of the cell cultures. *D*-Amino acid oxidase activity has been measured in mammalian (hog) CP homogenates (Yusko and Neims, 1973), and one can postulate that it is located in the epithelial layer. Accordingly, when used by several groups in primary cultures initiated from rat CPs (Esterle and Sanders-Bush, 1992; Villalobos et al., 1997; Rao et al., 1999; Chang et al., 2000; McGrew et al., 2002), *D*-valine–selective medium enabled control of fibroblast growth while epithelial cells proliferated. However, when used in bovine CPE cell cultures, the effect of *D*-valine–selective medium did not prove to be reproducible from one experiment to the next because of variability in cell density according to the authors (Crook et al., 1981).

Instead, this group observed in the same bovine cultures a more reliable effect of *cis*-4-hydroxy-*L*-proline (*cis*-OH proline), another classical growth inhibitor described by Kao and Prockop (1977). As mentioned previously, overgrowth of CPE cell cultures by fibroblasts is strongly facilitated on collagen-coated surfaces. Newly synthesized polypeptides of procollagen do not fold into the normal stable helical conformation when *cis*-OH proline is incorporated into collagen in replacement for proline. This leads to a marked decrease in the net production of extracellular collagen as a result of both a reduced rate of cellular extrusion and an enhanced susceptibility to protease degradation (Kao et al., 1979) and consequently limits the proliferation rate of fibroblasts. *Cis*-OH proline was also found to be active in rodent cultures (Zheng et al., 1998; Rao et al., 1999), but in all animal species used this inhibitor also affected the epithelial cells when added at the time of plating or rapidly after plating. Different groups have reported, for bovine and rat CPE cells, the importance of both *cis*-OH proline concentration and time of addition in achieving an optimal efficiency and selectivity toward fibroblasts (Crook et al., 1981; Zheng et al., 1998).

Supplementation of the medium with 20 μM cytosine arabinoside (Ara-C), a S-phase–specific blocking agent, appears as the method of choice to achieve purity of porcine CPE cells and has been used by independent groups (Gath et al., 1997; Peiser et al., 2000). By treating the cells with 20 μM Ara-C for the first five days of culture, Gath and colleagues reported a complete inhibition of all contaminating cells, without affecting the growth of epithelial cells. Yet the specificity of the compound toward fibroblasts remains unclear. The relative specificity of Ara-C for fibroblasts versus epithelial cells has often been assumed to depend mainly on the much faster rate of proliferation of the former cells, rather than on a strict cell-type specificity. Accordingly, inhibition by 50% of cell proliferation is obtained by similar doses of this compound for both corneal epithelial cells and conjunctive fibroblasts in culture (Mallick et al., 1985). As for CPE cell culture, the differential action of Ara-C has been attributed to the possible lack of recognition of Ara-C by choroidal nucleoside transporters (Gath et al., 1997; Haselbach et al., 2001). However, while Ara-C indeed has no affinity for the active nucleoside transport system in rabbit isolated CPs, it is taken up by this tissue via a saturable facilitating carrier and is phosphorylated (Spector, 1982), suggesting that the active metabolite is indeed synthesized within the CPE cells. Since Ara-C apparently did not affect the growth of porcine epithelial cells, it is possible that nucleoside transporters in these cells differ from those of the rabbit or are rapidly downregulated in culture.

23.2.2 Characterization of Cultured Choroidal Epithelial Cells

Characterization of the cells in culture aims at evaluating the extent to which cells in vitro retain their epithelial phenotype and the differentiated functions they exert in vivo in choroidal tissue.

23.2.2.1 General Morphology, Ultrastructure, and Epithelial Phenotype

The morphology and the structural organization of cells in culture assessed by light and electron microscopy are good indices of their epithelial nature. Following adhesion to the support and proliferation, clumps of epithelial cells consistently gave rise to monolayers with a classical tight pavement-like appearance. The epithelial cuboidal cells contrasted clearly with the elongated fibroblasts, which grew in the absence of appropriate inhibitors or on inadequate substratum (Gath et al., 1997; Strazielle and Ghersi-Egea, 1999).

Ultrastructure was studied by transmission electron microscopy on various rat and porcine cell monolayers cultured on permeable membranes, as well as on vesicles prepared in Matrigel from mouse embryos CPs (Thomas et al., 1992; Southwell et al., 1993; Ramanathan et al., 1996; Gath et al., 1997; Villalobos et al., 1997; Strazielle and Ghersi-Egea, 1999; Rao et al., 1999). All these works confirmed the single-layer structure and revealed several features characteristic of the choroidal epithelium in vivo and indicative of cell polarity: (1) junctional complexes located near the apical end of the lateral membrane domains, (2) complex invaginations of the basolateral membrane creating large intercellular lateral spaces, and (3) an apical membrane domain bearing microvilli and occasionally cilia. The number of microvilli displayed by the cells in vitro varies among the models, but does not appear to thoroughly match the abundance of this structure in vivo.

Because other cell types at high density may present a cobblestone epithelial-like pattern, reliable epithelial identification depends on the complementary recognition of specific markers. The most frequently used criteria in general culture of epithelial cells is the analysis of the epithelial-specific cytoskeletal intermediate filaments, namely the cytokeratins. The use of pan-cytokeratin antibodies, which recognize epitopes of many cytokeratins, is appropriate for that purpose. Cytokeratins 8, 18, and 19 have been demonstrated in human, rat, and mice adult CPs, where they stained exclusively the epithelial cells, in contrast to vimentin, the mesenchymal specific intermediate filament, which stained only stromal cells (Miettinen et al., 1986).

Importantly, there is a switch in intermediate filament expression in rodent CP epithelium around birth, from a vimentin positivity during fetal life, which is abolished by the fourth postnatal day, to a cytokeratin positivity that appears in some cells at embryonic day 18 and extends to most cells by the fourth postnatal day (Miettinen et al., 1986). Immunostaining with pan-cytokeratin antibodies demonstrated the epithelial origin of CP cells prepared from one- to two-day-old rats, with some variation in staining intensity from cell to cell (Strazielle and Ghersi-Egea, 1999). This variation most likely reflects the more or less advanced stage of individual cells in this intermediate filament switch process. Cytokeratins were also analyzed on CPE cell cultures from adult mice and allowed to define a 70 to 80% purity (Steffen et al., 1996). Desmosomal junctions that are specific to epithelial cells at confluence can be identified via desmosomal proteins such as desmoplakin. Immunostaining revealed the presence of this protein as discrete spots found only at the cell boundaries of cultured porcine choroidal cells, thus indicating their epithelial origin (Gath et al., 1997).

Demonstrating the lack of marker proteins exclusive for the possible contaminating cells constitutes a complementary approach to assess the purity of the epithelial cells in culture. Vimentin, constituting the specific mesenchymal intermediate filament, should in theory represent an appropriate marker for fibroblasts. However, it has been reported that most proliferating cultured cells contain vimentin-type filaments in addition to the tissue-specific intermediate (Virtanen et al., 1981). This was indeed observed in porcine CPE cells prepared from adult animals (Gath et al.,

1997). The typically elongated morphology of fibroblasts, contrasting with the cobblestone appearance of CPE cells, thus remains a useful criterion to detect the former cells. Several markers for endothelial cells, which can also display an elongated shape, have been analyzed in CPE cell cultures. The absence of cells of endothelial origin was established in rabbit and bovine cultures by the lack of staining with indocarboxythionine-conjugated acetylated low-density lipoprotein (Mayer and Sanders-Bush, 1993) and the lack of immunoreactivity of factor VIII-associated antigen (Crook et al., 1981). Staining with MESA 1 (mouse endothelial surface antigen-1) monoclonal antibody revealed the presence of endothelial cells in rodent choroidal cultures (Peraldi-Roux et al., 1990). Other specific endothelial markers include the adhesion molecule CD34 and flt-1, a receptor for vascular endothelial growth factor. These markers have been used to demonstrate the endothelial origin of a new choroid plexus cell line (Battle et al., 2000).

Immunostaining of specific markers for stromal myeloid cells has not been carried out in CPE cell cultures to our knowledge. These cells, however, are unlikely to be contaminating cells, particularly when differential attachment on plastic is included in the cell preparation protocol, as they will adhere very rapidly to the support. Furthermore, culture conditions selected to favor CPE cells do not support their in vitro proliferation.

23.2.2.2 Assessing Choroidal Differentiation

Investigation of tissue-specific features and functions of the choroidal epithelium is required to define the degree of choroidal differentiation achieved in vitro by the CPE cells. In addition, markers that are exclusive for the epithelial monolayer within the choroidal tissue provide further indication of epithelial purity when homogeneously expressed in cultured cells. The specific differentiation products can be assessed by biochemical, immunochemical, or molecular biological techniques in reference to their characteristics in the intact tissue.

23.2.2.2.1 Transthyretin

Transthyretin (TTR) is a carrier protein required for the transport of water insoluble thyroid hormones in biological fluids (see Chapter 11). In mammals, its synthesis is restricted to very few tissues: liver, retinal pigment epithelium, and within the brain to the meninges and the choroid plexus epithelium (Cavallaro et al., 1990; Blay et al., 1994). TTR expression in the hippocampus has been recently evidenced by microarray analysis and RT-PCR. However, this data requires confirmation using in situ hybridization, since the authors indicated that the individual hippocampus samples contained choroid plexus (Puskas et al., 2003). TTR is thus considered a highly specialized feature of CPs, and TTR mRNA is detected as early as day 12.5 of gestation in the mouse when this tissue starts differentiating from the neural ectoderm. The level increases 7- to 8-fold up to birth and thereafter varies only slightly (Thomas et al., 1992). Studies of protein secretion in adult sheep choroid plexus have shown that at least 80% of newly synthesized TTR is secreted into the cerebrospinal fluid through the apical surface of the epithelium (Schreiber et al., 1990).

With the exception of one report that revealed a loss of TTR immunoreactivity by CPE cells prepared from adult sheep after three to four days in culture (Holm et al., 1994), all CPE cells in primary culture tested for TTR demonstrated the maintenance of the choroidal specific expression. TTR mRNA was detected in fetal mouse CPE cells forming vesicles in Matrigel over the six-day period of culture (Thomas et al., 1992). The upregulation occurring in vivo between day 13 of gestation and birth, however, was not reproduced in vitro, suggesting the absence of regulatory factors, possibly of mesenchymal origin, in this cellular model. TTR mRNA or immunoreactive protein was also detected in rat CPE cell monolayers after 7 to 10 days of culture (Nilsson et al., 1996; Zheng et al., 1998; Strazielle and Ghersi-Egea, 1999). Newly synthesized TTR was immunoprecipitated from rat cells cultured on solid support, as well as in their conditioned medium, thus demonstrating that both synthesis and secretion of the protein persist during culture (Southwell et al., 1993; Zheng et al., 1999). Evidence for TTR secretion in conditioned medium of porcine cells was also obtained by Western blot analysis (Gath et al., 1997).

However, the in vivo polarity of TTR secretion is apparently not maintained in vitro. In mouse embryo epithelial cell vesicles prepared in Matrigel, equal amounts of TTR were immunoprecipitated from the intravesicular and extravesicular fluids despite a highly polarized morphology of the epithelial cells and the occurrence of polarized patterns of secretion observed for other proteins (Thomas et al., 1992). A nonpolarized TTR secretion was also observed in monolayers of rat CPE cells, contrasting with a preferentially apical secretion of the newly synthesized total proteins (apical to basolateral ratio of 2) (Southwell et al., 1993). Although not quantified, an apical polarity of TTR secretion was demonstrated by Western blot analysis of the media of porcine CPE cells cultured in serum-containing medium (Gath et al., 1997). Yet this result was not reproduced thereafter (Hakvoort et al., 1998b). Whether secretion is not yet polarized at early stages of fetal development or the normal pathways of TTR secretion are affected during culture has not been further investigated.

23.2.2.2.2 Na^+,K^+-ATPase

Na^+,K^+-ATPase, which maintains a Na^+ gradient across the apical cell membrane, is involved in secretory and absorptive processes in epithelial cells and plays a major role in the highly differentiated function of CP epithelium that is CSF production (see Chapter 6). This pump displays a specific distribution in the CP and retinal pigment epithelia, where it localizes to the apical membrane of the cells, which contrasts with other epithelia, in which it is found on the basolateral membrane. Peraldi-Roux and colleagues (1990) studied the distribution of Na^+,K^+-ATPase by ultrastructural immunochemistry in mouse cultures enriched in choroidal epithelial cells. They observed a high number of gold particles restricted to the surface of the apical microvilli. Na^+,K^+-ATPase distribution was also investigated by fluorescent immunocytochemistry in cultured cells prepared from rat and porcine CPs (Gath et al., 1997; Villalobos et al., 1997). The apical distribution of the pump was assessed in both models, providing evidence of choroidal differentiation and also of polarization in the CPE cells in culture.

23.2.2.2.3 Drug-Metabolizing Enzymes

Besides TTR, CPs share another specificity with the liver, in that they have a high capacity for drug metabolism, involving hydrolysis, reduction, or conjugation enzymes. Activities for several of these enzymes were measured in rat CPE cells in comparison to freshly isolated tissue. In contrast to cultured hepatocytes in which these enzymes are rapidly downregulated, the choroidal cells in culture for seven to nine days preserved all enzymatic activities at levels comparable to those determined in the initial choroidal tissue (Strazielle and Ghersi-Egea, 1999).

23.2.2.2.4 Insulin-Like Growth Factor-II and Insulin-Like Growth Factor Binding Protein-2

Insulin-like growth factor-II (IGF-II) and insulin-like growth factor binding protein-2 are essential for brain development and are widely expressed in this organ during fetal life. Their expression becomes more restricted in the adult brain and is limited to discrete regions, including the CP epithelium. The presence of these two choroidal markers in sheep CPE cells was demonstrated at the mRNA level by Northern blot, and secreted proteins were revealed in the conditioned medium of the cells on day 8 of culture (Holm et al., 1994). Interestingly, the same sheep cells had lost TTR immunoreactivity within a few days of culture (see above), indicating that different mechanisms regulate the synthesis of these various choroidal proteins.

23.2.2.2.5 Transferrin

Transferrin, a carrier protein involved in iron transport, is synthesized at high levels and secreted by the CP epithelium. In one of the early works, Tsutsumi and colleagues (1989) demonstrated that rat CPE cells cultured on solid supports synthesize and secrete this protein. It should be noted, however, that choroidal secretion of transferrin is species specific and may not constitute a criteria of choice for cell models established from other species (Tu et al., 1991). In particular, its mRNA was not detected in pig, sheep, cow, or human CPs.

23.2.2.3 Monitoring the Diffusion Barrier Properties

CPE cells function as a diffusion barrier for electrolytes and macromolecules requires apically located tight junctions (TJs) (*Zonula occludens*) to seal the paracellular pathway between apposing membranes of adjacent cells. Analyzing the diffusion barrier properties of CPE cells cultured in bicameral devices can be approached structurally by investigating the reconstitution of TJs and functionally through the measurement of transepithelial resistance and flux of inert paracellular markers.

23.2.2.3.1 Structural Investigations

TJs are formed by a network of branching and anastomizing pairs of strands contributed by two adjacent cells. Major progress has been made recently concerning the identification of the molecular components of the TJ and led to the identification of occludin and claudins as the integral constituents of TJ strands. Intracellularly, these transmembrane proteins interact with the cytoskeleton actin through cytoplasmic TJ-associated proteins such as the zonula occludens (ZO) proteins. Another integral membrane protein with a single spanning domain, named JAM, was shown to localize at TJs. However, unlike occludin and claudins, this protein has no ability to reconstitute TJ strands in fibroblasts (for a review of TJ structure and proteins, see Tsukita and Furuse, 2000, 2002; and Chapter 3, this volume). Claudins represent a multigene family of more than 24 members, several of which are commonly coexpressed in single cells. They not only are structural elements of TJ strands, but have also been identified as functional components of simple epithelium TJs. Different claudin species associate in a heterotypic manner within individual strands, but can interact as hetero- or homodimers between adjacent strands, and the combination and mixing ratios of claudin species appear to be important factors that determine the tightness of individual strands and their ion selectivity (Furuse et al., 1999; Tsukita and Furuse, 2002).

Investigation of the molecular composition of TJs in rat and mice CPs first indicated that occludin and ZO-1 are present in the epithelial junctions (Gath et al., 1997; Lippoldt et al., 2000a, 2000b). As for claudins, at least four different species are expressed, namely, claudin 1 or 3 or both (the antibody from Zymed originally developed against claudin 1 turned out to crossreact with claudin 3), claudin 2, claudin 5, and claudin 11 (Lippoldt et al., 2000b, Wolburg et al., 2001). Whereas claudins 1, 2, and 11 display a distribution restricted to the epithelial TJ, claudin 5 immunoreactivity is diffusely distributed in the vicinity of the junctional region. Expression and correct distribution of these TJ components have been demonstrated in rat CPE cells cultured on permeable membranes, for claudin 1/3, claudin 2 and for occludin, all of which formed honeycomb-shaped immunoreactivity patterns comparable to those in vivo (see Figure 23.2A,B) (Strazielle and Ghersi-Egea, 1999; Strazielle et al., 2003a). In porcine CPE cells in culture, ZO-1 immunoreactivity was observed with a similar pattern, and freeze-fracture studies confirmed the presence of TJ strands in these cells (Gath et al., 1997). Based on these results, TJ protein expression and TJ formation appear to be preserved in cultured cells.

Interestingly, the disaggregation of TJs after trypsin treatment and their in vitro reassembling during culture had been investigated much earlier in detail using the freeze etching technique (Dermietzel et al., 1977). This study was performed on CPE cells isolated from 11-day-old chick embryos, which at this stage display mature junctional complexes with fully developed TJs. Cells were cultured for up to seven days in 10% serum-containing medium. Trypsination induced first a degradation of the zonula occludens, via fragmentation of the junctional strands. While partial lysosomal digestion of TJ remnants occurred during the first couple of days of culture, other fragments of the former zonula occludens served as nucleation sites for the reconstruction of tight junctions. On the seventh day of culture, choroid plexus epithelial cells revealed well-differentiated junctional complexes similar to those in the starting tissue, displaying in particular an identical degree of branching in the strand network. The time course of in vitro reaggregation followed a pattern comparable to that of the in vivo formation of TJs studied in parallel in the embryo between the fifth and eleventh days.

23.2.2.3.2 *Functional Investigations*

While TJ protein expression and continuous pericellular distribution are prerequisite to recreate the diffusion barrier in cultured CPE cells, the latter needs be assessed

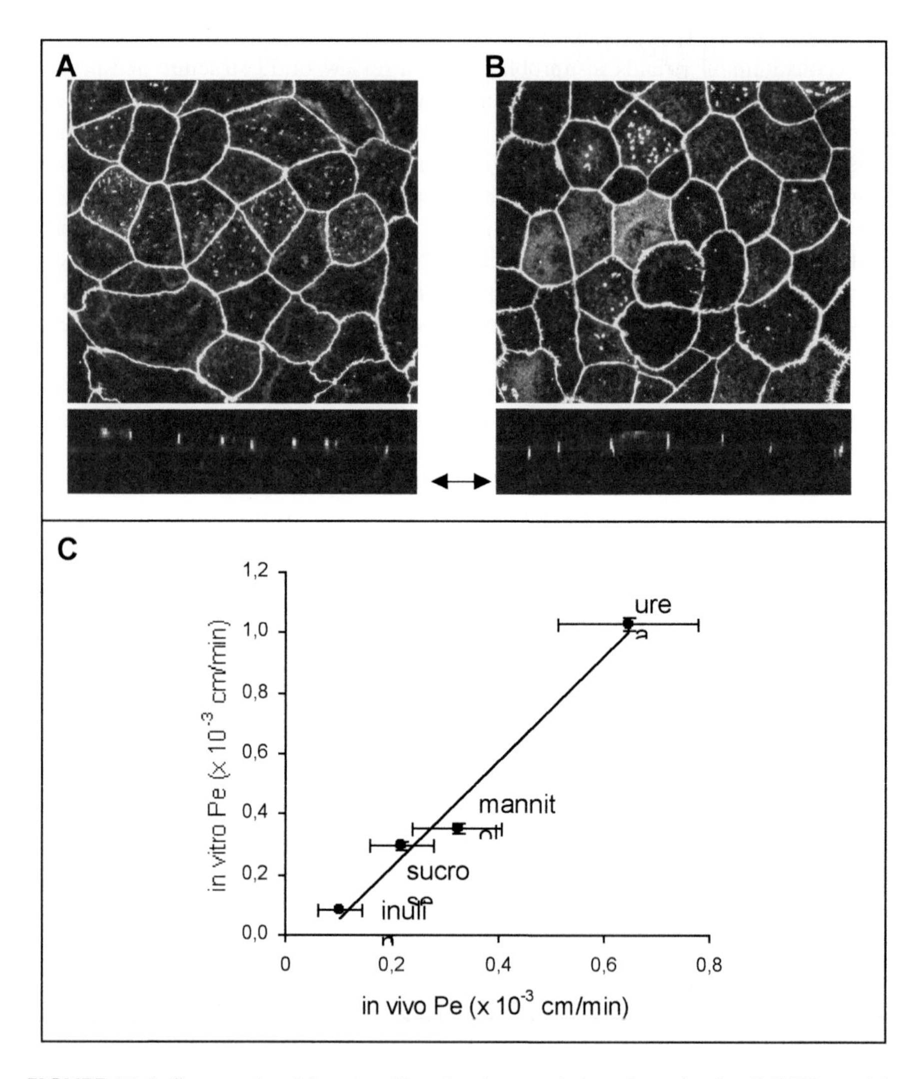

FIGURE 23.2 Structural and functional barrier characteristics of a rat in vitro BCSFB model. Immunocytochemical localization of TJ-associated proteins claudin 1 (A) and claudin 2 (B) in confluent cell monolayers, by confocal microscopy. The upper parts show the continuous circumferential distribution of both proteins when observed in the xy plan, and the lower parts show their apical (ap) localization when observed in the xz plan. The arrow indicates the position of the filter. (C) Correlation between in vitro and in vivo choroid plexus epithelium permeability coefficients for low-permeability polar molecules. The in vivo Pe values (mean ± SEM; n = 6–54) are the overall permeability coefficients measured across the rabbit choroidal epithelium by Welch and Sadler (1966). In vitro Pe values (mean ± SEM; n = 3–4) were measured on cell monolayers. Straight line: linear regression analysis; r = 0.98. (From Strazielle et al., 2003a.)

further by functional studies. This obviously will require the use of permeable filters and is particularly essential for subsequent transepithelial transport studies. It is classically achieved by evaluating certain criteria, namely, transepithelial electrical resistance (TEER) and the flux of inert paracellular markers across the cell monolayer, both of which should ideally be comparable to values measured in vivo, when available. It is worth mentioning that these two parameters are not equivalent in terms of their physiological significance and should not be used in an indifferent manner. While the former reflects the passive conductance of the TJ to small inorganic electrolytes, the latter represents the permeability of the cells to organic compounds with much higher molecular weights. They are influenced by different factors (see below), and consequently neither is strictly correlated nor varies concomitantly in response to cell treatment. Thus, the changes elicited by modifying the culture conditions on porcine CPE cells were of different magnitudes for the TEER and for the permeabilities to different paracellular tracers (Hakvoort et al., 1998b). Even more striking is the 40-fold decrease in TEER occurring in the absence of any effect on the permeability to dextran in MDCK cells upon exogenous claudin 2 expression (see next section).

23.2.2.3.3 In Vitro Transepithelial Electrical Resistance of the BCSFB—Does It Reflect In Vivo Resistance?

Measurement of the TEER of cell monolayers grown on permeable filters, using alternative current techniques, has become a nondestructive routine method for monitoring the growth of cell cultures. TEER measurement on CPE cell monolayers has been performed by impedance spectroscopy (Gath et al., 1997) or more often with commercially available resistance meter instruments, i.e., the Millicell-ERS™ or EVOM™ equipped with chopstick electrodes, which are widely used for their convenience. The more recent Endohm™ chambers, which provide a more uniform current and improved accuracy, may be preferred to the chopstick device. The resistance should be measured in parallel on blank filters (coated with matrix, but without cells) and subtracted from the resistance value of the cell-covered filter in order to determine the resistance of the cell monolayer proper. It should be noted that the electrical resistance is inversely proportional to the surface area covered by the cells. This latter is taken into account by calculating a resistance × surface area product, in $\Omega.cm^2$ (and not a resistance to surface area quotient, in $\Omega.cm^2$, as may sometimes be reported in the cell culture literature).

TEER values obtained for CPE cells cultured in serum-containing medium have been reported by several groups, and whichever the animal species from which the CP tissue originated, these values varied between 100 and 200 $\Omega.cm^2$ (Southwell et al., 1993; Gath et al., 1997; Ramanathan et al., 1996; Strazielle and Ghersi-Egea, 1999; Shu et al., 2002; Zheng et al., 1998). Comparison of this in vitro data with in vivo values is made difficult by the paucity of the latter, at least, for mammalian CPs. Indeed, for technical reasons, most TEER values have been determined on amphibian or elasmobranch posterior CPs. Electrophysiological studies revealed relatively low resistances in bullfrog (170 to 184 $\Omega.cm^2$), in sharks, and ray (100 $\Omega.cm^2$), but also in the cat (159 $\Omega.cm^2$) (Wright, 1972; Welch and Araki, 1975; Saito and Wright, 1983). This data designated the CP epithelium as a “leaky” epithelium, in contrast with “tight” epithelia such as the urinary bladder epithelium, which displays values of 1200 to 2000 Ω.cm^2 (Claude and Goodenough, 1973). These low in vivo electrophysiological values have been a matter of controversy, as they did not correlate with the rather high number of strands (4 to 7) commonly reported in CP TJs of various animal species, and which classifies this tissue as a tight epithelium according to the criteria of Claude and Goodenough (1973). A careful examination of rat CP junctions by freeze fracture using complementary replicas confirmed the high number of junctional strands (around 7.5), but also showed significant discontinuities within these strands (van Deurs and Koehler, 1979). According to these authors, these discontinuities may represent the hydrated pores previously suggested by electrophysiological data (Wright, 1972) and account for the high ionic conductance of the choroidal junction. In addition, other examples illustrating the lack of correlation between the complexity of the strand network revealed by freeze fracture and the permeability of the tight junctions to ions have been reported (Martinez-Palomo and Erlij, 1975), suggesting at that time that other features not revealed by electron microscopy such as the biochemical composition of the strand particles actually control the epithelial permeability to ions. This notion was supported by the recent molecular identification of the TJ integral proteins and their functional characterization, indicating that claudins cannot only increase but also decrease the tightness of TJ strands, depending on their combination. A comparative analysis of claudins expressed in two MDCK cell lines, the high resistance cell strain MDCK I (over 12000 Ω.cm^2), and the low resistance cell strain MDCK II (around 200 Ω.cm^2), which yet display similar TJ strand networks by freeze-etching electron microscopy (Furuse et al., 2001), showed that claudin 2, present in MDCKII, was the only TJ protein differentiating the strains. Upon transfection of MDCK I cells with dog claudin 2 (at expression levels comparable to that observed in MDCK II), the high-resistance phenotype was conversed into a low-resistance one, without any visible effect on TJ strand number and organization or on the permeability of the monolayers to 4 and 40 kDa-dextrans. Given these results, the established expression of claudin 2 in the TJ of the CP epithelium strongly supports the low resistance measured in vivo and validate the low TEER values determined in the various in vitro models.

Nonetheless, in vitro TEER values are difficult to interpret, as accumulating evidence is now stressing that many different factors can affect the measurement, such as the temperature at the time of measurement, the filter material, its pore size, and pore density. It should be remembered that TEER values measured on permeable filters result from both paracellular and transcellular parallel pathways. While the latter is assumed to be of much higher resistance than the former, and thus to be negligible, the former itself consists of several components, i.e., the tight junction resistance proper in series with the lateral intercellular resistance, which again is assumed to be low and negligible. More recently, a previously underestimated source of paracellular resistance was reported, namely the resistance contributed by the cell-substrate space, which is not taken into account by subtracting the TEER value of blank filters. It can represent as much as 90% of the (low) TEER measured for fibroblasts grown on polycarbonate filters or half of the TEER value measured for the epithelial cell line MDCK (Lo et al., 1999). Cell-substrate contact resistance is all the more important as the distance between the cell and the substrate is smaller and has been shown to be influenced highly by the material constituting the support, explaining that MDCK II cultivated on polycarbonate can exhibit up to 500 $\Omega.cm^2$ while the same cells grown on other materials display a low resistance below 100 $\Omega.cm^2$ (reviewed in Lo et al., 1999).

Electric cell-substrate impedance sensing (ECIS) was developed as a method to measure junctional resistance independently from other resistances (transcellular or contributed by the cell substrate interactions) (for details, see Lo et al., 1999). Impedance studies performed with confluent porcine CPE cell layers grown on porous polycarbonate filters were recently compared with ECIS measurements performed on monolayers of the same cells grown on gold-film electrodes (Wegener et al., 2000). Values were respectively $128 \pm 7\ \Omega.cm^2$ for transepithelial resistance and $97 \pm 1\ \Omega.cm^2$ for intercellular junctional resistance, indicating that about 25% of the TEER measured on polycarbonate filters represents a resistance contributed by cell substrate adhesion. Furthermore, the variations of overall TEER induced by cAMP analogs were shown to result from changes in both the junctional resistance and the cell substrate resistance, emphasizing the difficulty to correlate strictly TJ ion permeability with experimental TEER values. This data indicates that (1) the TEER values determined in vitro on CPE cell monolayers, approximating 100 to 200 $\Omega.cm^2$, may in fact be slightly overestimated (by no more than 25%), depending on the plating support; (2) yet, they remain close to the in vivo TEER values; and (3) care should be taken when relating TEER values to the trans-junctional electrolyte permeation.

23.2.2.3.4 Flux of Paracellular Markers

Small polar inert compounds such as radiolabeled mannitol, sucrose, inulin, polyethylene glycols, or fluorescent dextrans are useful markers of the paracellular pathway and have been used to monitor the establishment of barrier properties of CPE cells cultured on permeable supports. Transfer is measured on both filters with cells and blank filters (without cells), enabling one to generate permeability coefficients (Pe) for the epithelial monolayer proper. It should be noted that flux data expressed as percent values of the initial amount in the donor compartment should be avoided unless full information related to the initial tracer concentration, the surface area available for exchange, and the fluid volumes on both sides of the cells at the time of the experiment are available, as this flux data will not otherwise be fully interpretable. In addition, a number of factors such as agitation or fluid balance can influence the permeability measurement and should be standardized (for experimental conditions and calculation details see Strazielle and Preston, 2003).

Cultured CPE cells from rat or rabbit were shown to act as significant barriers to paracellular markers. They induced a 7- to 14-fold reduction of mannitol flux by comparison to blank filters (Ramanathan et al., 1996; Shu et al., 2002). A Pe value of $0.21 \times 10^{-3}\ cm.min^{-1}$ was calculated for the rat cells (Shu et al., 2002). Typical in vitro Pe values for different extracellular markers, measured on rat CPE cell monolayers, are presented in Figure 23.2C. As previously mentioned, determining in vivo permeability coefficients that reflect solely the choroidal pathway is difficult, and very little data is available. However, using rabbit, Welch and Sadler (1966) determined the permeability coefficients for a number of solutes of different molecular weights (including the paracellular markers mannitol, sucrose, and inulin) from CSF into blood across the CP, measuring their concentration in the cannulated choroidal vein when the solutes were included in the fluid bathing the exposed plexus. The permeability coefficients determined in vitro on the rat BCSFB model for this set of compounds not only were correlated to the in vivo data, they were also close to these values, indicating that the paracellular permeability of CPE cells reflects the in vivo pathway (see Figure 23.2C). Urea, whose transcellular diffusion is not negligible, accordingly displayed higher Pe values both in vitro and in vivo. It is conceivable that the cell-substrate contact space may influence to some extent the flux of paracellular tracers across cell-covered filters. Using rat CPE cells seeded on different membranes, we observed only slight differences in the clearance of sucrose between polycarbonate (known to promote higher cell-substrate resistance) (see Lo et al., 1999) and other membrane materials (N. Strazielle and J.F. Ghersi-Egea, unpublished observations). Thus, whichever type of insert is used, the apparent paracellular permeability estimated in vitro on these rat CPE cell monolayers will be comparable to that of the choroidal epithelium in vivo.

Besides their interest in assessing the reproducible barrier properties of successive cell preparations, paracellular markers may be included in transfer experiments for other purposes. The flux rate of a paracellular marker, chosen appropriately to match the size of the compound of interest, will provide an indication of the extent of transcellular transfer contribution to the overall permeability determined for this compound. They can also be used to monitor possible adverse effects occurring during transport experiments. Such effects may be caused (1) by the studied compound itself, or by a transporter inhibitor, (2) by an additive required to solubilize these molecules, or (3) by a specific medium required to investigate transmembrane transport or secretion mechanisms (e.g., a change in the ionic balance). Because a possible adverse effect will depend on the concentration of the toxic molecule, special attention must be paid to experiments in which high concentrations are needed to determine kinetic parameters or to mechanistic studies involving high concentrations of transporter inhibitors. One should also be cautious in carrying out transfer experiments for a long period without monitoring the paracellular pathway, as a deleterious compound may impair the barrier properties only after a time lag.

23.2.2.4 Other Choroidal Functions

CPs provide the brain with various micronutrients (e.g., amino acids, nucleosides, vitamins). Transport systems for these nutrients have been identified in the choroidal tissue, and for some a subcellular localization has been specified. Expression and functionality of these membrane proteins have been investigated in order to define further the degree of differentiation and polarization of the CPE cells.

23.2.2.4.1 Amino Acid Transport

As shown by uptake experiments on cells cultured on solid supports, both rabbit and rat CPE cells retained the Na^+-coupled transport system, known to mediate the accumulation of the neutral amino acid proline (Ramanathan et al., 1996; Villalobos et al., 1997). Primary cultures of rat cells also displayed a saturable transcellular transport process for phenylalanine from blood to CSF (Strazielle and Ghersi-Egea, 1999).

23.2.2.4.2 Nucleoside Transport

Nucleoside transport at the CP is mediated by both Na^+-dependent concentrative and Na^+-independent equilibrative systems. The Na^+-stimulated uptake of formycin-B or thymidine was not observed in primary cultures of rabbit CPE cells grown on solid supports (Ramanathan et al., 1996). A saturable blood-to-CSF transport of thymidine was, however, demonstrated in rat CPE cell monolayers grown on porous filters (Strazielle et al., 2003a). This difference may be related to the cell culture set-up. Cells from rabbit and rat may not maintain their specificities to the same extent, but it is also conceivable that cells grown on permeable supports achieved a more complete differentiation than cells grown on plastic. Alternatively, the nucleoside, added at the apical membrane of the rabbit cells cultured on solid supports, may not have had access to the relevant transporter, as suggested by these authors.

23.2.2.4.3 Glucose Transport

The expression and basolateral distribution of the Na^+-independent glucose transporter Glut-1 were demonstrated by confocal microscopy on rat CPE cell cultures (Villalobos et al., 1997).

In summary the studies reviewed above represent examples of some choroidal features only. This type of approach, aimed at characterizing cultured cells and verifying that they retain the specific properties of the cells in vivo, could be extended to a large number of other transport systems, as well as to receptors, enzymes, secreted proteins or polypeptides, adhesion molecules, and so on. Additional validation criteria should be chosen accordingly to one's personal experimental goal.

23.2.2.5 Effects of Serum Withdrawal on Cell Differentiation

While serum, classically used at a 10% concentration to supplement the culture medium, has proved crucial for CPE cell adhesion and proliferation in the days following plating, it has also been shown to impede cell differentiation and the establishment of certain specific properties of the CP epithelium in a porcine model. Following seven to nine days of propagation in the presence of serum, culture in serum-free medium (SFM) for a few days had morphological and functional consequences on porcine CPE cells (Hakvoort et al., 1998a,b). It induced, in particular, (1) a major increase in TEER (up to 1700 $\Omega.cm^2$), (2) a decrease in the paracellular flux of low molecular weight dextrans (i.e., a 4-fold reduction in the permeability coefficient of 4 kDa dextran), (3) a secretion of fluid into the apical chamber, and (4) a polarized apical protein secretion, illustrated in particular for TTR. According to the authors, the permeability changes elicited by withdrawing the serum are required in order to observe active transport processes visualized by the accumulation of phenol red (PR; phenolsulfonphthalein) contained in the culture medium, uphill from its building concentration gradient. Active transport of PR from the CSF to the blood was demonstrated in 1961 by Papenheimer using a sophisticated technique of ventriculocisternal perfusion in goats, and accordingly PR was transported across porcine CPE cells from the apical into the basolateral chamber of the cell culture bicameral device. Up to 70% of the compound initially present in the apical fluid (volume of 1.5 ml) was transported in 48 hours (Hakvoort et al., 1998b), which represents a clearance of 1050 μL for a cell surface of 4.7 cm^2.

These serum-dependent effects are in contrast with data obtained on rat CPE cell cultures by several groups. Similar to serum-deprived porcine cells, rat cells cultured for 10 to 12 days in medium supplemented with 10% serum displayed a clear polarity in the overall protein secretion, which occurred predominantly at the apical pole of the epithelial cells (apical-to-basolateral ratio of 2), although TTR was not following the rule, as it did for SFM-cultured porcine cells (Southwell et al., 1993). However, because polarized secretion of TTR by serum-fed porcine cells had been reported previously (Gath et al., 1997), the effect of different culture media on cells with respect to the in vitro polarity of TTR secretion remains to be clarified. PR active transport was also observed in rat CPE cell cultures grown for eight days with serum (see Figure 23.3). In this experiment, 188 μL were cleared by the 0.33 cm^2 surface area of cells over a 48-hour period, a volume higher than that reported above for porcine cells. Because on both porcine and murine cells PR active transport is not linear over the 2-day period, we compared the initial rates of transport. The initial clearance rates were respectively 31 $\mu L.cm^{-2}.h^{-1}$ for serum-starved porcine cells (deduced from Hakvoort et al., 1998a) (2-h time point at the non saturating concentration of 10 μM), and 20 ± 1 $\mu L.cm^{-2}.h^{-1}$ for serum-fed rat cells (calculated from three independent cell preparations, including the data presented in Figure 23.3, for a PR concentration of 22 μM). It should be noted that the latter value is underestimated since we observed that diffusion of PR across laminin-coated filters without cells is restricted by 30% in the presence of 10% serum, compared to serum-free medium as a result of PR binding to serum proteins (unpublished observations). In addition, the rate of clearance measured in rat cells was not maximal at the 22 μM concentration, as the apparent affinity constant of PR for the active apical to basolateral transport process in these cells was lower than the one reported for the porcine cells by one order of magnitude (N. Strazielle and J.F. Ghersi-Egea, unpublished observations). Consequently, as far as we can tell from these two criteria—polarity of total protein secretion and active transport processes—serum does not prevent differentiation of rat CPE cells as it does in porcine CPE cells. Furthermore, when active transport of PR by serum-fed rat cells was determined in serum-free medium over a 48-hour period, (i.e., a period sufficient to alleviate the inhibitory effect of serum in porcine cultures), only a small increase was observed, which was accounted for by the protein-binding component (see Figure 23.3 and Strazielle et al., 2003a). More importantly, the data collected from rat CPE cells demonstrate that investigation of choroidal active transport processes is compatible with the low, yet physiological, TEER of these primary cultures. It would be of interest to analyze the fluid secretion in the rat BCSFB models following the method described for the porcine cells (Hakvoort et al., 1998b).

The discrepancies between porcine and rat cell data could be related to the serum proper, as obviously different batches were used by the various investigator groups. Yet this is unlikely because the dedifferentiating effects observed on porcine cells could be elicited by multiple sera, originating from different animal species. Species-specific requirements could provide another explanation. Serum also displayed inhibitory effects on TJ formation in cultured retinal epithelial cells from rat

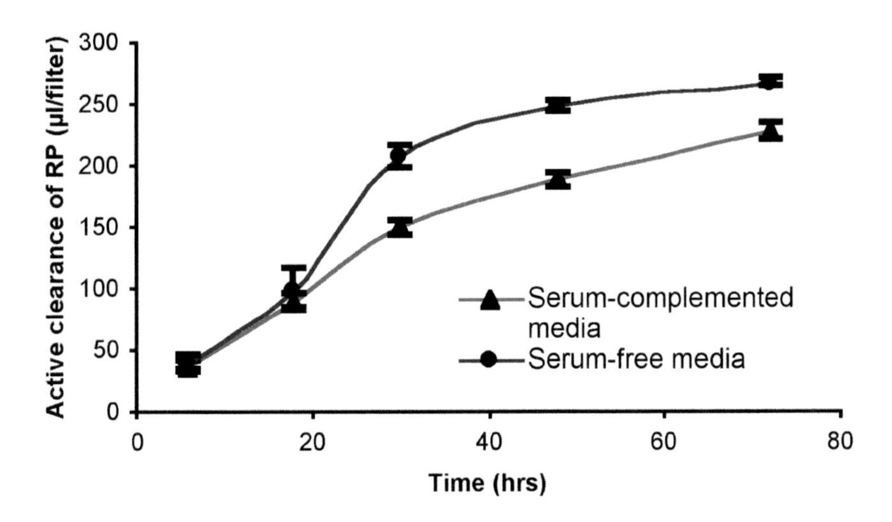

FIGURE 23.3 Apical to basolateral active clearance of RP across the choroidal epithelium in vitro: comparison between serum-free and serum-supplemented media. Following seven days of culture in the presence of serum, cell monolayers grown on Transwell Clear filters (surface area 0.33 cm^2) were incubated in fresh media, with or without serum, for 6, 18, 30, 48, or 72 hours. RP concentrations were measured by spectrophotometry in both apical and basolateral fluids, and active clearance was calculated as described in Strazielle et al., 2003b. Values are expressed as mean ± SD of three different monolayers.

(Chang et al., 1997), which would argue against this hypothesis. However, the preliminary characterization of the serum components involved in the modulation of retinal and porcine TJs indicated that they are different (Chang et al., 1997; Hakvoort et al., 1998b). In porcine cells, these characterization studies designated epidermal growth factor (EGF) as one possible culprit (Hakvoort et al., 1998b). Involvement of this growth factor in regulating TJ protein expression and TEER has been demonstrated in MDCK cells (Singh and Harris, 2004). However, in contrast to porcine cells, EGF induced a threefold increase in TEER, increased synthesis, and a membrane redistribution of claudins. EGF receptor-mediated responses are pleiotropic, and EGF signaling includes differentiation as well as dedifferentiation even in the same cell, depending on the context, i.e., cell density, type of matrix, and other factors (Wells, 1999). One could speculate that EGF exerts opposite effects on rat and porcine CPE cells, and this hypothesis is all the more conceivable as exogenous EGF was supplemented in the medium fed to both rat cell primary cultures (Southwell et al., 1993; Strazielle and Ghersi-Egea, 1999). Alternatively, since EGF alone did not fully reproduce the serum effects on the TEER of porcine cells, other factors are likely to be involved and may be species-specific. Finally, an important elevation in the cell-substrate space resistance may also contribute to the increase in TEER and decrease in 4 kDa dextran flux observed in serum-starved porcine cells. TEER measurement using the ECIS method should shed some light on the porcine-specific effects of serum.

23.3 CHOROIDAL EPITHELIAL CELL LINES

23.3.1 Spontaneous, Papilloma, and Carcinoma Cell Lines

So-called sheep choroid plexus (SCP) cells are being widely cultured in virology laboratories to support the propagation of Maedi-visna virus strains. These cells are produced from minced CPs, grown in medium supplemented with 14 to 20% serum. After a few subcultures, stocks of cells are stored in liquid nitrogen. They can then be further propagated by repeated subculture (up to 120 passages) (Thormar, 1963; Narayan et al., 1977; Torchio and Trowbridge, 1977). They display a fibroblastic appearance (Torchio and Trowbridge, 1977) and, accordingly, are also referred to as sheep or ovine choroid plexus fibroblasts (Chebloune et al., 1996). Although the mesenchymal origin of these SCP cells was not directly demonstrated, it is, in our opinion, more likely to be the case, given the protocol of cell isolation and maintenance.

The American Type Culture Collection (ATCC) offers a cell line also designated as SCP (CRL 1700). The product information sheet only states that it has been prepared from brain choroid plexus of *Ovis aries* and is susceptible to visna virus, strain 1514. Characterization of this cell line is scarce, and ATCC does not provide any information about its generation and phenotype. Indications in the current literature for its choroidal epithelial origin have been provided solely by the demonstration of mRNAs for TTR and IGF-II (Gee et al., 1993). These authors also presented evidence for mRNAs for neuropeptide-processing enzymes and neurosecretory proteins. The ATCC SCP cell line has been used as a model of choroidal epithelial cells by several groups (Thornwall et al., 1995; Angelova et al., 1996; Dickinson et al., 1998; Albert et al., 1999). It was shown in particular to possess mRNAs for various receptors whose presence in the CP tissue had been suggested by binding studies, such as growth hormone receptor and endothelin-1 receptor. From this little published data, the SCP cell line might represent a potential tool, at least for the study of certain neuroendocrine functions of the CP or some aspects of the regulation of CSF secretion. However, further characterization of the cell line would be required for any other specific study. The establishment of junctional complexes and diffusion barrier properties in particular remains undetermined.

Other cell lines were generated from human choroid plexus carcinoma and papilloma. Cells migrating from tissue fragments of a 4th ventricle choroid plexus papilloma resected from a 28-year-old male patient were shown to form a monolayer with a pavement-like appearance on plastic (Nakashima et al., 1983). On day 115 of culture, the cells still displayed ultrastructural features similar to that of the papilloma epithelial cells and retained in particular microvilli, cilia, and interdigitations. Interestingly, some spider-like cells present in the papilloma in situ in which they located mainly in the stroma and to a lesser extent between epithelial cells and on their apical surface, were also observed in the cultured cells on day 10. They showed an intense phagocytic activity for latex beads and stained positive with macrophage-specific dyes. Whether these spindle-like cells or macrophages survived or disappeared during further subculture was not reported. A choroid plexus carcinoma cell line was established from the resected tumor of a two-month-old boy. Although not characterized in details, it was shown to produce and secrete adrenomedullin, a potent vasodilatator peptide with possible neuromodulator functions, which has been detected by immunocytochemistry in human CP (Takahashi et al., 1997).

23.3.2 Laboratory Engineered Immortalized Cell Lines

Two approaches to cell immortalization have been applied to generate CPE cell lines. The first approach is to introduce the immortalizing gene by DNA transfection or by viral transduction in primary cultures. Both the small and large tumor (T) antigens of the simian virus (SV) 40 genome were transfected in primary cultures of rat CPE cells (Zheng and Zhao, 2002b). Upon cytotoxic selection and clone characterization, one cell line was generated and called Z310. It grows with a doubling time of 20 to 22 hours, and up to the 110th subculture it has not displayed any variability in TTR expression or growth rate.

The second approach takes advantage of transgenic mice for oncogenes, from which cell lines can be obtained by direct derivation of CP tissue. Enjoji and collaborators (1995) reported the generation of mouse CP cell lines from transgenic mice harboring the viral SV40 large T oncogene under the transcriptional control of an intronic enhancer region from the human immunoglobulin heavy-chain gene, which results in the restricted expression of the oncogene in lymphocytes and unexpectedly in CP. Two morphologically distinct cell lines, ECPC-3 and ECPC-4, have been cloned from the cultures of the resected CP carcinoma that developed in those mice (Enjoji et al., 1995, 1996). These cells proliferate with a doubling time of seven to nine hours and, at the time of report, had been maintained in culture for over a year with good stability. Recently, five rat CPE cell lines were produced from a transgenic rat harboring a temperature-sensitive mutant of the SV40 large T antigen, expressed ubiquitously under the control of its own promoter (Kitazawa et al., 2001). CPE cells were prepared using a trypsin-based protocol developed for primary cultures, and five clones, TR-CSFB 1 to 5, were selected as being epithelial clones. At the permissive temperature of 33°C, the TR-CSFB cell lines grow with a doubling time of 35 to 40 hours, and contact inhibition is observed after seven days.

For Z310 and TR-CSFB cells, clone selection was based on the typical polygonal shape of epithelial cells and pavement-like appearance of cell sheets. Yet, assessing the epithelial nature of immortalized cells using more definite criteria is as critical as it is for cells in primary culture. This is illustrated by the RCP cell line generated using a very similar strategy, i.e., primary cultures of CPE cells transduced with the TSOri minus adenovirus harboring the SV40 large T antigen (Battle et al., 2000). Patch-clamp experiments and immunopositivity for several endothelial markers indicated that the RPC cells are actually of endothelial origin.

Given the recent development of the CPE cell lines described above, extensive characterization data is not yet available. Choroidal tissue specificity was used as the first criteria, and all the epithelial cell lines were shown to express TTR (mRNA or protein) or to secrete this protein in the culture medium, which confirms their epithelial origin. Further evidence for choroidal differentiated status was provided by the expression of transferrin mRNA (Zheng and Zhao, 2002b) or by that of

α_2-macroglobulin (Enjoji et al., 1996). Immunopositivity for Na^+, K^+-ATPase (predominant in the apical domain of the cells) and saturable uptake processes for amino acids (system A) were reported for the TR-CSFB3 cell line (Kitazawa et al., 2001). mRNA expression of the organic anion transport protein Slc21A7 (oatp3) and the multidrug resistance associated protein ABCC1 (Mrp1), as well as a Slc21-mediated uptake process, have been demonstrated in one of the rat cell lines (Ohtsuki et al., 2003). This data indicates that the immortalized CPE cell lines retain differentiated features of the CP epithelium. However, they apparently do not maintain to a great extent the structural diffusion barrier properties that would be mandatory in order to study transepithelial transport processes and polarized phenomena. Indeed Terasaki et al, (2003) stated that TR-BCSFB cells cultured on permeable filters will not allow to generate reliable data in transcellular transport studies, unless prior reconstruction of the TJs is achieved. Occludin expression was observed in the Z310 cell line (J. Szmydynger-Chodobska and A. Chodobski, personal communication), and when seeded on permeable filters, these cells exerted some restriction to the diffusion of sucrose in comparison to filters without cells. This barrier effect was, however, limited (1.8-fold reduction in the flux of this compound estimated from a 2-h transfer experiment) (Zheng and Zhao, 2002b) and less than that of CPE cells in primary culture, which yield a 7- to 14-fold reduction in the flux of mannitol or sucrose through permeable filters (Ramanathan et al., 1996; Shu et al., 2002; and personal data). Although its expression was not investigated, the SV40 small T antigen transfected in Z310 cells may contribute to this relative TJ leakiness. In MDCK cells, this oncogene induced a deregulation and disorganization of the TJ proteins and actin cytoskeleton as well as an inhibition of TJ formation (Nunbhakdi-Craig et al., 2003) as a consequence of its interaction with the serine/threonine protein phosphatase A2. These acute changes were associated with an increased permeability of paracellular tracers.

23.4 CONTRIBUTIONS OF IN VITRO MODELS OF THE BCSFB TO THE UNDERSTANDING OF CP PHYSIOLOGY

An important purpose of establishing in vitro CPE cell cultures is to obtain reliable cellular models of the BCSFB which are complementary to more physiologic, but less mechanistic, methodological approaches in order to investigate CP functions. As said before, isolated perfused CP techniques or in vivo experiments based on CSF sampling are difficult to interpret or set up. Isolated CP remains a powerful model which can be used in a number of different ways, with the drawbacks that the experiments with the isolated tissue need be performed in a limited time frame, polarity studies (i.e., separate access to apical and basolateral membranes) are difficult to set up, and only uptake, but no quantitative transcellular transport experiments, can be performed in most cases (with the exception of some possible quantitation of fluorescent substrate concentration by video confocal microscopy) (Breen et al., 2002). Many of the studies involving newly developed cellular models of the BCSFB were aimed at characterizing the models by comparison to known properties of the CP epithelium. This aspect has been discussed in the previous sections and will not be reviewed here. The BCSFB cellular models, however, have already proven useful for a better understanding of some of the CP functions related to choroidal protein secretion, transport and enzymatic characteristics, or receptor-mediated signaling/regulation mechanisms at the CPE cells. Examples given below may not be exhaustive, but aim at illustrating this contribution.

23.4.1 Secretion of Biologically Active Polypeptides or Transport Proteins by the Choroidal Epithelium

23.4.1.1 Transferrin

Using rat CPE cells cultured on plastic support, Tsutsumi and colleagues (1989) showed that a four-day treatment by serotonin led to an increase of transferrin synthesis and secretion, while a treatment by factors activating the adenylate cyclase pathway reduced transferrin secretion. This data suggested a central control of choroidal transferrin secretion via the neurotransmitter serotonin released into the CSF. The authors subsequently showed that the serotonin effect on transferrin secretion was mediated by the $5HT_{2C}$ (formerly $5HT_{1C}$) receptor, whose expression is maintained in cultured cells (Esterle and Sanders-Bush, 1992).

23.4.1.2 Insulin-Like Growth Factor II

Using rat CPE cells cultured on solid support, Nilsson and colleagues (1996) showed that cells cultured for 48 hours in presence of an antibody raised against the mitogenic and trophic polypeptide IGF-II divided significantly less than cells grown in the absence of the antibody, suggesting an autocrine role of IGF-II in the regulation of CPE cell growth during development or repair. This data needs to be reinterpreted in the light of the apparent very low figure of division occurring in vivo in the choroidal epithelium (with the exception of the stalk), even following ischemia-induced injury (Johanson et al., 2000).

23.4.1.3 Transthyretin

Southwell and colleagues (1993) were the first to use a sophisticated model based on cultivated rat CPE cells onto permeable filters and investigated the relationship between T4 transport across the BCSFB and TTR secretion. TTR secretion was maintained in the cultured cells, but was not polarized in the experimental model, that is, a similar amount of TTR was secreted at the apical and basolateral membrane of the reconstituted epithelium. This led, however, to a higher concentration in the apical compartment, because the apical volume was half of the basolateral volume. As a result, the authors showed that T4 partitioning across the epithelium was influenced by the secretion of TTR, the hormone concentration in the apical compartment being higher than in the basolateral one at steady state. This suggests that T4 extent of transfer from bloodborne TTR to brain will be influenced by choroidal TTR synthesis, secretion in the CSF, and diffusion in the brain. Very similar *in vitro* results have been generated by Zheng and colleagues in the course of characterizing a rat BCSFB model (Zheng et al., 1998).

Exposure of the organism to lead (Pb) results in the accumulation of this metal in the CP and in a decrease in TTR concentration in the CSF. The effect of Pb on TTR synthesis and secretion as well as on T4 partitioning across the epithelial monolayer has been investigated using a rat CPE cell culture (Zheng et al., 1999). The authors showed that Pb exposure induced a decrease in TTR secretion from cells grown on solid supports and observed in parallel a reduction in the CSF-favoring polarity of T4 partition across the cells cultured in a bicameral device. These results provide a link between Pb accumulation in CPs during Pb intoxication and the decrease in CSF-borne TTR and hint at one possible T4-related mechanism to explain mental retardation observed in children exposed to this metal.

23.4.1.4 Matrix Metalloproteases

Matrix metalloproteases (MMPs) catalyze the proteolytic cleavage of basal lamina components and thus are involved in extracellular matrix remodeling and in cellular migration. They also participate in the maturation by cleavage of various membrane receptors and ligands. A dysregulation of the balance between MMPs and their endogenous inhibitors appears to play an important role in the pathophysiology of neuroinflammation. Basal secretions of MMP-2 and MMP-9 by intact CPs of both lateral and 4th ventricles have been demonstrated and can be reproduced by cultured rat CPE cells (Strazielle et al., 2003b). A strong upregulation and a polarization of MMP secretion, especially MMP-9, were induced upon pro-inflammatory cytokine exposure of the cultured cells. This data shows that the CP can contribute at least partially to the increased levels of MMP-9 observed in the CSF in both neuroinflammatory diseases in humans and in experimental neuroinflammation. It also suggests that the basolaterally secreted MMPs can be of relevance to activated immune cell invasion.

23.4.1.5 Vasoactive Polypeptides

The secretion of two polypeptides with opposite effects on microcirculation, endothelin-1, and adrenomedullin was demonstrated in cultured cells from a human CP carcinoma, suggesting their possible involvement in the tumoral pathophysiology. This secretion was induced by a 24-hour treatment with proinflammatory cytokines (Takahashi et al., 1997, 1998), but the significance of this data for normal CP functions and for neuroinflammatory diseases remains to be investigated.

23.4.2 Function and Regulation of Solute Ligand Carriers (SLC) and Multidrug Resistance Proteins (ABC Transporters) at the CP

Functional BCSFB cellular models obtained by cultivating cells in a bicameral device offer the advantages of generating true transcellular transfer measurement data and vectorial information, and, not least, of studying nonradioactive compounds, as culture medium can be easily analyzed by HPLC or by other analytical procedures. Furthermore, unlike transfected cell systems, which generally express one particular transporter, the relative contribution of the different native coexisting choroidal transport proteins to the transport of a given compound can be approached. Although less informative, more conventional uptake measurements on CPE cells cultured on solid supports or on permeable filters have also generated some new information with regard to transport mechanisms and regulation at the CP. Examples based on these different approaches are described below.

23.4.2.1 Na^+-K^+-$2Cl^-$ Cotransport

The Na^+-K^+-$2Cl^-$ cotransporter NKCC1 (Slc12a2) is an important component of transepithelial inorganic ion transport. The transporter is highly expressed in the choroid plexus, where its precise function in CSF formation remains unclear. The apical localization of the transporter, demonstrated in vivo, was maintained in rat CPE cells cultured on permeable membranes. The bumetanide-sensitive uptake of $^{86}Rb^+$ (an isotope that can replace K^+) at the apical side was much higher than at the basolateral side of the epithelium (Plotkin et al., 1997). This confirms by a functional test the presence of Slc12a2 at the brush border membrane and suggests for this protein an inwardly directed CSF-to-cell transport function.

23.4.2.2 Taurin

Taurin is a neuromodulator β-amino acid with osmoregulatory properties. An adaptive regulation of CP taurine transporter was demonstrated by measuring taurine uptake in rabbit CPE cells seeded on solid supports and exposed for 48 hours to high or no taurine-containing medium. The meaning of this uptake in terms of transepithelial transport directionality could not be addressed using this approach (Ramanathan et al., 1997).

23.4.2.3 Organic Cations

The transport of organic cations has been investigated using cultured rat CPE cells seeded on solid supports. Villalobos and collaborators (1999) showed that choline, the precursor of the neurotransmitter acetylcholine, accumulated at very high levels in cells, suggesting that CP participates in the maintenance of the constant choline concentration observed in the CSF despite large variations in the plasma. The use of cell culture enabled these authors to show that the uptake of choline was sensitive to plasma membrane electrical potential. Using the same model, the uptake and movement of the fluorescent organic cation Quinacrine was followed within the living cells by conventional and confocal microscopy (Miller et al., 1999). The authors showed the presence of the cation in acidic vesicles moving within the cytoplasm and possibly releasing their contents at the basolateral membrane. Interaction with microtubules could be shown. Although such an approach has now been set up by the same group on isolated CPs, incorporating the vascular compartment in the studies (Breen et al., 2002), this elegant imaging method applied to cultured cells will remain useful if long-term cell treatment must be achieved, the obvious drawback being that it can be used only with fluorescent markers.

23.4.2.4 Glutamine

Glutamine is the most common amino acid in the CSF and has multiple physiological roles. In an attempt to investigate the possible involvement of system N transporters (SN) at the BCSFB in the regulation of glutamine concentration, Xiang and colleagues (2003) used rat CPE cells cultured on porous filters. They confirmed the presence of transcripts for SN1 and SN2 in the cells and ascertained, by differential (apical vs. basolateral) uptake, the apical distribution of SN-like Na^+-dependent glutamine transporters at the choroidal epithelium.

23.4.2.5 5-Aminolevulinic Acid and PEPT2 Substrates

The implication of PEPT2 in vectorial CSF-to-blood transfer has been shown by transcellular transfer measurements of the model dipeptide substrate glycylsarcosine (GlySar) across the same rat CPE cell system. The results showed clear vectorial transport in favor of the apical-to-basolateral direction. Affinity constant determination showed that the apical uptake of GlySar was characterized by a higher affinity than that of the basolateral uptake (Shu et al., 2002). 5-Aminolevulinic acid (ALA), a precursor of porphyrins and heme, is involved in neuropsychiatric symptoms caused by hereditary porphyrias. Differential uptake experiments on rat CPE cells cultured on porous filters showed a much greater saturable uptake at the apical versus the basolateral membrane (Ennis et al., 2003). This suggested that an apical transport, possibly mediated by PEPT2, known to accept ALA as substrate and present at the apical membrane of the choroidal epithelium, keeps ALA concentration in the CSF at a low level. Overall, the data indicates the potential role of choroidal PEPT2 not only in endogenous peptide removal from CSF, but also in the biodisposition of peptidomimetic drugs.

23.4.2.6 Drugs and Other Bioactive Compounds

The implication of the BCSFB in the central bioavailability of pharmacological compounds has been investigated using in vitro cellular models. One particular study focused on the movement of antiretroviral compounds across the reconstituted epithelium. In AIDS, early suppression of the viral load in the CNS is critical for the efficacy of antiretroviral therapy in order to prevent the emergence of a reservoir of resistant viral strains and brain impairment. Targeting the CSF itself is important to enable drugs to reach ventricular, meningeal, and perivascular macrophages, which are the main infected and replicative cells in the brain. Using an in vitro cellular model of the BCSFB coupled to HPLC analysis, a classification based on the relative ability of clinically used antiretroviral nucleoside analogs to distribute into the CSF through the choroidal epithelium has been produced (Strazielle et al., 2003a). In accord with its higher lipophilicity, azidothymidine had the highest influx rate. The data also confirmed previous results suggesting that its influx was independent of thymidine transporters. Most compounds tested had an apical-to-basolateral efflux rate higher than the basolateral-to-apical influx rate. This efflux was shown to be governed by an apical transport system belonging to the solute carrier Slc22 (former organic anion transporter OAT) subfamily and could be reversed by therapeutic concentrations of uricosuric compounds (Strazielle et al., 2003a). Thus, therapeutic strategies aimed at increasing CSF delivery or decreasing brain elimination of drugs can be assessed by this in vitro BCSFB model.

A molecule can move across a cell membrane by diffusion, facilitated (equilibrative) transport, or active transport. An elegant way of investigating whether the major component of a transport is a unidirectional, active process is to add the compound of interest at the same concentration to both compartments of CPE cells cultured on porous filters. The appearance of an imbalance in the concentration between both compartments over time will indicate the involvement of an active transport system. In this setting, the tightness of the monolayer needs be ascertained (especially when very high doses of substrates are used for kinetic analysis purposes), as the backdiffusion via an altered paracellular pathway would counteract the active transport, which works uphill from the concentration gradient. Hakvoort and colleagues (1998a) used this approach with porcine CPE cells to demonstrate that drugs and endogenous bioactive compounds undergo vectorial active transport. The transepithelial active transport was in the apical-to-basolateral direction for benzylpenicillin, an antibiotic drug, and riboflavin (vitamin B2). It was in the basolateral-to-apical direction for myo-inositol (an inositol-phosphate precursor and an organic osmolyte) and ascorbic acid (vitamin C). Kinetic analysis revealed affinity constants close to those measured in isolated CPs, and provided evidence for several transport systems of riboflavin at the CP. Further investigation of ascorbic acid transport using the same model showed that the basolateral transporter was Na^+-dependent and, in all probability, was SVCT-2 (Angelow et al., 2003).

This "uphill transport" approach was also used on a rat CPE cell model with phenol red as a model substrate to demonstrate that active, vectorial apical-to-basolateral organic anion transport capacity of the BCSFB was impaired in an inflammatory environment (Strazielle et al., 2003b), indicating that some neuroprotective properties of the CP can be altered in pathological situations.

23.4.2.7 Enzymatic Barrier (Metabolism) and Multidrug Resistance

The CPs have a high metabolic capacity toward xenobiotic compounds, especially by conjugation pathways (see Chapter 7 for further discussion). The ability of the choroidal epithelium to act as an enzymatic barrier to lipophilic compounds has been addressed using an in vitro rat BCSFB model and performing HPLC analysis of the media. The cells had the ability to prevent the entry of phenolic compounds into the CSF compartment by conjugating the xenobiotics via a UDP-glucuronosyltransferase–dependent pathway (Strazielle and Ghersi-Egea, 1999). Furthermore, a strong basolateral, i.e., blood-facing, polarity of the efflux of the glucuronoconjugate was demonstrated. The involvement of the multidrug-resistant protein MRP1 (ABCC1), present at the basolateral membrane of the CPE cells, or other members of the MRP family was suggested by efflux inhibition studies. This data demonstrated an efficient neuroprotective function of the choroidal epithelium through coupled metabolic and transport processes. Expression of P-glycoprotein (Pgp, ABCB1), another multidrug-resistance protein known to be associated with the capillaries forming the BBB, has also been reported at the choroid plexus epithelium. Functional evidence that this protein mediates cellular extrusion from CPE cells was given by uptake experiments with nonspecific or specific Pgp substrates such as 99mTc-sestamibi or Taxol, respectively, in rat cell cultures grown on porous membranes (Rao et al., 1999). The authors showed that the accumulation of these substrates was enhanced in the presence of a Pgp-specific inhibitor. Immunofluorescent analysis of Pgp in the CPE cells revealed punctuate or granular staining throughout the cytoplasm with a predominantly subapical distribution. The in vivo polarity and functional significance (in terms of transport directionality) of the choroidal Pgp remain unclear at present. In the absence of paracellular permeability monitoring during transcellular transport studies carried out with Pgp inhibitors, it is difficult to ascertain whether the increase in the apical-to-basolateral flux of Taxol and ^{99m}Tc-sestamibi, occurring with a time lag of two hours, actually results from Pgp inhibition or rather from a delayed alteration of the barrier properties of the cell monolayers.

23.4.2.8 Immortalized Cell Line-Based Investigations

Immortalized cell lines have recently been developed as tools to study transport mechanisms at the blood–CSF barrier. So far, experiments performed on these cell systems have investigated known properties of the choroidal epithelium for validation purposes. Applications of these models relevant to new understandings of choroidal transport processes are not yet available. In the course of the characterization of a mouse immortalized cell line, Kitazawa and collaborators (2001) carried out uptake experiments on cells seeded on solid supports. They established the Na^+-dependent and Na^+-independent affinity constants for glutamic acid and proline uptake, some of which corresponded to previous values obtained using isolated CPs. They also described an additional high-affinity *L*-proline uptake, which could be related to the brain-specific transporter PROT. This data would be of interest for the understanding of BCSFB functions, as this transporter is also a carrier for neurotransmitters, osmolites, and metabolites, but the data requires careful evaluation as PROT mRNA has not yet been detected in the CP (Velaz-Faircloth et al., 1995). This group also described the characteristics of uptake by the immortalized cells of estrone-3-sulfate, a metabolite formed within the brain (Ohtsuki et al., 2003). The results were fairly consistent with known data generated using isolated CP and implicated both Slc21A5 (oatp2) and Slc21A7 (oatp3) in the uptake process. This indicates that the cell line may be a practical, easy-to-use tool to investigate uptake at the CP. However, vectorial information obviously cannot be drawn from this uptake-based approach as, in vivo, Slc21A7and Slc21A5 are located differently, i.e., on the apical and basolateral membranes, respectively.

23.4.3 Other Applications of Cellular BCSFB Models

23.4.3.1 $5HT_{2C}$ Receptor

Rat CPE cell cultures were used to demonstrate that serotonin increases intracellular calcium via activation of the $5HT_{2C}$ receptor in the choroidal epithelium, a study that could not have been completed in intact tissue due to high autofluorescence levels (Watson et al., 1995). Rat cells cultured on plastic supports were also used to identify the active G-protein subunits involved in the $5HT_{2C}$ receptor-mediated activation of phospholipase C and phospholipase D (Chang et al., 2000; McGrew et al., 2002). In studies unrelated to BCSFB physiology, the rat CPE cells were also used to investigate the paradoxical downregulation of $5HT_{2C}$ receptor-binding sites, which is induced in the cortex and spinal cord by 5HT antagonists, e.g., the antidepressant miancerin. A decrease in binding sites was also observed in CPE cells cultured on plastic after several days of miancerin exposure, enabling the authors to conclude that the effect observed in the brain tissue was not indirect, i.e., not due to a transsynaptic increase in 5HT production. They also showed that the effect was not associated to a decrease in the receptor mRNA level (Barker and Sanders-Bush, 1993).

23.4.3.2 Adhesion Molecules and Immune Interactions

Several cellular adhesion molecules (CAM) are present on the mouse choroidal epithelium (but not on the ependyma) and are upregulated during experimental acute encephalomyelitis. To study induction mechanisms of CAM in CPE cells, Steffen and colleagues (1996) cultured mouse CPE cells on solid support for five days, then pretreated them for up to 16 hours with pro-inflammatory molecules. Treatment with TNF-α, IL-1, IFN-γ or a lipopolysaccharide preparation led to an upregulation of ICAM-1 and VCAM-1 and a de novo expression of MadCAM-1 in some cells identified as epithelial cells. As CAM are involved in leukocyte adhesion and in cell–cell immune interactions, this data supports a role for the CP epithelium in neuroimmune regulation. Another indication of a relationship between the CP and the immune system was provided by Enjoji and colleagues (1996). They observed that transgenic mice harboring the viral oncogene SV40 large tumor antigen under transcriptional control of an enhancer region from the human immunoglobulin heavy-chain gene developed choroid plexus carcinoma and showed that transactivating factors with binding capacity toward the promoter of this immune-associated molecule were produced by TTR-expressing, i.e., epithelium-derived, cell lines originating from these tumors.

23.4.3.3 Trophic Function

The choroid plexus secretes different factors involved in neuronal growth, migration, and differentiation. The influence of a mixed culture from the whole rat CP on the neurite outgrowth of dorsal root neurons was investigated by direct plating of these neurons on CP cultures (Chakrabortty et al., 2000). Although the state of differentiation of the CPE cells kept for several weeks in culture prior to use has not been characterized, these cells appeared to induce faster initiation and promotion of neurite outgrowth than those induced by glial cells. This work provides a basis for more sophisticated cell culture studies on the influence of CP on neural cell growth and differentiation.

23.4.3.4 Cellular Communication

Using mixed cultures prepared from mouse embryo CPs, in which choroidal epithelial cells are maintained as aggregates, Bouillé and colleagues (1991) demonstrated the functionality of gap junctions whose incidence is high at the choroidal epithelium.

These examples, illustrating the contribution of cellular models of the BCSFB to current research, open new experimental strategies to investigate CP functions. They also show that the various BCSFB models may have different applications depending on their degree of cellular differentiation, a parameter particularly crucial when transepithelial transfer measurements or polarity of secretion are to be analyzed, as both endpoints require the presence of functional tight junctions.

23.5 CONCLUSION AND PERSPECTIVES

Over the last two decades, significant progress has been made in developing cultures of choroidal epithelial cells. Because of the refinement of cell-isolation protocols and the large selection of attachment factors and medium additives, pure or highly enriched populations of CPE cells have been established.

Different requirements for cultured cells have been demonstrated in relation to the animal species or the developmental stage, emphasizing the importance of newly established cultures to assess their choroidal phenotype. CPE cells in primary cultures in appropriate conditions exhibit a highly differentiated phenotype.

Many choroidal specific features investigated to date in the most differentiated cell models were stable in the course of culture. This observation cannot be considered as a general rule and will not preclude the need to assess the relevance of a given choroidal epithelial cell model for one's personal experimental goal.

A large advantage of the primary cultures is that they establish functional tight junctions and cell polarity, thus reproducing the barrier properties that are requisite to investigate transepithelial transport or polarized processes. Their relatively short life span, however, limits their use in long-term toxicity studies.

For that reason, choroidal epithelial cell lines were seen as a valuable alternative to primary cultures. They would also represent a more convenient system for high-throughput screening of pharmaceutical compounds with regard to their extent and mechanism of passage across the BCSFB. The various immortalized cell lines developed so far clearly retain choroidal features. However, a major drawback in their application to transport studies arises from their apparent loss of diffusion barrier properties. Introducing in these immortalized cells some TJ proteins, such as claudins, may provide a clue, but a challenge certainly lies in producing the pattern and ratio of the various claudins that will reliably reproduce the structural and functional characteristics of the choroidal junction in vivo.

Choroidal epithelial cell cultures have potential applications to study various aspects of CP function that involve or play a role in neuroendocrine and neuroimmune communication, neuroinflammation, neuroinfection, and transport, removal, and metabolism of xenobiotics. By allowing the researchers to gather reliable functional data, the choroidal cultures have already provided important mechanistic information on BCSFB. However, it should be born in mind that these data generated in vitro on cultured choroidal epithelial cells will be all the more valuable if their relevance can be appreciated in more integrated approaches.

23.6 ACKNOWLEDGMENTS

This work was supported by ANRS. We would like to thank Silvia Schanz for her contribution to transport experiments.

REFERENCES

Agnew, W. F., Alvarez, R. B., Yuen, T. G., Abramson, S. B. and Kirk, D. (1984) A serum-free culture system for studying solute exchanges in the choroid plexus. *In Vitro* 20:712–22.

Albert, O., Ancellin, N., Preisser, L., Morel, A. and Corman, B. (1999) Serotonin, bradykinin and endothelin signalling in a sheep choroid plexus cell line. *Life Sci* 64:859–67.

Angelova, K., Fralish, G. B., Puett, D. and Narayan, P. (1996) Identification of conventional and novel endothelin receptors in sheep choroid plexus cells. *Mol Cell Biochem* 159:65–72.

Angelow, S., Haselbach, M. and Galla, H. J. (2003) Functional characterisation of the active ascorbic acid transport into cerebrospinal fluid using primary cultured choroid plexus cells. *Brain Res* 988:105–13.

Barker, E. L. and Sanders-Bush, E. (1993) 5-Hydroxytryptamine1C receptor density and mRNA levels in choroid plexus epithelial cells after treatment with mianserin and (-)-1-(4-bromo-2,5-dimethoxyphenyl)-2-aminopropane. *Mol Pharmacol* 44:725–30.

Battle, T., Preisser, L., Marteau, V., Meduri, G., Lambert, M., Nitschke, R., Brown, P. D. and Corman, B. (2000) Vasopressin V1a receptor signaling in a rat choroid plexus cell line. *Biochem Biophys Res Commun* 275:322–7.

Blay, P., Nilsson, C., Hansson, S., Owman, C., Aldred, A. R. and Schreiber, G. (1994) An *in vivo* study of the effect of 5-HT and sympathetic nerves on transferrin and transthyretin mRNA expression in rat choroid plexus and meninges. *Brain Res* 662:148–54.

Bouille, C., Mesnil, M., Barriere, H. and Gabrion, J. (1991) Gap junctional intercellular communication between cultured ependymal cells, revealed by lucifer yellow CH transfer and freeze-fracture. *Glia* 4:25–36.

Breen, C. M., Sykes, D. B., Fricker, G. and Miller, D. S. (2002) Confocal imaging of organic anion transport in intact rat choroid plexus. *Am J Physiol Renal Physiol* 282:F877–85.

Cameron, G. (1953) Secretory activity of the chorioid plexus in tissue culture. *Anat Rec* 117:115–125.

Cavallaro, T., Martone, R. L., Dwork, A. J., Schon, E. A. and Herbert, J. (1990) The retinal pigment epithelium is the unique site of transthyretin synthesis in the rat eye. *Invest Ophthalmol Vis Sci* 31:497–501.

Chakrabortty, S., Kitada, M., Matsumoto, N., Taketomi, M., Kimura, K. and Ide, C. (2000) Choroid plexus ependymal cells enhance neurite outgrowth from dorsal root ganglion neurons *in vitro*. *J Neurocytol* 29:707–17.

Chang, C. W., Ye, L., Defoe, D. M. and Caldwell, R. B. (1997) Serum inhibits tight junction formation in cultured pigment epithelial cells. *Invest Ophthalmol Vis Sci* 38:1082–93.

Chang, M., Zhang, L., Tam, J. P. and Sanders-Bush, E. (2000) Dissecting G protein-coupled receptor signaling pathways with membrane-permeable blocking peptides. Endogenous 5-HT_{2C} receptors in choroid plexus epithelial cells. *J Biol Chem* 275:7021–9.

Chebloune, Y., Sheffer, D., Karr, B. M., Stephens, E. and Narayan, O. (1996) Restrictive type of replication of ovine/caprine lentiviruses in ovine fibroblast cell cultures. *Virology* 222:21–30.

Claude, P. and Goodenough, D. A. (1973) Fracture faces of zonulae occludentes from “tight” and “leaky” epithelia. *J Cell Biol* 58:390–400.

Crook, R. B., Kasagami, H. and Prusiner, S. B. (1981) Culture and characterization of epithelial cells from bovine choroid plexus. *J Neurochem* 37:845–54.

Dermietzel, R., Meller, K., Tetzlaff, W. and Waelsch, M. (1977) *In vivo* and *in vitro* formation of the junctional complex in choroid epithelium. A freeze-etching study. *Cell Tissue Res* 181:427–41.

Dickinson, K. E., Baska, R. A., Cohen, R. B., Bryson, C. C., Smith, M. A., Schroeder, K. and Lodge, N. J. (1998) Identification of [^{3}H]P1075 binding sites and P1075-activated K+ currents in ovine choroid plexus cells. *Eur J Pharmacol* 345:97–101.

Enjoji, M., Iwaki, T., Nawata, H. and Watanabe, T. (1995) IgH intronic enhancer element HE2 (mu B) functions as a cis-activator in choroid plexus cells at the cellular level as well as in transgenic mice. *J Neurochem* 64:961–6.

Enjoji, M., Iwaki, T., Hara, H., Sakai, H., Nawata, H. and Watanabe, T. (1996) Establishment and characterization of choroid plexus carcinoma cell lines: connection between choroid plexus and immune systems. *Jpn J Cancer Res* 87:893–9.

Ennis, S. R., Novotny, A., Xiang, J., Shakui, P., Masada, T., Stummer, W., Smith, D. E. and Keep, R. F. (2003) Transport of 5-aminolevulinic acid between blood and brain. *Brain Res* 959:226–34.

Esterle, T. M. and Sanders-Bush, E. (1992) Serotonin agonists increase transferrin levels via activation of 5-HT1C receptors in choroid plexus epithelium. *J Neurosci* 12:4775–82.

Furuse, M., Sasaki, H. and Tsukita, S. (1999) Manner of interaction of heterogeneous claudin species within and between tight junction strands. *J Cell Biol* 147:891–903.

Furuse, M., Furuse, K., Sasaki, H. and Tsukita, S. (2001) Conversion of zonulae occludentes from tight to leaky strand type by introducing claudin-2 into Madin-Darby canine kidney I cells. *J Cell Biol* 153:263–72.

Gabrion, J., Peraldi, S., Faivre-Bauman, A., Klotz, C., Ghandour, M. S., Paulin, D., Assenmacher, I. and Tixier-Vidal, A. (1988) Characterization of ependymal cells in hypothalamic and choroidal primary cultures. *Neuroscience* 24:993–1007.

Gabrion, J. B., Herbute, S., Bouille, C., Maurel, D., Kuchler-Bopp, S., Laabich, A. and Delaunoy, J. P. (1998) Ependymal and choroidal cells in culture: characterization and functional differentiation. *Microsc Res Tech* 41:124–57.

Gath, U., Hakvoort, A., Wegener, J., Decker, S. and Galla, H. J. (1997) Porcine choroid plexus cells in culture: expression of polarized phenotype, maintenance of barrier properties and apical secretion of CSF-components. *Eur J Cell Biol* 74:68–78.

Gee, P., Rhodes, C. H., Fricker, L. D. and Angeletti, R. H. (1993) Expression of neuropeptide processing enzymes and neurosecretory proteins in ependyma and choroid plexus epithelium. *Brain Res* 617:238–48.

Gilbert, S. F. and Migeon, B. R. (1975) *D*-valine as a selective agent for normal human and rodent epithelial cells in culture. *Cell* 5:11–7.

Hakvoort, A., Haselbach, M. and Galla, H. J. (1998a) Active transport properties of porcine choroid plexus cells in culture. *Brain Res* 795:247–56.

Hakvoort, A., Haselbach, M., Wegener, J., Hoheisel, D. and Galla, H. J. (1998b) The polarity of choroid plexus epithelial cells *in vitro* is improved in serum-free medium. *J Neurochem* 71:1141–50.

Hartter, S., Huwel, S., Lohmann, T., Abou El Ela, A., Langguth, P., Hiemke, C. and Galla, H. J. (2003) How does the benzamide antipsychotic amisulpride get into the brain?—An *in vitro* approach comparing amisulpride with clozapine. *Neuropsychopharmacology* 28:1916–22.

Haselbach, M., Wegener, J., Decker, S., Engelbertz, C. and Galla, H. J. (2001) Porcine choroid plexus epithelial cells in culture: regulation of barrier properties and transport processes. *Microsc Res Tech* 52:137–52.

Hoffmann, A., Gath, U., Gross, G., Lauber, J., Getzlaff, R., Hellwig, S., Galla, H. J. and Conradt, H. S. (1996) Constitutive secretion of beta-trace protein by cultivated porcine choroid plexus epithelial cells: elucidation of its complete amino acid and cDNA sequences. *J Cell Physiol* 169:235–41.

Holm, N. R., Hansen, L. B., Nilsson, C. and Gammeltoft, S. (1994) Gene expression and secretion of insulin-like growth factor-II and insulin-like growth factor binding protein-2 from cultured sheep choroid plexus epithelial cells. *Brain Res Mol Brain Res* 21:67–74.

Johanson, C. E., Palm, D. E., Primiano, M. J., McMillan, P. N., Chan, P., Knuckey, N. W. and Stopa, E. G., (2000) Choroid plexus recovery after transient forebrain ischemia: role of growth factors and other repair mechanisms. *Cell Mol Neurobiol* 20:197–216.

Kao, W. W. and Prockop, D. J. (1977) Proline analogue removes fibroblasts from cultured mixed cell populations. *Nature* 266:63–4.

Kao, W. W., Prockop, D. J. and Berg, R. A. (1979) Kinetics for the secretion of nonhelical procollagen by freshly isolated tendon cells. *J Biol Chem* 254:2234–43.

Kitazawa, T., Hosoya, K., Watanabe, M., Takashima, T., Ohtsuki, S., Takanaga, H., Ueda, M., Yanai, N., Obinata, M. and Terasaki, T. (2001) Characterization of the amino acid transport of new immortalized choroid plexus epithelial cell lines: a novel *in vitro* system for investigating transport functions at the blood-cerebrospinal fluid barrier. *Pharm Res* 18:16–22.

Lippoldt, A., Jansson, A., Kniesel, U., Andbjer, B., Andersson, A., Wolburg, H., Fuxe, K. and Haller, H. (2000a) Phorbol ester induced changes in tight and adherens junctions in the choroid plexus epithelium and in the ependyma. *Brain Res* 854:197–206.

Lippoldt, A., Liebner, S., Andbjer, B., Kalbacher, H., Wolburg, H., Haller, H. and Fuxe, K. (2000b) Organization of choroid plexus epithelial and endothelial cell tight junctions and regulation of claudin-1, -2 and -5 expression by protein kinase C. *Neuroreport* 11:1427–31.

Lo, C. M., Keese, C. R. and Giaever, I. (1999) Cell-substrate contact: another factor may influence transepithelial electrical resistance of cell layers cultured on permeable filters. *Exp Cell Res* 250:576–80.

Mallick, K. S., Hajek, A. S. and Parrish, R. K., 2nd (1985) Fluorouracil (5-FU) and cytarabine (ara-C) inhibition of corneal epithelial cell and conjunctival fibroblast proliferation. *Arch Ophthalmol* 103:1398–402.

Martinez-Palomo, A. and Erlij, D. (1975) Structure of tight junctions in epithelia with different permeability. *Proc Natl Acad Sci USA* 72:4487–91.

Mayer, S. E. and Sanders-Bush, E. (1993) Sodium-dependent antiporters in choroid plexus epithelial cultures from rabbit. *J Neurochem* 60:1308–16.

McGrew, L., Chang, M. S. and Sanders-Bush, E. (2002) Phospholipase D activation by endogenous 5-hydroxytryptamine 2C receptors is mediated by Galpha13 and pertussis toxin-insensitive Gbetagamma subunits. *Mol Pharmacol* 62:1339–43.

Miettinen, M., Clark, R. and Virtanen, I. (1986) Intermediate filament proteins in choroid plexus and ependyma and their tumors. *Am J Pathol* 123:231–40.

Miller, D. S., Villalobos, A. R. and Pritchard, J. B. (1999) Organic cation transport in rat choroid plexus cells studied by fluorescence microscopy. *Am J Physiol* 276:C955–68.

Nakashima, N., Goto, K., Tsukidate, K., Sobue, M., Toida, M. and Takeuchi, J. (1983) Choroid plexus papilloma. Light and electron microscopic study. *Virchows Arch A Pathol Anat Histopathol* 400:201–11.

Narayan, O., Griffin, D. E. and Silverstein, A. M. (1977) Slow virus infection: replication and mechanisms of persistence of visna virus in sheep. *J Infect Dis* 135:800–6.

Nathanson, J. A. (1979) Beta-adrenergic-sensitive adenylate cyclase in secretory cells of choroid plexus. *Science* 204:843–4.

Nilsson, C., Hultberg, B. M. and Gammeltoft, S. (1996) Autocrine role of insulin-like growth factor II secretion by the rat choroid plexus. *Eur J Neurosci* 8:629–35.

Nunbhakdi-Craig, V., Craig, L., Machleidt, T. and Sontag, E. (2003) Simian virus 40 small tumor antigen induces deregulation of the actin cytoskeleton and tight junctions in kidney epithelial cells. *J Virol* 77:2807–18.

Ohtsuki, S., Takizawa, T., Takanaga, H., Terasaki, N., Kitazawa, T., Sasaki, M., Abe, T., Hosoya, K. and Terasaki, T. (2003) In vitro study of the functional expression of organic anion transporting polypeptide 3 at rat choroid plexus epithelial cells and its involvement in the cerebrospinal fluid-to-blood transport of estrone-3-sulfate. *Mol Pharmacol* 63:532–7.

Papenheimer (1961) Active transport of Diodrast and phenolsulfonphtalein from cerebrospinal fluid to blood. *Am J Physiol* 200:1–10.

Peiser, C., McGregor, G. P. and Lang, R. E. (2000) Binding and internalization of leptin by porcine choroid plexus cells in culture. *Neurosci Lett* 283:209–12.

Peraldi-Roux, S., Nguyen-Than Dao, B., Hirn, M. and Gabrion, J. (1990) Choroidal ependymocytes in culture: expression of markers of polarity and function. *Int J Dev Neurosci* 8:575–88.

Plotkin, M. D., Kaplan, M. R., Peterson, L. N., Gullans, S. R., Hebert, S. C. and Delpire, E. (1997) Expression of the Na^+-K^+-Cl^- cotransporter BSC2 in the nervous system. *Am J Physiol* 272:C173–83.

Puskas, L. G., Kitajka, K., Nyakas, C., Barcelo-Coblijn, G. and Farkas, T. (2003) Short-term administration of omega 3 fatty acids from fish oil results in increased transthyretin transcription in old rat hippocampus. *Proc Natl Acad Sci USA* 100:1580–5.

Ramanathan, V. K., Hui, A. C., Brett, C. M. and Giacomini, K. M. (1996) Primary cell culture of the rabbit choroid plexus: an experimental system to investigate membrane transport. *Pharm Res* 13:952–6.

Ramanathan, V. K., Chung, S. J., Giacomini, K. M. and Brett, C. M. (1997) Taurine transport in cultured choroid plexus. *Pharm Res* 14:406–9.

Rao, V. V., Dahlheimer, J. L., Bardgett, M. E., Snyder, A. Z., Finch, R. A., Sartorelli, A. C. and Piwnica-Worms, D. (1999) Choroid plexus epithelial expression of MDR1 P glycoprotein and multidrug resistance-associated protein contribute to the blood-cerebrospinal-fluid drug-permeability barrier. *Proc Natl Acad Sci USA* 96:3900–5.

Saito, Y. and Wright, E. M. (1983) Bicarbonate transport across the frog choroid plexus and its control by cyclic nucleotides. *J Physiol* 336:635–48.

Sanders-Bush, E. and Breeding, M. (1991) Choroid plexus epithelial cells in primary culture: a model of 5HT1C receptor activation by hallucinogenic drugs. *Psychopharmacology (Berl)* 105:340–6.

Schreiber, G., Aldred, A. R., Jaworowski, A., Nilsson, C., Achen, M. G. and Segal, M. B. (1990) Thyroxine transport from blood to brain via transthyretin synthesis in choroid plexus. *Am J Physiol* 258:R338–45.

Shu, C., Shen, H., Teuscher, N. S., Lorenzi, P. J., Keep, R. F. and Smith, D. E. (2002) Role of PEPT2 in peptide/mimetic trafficking at the blood-cerebrospinal fluid barrier: studies in rat choroid plexus epithelial cells in primary culture. *J Pharmacol Exp Ther* 301:820–9.

Singh, A. B. and Harris, R. C. (2004) Epidermal growth factor receptor activation differentially regulates claudin expression and enhances transepithelial resistance in Madin-Darby canine kidney cells. *J Biol Chem* 279:3543–52.

Southwell, B. R., Duan, W., Alcorn, D., Brack, C., Richardson, S. J., Kohrle, J. and Schreiber, G. (1993) Thyroxine transport to the brain: role of protein synthesis by the choroid plexus. *Endocrinology* 133:2116–26.

Spector, R. (1982) Pharmacokinetics and metabolism of cytosine arabinoside in the central nervous system. *J Pharmacol Exp Ther* 222:1–6.

Stadler, E. and Dziadek, M. (1996) Extracellular matrix penetration by epithelial cells is influenced by quantitative changes in basement membrane components and growth factors. *Exp Cell Res* 229:360–9.

Steffen, B. J., Breier, G., Butcher, E. C., Schulz, M. and Engelhardt, B. (1996) ICAM-1, VCAM-1, and MAdCAM-1 are expressed on choroid plexus epithelium but not endothelium and mediate binding of lymphocytes *in vitro*. *Am J Pathol* 148:1819–38.

Strazielle, N. and Ghersi-Egea, J. F. (1999) Demonstration of a coupled metabolism-efflux process at the choroid plexus as a mechanism of brain protection toward xenobiotics. *J. Neurosci* 19:6275–89.

Strazielle, N., Belin, M. F. and Ghersi-Egea, J. F. (2003a) Choroid plexus controls brain availability of anti-HIV nucleoside analogs via pharmacologically inhibitable organic anion transporters. *AIDS* 17:1473–1485.

Strazielle, N., Khuth, S. T., Murat, A., Chalon, A., Giraudon, P., Belin, M. F. and Ghersi-Egea, J. F. (2003b) Pro-inflammatory cytokines modulate matrix metalloproteinase secretion and organic anion transport at the blood-cerebrospinal fluid barrier. *J Neuropathol Exp Neurol* 62:1254–64.

Strazielle, N. and Preston, J. E. (2003) Transport across the choroid plexuses *in vivo* and *in vitro*. *Methods Mol Med* 89:291–304.

Takahashi, K., Satoh, F., Hara, E., Murakami, O., Kumabe, T., Tominaga, T., Kayama, T., Yoshimoto, T. and Shibahara, S. (1997) Production and secretion of adrenomedullin by cultured choroid plexus carcinoma cells. *J Neurochem* 68:726–31.

Takahashi, K., Hara, E., Murakami, O., Totsune, K., Sone, M., Satoh, F., Kumabe, T., Tominaga, T., Kayama, T., Yoshimoto, T. and Shibahara, S. (1998) Production and secretion of endothelin-1 by cultured choroid plexus carcinoma cells. *J Cardiovasc Pharmacol* 31(suppl 1):S367–9.

Terasaki, T., Ohtsuki, S., Hori, S., Takanaga, H., Nakashima, E. and Hosoya, K. (2003) New approaches to *in vitro* models of blood-brain barrier drug transport. *Drug Discov Today* 8:944–54.

Thomas, T., Schreiber, G. and Jaworowski, A. (1989) Developmental patterns of gene expression of secreted proteins in brain and choroid plexus. *Dev Biol* 134:38–47.

Thomas, T., Stadler, E. and Dziadek, M. (1992) Effects of the extracellular matrix on fetal choroid plexus epithelial cells: changes in morphology and multicellular organization do not affect gene expression. *Exp Cell Res* 203:198–213.

Thormar, H. (1963) The growth cycle of visna virus in monolayer cultures of sheep cells. *Virology* 19:273–8.

Thornwall, M., Chhajlani, V., Le Greves, P. and Nyberg, F. (1995) Detection of growth hormone receptor mRNA in an ovine choroid plexus epithelium cell line. *Biochem Biophys Res Commun* 217:349–53.

Torchio, C. and Trowbridge, R. S. (1977) Ovine cells: their long-term cultivation and susceptibility to visna virus. *In Vitro* 13:252–9.

Tsukita, S. and Furuse, M. (2000) Pores in the wall: claudins constitute tight junction strands containing aqueous pores. *J Cell Biol* 149:13–6.

Tsukita, S. and Furuse, M. (2002) Claudin-based barrier in simple and stratified cellular sheets. *Curr Opin Cell Biol* 14:531–6.

Tsutsumi, M., Skinner, M. K. and Sanders-Bush, E. (1989) Transferrin gene expression and synthesis by cultured choroid plexus epithelial cells. Regulation by serotonin and cyclic adenosine 3.5'-monophosphate. *J Biol Chem* 264:9626–31.

Tu, G. F., Achen, M. G., Aldred, A. R., Southwell, B. R. and Schreiber, G. (1991) The distribution of cerebral expression of the transferrin gene is species specific. *J Biol Chem* 266:6201–8.

van Deurs, B. and Koehler, J. K. (1979) Tight junctions in the choroid plexus epithelium. A freeze-fracture study including complementary replicas. *J Cell Biol* 80:662–73.

Velaz-Faircloth, M., Guadano-Ferraz, A., Henzi, V. A. and Fremeau, R. T., Jr. (1995) Mammalian brain-specific *L*-proline transporter. Neuronal localization of mRNA and enrichment of transporter protein in synaptic plasma membranes. *J Biol Chem* 270:15755–61.

Villalobos, A. R., Parmelee, J. T. and Pritchard, J. B. (1997) Functional characterization of choroid plexus epithelial cells in primary culture. *J Pharmacol Exp Ther* 282:1109–16.

Villalobos, A. R., Parmelee, J. T. and Renfro, J. L. (1999) Choline uptake across the ventricular membrane of neonate rat choroid plexus. *Am J Physiol* 276:C1288–96.

Virtanen, I., Lehto, V. P., Lehtonen, E., Vartio, T., Stenman, S., Kurki, P., Wager, O., Small, J. V., Dahl, D. and Badley, R. A. (1981) Expression of intermediate filaments in cultured cells. *J Cell Sci* 50:45–63.

Watson, J. A., Elliott, A. C. and Brown, P. D. (1995) Serotonin elevates intracellular Ca^{2+} in rat choroid plexus epithelial cells by acting on 5-HT2C receptors. *Cell Calcium* 17:120–8.

Wegener, J., Hakvoort, A. and Galla, H. J. (2000) Barrier function of porcine choroid plexus epithelial cells is modulated by cAMP-dependent pathways *in vitro*. *Brain Res* 853:115–24.

Welch, K. and Sadler, K. (1966) Permeability of the choroid plexus of the rabbit to several solutes. *Am J Physiol* 210:652–60.

Welch, K. and Araki, H. (1975) Features of the choroid plexus of the cat studied *in vitro*. In Cserr H. F., Fenstermacher, J. D. and Fencl, V. (eds.). *Fluid Environment of the Brain*. New York: Academic Press, Inc.

Wells, A. (1999) EGF receptor. *Int J Biochem Cell Biol* 31:637–43.

Wolburg, H., Wolburg-Buchholz, K., Liebner, S. and Engelhardt, B. (2001) Claudin-1, claudin-2 and claudin-11 are present in tight junctions of choroid plexus epithelium of the mouse. *Neurosci Lett* 307:77–80.

Wright, E. M. (1972) Mechanisms of ion transport across the choroid plexus. *J Physiol* 226:545–71.

Xiang, J., Ennis, S. R., Abdelkarim, G. E., Fujisawa, M., Kawai, N. and Keep, R. F. (2003) Glutamine transport at the blood-brain and blood-cerebrospinal fluid barriers. *Neurochem Int* 43:279–88.

Yusko, S. C. and Neims, A. H. (1973) *D*-Aspartate oxidase in mammalian brain and choroid plexus. *J Neurochem* 21:1037–9.

Zheng, W., Zhao, Q. and Graziano, J. H. (1998) Primary culture of choroidal epithelial cells: characterization of an *in vitro* model of blood-CSF barrier. *In Vitro Cell Dev Biol Anim* 34:40–5.

Zheng, W., Blaner, W. S. and Zhao, Q. (1999) Inhibition by lead of production and secretion of transthyretin in the choroid plexus: its relation to thyroxine transport at blood-CSF barrier. *Toxicol Appl Pharmacol* 155:24–31.

Zheng, W. and Zhao, Q. (2002a) The blood-CSF barrier in culture. Development of a primary culture and transepithelial transport model from choroidal epithelial cells. *Methods Mol Biol* 188:99–114.

Zheng, W. and Zhao, Q. (2002b) Establishment and characterization of an immortalized Z310 choroidal epithelial cell line from murine choroid plexus. *Brain Res* 958:371–80.

24 *In Vivo* Techniques Used in Blood–CSF Barrier Research: Measurement of CSF Formation

Adam Chodobski
Brown University School of Medicine,
Providence, Rhode Island, U.S.A.

Malcolm B. Segal
King's College London,
London, United Kingdom

CONTENTS

24.1 INTRODUCTION

Various physiological and pathological conditions, such as aging of the central nervous system (CNS), injury, and hydrocephalus, are accompanied by changes in cerebrospinal fluid (CSF) dynamics. Therefore, measurement of the rate at which

CSF is generated provides important information needed to obtain a better mechanistic insight into the regulation of CSF formation and to enhance our understanding of the functional implications of altered CSF dynamics. In this chapter we will provide a detailed analysis of methodology relevant to the measurement of CSF formation in small rodents and will briefly review the techniques used to assess CSF production in humans.

24.2 MEASUREMENT OF CSF FORMATION IN EXPERIMENTAL ANIMALS

24.2.1 Ventriculo-Cisternal (VC) Perfusion Technique

The method of VC perfusion was developed by Royer and Leusen in 1950 (see Davson and Segal, 1995) and was elegantly adapted to the conscious goat by Pappenheimer and colleagues in 1962 (Heisey et al., 1962; Pappenheimer et al., 1962). VC perfusion is an indicator-dilution, constant infusion method based on the Stewart principle (Lassen and Perl, 1979). In this technique (see Figure 24.1), artificial CSF containing an indicator is infused into the lateral ventricle(s), while

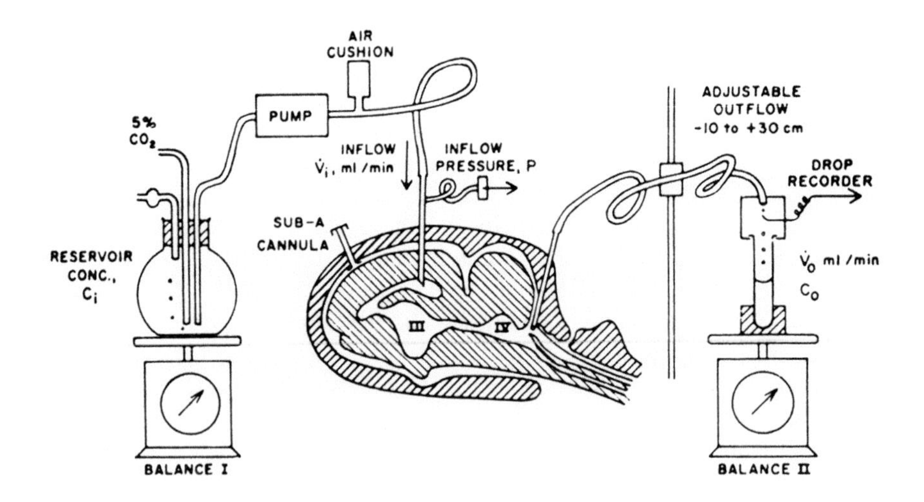

FIGURE 24.1 Schematic diagram of VC perfusion. In this method, artificial CSF containing an indicator is infused into the lateral ventricle(s), while the mixture of nascent and mock CSF is collected from the cisterna magna. The tip of an outflow line is usually positioned at or above the interaural line. During VC perfusion, the intraventricular CSF pressure is continuously monitored and maintained at a relatively constant level by adjusting the height of the outflow line. It is assumed that the indicator is eliminated from the CSF space only via bulk CSF absorption and its diffusion into the brain tissue is negligible. [Reprinted with permission from the American Physiological Society (*Am J Physiol* 203, 1962, 763–74).]

the mixture of nascent and mock CSF is collected from the cisterna magna. The tip of an outflow line is usually positioned at or above the interaural line. During the experiment, the intraventricular CSF pressure is continuously monitored and maintained at a relatively constant level by adjusting the height of the outflow line. In this method, an assumption is made that the indicator is eliminated from the CSF space only via bulk CSF absorption and its diffusion into the brain tissue is negligible. For this reason, moderate and high molecular weight markers, such as inulin and radiolabeled inulin (MW ~ 5,000), radiolabeled serum albumin (MW ~ 69,000), and Blue Dextran 2000 (MW ~ 2,000,000), have been used (Pappenheimer et al., 1962; Cutler et al., 1968, 1970; Davson and Segal, 1970; Lindvall et al., 1979; Chodobski et al., 1986b, 1992a,b, 1994, 1995, 1998a,b; Faraci et al., 1990; Nilsson et al., 1991; Seckl and Lightman, 1991; Maktabi et al., 1993). Lower molecular weight indicators (low molecular weight inulin, MW 1300–1800 and 2500–3500) are less desirable as their use can result in an overestimation of the CSF formation rate (Curran et al., 1970). The CSF formation rate is calculated using the following formula (Heisey et al., 1962):

$$V_f = V_i \times [(C_i - C_o)/C_o], \tag{24.1}$$

where C is the concentration of indicator (e.g., mg/mL), V is the rate of fluid flow (μL/min), and the subscripts f, i, and o denote formation, inflow, and outflow. In this equation, C_o represents the concentration of indicator in the outflow fluid measured after equilibrium in the system had been established; this is usually observed within two to three hours after the start of VC perfusion (Cutler et al., 1968; Chodobski et al., 1992b). If changes in CSF production resulting from experimental intervention are to be measured in the same animal, the system must attain a new equilibrium before the production rate can be assessed again. This necessitates the maintenance of stable physiological conditions for a relatively long period of time, which can be difficult to achieve. To avoid this problem, some authors used a separate group of control animals to compare with the animals subjected to experimental intervention (Chodobski et al., 1992b, 1994, 1995, 1998a).

In the original study by Pappenheimer et al. (1962), CSF formation was measured in conscious goats. The measurement of CSF formation in conscious animals poses several technical problems; for instance, head movement can cause sudden changes in CSF pressure and fluid flow, resulting in uneven mixing of mock and nascent CSF. For this reason, only a few studies have been reported where CSF production was measured in conscious animals (Seckl and Lightman, 1991; Chodobski et al., 1992a, 1998b). The following CSF formation rates were obtained for each species: 90–160 μL/min in goats (Heisey et al., 1962; Seckl and Lightman, 1991), 70–80 μL/min in sheep (Chodobski et al., 1992a, 1998b), 44–47 μL/min in dogs (Miller et al., 1986), ~30 μL/min in rhesus monkeys (Curran et al., 1970), 22 μL/min in cats (Chodobski et al., 1986b), 8–12 μL/min in rabbits (Davson and Segal, 1970; Lindvall et al., 1979; Faraci et al., 1990; Chodobski et al., 1992b; Maktabi et al., 1993), and 2.2–3.4 μL/min in rats (Nilsson et al., 1991; Chodobski et al., 1994, 1995, 1998a).

24.2.2 VC Perfusion in Rats

In this chapter we will provide a description of procedures for VC perfusion in rats. The reader will find more detailed information on the surgical preparation and protocols used in Chodobski et al. (1994, 1995). While surgical procedures in rat may appear technically more demanding compared to those in the larger laboratory animals, both the cost and the existence of ample physiological, biochemical, and molecular data make the rat an attractive and important animal model in which to study CSF dynamics. The experiments are performed on mechanically ventilated animals under general anesthesia (e.g., induced with intraperitoneal pentobarbital sodium, 50 mg/kg, and maintained by intravenous administration of the same anesthetic at 15–25 mg/kg/h). It is important to keep the arterial blood gas parameters within the physiological range because both metabolic alkalosis and hypoxia can influence CSF formation (Oppelt et al., 1963; Michael and Heisey, 1973). An increase in arterial CO_2 tension has been found either not to affect (Oppelt et al., 1963; Hochwald et al., 1973) or to increase CSF production (Davson and Segal, 1970), though in the latter studies this CO_2 effect was variable. Furthermore, CO_2 by itself is a potent vasodilator of cerebral vasculature (Chodobski et al., 1986a). Therefore, CO_2 has a profound effect on the CSF space, which may affect the mixing of artificial and nascent CSF. The use of pentobarbital sodium for general anesthesia is preferable because of the ease of induction and maintenance. Although some authors (Lindvall et al., 1979; Nilsson et al., 1991) have used a volatile anesthetic, halothane, this anesthetic may not be appropriate for VC perfusion since it has been shown to decrease CSF production (Maktabi et al., 1993). Special care has to be taken to replace the loss of fluids through intravenous infusion of lactated Ringer solution at a rate of 3 mL/kg/h and to maintain the core temperature of the animal at ~37°C.

The rat is mounted in a stereotaxic frame and 27-gauge stainless steel cannulas are introduced into one or both lateral ventricles (1 mm caudally and 1.5 mm laterally to the bregma; 2–3 mm below the level of the dura at a right angle to the surface of the skull). Artificial CSF (see Table 24.1), containing the indicator Blue Dextran

TABLE 24.1
Composition of Artificial Rat CSF

Constituent	Concentration (mM)
NaCl	125.0
KCl	2.5
$CaCl_2$	1.2
$MgCl_2$	0.9
$NaHCO_3$	25.0
Na_2HPO_4	0.5
KH_2PO_4	0.5
Glucose	3.7
Urea	6.5

2000 at a concentration of 5 mg/mL, is infused through these cannulas using a syringe infusion pump (e.g., Harvard Apparatus infusion pump). The total rate of infusion is 4 µL/min. The intraventricular CSF pressure is continuously monitored by means of T connectors inserted into the infusion lines and should be kept at 2–3 mmHg. CSF is collected at 20-minute intervals through a 27-gauge stainless steel cannula that is inserted into the cisterna magna. The concentration of Blue Dextran in these samples is determined colorimetrically by measuring absorbance at 620 nm.

One of the major technical problems encountered in the VC perfusion experiments is the difficulty of maintaining a patent outflow line during the two- to three-hour observation period. Access to the cisterna magna is usually gained by a puncture of the atlantooccipital membrane. However, successful drainage of the cisternal fluid at the beginning of VC perfusion does not guarantee a continuous outflow of CSF throughout the experiment. This problem was considerably minimized in sheep and rabbits when the cisterna magna was accessed via a metal cannula introduced through a hole drilled in the occipital bone (Chodobski et al., 1992a,b, 1998b). Although this could not be achieved in smaller animals, such as rats, during acute experiments, we were highly successful if the rats had their cannulas implanted 7–10 days before the experiment. For the implantation of the cisterna magna cannulas, rats were anesthetized intramuscularly with ketamine and xylazine (50 and 7.5 mg/kg, respectively). A 25-gauge thin-walled stainless steel guide cannula was then introduced 7 mm below the surface of the skull at an angle of 12° through a hole drilled medially at the junction of the interparietal and occipital bones (Greene, 1963). The guide cannula was fixed in place with dental acrylic and stainless steel screws. The scalp was closed with a silk suture, and the animals were allowed to recover for at least one week before the experiment. To drain the CSF from the cisterna magna, a 30-gauge stainless steel needle was inserted into the guide cannula. The inner needle protruded from the guide cannula by 0.5–1.5 mm. This setting allowed us to significantly increase the success rate in VC perfusion experiments (J. Szmydynger-Chodobska and A. Chodobski, unpublished observations).

Another serious problem specifically related to the use of Blue Dextran 2000 occurs when the CSF samples become contaminated with blood. The presence of plasma proteins in the samples affects the measurement of absorbance and, therefore, is a source of error in assessing the rate of CSF production (Hyman et al., 1987). For this reason, the surgical preparation associated with the exposure and the puncture of the atlantooccipital membrane should be done with the aid of an operating microscope to avoid damaging the small blood vessels that supply the area.

24.2.3 Measurement of CSF Absorption and Outflow Resistance

In VC perfusion experiments, the rate of CSF absorption can be calculated based on the following equation:

$$V_a = V_i + V_f - V_o \quad (24.2)$$

where V is the rate of fluid flow (μL/min), and the subscripts a, f, i, and o refer to absorption, formation, inflow, and outflow. Cutler et al. (1968) have demonstrated that the CSF absorption rate is linearly related to the outflow CSF pressure. Therefore, the resistance to CSF absorption or outflow resistance (R_{out}; mmHg·min·μL^{-1} or cmH_2O·min·μL^{-1}) is calculated using the following formula:

$$R_{out} = \Delta P_{CSF}/\Delta V_a, \quad (24.3)$$

where P_{CSF} (mmHg or cmH_2O) is the CSF pressure measured either close to the CSF outflow pathways or within the ventricular system. Thus, R_{out} represents the slope of the straight line (usually obtained by the least-squares fit) characterizing the relationship between V_a and P_{CSF} (see Figure 24.2). Acutely, CSF production does not depend on outflow CSF pressure ranging between 0 and 20 cmH_2O (Cutler et al., 1968). Accordingly, the relationship between V_a and P_{CSF} can be characterized experimentally by measuring V_a at various P_{CSF} levels. To change the outflow CSF pressure, the height of the outflow line is raised or lowered relative to the interaural line (Chodobski et al., 1998b). Although technically more cumbersome, it is preferable to measure the outflow CSF pressure rather than the intraventricular pressure because the changes in the latter not only depend on R_{out}, but also reflect the resistance to the CSF flow along the ventricular system. This is especially important if one needs to compare the R_{out} of normal and hydrocephalic animals with aqueductal stenosis.

24.3 MEASUREMENT OF CSF FORMATION IN HUMANS

24.3.1 Early Measurements of CSF Formation

In early human studies, the rate of CSF formation was estimated by measuring the time needed for the CSF pressure to return to its initial level after a certain volume of CSF (usually 20–50 mL) was removed by lumbar puncture. CSF production then equaled the volume of CSF removed divided by the time elapsed until the opening CSF pressure was restored. After removing 20–35 mL of CSF, Masserman (1934) assessed that the CSF formation rate in man is ~320 μL/min. Using a similar technique, Sjoqvist (1937) reported a CSF production rate of ~360 μL/min. It is of interest to note that despite several obvious drawbacks in this technique (Katzman and Pappius, 1973; Fishman, 1980), the rates of CSF formation found by Masserman and Sjoqvist are very similar to those obtained using a steady-state perfusion technique (Rubin et al., 1966; Cutler et al., 1968) (see below). Studies employing the CSF drainage method have also been done recently with useful results (May et al., 1990; Silverberg et al., 2001, 2002). Nevertheless, the Masserman method is subject to error, and, therefore, the data so obtained would be difficult to interpret in pathological or experimental situations associated with major changes in the pressure-volume relationship of the craniospinal axis (Fishman, 1980).

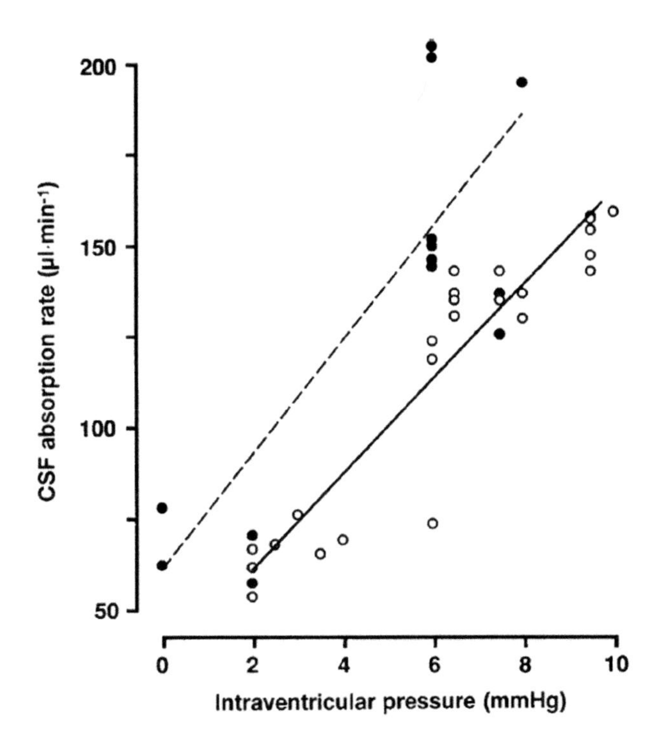

FIGURE 24.2 Relationships between CSF absorption rate (V_a) and intraventricular CSF pressure (IVP) in normally hydrated sheep (open circles and solid line) and in animals subjected to 48-hour dehydration (filled circles and broken line). IVP and V_a were measured in conscious sheep during VC perfusion experiments. Data points (IVP, V_a) on the bottom of the graph were obtained with the outflow cannula being positioned at the level of the interaural line, whereas those plotted on the top of the graph were collected when the outflow cannula was placed above the interaural line. The relationships between V_a and IVP are characterized by the following best-fit equations: $V_a = 36.0 + 12.8 \times IVP$ ($r = 0.93$; $p < 0.001$) and $V_a = 60.9 + 15.6 \times IVP$ ($r = 0.79$; $p < 0.001$) for control and dehydration, respectively. Although the difference between the water-replete and the dehydration states did not display statistical significance, IVP in water-deprived sheep tended to be lower than in controls, possibly because dehydration caused a decrease in extracellular fluid volume and central venous pressure. [Reprinted with permission from the American Physiological Society (*Am J Physiol* 275, 1998, F235–8).]

24.3.2 Ventriculo-Lumbar Perfusion and the Lumbar Infusion Technique

In humans, CSF production was measured using ventriculo-lumbar perfusion, in which artificial CSF was infused into the lateral ventricle and collected from the lumbar subarachnoid space (Rubin et al., 1966; Cutler et al., 1968). Both children and adult patients were studied, and the average CSF formation rates found in these two groups were 350 and 370 μL/min, respectively. In these studies, the patients were not healthy but suffered from intracranial diseases and, therefore, the CSF production rates reported may actually be lower than in normal subjects.

In 1970 Katzman and Hussey introduced the lumbar CSF infusion technique (Katzman and Hussey, 1970). In this method, after lumbar puncture, mock CSF is infused at a constant rate into the lumbar subarachnoid space and CSF pressure is measured in parallel. From this data one can determine the pressure-volume index, a measure of the compliance of the craniospinal axis, the resistance to CSF absorption, and, if one assumes that CSF production is constant, the CSF production rate (Czosnyka et al., 2001).

24.3.3 Noninvasive Measurement of CSF Formation

The implementation of ventriculo-lumbar perfusion or lumbar CSF infusion to measure CSF production in humans is limited because of the potential risk of medical complications. Therefore, more recent studies of CSF formation in humans have employed magnetic resonance imaging (MRI) (Nilsson et al., 1994). The rate of CSF production was calculated based on the measurements of linear velocity of CSF flow along the cerebral aqueduct and the estimation of the aqueduct's cross-section area. These studies have demonstrated circadian variations in CSF formation, with an average CSF production rate amounting to 340 μL/min during the time interval between 1500 and 1800 hours and 610 μL/min at night. The physiological significance of circadian variations in CSF production remains to be established. It is possible that the daily fluctuations in CSF dynamics play a role in central signaling involving the CSF pathways (see Chapter 10) and in the CSF-mediated removal of metabolites and other CSF constituents ("sink" action of CSF) (Oldendorf and Davson, 1967).

24.4 CONCLUSIONS

While VC perfusion is the technique of choice for measuring CSF production in small rodents, it is not appropriate for humans unless catheters are to be placed into the ventricles and subarachnoid space for medical reasons, e.g., ventriculo-lumbar perfusion for chemotherapy or administration of other therapeutic compounds. Accordingly, it is critical to develop a noninvasive, MRI-based method for the assessment of the CSF production rate in humans that would be applicable for both cohort and longitudinal studies. Preliminary data obtained by a Scandinavian group using MRI to estimate CSF production in humans is promising (Gideon et al., 1994a,b; Nilsson et al., 1994). However, their method will require further verification, as linear velocity of CSF flow through the aqueduct of Sylvius is difficult to measure and the rates of CSF formation reported by these authors are generally higher than those obtained earlier with ventriculo-lumbar perfusion.

24.5 ACKNOWLEDGMENTS

We are indebted to Dr. Gerald Silverberg for his comments on the measurement of CSF formation in humans. The authors also wish to thank Andrew Pfeffer for his help in the preparation of this manuscript. This work was supported by grant from NIH NS-39921 and by research funds from the Neurosurgery Foundation and Lifespan, Rhode Island Hospital.

REFERENCES

A. Chodobski, L. Królicki and K. Skolasinska, Chemical regulation of the cerebral blood flow in cats with rostro- or prepontine transection of the brainstem, *Acta Neurobiol Exp (Warsaw)* 46, 1986a, 47–56.

A. Chodobski, J. Szmydynger-Chodobska, A. Urbanska and E. Szczepanska-Sadowska, Intracranial pressure, cerebral blood flow, and cerebrospinal fluid formation during hyperammonemia in cat, *J Neurosurg* 65, 1986b, 86–91.

A. Chodobski, J. Szmydynger-Chodobska, E. Cooper and M.J. McKinley, Atrial natriuretic peptide does not alter cerebrospinal fluid formation in sheep, *Am J Physiol* 262, 1992a, R860–4.

A. Chodobski, J. Szmydynger-Chodobska, M.B. Segal and I.A. McPherson, The role of angiotensin II in regulation of cerebrospinal fluid formation in rabbits, *Brain Res* 594, 1992b, 40–6.

A. Chodobski, J. Szmydynger-Chodobska, M.D. Vannorsdall, M.H. Epstein and C.E. Johanson, AT_1 receptor subtype mediates the inhibitory effect of central angiotensin II on cerebrospinal fluid formation in the rat, *Reg Peptides* 53, 1994, 123–9.

A. Chodobski, J. Szmydynger-Chodobska, M.H. Epstein and C.E. Johanson, The role of angiotensin II in the regulation of blood flow to choroid plexuses and cerebrospinal fluid formation in the rat, *J Cerebr Blood Flow Metab* 15, 1995, 143–51.

A. Chodobski, J. Szmydynger-Chodobska and C.E. Johanson, Vasopressin mediates the inhibitory effect of central angiotensin II on cerebrospinal fluid formation, *Eur J Pharmacol* 347, 1998a, 205–9.

A. Chodobski, J. Szmydynger-Chodobska and M.J. McKinley, Cerebrospinal fluid formation and absorption in dehydrated sheep, *Am J Physiol* 275, 1998b, F235–8.

R.E. Curran, M.B. Mosher, E.S. Owens and J.D. Fenstermacher, Cerebrospinal fluid production rates determined by simultaneous albumin and inulin perfusion, *Exp Neurol* 29, 1970, 546–53.

R.W. Cutler, L. Page, J. Galicich and G.V. Watters, Formation and absorption of cerebrospinal fluid in man, *Brain* 91, 1968, 707–20.

M. Czosnyka, Z.H. Czosnyka, P.C. Whitfield, T. Donovan and J.D. Pickard, Age dependence of cerebrospinal pressure-volume compensation in patients with hydrocephalus, *J Neurosurg* 94, 2001, 482–6.

H. Davson and M.B. Segal, The effects of some inhibitors and accelerators of sodium transport on the turnover of 22.Na in the cerebrospinal fluid and the brain, *J Physiol (London)* 209, 1970, 131–53.

H. Davson and M.B. Segal, *Physiology of the CSF and Blood–Brain Barriers*. London: CRC Press, 1995, pp. 197–9.

F.M. Faraci, W.G. Mayhan and D.D. Heistad, Effect of vasopressin on production of cerebrospinal fluid: possible role of vasopressin (V_1)-receptors, *Am J Physiol* 258, 1990, R94–8.

R.A. Fishman, *Cerebrospinal Fluid in Diseases of the Nervous System.* London: W.B. Saunders Company, 1980, p. 20.

P. Gideon, F. Ståhlberg, C. Thomsen, F. Gjerris, P.S. Sørensen and O. Henriksen, Cerebrospinal fluid flow and production in patients with normal pressure hydrocephalus studied by MRI, *Neuroradiol* 36, 1994a, 210–5.

P. Gideon, C. Thomsen, F. Ståhlberg and O. Henriksen, Cerebrospinal fluid production and dynamics in normal aging: a MRI phase-mapping study, *Acta Neurol Scand* 89, 1994b, 362–6.

E.C. Greene, *Anatomy of the Rat.* New York: Hafner Publishing Company, 1963, p. 12.

S.R. Heisey, D. Held and J.R. Pappenheimer, Bulk flow and diffusion in the cerebrospinal fluid system of the goat, *Am J Physiol* 203, 1962, 775–81.

G.M. Hochwald, C. Malhan and J. Brown, Effect of hypercapnia on CSF turnover and blood-CSF barrier to protein, *Arch Neurol* 28, 1973, 150–5.

S. Hyman, J.G. McComb, L. Megerdichian and M.H. Weiss, Blood-cerebrospinal fluid barrier alteration following intraventricularly administered cholera toxin, *Brain Res* 419, 1987, 104–11.

R. Katzman and F. Hussey, A simple constant-infusion manometric test for measurement of CSF absorption. I. Rationale and method, *Neurology* 20, 1970, 534–44.

R. Katzman and H.M. Pappius, *Brain Electrolytes and Fluid Metabolism.* Baltimore: Williams and Wilkins Company, 1973, pp. 18–19.

N.A. Lassen and W. Perl, *Tracer Kinetic Methods in Medical Physiology.* New York: Raven Press, 1979, pp. 1–13.

M. Lindvall, L. Edvinsson and C. Owman, Effect of sympathomimetic drugs and corresponding receptor antagonists on the rate of cerebrospinal fluid production, *Exp Neurol* 64, 1979, 132–45.

M.A. Maktabi, F.F. Elbokl, F.M. Faraci and M.M. Todd, Halothane decreases the rate of production of cerebrospinal fluid. Possible role of vasopressin V_1 receptors, *Anesthesiology* 78, 1993, 72–82.

J.H. Masserman, Cerebrospinal hydrodynamics. IV. Clinical experimental studies, *Arch Neurol Psychiatry* 32, 1934, 523–53.

C. May, J.A. Kaye, J.R. Atack, M.B. Schapiro, R.P. Friedland and S.I. Rapoport, Cerebrospinal fluid production is reduced in healthy aging, *Neurology* 40, 1990, 500–3.

D.K. Michael and S.R. Heisey, Effects of brain ventricular perfusion and hypoxia on CSF formation and absorption, *Exp Neurol* 41, 1973, 769–72.

T.B. Miller, H.A. Wilkinson, S.A. Rosenfeld and T. Furuta, Intracranial hypertension and cerebrospinal fluid production in dogs: effects of furosemide, *Exp Neurol* 94, 1986, 66–80.

C. Nilsson, M. Lindvall-Axelsson and C. Owman, Simultaneous and continuous measurement of choroid plexus blood flow and cerebrospinal fluid production: effects of vasoactive intestinal polypeptide, *J Cerebr Blood Flow Metab* 11, 1991, 861–7.

C. Nilsson, F. Ståhlberg, P. Gideon, C. Thomsen and O. Henriksen, The nocturnal increase in human cerebrospinal fluid production is inhibited by a β_1-receptor antagonist, *Am J Physiol* 267, 1994, R1445–8.

W.H. Oldendorf and H. Davson, Brain extracellular space and the sink action of cerebrospinal fluid. Measurement of rabbit brain extracellular space using sucrose labeled with carbon 14, *Arch Neurol* 17, 1967, 196–205.

W.W. Oppelt, T.H. Maren, E.S. Owens and D.P. Rall, Effects of acid-base alterations on cerebrospinal fluid production, *Proc Soc Exp Biol Med* 114, 1963, 86–9.

J. Pappenheimer, S.R. Heisey, E.F. Jordan and J.Dec. Downer, Perfusion of the cerebral ventricular system in unanesthetized goats, *Am J Physiol* 203, 1962, 763–74.

R.C. Rubin, E.S. Henderson, A.K. Ommaya, M.D. Walker and D.P. Rall, The production of cerebrospinal fluid in man and its modification by acetazolamide, *J Neurosurg* 25, 1966, 430–6.

J.R. Seckl and S.L. Lightman, Intracerebroventricular vasopressin reduces CSF absorption rate in the conscious goat, *Exp Brain Res* 84, 1991, 173–6.

G.D. Silverberg, G. Heit, S. Huhn, R.A. Jaffe, S.D. Chang, H. Bronte-Stewart, E. Rubenstein, K. Possin and T.A. Saul, The cerebrospinal fluid production rate is reduced in dementia of the Alzheimer's type, *Neurology* 57, 2001, 1763–6.

G.D. Silverberg, S. Huhn, R.A. Jaffe, S.D. Chang, T. Saul, G. Heit, A. Von Essen and E. Rubenstein, Downregulation of cerebrospinal fluid production in patients with chronic hydrocephalus, *J Neurosurg* 97, 2002, 1271–5.

O. Sjoqvist, Beobachtungen über die Liquorsekretion beim Menschen *Zentralbl für Neurochir* 2, 1937, 8–17.

Index

B

C

D

E

F

G

I

J

K

L

M

N

O

P

Q

R

S

U

V

W

X

Z

T - #0017 - 121019 - C2 - 234/156/29 - PB - 9780367393106